Clifford

REPORT OF INVESTIGATIONS NO. 89

Cretaceous Carbonates of Texas & Mexico
Applications to Subsurface Exploration

D. G. BEBOUT AND R. G. LOUCKS, EDITORS

BUREAU OF ECONOMIC GEOLOGY
THE UNIVERSITY OF TEXAS AT AUSTIN
AUSTIN, TEXAS
W. L. FISHER, DIRECTOR
1977

Second Printing, July 1981

Papers presented at a Cretaceous symposium held in conjunction with the 27th annual meeting of Gulf Coast Association of Geological Societies and Gulf Coast Section/Society of Economic Paleontologists and Mineralogists.

CONTENTS

CRETACEOUS CARBONATES OF TEXAS AND MEXICO: APPLICATIONS TO SUBSURFACE EXPLORATION—OBJECTIVES AND SCOPE

D. G. Bebout[1] and R. G. Loucks[1]

The objectives of this symposium are (1) to bring together individuals from universities and industry who have conducted research on Cretaceous carbonates, in order to accelerate communication, exchange ideas, and discuss differences of opinion; (2) to emphasize subsurface carbonate studies but to include outcrop investigations which provide insight into interpreting the subsurface; (3) to cover all aspects of carbonate studies—regional geological setting, facies and depositional environments, paleontology, and diagenesis.

The following papers and abstracts include studies of regional stratigraphy, facies and depositional environments, diagenesis, geochemistry, paleontology, seismic stratigraphy, porosity, and hydrocarbon production and economics (fig. 1). Some consider the subsurface exclusively; others, the outcropping Cretaceous; and still others describe both the outcrop and the subsurface (fig. 1). The first extensive paper on Texas rudists, the dominant fossil of the Cretaceous, is included in this volume.

A majority of the papers cover Cretaceous carbonates of South and Central Texas, while a few deal with

[1]Bureau of Economic Geology, The University of Texas at Austin.

Author	Outcrop	Sub-surface	Regional-Stratigraphic	Facies-Depositional Environments	Diagenetic-Geochemical	Paleontological	Seismic-Stratigraphic	Porosity	Economic
1 Bebout-Schatzinger		X	X						
2 McFarlan		X	X	X					
3 Bay	X	X	X	X			X		
4 Cook		X							X
5 Coogan	X	X				X			
6 Wooten-Dunaway		X					X		X
7 Bebout		X		X				X	
8 Loucks		X		X	X			X	
9 Achauer	X	X		X	X			X	
10 Petta	X			X	X				
11 Cleaves	X			X					
12 Scott-Kidson	X		X	X		X			
13 Jacka	X			X	X			X	
14 Prezbindowsky		X			X				
15 Land	X	X			X				
16 Halley-Rose	X	X		X	X				
17 Kerr	X			X	X			X	
18 Bebout-Schatzinger-Loucks		X		X				X	
19 Scholle	X	X			X			X	X
20 Luttrell		X		X	X			X	
21 Wilson-Pialli Barrier	X		X	X		X		X	
22 Stabler-Marquez	X			X					
23 Conklin-Moore	X			X	X				
24 Enos		X		X	X			X	
25 Young	X		X			X			

Figure 1. Subject classification of papers in the symposium on Cretaceous carbonates of Texas and Mexico.

the extensive but less well known Cretaceous carbonates of West Texas (fig. 2). Studies of the Mexican Cretaceous are also included, with much attention given to outcropping facies that are similar to subsurface facies in Texas.

Most of the papers are concerned with carbonates of Aptian, Albian, and Cenomanian age which are widespread in the subsurface and on the outcrop of Texas and Mexico (fig. 3), although several cover older carbonates (Berriasian-Berremian) from the outcrop in Mexico. Two papers discuss the younger subsurface Texas carbonates of Coniacian to Campanian age which are presently of considerable economic importance.

Several Cretaceous symposia have been published previously (Lozo and others, 1959; Hendricks and Wilson, 1967; Perkins, 1974). Evidence of the widespread use of these valuable contributions is shown by the numerous references to papers from them in the present symposium volume. It is hoped that this new volume will prove to be of equal significance.

REFERENCES

Hendricks, L., and Wilson, W. F., eds., 1967, Comanchean (Lower Cretaceous) stratigraphy and paleontology of Texas: SEPM, Permian Basin Section, Special Pub. 67-8, 412 p.

Lozo, F. E., Nelson, H. F., Young, Keith, Shelburne, O. B., and Sandidge, J. R., 1959, Symposium on Edwards Limestone in Central Texas: Univ. Texas Pub. 5905, 235 p.

Perkins, B. F., ed., 1974, Aspects of Trinity Division geology—A symposium on the stratigraphy, sedimentary environments, and fauna of the Comanche Cretaceous Trinity Division (Aptian and Albian) of Texas and northern Mexico: Baton Rouge, Louisiana State Univ., Geoscience and Man, v. VIII, 228 p.

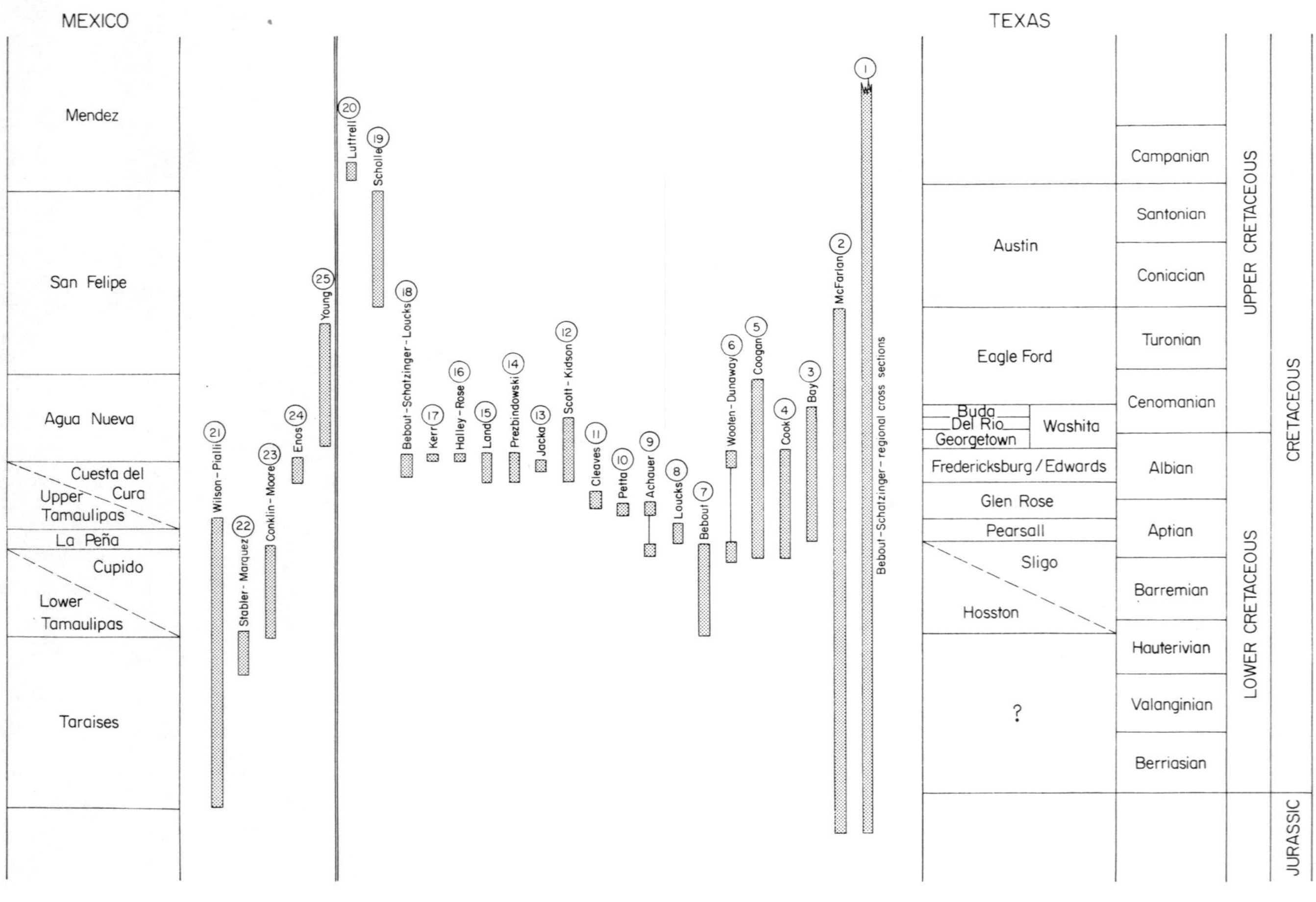

Figure 2. Stratigraphic and age distribution of study areas.

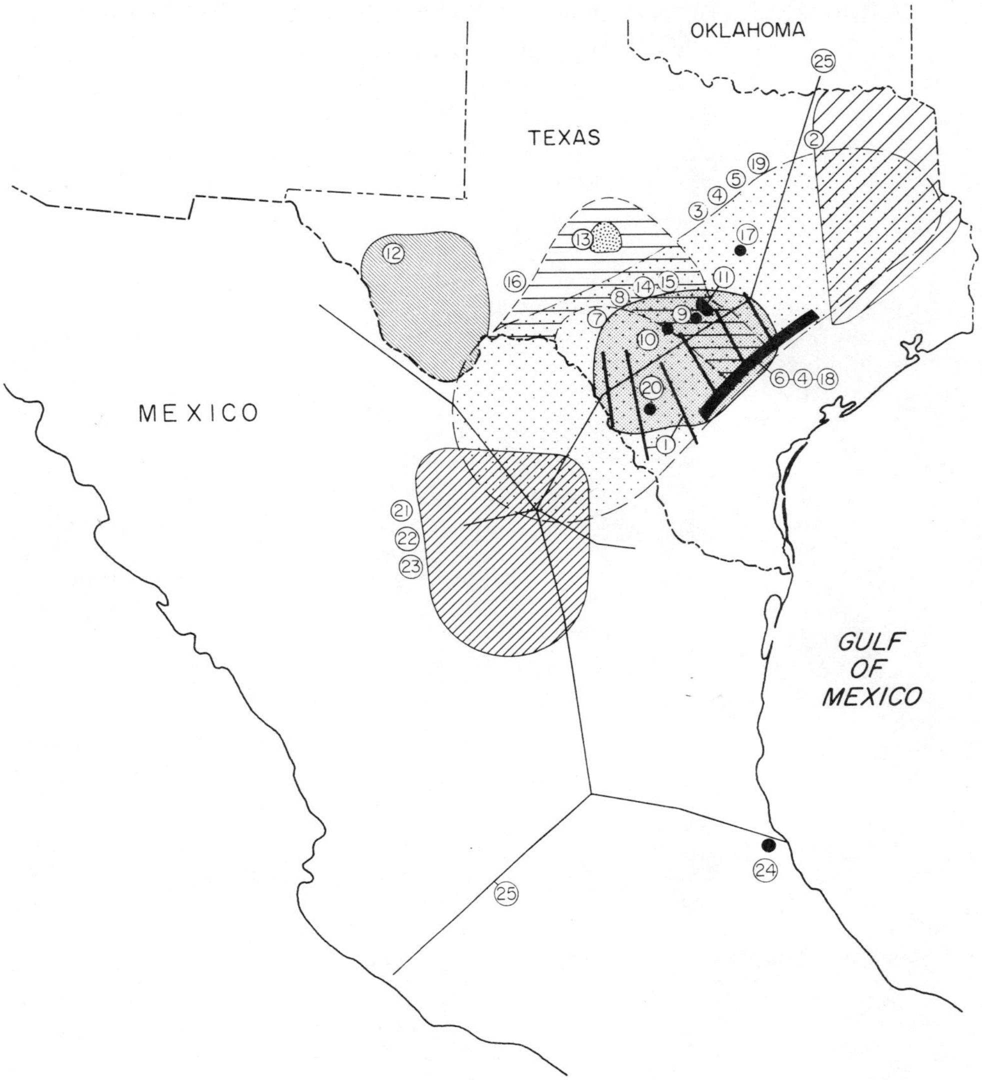

Figure 3. Geographic distribution of study areas.

REGIONAL CRETACEOUS CROSS SECTIONS—SOUTH TEXAS

D. G. Bebout[1] and R. A. Schatzinger[1]

Regional studies at the Bureau of Economic Geology of facies distribution and depositional environments of Lower Cretaceous carbonate strata have emphasized the need for a network of regional cross sections. The following sections were constructed to meet this need and are meant only to show the general changes in lithology and thickness within and between formations. In order to show these changes more clearly, faults have been omitted from the sections. Similar cross sections which extend farther downdip and include primarily Tertiary formations have been published by the Bureau of Economic Geology (Bebout and others, 1976).

Formation tops have been taken from many sources. The Tertiary tops are mainly from unpublished cross sections constructed for the Bureau of Economic Geology's geothermal project and from the Tertiary cross sections cited above. The tops of Cretaceous units have been picked with the aid of published reports by Tucker (1962), Winter (1961), and Flawn (1961). Also consulted were unpublished cross sections provided by Exxon Company U.S.A., Tenneco Oil Company, and additional Bureau of Economic Geology sections.

[1]Bureau of Economic Geology, The University of Texas at Austin.

REFERENCES

Bebout, D. G., Luttrell, P. E., Seo, J. H., 1976, Regional Tertiary cross sections - Texas Gulf Coast: Univ. Texas, Austin, Bureau of Economic Geology Geological Circular 76-5, 10 p.

Flawn, P. T., 1961, Appendix: Part 1. Summary reports on wells penetrating rocks of the Ouachita belt and immediately adjacent foreland in Texas, *in* Flawn, P. T., Goldstein, A., Jr., King, P. B., and Weaver, C. E., The Ouachita System, Univ. Texas, Austin, Bureau of Economic Geology Pub. No. 6120.

Tucker, D. R., 1962, Subsurface Lower Cretaceous stratigraphy, Central Texas: Univ. Texas, Austin, Ph. D. dissertation, 137 p. (unpublished).

Winter, J. A., 1961, Fredericksburg and Washita strata (subsurface Lower Cretaceous), Southwest Texas: Univ. Texas, Austin, Ph. D. dissertation, 135 p. (unpublished).

See appendix at end of text for cross sections.

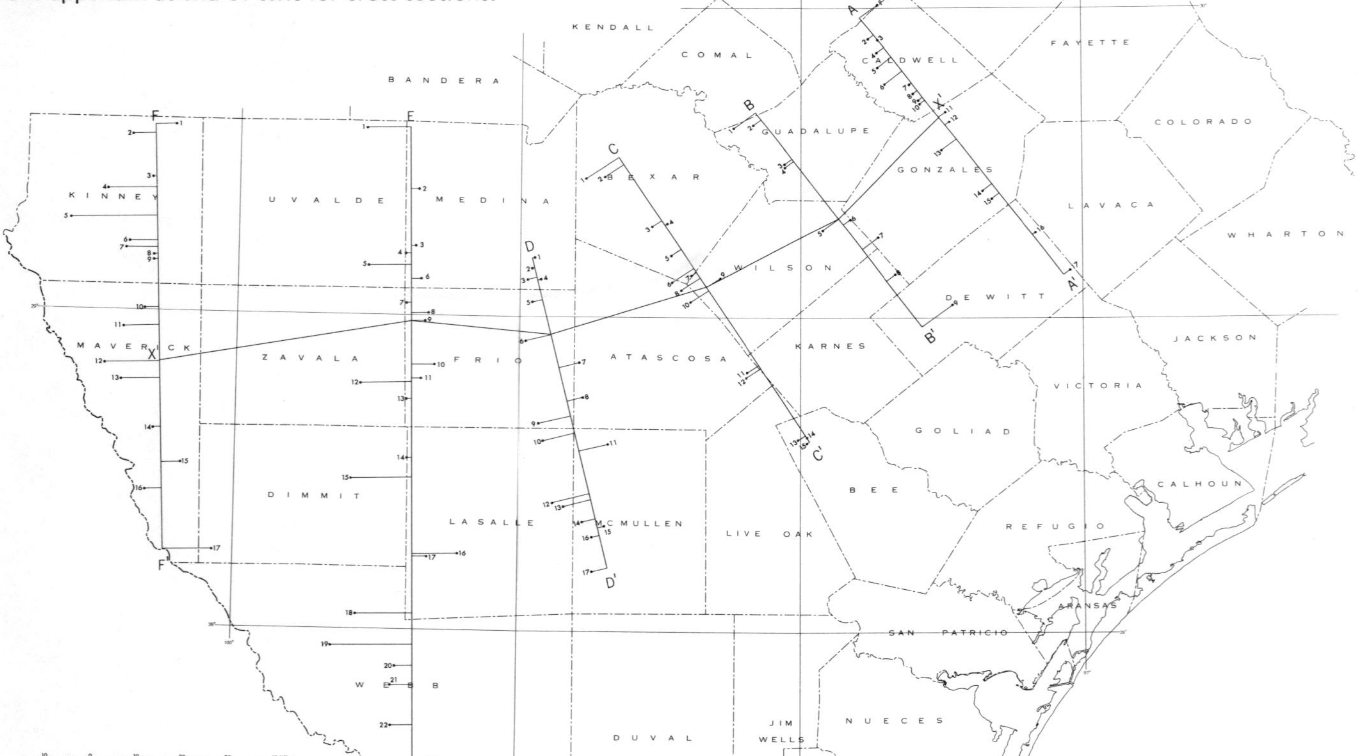

Location of cross sections.

LOWER CRETACEOUS SEDIMENTARY FACIES AND SEA LEVEL CHANGES, U. S. GULF COAST

Edward McFarlan, Jr.[1]

ABSTRACT

In the northern United States Gulf Coast, Lower Cretaceous sediments form an arcuate prism which thickens from a few hundred feet updip to more than 10,000 feet along the ancient shelf margin 100 to 300 miles downdip. This prism was divided into 11 time-stratigraphic units using hundreds of control wells with lithologic and faunal data. This information led to the recognition and mapping of major depositional facies including alluvial valley, delta, prodelta, inner shelf, middle shelf, outer shelf, and basin within each time-stratigraphic unit. During continuous deposition in Early Cretaceous time, these major facies units have transgressed and regressed many times across the broad subsiding shelf areas. The transgressions in upper Cotton Valley, lower Hosston through James, and Mooringsport through Washita times are thought to be controlled primarily by eustatic relative rise in sea level. Regressions in upper Hosston, Rodessa, Glen Rose, and Paluxy times are probably controlled by a decreased rate of subsidence and an increase in the supply of clastic sediments from rising uplands inland.

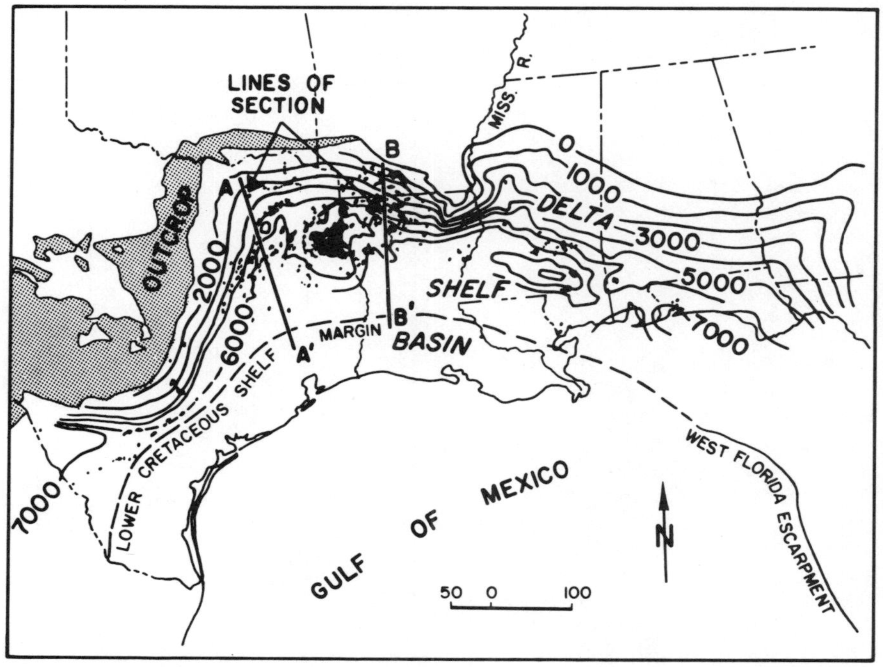

Figure 1. Lower Cretaceous thickness patterns, producing fields, and major features in the U. S. Gulf Coast (after E. H. Rainwater, 1971; this figure reproduced with permission of AAPG).

INTRODUCTION

This investigation was undertaken to determine the extent to which tectonic movements inland, shelf subsidence, and eustatic changes in sea level were factors controlling the entire sequence of sedimentary transgressions and regressions across the Early Cretaceous shelf in the northern U. S. Gulf Coast.

In this area, the Lower Cretaceous sediments form an arcuate prism which thickens from less than several hundred feet along the outcrop in Texas to more than 10,000 feet along the ancient shelf margin 100 to 300 miles downdip (fig. 1). From East Texas to northern Florida, the updip parts of the prism consist primarily of clastic sediments which were deposited in deltaic and alluvial valley environments. On the broad shelf, carbonates accumulated in shelfal environments. Seaward of the shelf margin, thick basinal deposits of micritic limestone and shale were laid down in the ancient Gulf basin. During Early Cretaceous time, coastal plain and shelf sediments in these major environmental settings transgressed and regressed many times across the broad shelf areas adjacent to the ancient Gulf basin.

These Lower Cretaceous transgressions and regressions of major depositional facies record continuous sedimentation and the principal depositional events in the geologic history of the U. S. Gulf Coast. The landward and seaward migration of major facies have been interpreted in different ways by many geologists using data from subsurface control wells. Murray (1961) and Rainwater (1971) in their reviews of U. S. Gulf Coast Mesozoic geology attributed the transgressions and regressions to varying rates of tectonic uplift inland and shelfal subsidence basinward. In a study of the Lower Cretaceous Trinity sedimentary facies, Forgotson (1963) developed a tectofacies model in which the area of Lower Cretaceous deposition was divided into linear belts including a stable shelf, an unstable shelf, and an intracratonic basin, each with "varying degrees of stability." Transgression and regression of facies were caused by changes in sediment supply and the rate of subsidence on the unstable shelf.

In a study of selected Lower Cretaceous sedimentary facies in the East Texas basin, Bushaw (1968) interpreted a major transgression in Hosston through James times and a regression in the Rodessa and Ferry Lake intervals in terms of a eustatic rise in sea level followed by a high stillstand. Rose (1972) in his study of the Edwards Group in Central Texas described varying rates of subsidence, tectonic uplift inland, and sea-level rise to account for changes in the type of carbonate and clastic sediment laid down in different areas.

[1]Exxon Company, U.S.A., Houston, Texas.

METHOD

The study was initiated by constructing a series of regional stratigraphic cross sections with deep control wells in key areas. In places, seismic sections were used to determine the configuration of mapping surfaces beyond well control. The relationships demonstrated on the cross sections were carried laterally with additional infill control wells. Many hundreds of wells were analyzed from Texas to Florida; for each well, descriptions of lithologies, fauna, and flora together with electric logs were used to identify time-stratigraphic units and major depositional facies. The regional pattern of the depositional facies throughout the Lower Cretaceous prism of sediments was interpreted in terms of major controls on sedimentation.

REGIONAL STRATIGRAPHY

The Lower Cretaceous prism of sediments was divided into 13 mapping units, shown by the heavy lines on the correlation chart (fig. 2). The conformable boundaries of each mapping unit approximate geologic-time surfaces; hence, each unit can be considered an informal time-stratigraphic unit. The ages of these units were determined from index fossils identified in the downdip marine parts of the sections. In general, these mapping units were named for well-known lithostratigraphic units in the U. S. Gulf Coast, such as the shelf limestones of the Fredericksburg or the sandstones and shales of the Paluxy. In places, the Fredericksburg unit changes to clastic sediments updip and the Paluxy unit changes to carbonate rocks downdip. Where these changes occur, the common lithostratigraphic name was applied for mapping purposes to rocks of all lithologies interpreted to have been deposited during the time assigned to the stratigraphic unit at the type locality.

The time-stratigraphic units with index fossils in the marine facies were correlated updip into nonmarine facies using an interconnected network of control wells. The correlation of mapping surfaces in the nonmarine facies was based on the interpretation of time-equivalent facies and associated lithologies in control wells. In places, the Ferry Lake Anhydrite served as a marker surface for correlation. In other places, erosional unconformities were recognized and correlated along surfaces associated with the angular truncation of older beds.

In general, the time-stratigraphic units of the Lower Cretaceous thicken basinward (fig. 3, dip section B-B′). This basinward thickening reflects continuous sedimentation and regional subsidence during Early Cretaceous time. In East Texas, the Mooringsport through Washita units thicken in a large shelf basin in response to a more rapid subsidence in this area compared with the shelf margin (fig. 3, section A-A′). The prominent unconformities at the top of the upper Cotton Valley and Washita-Fredericksburg units were recognized by the truncation of underlying beds and the onlap of post-unconformity deposits. The marked changes in slope at the shelf margin in each unit were taken from selected seismic sections.

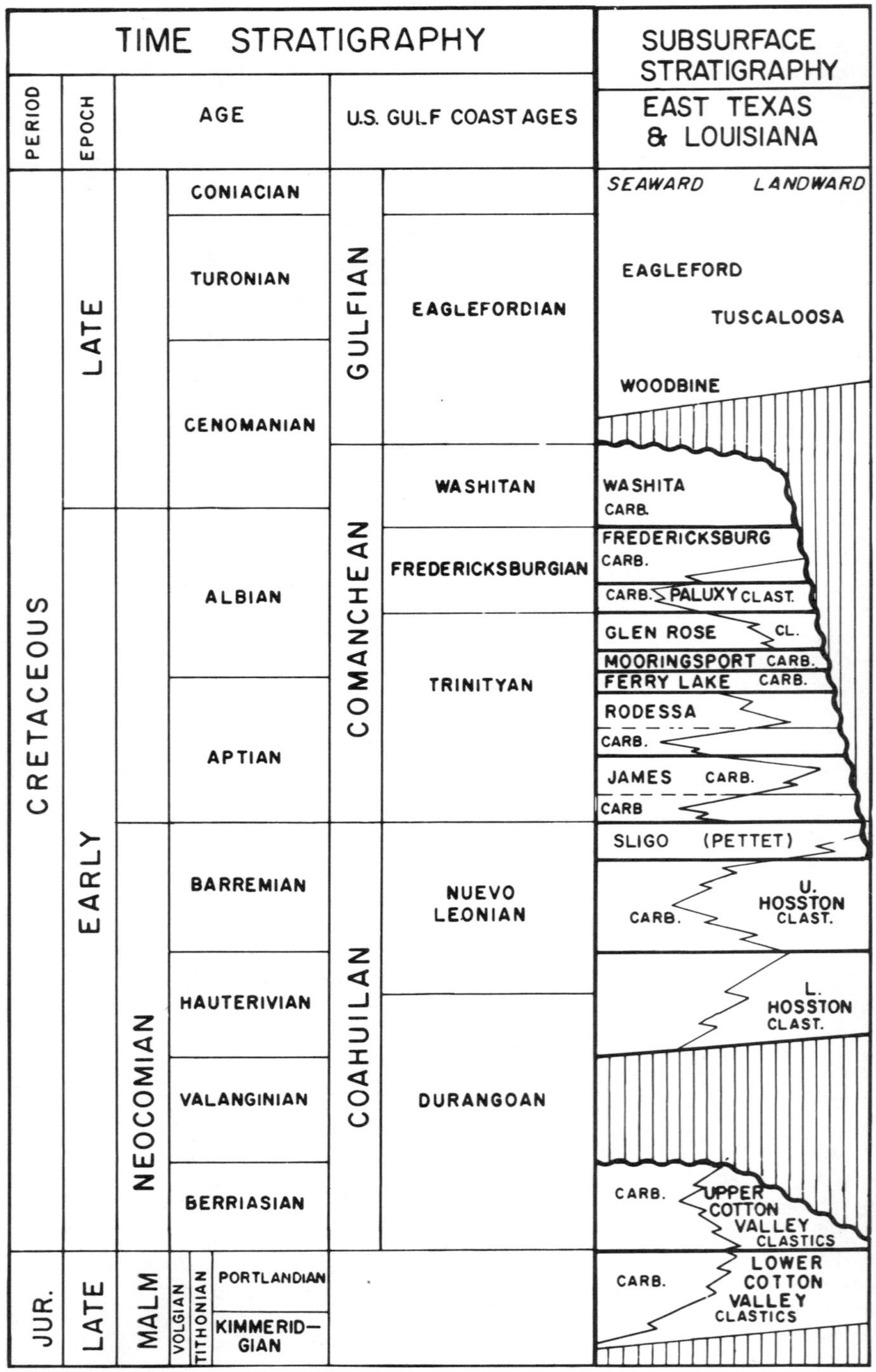

Figure 2. Correlation chart for Lower Cretaceous sediments in U. S. Gulf Coast.

MAJOR DEPOSITIONAL FACIES

Within regional Lower Cretaceous mapping units, sequences of major depositional facies were recognized as illustrated on the regional cross sections (fig. 3). These facies developed as a result of simultaneous deposition of clastic and carbonate sediments in continental and marine environments extending from an alluvial valley and deltaic complex in the coastal plains across a broad shelf to the deeper ancient Gulf basin. For each mapping unit, the shelf facies can be subdivided into inner, middle, and outer units as shown diagrammatically on figure 4. The shelfal facies are replaced updip by prodelta, delta, and alluvial valley facies in East Texas, Louisiana, Mississippi, Alabama, and northern Florida.

Each of these major facies units in the Lower Cretaceous sedimentary prism contains distinctive lithology, fauna, flora, and key minerals used to identify the depositional environments. These distinctive criteria are listed on the facies model (fig. 4). The lithologic and faunal characteristics of each facies were taken from sample descriptions for control wells. The identification of the major facies using these criteria was made with the application of well-known carbonate and clastic environmental facies descriptions and interpretations (Bebout, 1972; Bebout and Loucks, 1974; Coleman and Wright, 1975; Fisk, 1944; Kauffman, 1974; Maxwell, 1968; Newell and Rigby, 1957; Wilson, 1969).

In general, from East Texas to northern Florida alluvial valley and deltaic sandstones and gravels interfinger seaward with prodelta sandstones and red shales containing oyster shells, wood fragments, and lignite. These clastic sediments are replaced seaward by inner-shelf facies consisting of skeletal, oolitic, pelletal, and micritic limestones interbedded with thin gray and red shales containing marine mollusks, gastropods, and benthonic foraminifers. The fauna are typical of nearshore, high-energy, shallow-water conditions. Inner-shelf facies interfinger basinward with middle-shelf facies composed of fine- to coarse-grained limestones and dolomites, with open-marine fauna including corals, echinoids, and miliolids. These sediments change seaward to rudistid-bank limestones and backbank oolitic and skeletal limestones interbedded with micritic limestones. The outer-shelf fauna, including rudistids, corals, encrusting algae, and stromatolites, occurs in abundance. Beyond the shelf margin, widespread basinal facies are composed of fine-grained limestones interbedded with dark-gray shales with abundant open-marine pelagic fauna.

LOWER CRETACEOUS TRANSGRESSIONS AND REGRESSIONS

In general, major Lower Cretaceous depositional facies shifted landward and basinward on the ancient shelf in a regular pattern which can be correlated throughout the epoch (fig. 3). Major facies patterns are transgressive or regressive in any particular time-stratigraphic unit wherever it occurs throughout the U. S. Gulf Coast. Accordingly, the changes in direction of facies movement occurred simultaneously throughout the area of study.

A summary of these transgressions and regressions records the major depositional events in Early Cretaceous time. At the onset of the epoch in upper Cotton Valley time, major facies units shifted landward 40 to 60 miles on the ancient shelf as shown by the transgression of inner shelf facies on prodelta facies (fig. 3).

Following upper Cotton Valley time, a period of erosion occurred; the upper Cotton Valley sediments on the broad shelf were exposed to subaerial weathering and scouring. On the inner part of the ancient shelf, previously deposited sediments were removed. The sites of deposition shifted to the outer shelf margin and the ancient Gulf basin.

During Valanginian time, deposition is presumed to have continued on the outer shelf margin and in the Gulf basin seaward of well control.

In lower Hosston through James times, major depositional facies transgressed 40 to 60 miles in a landward direction (fig. 3). The seaward boundary of the middle-shelf facies moved from the shelf margin near the beginning of lower Hosston time to a position about 60 miles inland at the end of James time (fig. 3, section B-B′). This transgression was interrupted in the early part of upper Hosston time with 10 to 20 miles of basinward progradation of deltaic and associated shelf facies.

An abrupt regression of facies was recorded in the sediments deposited during basal Rodessa time. Middle-shelf facies were shifted seaward over outer-shelf facies for a distance of 10 to 20 miles in a relatively short interval of geologic time (fig. 3, sections A-A′ and B-B′).

From the end of James time and through all of Rodessa time, shelf facies moved basinward 20 to 40 miles (fig. 3, sections A-A′ and B-B′) in response to progradational sedimentary processes.

During the ensuing Ferry Lake interval, the entire shelf from East Texas across Louisiana and Mississippi became an area of evaporite deposition in a restricted shelf environment. Shelf-margin rudistid-bank buildups (not shown on cross sections) continued to develop during this time.

In the following Mooringsport through Washita intervals, a 40- to 50-mile transgression of major facies is recorded in the shelfal area. The transgression was interrupted in Glen Rose and Paluxy time by 20- to 30-mile basinward progradation of deltaic and shelfal facies. In the post-Ferry Lake interval on the shelf margin, the bank complex within the outer-shelf facies migrated seaward 10 to 20 miles in response to continued rapid sedimentation in this environment.

The interval ended with the post-Washita development of an erosional unconformity on Lower Cretaceous sediments exposed to subaerial weathering and erosion on the ancient shelf.

GEOLOGIC HISTORY

In the northern U. S. Gulf Coast, these transgressions and regressions of major depositional facies within the Lower Cretaceous prism of sediments were recognized from Texas to Florida. Although these facies patterns are thought to reflect eustatic sea-level changes, they could be due in part to inland uplift, shelf subsidence, and associated sedimentation. However, the extensive area containing the transgressions and regressions and the correspondence of geologic ages for the major shifts in facies migration throughout the area suggest that the primary control could be eustatic changes in sea level. Hence, these transgressions and regressions were compared with the global sea-level chart developed by Vail and others (in press (a)) and shown with the facies patterns on figure 5. By comparison of the generalized facies plot with the sea-level curve, the major breaks can be accounted for. Therefore, it is concluded that eustatic changes in sea level were a major control of facies patterns in the U. S. Gulf Coast. Vail and others (in press (b)) point out that a facies transgression indicates a relative rise in sea level, but it may not indicate the entire rise.

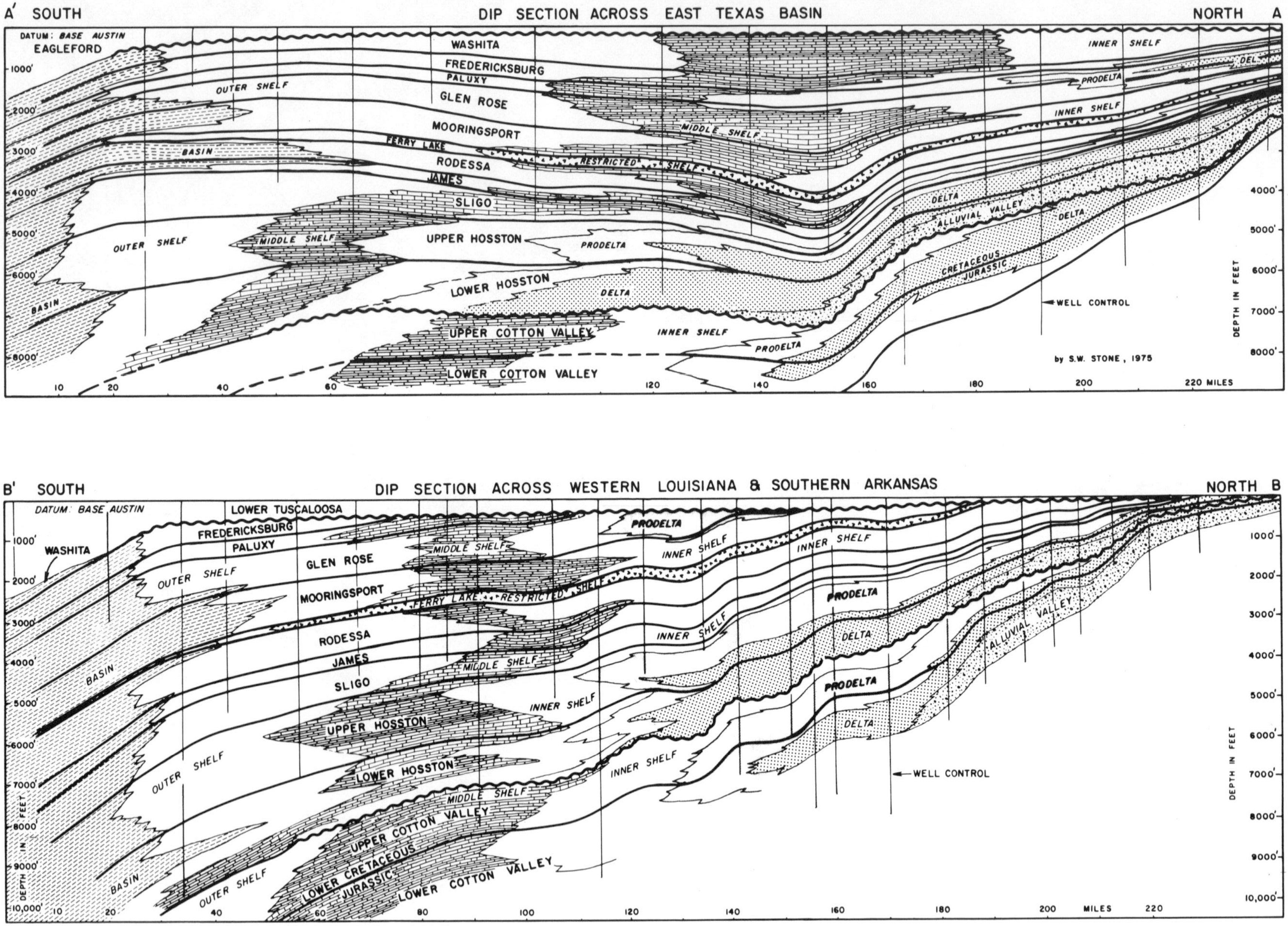

Figure 3. Representative stratigraphic dip sections across East Texas and western Louisiana.

The oldest Early Cretaceous event was the eustatic rise in sea level during upper Cotton Valley time which was marked by a landward shift of the shoreline and associated alluvial valley, deltaic, prodelta, and shelfal facies. In a study of the Cotton Valley, Todd and Mitchum (1975) used well and seismic-stratigraphic data and observed the landward onlap of seismic reflection patterns. They interpreted this pattern to be related to a eustatic rise in sea level, and the top of the Cotton Valley was judged to be the end of a supercycle marked by a highstand of sea level.

The ensuing period of erosion and the truncation of previously deposited upper Cotton Valley beds in the updip areas are evidence for the worldwide drop in sea level in Valanginian time. The seismic evidence for erosion was observed by Todd and Mitchum (1975) and attributed to a lowering of sea level.

The sediments presumed to be deposited during Valanginian time were located seaward of well control. These sediments were identified on seismic sections and were interpreted to have been deposited during a lowstand of sea level (Todd and Mitchum, 1975).

The general transgression of major facies noted in lower Hosston, upper Hosston, Sligo, and James time-stratigraphic units is interpreted to be related to the rise in sea level noted on the global cycle chart (Vail and others, in press (a)) and originally suggested by Bushaw (1968). The upper Hosston regression was probably a Gulf Coast event involving uplift in the inland areas including the Appalachian and Ouachita uplands. Streams draining these uplands carried an unusually large amount of clastic sediments onto the ancient shelf where deposition occurred in a prograding sequence of depositional environments. A decrease in the rate of regional subsidence may have been an additional factor enhancing this regression.

The abrupt regressive shift in facies in the basal part of the Rodessa unit probably reflects a lowering of sea level in post-James time (fig. 5). Sea level continued to rise through Rodessa time but inland uplift and/or reduction in the rate of subsidence resulted in a continued regression of major facies. The resulting progradation of shelf sediments ended in the development of a restricted shelf from East Texas to South Florida. The restricted shelf facies of Ferry Lake age are primarily anhydrite evaporites interbedded with fine- to coarse-grained carbonates. The geologic data suggest that the shelf-margin barrier bank continued to grow and was enlarged sufficiently to restrict the flow of marine water onto the shallow shelf. Thus, the restricted shelf area became a site for evaporite deposition. After Ferry Lake time, sea level continued to rise, and the restricted shelf facies were replaced by inner-, middle-, and outer-shelf facies which migrated inland in Mooringsport time.

The next major depositional event was progradation of Glen Rose and Paluxy delta and prodelta facies units onto the shelf probably in response to inland uplift and/or reduced rate of subsidence across the entire northern Gulf Coast.

In the following Fredericksburg and Washita intervals, the major facies units maintained their positions on the shelf or shifted toward the ancient Gulf basin where large delta systems were active in Mississippi. The rate of

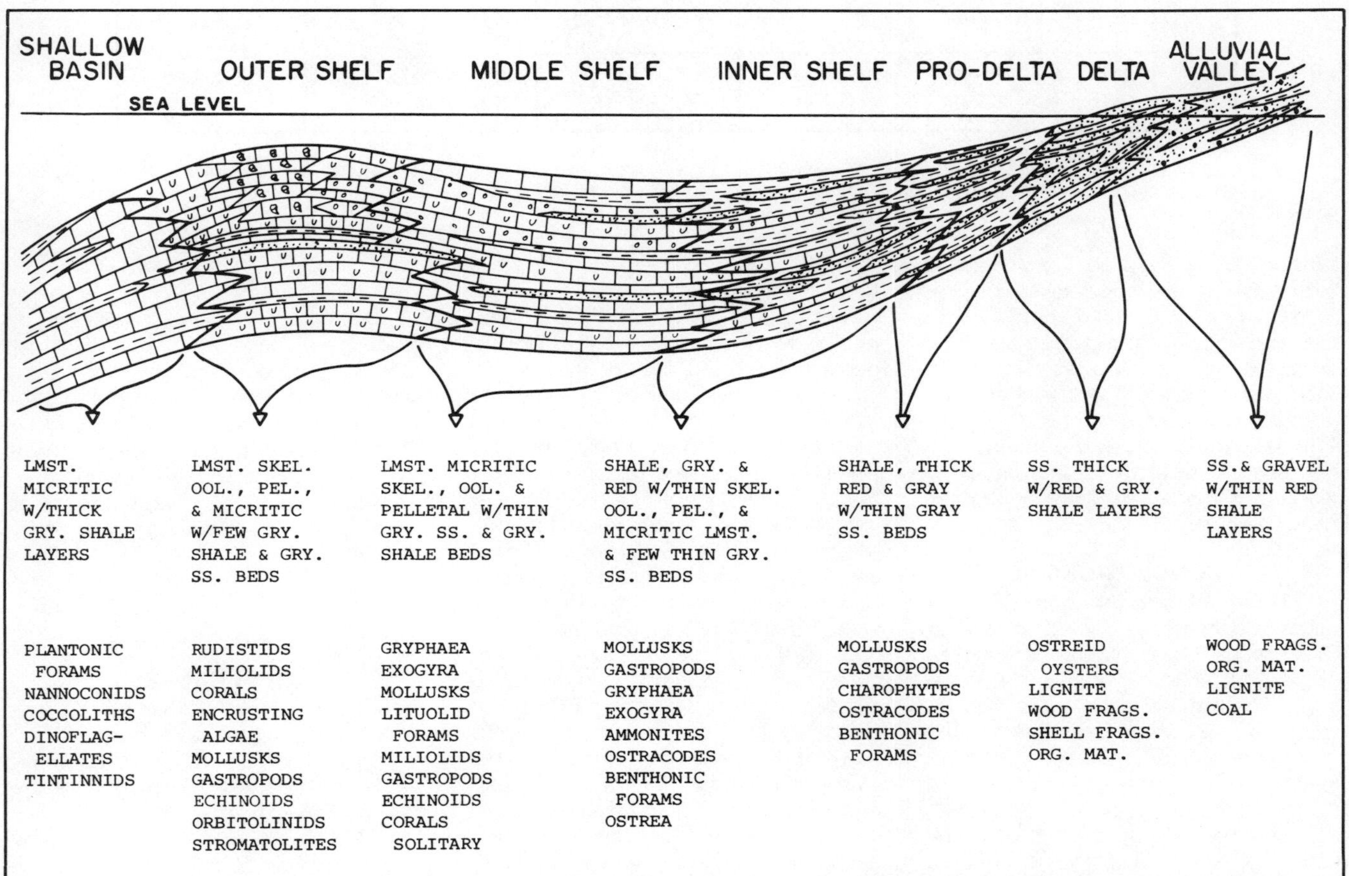

Figure 4. Criteria for recognition of major depositional facies in Lower Cretaceous deposits.

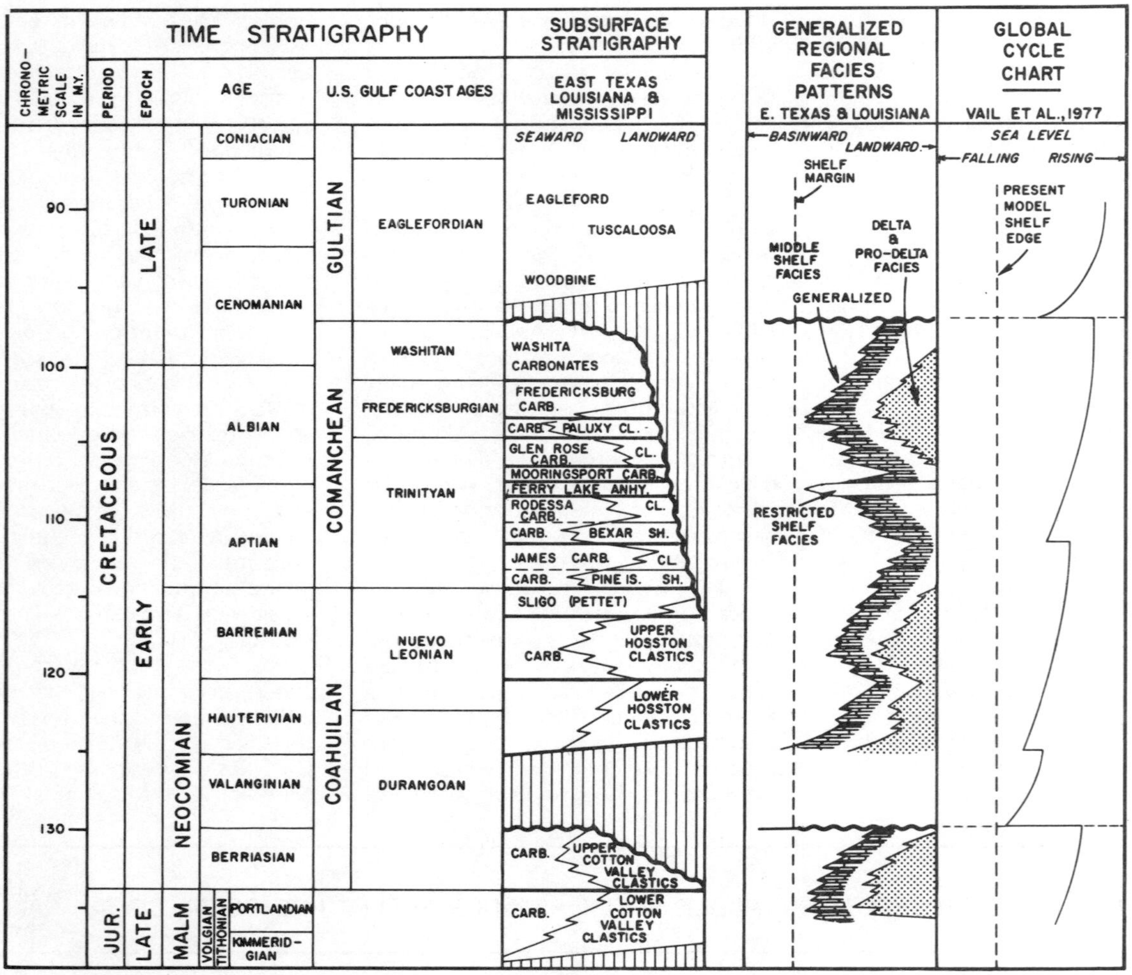

Figure 5. Lower Cretaceous correlation chart of the U. S. Gulf Coast with major regional facies patterns and major eustatic cycles of sea-level changes.

subsidence was probably balanced with the rate of sea-level rise as suggested by Rose (1972). Finally at the end of Washita time, sea level dropped, and the Lower Cretaceous shelf was exposed to subaerial erosion, noted on the cross sections (fig. 3). This is a major break worldwide and has been described by a number of workers (Gussow, 1972, 1973; Coogan and others, 1972).

CONCLUSIONS

Three of the transgressions of major depositional facies in Lower Cretaceous beds were controlled primarily by eustatic relative rise in sea level. These three facies transgressions occurred in the following units: (1) upper Cotton Valley, (2) lower Hosston through James, and (3) Mooringsport through Washita.

The marked regressive shift in the basal Rodessa facies patterns reflects a slight drop in sea level followed by a gradual rise.

All other regressions including those of upper Hosston, upper Rodessa, Glen Rose, and Paluxy were related to widespread events involving decreased rate of subsidence and increased supply of clastic sediments from rising uplands inland across the northern U. S. Gulf Coast. Possibly some of these regressions may be associated with small sea-level lowerings during the overall cycle of relative sea-level rise. Perhaps some of them could be caused by pauses in the rise in sea level when progradation could occur. Because Vail and others (in press (a)) did not discuss all specific cycles identified in the Early Cretaceous supercycle, there may be a closer correlation to eustatic cycles than is evident from their curve. Additional study of U. S. Gulf Coast geology and worldwide eustatic cycles of sea level will be required for a detailed correlation of regional facies patterns with small sea-level changes that is not possible at this time.

ACKNOWLEDGMENTS

The writer thanks Exxon Company, U.S.A. for permission to publish this paper. In the study of the Lower Cretaceous sediments in Texas and in the construction of section A-A′ on figure 3, the assistance of Dr. S. W. Stone is acknowledged with appreciation. Special thanks are extended to R. G. Todd and R. M. Mitchum, Jr. (Exxon Production Research Company) for their critical reading of the manuscript and their constructive suggestions included in this paper.

REFERENCES

Bebout, D. G., 1972, Sand deposition in continental slope basins off the Magdalena River, southern Caribbean Sea, Part II: Sand facies and their distribution: Exxon Production Research Company, EPR.55EX.72.

______ , and Loucks, R. G., 1974, Stuart City Trend, Lower Cretaceous, South Texas—A carbonate shelf-margin model for hydrocarbon exploration: Univ. Texas, Austin, Bureau of Economic Geology Report of Investigations 78, 80 p.

Bushaw, Dewey J., 1968, Environmental synthesis of the East Texas Lower Cretaceous: GCAGS, Transactions, v. XVIII, p. 416-438.

Coleman, J. M., and Wright, L. D., 1975, Modern river deltas: Variability of processes and sand bodies, Deltas—Models for exploration: Houston Geological Society, p. 99-149.

Coogan, A. H., Bebout, D. G., and Maggio, Carlos, 1972, Depositional environments and geologic history of Golden Lane and Poza Rica Trend, Mexico, An alternative view: AAPG Bull., v. 56, p. 1419-1447.

Fisk, Harold N., 1944, Geological investigation of the alluvial valley of the lower Mississippi River: U. S. Army, Corps of Engineers, War Dept.

Forgotson, James M., Jr., 1963, Depositional history and paleotectonic framework of Comanchean Cretaceous Trinity stage, Gulf coast area: AAPG Bull., v. 47, no. 1, p. 69-103.

Gussow, W. C., 1972, East Texas field—A new interpretation: 24th International Geological Congress, Session Proceedings, Section 5, p. 82-91.

______ 1973, East Texas field—The clastic trap has new interpretation: Oil and Gas Journal, v. 71, no. 7, p. 134-136.

Kauffman, Erle G., 1974, Cretaceous assemblages, communities, and associations: Western interior United States and Caribbean Islands, Principles of benthic community analysis (Notes for a short course): Univ. Miami, Sedimenta IV Comparative Sedimentology Laboratory, p. 12.1-12.27.

Maxwell, W. G. H., 1968, Atlas of the Great Barrier Reef: Amsterdam-London-New York, Elsevier Publishing Company.

Murray, Grover E., 1961, Geology of the Atlantic and Gulf Coastal Province of North America: New York, Harper and Brothers, Harper's Geoscience Series.

Newell, Norman D., and Rigby, J. Keith, 1957, Geological studies on the Great Bahama Bank, Regional aspects of carbonate deposition—A symposium with discussions: Society of Economic Paleontologists and Mineralogists, Special Pub. No. 5.

Rainwater, E. H., 1971, Possible future petroleum potential of Lower Cretaceous western Gulf basin, *in* Cram, Ira H., ed., Future petroleum provinces of the United States—Their geology and potential: AAPG Memoir 15, p. 901-926.

Rose, Peter R., 1972, Edwards Group, surface and subsurface, Central Texas: Univ. Texas, Austin, Bureau of Economic Geology Report of Investigations 74, 198 p.

Todd, R. G., and Mitchum, R. M., Jr., 1975, Seismic stratigraphic identification of eustatic cycles in Late Triassic, Jurassic, and Early Cretaceous rocks, Gulf of Mexico and West Africa: GCAGS, Transactions, v. XXV, p. 41-43.

Vail, P. R., Mitchum, R. M., Jr., and Thompson, S., III, in press (a), Global cycles of relative changes of sea level: AAPG Memoir.

______ , Mitchum, R. M., Jr., and Thompson, S., III, in press (b), Relative changes of sea level from coastal onlap: AAPG Memoir.

Wilson, James Lee, 1969, Microfacies and sedimentary structures in "deeper water" lime mudstones, Depositional environments in carbonate rocks: SEPM, Special Pub. No. 14, p. 4-7.

LOWER CRETACEOUS STRATIGRAPHIC MODELS FROM TEXAS AND MEXICO

T. A. Bay, Jr.[1]

ABSTRACT

Carbonate ramp and carbonate shelf depositional models are utilized in interpreting Lower Cretaceous carbonate stratigraphy in the Gulf of Mexico province. The lower Glen Rose (upper Trinity) high-energy, rudist reef and grainstone complex began on a low-relief carbonate ramp profile and prograded seaward over slightly deeper water, low-energy, lime mudstones and wackestones with a resultant growth of some 1,250 feet vertically over a horizontal distance of 30 miles. During late Glen Rose, Fredericksburg, and Washita time, this same high-energy facies complex built about 1,300 feet vertically in a horizontal distance of 6 miles, creating a marked break in slope between the shallow water of the shelf and shelf margin and the adjacent deeper water of the ancestral Gulf of Mexico. This topographic break in slope changed a carbonate ramp to a carbonate shelf profile of deposition. A faster rising sea level, perhaps a result of a more rapid rate of sea-floor spreading, probably accounts for the pronounced vertical buildup.

Regional cross sections of Fredericksburg and Washita strata show the shallow-water depositional attributes of carbonates on the Central Texas platform. High-energy grainstone and rudist reef complexes separate these shallow-water carbonates and evaporites from deeper water strata in the East Texas basin and, during Washita time, in the McKnight basin.

Density and acoustical contrasts present within the prograding lower Glen Rose strata deposited on a ramp profile and in the upper Glen Rose, Fredericksburg, and Washita strata deposited at the shelf margin are sufficient to recognize the character of these buildups on reflection seismic profiles.

INTRODUCTION

Several major oil companies began extensive geological research programs on sedimentology following World War II. Intensive studies of Recent carbonates revealed that geometry of facies zones and identification of these facies with depositional environments enabled prediction of the distribution of potential reservoir rocks and more dense lateral seals. Recent carbonate models mainly show lateral facies relationships because not enough time has evolved nor have sufficient sea-level variations occurred to show vertical attributes. Criteria derived from Recent carbonate studies, led by R. N. Ginsburg for Shell, were simultaneously applied to ancient carbonate outcrop investigations where three-dimensional relationships are more easily recognized; subsequently outcrop correlations and other data were extended into the subsurface.

Most carbonate sediments are produced and deposited in a marine basin. The marine environment begins at a shoreline and extends into shallow and ultimately deep water; deep-water deposits may be euxinic and extend to either bathyl or abyssal depths. According to Laporte (1968), the distribution of land and water masses, hydrography, and water-energy level "are the dominant controls in carbonate facies genesis and differentiation" in Recent shallow-water carbonate environments. Each depositional basin is unique in that it has specific tides, marine currents, and local wind-driven currents. Each basin also has unique geography and hydrography which is quite complex, but many basins have some common attributes. Some of these common attributes may be expressed by two basic, genetic profiles: (1) the carbonate shelf and (2) the carbonate ramp.

The carbonate shelf-margin model (fig. 1) may be simply defined as an essentially flat region of shallow-water sedimentation bounded on one side by land and on the other by a break in slope which separates it from adjacent areas of deeper, open-marine circulation. The break in slope may circumscribe the shallow flat area in the form of a submerged plateau surmounted by an island chain. This definition is intended to include features with some regional extent. The carbonate ramp profile (fig. 2), originally defined in the literature by Ahr (1973) as "a sloping surface connecting two levels," is here defined as a marine depositional surface that begins at a shoreline and descends into deeper water with

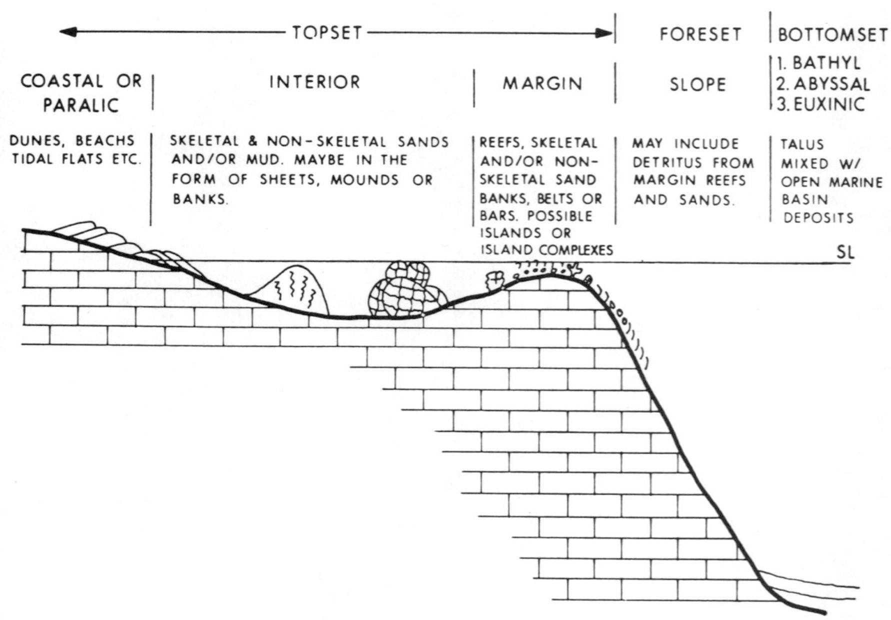

Figure 1. Carbonate shelf profile.

[1]Shell Development Company, Houston, Texas

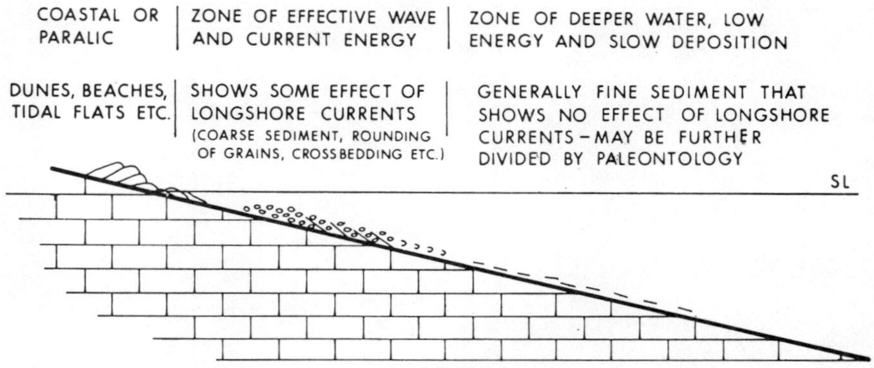

Figure 2. Carbonate ramp profile.

no marked break in slope in the zone of effective wave and current energy.

An understanding of facies relationships, depositional environments, and the carbonate shelf and ramp profiles concept is essential in interpreting Lower Cretaceous carbonate stratigraphy in the Gulf of Mexico province.

TRINITY DIVISION

Stratigraphy

One of the early applications using Recent carbonate models as a guide to interpreting facies zonations and depositional environments of Cretaceous carbonates was reported by Stricklin and others (1971); their work during the 1950's concentrated on a study of Lower Cretaceous Trinity division strata in outcrops between the Llano uplift and the Balcones fault zone in Central Texas (fig. 3). Included among the environmentally diverse strata reported are blanketlike beach deposits, rudist reefs, widespread tidal-flat deposits, shallow-water evaporites with an association of unusual diagenetic features, and shore deposits of caliche and alluvium.

Trinity strata are conveniently divided into three "clastic-carbonate couplets" reflecting cyclic sedimentation during an overall transgression of the sea upon the land. The three couplets in ascending order are designated the lower (Sycamore Sandstone and Sligo Limestone), middle (Hammett Shale and Cow Creek Limestone), and upper Trinity (Hensel Sandstone and Glen Rose Limestone) (fig. 4).

Upper Trinity Depositional History

The upper Trinity Glen Rose Limestone is the thickest and most extensively exposed Trinity Formation on the San Marcos arch and, together with the underlying Hensel Sandstone, constitutes a massive wedge that thins from about 1,000 feet near the Balcones fault zone to about 50 feet at the updip edge near the Llano uplift. Because the Glen Rose Formation is so well exposed and has been studied in considerable detail, this stratigraphic unit provides a foundation for discussion of the carbonate ramp-carbonate shelf concept from upper Trinity through Fredericksburg and Washita time.

The Glen Rose Formation is separated into upper and lower members by the ***Corbula*** bed, a widespread marker datum initially recognized by F. L. Whitney. The lower Glen Rose is exposed only in deeply incised river and creek valleys from the vicinity of Austin to some 75 miles northwest of San Antonio. Outcrops of the upper Glen Rose are more extensive and exposed all the way to the south end of the Llano uplift. Interpretations of depositional environments for these deposits are based on inferred degrees of water depth, energy, and salinity. Abrupt fluctuations in water circulation over a broad, almost flat, low-relief shelf resulted in abrupt changes in rock types.

In a more detailed report, Stricklin and Amsbury (1974) recognized a sequence in part of the lower Glen Rose (fig. 5) consisting of low-energy, shallow-water deposits identified with an immense tidal flat, followed by slightly deeper water with good circulation, which gave way to restricted circulation during the deposition of the lagoonal ***Corbula*** bed. The lowermost part of the upper Glen Rose immediately above the ***Corbula*** bed consists of evaporite stringers with ***Corbula*** interbeds, suggesting an even more restricted circulation environment.

Evidence for the tidal-flat origin of the rocks at the base of the sequence consists of the close association of stromatolitic algae with mud cracks, pholad-bored surfaces, dinosaur tracks, and reptilian bones and scratch marks in a supratidal unit some 6 to 18 inches thick immediately overlying the stromatolites. A rich, varied fauna is present in the nodular limestone above the supratidal unit, indicating slightly deeper water with more open circulation. Finally, increasing salinity at the end of early Glen Rose time resulted in the proliferation of the small clam ***Corbula***, the only organism able to tolerate the saline water. Eventually even ***Corbula*** became extinct. Behrens (1965) lists additional evidence confirming Stricklin's interpretation of these depositional environments.

Caprinid "reefs" have been studied in detail by Perkins (1974) in an interval in the upper 100 to 180 feet of the lower Glen Rose (fig. 6). Most of the reefs occur in this interval, however one composed dominantly of corals has been recognized near the base of the Glen Rose at the Narrows locality in Blanco County (Stricklin and others, 1971). Within the reef complex, Perkins recognized "four principal laterally equivalent facies that include (1) oyster biostromes, (2) monopleurid biostromes, (3) caprinid reefs, and (4) plant fragment beds." (figs. 7, 8).

Perkins (1974) states, "the oyster, monopleurid, and caprinid facies are aligned in northeast-trending bands paralleling the early Glen Rose shoreline of the southeastern side of the Llano uplift. The oyster biostrome facies lies nearest the ancient shoreline and is transitional seaward (southeastward) into the monopleurid biostrome facies. The most seaward band exposed is the caprinid reef facies. The plant-fragment facies occupies a small oval area in the center of the caprinid reef facies that surrounds the plant fragment facies on all sides except the west."

Stricklin and others (1971) comment on the high-energy reef deposits of the lower Glen Rose "resulting from a high degree of water circulation" as evidenced by the "varied fauna containing large rudists and corals that depended on efficient water circulation for food supply, the abundance of coarse, abraded shell debris in the larger reefs, and shingle beach-like accretions fringing or overlying the larger reefs."

According to Stricklin and others (1971), the upper Glen Rose member is primarily a sequence of nonresistant beds of calcareous clay alternating with resistant beds of either dolomite, lime mudstones, or fine- to medium-grained limestones. Eight distinct stratigraphic units, including two evaporite intervals, are recognized

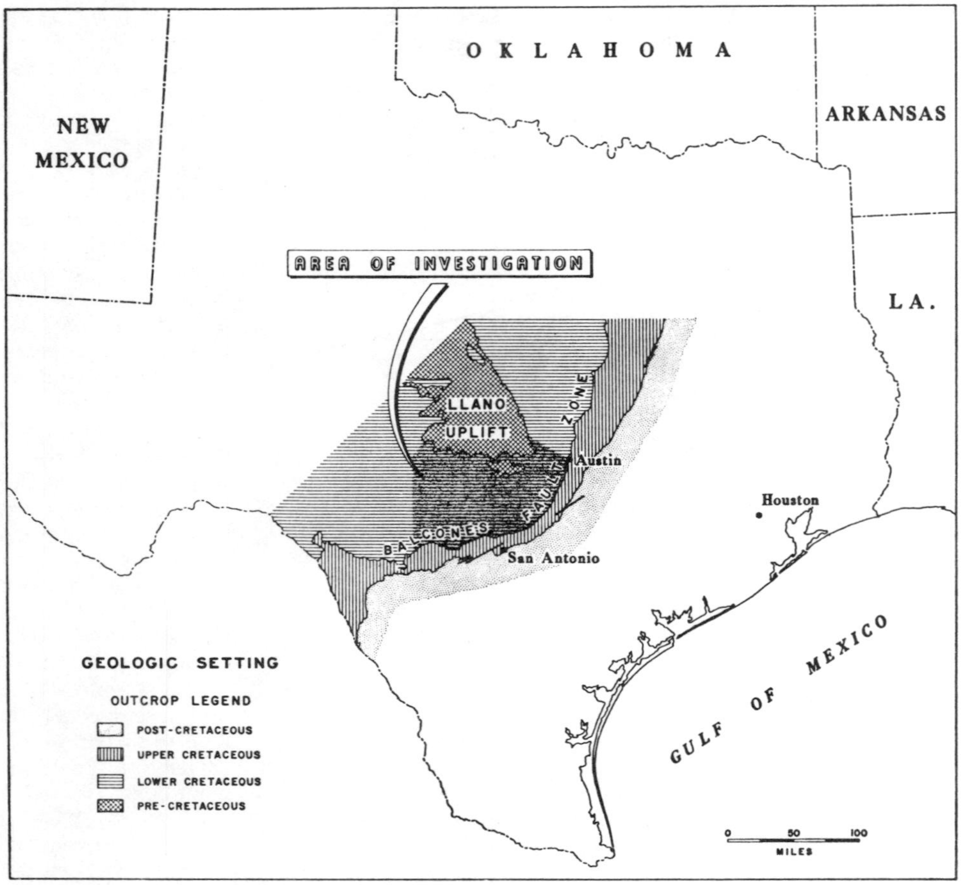

Figure 3. Geologic setting of Central Texas (from Stricklin and others, 1971).

within the upper Glen Rose. All these units have a high degree of lateral continuity. The carbonate beds apparently reflect deposition on a broad, gently sloping shallow sea floor that was frequently combined with an influx of clastic sediment (mainly clay) from the nearby Llano uplift. The evaporite beds indicate that the sea was at times hypersaline.

Stricklin and others (1971) state that Trinity strata in Central Texas were deposited "on a broad shelf with gentle seaward slope." Perkins reiterates the concept of a gently sloping featureless shelf specifically for the lower Glen Rose. According to Perkins (1974), "The lower Glen Rose deposits originated on a nearly featureless shelf that curved around the southeast side of the emergent Llano uplift and gently sloped away from the shoreline to the shelf edge at least 150 miles to the south and southwest. Although the most seaward parts of the area studied were within 50 miles of the early Glen Rose shoreline, most of the area lay within 25 miles of the shoreline.

"The geologic setting of this outcrop Glen Rose reef complex (north of the Balcones fault zone) differs from that of the "Deep Edwards" subsurface reefs of South Texas and the El Abra reefs of Mexico chiefly in its remoteness from the shelf edge. All of these reefs have similar rock types, fossil assemblages, facies relationships, and early diagenetic features, indicating similar environments and early geologic histories. The areas of reef growth nearer the shelf edge probably subsided more rapidly than did the areas closer to the shoreline, as is reflected in the much thicker local reef sequences in the Deep Edwards and El Abra reef trends. The individual reefs in these thicker sequences are not necessarily thicker than the more shoreward reefs, but commonly near the shelf edge, reef complexes of successive ages are superposed in an area to form a local reef sequence several hundred feet thick."

Perkins suggests that the lower Glen Rose reefs exposed at the surface on the San Marcos arch are on the shelf some 150 miles from the shelf edge. The term shelf edge implies a break in slope. Surface and subsurface studies indicate that a marked break in slope did not exist during lower Glen Rose time. Consequently, the outcrop reefs in Central Texas represent the highest energy, shallow-water sedimentation occurring during lower Glen Rose deposition. Seaward of these reefs, subsurface data indicate the presence of a lower energy, slightly deeper water mudstone and wackestone facies. This facies relationship has also been observed in isolated outcrops. Contrary to Perkins interpretation, a lower Glen Rose shelf margin is not known either from surface outcrops or from the subsurface. The deeper water sediments known seaward from the outcrop reefs suggest that these reefs were deposited on a ramp profile.

The ramp profile concept for the lower Glen Rose and the lower part of the upper Glen Rose has been interpreted in numerous north-south surface to subsurface cross sections in South and East Texas. This concept is illustrated in an outcrop cross section from northern Coahuila, Mexico (Smith and Bloxsom, 1974) (figs. 9, 10).

Smith and Bloxsom (1974) correlate the outcrop sequence in Coahuila with middle and upper Trinity rocks in

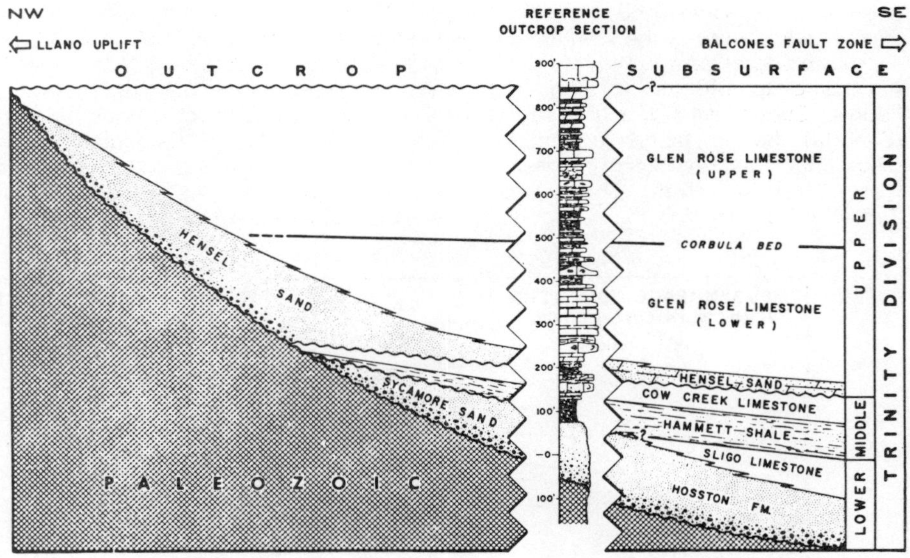

Figure 4. Stratigraphic diagram of the Trinity division (from Stricklin and others, 1971).

Central Texas and in the subsurface of South Texas. The top of the Trinity is well established paleontologically to coincide with the lowermost position of the lituolid foraminifer ***Barkerina barkerensis*** in shallow-water Glen Rose carbonate strata and at the top of the tintinnid ***Colomiella recta*** zone in the deeper water, upper Tamaulipas lime mudstones which have a pelagic fauna.

According to Smith and Bloxsom (1974), all of the Trinity rocks above the La Pena shales north of Cerro El Palomo are soft marls alternating with more resistant lime wackestones that produce the stairstep topographic profile that is typical of the upper Glen Rose in Central Texas. Bloxsom (1972) interprets these alternating beds as a platform-interior facies reflecting deposition in shallow, poorly circulated waters that was periodically interrupted by an influx of argillaceous material from land areas to the north, presumably the Llano uplift. A tongue of the upper Tamaulipas basinal facies splits the Glen Rose into a thinner, unnamed, lower shelf carbonate unit and a much thicker upper shelf carbonate unit at Cerro El Palomo.

The terrigenous facies decreases south of Cerro El Palomo and the lower part of the upper unit at the El Cedral section is a massive-bedded, sponge, carpinid, oyster, coated-grain, boundstone and wackestone that thickens and thins from about 100 to 75 feet in biohermal-biostromal alternations. The contact with the upper Tamaulipas is very abrupt. The biohermal-biostromal facies (called platform-margin facies by Smith and Bloxsom, 1974) is overlain by some 1,280 feet of strata alternating between platform-interior and platform-bank facies. The platform-bank facies consists of lime grainstones, packstones, and wackestones with miliolids, green algae, ***Dictyoconus***, and ***Toucasia*** thought to have been deposited in a shallow-water environment with unrestricted circulation.

The high-energy biohermal-biostromal facies, followed by an onlap sequence of the platform-bank facies, progrades seaward over upper Tamaulipas lime mudstones and fore-reef detritus. It climbs vertically 1,250 feet in 30 miles between the El Cedral section to the north and the San Rafael section to the south. The lime boundstone, wackestone, and grainstone biohermal facies is about 300 feet thick and occurs at the top of the Glen Rose at San Rafael; just below the massive biohermal buildup is a 200-foot sequence of grainstones interbedded with upper Tamaulipas pelagic mudstones.

The lower and part of the upper Glen Rose Formation in South Texas and northeastern Mexico was deposited during a slow, continuous rise in sea level characterized by a more pronounced horizontal progradation of high-energy deposition rather than by vertical buildup. The upper Tamaulipas Formation was deposited in a low-energy environment and below wave base and seaward of the high-energy biohermal facies. Sea-level rise must have been sufficiently slow during early Glen Rose time for Tamaulipas sedimentation to have essentially kept pace with that of the biohermal facies. The thicker buildup in the uppermost Glen Rose at the San Rafael section suggests that sea level may have been rising more rapidly at this time and the biohermal facies began to grow vertically with less horizontal progradation. Although sea level may have been rising at a faster rate, the ancestral Gulf of Mexico basin was also continuously subsiding.

Smith and Bloxsom's (1974) cross section in Mexico (fig. 9) shows the lower Glen Rose reefs connecting with upper Glen Rose reefs; the connection indicates a continuous progradation of younger reefs building on older reefs. The outcrops of these reefs in Central Texas (fig. 7) do not substantiate continuity of connecting reefs, rather they are distributed sporadically on a low-relief profile and do not form a continuous barrier. According to Heckel (1974), "Certain portions of a sea bottom may be more favorable than others for prolific growth of various types of carbonate secreting organisms. Favorable factors for organic proliferation normally include good nutrient and oxygen replenishment and may involve firm substrate, lack of terrigenous dilution, hydrodynamics and other factors." A surface lithofacies map (fig. 7) of these strata across the San Marcos platform shows the horizontal separation of lower Glen Rose reefs. McNamee (1969) also shows the separation of lower and upper Glen Rose reefs in the subsurface in East Texas and Louisiana (fig. 11).

FREDERICKSBURG-WASHITA DIVISIONS

Stratigraphy

During Fredericksburg and Washita time, the biohermal facies built upward some 1,300 feet in only 6 miles between the San Rafael and Los Ojos measured sections, as exemplified in cross section (fig. 12). The biohermal facies was able to keep up with a more rapid rising sea level, whereas the basinal facies did not receive sufficient sediments to keep pace; consequently, a depositional break in slope was created, which changed the profile of deposition from that of a ramp to a shelf. According to Smith and Bloxsom (1974), "the upper Trinity-Fredericksburg-Washita biohermal facies complex has been recognized in the subsurface east of the outcrop sections in Pemex wells on the Peyotes anticline and these same facies extend into south Texas as the Stuart City reef trend." The Stuart City (Edwards) reef trend developed as an almost continuous barrier around the ancestral Gulf of Mexico (fig. 13) during Fredericksburg and part of

Washita time; some 1,500 to 2,000 feet of relief between the shelf margin and basin was present in Washita time. According to Bloxsom (1972), the basinal Cuesta del Cura Formation (Washita) has a faunal assemblage suggesting "a significantly deeper water facies than the Upper Tamaulipas facies."

Neither outcrop nor subsurface stratigraphic and structural data nor seismic structural data suggest that the abrupt vertical shelf margin buildup was induced by a directly underlying structural hinge line or topography. A more accelerated rise in sea level seems a logical answer to the abrupt rapid vertical buildup. According to Hays and Pitman (1973), only 10 percent of the present land surface was covered by seas 130 million years ago, which is approximately coincident with the beginning of lower Glen Rose time. Twenty percent of the present land surface was covered by seas 110 million years ago and a maximum transgression occurred about 70 mil-

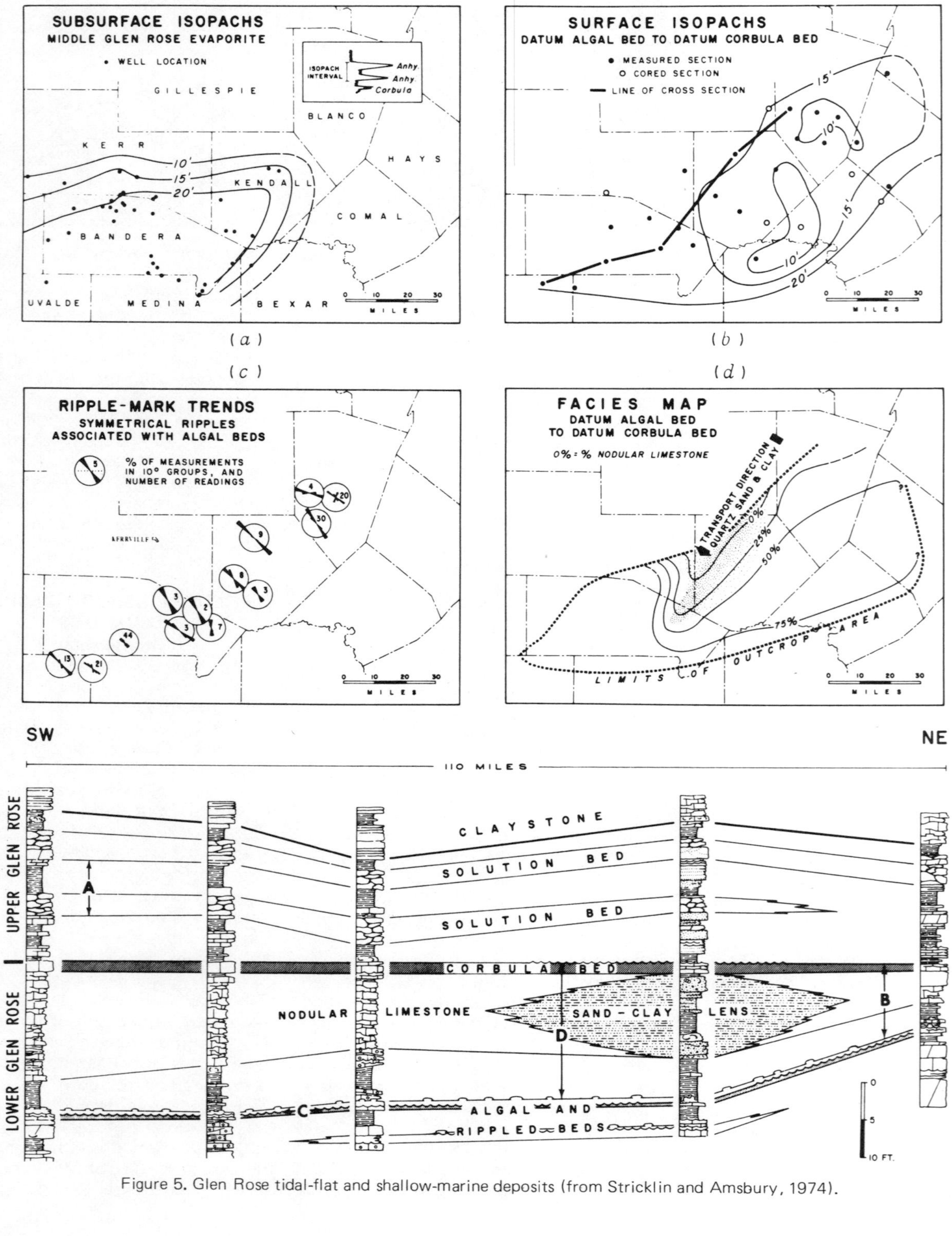

Figure 5. Glen Rose tidal-flat and shallow-marine deposits (from Stricklin and Amsbury, 1974).

lion years before present (MYBP) covering 45 percent of the land. The curve from Hays and Pitman (fig. 14) plotting age versus land area covered becomes much steeper between 110 MYBP and 70 MYBP, suggesting a more rapid rise in sea level. Gussow (1976) published charts by Vail and Mitchum (fig. 15) that also indicate a more accelerated rise in sea level at about 110 MYBP. Upper Trinity - lower Fredericksburg deposition began about 110 MYBP, and the more rapid rise in sea level roughly coincides with the vertical growth of the Edwards - Stuart City reef during upper Trinity, Fredericksburg, and Washita time. Hays and Pitman attribute rising sea level to more rapid sea-floor spreading with attendant expansion of mid-oceanic ridges, thus displacing more water onto land areas.

The high-energy shelf-margin caprinid reefs were able to maintain growth concomitant with a rapidly rising sea level and basin subsidence; because the basin received sediment at a more constant rate and in lesser amount, the reefs could only build upward and slightly basinward over their own debris. The reefs across South Texas were eventually drowned at the end of Washita time with continued sea level rise and were transgressed by deeper water Eagle Ford Shale. Caprinid reefs continued to grow in central Mexico through most of Eagle Ford time but were ultimately drowned by Aqua Nueva Shale (upper Eagle Ford) deposition.

The Comanche shelf was named by Rose (1972) for the vast shallow-water area of Fredericksburg-Washita carbonate deposition in Texas (fig. 16). The Central Texas platform, occupying a part of the ancestral positive area Concho arch - Llano uplift, separated the Maverick basin (Winter, 1962) on the southwest and the North Texas - Tyler (Fisher and Rodda, 1967) or East Texas basin on the north and northeast. The southeastern plunge of the Central Texas platform is known as the San Marcos platform (Adkins, 1933). The Stuart City reef trend separates the platform from the deeper waters of the ancestral Gulf of Mexico to the south. Tidal-flat carbonates and evaporites were deposited on the Central Texas platform during Fredericksburg time. The Antlers and Paluxy (lower Fredericksburg) sandstones were deposited from a northerly source in part of the East Texas basin and around the north, northeastern, and northwestern flanks of the Central Texas platform.

Fredericksburg and Washita stratigraphy on the Central Texas platform has been studied and described by Rose (1972) and correlated with the subsurface on the San Marcos platform. The Edwards Plateau area west of the platform and in the Maverick basin of Texas has been studied and described by Smith and Lozo (1964) and Freeman (1968);

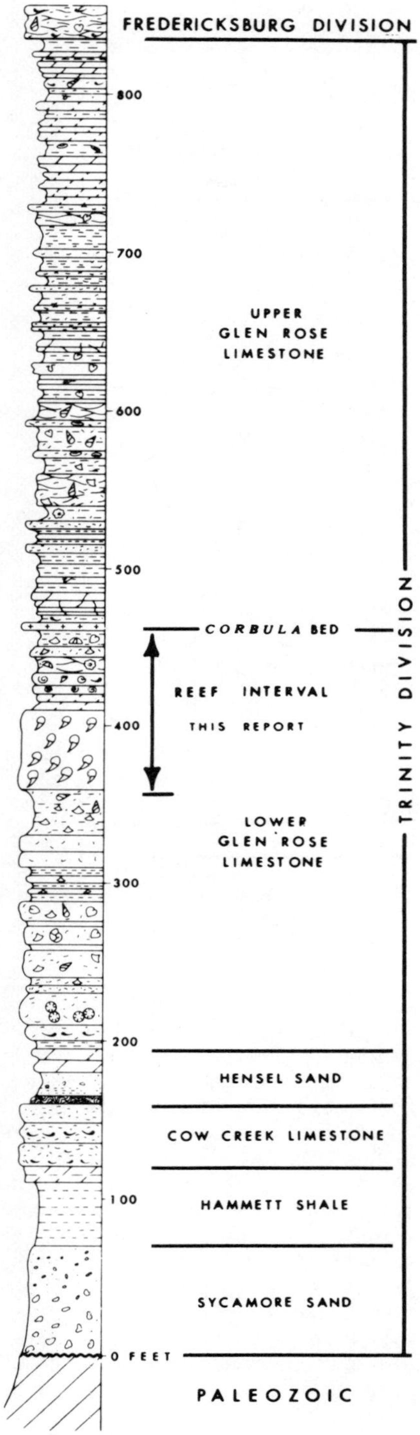

Figure 6. Generalized graphic section of the Trinity division, Central Texas (from Perkins, 1974).

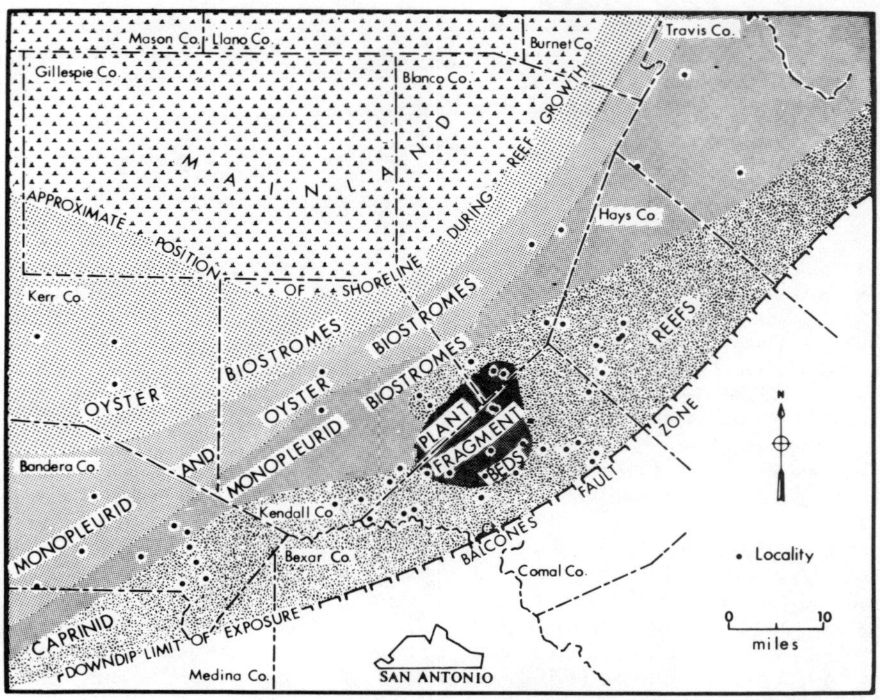

Figure 7. Distribution of the principal facies in the lower Glen Rose reef interval (from Perkins, 1974).

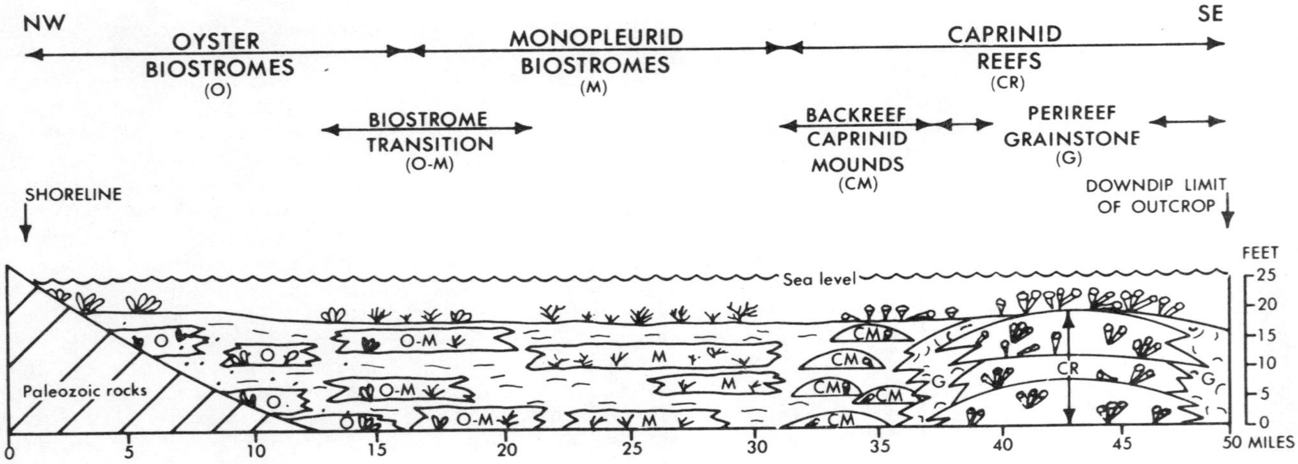

Figure 8. Schematic northwest-southeast cross section during lower Glen Rose reef interval time showing lateral zonation, bottom relief, and water depths (from Perkins, 1974).

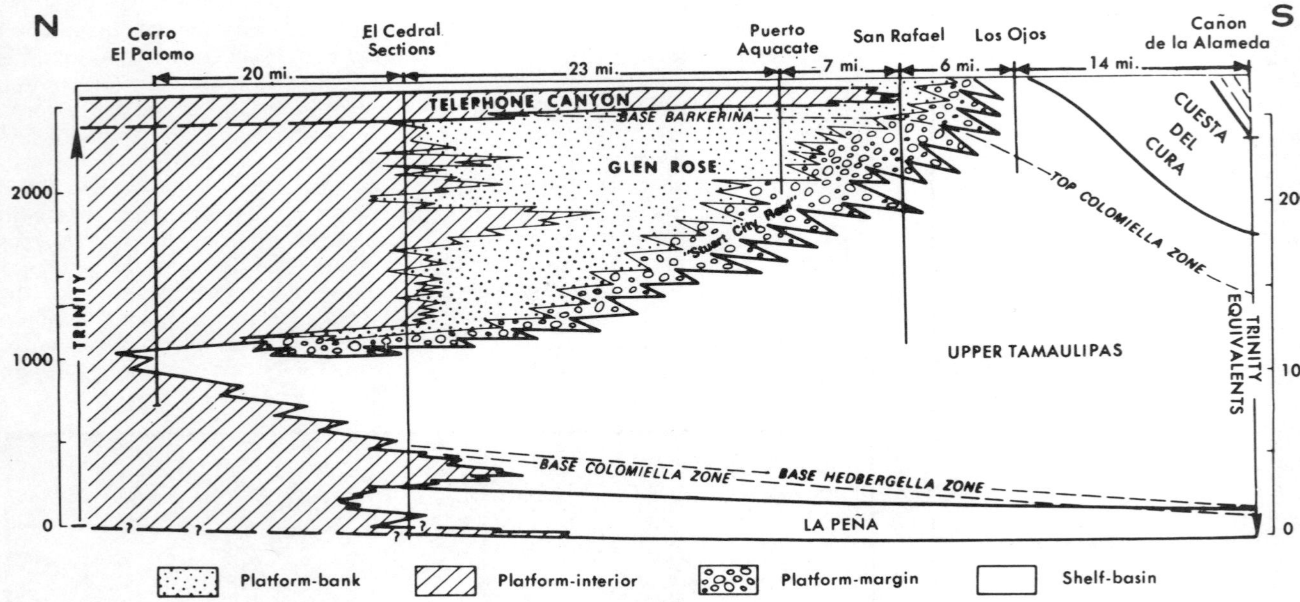

Figure 9. Upper Trinity stratigraphic cross section, northern Coahuila, Mexico (from Smith and Bloxsom, 1974).

these strata have been studied in Mexico by Smith (1970), Bloxsom (1972), and Smith and Bloxsom (1974). The area east of the platform was studied and reported on by Nelson (1959), Lozo (1949, 1959), Tucker (1962), Moore (1964), and Amsbury and others (1973).

Lower and middle Fredericksburg strata were studied on the northern part of the Edwards Plateau in the vicinity of Abilene, Sweetwater, and San Angelo in west-central Texas at the northwest end of the Central Texas platform by Moore (1967, 1969), Boutte (1969), E. Marcantel (1969), and J. Marcantel (1969). All these studies utilized detailed, measured and described outcrop sections to correlate and construct the regional stratigraphic framework.

Two regional cross sections based on work of the previously cited authors have been constructed to show the generalized stratigraphy of Fredericksburg and Washita divisions carbonates in parts of Texas and Mexico. Cross section A-A′ (fig. 17) begins at the northwest end of the Central Texas platform near Abilene and terminates at the Stuart City reef trend to the southeast. Cross section B-B′ (fig. 18) begins at the outcrop shelf margin in Coahuila, Mexico, crosses the Maverick basin and the Central Texas platform, and terminates on the northeast side of the platform adjacent to the East Texas basin.

High-energy, rudist-rich sedimentation of the Stuart City reef trend (Bebout and Loucks, 1974) continued at the shelf margin throughout Fredericksburg and Washita time. Sedimentation on the shelf and in adjacent basins was also more or less continuous but reflects a variety of depositional environments and consequently different rock types.

Rose (1972) referred to the shallow-water Fredericksburg and Washita carbonate strata on the eastern Edwards Plateau as the "Edwards" group and divided it into two formations. He named the lower formation Fort Terrett and the upper formation Segovia. The Fort Terrett Formation is subdivided into four informal members. These members in ascending order are (1) Basal nodular, (2) Burrowed, (3) Dolomitic, and (4) Kirschberg evaporite. The Fort Terrett Formation is all within the

Fredericksburg division; the overlying "Dr. Burt Ammonite" zone contains the youngest known Fredericksburg ammonites. Rose named the basal unit of the Segovia Formation the Burt Ranch member (the "Dr. Burt Ammonite Zone" is at the base of this member). Correlations were made in the upper part of the Segovia Formation by the recognition of four key beds which, in ascending order, are: (1) Allen Ranch breccia, (2) Gryphaea bed, (3) Orr Ranch bed, and (4) Black bed.

According to Rose (1972) the subsurface Kainer Formation of the San Marcos platform comprises the entire Fredericksburg division there, and the subsurface Person Formation is within the Washita division.

Fredericksburg Depositional History

The basal transgressive unit of the Fort Terrett Formation is a nodular, argillaceous lime wackestone and mudstone. The nodular appearance is the result of differential weathering caused by thorough burrowing and churning of the sediments by organisms and the argillaceous content of the rock. The unit, deposited below effective wave base in a shallow-water, low-energy, marine environment, is recognizable in the western Maverick basin (Smith, 1970) where it is named the Telephone Canyon Formation and on the southern and eastern Edwards Plateau and on the San Marcos platform where it is named the Basal Nodular member (Rose, 1972). The Basal Nodular member is laterally equivalent to the upper part of the Antlers Sandstone on the northwest side of the Central Texas platform; the upper part of the Antlers is equivalent to the Paluxy Sandstone on the northeast side of the platform, and the Paluxy interfingers to the south and is ultimately replaced by the Bull Creek member of the Walnut Formation (Moore, 1964; Moore and Martin, 1966).

The Burrowed member consists of lime mudstones with a sparse fauna of miliolids and mollusk fragments. These sediments were deposited below effective wave base in a shallow-water marine environment devoid of any argillaceous influx on the eastern Edwards Plateau. The Burrowed member becomes dolomitic in the subsurface to the south and interfingers with the more argillaceous Comanche Peak - Walnut Formations to the north (cross section B-B′, fig. 18). This unit changes facies to grainstones around the periphery of the Maverick basin (cross section A-A′, fig. 17) where it becomes part of the Devils River Formation and is equivalent to the Walnut and Comanche Peak Formations on the northeast side of the platform.

The Dolomitic member is recognizable over the entire platform and consists of beds of dolomitized shallow-marine grainstones and packstones alternating with dolomitized tidal-flat and supratidal mudstones. This entire unit changes facies from dolomitized mudstones to lime grainstones around the periphery of the Maverick basin, and the lower part of the unit consists of grainstones at the northwestern end of the Central Texas platform in the vicinity of Abilene. The upper part of the Dolomitic member and the overlying Kirshberg evaporites are replaced by grainstones at Moffatt mound on the northeast side of the platform.

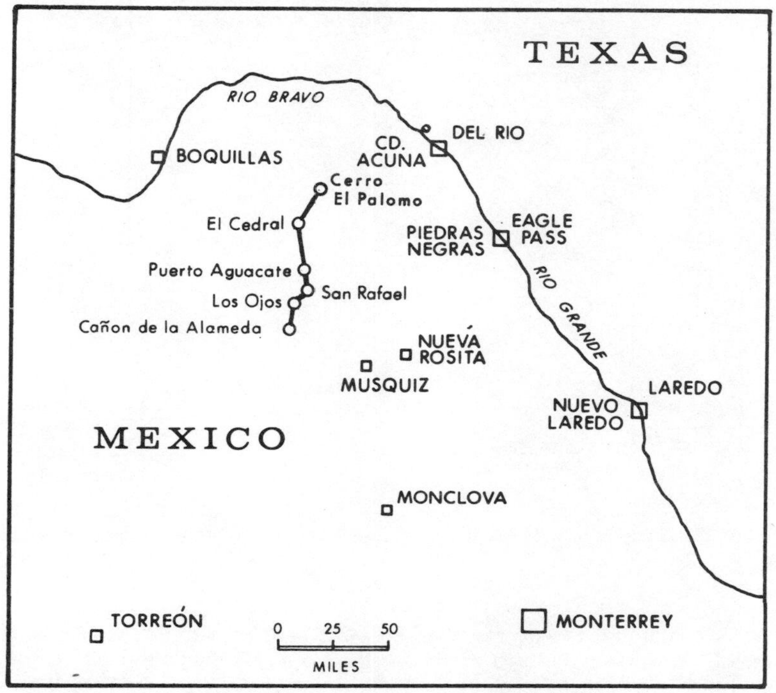

Figure 10. Location map with section localities (from Bloxsom, 1972).

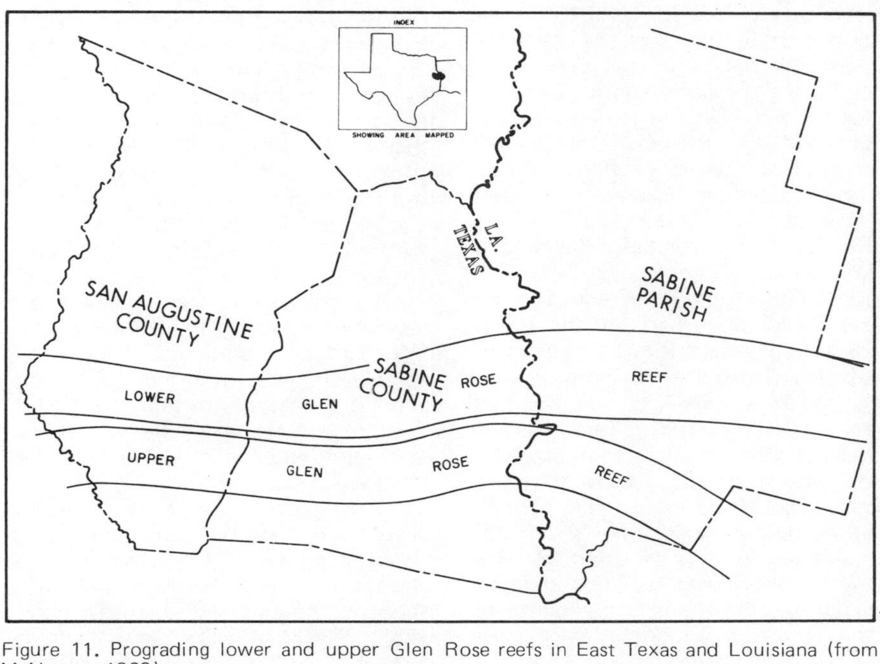

Figure 11. Prograding lower and upper Glen Rose reefs in East Texas and Louisiana (from McNamee, 1969).

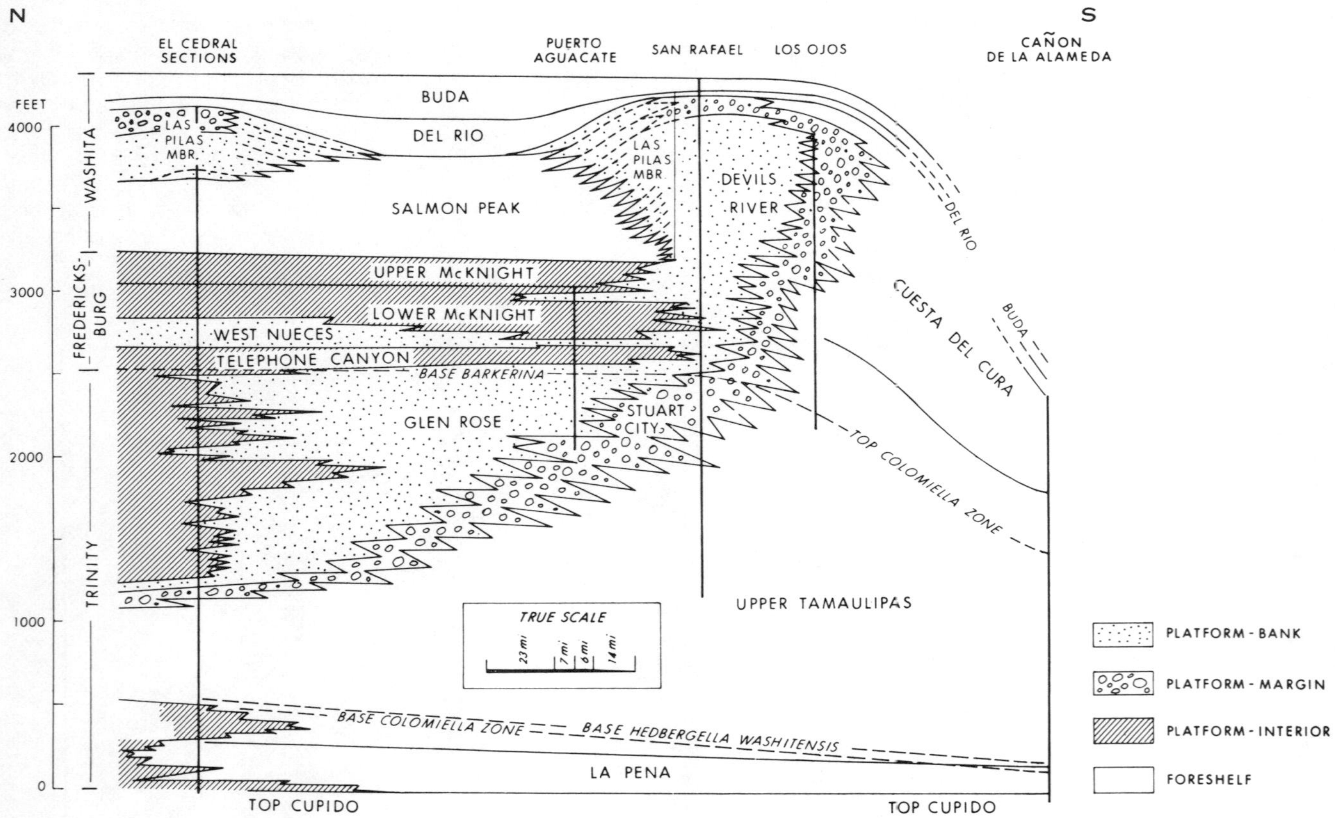

Figure 12. Upper Trinity, Fredericksburg, and Washita stratigraphic cross section, northern Coahuila, Mexico (from Bloxsom, 1972).

The Kirshberg evaporite member was deposited over a large area of the Central Texas platform. A paleoenvironment map (fig. 19) restored at the top of the Kirshberg and its lateral equivalents shows the probable original extent of the evaporite member. Gypsum is found presently only in the vicinity of Fredericksburg, Gillespie County, and Menard, Menard County (Barnes, 1944). Elsewhere, the formation is recognized as a collapse breccia or as recrystallized, box-work-textured limestone. The lower part of the McKnight Formation began forming in toxic, euxinic waters in the Maverick basin during middle Fredericksburg time. The upper part of the lower McKnight is anhydritic and appears to be equivalent to the Kirshberg collapse breccia as exposed at the Leakey section (fig. 18). The upper Fredericksburg Moffatt mound trend along the northeast side of the platform (figs. 19, 20) formed a high-energy, linear grainstone shoal approximately 5 miles wide and 50 miles long; the shoal separated tidal-flat and evaporite sediments on the platform from low-energy, argillaceous lime mudstone deposited below effective wave base in slightly deeper water as the upper Comanche Peak - Goodland Formations in the East Texas basin.

Sedimentation on the Central Texas platform changed after Kirshberg evaporite deposition. Lime mudstones were deposited in normal-marine water. Shortly after the mudstones were lithified, the underlying evaporites were partially dissolved, creating collapse breccia of the carbonate interbeds within the evaporites and deforming the overlying mudstones. Nodular chert beds present in the collapse breccia formed very early. The seas subsequently deepened and Fredericksburg time was concluded by deposition of marly mudstones of the "Dr. Burt Ammonite Zone" over the northwest end of the platform, miliolid grainstones in topographically high areas in the lee of the Stuart City reef trend, and rudist reefs around the Maverick basin.

According to Rose (1972), "Eventually a brief still-stand produced a broad, flat surface of by-passing, submarine exposure and alteration on the axis and northeast flank of the San Marcos Platform. This flat surface of by-passing was then covered by open-shelf marine seas of shallow to moderate depth, signaling the end of the Fredericksburg and the beginning of Washita time."

Washita Depositional History

Rudist reefs and bioclastic grainstone continued to develop in the Stuart City reef trend and around the periphery of the Maverick basin. Argillaceous lime mudstones were deposited on the Central Texas platform reflecting a low-energy, slightly deeper water environment. These mudstones are named the Regional Dense member of the Person Formation in the subsurface and the upper part of the Burt Ranch member of the Segovia Formation on the outcrop. An exact time correlation cannot be made because of the lack of a diagnostic ammonite fauna in these strata, but physical stratigraphic evidence suggests at least a partial time equivalency of the Regional Dense member, the upper part of the Burt Ranch member, the Kiamichi Shale in the East Texas basin, and thinly laminated, organic-rich lime mudstones ("oil shales") in the lower part of the upper McKnight in the Maverick basin. A pelagic fauna found in the McKnight mudstones strongly suggests that the Maverick basin was deepening.

According to Rose (1972), "Several related tectonic and depositional events now followed:

(1) The Karnes and Atascosa troughs began to subside actively.

(2) The subsidence rate of the southwest flank of the Central Texas Platform began not only to increase but also to accelerate progressively southwestward.

(3) The axis of the San Marcos Platform began to rise very gently.

(4) Clinoforms were established on both flanks of the Central Texas Platform at the margins of the Maverick and North Texas - Tyler basins.

(5) The gentle uplift of the Devils River belt that began during late Fredericksburg continued, insuring the permanent shift of the primary belt of rudist-bank deposition across the Maverick basin to the Devils River belt. Some rudist-bank deposition continued, however, on the slowly submerging southwestern end of the Stuart City Reef.

(6) Despite the overall retreat of shallow-water facies tracts toward the axis of the Central Texas Platform, a considerable area of tidal flats remained on the axis of the San Marcos Platform in the lee of the Stuart City Reef and a large sinuous area of restricted shallow marine deposits followed the Central Texas Platform northwestward from the Stuart City to the Llano Uplift, where it deflected southwestward, faithfully reflecting the influence of the Medina axis. Dolomitization was an active contemporary process in both facies tracts."

The Central Texas platform became shallow and partially emergent following the deposition of the Burt Ranch member. The Allen Ranch evaporites (now collapse breccia) and associated tidal flats probably reflect sabkha conditions as does the collapse breccia member in the subsurface part of the San Marcos platform. Grainstones and tidal-flat dolomite sepa-

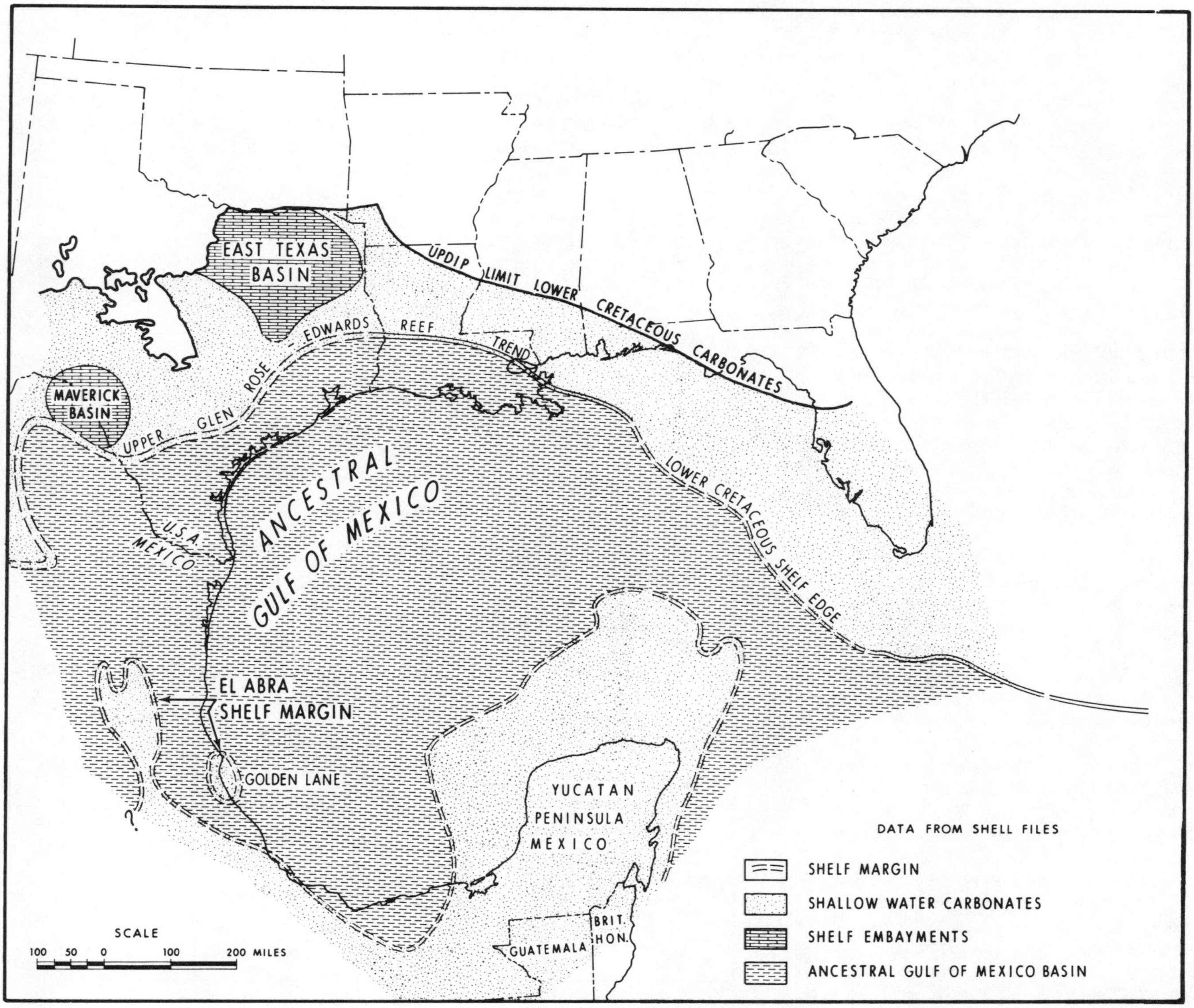

Figure 13. Edwards-El Abra carbonate shelves and basins.

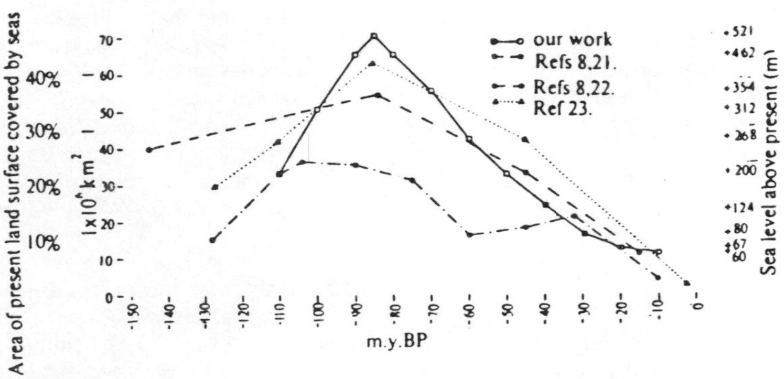

Figure 14. Area of present-day land covered by former seas plus the height of these seas above present levels plotted against age (from Hays and Pittman, 1973).

rated the evaporites from marly mudstones with ammonites to the north. The upper McKnight evaporites were forming at the same time, but again associated pelagic faunas suggest that these strata may have been deposited in deeper euxinic waters of the subsiding Maverick basin. Argillaceous lime mudstones of the lower part of the Georgetown Formation were deposited in the East Texas basin, and the ever-present rudist and grainstone banks continued to form along the Stuart City reef trend and in the Devils River Formation around the periphery of the Maverick basin.

Pelagic lime mudstones of the Salmon Peak Formation were subsequently deposited in the rapidly subsiding Maverick basin, while similar but argillaceous lime carbonates of the Georgetown Formation were deposited in the East Texas basin. Rudist wackestones and grainstones of the Los Pilas member of the Devils River Formation began to prograde over Salmon Peak mudstones, subsequently reducing the areal extent of the Maverick basin. Tidal-flat sediments continued to form in the lee of the Stuart City reef as the San Marcos platform remained emergent.

According to Rose (1972), the San Marcos platform eventually became almost completely emergent and the Cyclic and Marine members of the Person Formation were eroded from the axis of the platform but were preserved in the subsiding Karnes trough. Concurrently, a soil zone was forming on the northwestern part of the Edwards Plateau. Ultimately, as sea level continued to rise, pelagic lime mudstones were deposited first on the flanks and finally covered the southeastern part of the platform. The Stuart City reef trend probably became an isolated group of islands during the remainder of Washita time. Meanwhile the Del Rio Clay covered most of the platform and the Maverick basin and Washita sedimentation culminated with the deposition of a sheetlike lime mudstone, the Buda Limestone. The entire shelf, including the shelf-margin Stuart City reef was covered by the Eagle Ford Shale.

RECOGNITION OF STRATIGRAPHIC MODELS FROM REFLECTION SEISMIC PROFILES

Two regional stratigraphic models have been described: (1) the lower Glen Rose, and (2) the upper Glen Rose, Fredericksburg, and Washita shelf margin. Seismic data indicate that sufficient density and acoustical contrasts are present within these strata to differentiate these models on the basis of reflection character.

The lower Glen Rose reef progradations appear to be separated both horizontally and vertically. The internal zonation of an outcrop reef complex (fig. 21a) and the envisioned prograding character of these reefs (fig. 21b) are shown diagrammatically. An interpreted seismic profile of Lower Cretaceous carbonates (fig. 22) shows a series of prograding carbonate strata similar to the model described. The geometry of a specific buildup is indicated by abrupt changes in slope between the more or less horizontal reflectors of landward shallow-water carbonates and the slightly deeper water, seaward carbonates. The seaward termination of each slope should indicate the seaward limit of rudist mound complexes.

The more vertical character of the Stuart City shelf margin is shown in outcrop section (fig. 23a) and depicted diagrammatically (fig. 23b); the vertical character is attributed to both a more rapid rise in sea level and a subsiding basin.

The Stuart City shelf margin has long been recognized on seismic profiles because of the abrupt change in slope of the reflector emanating from the top of the Buda Limestone. A

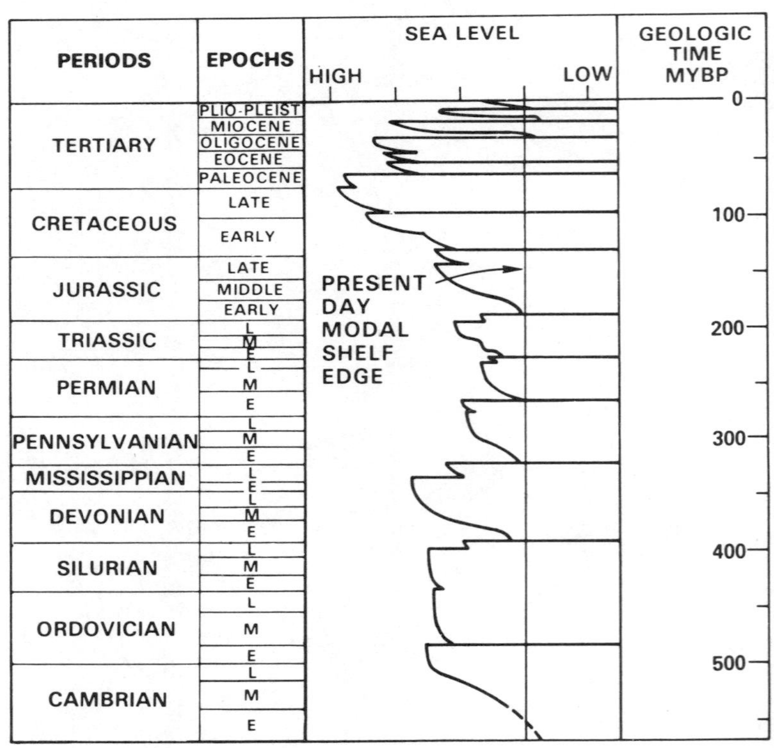

Figure 15. Eustatic supercycle chart of the Phanerozoic (from Gussow, 1976; chart by Vail and Mitchum).

high-resolution seismic profile (fig. 24) shows a continuous reflector from the shelf, across the shelf margin, and into the basin; it also shows more or less parallel deeper reflections. The interval between the top Buda and top Pearsall seismic events is thicker at the margin than either on the shelf or in the basin.

Although cross sections usually indicate one continuous reef buildup at the shelf margin, outcrop data from the Ciudad Valles area, San Luis Potosi, Mexico, suggest that the complex consists of a series of rudist mounds and related grainstones some 30 to 40 feet thick growing on top of older mound complexes. Ultimately these smaller mound complexes when stacked may attain a thickness of several thousand feet. The subsurface Stuart City shelf-margin facies zonation is vividly portrayed in a series of shelf - shelf-margin cross sections by Bebout and Loucks (1974). Although the seismic profile across the Stuart City reef (fig. 23) shows only one reef edge, the internal structure of the shelf margin is envisioned to be similar to that portrayed by these authors and to the El Abra margin of approximately the same age in the Golden Land Trend in Mexico (fig. 25). Internal reflectors within the El Abra margin complex, labeled "fossil reef fronts" by Rockwell and Rojas (1953) indicate the presence of several smaller slope breaks within the shelf margin. The shelf and basinal strata are recognized by zones of more or less parallel, horizontal reflectors on each side of the margin. The onlap of Tertiary reflectors onto the slope is also very pronounced. This entire profile has been documented by well control.

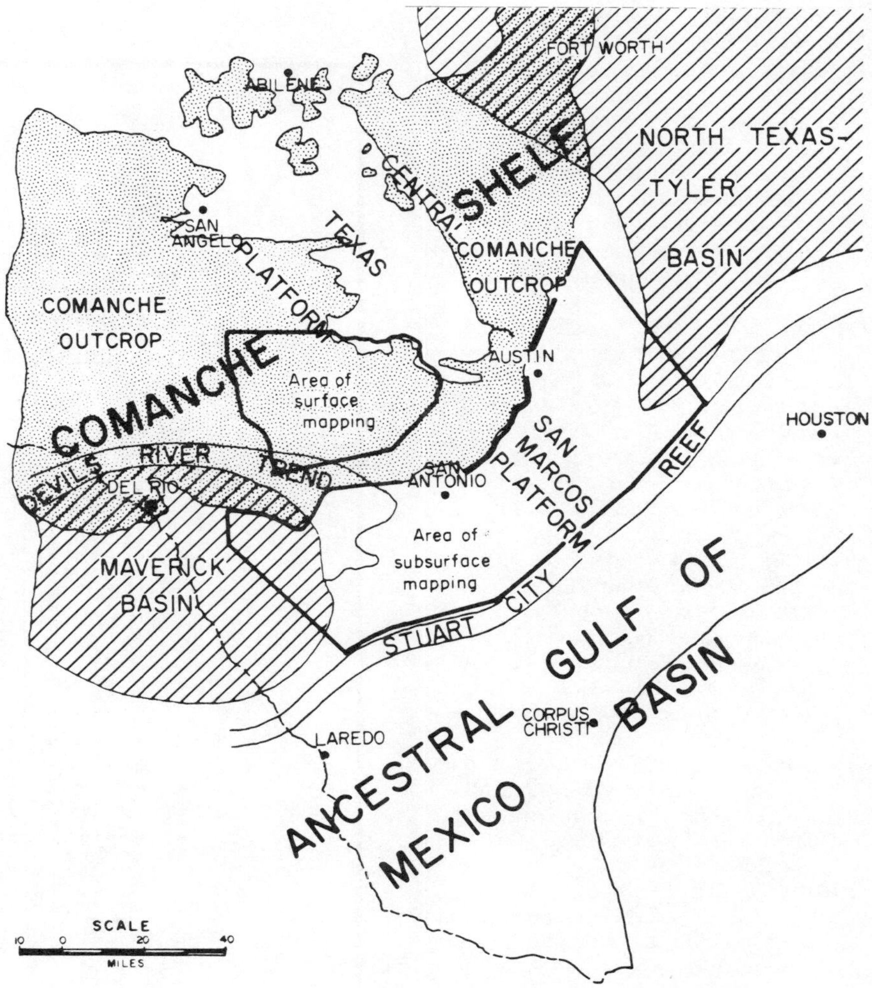

Figure 16. Regional elements and areas of investigation (from Rose, 1972).

ACKNOWLEDGMENTS

This report is a synthesis of data from published and unpublished reports of many people. The relating of stratigraphic models to seismic reflection profiles is the author's interpretation, also based in part on published and unpublished data of others.

Dr. W. S. Adkins should be acknowledged first because he encouraged and supported Cretaceous stratigraphic studies of Dr. F. E. Lozo, Shell Development Company. Frank Lozo in turn followed a planned program of research that included the systematic mapping of the regional stratigraphy of Lower Cretaceous carbonate strata where they crop out in Texas and northern Mexico. Among the workers on the Shell Development Company project were F. L. Stricklin, Jr., C. I. Smith, B. F. Perkins, D. L. Amsbury, C. H. Moore, Jr., K. G. Martin, and J. B. Brown. P. R. Rose related the surface stratigraphy in a part of Central Texas with the subsurface in South Texas. W. E. Bloxsom made similar detailed studies in northern Mexico. I have drawn freely on the work of all the forenamed geologists and gratefully acknowledge their contribution to this synthesis.

Sincere appreciation is expressed to D. S. Haglund, F. E. Lozo, R. L. Nicholas, F. L. Stricklin, Jr., and R. H. Waite who edited the manuscript and to Shell Oil Company for permission to publish.

REFERENCES

Adkins, W. S., 1933, The Mesozoic System in Texas, *in* The geology of Texas, Volume 1, Stratigraphy: Univ. Texas Bull. 3232, p. 239-517.

Ahr, W. M., 1973, The carbonate ramp: An alternative to the shelf model: GCAGS Transactions, v. 23, p. 221-225.

Amsbury, D. L., Bay, T. A., and Lozo, F. E., 1973, Lower Cretaceous strata in Central Texas, *in* A field guide to the Moffatt Mound Area near Lake Belton, Bell County, Texas: Houston Geological Society, 21 p.

Barnes, V. E., 1944, Gypsum in the Edwards Limestone of Central Texas: Univ. Texas Bull. 4301, p. 35-46.

Bebout, D. G., and Loucks, R. G., 1974, Stuart City Trend, Lower Cretaceous, South Texas, A carbonate shelf-margin model for hydrocarbon exploration: Univ. Texas, Austin, Bureau of Economic Geology Report of Investigations 78, 80 p.

Behrens, E. W., 1965, Environment reconstruction for a part of the Glen Rose Limestone, Central Texas: Sedimentology, v. 4, p. 65-111.

Bloxsom, W. E., 1972, A Lower Cretaceous (Comanchean) prograding shelf and associated environments of deposition, northern Coahuila, Mexico: Univ. Texas, Austin, Master's thesis (unpublished).

Boutte, A. L., 1969, Callahan carbonate-sand complex, west-central Texas, *in* A guidebook to the depositional environments and depositional history of Lower Cretaceous shallow shelf carbonate sequence, west-central Texas: Dallas Geological Society, p. 40-74.

Dunham, R. J., 1962, Classification of carbonate rocks according to depositional texture, *in* Ham, W. E., Classification of carbonate rocks: AAPG Memoir 1, p. 108-121.

Enos, P., 1975, Tamabra Limestone of Poza Rica Trend, *in* Annual Meeting Abstracts: AAPG and SEPM, v. 2, p. 91.

Fisher, W. L., and Rodda, P. U., 1969, Edwards Formation (Lower Cretaceous) Texas: Dolomitization in a carbonate platform system: AAPG Bull., v. 53, p. 55-72. *Reprinted as* Univ. Texas, Austin, Bureau of Economic Geology Geological Circ. 69-1.

Freeman, V. L., 1968, Geology of the Comstock - Indian Wells Area, Val Verde, Terrell, and Brewster Counties, Texas: USGS Professional Paper 594-K, p. 1-26.

Griffith, L. S., Pitcher, M. G., and Rice, G. W., 1969, Quantitative environmental analysis of a Lower Cretaceous reef complex, *in* Friedman, G. M., ed., Depositional environments in carbonate rocks, A symposium: SEPM Special Pub. 14, p. 120-138.

Gussow, W. C., 1976, Sequence concepts in petroleum engineering: Geotimes, v. 21, no. 9, p. 17.

Hays, J. D., and Pitman, W. C., III, 1973, Lithospheric plate motion, sea level changes and climatic and ecological consequences: Nature, v. 246, no. 5427, p. 18-22.

Heckel, P. H., 1974, Carbonate buildups in the geologic record: A review, *in* Laporte, L. F., ed., Reefs in time and space, selected examples from the recent and ancient, A symposium: SEPM Special Pub. 18, p. 90-154.

Hendricks, L., ed., 1967, Comanchean (Lower Cretaceous) stratigraphy and paleontology of Texas: SEPM Permian Basin Section, Pub. no. 67-8, 410 p.

Kerr, R. S., 1976, Development and diagenesis of a Lower Cretaceous bank complex Edwards Limestone, north-central Texas: Univ. Texas, Austin, Master's thesis, 203 p. (unpublished).

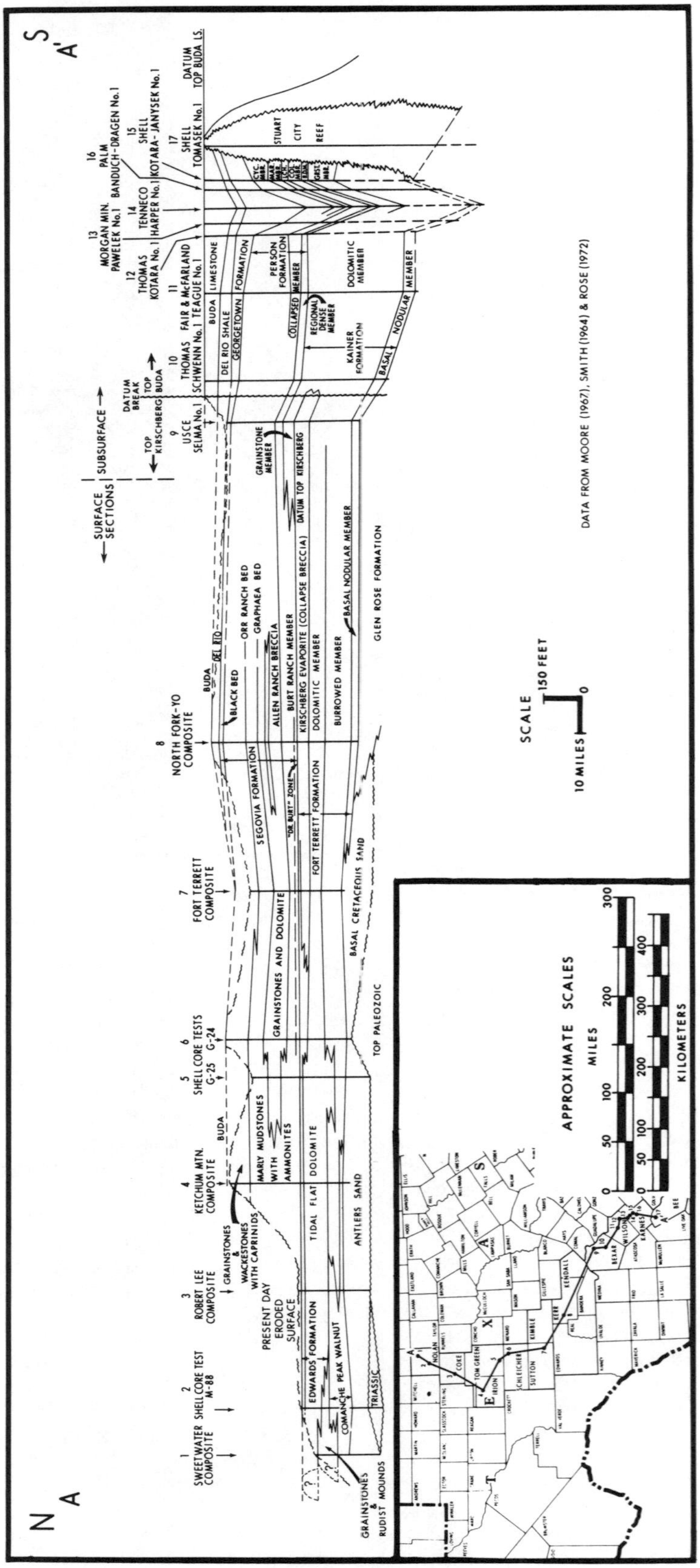

Figure 17. Cross section A-A′, Fredericksburg and Washita surface to subsurface stratigraphy, Nolan County to Bee County, Texas.

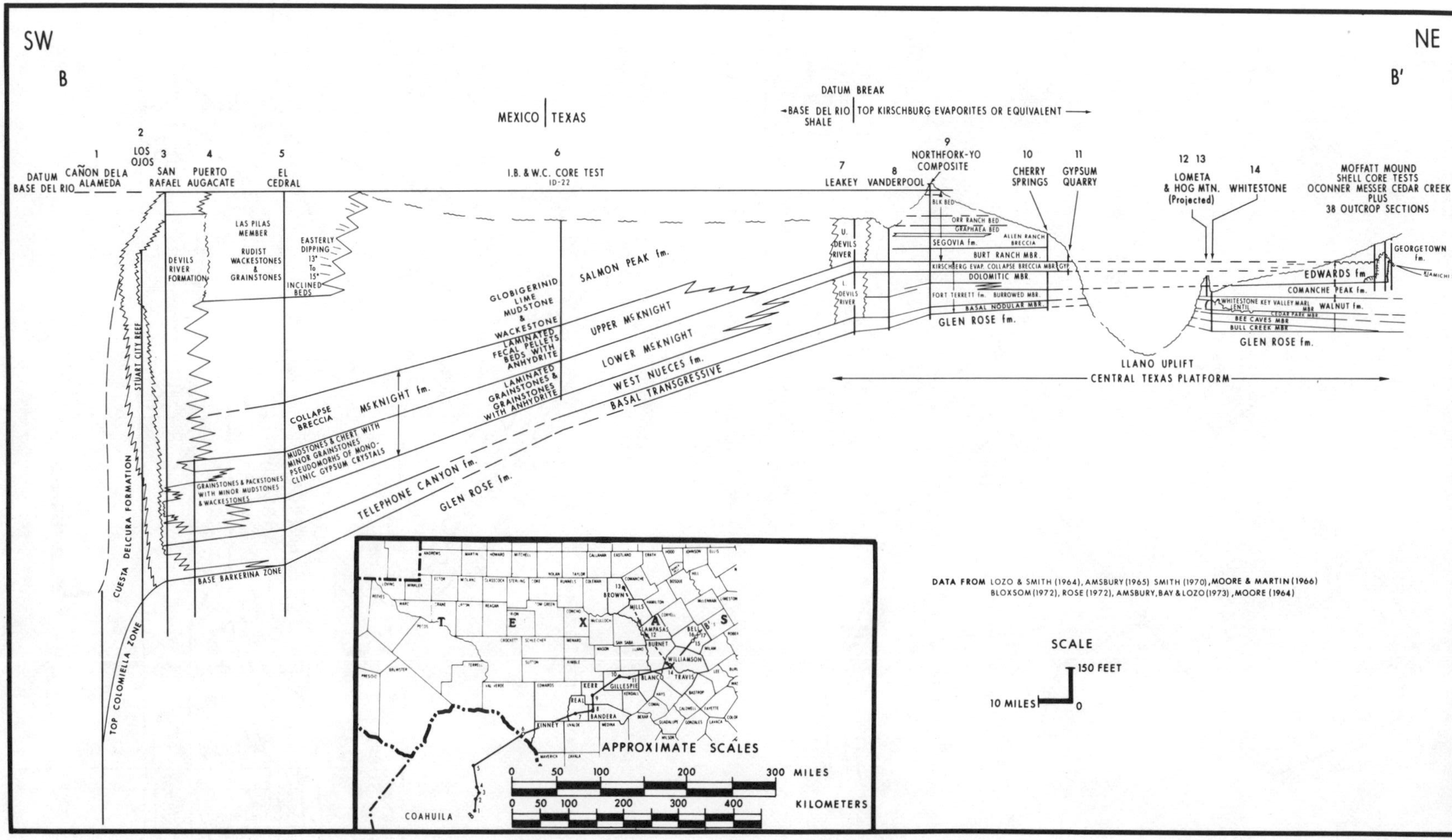

Figure 18. Cross section B-B', Fredericksburg and Washita stratigraphy, Coahuila, Mexico, to Bell County, Texas.

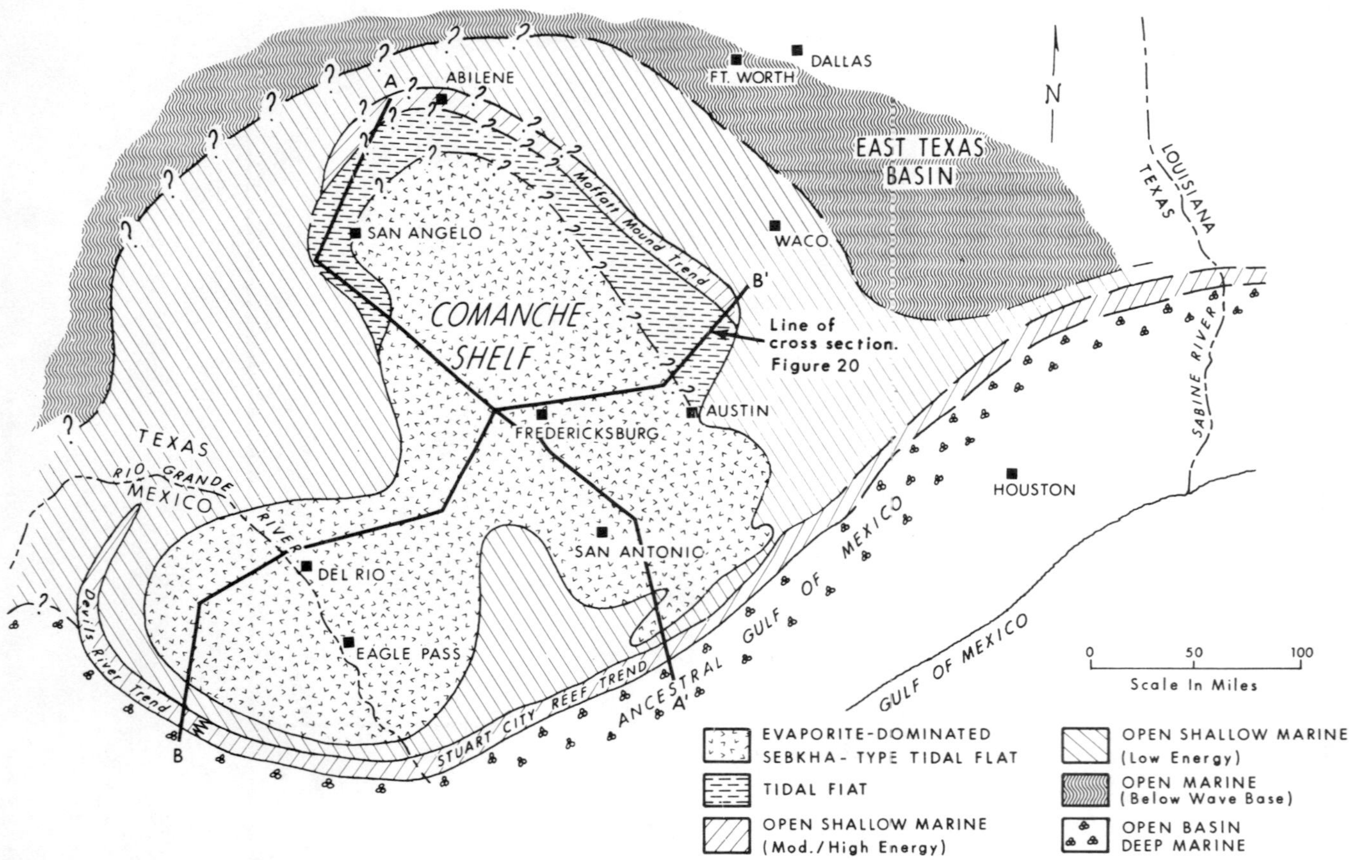

Figure 19. Generalized, restored paleoenvironment map near top of Fredericksburg-Kirschberg and lateral equivalents.

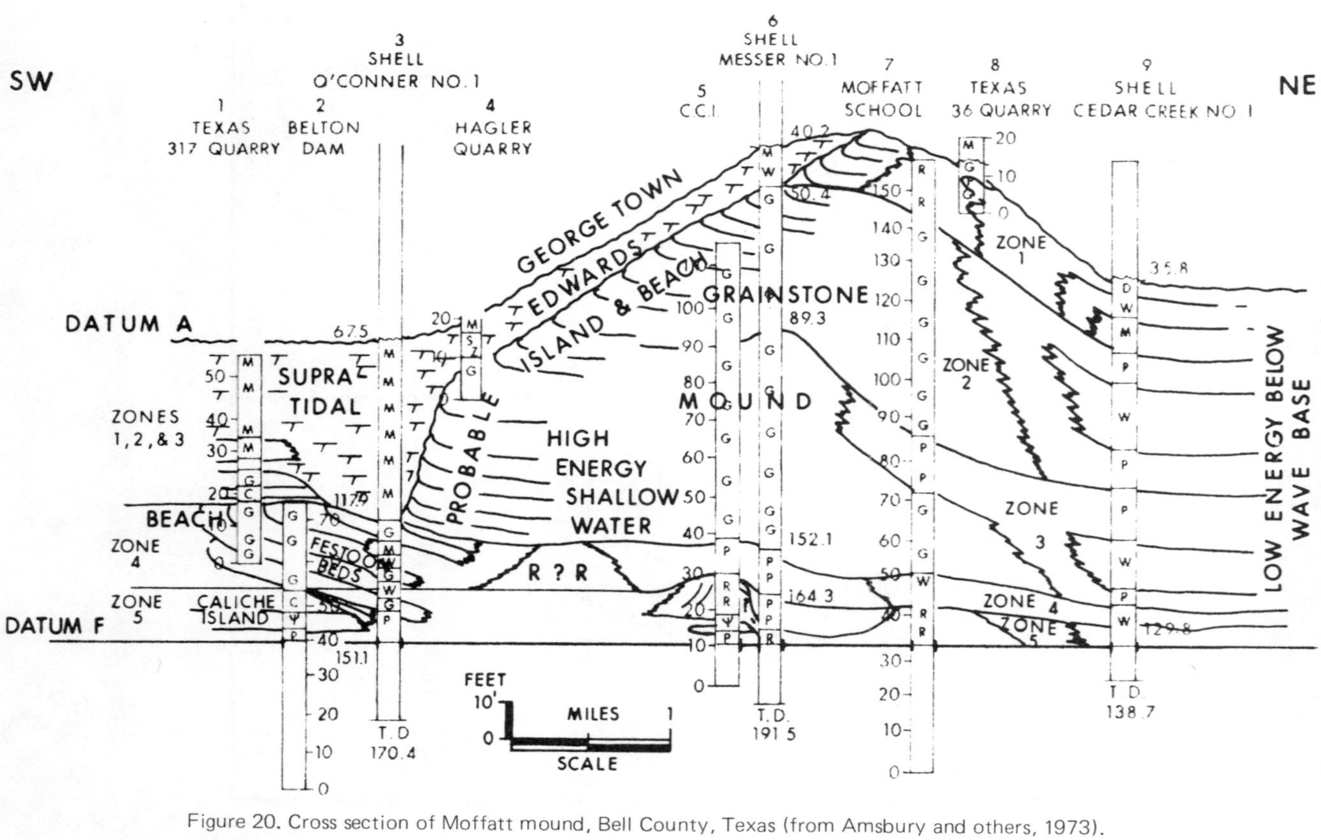

Figure 20. Cross section of Moffatt mound, Bell County, Texas (from Amsbury and others, 1973).

Laporte, L. F., 1968, Recent carbonate environments and their paleoecologic implications, *in* Drake, E. T., ed., Evolution and Environment: Yale Univ.

Lozo, F. E., 1949, Stratigraphic relations of Fredericksburg Limestones, north-central Texas: Shreveport Geological Society Guidebook, 17th Annual Field Trip (Cretaceous of Austin, Texas, area), p. 85-91.

______ 1959, Stratigraphic relations of the Edwards Limestone and associated formations in north-central Texas, *in* Symposium on Edwards Limestone in Central Texas: Univ. Texas Pub. 5905, p. 1-19.

______ , and Smith, C. I., 1964, Revision of Comanche Cretaceous stratigraphic nomenclature, southern Edwards Plateau, Southwest Texas: GCAGS Transactions, v. 14, p. 285-307.

______ , and Stricklin, F. L., Jr., 1956, Stratigraphic notes on the outcrop basal cretaceous, Central Texas: GCAGS Transactions, v. 6, p. 67-78.

Marcantel, E. L., 1969, The Skelly-Hobbs rudist reef complex, *in* A guidebook to the depositional environments and depositional history of Lower Cretaceous shallow shelf carbonate sequence, west-central Texas: Dallas Geological Society, p. 19-39.

Marcantel, J. B., 1969, The origin and distribution of Lower Cretaceous dolomites in west central Texas, *in* A guidebook to the depositional environments and depositional history of Lower Cretaceous shallow shelf carbonate sequence, west-central Texas: Dallas Geological Society, p. 75-104.

McNamee, D. F., 1969, The Glen Rose reef complex of East Texas and Central Louisiana: GCAGS Transactions, v. 19, p. 11-21.

Moore, C. H., Jr., 1964, Stratigraphy of the Fredericksburg division, south-central Texas: Univ. Texas, Bureau of Economic Geology Report of Investigations 52, 48 p.

______ 1967, Stratigraphy of the Edwards and associated formations, west-central Texas: GCAGS Transactions, v. 17, p. 61-75.

______ 1969, Stratigraphic framework, Lower Cretaceous, west-central Texas, *in* A guidebook to the depositional environments and depositional history of Lower Cretaceous shallow shelf carbonate sequence, west-central Texas: Dallas Geological Society, p. 1-18.

______ , and Martin, K. G., 1966, Comparison of quartz and carbonate shallow marine sandstones, Fredericksburg Cretaceous, Central Texas: AAPG Bull., v. 50, no. 5, p. 981-1000.

Nelson, H. F., 1959, Deposition and alteration of the Edwards Limestone, Central Texas, *in* Symposium on Edwards Limestone in Central Texas: Univ. Texas Pub. 5905, p. 21-86.

Perkins, B. F., 1969, Rudist faunas in the Comanche Cretaceous of Texas, *in* The Comanchean stratigraphy of the Fort Worth-Waco-Belton area, Texas: Shreveport Geological Society Guidebook 23d Annual Field Trip, p. 121-137.

______ 1974, Paleoecology of a

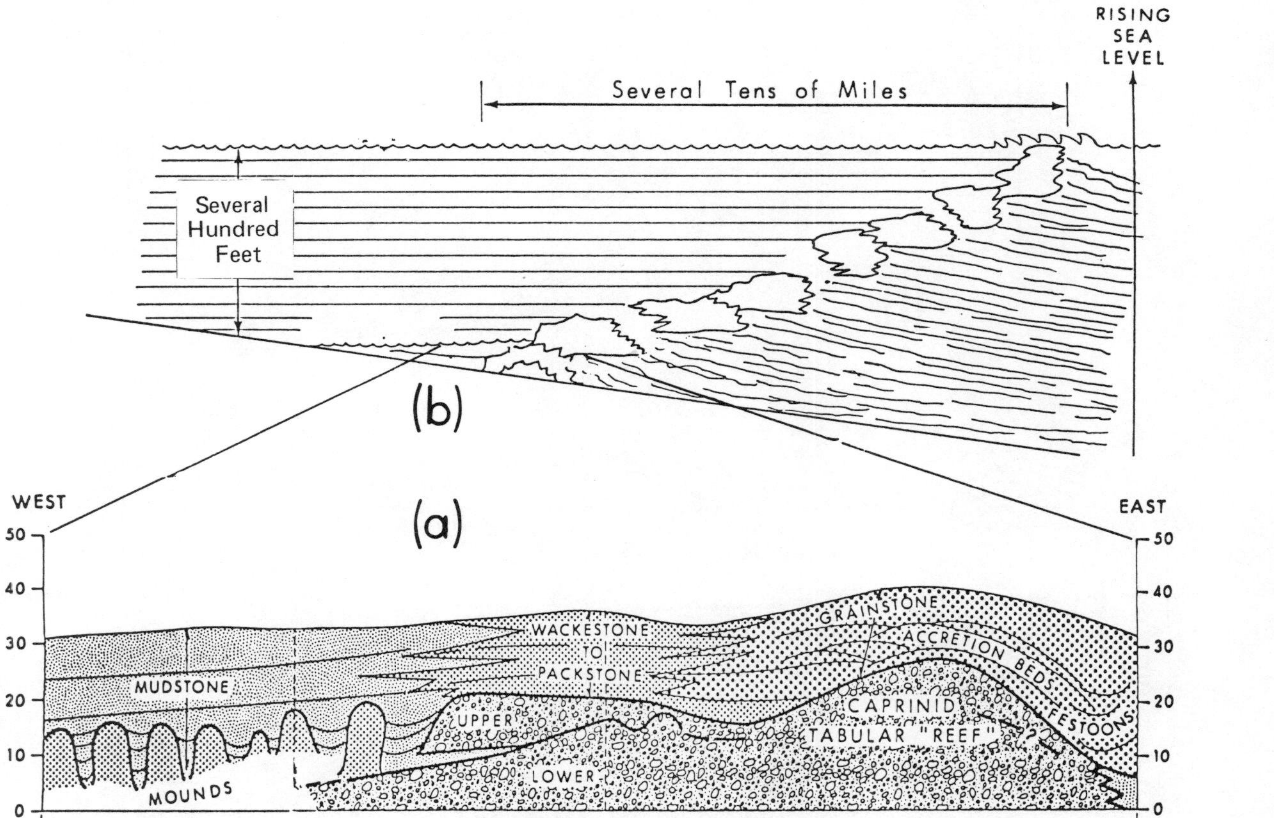

Figure 21. (a) Facies cross section of Glen Rose reef at Pipe Creek locality, Bandera County, Texas. (b) Diagrammatic cross section of prograding lower Glen Rose rudist reef complex.

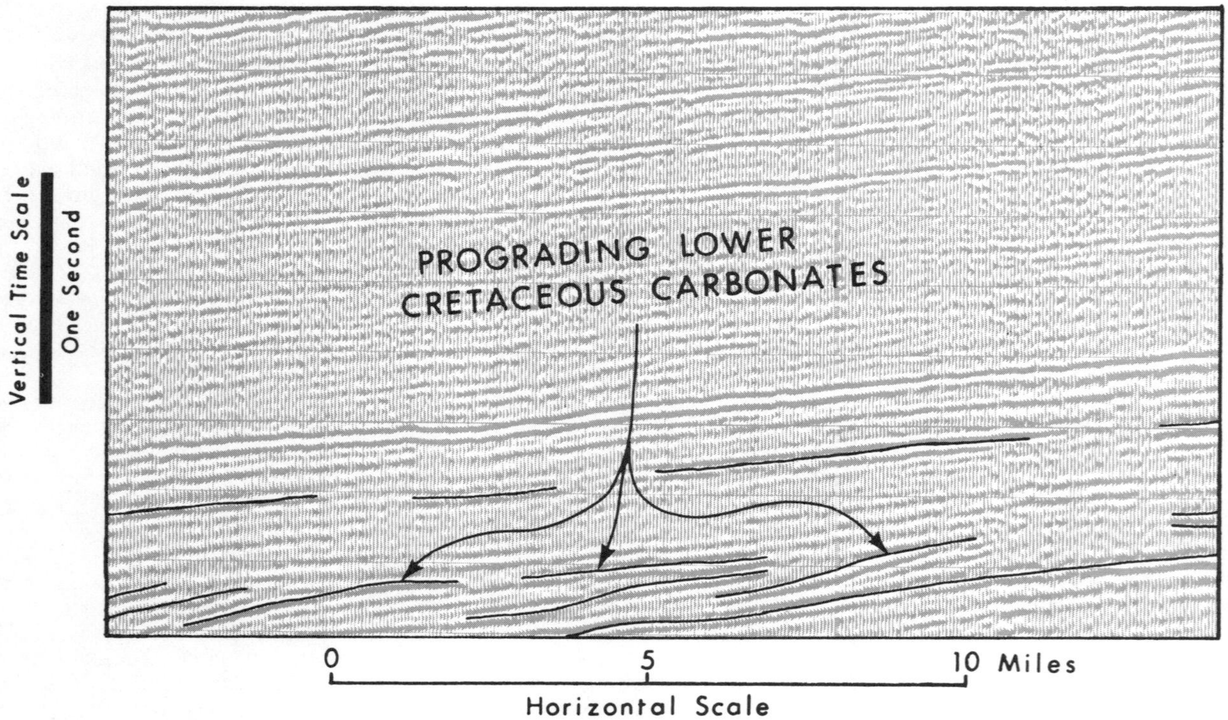

Figure 22. Interpreted reflection seismic profile showing prograding lower Glen Rose rud st reef complex and Edwards shelf margin.

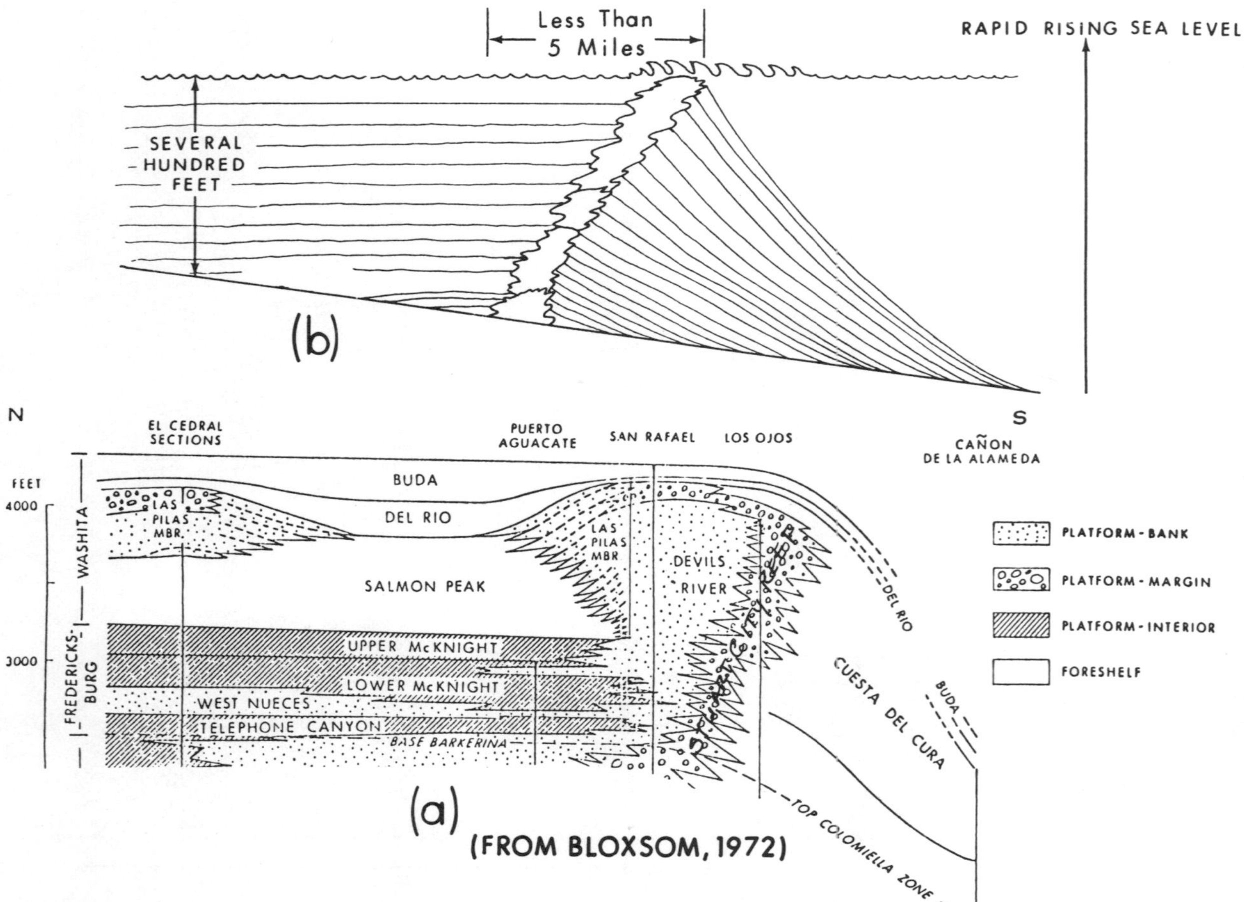

Figure 23. (a) Cross section showing vertical buildup of Fredericksburg-Washita strata (northern Coahuila, Mexico) creating topography that forms a break in slope. (b) Diagrammatic cross section depicting a vertical buildup with a rapid rising sea level.

rudist reef complex in the Comanche Cretaceous Glen Rose Limestone of Central Texas, *in* Aspects of Trinity division geology: Geoscience and Man, v. VIII, p. 131-173.

Petta, T. J., 1976, Diagenesis and paleohydrology of a rudist reef complex (Cretaceous), Bandera County, Texas: Baton Rouge, Louisiana State Univ., Ph. D. dissertation, 212 p. (unpublished).

Rockwell, D. W., and Rojas, A. G., 1953, Coordination of seismic and geologic data in Poza Rica - Golden Land area, Mexico: AAPG Bull., v. 37, no. 11, p. 2551-2565.

Rose, P.R., 1963, Comparison of type El Abra of Mexico with "Edwards Reef Trend" of South-Central Texas, *in* Geology of Peregrina Canyon and Sierra de El Abra: Corpus Christi Geological Society Annual Field Trip, p. 57-64.

______ 1972, Edwards Group, Surface and subsurface Central Texas: Univ. Texas, Austin, Bureau of Economic Geology Report of Investigations 74, 198 p.

Smith, C. I., 1970, Lower Cretaceous stratigraphy, Northern Coahuila, Mexico: Univ. Texas, Austin, Bureau of Economic Geology Report of Investigations 65, 101 p.

______ , and Bloxsom, W. E., 1974, The Trinity Division and equivalents of northern Coahuila, Mexico, *in* Aspects of Trinity division geology: Geoscience and Man, v. VIII, p. 67-76.

______ , Charleston, A. S., and Brown, J. B., 1974, Lower Cretaceous shelf, platform reef and basinal deposits, southwest Texas and northern Coahuila: West Texas Geological Society - SEPM Permian Basin Section, 1974 Field Trip Guidebook, Pub. No. 74-64.

Stricklin, F. L., Jr., and Amsbury, D. L., 1974, A low-relief carbonate shelf, middle Glen Rose deposits, Central Texas, *in* Aspects of Trinity division geology: Geoscience and Man, v. VIII, p. 53-66.

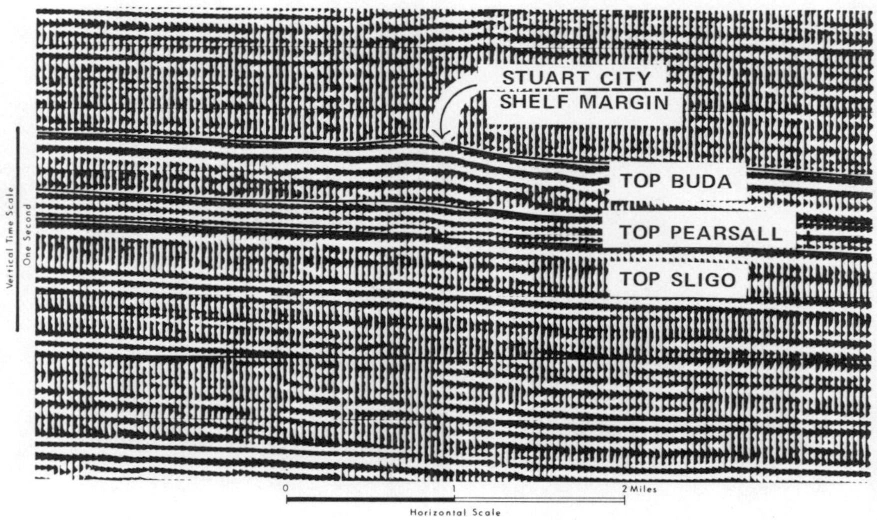

Figure 24. Interpreted reflection seismic profile, Stuart City reef shelf margin.

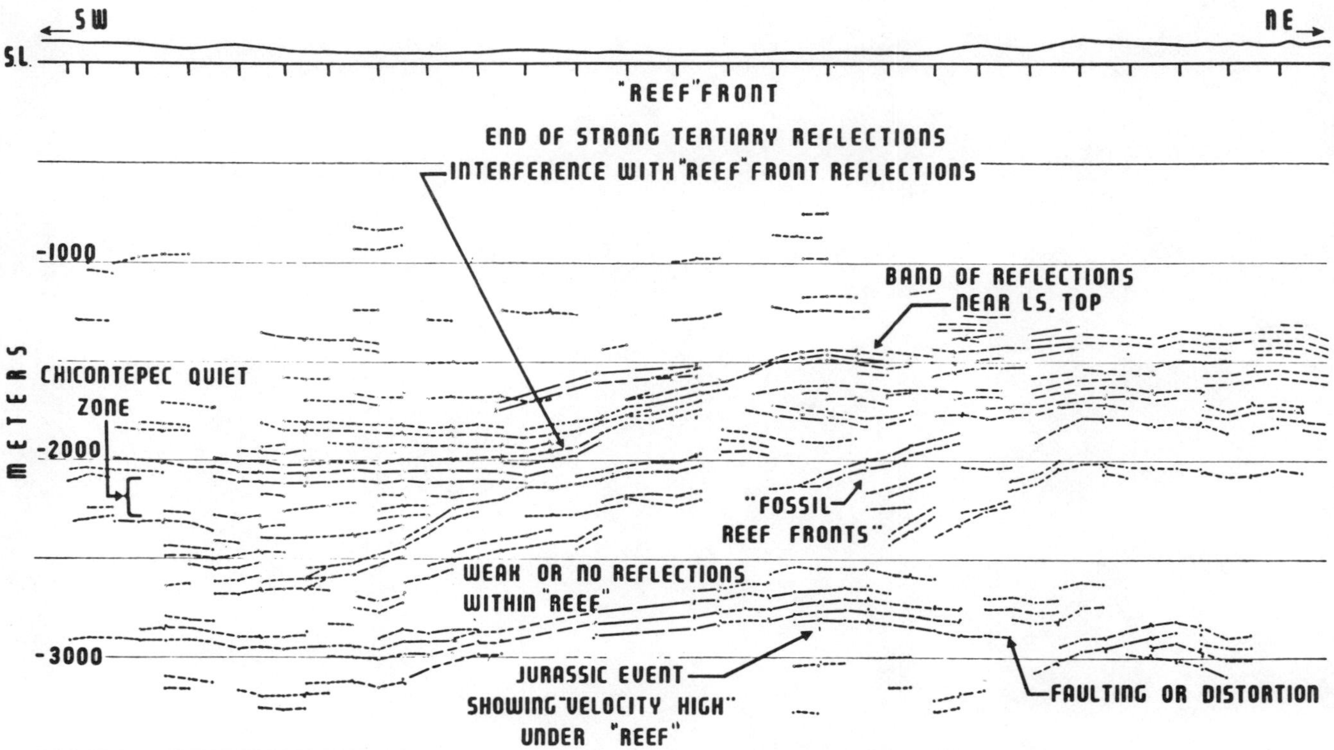

Figure 25. Reflection seismic profile showing internal structure of the El Abra shelf margin (from Rockwell and Rojas, 1953).

________, Smith, C. I., and Lozo, F. E., 1971, Stratigraphy of Lower Cretaceous Trinity deposits of Central Texas: Univ. Texas, Austin, Bureau of Economic Geology Report of Investigations 71, 63 p.

Tucker, D. R., 1962, Subsurface Lower Cretaceous stratigraphy, Central Texas, *in* Contributions to the geology of South Texas: San Antonio, South Texas Geological Society, p. 177-217.

Winter, J. A., 1962, Fredericksburg and Washita strata (subsurface Lower Cretaceous), southwest Texas, *in* Contributions to the geology of South Texas: South Texas Geological Society, p. 81-115.

Young, K. P., 1966, Texas Mojsisovicziinae (Ammonoidea) and the zonation of the Fredericksburg: GSA Memoir 100, 225 p.

Zink, E. R., 1957, Resume of the Lower Cretaceous of South Texas: GCAGS Transactions, v. 7, p. 13-22.

EXPLORATION HISTORY OF THE SOUTH TEXAS LOWER CRETACEOUS CARBONATE PLATFORM

T. D. Cook[1]

ABSTRACT

The search for hydrocarbons in reservoirs of the Lower Cretaceous of south-central Texas has been continuous for more than 60 years. To date accumulations have been found in significant quantities in but four areas: (1) the very shallow fault traps high on the San Marcos arch in Caldwell and Guadalupe Counties, (2) a fault trend stretching across central Atascosa County, (3) a fault trend extending from southeastern Atascosa County to southern Gonzales County, and (4) a narrow, elongate band extending across the entire area known as the Stuart City reef trend.

Reservoirs which contain the hydrocarbons were deposited in a myriad of environments all related to a broad carbonate shelf covered by an extremely shallow sea. The sea deepened dramatically at the shelf margin parallel to the reef trend. Dolomites contain the accumulations in the fault trends, and porosity and permeability are reasonably good. Limestones in the reef trend were seldom extremely porous initially, and late cementation has diminished even that porosity making the reservoir of lesser quality.

Oil is the dominant hydrocarbon in the shallow fields, is less dominant in the other fault trends, and is nonexistent in the reef reservoirs. Proved ultimate for the fault trends is about 350 million barrels of oil and 1½ trillion cubic feet of gas. Reserves for the reef trend's dry gas reservoirs are difficult to estimate because of highly variable reservoir conditions but should fall between 1 and 1½ trillion cubic feet.

Intensity of exploration decreases from late Lower Cretaceous to older rocks. The Sligo Formation still holds the promise of success but lies at considerable depths over much of the area. Edwards and Glen Rose rocks are more densely explored, but there are ample opportunities for new plays even in these beds. Geologists who examine cores and cuttings, determine depositional patterns, understand modern carbonate sedimentology, and study patterns of diagenesis will have an advantage in developing new concepts for exploration.

[1]Shell Oil Company, Stratigraphic Services, Houston, Texas

EARLY AND MIDDLE CRETACEOUS HIPPURITACEA (RUDISTS) OF THE GULF COAST

Alan H. Coogan[1]

ABSTRACT

A rich and diverse Early and Middle Cretaceous (Aptian-Cenomanian) rudist fauna of five stocks—the requieniids, monopleurids, caprotinids, radiolitids, and coalcomaninid caprinids—built the widespread and often thick bank and reef deposits of the Gulf Coast area of the United States, Mexico, and the Caribbean islands.

Provincialism in the Early and Middle Cretaceous fauna, especially among the coalcomaninid caprinids in the Western Hemisphere, limits the use of caprinid genera for interhemispherical correlation. Late Cretaceous (Turonian-Maastrichtian) provincialism is more pronounced in both hemispheres among the rudists involving caprinid, radiolitid, and hippuritid genera.

Biostratigraphically, the coalcomaninid caprinids are as useful for dating the Early and Middle Cretaceous strata of the Western Hemisphere reef and bank facies as are the large arenaceous foraminifers. Establishment of an interrelated stratigraphic framework based on surface and subsurface control allows the recognition of Aptian, early, middle and late Albian, and Cenomanian stages using genera and species of rudists and arenaceous foraminifers of the reef and bank facies. The rudist faunas, described here, are correlated to the Gulf Coast Cretaceous stages and less precisely to the established ammonite zonation.

Further biostratigraphic and paleontologic work is necessary to elucidate fully the composition of Early and Middle Cretaceous rudist faunas. Nevertheless, at this time, morphologically distinct and widespread genera can be used.

INTRODUCTION

Members of the Hippuritacea, collectively called rudists, were a group of solitary and gregarious bivalve mollusks which constituted an extremely varied and abundant component of the tropical and subtropical Cretaceous reef and associated carbonate lagoonal faunas. In addition to their occurrence in small biostromes and bioherms, the rudists, together with their associated reef biota of corals, hydrozoans, algae, gastropods, and foraminifers, built the reservoir rocks of oil fields in Iran, Saudi Arabia, Mexico, Texas, Louisiana, and Mississippi (fig. 1). Although much work needs to be done to understand completely the variety, range, and paleoecologic setting of individual genera of Early and Middle Cretaceous rudists, it is possible now to sketch the major components of the common rudist groups during this period and to highlight those particularly useful for biostratigraphic correlation and dating.

A brief survey of the Early and Middle Cretaceous rudists of the Gulf Coast shows that there are five major paleontologic groups that are readily recognizable and which correspond in the systematic classification to five rudist families. These are the requieniids, monopleurids, caprotinids, radiolitids, and caprinids. The purpose of this paper is to describe and illustrate the commonly found genera of these five groups from wells and outcrop, to summarize their known stratigraphic distribution from the Barremian (Early Cretaceous) to Cenomanian (Middle Cretaceous) on the Gulf Coast, and to discuss their geographic distribution in the carbonate seas of those times.

The literature on American Early and Middle Cretaceous rudists until recently has been difficult to manage. The descriptions were, and to an extent still are, scattered mainly in old European journals. The literature is confusing to anyone who lacks facilities to assemble systematically all of it for review. The caprinids were especially difficult to decipher from the literature until the 1960's because the North American forms, although quite distinct from the European ones, were identified in part as European genera and in part as new Western Hemisphere genera.

In the last century a few species of rudists were described from the Edwards Limestone in Texas by the German geologist Ferdinand Roemer (1849, 1852, 1888) and by the U. S. Geological Survey geologists C. A. White (1844), and R. T. Hill (1893). T. A. Conrad described several rudists from the Devils River Formation (1855) collected during the exploration of the United States - Mexican boundary. Later in the 1920's, W. S. Adkins' interest in the Texan and Mexican Cretaceous resulted in a summary of the then-named rudists in his catalog of Cretaceous fossils (Adkins, 1930). In the meantime, varied faunas of caprinids and radiolitids from Mexico had been described by Georg Boehm (1898, 1899) in Germany, by Henri Douvillé (1900) in France, and by Robert Palmer (1928) in Mexico. Harris and Hodson (1922) who worked on Trinidad rudists pointed out the differences between Western and Eastern Hemisphere caprinids and named several new genera. In the 1930's several Dutch geologists, A. A. Thiadens (1936), L.A.H. Bouwman (1937), and H. J. MacGillavry (1937), described Albian and Cenomanian rudists from Caribbean islands, mainly Cuba. The work in Jamaica by Chubb (1956, 1971), although touching on Early Cretaceous rudists, did not add significantly to an understanding of their stratigraphy.

The first attempt to synthesize the literature on American Cretaceous rudists was done by MacGillavry in 1937. He examined the convergence and superficial similarity of many American and Eastern Hemisphere caprinids, illustrated several lines of caprinid evolution, and suggested the significance of these groupings for biostratigraphic correlation. Two decades later, MacGillavry (1959) reviewed and expanded on his earlier conclusions. Shortly thereafter Perkins (1960) described a new caprinid species of significance from the late Albian of Mexico. The study of radiolitids, the other important rudist group for stratigraphic correlation, languished during this period, although several genera and species had been described by Hill (1893), Roemer

[1]Department of Geology, Kent State University

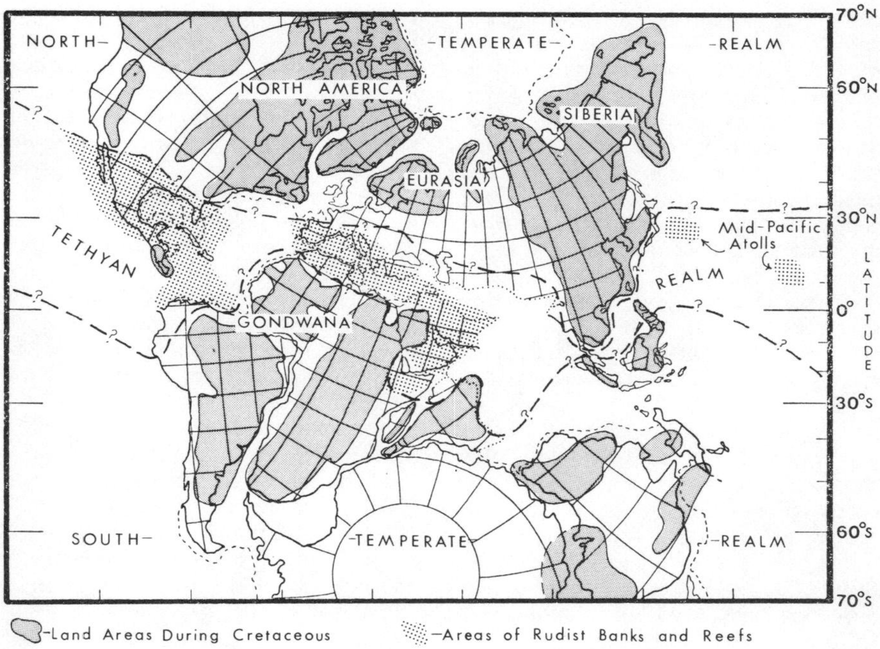

Figure 1. Distribution of rudist banks and reefs during the Cretaceous in the Tethyan Realm displayed on a predrift map (Jurassic-Early Cretaceous) modified from "Organisms and Continents through Time" (see Coates, 1973).

(1849), Adkins (1930), and others. The first modern, comprehensive but brief description of the rudists of the Gulf Coast is contained in "Treatise on Invertebrate Paleontology, Part N" (Moore, 1969). In that volume, Perkins and Coogan summarized much of what was known of the Western Hemisphere caprinids and radiolitids, clarified generic descriptions, and revised the ranges of some genera. Subsequently, Coogan (1973) described new genera of coalcomaninid caprinids from Mexico and Texas and related the stratigraphic distribution of members of this new subfamily, associated radiolitid genera, and larger foraminifers commonly occurring with them to the standard stage divisions of the Early and Middle Cretaceous.

Paleoecologic work on rudists has moved even more slowly. The theoretical base for establishing the paleoecologic setting of various rudist groups was delayed until the widespread application of standard sedimentological tools of carbonate petrography and environmental analysis developed in the 1960's. As part of a symposium on the Edwards Limestone, Young (1959) described the rudist zonation of a middle Albian Texas patch reef. Many other authors have since described both the general and specific setting of various rudist bodies as part of regional or local carbonate facies analyses (Coogan and others, 1972; Bebout and Loucks, 1974; Perkins, 1974; Rose, 1972; Fisher and Rodda, 1969; Keith, 1963; Rose, 1963; Robertson, 1972; and others). Of particular interest is the work of Perkins (1974) on the paleoecology of a rudist reef complex in the Glen Rose Limestone of Central Texas. The paper emphasizes the paleoecologic setting of early Albian requieniid and monopleurid biostromes and caprinid biostromes and bioherms.

The attempt here will be to bring into focus a current evaluation of three aspects of our understanding of the rudists of the Early and Middle Cretaceous—morphologic characteristics necessary for identification of group and genus, potential for biostratigraphic correlation, and provincialism of faunas. The paleoecologic significance of rudist distribution is left for others or another time.

PROVINCIALISM AND CORRELATION OF RUDIST FAUNAS

Early Cretaceous

The earliest Cretaceous (pre-Albian) rudist faunas of the requieniid and monopleurid stocks were presumably cosmopolitan generically. This conclusion differs from that of Coates (1973) who shows a 40 percent endemism for the Mediterranean province in pre-Albian time. However, little is known of the rudists of the earliest Cretaceous of the Western Hemisphere outside of Trinidad (Harris and Hodson, 1922) because the many physically accessible facies do not contain rudists and the facies with rudists are unavailable, unstudied, or the rocks are too poorly known to be used as a datable framework.

Middle Cretaceous

By Middle Cretaceous (Albian-Cenomanian), in fact by the preceding Aptian, five of the six stocks of rudist families had evolved; they are distributed in both the Eastern and Western Hemispheres. By my estimate, there are 40 genera of Middle Cretaceous rudists: 6 requieniids, 6 monopleurids, 7 caprotinids, 6 radiolitids, and 15 caprinids. The caprinids and radiolitids were rapidly evolving plastic groups (fig. 2) and hence generally more useful for dating (Coogan, 1969). Of these 40 genera, 26 (65 percent) occur in the Western Hemisphere and 31 (77 percent) in the Eastern Hemisphere. The bulk of the Middle Cretaceous fauna is cosmopolitan (fig. 3), although provincialism is apparent in a few groups. Of the Western Hemisphere rudists, seven genera (18 percent), all coalcomaninid caprinids, occur only in the Western Hemisphere and on Pacific guyots (Matthews and others, 1974; Hamilton, 1956). Of the Eastern Hemisphere fauna, an estimated 6.5 genera (16 percent), mainly caprinids, occur only in the Eastern Hemisphere. These percentages will doubtless change with further discoveries and further differentiation among caprinids and caprotinids. It is important to note that the requieniids, monopleurids, caprotinids, and radiolitids are essentially cosmopolitan at the generic level in the Middle Cretaceous and that provincialism is confined to the different caprinid stocks (mainly subfamilies). This provincialism must be a result of biogeographic and ecological controls rather than the result of transoceanic distance barriers, since coalcomaninid caprinids are found on western Pacific Ocean sunken atolls (Matthews and others, 1974).

The significance of provincialism for dating is that the conservative requieniids and monopleurids which are of little use for dating are cosmopolitan. Members of these groups are best used for paleoecologic information (Perkins, 1974) and composite faunal dating. The radiolitids are useful for local, regional, and intercontinental dating, but radiolitid genera encompass several Cretaceous stages or substages and, therefore, provide less precise dates than desired. The caprinids, especially the coalcomaninids, are known to have been rapidly evolving in the Middle Cretaceous coincident with the development of the great reef and bank complexes of the Sligo, Glen Rose, Sunniland, Edwards, Washita, Stuart City, Devils River, Golden Lane, and Poza Rica deposits. These coalcomaninids are as useful and precise for dating as are the accompanying arenaceous foraminifers found in similar environments, for example, ***Orbitolina, Dictyoconus, Coskinolinoides***, and ***Coskinolina*** (Coogan, 1973). Caprinid genera are useless for Middle Cretaceous interhemisphere correlation, as are the foraminifers, without first locally correlating across facies to the ammonite or comparable zonation.

The establishment of the necessary correlation framework partially has been done for the Middle Cretaceous by using intertonguing physical relationships, associated arenaceous and planktonic foraminifers, ammonites, and well-log correlations (Coogan, 1973). The stages of the Aptian, early, middle, and late Albian, and Cenomanian, and probably the Barremian and Turonian as well, are recognizable in the Gulf Coast region in the reef and bank facies on the basis of rudists alone.

Late Cretaceous

In comparison, the Late Cretaceous (Turonian-Maastrichtian) rudist faunas are estimated at more than 65 genera. Deliberately excluded from this tabulation are many poorly described rudists, those with unique and singular occurrences, and a few described since 1970 in the Eastern Hemisphere which are known to occur at only one locality. The inclusion of these would increase the percentage of provincialism in the Late Cretaceous.

Of the 65 genera, 31 (47 percent) occur in Western Hemisphere localities and 53 (81 percent) occur in the Eastern Hemisphere (fig. 4). In the Western Hemisphere fauna, only 15 percent are provincial. This 15 percent consists of several genera of ***Barretia***-like hippuritids (Moore, 1969; Chubb, 1971), several genera of sauvagesiid radiolitids, and several caprinids.

Of the Eastern Hemisphere fauna, a startling 47 percent of the genera are not found in the Western Hemisphere. Most of the previously cosmopolitan radiolitid fauna is now separate in the Eastern Hemisphere, including eight genera of lapeirousiines, all the Late Cretaceous radiolitine genera except the two carry-overs from the Middle Cretaceous (***Radiolites*** and ***Praeradiolites***), and three sauvagesiinids including two carry-overs (***Sauvagesia*** and ***Durania***). The biradiolitine stock of radiolitids is cosmopolitan. Other provincial elements include 7 caprinid genera and 4 of the 12 hippuritid genera. The hippuritid genera, which first appear in the Turonian, are divided equally into four cosmopolitan genera (including ***Hippurites, Hippuritella***, and ***Vaccinites*** all appearing in the Turonian), four Western Hemisphere and four Eastern Hemisphere genera.

Thus, the Late Cretaceous rudist faunas are far more provincial than those of the Middle Cretaceous, as previously recognized (Coates, 1973), and the endemism is displayed in the caprinids, radiolitids, and Late Cretaceous hippuritids. Late Cretaceous rudists are important for dating in the rudist bank and reef facies in Jamaica, Cuba, Puerto Rico, and Mexico but have not been used much along the United States Gulf Coast. A well-founded sequence of rudist faunas upon which one can rely for correlation is yet to be established for the Late Cretaceous of the Western Hemisphere, although the systematic descriptions by Chubb (1971) and MacGillavry (1937) serve as a starting point. It appears that in the Turonian to Santonian interval, a combination of regular hippuritids, caprinids, and foraminifers should provide a suitable basis for correlation and dating. In the Campanian-Maastrichtian interval, the bizarre ***Barretia***-like hippuritids and ***Antillocaprina***-like caprinids should be useful in the Western Hemisphere dating based on rudists alone, even where there is a lack of datable foraminifers and rich and unique radiolitid fauna of the Eastern Hemisphere.

Increased provincialism among rudists in the Late Cretaceous is attributed by Coates (1973) to the effect of development of the oceanic ridge system as a migration barrier with endemism increasing through the Maastrichtian because of the spreading of the Atlantic Ocean and its consequent effect in isolating the Mediterranean province. Coates (1973) also recognizes the provincialism of the Caribbean region in the Aptian-Albian interval, remarking that the endemic trend is more pronounced for rudists than for corals but that both follow similar trends.

The reader should be aware that, in the use of the Treatise, Part N (Moore, 1969), on rudists, lack of recognition of subfamilies among the caprinids coupled with the inclusion of somewhat suspect geographic distributional data for genera in certain cases results in the superficial appearance that the caprinid genera are less recognizable and less useful for dating than in fact is the case.

BIOSTRATIGRAPHY

Rudists

Biostratigraphic use of Early and Middle Cretaceous rudists in the Western Hemisphere was limited in the past by lack of adequate stratigraphic control, lack of systematic descriptions, and lack of a catalog of rudist faunas interval by interval. This problem has been partially corrected in the last few years (Moore, 1969; Coogan, 1973) so that at the generic level some rudists, especially genera of the coalcomaninid caprinids, are

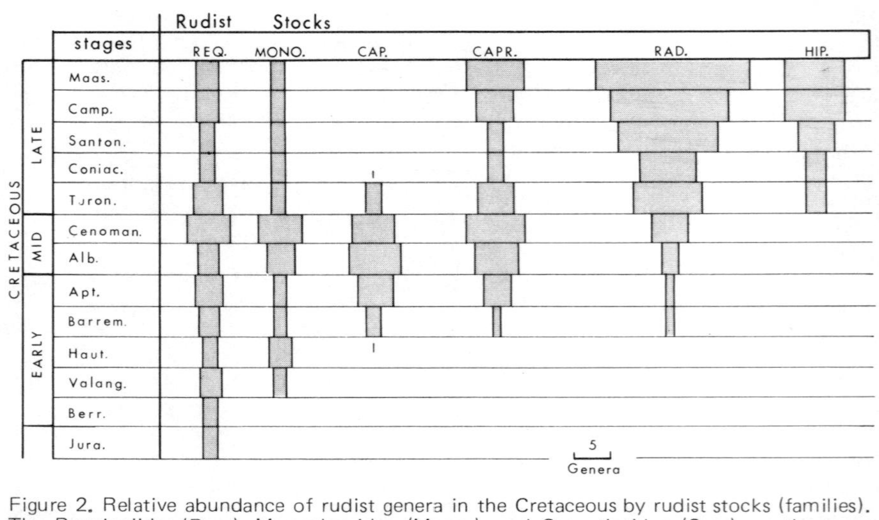

Figure 2. Relative abundance of rudist genera in the Cretaceous by rudist stocks (families). The Requieniidae (Req.), Monopleuridae (Mono.), and Caprotintidae (Cap.) are the more conservative, less rapidly evolving stocks. The Caprinidae (Capr.), Radiolitidae (Rad.), and Hippuritidae (Hip.) are the more plastic, diverse, and rapidly evolving stocks.

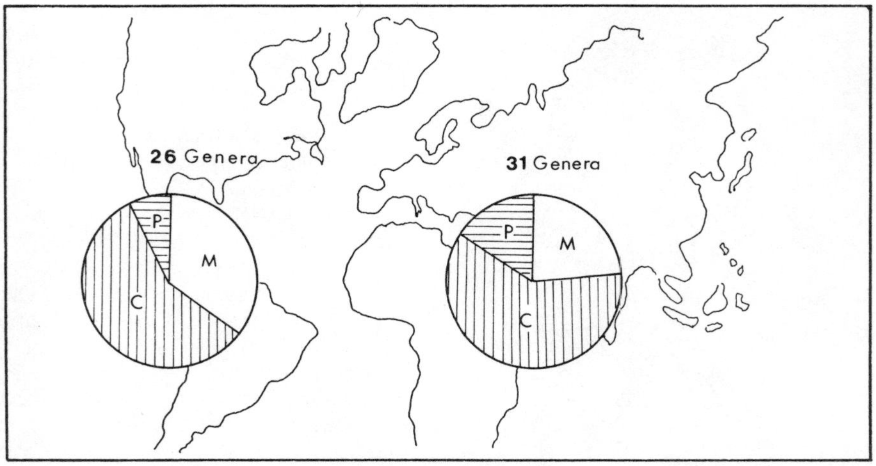

Figure 3. Middle Cretaceous (Albian-Cenomanian) provincialism in rudist faunas by hemisphere. Over half the rudist fauna of both hemispheres is cosmopolitan at the generic level. Circle=40 genera or total world fauna. P=provincial elements, C=cosmopolitan elements, M=elements missing in the particular hemisphere of the total world fauna at the generic level.

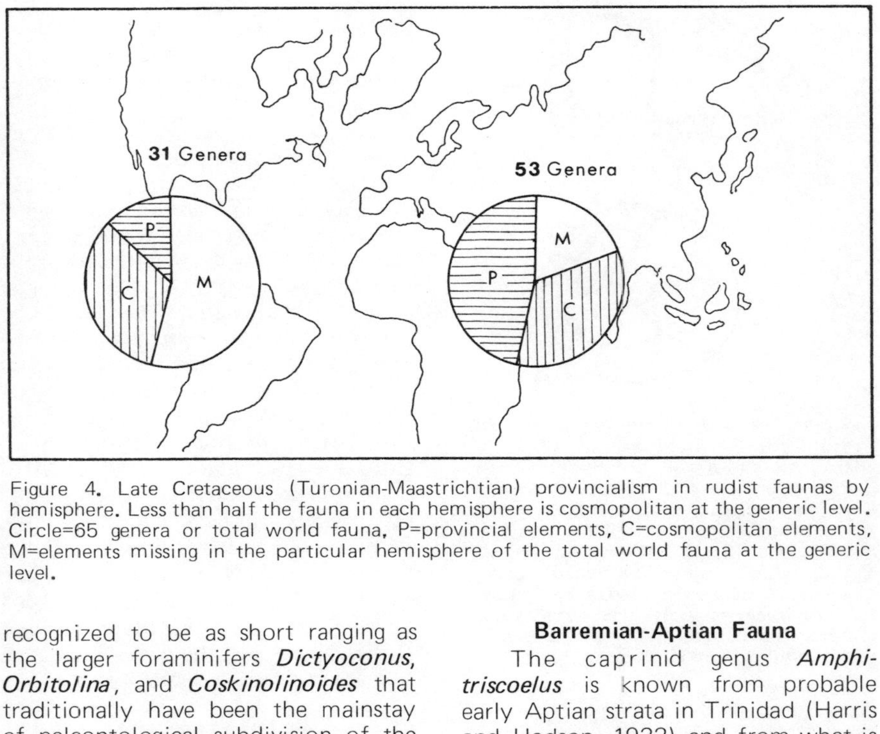

Figure 4. Late Cretaceous (Turonian-Maastrichtian) provincialism in rudist faunas by hemisphere. Less than half the fauna in each hemisphere is cosmopolitan at the generic level. Circle=65 genera or total world fauna, P=provincial elements, C=cosmopolitan elements, M=elements missing in the particular hemisphere of the total world fauna at the generic level.

recognized to be as short ranging as the larger foraminifers ***Dictyoconus, Orbitolina***, and ***Coskinolinoides*** that traditionally have been the mainstay of paleontological subdivision of the Early and Middle Cretaceous reef facies.

A chart of the caprinid genera from Barremian to middle Albian (fig. 5) shows the effect of evolution in the main coalcomaninid line of ***Amphitriscoelus, Planocaprina, Coalcomana***, and ***Caprinuloidea***. Combining outcrop age determinations, correlation to the ammonite zones, foraminiferal zonation (fig. 6), and some subsurface control, it is possible to develop an interrelated sequence in which rudists, especially caprinids, can be used for dating.

Barremian-Aptian Fauna

The caprinid genus ***Amphitriscoelus*** is known from probable early Aptian strata in Trinidad (Harris and Hodson, 1922) and from what is probably the upper Barrenquin Formation of Barremian or early Aptian age in Venezuela (Imlay, 1944). ***Planocaprina*** is recorded from outcrops in Jalisco and Veracruz, Mexico, and from the Sligo Formation in Texas wells where it occurs with ***Choffatella decipens*** and other Aptian fossils. The Cenomanian age attributed to ***Planocaprina*** by Palmer (1928) is incorrect.

Early Albian Fauna

The key early Albian coalcomaninid caprinid genus ***Coalcomana*** occurs with ***Orbitolina texana*** in outcrops of the Glen Rose Limestone in Central Texas. ***Coalcomana ramosa*** is known from Texas, Mexico, and Cuba. The Mexican occurrences are recorded from Boehm's (1898) descriptions from the Cerro Escamela near Orizaba, Veracruz, and Palmer's (1928) descriptions from Soyatlan de Adentro, Jalisco. In Cuba, Thiadens (1936) assigned several specimens from near Formento and Cruces in southern Las Villas Province to ***C. ramosa***. In Texas, species of ***Coalcomana*** occur in the middle Glen Rose reef in Hays County (Perkins, 1974) together with species of the radiolitid ***Sphaerulites***? sp., the requieniid rudist ***Toucasia***, the monopleurid rudist ***Monopleura***, and the larger foraminifers ***Orbitolina texana*** and ***Coskinolina sunnilandensis*** (Perkins, 1974). ***Coalcomana*** is also found in cores of the Glen Rose Limestone in a well in Trinity County, Texas, with ***Orbitolina texana*** and in a well in Sabine Parish, Louisiana, with the radiolitid ***Eoradiolites plicatus.***

The coalcomaninid genus ***Caprinuloidea*** begins late in the early Albian and ranges into the Cenomanian. Species of the ***Caprinuloidea perfecta*** group are recorded from the early Albian subsurface Glen Rose Limestone where they occur with both ***Orbitolina texana*** and ***Dictyoconus walnutensis*** in the Mobil No. 1 Kahanek (Lavaca County, Texas) in an interval probably correlative with a portion of the Paluxy Sandstone.

Middle Albian Fauna

Radiolitid, coalcomaninid caprinid, caprotinid, and requieniid rudists are abundant in middle Albian rocks which outcrop in Central Texas. Biostratigraphic control there is excellent owing to the partly interbedded occurrence of ammonite and rudist-bearing strata.

The radiolitid fauna contains ***Eoradiolites angustus, E. davidsoni, E. quadratus, E. liratus, E. robustus***, and other unnamed species of ***Eoradiolites*** (Davis, 1976). ***Praeradiolites edwardensis*** begins in the Albian and ranges into the latest Cretaceous. It is recorded from the Edwards Limestone outcrop in Central Texas. ***Eoradiolites robustus*** is also known from Huescalpa, Jalisco, Mexico, and numerous specimens of ***Eoradiolites*** are recorded from middle Albian intervals in wells in Texas and elsewhere (Bebout and Loucks, 1974; Coogan and others, 1972). A possible occurrence of ***Radiolites***, designated as ***Radiolites*** (?) sp., is recorded here from the subsurface middle Albian (pl. 9).

The caprinid fauna includes a

number of species of *Caprinuloidea* and *Texicaprina.* The species of *Caprinuloidea* include those of the *C. gracilis* group (*C. gracilis, C. anguis*), *C. bisulcatus*, the *C. multitubifera* group (*C. multitubifera, C. felixi*), and the *C. perfecta* group (*C. perfecta, C. lenki*). *Caprinuloidea lenki*, named from the Cerro Escamela, Veracruz, occurs in a well in the Stuart City field which is the reefal equivalent of the surface Edwards Limestone. *Caprinuloidea gracilis*, named from Jalisco, occurs in the Edwards Limestone in Central Texas and in wells in the Stuart City trend. Its close relative, *C. anguis*, named from the outcrop Edwards Limestone near Austin, is known from wells in Live Oak and San Jacinto Counties, Texas, and in the Tamabra Formation in the San Andrés field, Veracruz, Mexico. *C. bisulcata* is named from Jalisco, Mexico. *Caprinuloidea multitubifera*, named from Jalisco, is known from the Tenneco No. 1 Schulz, Live Oak County, Texas (Keith, 1963; Bebout and Loucks, 1974). The species *C. felixi* from the Cerro Escamela, Veracruz, is very similar to *C. multitubifera*. The recently reported occurrence of *ichthyosarcolites* by Davis (1976) is questioned.

Species of the coalcomaninid genus *Texicaprina* are abundant in middle Albian reef and bank deposits of the Edwards outcrop in Central Texas (Coogan, 1973; Davis, 1976); they also occur in Albian strata from Cuba (MacGillavry, 1937) and Jamaica (Chubb, 1971) and in younger strata. The type specimen comes from Paso del Rio, Colima, Mexico (Palmer, 1928).

Other middle Albian rudists which are abundant to scarce in Edwards Limestone outcrops in Central Texas and equivalent subsurface strata include the caprotinid genera *Caprotina*, *Sellaea*, and *Pachytraga* (see Davis, 1976) and species of the genera *Monopleura* and *Toucasia*.

Ammonites such as *Oxytropidoceras* provide intercontinental correlation of outcropping Edwards Limestone, and the foraminifer *Dictyoconus walnutensis* provides a convenient regional stratigraphic marker for approximately the middle Albian.

An interesting occurrence of middle Albian rudists is in the Humble No. 1 Ben Ogletree, San Jacinto County, Texas, from a depth of about 13,690 to 13,700 feet. The interval includes specimens of *Caprotina* sp. (pl. 2, fig. 8) and *Texicaprina* sp. (pl. 15, fig. 4). *Eoradioites* sp. (pl. 4, fig. 4) occurs deeper at 13,739 feet. *Radiolites* (?) sp. (pl. 9, fig. 1), if correctly identified, would be the oldest occurrence of the genus. In addition, a caprinid similar to *Immanites rotunda* (Palmer), designated *Immanites* (?) sp., cf. *I. rotunda* (Palmer) (pl. 15, fig. 7), may be more common in middle Albian strata than is usually recognized. This assemblage is typical of those found in the outcropping Edwards Limestone banks and patch reefs in Central Texas and is unlike the predominately *Caprinuloidea*-coral-requieniid assemblage found in the cores of wells from the Stuart City Formation of the "Deep Edwards Trend" in equivalent middle Albian strata (Bebout and Loucks, 1974).

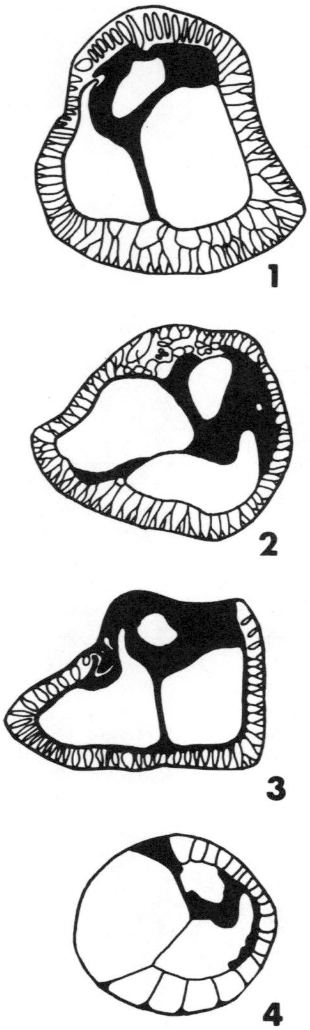

Figure 5. Progressive changes in marginal radial plate structure in the main coalcomaninid caprinid line from the simple unbranched radial plates of the Barremian-Aptian genus *Amphitriscoelus* (4), to bifurcating plates in the Aptian genus *Planocaprina* (3), to trifurcating plates in the early Albian genus *Coalcomana* (2), and to trifurcating plates crossed by radial plates to form marginal polygonal canals in the middle Albian and younger species of *Caprinuloidea* (here *C. perfecta*, 1).

Late Albian Fauna

The age assignment of rudists to the late Albian is based on a study of the Washita-age equivalent Segovia Formation of the Edwards Plateau, the Devils River Formation along the Pecos River (Rose, 1972), and the outcrops of the Fort Stockton area (Coogan, 1973). The late Albian fauna consists of the radiolitid rudists *Eoradiolites* and possibly *Sauvagesia. Eoradiolites*, which is abundant in middle Albian strata, ranges at least into the late Albian along the Gulf Coast. Species of the genus are found in late Albian rocks of the Devils River Formation, in a well in Hancock County, Mississippi, and one in San Jacinto County, Texas. The genus *Sauvagesia* evolved during the late Albian. Specimens of *S. texana* from Texas (Roemer, 1849) may be as old as late Albian and occur in younger strata.

Coalcomaninid caprinids are abundant in late Albian strata and include species of the genera *Kimbleia, Texicaprina*, and *Caprinuloidea* which are known to occur on the outcrop above ammonites and nautiloids of the lower Washita strata (Duck Creek equivalent, in part). The key coalcomaninid genus *Kimbleia* has two species: *K. albrittoni* found in the upper Aurora Formation, Mexico (Perkins, 1960) and *K. capacis* in the Segovia Formation (Rose, 1972; Coogan, 1973) on the Edwards Plateau. These genera are accompanied by species of requieniid rudists, probably *Toucasia*. Based on the occurrence of *Mexicaprina* 117 feet below the base of the Buda Limestone in the Stuart City Formation, Mobil No. 1 Zula Boyd, the genus may have appeared as early as late Albian.

Cenomanian Fauna

The assignment of rudists to the Cenomanian fauna is based first on analysis of the faunal and stratigraphic relationships of the El Abra Limestone near Valles in San Luis Potosí, Mexico (Coogan, 1973). The rationale for dating the Taninul reef-bearing rudists and equivalent backreef El Abra Limestone as Cenomanian is that the fauna does not contain the characteristic elements of the older middle or late Albian rudist or foraminifers. The El Abra does contain three fossils (*Radiolites, Dicyclina, Pecten roemeri*) not previously known from strata older than Cenomanian, along with new elements; finally, it lacks the Turonian rudist fauna.

The El Abra radiolitids include ***Radiolites abraensis*** and ***Sauvagesia texana. Radiolites*** is not known from strata older than Cenomanian (unless the ***Radiolites*** (?) sp. in the Ben Ogletree is properly identified). ***Sauvagesia*** ranges up from the late Albian.

Among the El Abra caprinids are ***Caprinuloidea multitubifera***, other undescribed species of ***Caprinuloidea***, and several species of the genus ***Mexicaprina***. No older coalcomaninids such as ***Kimbleia*** or ***Texicaprina*** have been recorded in the El Abra section. Requieniids are common.

A distinctive nonrudist component of the El Abra fauna is the hydrozoan ***Parkeria sphaerica*** which is also known from the Vraconian and early Cenomanian of England, France, and Italy (Brown and Coogan, 1968). Also present is a species of the foraminifer ***Dicyclina*** which ranges from the Cenomanian to Maastrichtian. The older foraminifers ***Coskinolinoides texana*** and ***Dictyoconus*** are absent. ***Pecten roemeri*** is also present.

Large Arenaceous Foraminifers

According to Brown (personal communication, 1966)[2], there is a well-known developmental sequence from ***Coskinolina sunnilandensis*** (early Albian) to ***Dictyoconus walnutensis*** (middle Albian) (fig. 6). In addition, based on samples collected by Coogan for late Albian rudists, there is another recognizable sequence which begins with ***Coskinolinoides texanus*** (middle Albian), leads to a primitive form of ***Coskinolina*** n. sp., and then to an advanced form of ***Coskinolina*** sp., both of which are found in the late Albian (fig. 7).

Coskinolina n. sp. in its primitive and advanced forms occurs with ***Coskinolinoides texanus*** in the late Albian beds of the Devils River Formation in Val Verde County, Texas (pl. 18). The advanced form is larger and has more pillars than the primitive one. The advanced form is very similar to the early Albian species ***Coskinolina sunnilandensis*** Maync (pl. 18) from the early Albian, upper Glen Rose. There the transitional form of ***C. sunnilandensis*** has no horizontal plates between septa, a feature which differentiates it from ***Dictyoconus walnutensis.***

Coskinolinoides texanus Keijzer, which ranges from middle to late Albian, is found with primitive ***Coskinolina*** sp. in the late Albian beds in the caprock at Twelve-Mile Mesa, Pecos County, Texas (Coogan, 1973). It differs from ***C. texanus*** in being slightly larger and having fewer vertical pillars.

In addition, a single specimen of ***Orbitolina*** sp. was found in the Devils River Formation (pl. 18). This is the latest known occurrence of ***Orbitolina*** in the Western Hemisphere where almost all orbitolinas belong to the group of ***O. texana*** (Roemer) (pl. 18) of early Albian age. In the Eastern Hemisphere, ***Orbitolina*** ranges from Barremian to Cenomanian. The specimen of ***Orbitolina*** sp. is smaller than the group of ***O. texana***; its height about equals its diameter.

Further study of these late Albian large arenaceous foraminifers seems warranted.

MORPHOLOGIC CHARACTERISTICS AND GENERIC COMPOSITION OF IMPORTANT EARLY AND MIDDLE CRETACEOUS RUDIST GROUPS

Introduction

Of the five groups important to geologists working in the Gulf Coast, three—requieniids, monopleurids, and caprotinids—fall into those families which are the conservative, little changing, and, for the most part, older families of rudists (Coogan, 1969). The more plastic, rapidly changing, and adaptive families of importance in this interval are the radiolitids and caprinids (fig. 2). In the conservative group, distinctions between families involve slight differences in dentition and musculature, different modes of attachment, variations in shape and size, and variations in development of internal shell structures. In the more plastic families, these same differences occur at the subfamily or generic level; in addition, dramatic changes are observed in the outer wall structure, the mechanism of opening the valves, the development of structures for lightening and strengthening the shell, and the support of the siphons. Parallel evolutionary trends in one or more of these features are known from several different rudist families and these provide, in part, an evolutionary basis for biostratigraphic correlation once the sequence of evolutionary change is placed firmly in an overall biostratigraphic framework.

Requieniids
(Plate 1)

The rudist mollusks of this group are equivalent to the family Requieniidae. They have two inequivalve nontabulate valves. The shell is fixed by the left (attached) valve, which is invariably larger than the operculiform right (free) valve. A single tooth is in the attached valve and two teeth are in the free valve. The attached valve is spirally coiled; the free valve is a low, twisted, weakly inflated cone or a concentric or disk-like lid. The genera of this group are distinguished on the basis of the shell shape, the placement of the teeth and sockets, the muscle attachment areas (myophores), the external ornamentation, and the wall structure. Externally, the shell is generally smooth or has fine longitudinal ribs. The requieniid shell wall is thin, compact, and consists of two layers. The inner shell wall is usually the thinner of the two and consists of coarse-crystalline calcite. The original microstructure generally is not preserved. The outer wall is composed of very closely spaced, thin, shallow laminae which are typically dark brown to pale amber. The color is distinctive unless the shell is replaced.

The requieniids are recognizable in Gulf Coast strata in this interval by their nontabulate, oval body cavity, subequal valves, two-layered shell wall, and especially the outer, brown, laminated wall.

Occurrence: The requieniids appeared in the Late Jurassic (Tithonian) and as a group extend to the Late Cretaceous (Maastrichtian). Individual genera have shorter ranges. Although several Early Cretaceous genera have been described (Palmer, 1928), only the most common, ***Toucasia*** is shown here.

Toucasia Munier-Chalmas, 1873
(Plate 1)

Toucasia has keeled valves which are carinate or frilled with shallow siphonal bands on the posterior side of the attached valve. Species of this genus differ from ***Requienia*** in having the areas of muscle insertion in the attached valve on plates and platforms which extend from the shell wall. The characters which clearly distinguish ***Toucasia*** from ***Requienia***, related as they are to patterns of muscle insertion and tooth location, require knowledge of rudist dentition and musculature patterns beyond the scope of this paper. Readers are advised to consult the Treatise, Part N (Moore, 1969).

Named species from the Early and Middle Cretaceous include ***T. hancockensis*** Whitney from the Glen Rose Limestone, Bandera County, ***T. patagiata*** (White) from the Edwards Limestone at Austin, and ***T. texana***

(Roemer) from the Edwards Limestone in Comal, Hays, Crawford, and McLennan Counties, Texas.

Type Species: ***Requienia carinata*** Matheron, 1843.

Occurrence: Early Cretaceous (Barremian) to Middle Cretaceous (Cenomanian) of Europe, North Africa, and North America.

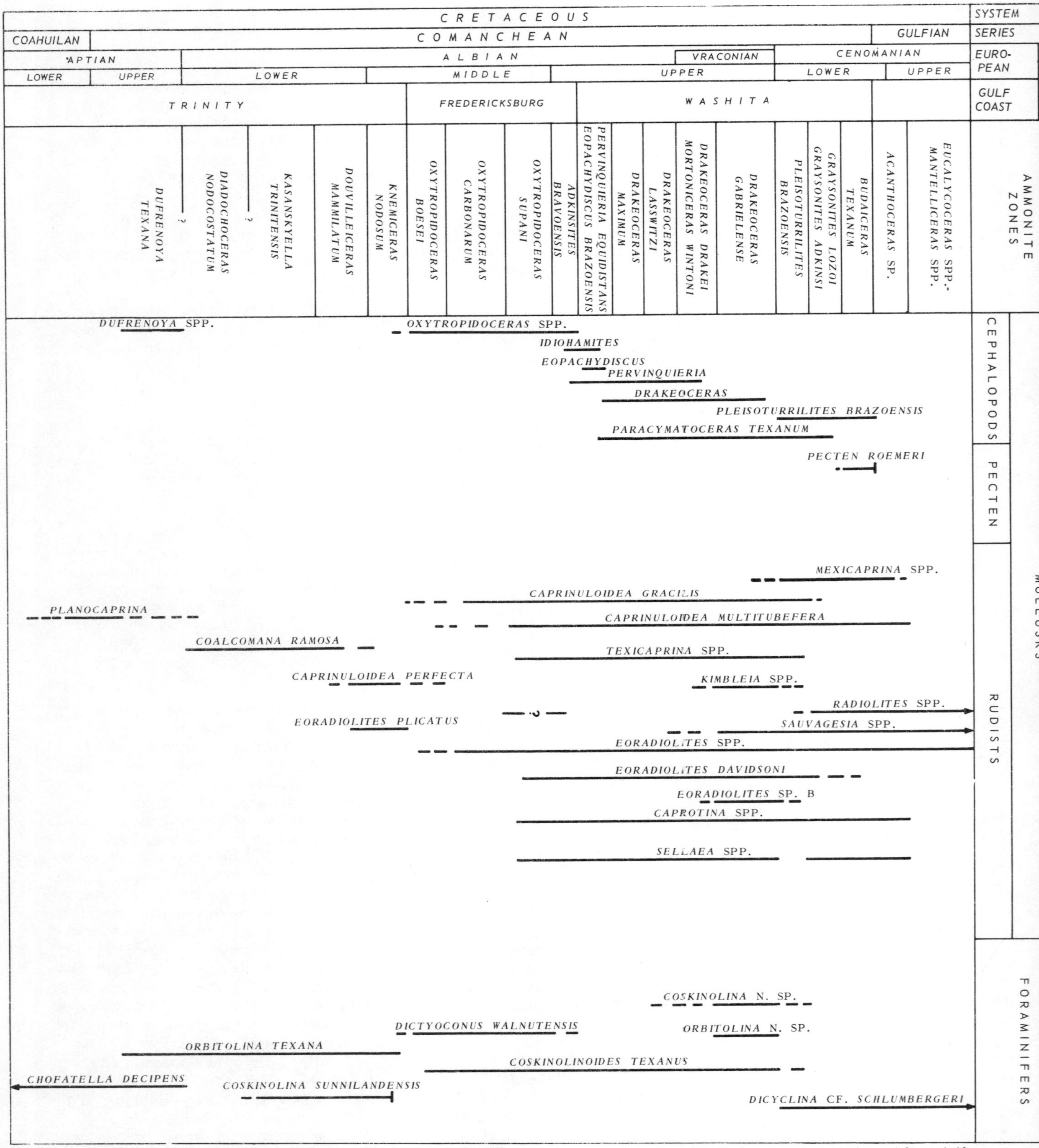

Figure 6. Zonation of Aptian to Cenomanian strata of the Gulf Coast and ranges of stratigraphically important mollusks and foraminifers.

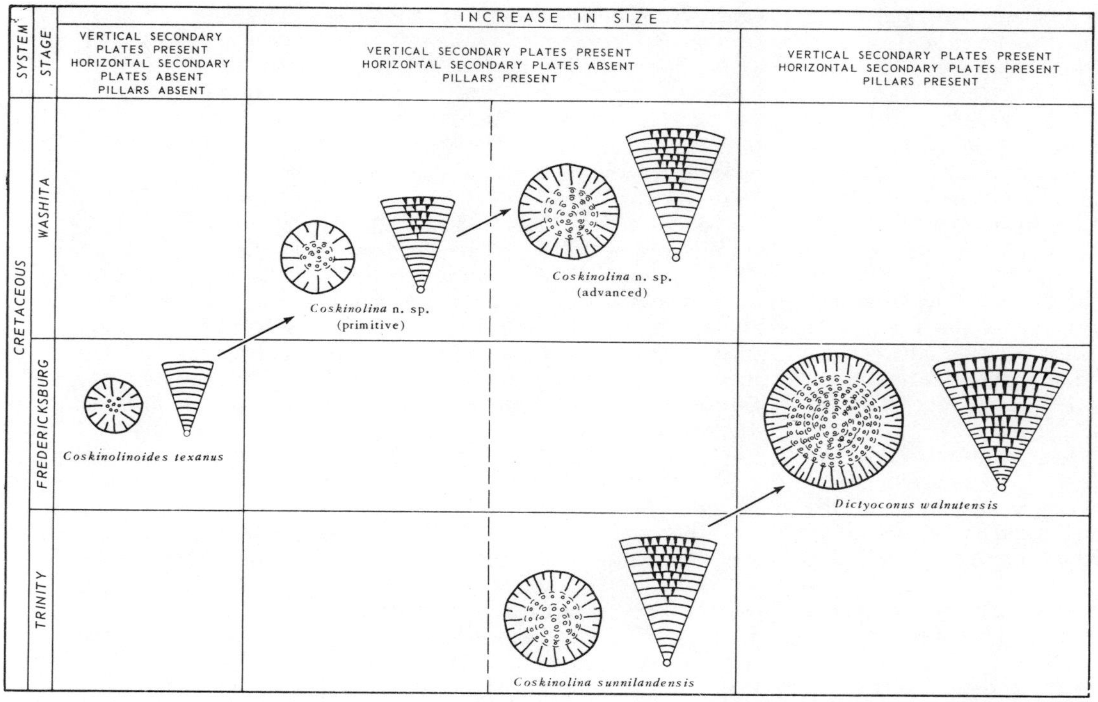

Figure 7. Diagrammatic illustration of progressive changes in size and development of pillars and plates in two evolutionary lines of large arenaceous foraminifers during the Middle Cretaceous.

Monopleurids
(Plate 2)

The monopleurids are the simplest rudists. They are attached on the right valve (in contrast to the requieniid attachment pattern). The valves are unequal. The free (left) valve has two teeth of equal or unequal size which fit into sockets of the attached valve; the attached valve has one tooth. The muscle attachment areas generally are on extensions of the cardinal platform. The attached valve is usually conical or spirally coiled. The smaller free valve is low, operculiform, and slightly coiled.

The ligament which opened the valves was external in the monopleurids, in contrast to internal ligament in other rudists attached by the right valve, such as the radiolitids. It is marked by a shallow external groove on the dorsal surface.

In cross section the body cavity of the attached valve is oval. The tooth sockets are small and sharply divided from the body cavity. The external portion of the shell is relatively smooth or may have fine growth lines and fine vertical strata.

The wall of monopleurid shell is thin, compact, and two layered. The thickness of the mature attached valve is from 1 to 5 mm, and the free valve wall is usually thinner. The inner shell wall, like the requieniids, is of coarse-crystalline calcite without the original structure preserved. The outer wall consists of fine lamellae (funnel plates) which extend from the inner wall to the external wall surface. These may be traversed by discontinuous irregular vertical radial plates similar to the wall structure of ***Sphaerulites*** or ***Radiolites***. The wall structure is generally compact.

The monopleurids lack the accessory cavities, strong siphonal bands, complicated wall structure, or specialized shell parts which are common and distinguishing features of the caprotinids, radiolitids, and caprinids.

In general, the monopleurids in the Early and Middle Cretaceous can be recognized by the smoothly circular to oval interior outlines of the attached valve which is nontabulate, by the compact outer shell wall, by external ligament, and by the small size of the shell.

A number of inadequately described and poorly dated genera of the family Monopleuridae are reported from North America, including ***Monopleura, Gyropleura, Himeraelites***, and ***Petalodontia***. Monopleurid biostromes are known from the early Albian Glen Rose Limestone (Perkins, 1974) and individual species of ***Monopleura*** have been described from the middle Albian Edwards Limestone.

Stratigraphic Range: The monopleurids range from the Early Cretaceous (Valanginian) to the Late Cretaceous (Maastrichtian) in central and southern Europe, the Gulf Coast, and Caribbean Islands.

Monopleura Matheron, 1843
(Plate 2)

The attached valve is conical and spirally twisted; the free valve is operculiform or coiled in a low spiral. The shell exterior has ribs from the apex to the commissure. The two teeth of the free valve are conical and subequal, the single tooth of the attached valve is oval in cross section. Muscle attachments in both valves are on extensions of the cardinal platform.

Two species described from Texas are ***M. marcida*** White and ***M. pinguiscula*** White from the Edwards Limestone near Austin.

Type Species: ***M. varians*** Kutassy, 1934.

Occurrence: Early Cretaceous (Valanginian) to Late Cretaceous (Maastrichtian) of Europe and North America. In Gulf Coast strata the occurrence of species of ***Monopleura*** is confirmed from outcropping Glen Rose and Edwards Limestone (early and middle Albian) strata.

Caprotinids
(Plates 2, 3)

The caprotinids, a group of rudists essentially synonymous with the genera of the family Caprotinidae (Moore, 1969), are intermediate between the monopleurids and other rudists attached by the right valve in the complexity of their shell development. The whole shell is strongly inequivalve. The attached valve is straight, curved, or slightly spirally coiled. The free valve is operculate or slightly inflated or may be in the form of an overhanging often coiled cone. The principal differences between the monopleurids and the caprotinids are that the caprotinids have elevated areas (myophores) for the attachment of the adductor muscles which close the valves and accessory cavities on the posterior, and in some species on the anterior, side of the shell wall.

The two teeth of the free valve are subequal or slightly curving and conical. The single tooth of the attached valve is large, oval, subtriangular or round in cross section, and commonly erect.

The accessory cavities are located between the elevated myophores and the outer shell wall. They may be subdivided in some genera by vertical plates. The body cavity is deep, nontabulate and generally oval or round in cross section. The ligament is impressed into the shell and is internal but has an external expression. The attachment area of the attached valve varies from a small spat to a large broad surface of the cone-shaped valve.

The external surface of the shell is smooth to longitudinally ribbed. The siphons are marked by folds on the exterior of some shells. The operculate type of free valve may have faint ribs.

The shell wall is two layered, thick, and compact; the inner wall is usually thinner. The compact outer shell wall is made of closely spaced laminae. The inner wall is usually recrystallized calcite but may have fine laminae preserved.

Caprotinids are distinguished from radiolitids which have a coarse reticulate outer shell wall of intersecting vertical radial and funnel plates and have a tabulate body cavity. Caprinids also have a tabulate body cavity and commonly have tabulate accessory cavities. In addition, all but the oldest caprinid genera commonly have marginal radial plates between the body cavity and the outer shell wall which bifurcate or form a pattern of polygonal canals.

Nine genera are assigned to the Caprotinidae including ***Caprotina, Chaperia, Horiopleura, Polyconites, Pachytraga, Praecaprotina, Retha, Baryconites***, and ***Sellaea***. Those reported from the Western Hemisphere are ***Caprotina, Sellaea, Chaperia, Horiopleura, Pachytraga***, and ***Baryconites. Baryconites*** and ***Chaperia*** are poorly described and not well understood. ***Caprotina*** and ***Sellaea*** are now well known in Texas. ***Pachytraga*** has recently been reported from the outcrop Edwards Limestone in Central Texas (Davis, 1976).

Occurrence: Caprotinids range from the Neocomian to the Turonian. They commonly are found in Europe in Aptian, Albian, and Cenomanian strata. In the Western Hemisphere their range is not well established, but they are well known from the middle Albian Edwards Limestone in Central Texas.

Caprotina d'Orbigny, 1842
(Plate 2)

The shell is inequivalve. The attached valve is elongate or curved and is ornamented with weak to strong costae. It lacks accessory cavities or pallial canals. The free valve has two or three accessory cavities. The posterior myophore is on an erect plate which parallels the shell margin. The anterior myophore is an extension of the cardinal hinge.

No species from North America definitely attributable to ***Caprotina*** have been formally named. Worldwide, about 20 species are listed for the genus. Ten are from the Cenomanian of Bohemia, Czechoslovakia, described by Počta (1899) in an obscure article in Czech. Five were described by Distefano (1888) and Parona (1909) from Italy, and two are from France. Several named species are invalid. The Italian species are adequately described and figured. The criteria used for species differentiation by Italian authors include external shape, size, and coiling, number and depth of accessory cavities, size of the body cavity, and position and size of the myophores. Davis (1976) used additional criteria. The specimens illustrated here (pl. 2) from the Edwards Limestone are simply designated ***Caprotina*** sp.

Type Species: ***Caprotina striata*** d'Orbigny, 1842.

Occurrence: ***Caprotina*** is reported

PLATE DESCRIPTIONS
KEY TO MORPHOLOGICAL FEATURES ON PLATES

B	Body Cavity	VRP	Vertical radial plate
AC	Accessory Cavity	FP	Funnel plate
L	Ligament	S	Excurrent siphonal band in radiolitids
C	Canals	E	Incurrent siphonal band in radiolitids
O-O′	Accessory cavities in caprotinids	S	Socket for tooth
N	Tooth of attached valve	OW	Outer wall
T-T′	Teeth of free valve	IW	Inner wall
ma	Anterior myophore (muscle attachment area)	V	Pedal band in radiolitids
mp	Posterior myophore	I	Interband in radiolitids

PLATE 1
EARLY AND MIDDLE ALBIAN REQUIENIIDS
Species of *Toucasia* from Texas.

Figure

1. ***Toucasia texana*** (Roemer) from the Edwards Limestone outcrop, Middle Bosque River, near Crawford, McLennan County, Texas, xl.
   (a) Upper surface of attached valve and upper free valve.
   (b) Apertural view.
2. ***Toucasia*** sp., cf. ***T. hancockensis*** Whitney from the Glen Rose Limestone, Bandera County, Texas, xl.
   (a) Anterior view of both valves. Beginning whorls of upper large valves broken off. Lower valves serve as cap.
   (b) View of attached valve.
3. ***Toucasia*** sp., cf. ***T. hancockensis*** Whitney from the same area, xl.
   (a) Adult specimen having strong twist to shell.
   (b) View of attached side. Surface of large valve strongly ribbed.
4. ***Toucasia texana*** (Roemer) from the Edwards Limestone or Glen Rose Limestone, collected by Roemer between New Braunfels and Fredericksburg. Both valves of a specimen from Roemer (1852), xl.
5. ***Toucasia patagiata*** (White) from the Edwards Limestone near Austin, Texas, xl. Figures from White (1884).
   (a) Both valves of a specimen.
   (b) Smaller specimen from same locality as 5.
6. ***Toucasia patagiata*** (White) from the Edwards Limestone, road cut on Red Bud Trail, top of hill west side of Tom Miller Dam, Austin, Texas, xl.
7. ***Toucasia***, cf. ***T. hancockensis*** Whitney, from Bandera County, Texas. Cross section through a single specimen in rock. Successive whorls are larger and appear to be different shells. Siphonal notch is distinctive for ***Toucasia***, xl.

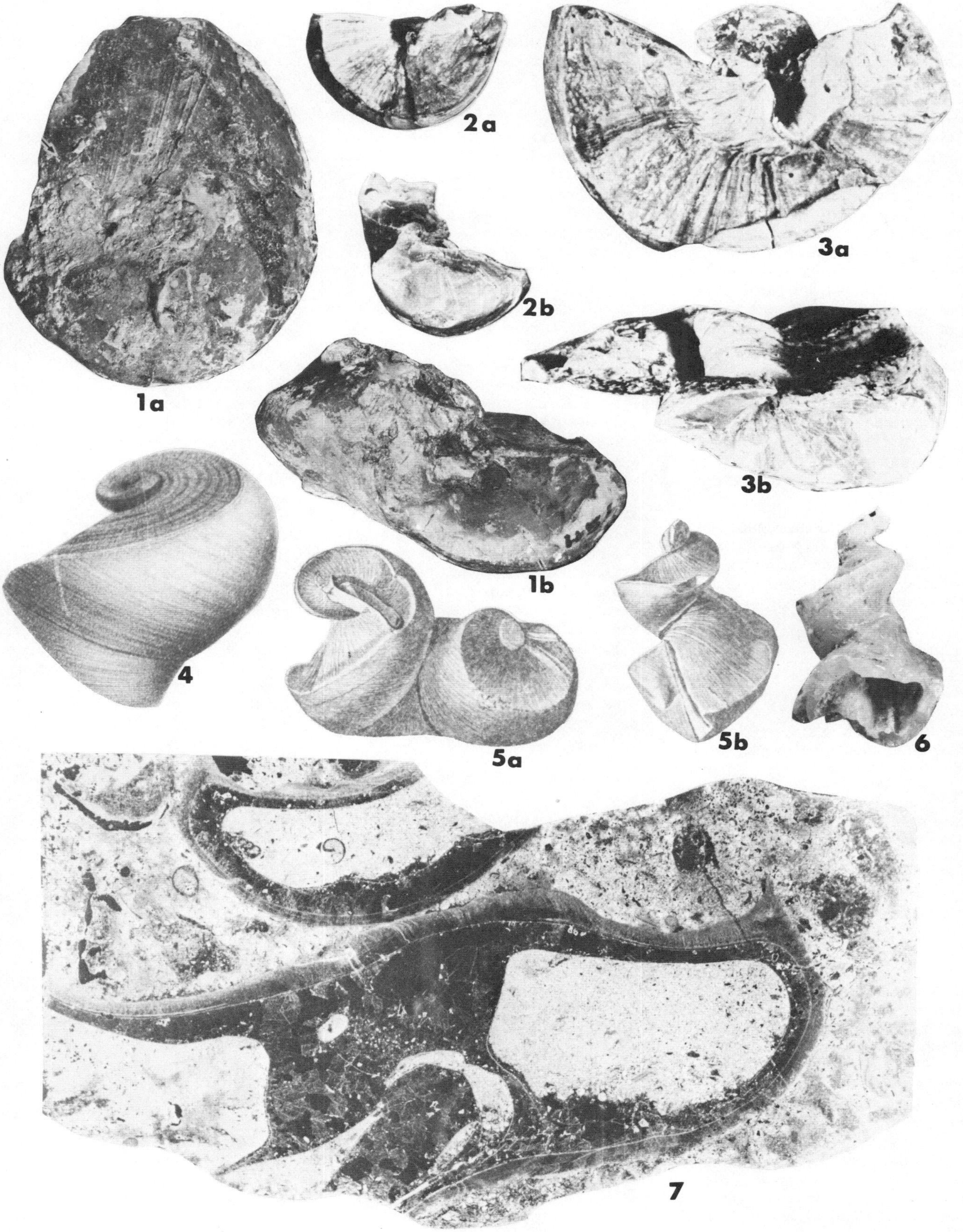

PLATE 1

from Neocomian to Turonian strata in Europe and North America, but because many occurrences are based on the earlier broad use of the name, the range of the genus needs revision. For example, the Edwards Limestone was once called the *Caprotina* Limestone. Early American paleontologists, including Meek and Conrad, named species of *Caprotina* over a hundred years ago. In 1937, however, McGillavry could not find a single species from North America correctly assigned to the genus. Since that time, however, species of *Caprotina* have been identified from North America, but no new species names have been published. Verified occurrences based on well-illustrated species show that *Caprotina* is common in rudist deposits of the Cenomanian of Sicily, Italy, and France and now is recorded in the middle Albian Edwards Limestone of Central Texas, and probable Albian deposits on the sunken guyots of the western Pacific (Matthews and others, 1974).

Sellaea Distefano, 1888
(Plates 2, 3)

Sellaea has the shape and general internal and external structure of *Caprotina* but differs from *Caprotina* by the development in the attached valve of irregular, elongate, subquadrangular, or subelliptical canals between the anterior myophore and the shell wall. The canals are the result of the subdivision of an accessory cavity by radial plates which, in turn, may be subdivided into small canals by two sets of vertical plates lying parallel to the shell wall.

Type Species: *Caprotina (Sellaea) cespitosa* Distefano (1888).

Sellaea was proposed by Distefano as a subgenus of *Caprotina* and later in 1899 raised to generic rank. The species *S. zitteli* is listed as the type designated by Kutassy (1934) in the Treatise, Part N (Moore, 1969). Distefano (1888) apparently did not designate a formal type, but *Sellaea cespitosa* is discussed under the 1888 subgeneric description; in 1899, Distefano described it as characteristic of the new group.

About 12 species are assigned to *Sellaea*. They are distinguished on the same basis as species of *Caprotina* as well as by differences in the subdivision of the accessory cavities. New species have been distinguished (Davis, 1976) from the Edwards Limestone. The two specimens illustrated are designated *Sellaea* sp. A and *Sellaea* sp. B. In addition, *Plagiotychus cordatus* Roemer (1888) may be a specimen of *Sellaea*.

Occurrence: Species of *Sellaea* are known from the Cenomanian "Beds with *Caprotina*" at Rupe del Castillo, Termini-Imerese, Sicily, and sparsely from the Albian(?) "Beds with *Polyconites*" at the same locality. They are also described from probable Cenomanian strata of Monti d'Ocre, Italy, by Parona (1909). The occurrence of *Sellaea* in the middle Albian Edwards Limestone is apparently the oldest verified occurrence. The Edwards Limestone specimens are abundant in Hamilton County near Cranfills Gap and south of Meridian in Bosque County. *Sellaea ? cordata* (Roemer) is from the Edwards Limestone on Barton Creek in Austin, Texas. A possible occurrence of *Sellaea* is on a sunken Japanese guyot in the western Pacific (Matthews and others, 1974).

Sellaea sp. (pl. 3) is similar to *Sellaea himerensis* Distefano in size, shape, and internal arrangement of teeth and muscle attachments. In the free valve in both species, the posterior accessory cavities are shallow and the anterior ones are broad. The different anterior canal pattern in the attached valve is distinctive. *Sellaea* sp. has four subquadrangular and one elongate canal; *S. himerensis* has seven subquadrangular anterior canals and two smaller dorsal canals. *Sellaea cespitosa* Distefano has seven canals arranged vertically on the anterior face. *Sellaea* sp. B. (pl. 3) is represented by one specimen of an attached valve which has a maze of small canals on the dorsal anterior margin and what appears to be canals around the whole ventral edge of the valve. *Sellaea plagiotychoides* Distefano has a similar grouping of anterior canals. Lack of discovery of a free valve prevents further comparison. *Sellaea ? cordatus* is presumed to be a specimen of *Sellaea* by inference because Roemer included it in *Plagiotychus.*

Pachytraga Paquier, 1900
(Plate 2)

The shell is inequivalve. The attached valve is cap shaped, curved, or slightly coiled. The anterior tooth is large and projecting; the posterior tooth is small. The anterior myophore is on the face of a plate connecting the anterior tooth with ventral shell wall. An anterior accessory cavity may be present and may be divided into canals by radial plates. The posterior myophore is inserted into the shell wall but at its outer edge may be separated from the wall by a small accessory cavity. A vertical radial plate runs from the anterior tooth to the ventral-posterior margin, forming a large cavity into which the posterior myophore slopes. The free valve is conical, spirally twisted to more or less straight or curved. The ligamental fold is shown by a groove on the dorsal side. The single tooth is robust and strongly projecting. The anterior socket is large; the posterior one is small. The posterior myophore is separated from the wall by an elongate

PLATE 2
MONOPLEURIDS AND CAPROTINIDS
Species of *Monopleura, Caprotina*, and *Sellaea* from the middle Albian Edwards Limestone in Texas and of *Pachytraga* from the Jubulee Limestone in Jamaica.

Figure
1-4. *Monopleura marcida* White.
(1) Free valve with two equally developed teeth, from White (1884), xl.
(2) Both valves. Attached valve is twisted and marked exteriorly with a ligamental scar. From the "Deep Eddy Bluff," U.S. 290 crossing of the Colorado River, Austin, Texas, after White (1884), xl.
(3) Free valve completely silicified from same locality as 2, xl. 5.
(4) Lateral view of whole specimen from the same locality as 2, xl. 5.
5. *Monopleura pinguiscula* White. Both valves. Attached valve is coiled and distorted, free valve has smaller individual growing on it, from White (1884), xl.
6. *Monopleura marcida* White, colony of individuals from same locality as 2, xl. 5.
7. *Caprotina* sp. from the Edwards Limestone outcrop 5.5 miles south of Cranfills Gap on Texas 22, Hamilton County, Texas.
(a) Cross section of attached valve, xl.
(b) Drawing of (a) illustrating internal morphological features, xl. 5.
(c) Cross section of free valve, xl.
(d) Drawing of (c) illustrating internal morphological features, xl.
(e) Lateral view of both valves showing elongate free valve and smaller, curved, attached valve, xl.
8. *Caprotina* sp., attached valve from the Stuart City Formation at a depth of 13,694 feet in the Humble No. 1 Ben Ogletree, San Jacinto County, Texas, xl. 5.
9. *Pachytraga jubilensis* Chubb from the Jubilee Limestone, Benbow Inlier, Jamaica, after Chubb (1971), xl.
(a) Transverse section of free valve showing ligamental fold, large centrally located tooth, body cavity, and posterior accessory cavity.
(b) Oblique section of attached valve showing elongate body cavity, elongate accessory cavity, and elongate tooth of free valve set into socket of attached valve.
10. *Sellaea* sp. from the Edwards Limestone, 6 miles south of Meridian on Texas 6, Bosque County, Texas. Both valves showing the outer surface of the shell, xl.

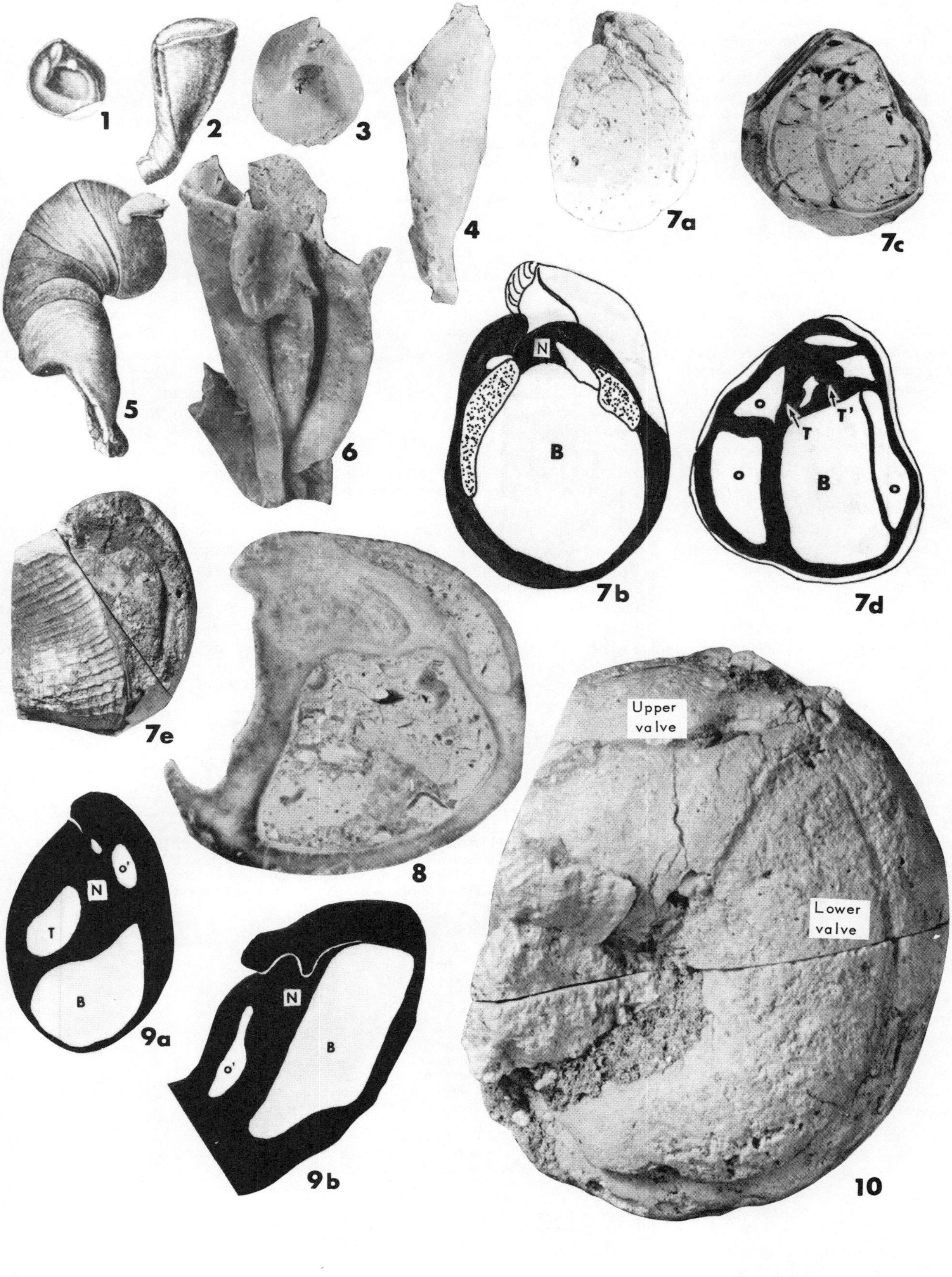

PLATE 2

accessory cavity. An anterior accessory cavity may be present.

Type species: ***Sphaerulites paradoxa*** Pictet and Campiche, 1869.

Occurrence: ***Pachytraga*** is known from the Early Cretaceous Urgonian facies of central Europe. In the Western Hemisphere, a species of ***Pachytraga, P. jubilensis*** is described by Chubb (1971) from the Early Cretaceous Jubilee Limestone of the Benbow Inlier, Jamaica. The Jubilee Limestone lies beneath the Seafield Limestone, which is believed to be middle Albian (Chubb, 1971). ***P. jubilensis*** has only a posterior accessory cavity which, like the body cavity, extends to the base of the shell. A massive concentric tooth encloses a large oval anterior tooth socket. The smaller posterior tooth socket is separated from the accessory cavity.

Recently, Davis (1976) has identified a species of ***Pachytraga*** from the Edwards Limestone in Central Texas.

Radiolitids

The Radiolitids are a diverse group of rudists encompassed in the Family Radiolitidae, Gray, which consists of mollusks with inequivalve shells attached by a large conical right valve with a smaller, operculiform free valve. The structure of the outer wall consists of coarse vertical radial and funnel plates, the intersection of which produces a reticulate pattern in cross section.

The family is divided into several subfamilies including the Radiolitinae, Biradiolitinae, Sauvagesinae, Lapeirousinae, and possibly an additional subfamily of Late Cretaceous radiolitids depending on the placement within the family of several genera of disputed affinities.

In general, the body cavity of the attached valve is tabulate. The ligamental ridge is prominent in those genera containing this feature. Commonly siphonal bands mark the position of the excurrent and incurrent siphons. In the free valve of some genera, there are openings called oscules. Where the shell lacks siphonal bands, some genera develop broad longitudinal projections of the shell wall termed pseudopillars. The teeth of the free valve and corresponding sockets of the attached valve are prominent, as are the myophores of the attached valve in many genera.

The major features of the radiolitids, the siphonal bands and their ornamentation, the presence or absence of a ligamental ridge and myophores, and the structure of the thick outer wall are characteristics by which different genera of the family are distinguished.

About 40 genera of radiolitids are known from the Early Cretaceous (Barremian) to the Late Cretaceous (Maastrichtian) strata.

Subfamily Radiolitinae Gray, 1848

The subfamily consists of radiolitids which have attached valves with a posterior ligament and an outer wall structure which is rectangular in transverse section. Genera of the subfamily include ***Radiolites, Agriopleura, Eoradiolites, Praeradiolites***, and ***Sphaerulites*** in the Early-Middle Cretaceous. The fauna is essentially cosmopolitan at the generic level.

Eoradiolites Douvillé, 1909
(Plates 4 through 7)

The attached valve is conical, large, and characterized by smooth, jutting siphonal bands, the presence of a ligament, and an outer wall which is coarsely reticulate. Large ribbed sockets in the dorsal inner wall are separated from the shell body cavity by a lateral extension of the cardinal platform from which the lower valve tooth rises. The free valve is small, operculiform, concave to flat, and fits tightly into the attached valve. Myophores are present.

The named species of ***Eoradiolites*** identified in the Western Hemisphere include ***E. davidsoni, E. robustus, E. perforatus, E. inflatus, E. angustus, E. quadratus, E. plicatus, E. liratus***, and additional unnamed species are recognizable (Davis, 1976).

Type Species: ***E. davidsoni*** (Hill), 1893.

Eoradiolites davidsoni group

E. davidsoni is a colonial species characterized by a somewhat straight, elongate, heavy, attached valve with two prominent smooth siphonal bands and, in mature specimens, a flaring oral end. Several species, similar to ***E. davidsoni*** in cross-sectional view, probably cannot be distinguished from it when the specimens are imbedded in rock or viewed in sawed cores. The ***E. davidsoni*** group includes part of Adkin's ***E. quadratus, E. robustus*** (Palmer) from Mexico, and specimens referred to ***E. liratus***, a species named from North Africa.

Eoradiolites robustus is characterized by the flaring oral end of the attached valve and by an external ornamentation of jutting growth ridges which give the shell a foliaceous appearance that distinguishes it from ***E. davidsoni***. Comparison of the types of ***E. robustus*** with other specimens of ***Eoradiolites*** described by Palmer as ***E. perforatus*** and ***E. inflatus*** leads to the conclusion that these two later species are indistinguishable from ***E. robustus***.

Adkins (1930) proposed the species ***E. quadratus*** for shells which differ slightly in cross-sectional view and ribbing from ***E. davidsoni***. Restudy of his types specimens (The University of Texas at Austin, Bureau of Economic Geology nos. 20547-20650) reveals that the speci-

PLATE 3
MIDDLE ALBIAN CAPROTINIDS AND EARLY ALBIAN RADIOLITIDS
Species of the caprotinid *Sellaea* from the middle Albian Edwards Limestone, Texas, and from Italy and the radiolitid *Sphaerulites*? from the early Albian middle Glen Rose Limestone of Hays County, Texas.

Figure
1. ***Sellaea*** sp. A. from the Edwards Limestone outcrop 5.5 miles south of Cranfills Gap on Texas 22, Hamilton County, Texas, xl.
   (a) Cross section of attached valve.
   (b) Cross section of free valve.
   (c) Lateral anterior view of both valves.

2-3. ***Sellaea zitteli*** Distefano from the "Calcari con ***Caprotina***", Rupe del Castillo, Termini-Imerese, Sicily of Cenomanian age. Drawings after Distefano (1888), xl.
   (2) Attached valve showing typical accessory cavities (o-o').
   (3) Free valve of another specimen showing marginal anterior canals.

4-5. ***Sellaea ? cordata*** (Roemer) from a cave (Edwards Limestone) on Barton Creek, 2 miles from the confluence with the Colorado River, Austin, Texas, after Roemer (1888).
   (4) Attached valve of small specimen showing internal morphological features of the shell as a caprotinid.
   (5) Lateral view of both valves.

6. ***Sellaea*** sp. B. from the Edwards Limestone, 6 miles south of Meridian on Texas 6, Bosque County. Cross section of free valve showing many anterior canals, xl.

7-8. ***Sphaerulites***? sp. from the middle Glen Rose Limestone at the Blanco River, 5.1 miles north of Fischer on the Fischer-Wimberly Road on the Rod-E Ranch (old Kreh Ranch locality), Hays County, Texas.
   (7) Cross section of attached valve.
      (a) Longitudinal cross section of attached valve, xl. 2.
      (b) Fine reticulate outer wall structure, cross section, x40.
      (c) Fine radial plates, longitudinal cross section x40.
   (8) Transverse cross section of two valves. Part of free valve is also shown in oral cavity of attached valve, left specimen, xl. 5.

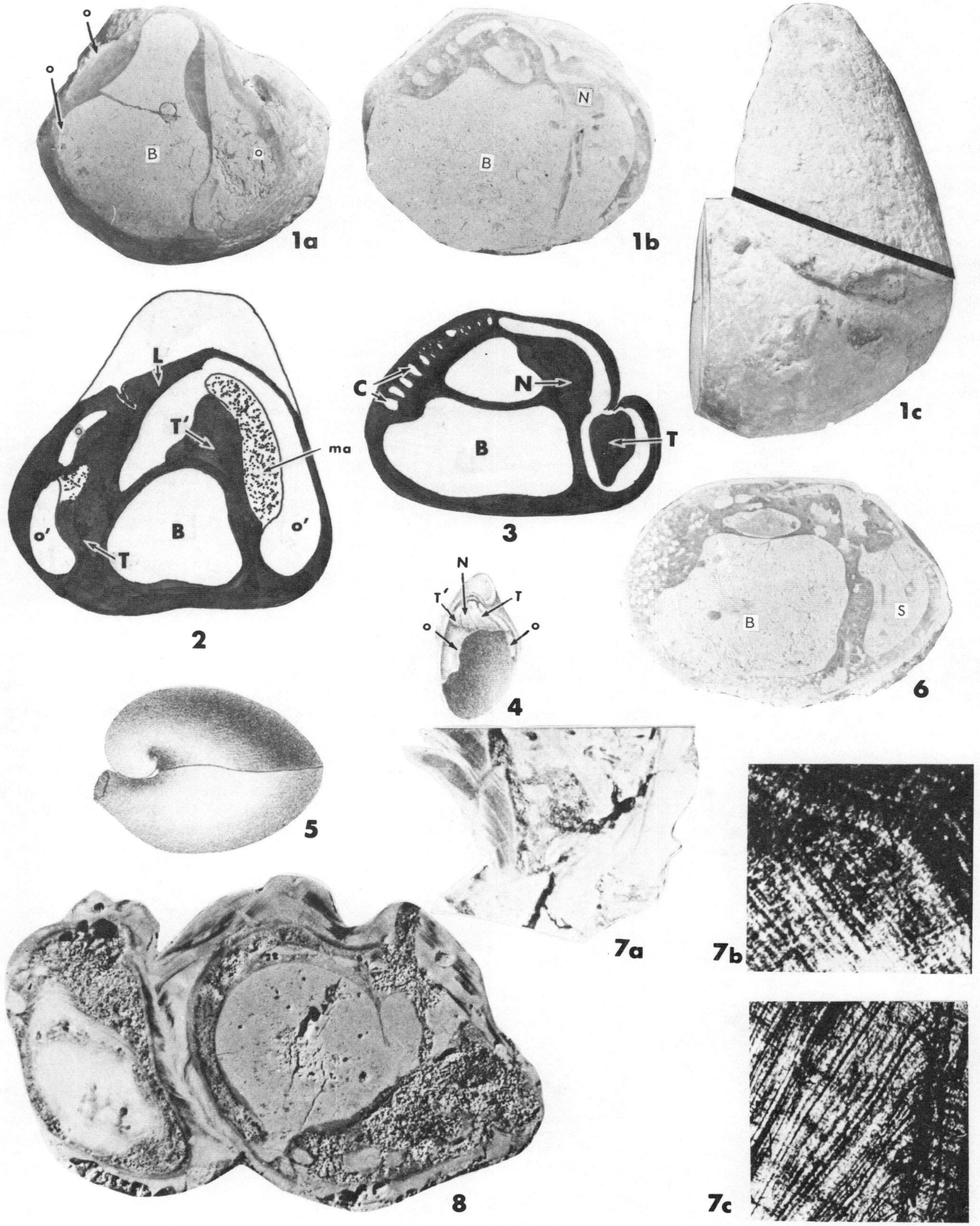

PLATE 3

mens are of two sorts. The type specimen (pl. 6, fig. 1) and two others differ from *E. davidsoni* in the quadrate outline. However, most of the differences listed by Adkins do not exist and others, such as the presence of four to five ribs instead of six, are of minor importance. The types of *E. quadratus* listed by number above may be well within the range of variation of *E. davidsoni*.

The specimens (nos. 20647-20648) as paratypes of *E. quadratus* differ from those here attributed to *E. davidsoni* and *E. quadratus* in having a shorter, more strongly curved lower valve which expands very rapidly, and are recognized as *Eoradiolites* sp. A (pl. 7, figs. 4a-b).

The group occurs in the Edwards Limestone of Texas. It has been reported from Iran (Douvillé, 1910) and from near Trieste, Italy (Wiontzek, 1934). Radiolites of the *E. davidsoni* group are common in middle and late Albian strata (Edwards, Washita) in wells in South Texas, Louisiana, and Mississippi and in outcrop in Mexico (Palmer, 1928).

Other Species of Eoradiolites

(1) The shells described by Adkins from the middle Albian Edwards Limestone in Bell County, Texas, as *E. angustus* have extremely slender attached valves with few ribs, smooth siphonal bands, nearly quadrate cross sections, and flattened anterior face. The reticulate wall structure is correspondingly finer. The species is clearly distinguished from others described here.

(2) In another species, designated here as *Eoradiolites* sp. B. from the Edwards Limestone, the attached valve is broad and thick and expands rapidly from the spat to near mature width (pl. 7, fig. 1). The siphonal bands are smooth. The pedal fold is moderately strong; the attached valve lies on its dorsal side. The dorsal surface is folded and irregular as is the anterior surface. The posterior surface has about six strong costae. Internally the body cavity is elongate in the posterior-anterior direction. The structure of the outer wall is typical for *Eoradiolites*. The free valve is unknown.

The species differs from *E. davidsoni* and related species in having a thicker, more rapidly expanding lower valve, an irregularly shaped body cavity, and a folded dorsoanterior surface. Other new species of *Eoradiolites* may be recognizable in the Edwards Limestone.

(3) An older, early Albian *Eoradiolites* is recognized from the Glen Rose Limestone as *E. plicatus* (pl. 4, figs. 1-3) in which the attached valve is subquadrangular in cross section. The dorsal edge has two rounded folds and a depression. The siphonal bands are not prominent and the interband area is shallow. Ribs are almost lacking on all but the ventral side. The wall structure of the species is distinctive. Where the shell is thin, the wall is composed of compact laminar layers. Where it is wide, the characteristic coarse reticulate texture of the *Eoradiolites* outer wall is present.

E. plicatus differs from *E. davidsoni* and similar species which are costate and have a coarsely reticulate wall. Other early Albian species are not well known. *E. choffati* has a circular cross section, smooth siphonal bands, and no plications; its wall structure is unknown. *E. rousseli* is externally smooth. It is a small shell which expands rapidly. Other details are lacking. *E. plicatus* could be confused with *Monopleura*. The latter, however, lacks a tabulate body cavity.

More complete information for this species in North America will require additional collections. Whitfield (1891) searched for Conrad's types and was not able to locate them at any of the museums where he would likely have deposited them.

E. plicatus is known from the middle Albian (zone of *Knemiceras syriacum*) with *Orbitolina lenticularis* at Nahr el Kelb, Libya, and at other sites in Libya and northern Iran. It is reported and figured here for the first time from the upper Glen Rose Limestone at several intervals in the Hodges No. 1 Fee, Sabine Parish, Louisiana.

Occurrence of *Eoradiolites*: Species of *Eoradiolites* are known to range from the early Albian (middle Glen Rose with *Orbitolina texana*) to the late Albian (Washita) strata of the Devils River Formation and in equivalent strata in the subsurface along the Gulf Coast. In addition, Eastern Hemisphere records place *Eoradiolites* as high as Turonian.

Sphaerulites Lamarck, 1819
(Plate 3)

The attached valve is broadly depressed, expands rapidly, and has a mushroom-like appearance. External folds are foliaceous, slack, and undulating. The siphonal zones are marked by sinuses and separated by folds. Internally two sinusal folds indent the shell cavity. The structure of the outer wall is probably finely to coarsely reticulate. The free valve is flatted or convex. The ligament is well developed.

Sphaerulites, one of the first named radiolitid genera, was named for flattened, spherical, mushroom-like shaped specimens from the Cenomanian of Ile d'Aix, near Rochefort, France. Younger species have a coarsely reticulate wall judging from Astre's (1954) illustrations of Turonian specimens.

Type Species: ***Sphaerulites foliaceous*** Lamarck, 1819. The earlier description by Deleméthérie (1805) is considered to have been a non-Linnean designation.

Several specimens of a radiolitid which occurs with *Coalcomana ramosa* in the Glen Rose Limestone are questionably referred to the genus *Sphaerulites*. The external shape and envelopment of external and internal siphonal folds are similar to *Sphaerulites*, but the coarsely imbricating foliaceous wall of younger species is lacking. The detailed wall structure of

PLATE 4
EARLY AND MIDDLE ALBIAN RADIOLITIDS
Species of *Eoradiolites* from the Glen Rose and Edwards Formations.

Figure

1-3. *Eoradiolites plicatus* Conrad from the upper Glen Rose Limestone, Hodges No. 1 Fee, Sabine Parish, Louisiana.
- (1) Cross section of attached valve near commissure, from a depth of 9,697 ft, xl. 5.
- (2) Cross section of valve showing ligament, from a depth of 9,670-77 ft, x5.
- (3) Anterior portion of another attached valve from a depth of 9,670-77 ft.
  - (a) Coarse reticulate wall, irregularly developed from the fine laminar structure, x5.
  - (b) Laminar wall develops into reticulate wall by "compressing" shell material to form funnel plates, x40.

4. *Eoradiolites* sp., longitudinal section of attached valve from the Stuart City Formation, Humble No. 1 Ben Ogletree, San Jacinto County, Texas, from a depth of 13,739 feet (middle Albian), x2.

5. *Eoradiolites*, cf. *E. perforata* (Palmer), from 5.5 miles south of Cranfills Gap on Texas 22, Hamilton County, Texas, xl.
- (a) Lateral posterior view.
- (b) Lateral view, attached valve, showing siphonal bands.
- (c) Oral view showing body cavity and ligament.

6. *Eoradiolites* sp. from Meridian State Park, Bosque County. Diagrammatic longitudinal cross section showing body cavity, sockets, attached valve tooth, tabulae, ligament, and outer reticulate wall structure, x2.

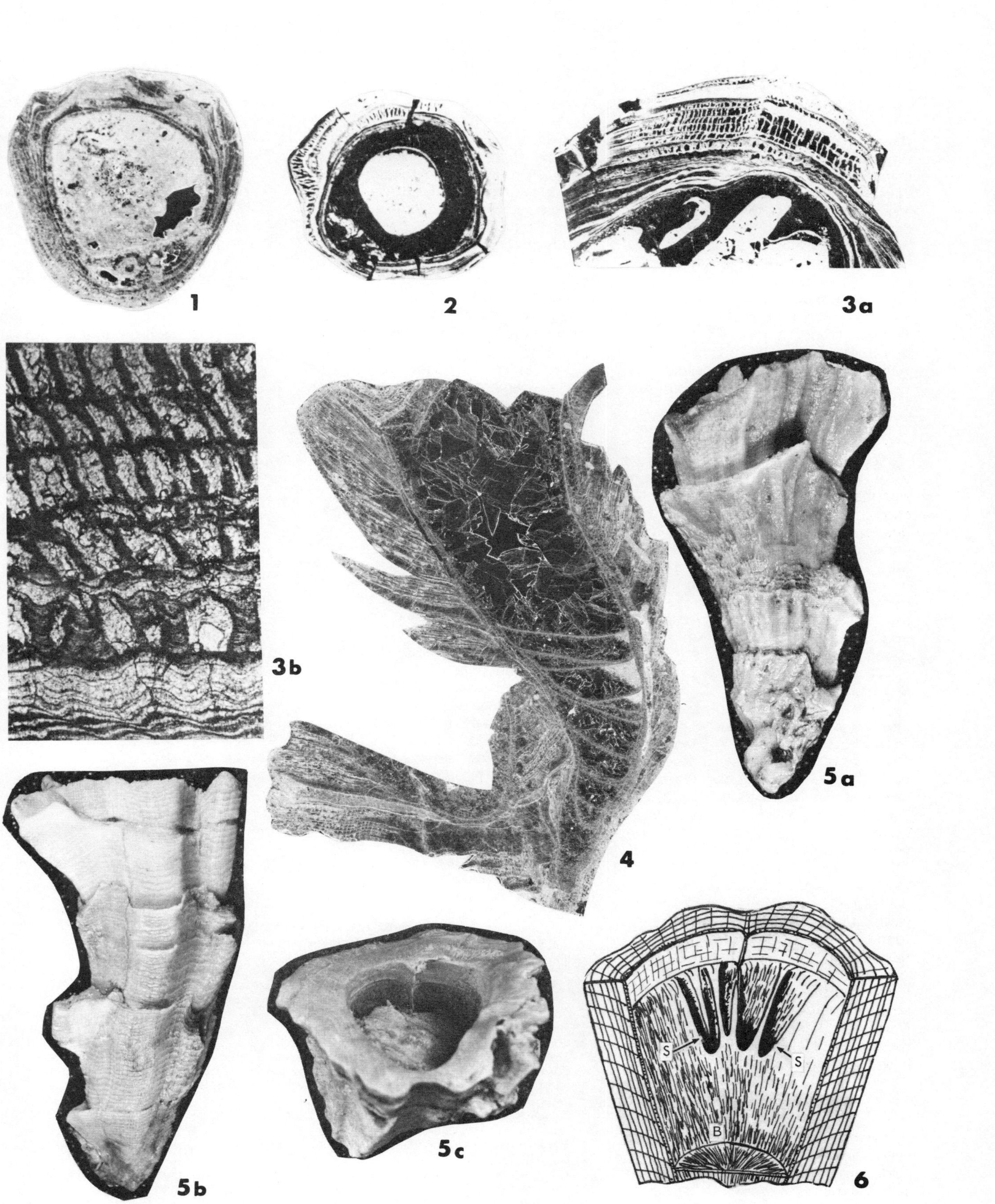

PLATE 4

the type species has not been described. Specimens from the Glen Rose have a finely reticulate outer wall. The free valve is convex.

The occurrence of this species at the Old Kreh Ranch locality is the first report of *Sphaerulites* or any specimens similar to *Sphaerulites* in the Western Hemisphere. The Glen Rose species is closest to *S. cantabricus* Douvillé (Toucas, 1907-1909) from the Albian of Portugal and the Pyrenees but is smaller, less foliaceous, and more flattened.

Occurrence: The genus ranges from the Albian to Maastrichtian in France. Many species in the early literature may belong to other genera, but approximately 10 species still can be reasonably assigned to *Sphaerulites*.

Praeradiolites Douvillé, 1902
(Plate 7)

Praeradiolites has an attached valve which is conical with smooth siphonal bands in the form of inverted cones. In younger species, the cones developed into an undulating chevron pattern around the whole valve. The outer wall is coarsely reticulate. The ligament is present. The free valve is operculiform and convex to flat; myophores are present. The stacked, undulating siphonal-band folds are distinctive of the genus.

Type Species: *Radiolites fleuriaui* d'Orbigny (1842)

About 40 species presently are assigned to *Praeradiolites*. The type is from the Cenomanian of France and has one of the least well-developed patterns of chevron folding. *Praeradiolites edwardensis* (Adkins, 1930) has an attached valve which is conical, squarish, and thick-walled. The siphonal bands are blunt, outward-folded horns. In cross section, the incurrent siphonal band is in a depression and not a projection of the valve as in *Eoradiolites*. The species is found in the Edwards Limestone near Belton, Bell County, Texas.

Occurrence: *Praeradiolites* is reported from strata of Albian to Maastrichtian age in central Asia, North Africa, and southern Europe. In the Western Hemisphere, the genus has been found in Texas and Jamaica.

Radiolites Lamarck, 1801
(Plates 8, 9)

Radiolites has a conical attached valve ornamented with longitudinal folds around the whole valve. The siphonal bands are smooth accentuations of the regular folds. The outer wall is coarsely reticulate. The free valve is small, conical, strongly convex or, rarely, flat. Myophore apophyses are present.

Type Species: *Ostracites angeiodes* Picot de Lapeirouse, 1781.

More than 50 species have been referred to the genus. The type comes from the Campanian of Aude, France. *Radiolites* differs from *Eoradiolites* in lacking strong, smooth siphonal bands and pedal fold. *Sphaerulites* has a broad foliaceous outer wall and mushroom-like shape. *Praeradiolites* has rounded siphonal bands in the form of collared horns.

Of the nine named species of *Radiolites* described from North America, only *Radiolites abraensis* (pl. 8) and two others from Mexico are clearly assigned to the genus. Two species named by Bauman (1958), *Radiolites mullerriedi* and *R. newelli* from the Turonian Limestones near Cuernavaca, Morelos, Mexico, are less than half the size of *R. abraensis*. *Radiolites maximus* Logan (1898) from Kansas is probably a species of *Biradiolites*. *Radiolites rugosa* Giebel (1853) from the Edwards Limestone of Texas is probably either *Eoradiolites* or *Praeradiolites* but is too poorly described to be given any generic assignment. The four species of Toumey (1854), *Radiolites undata, R. aimesii, R. lamellosis*, and *R. ormondii* from the Late Cretaceous of Alabama, are unfigured and too poorly described for placement even in subfamilies. The radiolitid described by Müllerried (1947) from faces of quarried building stone from Peñuela, Veracruz, Mexico, as *Neoradiolites ordonezi* is not a *Neoradiolites*. No ligamental ridge was found, so it is not known whether it belongs to the radiolitines or biradiolitines.

Occurrence: Species of Radiolites occur in strata of Cenomanian to Maastrichtian age in North America, Europe, North Africa, and Asia. The oldest confirmed North American species, *R. abraensis*, is from the Cenomanian Taninul Member of the El Abra Limestone near Valles, San Luis Potosi, Mexico. In addition, a possible specimen of *Radiolites* sp. (pl. 9, fig. 1) is recorded from a depth of 13,699 feet in the Humble No. 1 Ben Ogletree, San Jacinto County, Texas, in probably middle Albian strata.

Subfamily Sauvagesiinae Douvillé, 1908

This subfamily of radiolitid rudists is characterized by species with a wall structure in the attached valve composed of cells of intersecting vertical radial and funnel plates which appear prismatic in longitudinal section but polygonal in transverse section. Of the seven genera in the subfamily, four are from the Western Hemisphere but only one of these, *Sauvagesia*, occurs in Middle Cretaceous strata.

Sauvagesia Choffat, 1886
(Plate 9)

The attached valve is conical to cylindro-conical and is ornamented with longitudinal ribs. The siphonal bands are finely costulate folds separated by an interband. The ligamental ridge is present. The free valve is operculiform with radial folds.

Type species: *Sphaerulites sharpei* (Bayle, 1857).

Approximately 20 species of *Sauvagesia* described to date are about equally distributed between the Eastern and Western Hemispheres. They are associated with both reef and open-shelf sediments in the Late Cretaceous. Only one species of the genus, *Sauvagesia texana*, is known

PLATE 5
MIDDLE ALBIAN RADIOLITIDS
Species of *Eoradiolites* from the middle Albian Edwards Limestone, Central Texas.

Figure

1. *Eoradiolites davidsoni* (Hill) from one-half mile east of Meridian, Bosque County, Texas. Longitudinal section of both valves showing body cavity of last living chamber floored by tabular portion and roofed by free valve with downward extended tooth, xl. 5.
2. *Eoradiolites* sp. Diagrammatic view of free valve showing teeth and posterior muscle attachment area (mp), x2.
3. *Eoradiolites davidsoni* (Hill) from same locality as 1. Two individuals grown together. Siphonal band of right specimen strongly indents anterior side of other specimen, xl. 5.
4. *Eoradiolites davidsoni* (Hill) from 6 miles south of Meridian, Bosque County, Texas. Cross section of attached valve, outer reticulate wall showing bifurcating radial and funnel plates, x40.
5. *Eoradiolites davidsoni* (Hill) from the same locality as 1. Posterior-ventral external view showing siphonal bands, xl.
6. *Eoradiolites davidsoni* (Hill) from 3 miles northwest of Gatesville School for Boys, Coryell County, xl.
   (a) Oral view, free valve recessed as tightly fitting plug into body of an attached valve.
   (b) Ventral side showing smooth fold of incurrent siphonal band.
7. *Eoradiolites* sp. Cross section of valve from same locality as 1. Section shows two large teeth, myophores, and dorsal portion of free valve set at attached valve, xl. 5.
8. *Eoradiolites davidsoni* (Hill) from Belton, Bell County, Texas, xl. Large specimen designated as paratype of *E. quadratus* by Adkins (1930).

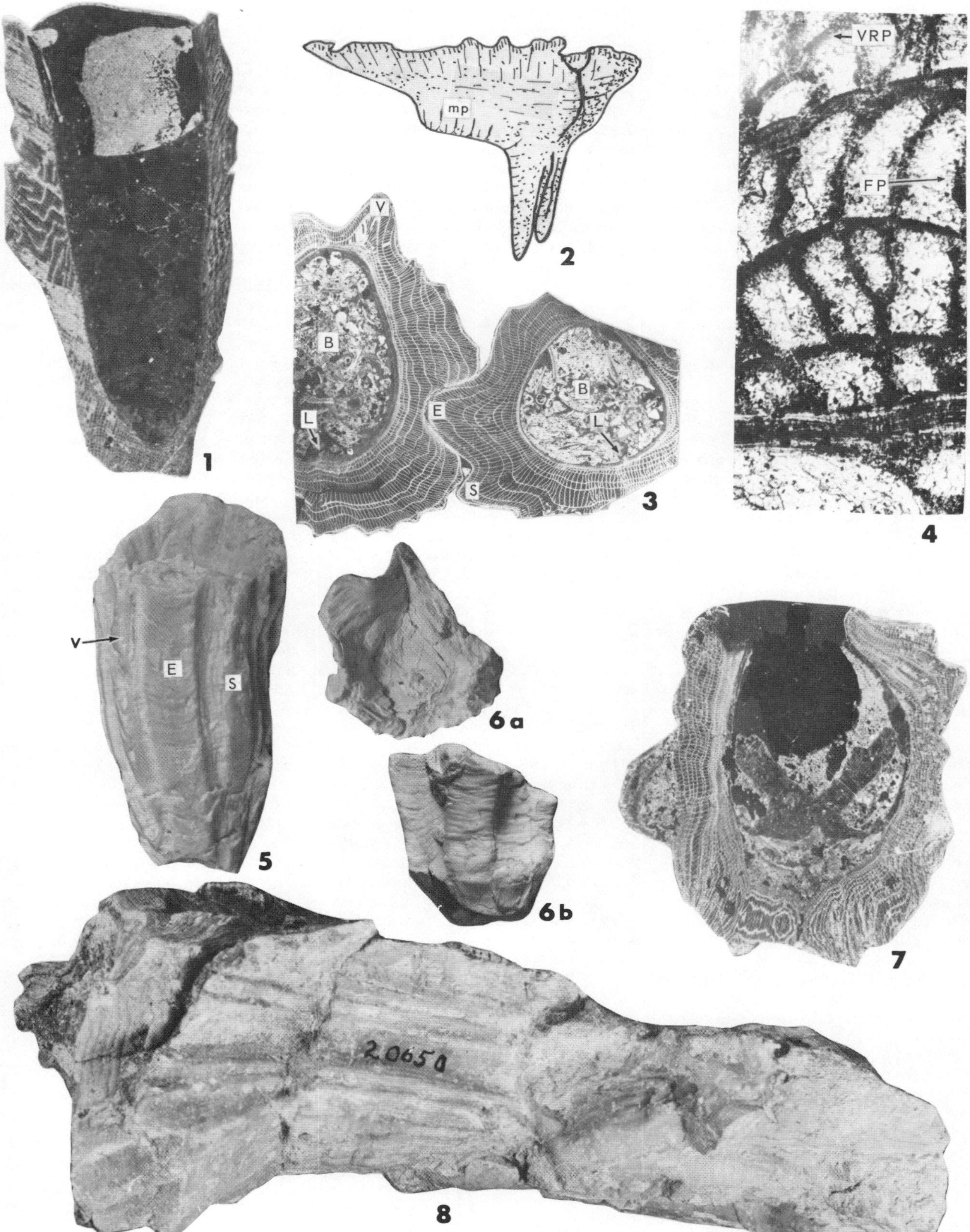

PLATE 5

from Middle Cretaceous strata. ***S. texana***, one of the oldest named species, was described first by Roemer (1849) from float in the Guadalupe River near New Braunfels, Texas. Although generally listed as part of the Edwards Limestone fauna, its true occurrence in Texas remains unknown. It is recorded from Cenomanian El Abra Limestone near Valles, San Luis Potosí, Mexico (Coogan, 1973). The species is characterized by its regular conical shape and the nearly 12 equal, strong rounded ribs of the attached valve. The closest related species, ***S. nicaisei*** Coquand from the Cenomanian of Algeria and France, has more numerous less-pronounced ribs and clearly differentiated incurrent and excurrent sinuses.

Occurrence: The genus ranges from possibly late Albian (float from undifferentiated Edwards-Washita limestones near San Marcos, Texas) to Maastrichtian and is found in Europe, North Africa, the Antilles, and the Gulf Coast. ***S. texana*** is also reported from the early Cenomanian strata north of Trieste in Yugoslavia (Wiontzek, 1934). Other North American genera of the subfamily, including ***Durania, Chiapasella***, and ***Tampsia***, occur in Turonian or younger strata.

Family Caprinidae Fischer, 1887

The shells are inequivalve. The attached valves can attain large size. Teeth and myophores are well developed, massive in some species. Accessory cavities and pallial canals formed of bifurcating radial plates may occur in one or both valves.

The Caprinidae are an extremely diverse group of rudists with over 25 genera. Controversy exists regarding the natural affinities of various genera, and, as a consequence, there are several methods of classification of caprinid genera within the family. One method, adopted by the authors of the Treatise, Part N (Moore, 1969), groups all caprinid genera together without further differentiation, on the rationale that many genera need restudy before a sound phylogenetically based classification can be derived. On the other hand, several subfamilies of caprinids have been recognized in the last century and, even if subject to revision, are useful now for recognizing biogeographically and stratigraphically distinct groups of caprinid genera.

In the Treatise, Part N (Moore, 1969), Cox recognized four groups of caprinids:

(1) forms typical of the family including ***Caprina, Caprinula, Neocaprina, Orthoptychus, Paracaprinula, Plagioptychus, Praecaprina*** and ***Sphaerucaprina***,

(2) variants possibly representing expression of regional differentiation (***Amphitriscoelus, Antillocaprina, Caprinuloidea, Coalcomana, Kipia, Mitrocaprina, Offneria, Planocaprina, Sabinia***, and ***Schiosia***),

(3) forms similar in morphology to radiolitids (***Coralliochama, Dictyoptychus, Ichthyosarcolites***, and ***Titanosarcolites***), and

(4) imperfectly known forms assignable to the Caprinidae only because of the presence of accessory cavities.

In contrast, a number of subfamilies have been named for various groupings of caprinid genera. These may be summarized as follows:

(1) Subfamily Caprininae Fischer, 1887, including ***Caprina, Caprinula, Neocaprina, Schiosia***, and possibly ***Sabinia, Orthoptychus, Paracaprinula, Praecaprina***, and ***Sphaerucaprina***. The subfamily as so defined is restricted to the Eastern Hemisphere. The species assigned to ***Sphaerucaprina*** in Jamaica by Chubb (1971) appears to be a coalcomaninid.

(2) Subfamily Antillocaprininae MacGillavry, 1937, including ***Antillocaprina*** and ***Titanosarcolites***. As defined, the subfamily is restricted to Late Cretaceous strata of the Western Hemisphere.

(3) Subfamily Plagioptychinae Douvillé, 1887, including the genera ***Plagioptychus, Mitrocaprina***, and perhaps ***Coralliochama***. As defined, the subfamily is restricted to the Late Cretaceous but is cosmopolitan.

(4) Subfamily Coalcomaninae Coogan, 1973, including ***Amphitriscoelus, Planocaprina, Coalcomana, Caprinuloidea, Kimbleia, Texicaprina, Mexicaprina***, and possibly ***Kipia***. As defined, the subfamily is restricted to Barremian-Cenomanian strata of the Western Hemisphere.

Other named subfamilies of less clear identity are Ichthyosarcolitinae Douvillé, Rousselinnae MacGillavry, and Trechmannellinae, the latter for ***Trechmanella***, an objective synonym of ***Dictyoptychus***.

Of these, the genera of the subfamily Coalcomaninae are considered here mainly because they are the great reef and bank builders of the Aptian, Albian, and Cenomanian of the Gulf Coast and are useful stratigraphically.

Subfamily Coalcomaninae Coogan, 1973

The attached valve is elongate, curved, or slightly coiled. The free valve is horn shaped and tightly or loosely coiled. In the attached valve, the accessory cavity adjacent to the posterior myophore is wide, short, and connected with the posterior tooth socket. A cross section of this combined tooth plus accessory cavity is commonly shaped like a small yellow squash or bean. The accessory cavity is set off from the peripheral vertical radial plates. Both valves have similar or identical peripheral vertical plate arrangement. The ligament is marked by an external groove; internally it forms an anchor-shaped deep fold in the valve wall. The anterior tooth of the free valve is large. The genera differ mainly in the complexity of the vertical plate development.

Occurrence: Barremian to Cenomanian in the Western Hemisphere and on the sunken Japanese and mid-Pacific guyots (Matthews and others, 1974; Hamilton, 1956).

Amphitriscoelus Harris and Hodson, 1922 (Plate 10)

Coalcomaninid rudist in which the vertical radial plates do not bifurcate.

Type species: ***Amphitriscoelus waringi*** Harris and Hodson, 1922.

PLATE 6
MIDDLE AND LATE ALBIAN RADIOLITIDS
Species of *Eoradiolites* from the Edwards, Stuart City and Devils River Formations.

Figure

1. ***Eoradiolites quadratus*** Adkins. Attached valve designated by Adkins (1930) as type of ***E. quadratus*** from the Edwards Limestone, Nolan Creek, southeast of Belton, Bell County, Texas, xl.
   (a) Anterior external view.
   (b) Cross section showing deep interband and thin outer wall.
2. ***Eoradiolites*** sp. from the upper few feet of the Devils River Formation in a road cut on U. S. 90, just east of the high bridge on the Pecos River, Val Verde County, Texas, late Albian. Transverse section of attached valve, x2.
3. ***Eoradiolites*** sp. from the Edwards Limestone, 5.5 miles south of Cranfills Gap on Texas 22, Hamilton County, Texas. Colony of individuals showing overlap of younger shells on older shells, xl.
4. ***Eoradiolites*** sp. from the Stuart City Formation, middle Albian, in the Humble No. 1 Ben Ogletree, San Jacinto County, Texas, at a depth of 13,699 feet, x2.5.
5. ***Eoradiolites*** from the Stuart City Formation (late Albian) in the Humble No. 1 Weston Lumber Company, and others, Hancock County, Mississippi.
   (a) Transverse section of attached valve, x3.
   (b) Detailed section of outer wall showing vertical radial and funnel plates, x40.

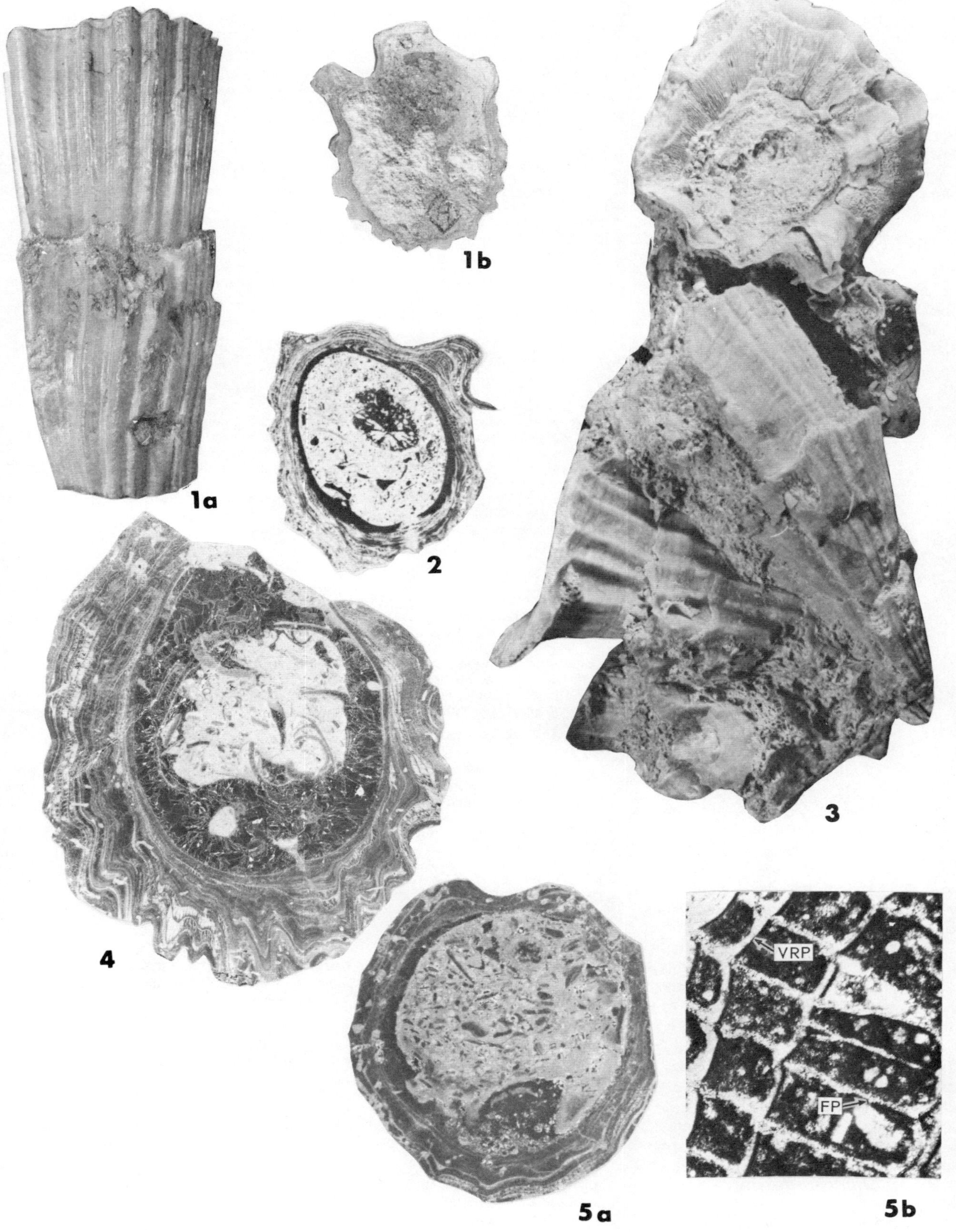

PLATE 6

Occurrence: The genus and species are known from strata probably assignable to the Barremian or Aptian age in Trinidad and Venezuela.

Planocaprina Palmer, 1928
(Plate 10)

Coalcomaninid rudist in which the vertical radial plates bifurcate ventrally.

Type species: *Planocaprina trapezoides* Palmer, 1928.

The genus was proposed for caprinids in which the bifurcating plates ventrally form pyriform canals in both valves. Two species belong to the genus, ***P. trapezoides*** and the more common form described as ***"Caprinia* cf. *adversa"*** by Boehm (1898). ***Planocaprina*** is intermediate between ***Amphitriscoelus*** in which the plates do not bifurcate and ***Coalcomana*** in which they trifurcate along the ventral side.

Occurrence: ***Planocaprina*** is known from outcrops in Jalisco and Veracruz, Mexico, and from limestones correlated with the Sligo Formation in wells in Texas. It occurs with ***Chofatella*** and other Aptian fossils. The Cenomanian age attributed to ***Planocaprina*** by Palmer (1928) is incorrect.

Coalcomana Harris and Hodson, 1922
(Plate 11)

Coalcomaninid rudist in which the vertical radial plates trifurcate along the ventral side, peripheral to the body cavity and to the posterior accessory cavity. The plates are not connected by tangential plates to form circular or polygonal canals.

Type species: ***Caprina ramosa*** Boehm, 1898.

Two named species of ***Coalcomana, C. ramosa***, and ***C. texana*** occur in Mexico and Texas, respectively. ***C. texana*** from the middle Glen Rose Limestone is either assignable to ***C. ramosa*** or differs from it in external form. ***C. ramosa*** is known from Orizaba, Veracruz, and Coalcoman, Michoacan, Mexico, and from Las Villas Province, Cuba (Thiadens, 1936). The specimen designated by Palmer (1928) as ***Caprinuloidea costata*** from Jalisco is referable to ***Coalcomana ramosa.*** The age recognized here is based on the Texas occurrences.

Occurrence: Early Albian strata of Texas, Mexico, and Cuba, commonly found with ***Orbitolina texana.***

Caprinuloidea Palmer, 1928

Coalcomaninid rudist in which the vertical radial plates trifurcate and are connected laterally to adjacent radial plates by tangential ones to form a set or sets of polygonal canals. These polygonal canals develop inside a row of pyriform canals similar to those present in ***Coalcomana*** and ***Planocaprina***.

Type species: ***Caprinuloidea perfecta*** Palmer, 1928.

About 12 species of coalcomaninids possibly referable to ***Caprinuloidea*** have been named in the last century. Four of these, ***Caprina crassifibra*** Roemer (1849), ***Caprina***

PLATE 7 *(page 53)*
MIDDLE ALBIAN RADIOLITIDS
Species of *Eoradiolites* and *Praeradiolites* from the Edwards Limestone.

Figure
1. ***Eoradiolites*** sp. B. from a quarry on the Middle Bosque River, 3 miles southeast of Crawford, McLennan County, Texas.
   (a) Lateral view of ventral side. Smooth siphonal bands are prominent, xl.
   (b) Cross section showing outer wall, siphonal bands, and body cavity, xl. 5.
2-3. ***Eoradiolites angustus*** Adkins from Nolan Creek, southeast of Belton, Bell County, Texas, external views, xl.
   (2a) Type specimen, ventral view.
   (2b) Anterior view.
   (3) Ventral view of Adkins' specimen No. 20653.
4. ***Eoradiolites*** sp. A., attached valve of paratype of ***E. quadratus*** Adkins (1930) from Belton, Bell County, Texas, xl.
   (a) Anterior lateral view showing curvature and rapid expansion of valve.
   (b) Oral view showing quadrate shape.
5. ***Praeradiolites edwardensis*** Adkins from Belton, Bell County, Texas, xl.
   (a) Oral view of attached valve, siphonal bands at 55° to ligamental axis.
   (b) Ventrolateral view showing external folds of ***Praeradiolites.***
6. ***Praeradiolites*** sp. from a quarry on the Bosque River, 3 miles southeast of Crawford, McLennan County, Texas, xl.
   (a) Longitudinal section of both valves. Free valve sits in body cavity of attached valve. Lower portion of body cavity tabulate. Edges show imbricated folds.
   (b) Lateral external view showing imbricated folds.

PLATE 8 *(page 54)*
CENOMANIAN RADIOLITIDS
***Radiolites abraensis* Coogan from the Cenomanian El Abra Limestone, Taninul member, 14 km east of Valles, San Luis Potosí, Mexico.**

Figure
1. Longitudinal section of attached valve showing thin inner wall of dense calcite merging with body tabula. Outer wall of fine vertical radial and funnel plates, x5.
2. Anterior-posterior longitudinal section of both valves showing the large muscle attachment areas of the free valve and dark outer wall and light inner wall of the attached valve, xl. 5.
3. Colony of individuals cut at three different levels below the commissure, xl.
   (a) Highest cut showing free and attached valves. Muscle attachment areas of free valve, ligamental crest, and stringy calcite supports for the retractor (?) muscles show clearly.
   (b) Cut below level of (c) shows siphonal inflexions on attached valve border of shell, broad interband, and curved and notched inner marginal muscle attachment areas in attached valve.
   (c) Cut between level of (a) and (b) showing free valve myophores and teeth set in attached valve.
4. Free valve upper surface showing regular hemiperipheral growth, siphonal notches, and intersiphonal band, xl.
5. Exterior of attached valve with evenly ribbed surface, xl.

PLATE 9 *(page 55)*
ALBIAN AND CENOMANIAN RADIOLITIDS
Species of *Radiolites* (?) and *Sauvagesia* from the Albian and Cenomanian of Texas and Mexico.

Figure
1. ***Radiolites*** (?) sp. from a depth of 13,699 feet in the Humble No. 1 Ben Ogletree, San Jacinto County, Texas, Stuart City Formation, middle Albian.
   (a) Longitudinal section of attached valve showing body cavity and outer wall, xl. 5.
   (b) Attached valves of two individuals. Specimen on the upper left shows ligament, large myophores and teeth of free valve, and outer wall of attached valve.
   (c) Enlargement of outer wall of attached valve showing finely reticulate structure, x40.
2. ***Sauvagesia texana*** (Roemer) from the late Albian (?) Edwards Limestone (?) bound by Roemer as a cobble in the Guadalupe River near New Braunfels, Comal County, Texas. From Roemer (1852), xl.
   (a) Attached valve, lateral view.
   (b) Attached valve, oral view.
3. ***Sauvagesia texana*** (Roemer) from the back-reef calcarenite of the El Abra Limestone, 14 km east of Valles, San Luis Potosí, Mexico, Cenomanian.
   (a) Transverse section of attached valve wall showing polygonal canals, x5.
   (b) Longitudinal section of attached valve wall showing rectangular cells, x5.
   (c) Transverse section of partial attached valve, xl.

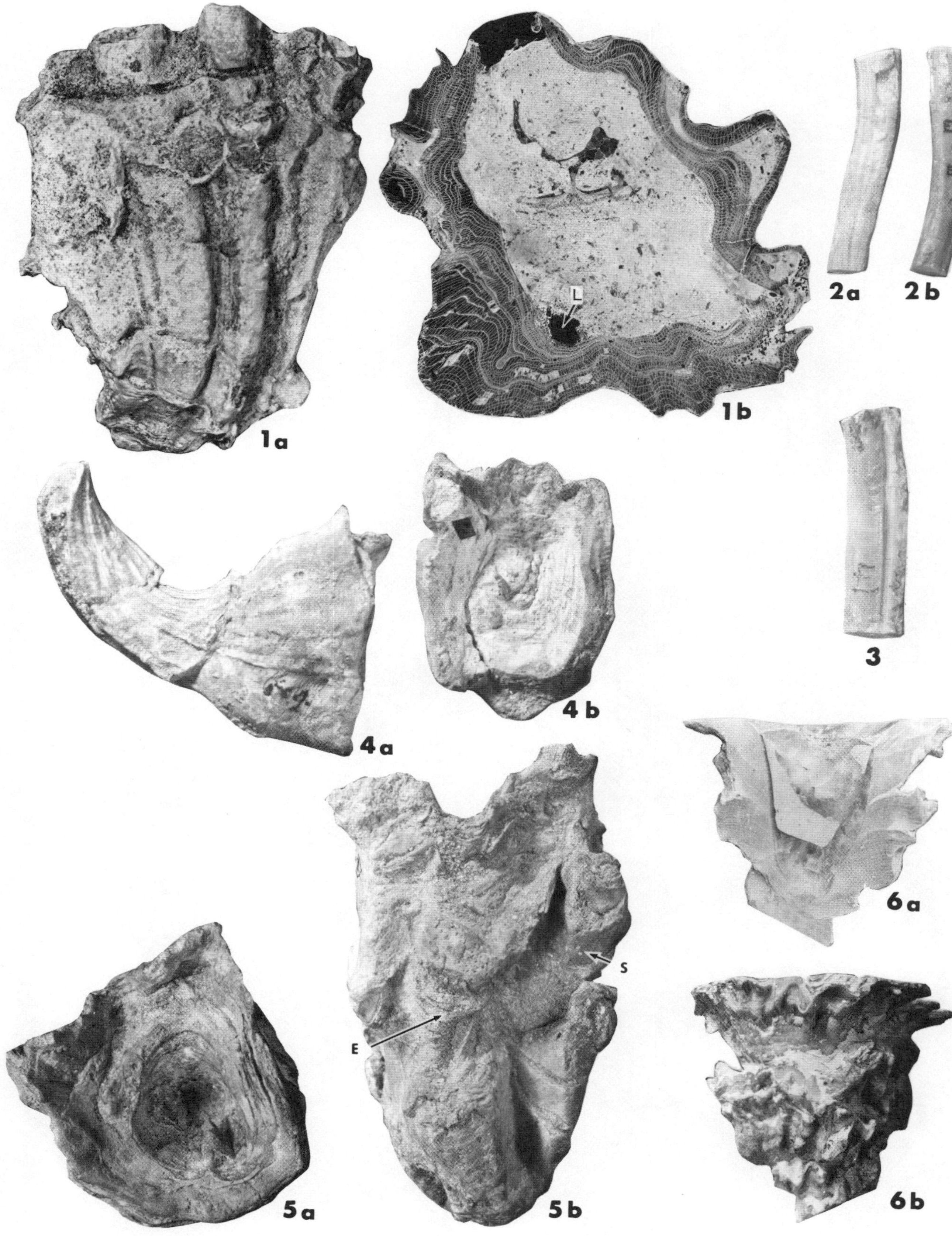

PLATE 7

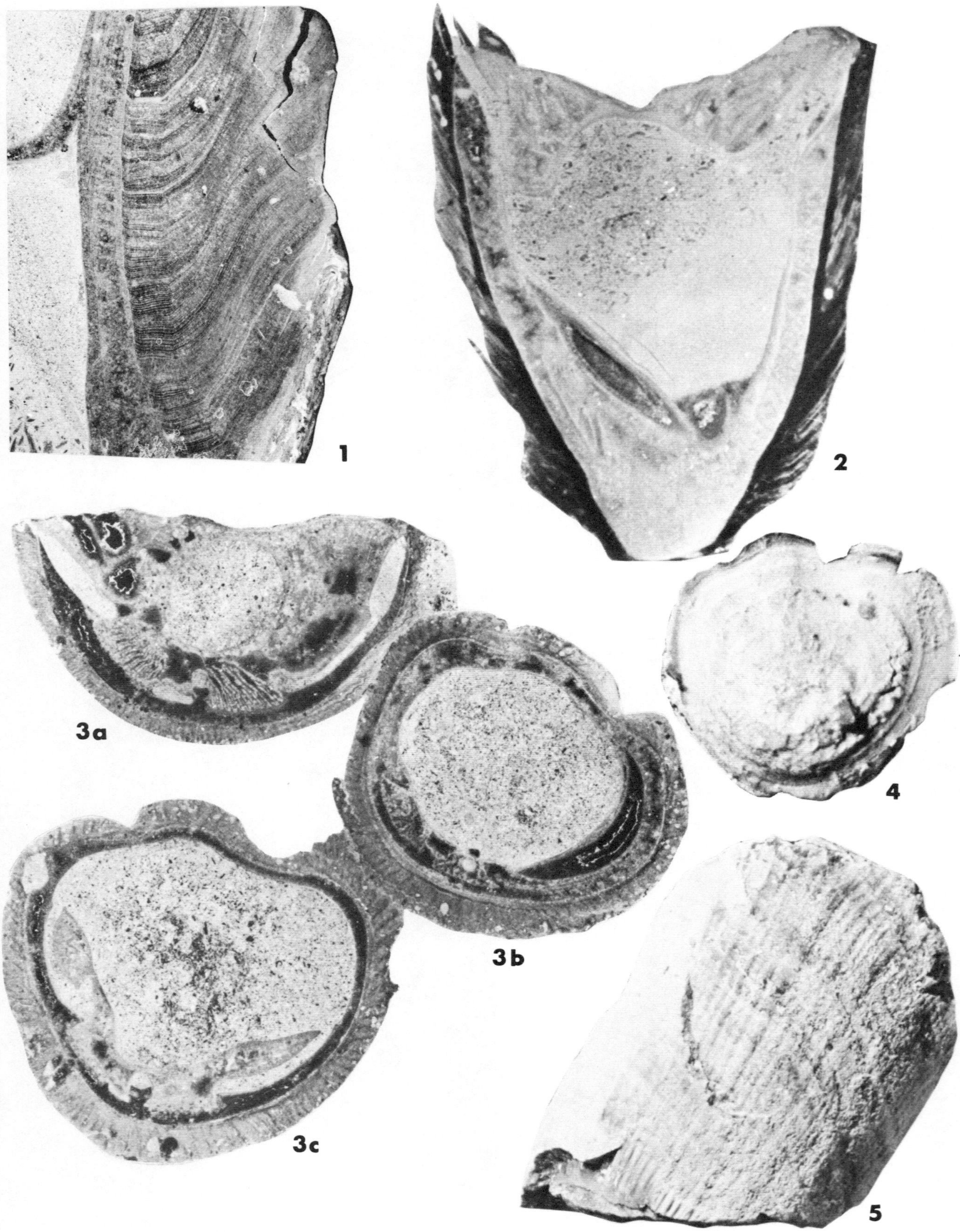

PLATE 8

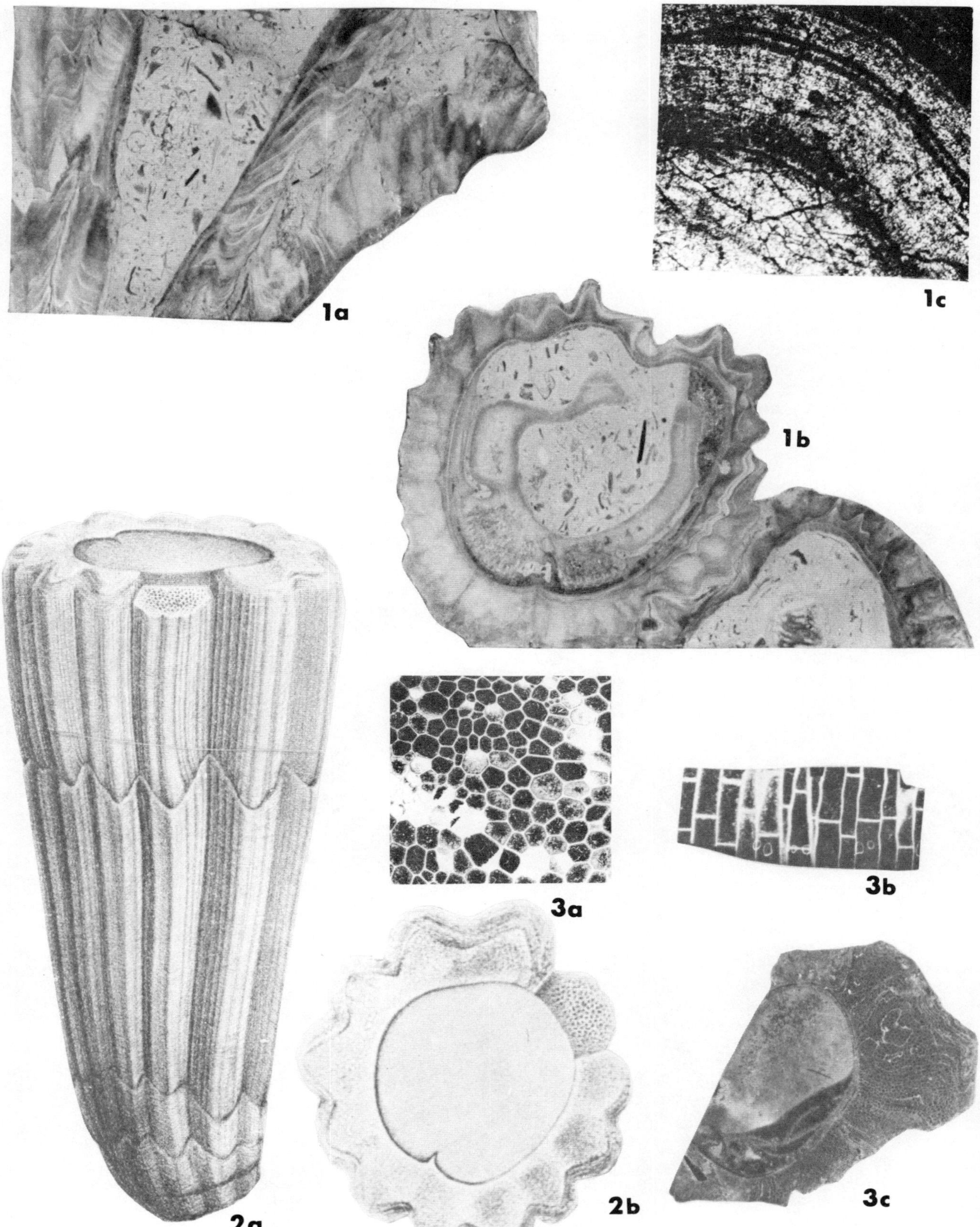

PLATE 9

guadalupae Roemer (1849), *Caprina occidentalis* Conrad (1855), and *Caprina planata* Conrad (1855), although named for specimens from the Texas Albian strata, are too poorly described and figured to be recognizable. In addition, Roemer (1888) also named *Ichthyosarcolites anguis*, a Texas species assignable to *Caprinuloidea*. Boehm (1898) named two species of *Caprinuloidea* from Mexico as *Sphaerucaprina lenki* and *S. felixi*. In 1928, Palmer named *Caprinuloidea perfecta, C. perfecta gracilis, C. multitubifera, C. septata*, and *C. bisulcata* from Mexico. Restudy of the literature and materials suggests that there now are eight available names for species of *Caprinuloidea*, but only about four distinct morphologic types. These are the:

1. *Caprinuloidea perfecta* group (pls. 12-13), including *C. perfecta, C. septata,* and *C. lenki* which have simple tangential plates, a large shell, and one to two rows of polygonal canals.
2. *C. gracilis* group (pls. 12-13), including *C. gracilis* and *C. anguis* which have two or more rows of polygonal canals and a small shell.
3. *C. bisulcata* with two or more rows of polygonal canals, large shell, and external sulcus. Restudy may place this in the next group.
4. *C. multitubifera* group (pl. 13) including *C. multitubifera, C. felixi,* and possibly other unnamed specimens. The shell has four or more rows of polygonal canals; the canals are small.

The *Caprinuloidea perfecta* group deserves special note. The specimens of *C. perfecta, C. septata*, and *C. lenki* may simply represent one species. In any case, all have a free valve with ventrally trifurcating plates and a single row of polygonal canals. The attached valve has two rows of polygonal canals. The *C. perfecta* group is older than the other species groups of *Caprinuloidea* congruent with its less well developed polygonal canal system which is intermediate between that of *Coalcomana* and *C. gracilis. C. perfecta* occurs in Glen Rose Limestone in Trinity County, Texas, and a free valve was found in the "Edwards Limestone" in the Mobil No. 1 Kahanek, Lavaca County, Texas, with both *Dictyoconus walnutensis* and *Orbitolina texana* in strata probably correlative with the Paluxy Sandstone of north Texas.

Caprinuloidea gracilis is common in middle and late Albian surface and subsurface strata in Texas and Mexico as are *C. anguis, C. bisulcata*, and *C. multitubifera*. The latter is also found in Cenomanian strata of the Taninul member, El Abra Limestone.

Occurrence: Species of *Caprinuloidea* range in age from early Albian to Cenomanian along the Gulf Coast. The *C. perfecta* group is the older less well-developed early Albian group of species.

Texicaprina Coogan, 1973
(Plates 14, 15, 17)

A coalcomaninid caprinid in which the attached valve is elongate, subrounded, or flattened and slightly curved. The free valve is shorter and curved. In the attached valve, the accessory plus tooth socket is shaped

PLATE 10 *(page 57)*
BARREMIAN-APTIAN COALCOMANINID CAPRINIDS
Species of *Amphitriscoelus* and *Planocaprina* from Trinidad, Mexico, and Texas.

Figure
1. *Amphitriscoelus waringi* Harris and Hodson from the Aptian of Trinidad, from Harris and Hodson (1922, pl. 4), xl.
   (a) Cap-shaped free valve with external trace of ligament.
   (b) Cross section of attached valve showing body cavity, accessory cavity, and canals formed of marginal radial plates.
   (c) Cross section of attached valve showing unbranched radial plates.
2. *Planocaprina trapezoides* Palmer from Soyatlan de Adentro, Jalisco, Mexico, from Palmer (1928, p. 65-58).
   (a) Both valves, upper one restored, x0.5.
   (b) Free valve showing trapezoidal outline of species. Note bifurcating radial plates marginal to outer wall.
3. *Planocaprina* sp. from the Cerro Escamela, Orizaba, Veracruz, Mexico, from Boehm (1898, p. 326-327) designated there as *Caprina adversa*, xl.
   (a) Cross section of free valve showing bifurcating radial plates.
   (b) Cross section of attached valve.
4. *Planocaprina* sp. from the Sligo Formation, Pan American No. 1 Lutcher-Moore, Newton County, Texas, Aptian. Attached valve from a depth of 15,143-46 feet, xl.5.
5. *Planocaprina* sp. from the Sligo Formation, Mobil No. 1 DD Spanihal, De Witt County, Texas. Oblique section of free valve showing bifurcating radial plates along ventral margin. From a depth of 16,495-98 feet, xl.5.

PLATE 11 *(page 58)*
EARLY ALBIAN COALCOMANINID CAPRINIDS
Species of *Coalcomana* from the early Albian strata of Texas, Louisiana, and Mexico.

Figure
1. *Coalcomana ramosa* (Boehm) from Coalcoman, Michoacan, Mexico, from Douvillé (1900, figs. 1-4), slightly less than xl.5.
   (a) Free valve with massive teeth, pyriform canals on the ventral side of shell margin, and polygonal canals on the dorsal and anterior margin formed of bifurcating radial plates.
   (b) Attached valve with same pattern of canals as free valve.
2. *Coalcomana texana* Whitney from about 235 feet above the base of the Glen Rose Limestone, 3 miles north of Hancock, Comal County, Texas, from Whitney (1952, pl. 87). Length about 55 cm, width about 5 cm.
3. *Coalcomana ramosa* (Boehm) from the middle Glen Rose Limestone at the Blanco River, 5.1 miles north of Fischer (Store) on the Rod-E Ranch (Old Kreh Ranch locality), Hays County, Texas. Free valve showing pyriform ventral marginal radial plates, xl.5.
4. *Coalcomana* sp. from the middle Glen Rose Limestone, Carter No. 1 Mansfield, Sabine Parish, Louisiana. Oblique cross section of free valve from a depth of 5,697 feet, xl.5.

PLATE 12 *(page 59)*
EARLY AND MIDDLE ALBIAN COALCOMANINID CAPRINIDS
Species of *Caprinuloidea* from Texas and Mexico.

Figure
1. *Caprinuloidea perfecta* Palmer from the upper Glen Rose (early Albian) strata in the Pauley No. 1 Cameron Heirs, Trinity County, Texas. Cross section of attached valve showing sets of trifurcating marginal radial plates. From a depth of 11,727 feet, xl.5.
2. *Caprinuloidea perfecta* Palmer from Soyatlan de Adentro, Jalisco, Mexico. Type specimen from Palmer (1928, pl. 11), xl.5.
   (a) Free valve, drawing of (b) to show development of polygonal canals.
   (b) Free valve turned to show internal details. Note massive projecting teeth.
   (c) Elongate attached valve and smaller free valve.
3. *Caprinuloidea anguis* (Roemer) from the Edwards Limestone (middle Albian) on Barton Creek about 2 miles from its confluence with the Colorado River, Austin, Texas, from Roemer (1888), xl.
   (a) Transverse section showing shape of attached valve, size, and marginal canals. Accessory cavity not shown in rendering.
   (b) Cap-shaped free valve of small specimen showing projecting teeth and marginal canals.
   (c) Specimen from Roemer's collection at Breslau, Germany, described and figured by Douvillé (1900), probably the same as illustrated in (b), showingfree valve with accessory cavity.
   (d) Complete specimen, from Roemer (1888), showing coiled free valve and elongate attached valve.
4. *Caprinuloidea* sp., cf. *C. anguis* (Roemer), attached valve from the middle Albian strata in the Humble No. 1 Ben Ogletree at a depth of 13,698 feet, xl.5.

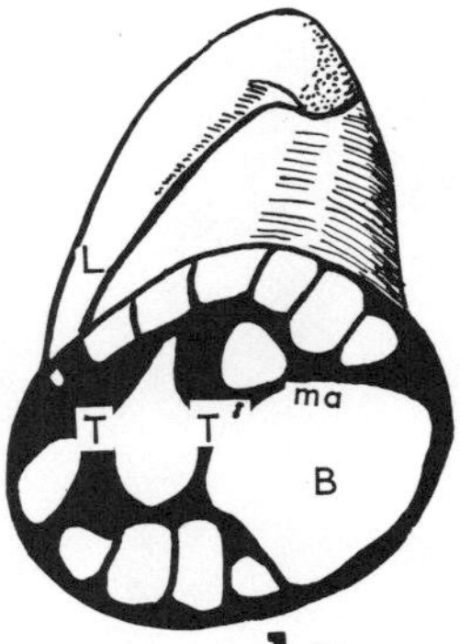

1a

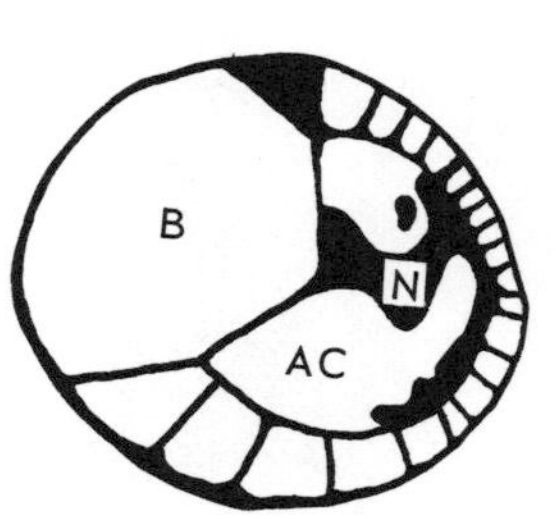

1b

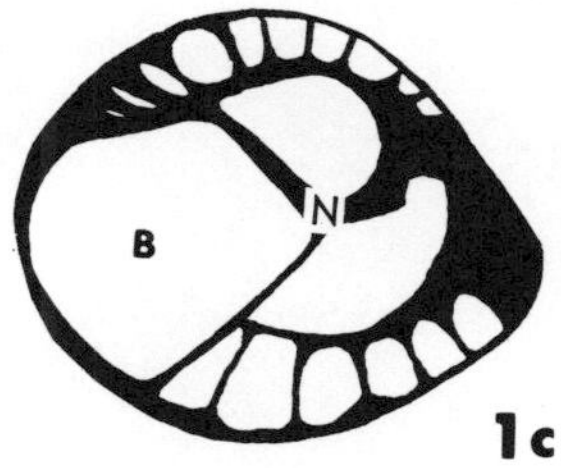

1c

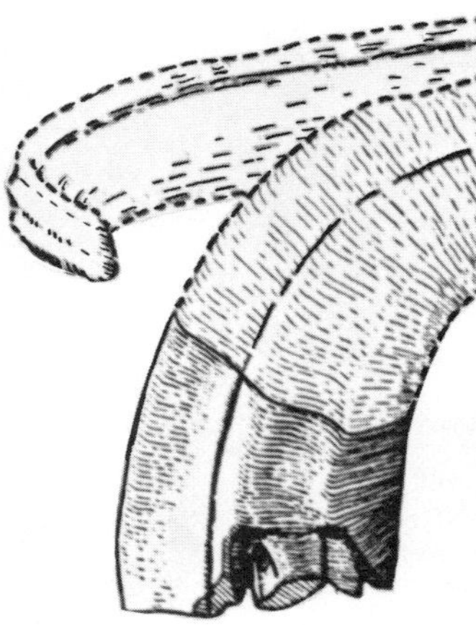

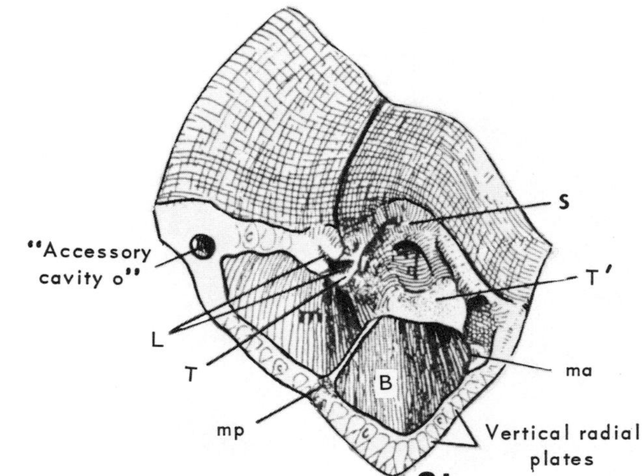

2b

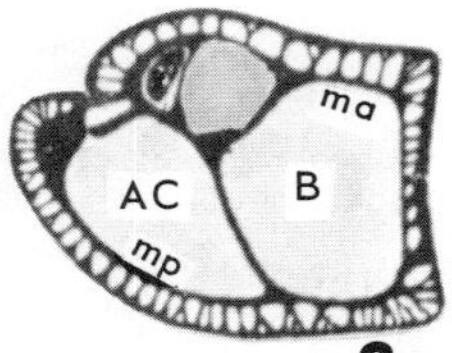

3a

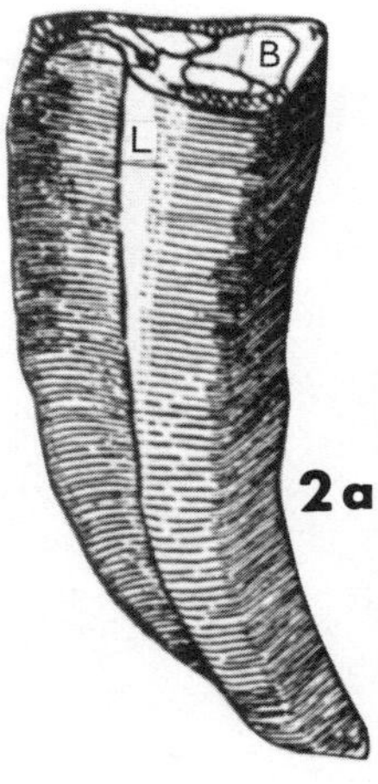

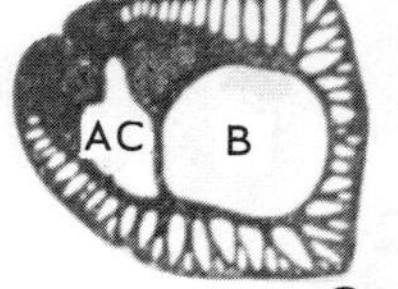

3b

2a

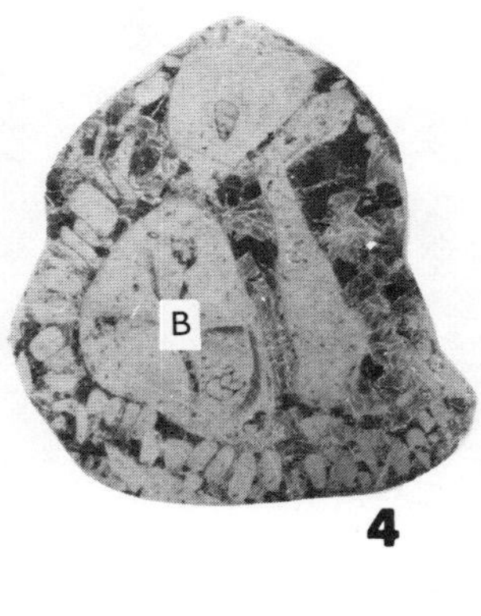

4

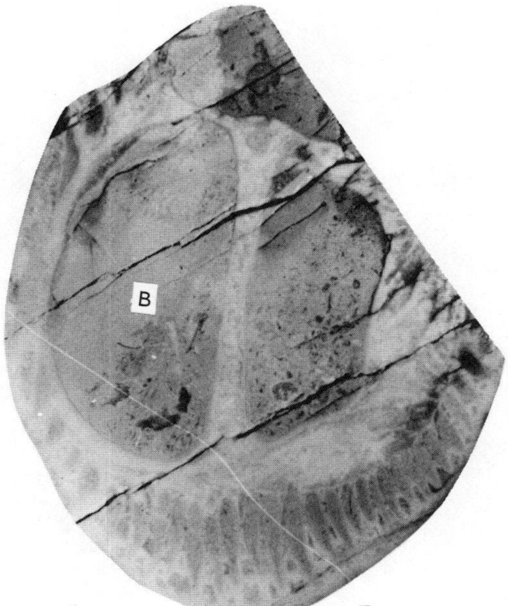

5

PLATE 10

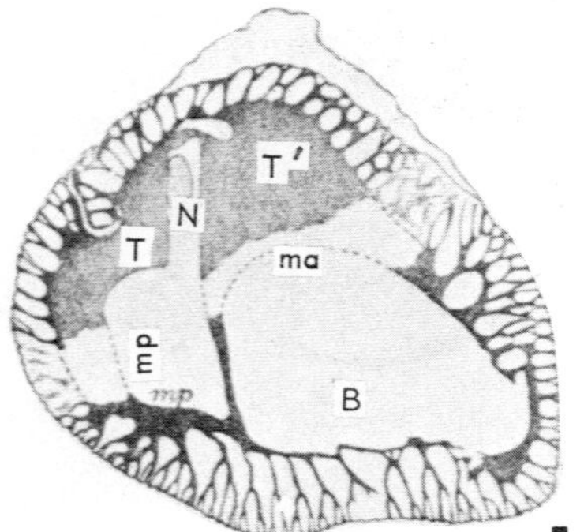

1a

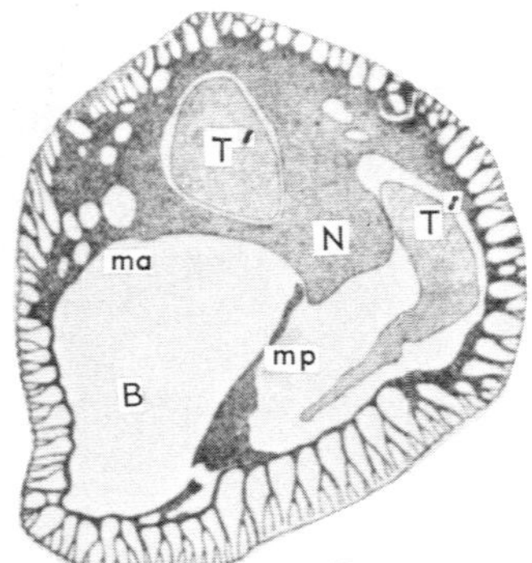

1b

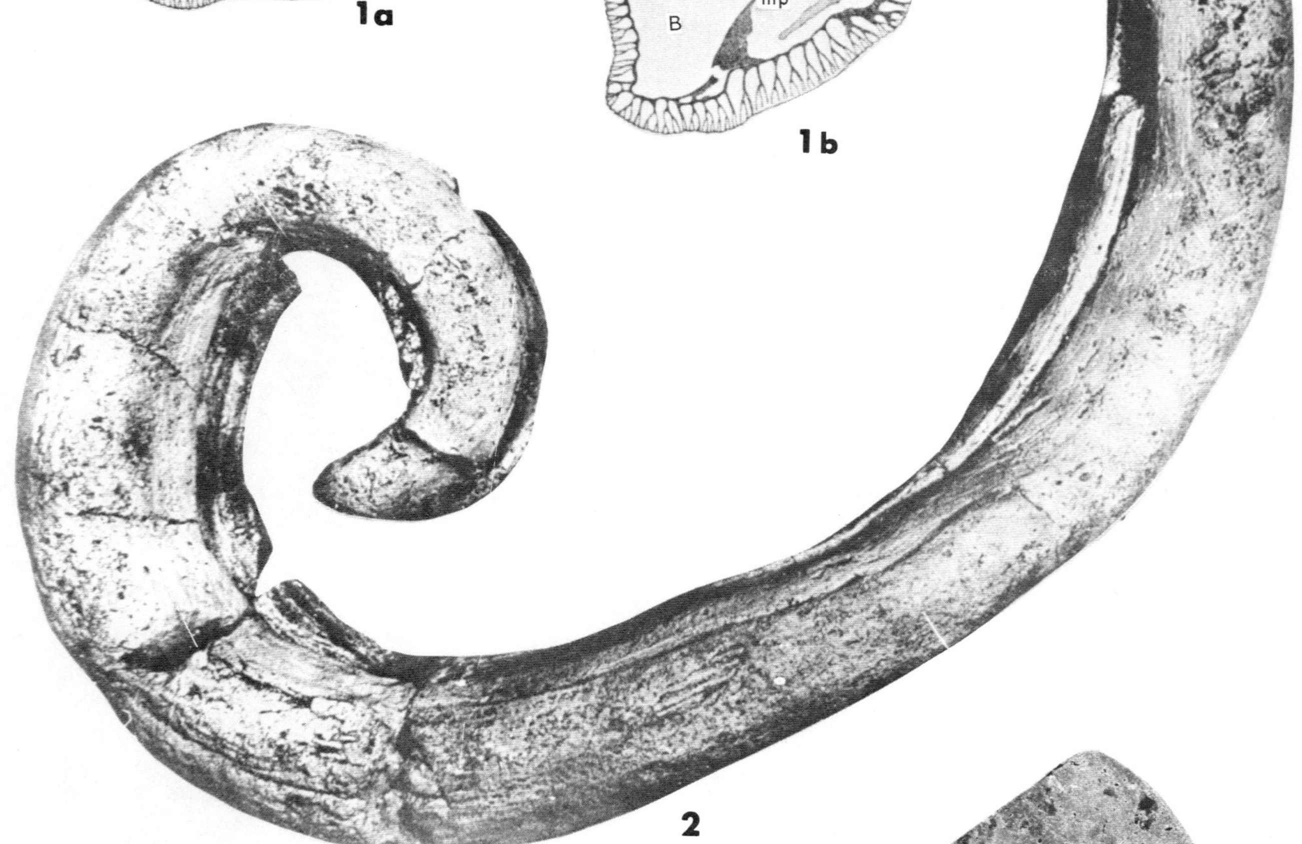

2

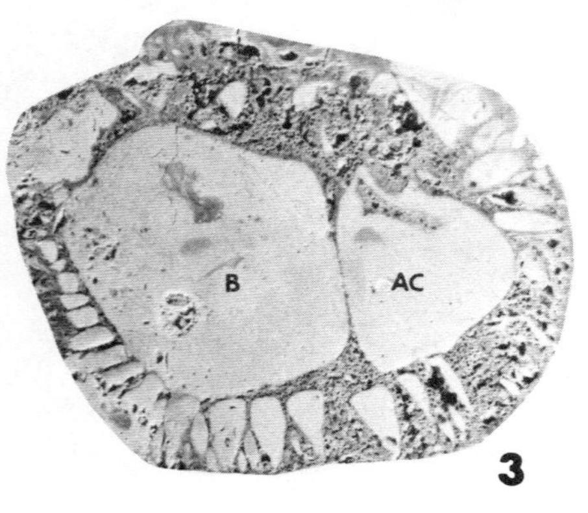

3

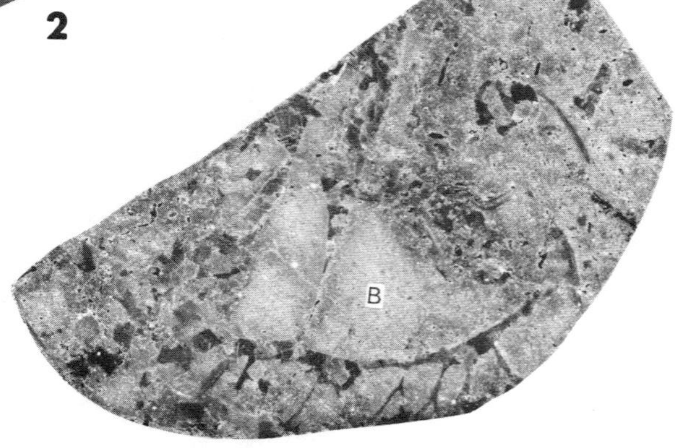

4

PLATE 11

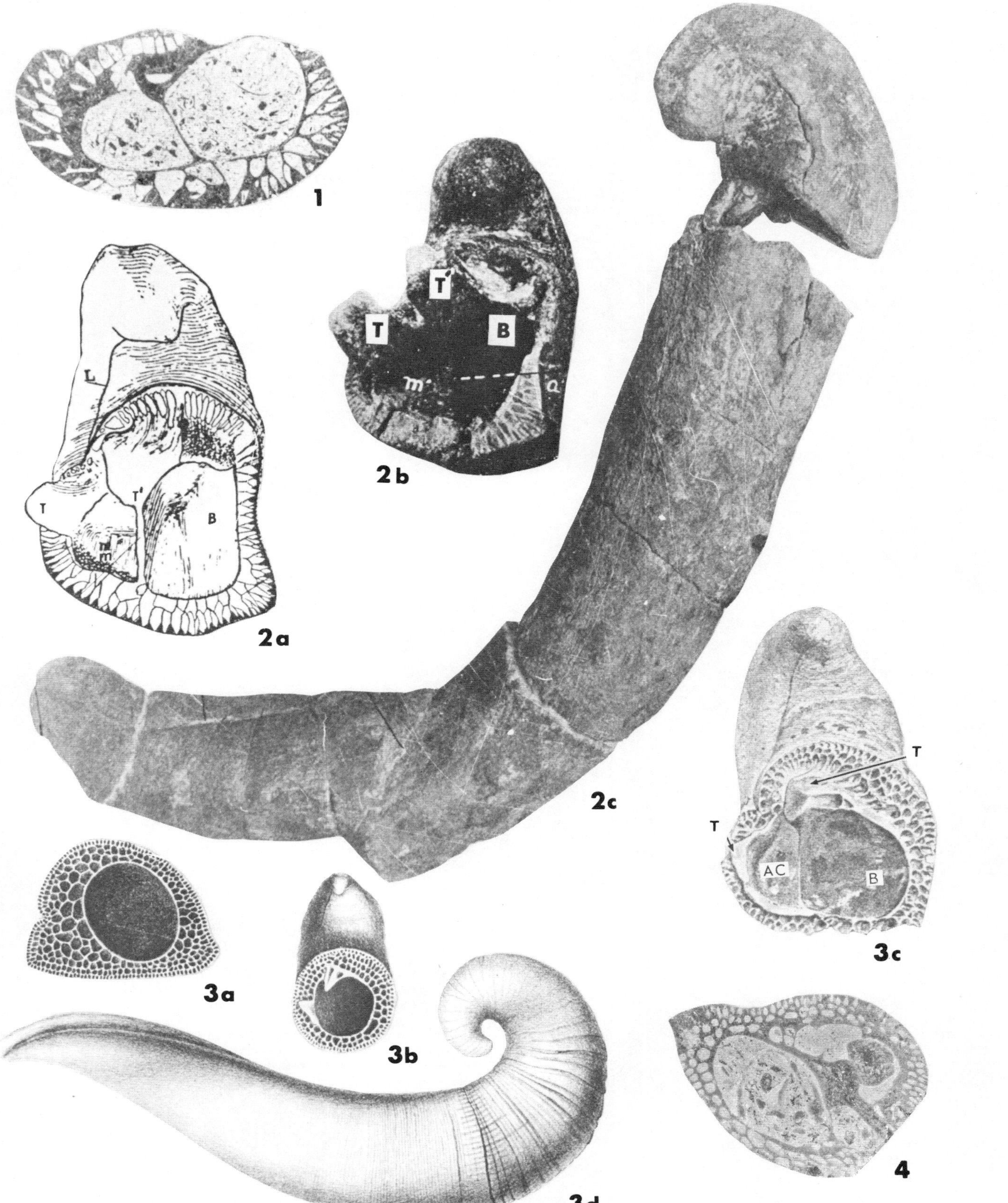

PLATE 12

like a broad low letter "W". The body cavity is moderately large but becomes smaller as it grows and is filled with tabulae. The shell may change its cross-sectional outline from subcircular to flattened subquadrangular or irregular as it grows. The shell wall is filled with polygonal canals which are slightly larger near the margin. A row of pyriform canals is present around the shell margin. The polygonal canals have widely spaced tabulae. The ligament is marked by a deep furrow externally; internally it is shaped like a fishhook. The accessory cavity and tooth sockets are small and appear to be tabulate. The teeth are large, and in the free valve the anterior one is larger. Both teeth and muscle attachment areas are invaded by canals and are generally difficult to distinguish.

Type species: *Sabinia vivari* Palmer, 1928.

The genera *Sabinia* Parona, *Ichthyosarcolites* Desmarest, and *Immanites* Palmer are apparently similar to *Texicaprina* in having the shell wall filled with canals. *Sabinia*, from the Senonian and Maastrichtian of Italy, is usually considered a caprinid, but it may be a radiolitid. It has siphonal bands on the exterior of both valves, a long straight radiolitid-like internal ligament, muscle attachment areas next to the body cavity, and small nonporous teeth.

Ichthyosarcolites, supposedly from strata of Urgonian (early Cretaceous), Cenomanian, and Turonian age in Europe, has a radiolitid-like dentition and muscle attachment areas set in slots in the body cavity walls; it lacks both the external ligamental furrow and internal ligament of *Texicaprina*. Early descriptions of Texas caprinids (Roemer, 1888; Douvillé, 1900) attributed caprinid specimens to *Ichthyosarcolites,* for example, *Ichthyosarcolites anguis* Roemer (now properly assigned to *Caprinuloidea*, cf. pl. 12, fig. 3). Recently Davis (1977) has suggested that *Ichthyosarcolites* may occur in the Edwards Limestone in Central Texas. One of the difficulties with *Ichthyosarcolites* is that the illustrations of early described specimens are inadequate to compare to morphology of this taxon with those being described today. Superficially, the genus is like *Texicaprina*. Davis has attributed a specimen to *Ichthyosarcolites* which is nearly identical with that shown on plate 15, figure 2 as *Texicaprina* sp. For comparison, an illustration of *Ichthyosarcolites triangularis* Desmarest made by Montagne (1938) from middle Dalmatia is shown (pl. 15, fig. 6).

Immanites, especially *I. rotunda* Palmer (1928), is a mystery and requires restudy. If it is related to *Texicaprina*, the relationship is not obvious. Palmer, who had examples of both on hand, did not compare or contrast them.

Texicaprina differs from other coalcomaninids in having the shell wall filled with canals. Several species assigned by Palmer (1928), Thiadens (1936), and Bouwman (1937) to *Sabinia* were recognized by MacGillavry (1937) as separate from

PLATE 13 *(page 61)*
ALBIAN COALCOMANINID CAPRINIDS
Species of *Caprinuloidea* from Texas and Mexico.

Figure
1. ***Caprinuloidea anguis*** (Roemer) from the middle Albian strata in the Tenneco No. 1 Schultz, Live Oak County, Texas, at a depth of 13,796 feet, x2.5.
2. ***Caprinuloidea lenki*** (Boehm) from the Cerro Escamela, Veracruz, Mexico, after Boehm (1898), free valve, xl.2.
3. ***Caprinuloidea perfecta*** Palmer (= ***C. septata*** Palmer), attached valve from probable Albian strata at Soyatlan de Adentro, Jalisco, Mexico, from Palmer (1928, pl. 11), xl.5.
4. ***Caprinuloidea lenki*** (Boehm) from the middle Albian strata in the Stanolind No. 1 Martin, La Salle County, Texas. Free valve from a depth of 10,343 feet, xl.5.
5. ***Caprinuloidea gracilis*** Palmer from the middle Albian strata of the Stuart City Formation in the Tenneco No. 1 Schultz, Live Oak County, Texas, at a depth of 13,814 feet, x2.
6. ***Caprinuloidea gracilis*** Palmer, transverse section of attached valve from the same locality as 5, x2.
7. ***Caprinuloidea felixi*** (Boehm) from the Cerro Escamela, Orizaba, Veracruz, Mexico, from Boehm (1898). Free valve showing multiple rows of marginal canals, xl.2.
8. ***Caprinuloidea multitubifera*** Palmer, free valve (type specimen) from Palmer (1928, pl. 10), x1.5 from probable middle Albian strata at Soyatlan de Adentro, Jalisco, Mexico.
9. ***Caprinuloidea*** sp. (***C. gracilis*** group), attached valve of worn specimen from the late Albian, Stuart City Formation in the Mobil No. 1 Zula Boyd, Lavaca County, Texas, at a depth of 14,212 feet, xl.

PLATE 14 *(page 62)*
MIDDLE ALBIAN COALCOMANINID CAPRINIDS
Species of *Texicaprina* from Texas, Mexico, and Trinidad.

Figure
1. ***Texicaprina vivari*** (Palmer) from Paso del Rio, Colima, Mexico, from Palmer (1928).
   (a) Transverse section of attached valve, approximately xl.
   (b) Another specimen showing strongly compressed attached valve, x0.7.
2. ***Texicaprina kugleri*** (Bouwman) from a conglomerate at Point à Pierre, Trinidad, after Bouwman (1937), x0.5.
   (a) Attached valve, lateral view, showing body cavity and protruding teeth.
   (b) Transverse section of free valve showing pervasive canals.
3. ***Texicaprina vivari*** (Palmer) from the Edwards Limestone, 5.5 miles south of Cranfills Gap on Texas 22, Hamilton County, Texas, xl.5.
   (a) Transverse section of free valve showing small body cavity filled with tabulae.
   (b) Thin section of same specimen showing shell invaded by canals.
   (c) Opposite end, same shell, attached valve.
   (d) Cross section of valve with irregular anterior underside.
4. ***Texicaprina vivari*** (Palmer) from the same locality as 3, xl.5.
   (a) Transverse section of attached valve showing posterior coalcomaninid tooth plus accessory cavity just below ligament, upper left.
   (b) Transverse section of another specimen showing nearly circular cross section of valve with internal canals.

PLATE 15 *(page 63)*
MIDDLE-LATE ALBIAN COALCOMANINID AND OTHER CAPRINIDS
Species of *Texicaprina* and *Immanites* (?) from Texas and of *Ichthyosarcolites* from Europe.

Figure
1. ***Texicaprina*** sp. from 5.5 miles south of Cranfills Gap on Texas 22, Hamilton County, Texas, showing another aspect of internal structure of valve, xl.5.
2. ***Texicaprina*** sp. from the upper Devils River Formation at Seminole Creek crossing of U. S. 90, 8 miles west of Comstock, Val Verde County, Texas, showing multiple rows of large polygonal canals, x2.
3. ***Texicaprina vivari*** from same locality as 1. showing well-preserved internal ligament on upper right (posterior) side of specimen, xl.5.
4. ***Texicaprina*** sp., oblique section from the Humble No. 1 Ben Ogletree, San Jacinto County, Texas, at a depth of 13,699 feet. Note ligament and marginal canals, x2.
5. ***Texicaprina*** sp. from same locality as 1. showing tabulate body cavity and coalcomaninid tooth plus accessory cavity.
6. ***Ichthyosarcolites triangularis*** (Desmarest), from Montagne (1938), x2.
7. ***Immanites*** sp. ?, cf. ***I. rotunda*** Palmer, from a depth of 13,747 feet in the Humble No. 1 Ben Ogletree, San Jacinto County, Texas, xl.5.
   (a) Longitudinal section showing caprinid body cavity filled with tabulae, septate outer wall, and accessory cavity.
   (b) Transverse section.

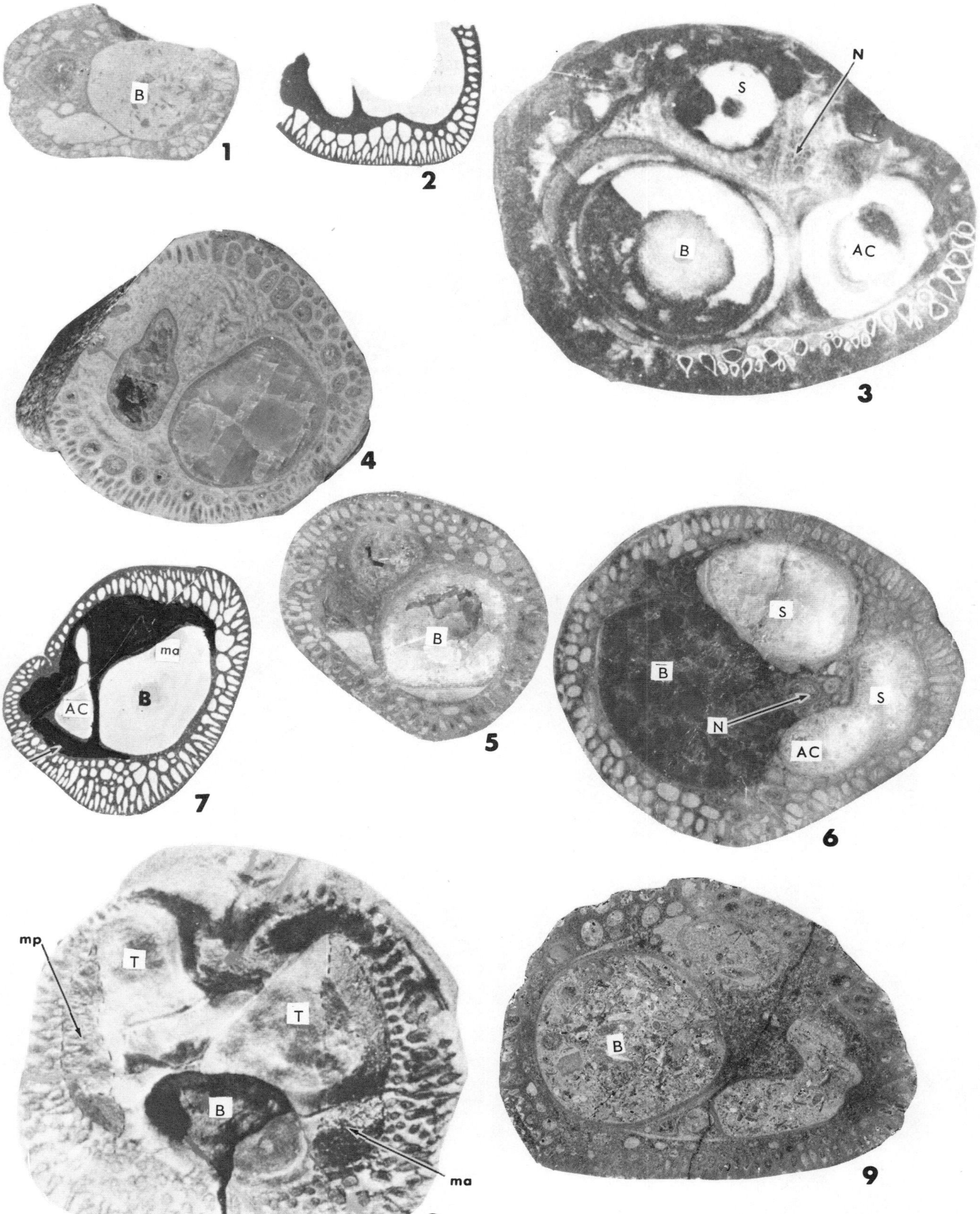

PLATE 13

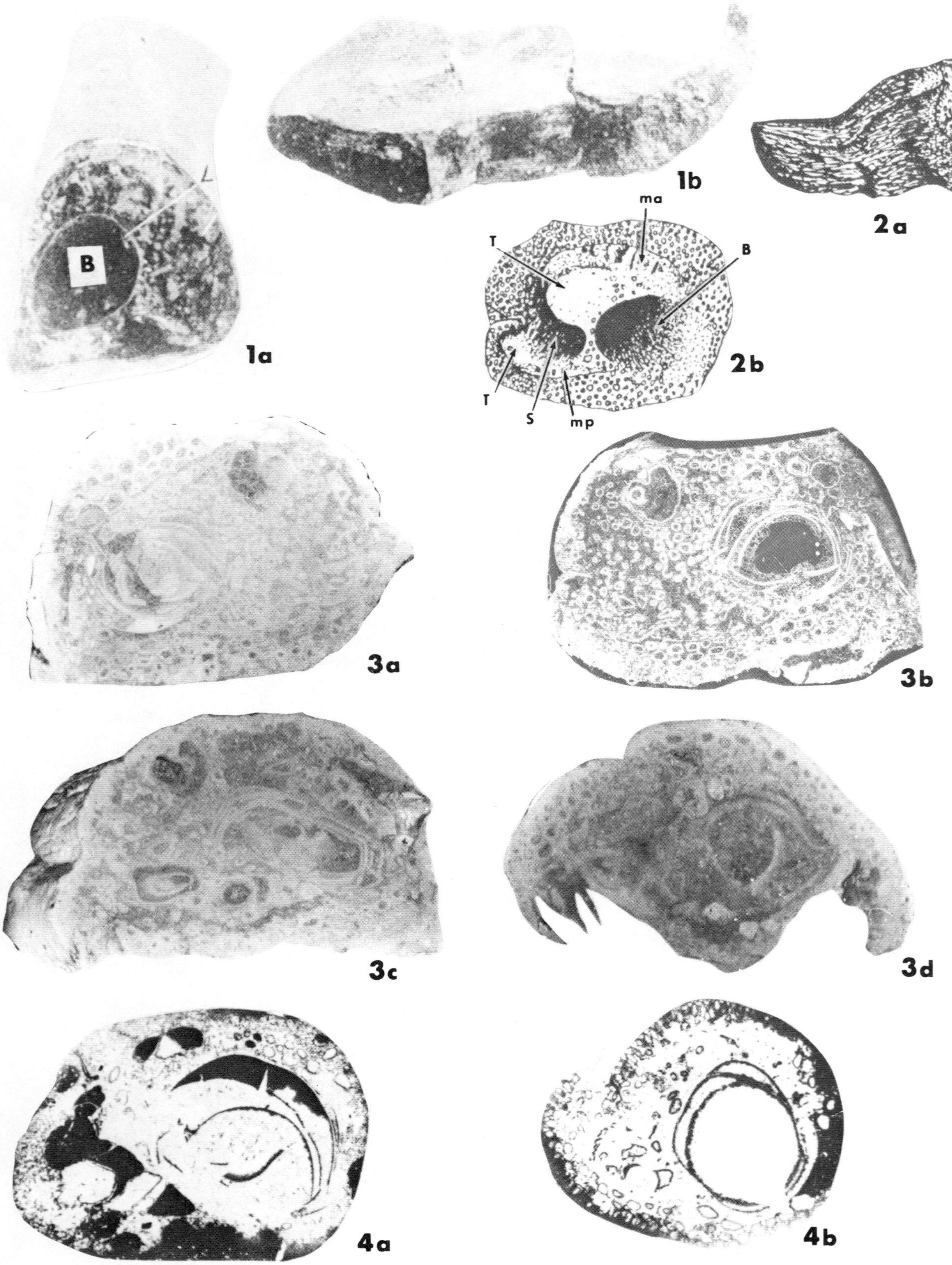

PLATE 14

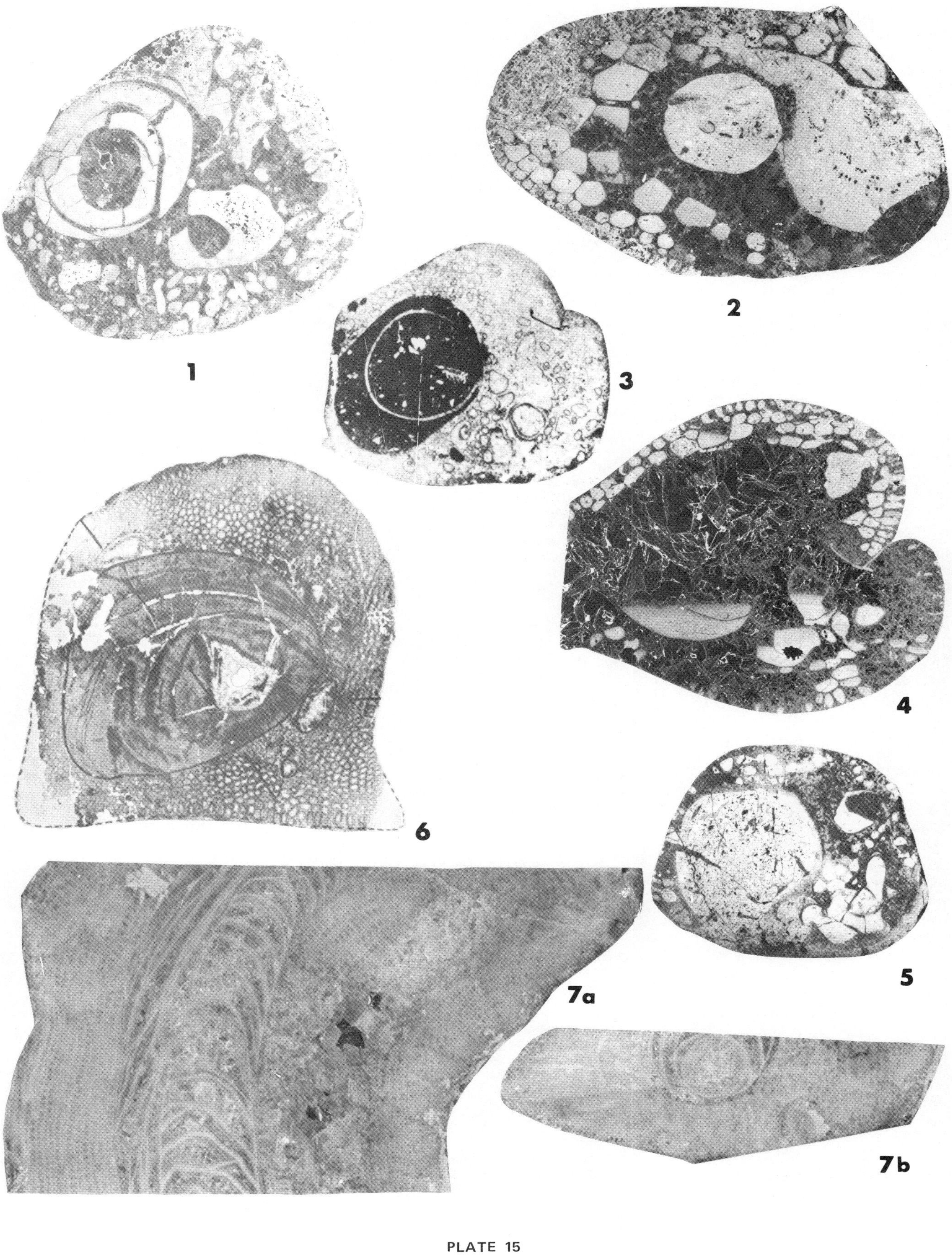

PLATE 15

that genus and closely related to *Caprinuloidea*. These species, *S. kugleri* Bouwman, *S. vivari* Palmer, *S. orbiculata* Palmer, and *Sabinia* sp. Thiadens, belong to the genus *Texicaprina*. In addition, specimens illustrated as *S. totiseptata* by Chubb (1971) from the Albian Seafield Limestone of Jamaica may belong to *Texicaprina*.

Occurrence: Species of *Texicaprina* are known from middle and late Albian strata in Texas and from limestones of less precisely dated strata in Mexico, Cuba, Jamaica, Trinidad, and on the sunken Japanese guyots of the western Pacific Ocean (Matthews and others, 1974). In Texas, *Texicaprina* is recorded from the Edwards Limestone outcrop in Central Texas, from the Devils River Formation along the Pecos River, and at Twelve-Mile Mesa in Pecos County near Fort Stockton. The Cenomanian age listed by Palmer (1928) for the Paso del Rio, Colima, Mexico, occurrence is not confirmed.

Kimbleia Coogan, 1973
(Plate 16)

A coalcomaninid caprinid similar to *Caprinuloidea* but differing in that the attached valve has the porous dorsal tooth closer to the ligament leaving the anterior tooth socket separated from the posterior tooth plus accessory cavity by a septum and in having straight rather than curved myophores. The free valves found with *K. capacis* and *K. albrittoni* (pl. 16) do not seem to match the attached valves of either but may nevertheless represent the genus.

Type species: *Kimbelia capacis* Coogan, 1973.

Two species of *Kimbleia* are now known, *K. capacis* and *K. albrittoni. K. capacis* differs from *K. albrittoni* in having much larger, suboval to subrectangular canals next to the body cavity and tooth plus accessory cavity in the inner row of marginal canals, and in having two to three rows of polygonal canals.

Occurrence: The genus *Kimbleia* is found in late Albian strata of the upper Aurora Limestone, Sierra de Tlahualillo, Coahuila, Mexico, in the limestones of the Edwards Group on the Edwards Plateau called the Segovia Formation (Rose, 1972), and in the Devils River Formation and its equivalents in Kimble, Menard, Pecos, and Val Verde Counties, Texas (Coogan, 1973).

Mexicaprina Coogan, 1973
(Plate 17)

A coalcomaninid rudist with a small shell for a full grown caprinid.

PLATE 16
(page 65)

LATE ALBIAN COALCOMANINID CAPRINIDS
Species of *Kimbleia* from the late Albian Segovia Formation of Kimble County, Texas, on the Lopez Ranch, one-quarter mile south of the Menard-Kimble County line off FM 1674.

Figure

1. *Kimbleia albrittoni* (Perkins), transverse section of attached valves, xl.
   (a) Specimen showing coalcomaninid tooth plus accessory cavity.
   (b) Another specimen showing nearly even-sized marginal canals.
2. *Kimbleia* sp., free valves, xl.
   (a) Oral view showing teeth and body cavity.
   (b) Oral view of coiled, moderate-sized specimen showing broken but still protruding teeth.
3. *Kimbleia capacis* Coogan, attached valves, xl.
   (a) Transverse section showing silicified portion of dorsal tooth (upper left) marginally set and marginal canals.
   (b) Transverse section showing large canals next to the accessory cavity and body cavity.
4. *Kimbleia albrittoni* (Perkins), attached valve, longitudinal section of weathered specimen showing tabulate body cavity (right) and accessory cavity (left), xl.

PLATE 17
(page 66)

LATE ALBIAN AND CENOMANIAN COALCOMANINID CAPRINIDS
Species of *Caprinuloidea* ?, *Kimbleia*, and *Mexicaprina* from Texas and Mexico.

Figure

1. *Kimbleia* sp. from the late Albian cap rock at Twelve-Mile Mesa, Pecos County, Texas. Transverse section of posterior-dorsal part of compressed valve. Position of tooth indicates specimen belongs to *Kimbleia*, xl.
2. *Kimbleia* ? sp., partial section of a specimen from the late Albian strata of the Stuart City Formation at a depth of 13,550 feet in the Tenneco No. 1 Schultz, Live Oak County, Texas, xl.5.
3. *Caprinuloidea* ? sp. from the El Abra Limestone, Teninul member, 14 km east of Valles, San Luis Potosí, Mexico, Cenomanian. Transverse section of partial specimen showing multiple rows of polygonal canals similar to *C. multitubifera*, xl.5.
4. *Mexicaprina ? whitei* (Boehm), oblique section from the Sierra de la Boca del Abra near El Chey (Choy), El Abra Limestone, San Luis Potosí, Mexico, from Boehm (1898), xl.1.
5. *Mexicaprina* sp., cf. *M. cornuta*, transverse section of free valve from the late Albian strata of the Stuart City Formation at a depth of 14,240 feet (117 feet below the base of the Buda Limestone) in the Mobil No. 1 Zula Boyd, De Witt County, Texas, xl.5.
6. *Mexicaprina cornuta* Coogan from the El Abra Limestone, Taninul member, 14 km east of Valles on Highway 110, San Luis Potosí, Mexico.
   (a) Transverse section of attached valve showing detail of polygonal canals in the projecting ridge and two bifurcating plates in the anterior accessory cavity, x2.
   (b) Transverse section of attached valve showing posterior and anterior accessory cavities subdivided by several plates, x2.
   (c) Transverse section of partly recrystallized specimen showing canal pattern near the ligament and base of the projecting ridge, xl.5.
   (d) Slightly oblique longitudinal section showing tabulate body cavity, spetate polygonal canals, and projecting ridge, xl.
7. *Mexicaprina minuta* Coogan, from the same locality as 6.
   (a) Transverse section of attached valve showing development of canals around the entire periphery of valve, x2.
   (b) Transverse section of attached valve showing posterior accessory cavity divided by three plates, x2.
   (c) Transverse section showing anterior accessory cavity with two radial plates, xl.
   (d) Transverse section of free valve (?), xl.
8. *Mexicaprina* sp. from the Taninul quarry showing development of two projecting ridges, xl.

PLATE 18
(page 67)

SPECIES OF ALBIAN LARGER ARENACEOUS FORAMINIFERS

Figure

1. *Orbitolina* sp. from the Devils River Formation, top 50 to 60 feet, along the road cut on old U. S. 90, west side of the Pecos River, Val Verde County, Texas. The specimen shows upward-pointing triangles, an aspect of the main partitions of *Orbitolina* cut normal to their length, x40.

2-5. Topotypes of *Coskinolinoides texanus* (= *Coskinolina adkinsi* Barker) from Barker (1944) and Maync (1955).
   (2) Axial section, x84.
   (3-4) Axial sections, x100.
   (5) Transverse section, x100.

6-8. *Coskinolina sunnilandensis* Maync, from Maync (1955), x40.
   (6-8) Axial sections.
   (7) Transverse section of topotype.

9-13. *Coskinolina*, n. sp., "advanced form" from the upper 50 to 60 feet of the Devils River Formation along road cut on the west side of the Pecos River, old U. S. 90, Val Verde County, Texas, x40.
   (9, 11, 13) Axial sections.
   (10) Transverse section.
   (12) Oblique section.

14. Axial section of *Coskinolina* n. sp., "primitive form" from the caprock at Twelve-Mile Mesa, Pecos County, Texas, x60.

15-19. Topotypes of *Dictyoconus walnutensis* (Carsey) from Cole (1942) and Barker (1944).
   (15) Axial section, x20.
   (16, 18) Axial sections, x40.
   (17, 19) Transverse sections, x40.

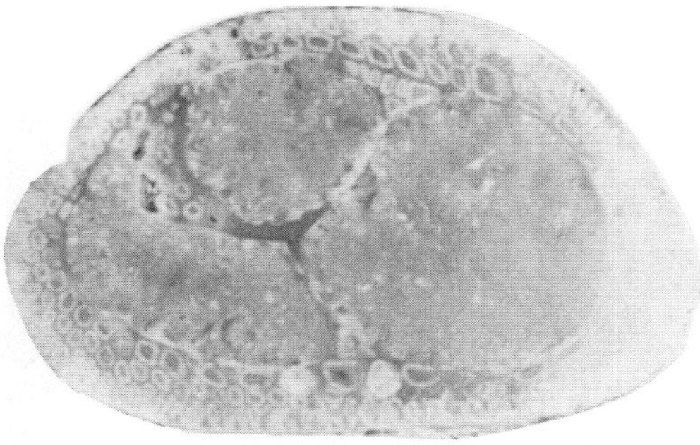

1a

2a

2b

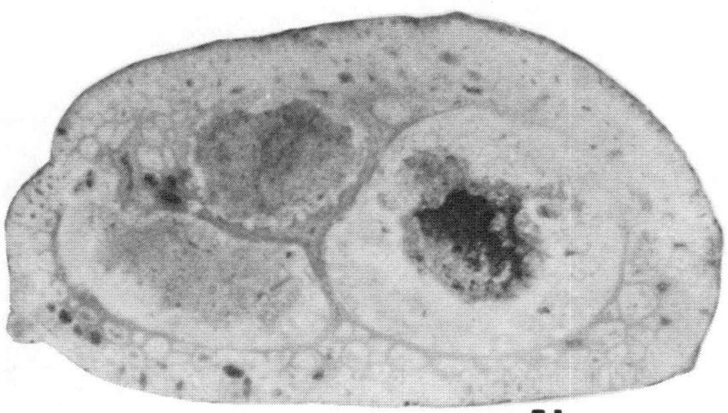

1b

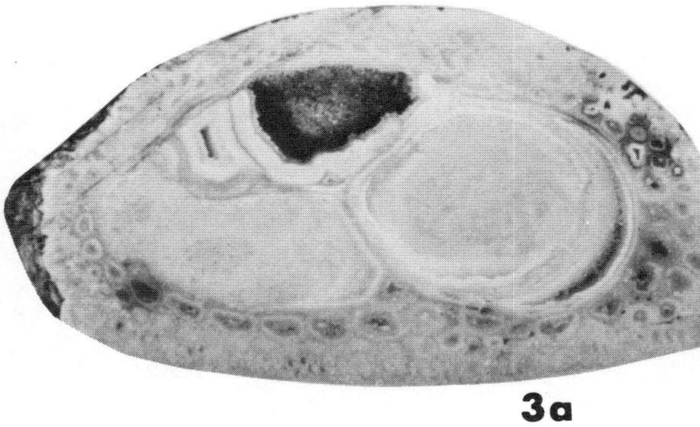

3a

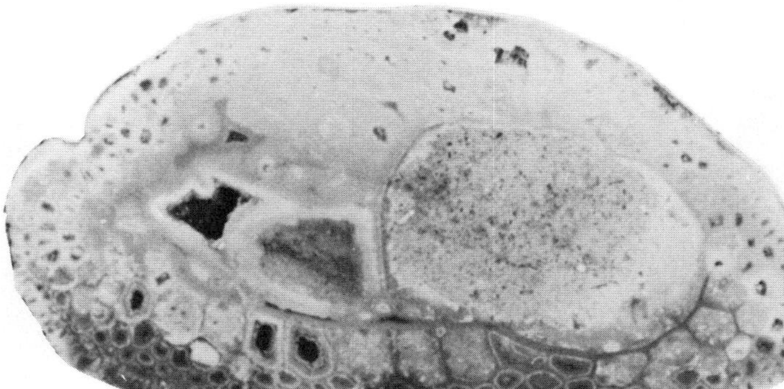

3b

4

PLATE 16

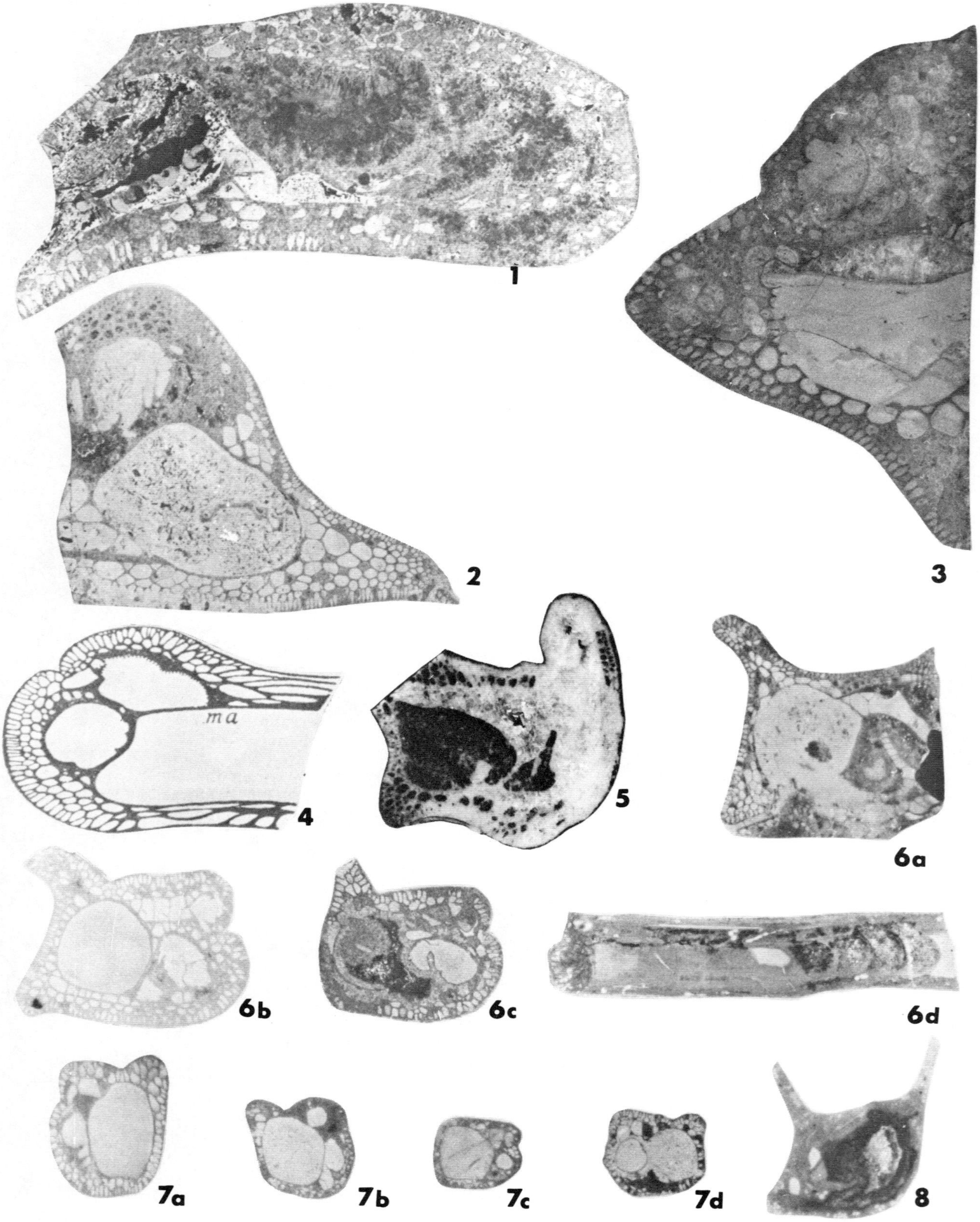

PLATE 17

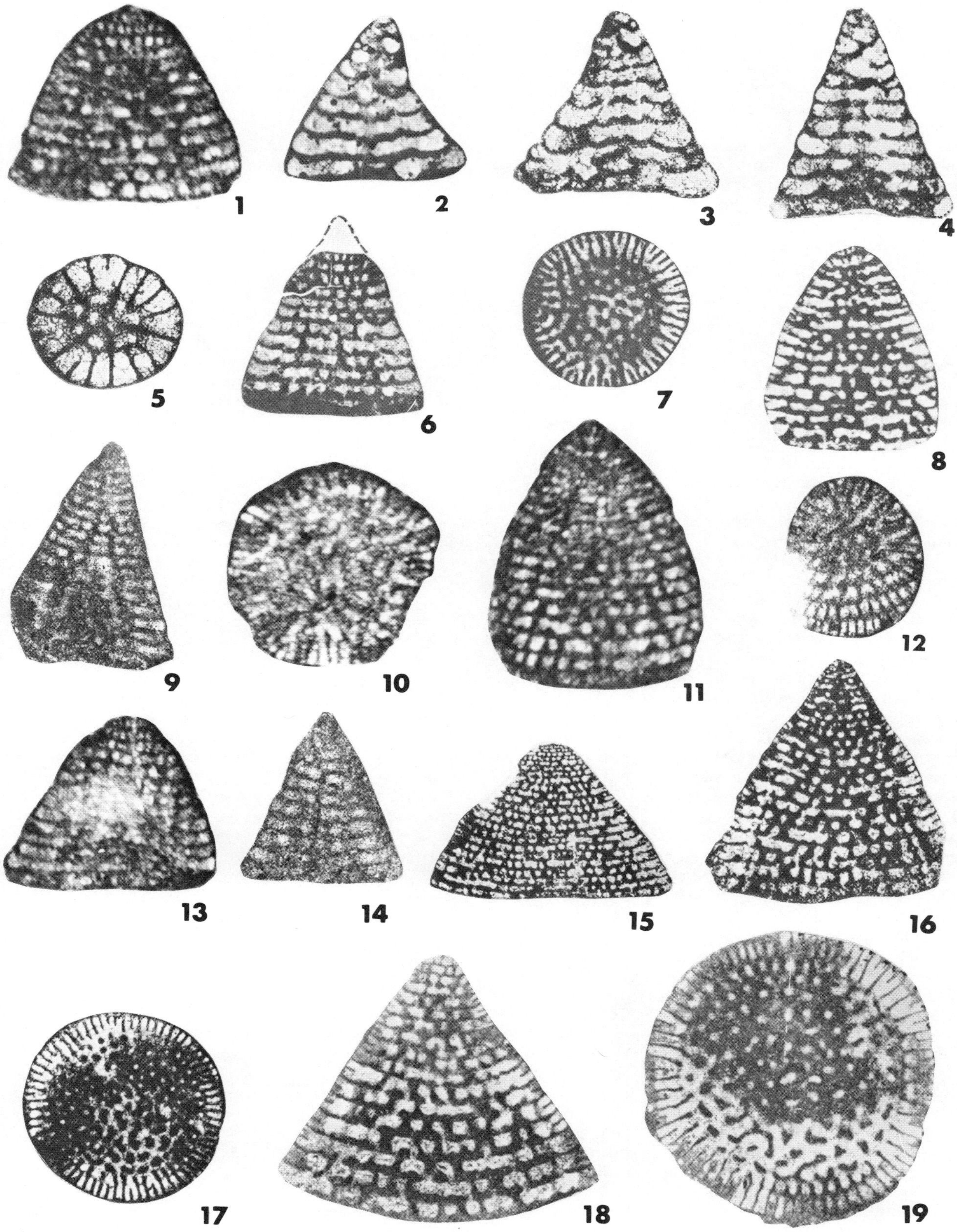

PLATE 18

The attached valve has a posterior accessory cavity separated by an oblique plate from the tooth cavity, which itself is divided by one or two oblique plates as in *Caprinula*. The single tooth of the attached valve is small. To the dorsal and posterior side of this tooth is an accessory cavity divided by vertical plates which is inside the rows of marginal polygonal canals. One row of pyriform canals is present around the circumference of the valves; two or three rows of polygonal canals are developed ventrally. The free valve is not known.

Type species: *Mexicaprina cornuta* Coogan, 1973.

Mexicaprina differs from *Caprinula* in having a much larger posterior tooth socket and a subdivided accessory cavity inside a row of polygonal canals anterior of the tooth in the lower valve, and in lacking a simply divided cavity dorsoanteriorly. It is similar to *Caprinula* in having a subdivided posterior accessory cavity. *Mexicaprina* differs from *Caprinuloidea* in having a subdivided posterior accessory cavity separate from the tooth socket, an anterior cavity, and a much smaller tooth in the attached valve. It is similar to many species of *Caprinuloidea* in the appearance of the marginal canals. The several species of *Mexicaprina* differ in size, development of the plates in the anterior accessory cavity, and development of a projecting ridge.

Mexicaprina cornuta is characterized by its anterior-projecting ridge, transversely set anterior accessory cavity divided into rectangular canals, and medium size (for the genus). *Mexicaprina minuta* differs from *M. cornuta* in being smaller, in lacking a strongly projecting thumblike ridge, and in having a more squarish anterior accessory cavity subdivided by fewer plates. One specimen of *Mexicaprina* with dual matching hornlike anterior and posterior projecting ridges is too poorly preserved internally to describe further. An obliquely cut specimen described by Boehm (1898) as *Sphaerucaprina occidentalis* appears to belong to *Mexicaprina* but is difficult to compare because of its orientation. Boehm reserved the name *"white"* as a specific designation should the specimen not belong to *Sphaerucaprina*. MacGillavry (1937) validated the name and assigned the species to *Caprinuloidea*.

The discovery of a specimen assignable to *Mexicaprina* in the Mobil No. 1 Zula Boyd (pl. 17, fig. 5) probably in strata in late Albian age, extends the range of the genus to strata older than Cenomanian (Coogan, 1973).

Family Uncertain
(Plate 15)
Immanites Palmer, 1928

The genus *Immanites*, assigned to Family Uncertain by Perkins and Coogan (Moore, 1969), is superficially similar to members of the Family Caprinidae. Based on the genotype, its shell is equivalve (Palmer, 1928). It is large, irregular, and curved. The shell wall is filled with round or polygonal canals which are commonly septate. The body cavity is small and tabulate. No myophores were observed. The teeth are located in the shell wall rather than separate and adjacent to the body cavity.

The recorded size is at least 14 cm in width and 62 cm in length.

Type species: *Immanites anahuacensis* Palmer, 1928.

The species *I. rotunda* has the whole shell filled with small canals similar to *Texicaprina.* A W-shaped fold or accessory cavity is present in the shell. The body cavity is small and tabulate. The specimen of *I. rotunda* illustrated here (pl. 15, fig. 7) is only tentatively assigned to the genus.

Immanites appears to lack the clear coalcomaninid tooth plus accessory cavity of *Texicaprina* but has a similar shell filled with canals. Until the genus is restudied, it remains a partial enigma. However, like many of the taxa named by Palmer (1928), it may be common in the Albian rudist facies and should be looked for with care.

Occurrence: The genus is named from the Paso del Rio outcrops at Colima, Mexico, attributed to the Cenomanian by Palmer. Another Paso del Rio rudist described by Palmer, *Texicaprina vivari* is definitely Albian, especially middle and late Albian, in age.

ACKNOWLEDGMENTS

Appreciation is due my former colleagues at Esso Production Research Company, D. G. Bebout, H. H. Beaver, R. M. Jeffords, R. Woods, and the late H. N. Fisk, Chief of the Geologic Research Section, Exploration Department, Humble Oil and Refining Company, all of whom provided assistance and encouragement in the beginning phases of my study of Cretaceous rudists.

Many other colleagues in petroleum companies, with governmental agencies, in Mexico and Jamaica, and at universities provided core samples, well and outcrop information, their counsel on stratigraphic and paleontological relationships, and other assistance. A list of them would be too long for this space, but I want to thank them at least in this way.

The Esso Production Research Company (EXXON) financed the early part of this work in the 1960's.

REFERENCES

Adkins, W. S., 1928, Handbook of Texas Cretaceous fossils: Univ. Texas, Austin, Bull. 2838, 385 p.

Astre, Gaston, 1954, Radiolitidés Nord-pyrénéens: Société Géologique de France, Mémoires, n. sér., v. 33, no. 71, 140 p.

Baker, R. W., 1944, Some larger Foraminifera from the Lower Cretaceous of Texas: Journal of Paleontology, v. 18, p. 204-209.

Bauman, C. F., 1958, Dos radiolitidos nuevos de la región de Cuernavaca, Morelos: México Univ. Nacional Aútonoma Instituto de Geologiá Paleontologia Mexicana, no. 3, 8 p.

Bauwman, L. A. H., 1937, Sur une espèce nouvelle du genre *Sabinia* (Caprinines): Koninklijke Nederlandse Akademie van Wetenschappen Proceedings, v. 40, p. 1449-1453.

Bebout, D. G., and Loucks, R. G., 1974, Stuart City Trend, Lower Cretaceous, South Texas: Univ. Texas, Austin, Bureau of Economic Geology Report of Investigations 78, 80 p.

Boehm, Georg, 1898, Über Caprinidenkalke aus Mexiko: Zeitschrift Deutschen der Geologishen Gesellschaft, v. 50, p. 323-333.

_______ 1899, Beiträge zur Kenntnis mexicanischer Caprinidenkalke, *in* Felix, J. and Lenk, H. (1889-1899), Beiträge zur Geologie und Paläontologie der Republik Mexico, II Theil, 3 Heft, p. 143-154, Leipzig, Arthur Felix Verlag.

Brown, N. K., Jr., and Coogan, A. H., 1968, The hydrozoan genus *Parkeria* Carpenter, 1869, and its occurrence in the Cretaceous of Mexico: GSA, 1968 Annual Meeting Program Abstracts, Special Paper 121, p. 40.

Chubb, L. J., 1956, Some rarer rudists from Jamaica, B. W. I.: Paleontographica Americana, v. 4, p. 1-30.

_______ 1971, Rudists of Jamaica: Palaeontographica Americana, v. 7, no. 45, p. 161-222.

Coates, A. G., 1973, Cretaceous Tethyan coral-rudist biogeography related to the evolution of the Atlantic Ocean, *in* Organisms and continents through time: London, The Palaeontological Association,

Special Papers in Palaeontology, no. 12, p. 169-174.

Cole, W. S., 1942, Stratigraphic and paleontologic studies of wells in Florida—No. 2; Suwannee Petroleum Corporation Sholtz No. 1, Florida Oil Discovery Company Cedar Keys No. 2: Florida Geological Survey Bull. 20, 89 p.

Conrad, T. A., 1855, Description of one Tertiary and eight new Cretaceous fossils from Texas, in the collection of Major Emory: Academy of Natural Sciences of Philadelphia, Proceedings, v. 7, p. 268-269.

Coogan, A. H., 1969, Evolutionary trends in rudist hard parts, *in* Moore, R. C., ed., Treatise on invertebrate paleontology, Part N, Vol. 2: GSA and Univ. Kansas, p. N766-776.

_______ 1973, New rudists from the Albian and Cenomanian of Mexico and adjacent South Texas: Revista del Instituto Mexicano del Petróleo, v. 5, p. 51-83.

_______ , Bebout, D. G., and Carlos Maggio, 1972, Depositional environments and geologic history of Golden Lane and Poza Rica Trend, Mexico, an Alternative View: AAPG Bull., v. 56, p. 1419-1447.

Cox, L. R., 1969, Family Caprinidae—Systematic descriptions, part of Dechaseaux, C. and Perkins, B. F., *in* Moore, R. C., ed., Treatise on invertebrate paleontology, Part N, Vol. 2: GSA and Univ. Kansas, p. N787.

Davis, E. R., 1976, Paleoecology and distribution of Albian rudists of north-central Texas with special emphasis on the Edwards Formation in Bell, Bosque, McLennan and Coryell Counties: Baylor Univ., Master's thesis, 161 p.

Delamétherie, J. C., 1805, De La Sphérulite: Journal de Physique, v. 61, p. 396-399.

Distefano, Giovanni, 1888, Studi stratigrafici e paleontologici sul sistema cretaceo della Sicilia; I. Gli strati con *Caprotina* di Termini-Imerese: Roy. Accad. Sci. Let. Art., Palermo, Atti., v. 10, p. 1-44.

_______ 1899, Studi stratigrafici e paleontologici sul sistema cretaceo della Sicilia: II, I calcari con ***Polyconites*** di Termini-Imerese: Paletontografia Italica, v. 4, p. 1-46.

Douvillé, Henri, 1887, Sur quelques formes peu connues de la familie des Chamides: Société Géologique de France, Bull., sér. 3, v. 15, p. 756-802.

_______ 1900, Sur quelques rudistes americains: Société Géologique de France, Bull., sér. 3, v. 28, p. 205-221.

_______ 1910, Sur le genre ***Eoradiolites***: Société Géologique de France, Bull., sér. 4, v. 9, p. 76.

Fisher, W. L., and Rodda, P. U.,1969, Edwards Formation (Lower Cretaceous), Texas: Dolomitization in a carbonate platform system: AAPG Bull., v. 53, p. 55-72.

Giebel, C. G., 1853, Beitrag zur Paläontologie des Texanischen Kreidegebirges: Naturw. Ver. Sachsen u. Thuringen in Halle, Jahresber., 5, p. 358-375.

Hamilton, E. L., 1956, Sunken islands of the mid-Pacific mountains: GSA, Memoir 64, 97 p.

Harris, G. D., and Hodson, Floyd, 1922, The rudistids of Trinidad: Paleontographica Americana, v. 1, p. 119-162.

Hill, R. T., 1893, The paleontology of the Cretaceous formations of Texas; The invertebrate fossils of the *Caprina* limestone beds: Biological Society of Washington Proceedings, v. 8, p. 97-108.

Imlay, R. W., 1944, *in* Hedberg, H. D., and Pyre, Augustin, Stratigraphy of northeastern Anzoátegui, Venezuela: AAPG, Bull. v. 28, p. 1-28.

Keith, J. W., 1963, Environmental interpretation of subsurface Washita-Fredericksburg limestones, northern Live Oak County, South Texas, *in* Geology of Peregrina Canyon and Sierra de El Abra: Corpus Christi Geological Society Annual Field Trip, p. 72-78.

Kutassy, A., 1934, Pachydonta mesozoica (Rudistes exclusis): Fossilium catalogus: I, Animalia, Pars 68: Berlin, W. Junk, 189 p.

Logan, W. N., 1898, The invertebrates of the Benton, Niobrara, and Fort Pierre Groups: Kansas Geological Survey Bull., v. 4, p. 431-518.

MacGillavry, H. J., 1937, Geology of the province of Camaguev, Cuba with revisional studies in rudistid paleontology: Utrecht Rijksuniv. Geog. Inst., Geog. en Geol. Mededell., Physiog.-geol. Reeks, no. 14, 168 p.

_______ 1959, Rudistids from North America and their significance for a correlation of the Cretaceous, *in* El Sistema Cretacico, un symposium sombre el Cretacico en el hemisferio occident y su correlatcion mundial, v. 1: International Geological Congress, 20th, Mexico, 1956, p. 39-43.

Matthews, J. L., Heezen, B. C., Catalano, R., Coogan, A., Tharp, M., Natland, J., and Rawson, M., 1974, Cretaceous drowning of reefs on Mid-Pacific and Japanese guyots: Science, v. 184, p. 462-464.

Maync, Wolf, 1955, ***Coskinolina sunnilandensis***, n. sp., a Lower Cretaceous (Urgo-Albian) species: Cushman Foundation for Foraminiferal Research, Contributions, v. 6, p. 105-111.

Montagne, D. G., 1938, Einige Rudisten und Nerinen aus Mittle-Dalmatien: Koninklijke Nederlandse Akademie van Wetenschappen Proceedings, v. 41, no. 9, p. 979-986.

Moore, R. C., ed., 1969, Treatise on invertebrate paleontology, Part N, Vol. 2, Mollusca 6, Bivalvia: GSA and Univ. Kansas, p. N491-951.

Müllerried, F. K. G., 1947, Paleobiologia de la caliza de Cordoba y Orizaba, Veracruz: Anales Inst. Biol., v. 18, p. 361-462.

Palmer, R. H., 1928, The rudistids of southern Mexico: California Academy of Science, Occasional Papers, no. 14, 137 p.

Parona, C. F., 1909, La fauna corallegena del Cretaceo del Monte d'Ocre nell Abruzzo Aquilano: Mem. Serv. Carta Geol. d'Italia, Roy. Com. Geol. del Regno. v. 5, pt. 1, 242 p.

Perkins, B. F., 1960, Biostratigraphic studies in the Comanche (Cretaceous) series of northern Mexico and Texas: GSA Memoir 83, 138 p.

_______ 1974, Paleoecology of a rudist reef complex in the Comanche Cretaceous Glen Rose Limestone of Central Texas: Geoscience and Man, v. 8, p. 131-173.

_______ , and Coogan, A. H., 1969, Systematic descriptions. Family uncertain; *in* Moore, R. C., ed., Treatise on invertebrate paleontology, Part N, Vol. 2: GSA and Univ. Kansas, p. N817.

Počta, Filip, 1889, O Rudistechı vymrele celedi mlzu z ceskeho kridoveho utvaru (On rudists, an extinct family of lamellibranchs from the Bohemian Cretaceous): Rozp. K. Ceske splecnosti nauk, v. 7, no. 3, 78 p. (German summary, p. 79-92).

Roemer, Ferdinand, 1849, Texas, mit besonderer Rücksicht auf deutsche Auswanderung und die physikalischen Verhältnisse des Landes nach eigener Beobachtung geschildert: Bonn, 464 p.

______ 1852, Die Kreidebildung von Texas und ihre organischen Einschlüsse: Bonn, Adolph Marcus, 100 p.

______ 1888, Über eine durch die Häufigkeit Hippuriten-artiger Chamiden ausgezeichnete Fauna der oberturonen Kreide von Texas: Palaeont. Abh. (Dames u. Kayser), v. 4, p. 1-18.

Rose, P. R., 1963, Comparison of type El Abra of Mexico with "Edwards Reef Trend" of South-Central Texas, *in* Geology of Peregrina Canyon and Sierra de El Abra: Corpus Christi Geological Society Annual Field Trip p. 57-64.

______ 1972, Edwards Group, surface and subsurface, Central Texas: Univ. Texas, Austin, Bureau of Economic Geology, Report of Investigations 74, 198 p.

Thiadens, A. A., 1936, On some caprinids and a monopleurid from southern Santa Clara, Cuba: Koninklijke Nederlandse Akademie van Wetenschappen Proceedings, v. 39, p. 1132-1141.

Toucas, A., 1907-1909, Études sur la classification et l'évolution de Radiolitidae: Société Géologique de France Memoires 36, 132 p.

Toumey, Michael, 1854, Description of some new fossils from the Cretaceous rocks of the southern States: Academy of Natural Sciences of Philadelphia, Proceedings, v. 7, p. 167-172.

White, C. A., 1884, Description of certain aberrant forms of the Chamidae from the Cretaceous rocks of Texas, *in* On Mesozoic fossils: USGS Bull. 4, p. 5-9.

Whitfield, R. P., 1891, Observations on some Cretaceous fossils from the Beirut District of Syria, in the collection of the American Museum of Natural History, with descriptions of some new species: American Museum of Natural History Bull., v. 3, p. 381-441.

Whitney, M. I., 1952, Some new Pelecypoda from the Glen Rose Formation of Texas: Journal of Paleontology, v. 26, p. 697-707.

Wiontzek, Hebert, 1934, Rudisten aus der oberen Kreide des mittleren Isonzogebietes: Palaeontographica, v. 80, pt. A, p. 1-40.

Young, Keith, 1959, Edwards fossils as depth indicators, *in* Lozo, F. E., and others, Symposium on Edwards Limestone in Central Texas: Univ. Texas, Pub. 5905, p. 97-104.

[2] In the middle 1960's, Noel K. Brown and the author reported on the Vraconian and Cenomanian occurrence of the hydrozoan ***Parkeria*** (Brown and Coogan, 1967). The journal article was never published owing to the untimely death of Noel Brown in 1967. A few illustrations of ***Parkeria*** were included in the description of the Cenomanian rudist fauna from the El Abra Limestone (Coogan, 1973). As envisioned in 1967, the study would have reported on the distribution of large arenaceous foraminifers found with ***Parkeria*** and on the rudists described here by Coogan (1973). A portion of that foraminiferal work by Brown is discussed here and illustrated to encourage further study of late Albian forms.

LOWER CRETACEOUS CARBONATES OF CENTRAL SOUTH TEXAS: A SHELF-MARGIN STUDY

J. W. Wooten[1] and W. E. Dunaway[1]

ABSTRACT

Recent research on the Albian and lower Cenomanian Edwards shelf margin along the Gulf of Mexico has disclosed a narrow band of biogenic growth consisting of reefs, banks, bars, and islands. For the purpose of this paper, in light of evidence obtained from Sligo tests within the study area, the shelf-margin theory derived from this research is assumed to apply to the Aptian Sligo shelf margin as well.

The objectives of this paper are to show the application of the shelf-margin exploration model in light of new information obtained from the recent surge of drilling activity. We have attempted to exemplify the major facies and structural components of the shelf-margin model and to demonstrate the effect that these have on accumulation of hydrocarbons.

Utilizing good quality modern CDP seismic data in areas of adequate well control, we can relate the major facies and structural components of the model to certain geophysical criteria common to several hundred miles of seismic dip control. Selected seismic sections demonstrate the validity of these criteria in defining the shelf, shelf margin, and shelf slope of each formation, as well as their relationship to one another. Through the correlation of these geophysical criteria with existing well control, we obtain a detailed structural configuration and facies distribution as well as a prime fairway for hydrocarbon accumulation for both the Edwards and Sligo.

INTRODUCTION

General

In recent years an abundance of subsurface information previously held confidential has been released and utilized by researchers in their efforts to understand and reconstruct paleodepositional environments. Data from detailed analyses of modern deposi-

[1]Dixel Resources, Inc., Houston, Texas

			MEXICO	STUART CITY TREND	
CRETACEOUS	UPPER	MAESTRICHTIAN	MENDEZ	NAVARRO	
				TAYLOR	
		CAMPANIAN			
		SANTONIAN	SAN FELIPE	AUSTIN	
		CONIACIAN			
		TURONIAN	AGUA NUEVA	EAGLE FORD	
		CENOMANIAN		WOODBINE	
				BUDA	
			EL ABRA, TAMABRA, U. TAMAULIPAS	DEL RIO	
				GEORGETOWN	
	LOWER	ALBIAN		EDWARDS	STUART CITY
				GLEN ROSE	
		APTIAN	OTATES	PEARSALL	JAMES PINE ISLAND
			LOWER TAMAULIPAS	SLIGO	
				HOSSTON	
		NEOCOMIAN		COTTON VALLEY	

Figure 1. Correlation chart of Lower Cretaceous formations, Central South Texas to Mexico (after Bebout and Loucks, 1974).

tional environments and from outcrops of paleodepositional environments, when correlated with the ever increasing quantity of available subsurface information, make possible an accurate reconstruction of the paleodepositional environments of many of the producing trends of the world today.

The carbonate deposition which took place along the shelf of the Gulf of Mexico during Aptian, Cenomanian, and Albian time (fig. 1) is one of these trends. Until recently, it had not been well understood and had undergone a sporadic history of exploration. Owing to the concern of those in the industry who have released vital subsurface information, as well as to the initiative, imagination, and persistence of researchers, the Lower Cretaceous carbonates of central South Texas are better understood than ever before and are presently receiving a generous portion of the industry's attention and risk capital.

This attention and activity centers in Texas along what are defined as the Edwards and Sligo shelf-margin trends (fig. 2). The Edwards shelf-margin trend (Edwards "Reef" Trend or "Stuart City" Trend) underwent a surge of exploration activity in the early 1950's, when the industry attempted to duplicate the prolific production found in rocks of the same age in the Poza Rica and Golden Lane Trends on the Gulf Coast of central Mexico. This surge led to the discovery of the Stuart City field in La Salle County, Texas, which further stimulated activity and led to the discovery of 16 more fields by 1962. With the exception of the up-to-the-coast fault line fields (for example, Panna Maria, Big John, Fashing), which produce from a dolomitized lagoonal facies on the Edwards shelf, the majority of production from the Edwards shelf-margin trend during this period was marginal to subcommercial. From 1962 until the early 1970's activity diminished and, of the few wells drilled, several were drilled deeper to test the Sligo. A timely increase in the price of natural gas, a broader understanding of shelf-margin model, and advancements in drilling, completion, and stimulation technology have spurred a recent surge in activity along both the Edwards and Sligo Trends. Along the Edwards shelf margin, in 1975 and 1976, one operator alone drilled and completed 27 wildcat and development wells of which only two have been plugged and abandoned to date. Within the study area, three existing fields were expanded and seven new fields discovered. A total of

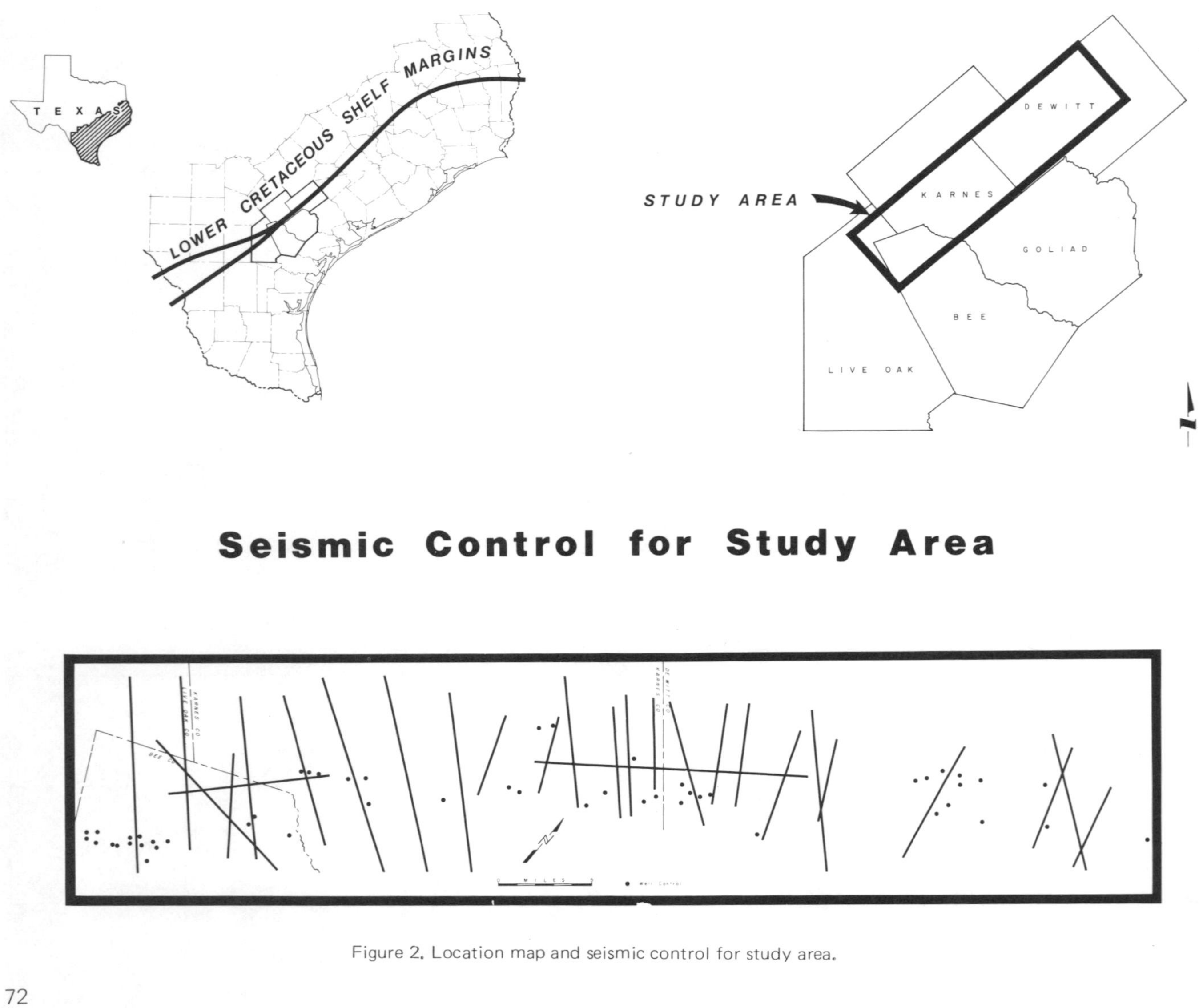

Figure 2. Location map and seismic control for study area.

six wildcat Sligo wells were drilled with one discovery. Two of the dry holes were offsets to production or shows and, although dry, serve well to exemplify the application of shelf-margin model to the sparsely controlled Sligo trend.

Paleogeography and Derivation of the Shelf-Margin Model

Lower Cretaceous carbonates accumulated along a broad shelf which completely encircled the Gulf of Mexico. Research shows that water depths along this shelf area ranged from a few feet to one or two hundred feet providing an excellent environment in which warm-water marine organisms flourished. The majority of these carbonates represent an open-shelf environment with little to no prospective merit for hydrocarbon exploration. A recent publication on the Stuart City Trend of South Texas (Bebout and Loucks, 1974) describes the environmental, depositional, diagenetic, and facies relationships of the Lower Cretaceous Edwards along an extremely narrow band on the basinward edge of this broad shelf. Along this shelf margin, biogenic growth climaxed in a heterogeneous assemblage of carbonate reefs, banks, bars, and islands which almost completely encircled the Gulf of Mexico and attained thicknesses in excess of 2,000 feet. The shelf-margin model described by Bebout and Loucks (fig. 3) provides the basis for a concentrated exploratory effort as well as the basic concept upon which this study is based.

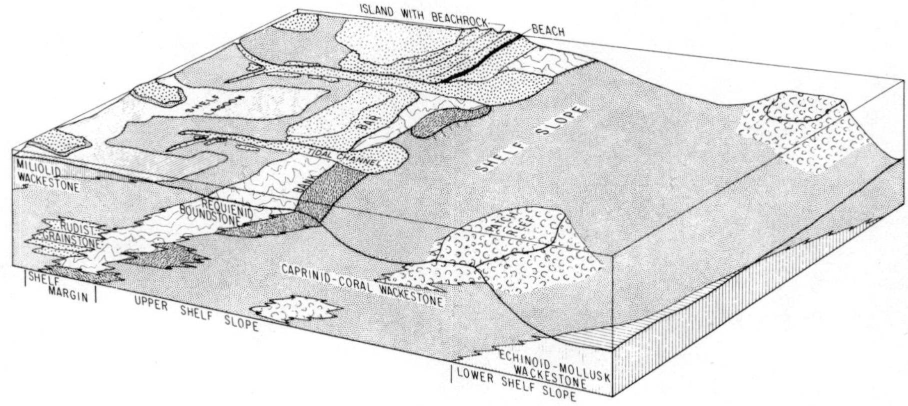

FACIES	ENVIRONMENTS OF DEPOSITION AND PROBABLE WATER DEPTH	
Miliolid wackestone Mollusk wackestone Toucasid wackestone Mollusk-miliolid grainstone	Shallow-water shelf lagoon Water depth less than 20 feet	
Algae-encrusted miliolid-coral-caprinid packstone	Stable grain flat Water depth 1-5 feet	SHELF MARGIN
Rudist grainstone	Beaches, tidal bars, spits, channel fill Water depth less than 10 feet	SHELF MARGIN
Requienid boundstone Coral-caprinid boundstone	Reefs and banks—low relief and discontinuous Water depth 5-15 feet	SHELF MARGIN
Caprinid-coral wackestone Coral-stromatoporoid boundstone	Upper shelf slope Water depth 10-30 feet	
Intraclast grainstone Echinoid packstone Echinoid-mollusk wackestone	Lower shelf slope Water depth 30-60 feet	
Planktonic foraminifer wackestone	Open marine Water depth greater than 60 feet	

Figure 3. Facies and interpreted depositional environments across the Edwards shelf margin (modified from Bebout and Loucks, 1974).

Objectives

The shelf-margin model—its depositional environments, facies relationships, diagenetic history, and structural framework—provides the explorationist with an accurate and meaningful tool for hydrocarbon exploration. Although research shows that diagenetic loss of porosity is the greatest detriment to carbonate exploration, the narrow nature of the shelf-margin complex averaging less than 2 miles in width also serves to increase the risk of drilling in the much less prospective shelf-lagoon or shelf-slope facies. To date, production within the study area (fig. 2) has been restricted to lithologies of the shelf-margin complex as described by Bebout and Loucks (1974) and, to a lesser extent, to the grainstone facies fringing the small shelfal islands in the lagoon immediately behind the shelf margin.

The model describes the shelf-margin complex as a progradational rigid framework of coalescing reefs, banks, and bars overlain by shelf-lagoon muds. The objective of this study is to show that geophysics can establish certain criteria indicative of the shelf-margin complex which can be used to define the more apparent shelf-margin - shelf-slope interface as well as the less apparent shelf-lagoon - shelf-margin interface. Establishing the existence of these criteria, finding them common to several hundred miles of seismic dip control (fig. 2), and integrating them with existing well control prove the ability of the seismic tool to define the major structural components and facies relationships of the shelf-margin complex.

Study of the Lower Cretaceous Sligo carbonates indicates that the depositional environment along the Sligo shelf margin was similar to that along the Edwards shelf margin. One well penetrated over 5,000 feet of Sligo shelf-margin facies. The shelf-margin model and the geophysical criteria derived from this study are therefore assumed to apply to the Sligo.

GEOPHYSICAL CRITERIA

Shelf

The average length of each seismic dip line within the study area is 7 to 10 miles. Each line therefore covers 4 to 6 miles of shelf, 1 to 2 miles of shelf margin, and 1 to 3 miles of shelf slope. Upon close examination of dip lines in areas of adequate well control, certain seismic relationships become evident. North of the Edwards shelf margin (fig. 4a, 4b) the section immediately overlying the Edwards consists of the Eagle Ford, Buda, Del Rio, and Georgetown as discernible lithologic units; here, the upper portion of the Edwards Formation comprises varying thicknesses of miliolid wackestone and other lithologies indicative of shelf and shelf-lagoon deposition.

Although less controlled, a similar facies distribution exists in the deeper Sligo (fig. 5a, 5b) where some wells drilled north of the shelf margin encounter miliolid wackestone for the first 300 to 400 feet below the top of the Sligo. Geophysically this facies distribution is reflected more in the Edwards than in the Sligo as a velocity

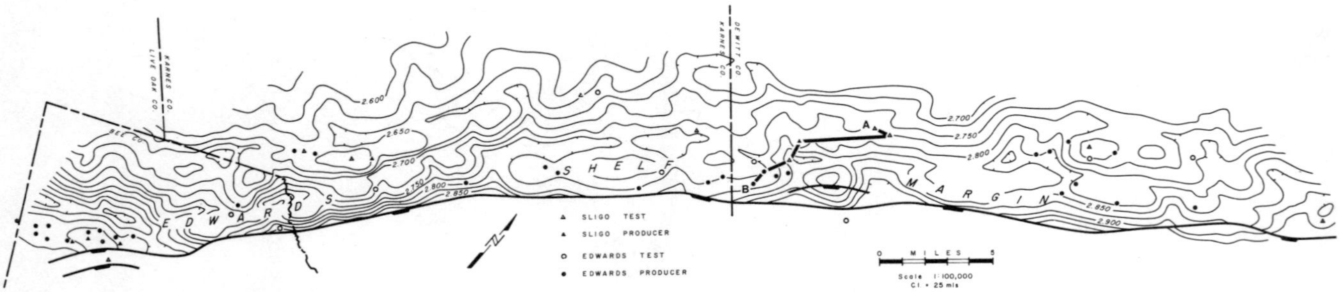

Figure 4a. Edwards seismic structure map.

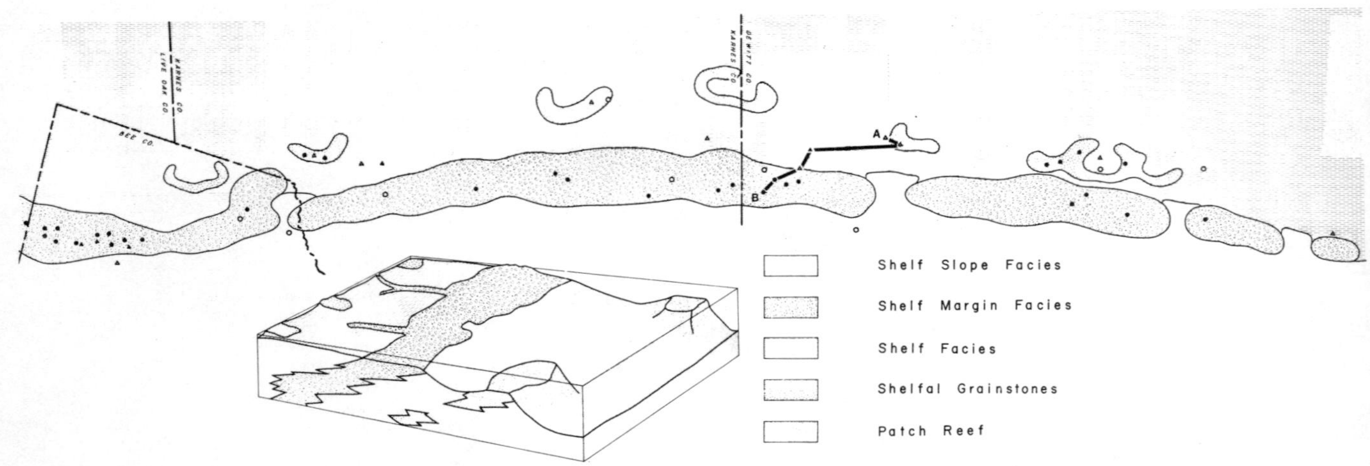

Figure 4b. Generalized facies and interpreted depositional environments across the Edwards shelf margin (diagram modified from Bebout and Loucks, 1974).

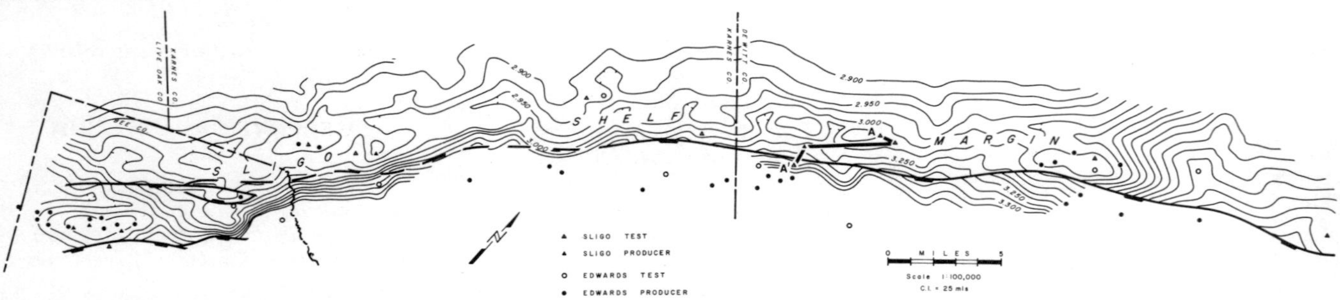

Figure 5a. Sligo seismic structure map.

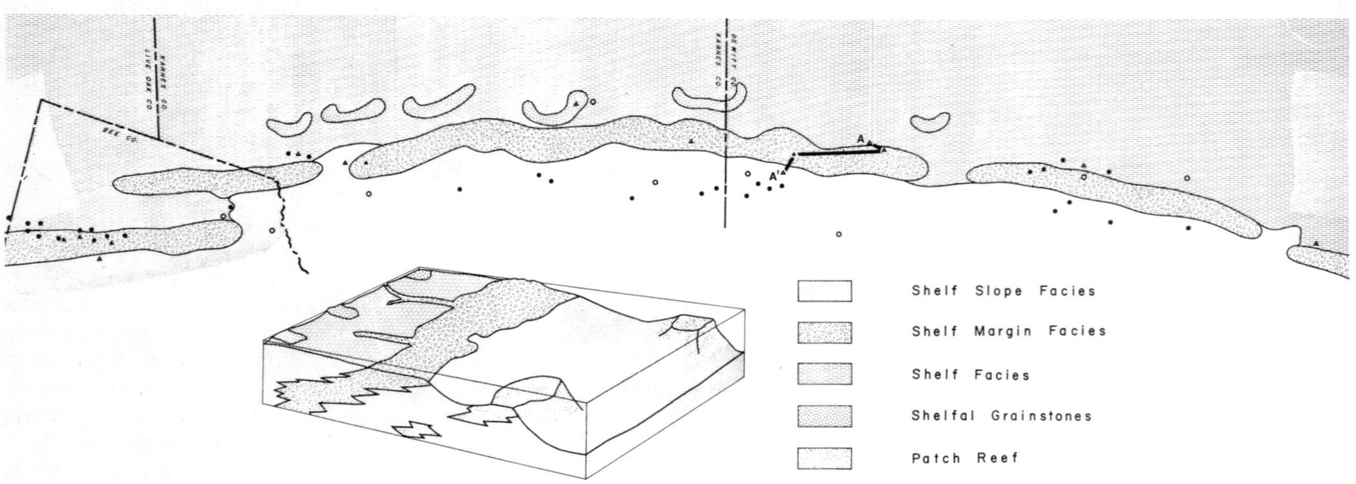

Figure 5b. Generalized facies and interpreted depositional environments across the Sligo shelf margin (diagram modified from Bebout and Loucks, 1974).

controlled reflection overlain and/or underlain by a series of tabular reflections with little or no isochron variations. This relationship exists in the overwhelming majority of lines utilized in this study (figs. 6-8). It persists in a downdip direction to a point where, from what is assumed to be a compaction phenomenon, it is terminated. Since the shelf-margin complex is primarily composed of skeletal debris and organisms in growth position, the rate of sediment accumulation is much greater and the compactibility much less than that of the shelf or shelf-lagoon environment where the opposite occurs. This shelf or shelf-lagoon - shelf-margin interface is therefore reflected seismically as a sharp change or flattening in the seismic dip rate, which is termed the inflection point. It is further reflected in both the Edwards and Sligo by an isochron thinning within the reflection events immediately overlying and/or underlying the Edwards and Sligo. Along some lines there is complete extinction of one or more of these reflections approaching the axis of the shelf-margin complex. This inflection point, thus defined and found common to the majority of lines utilized, represents the northern limit to which one can expect to encounter the favorable facies of the shelf-margin complex. The porous and prospective grainstone facies fringing the shelfal islands immediately behind the shelf margin have heretofore been found productive only in the Edwards. Currently, Sligo well control is extremely sparse; commercial accumulations may yet be found to exist. Within the study area, the facies is found to carry hydrocarbons at three locations where it is encountered on the fringe of Edwards features behind the shelf margin and superimposed over deeper Sligo structures. This facies, completely separate from the shelf-margin trend, is currently emerging as a viable exploratory objective. Anomalous Sligo structural features commonly affect the structure of the Edwards so that a well drilled to test the Sligo has an excellent opportunity to encounter the grainstone facies within the Edwards.

Prior to discussing the criteria associated with the shelf-margin complex and the shelf slope, it is well to note the structural relationship between the Edwards and Sligo shelf margins (fig. 9a, 9b). With one exception in the study area, the Edwards shelf margin is consistently basinward of the Sligo; the inflection point at the Edwards is precisely superimposed

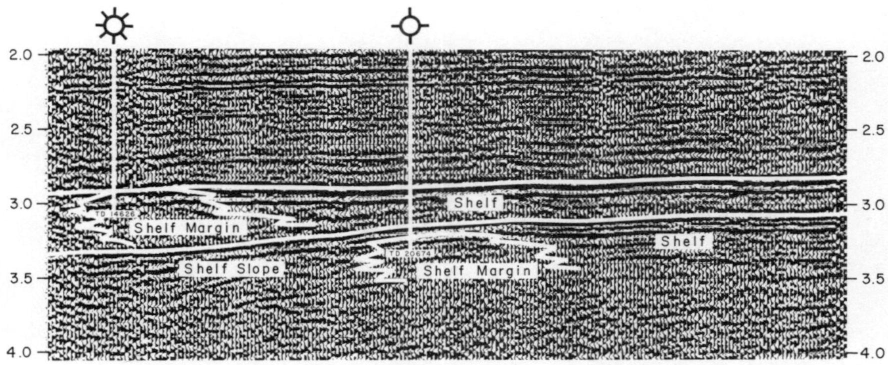

Figure 6. Dip-seismic line across Sligo and Edwards shelf margins.

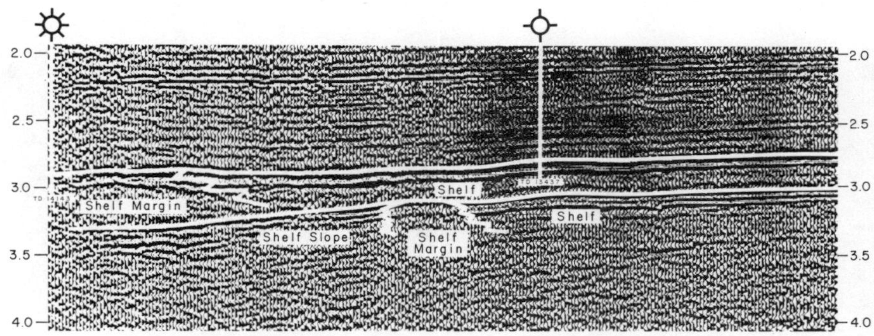

Figure 7. Dip-seismic line across Sligo and Edwards shelf margins.

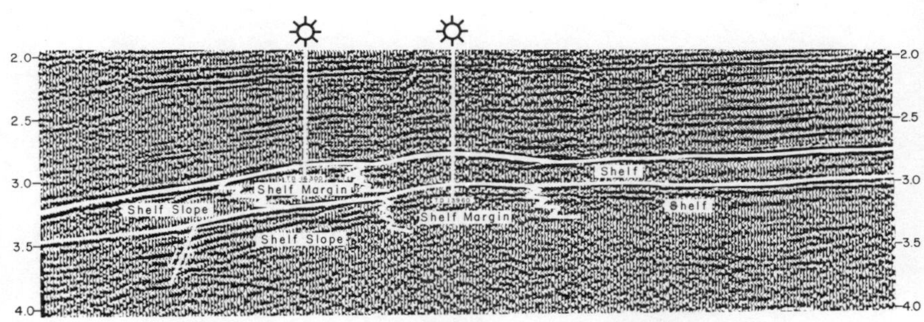

Figure 8. Dip-seismic line across Sligo and Edwards shelf margins.

over the shelf-margin - shelf-slope interface at the Sligo.

Shelf Margin

Basinward of the inflection point, data quality within the Edwards and Sligo as a rule becomes very poor. Density variation between the shelf-lagoon sediments and shelf-margin sediments might possibly be sufficient to provide a reflection event. Within the shelf-margin sediments, however, their homogeneous mineralogical composition as well as their heterogeneous morphology strongly negate the probability of the existence of reliable mappable events. Short discontinuous segments of mappable data which occur near the shelf-lagoon - shelf-margin interface reflect a pronounced flattening and, commonly, north dip.

It is important here to reevaluate the structural morphology and facies relationships of the shelf-margin complex. An examination of the model (fig. 3) and detailed correlation of well control demonstrate that stratigraphic north dip has been an integral part of the prograding shelf-margin complex throughout time. Any present-day structures with antiregional dip represent local areas along the shelf margin where biogenic growth was extremely high. Differential compaction between these and the shelf-lagoon sediments served to accentuate these anomalies, and basinal subsidence served to rotate their axes and diminish their structure. Within the study area the majority of Edwards production established along the shelf-margin trend is associated with structural closure. Along the Sligo Trend, however, production has been established from the shelf-margin complex on south dip as well as on structure. Whether north dip is a requirement for hydrocarbon accumu-

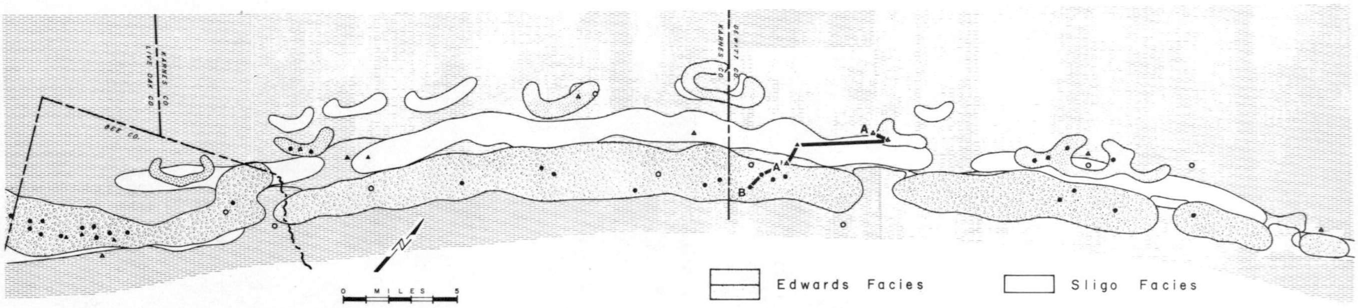

Figure 9a. Structural relationship of the shallow Edwards shelf-margin complex to the deeper Sligo shelf-margin complex.

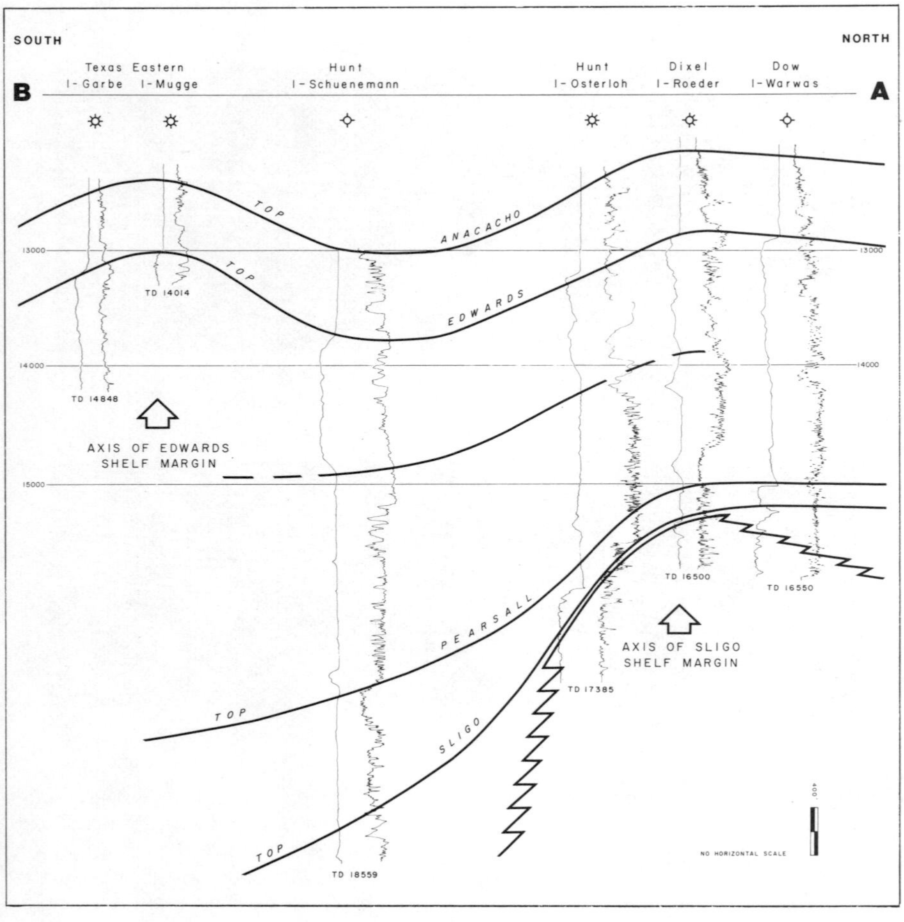

Figure 9b. Cross section A-B.

lation is a debatable matter and not in the scope of this study. It appears obvious from the model that the stratigraphic north dip created by the prograding shelf-margin complex is ample for hydrocarbon entrapment as these porous carbonates are overlain by the nonporous shelfal wackestones which form an updip impermeable barrier.

Although it is not pertinent to the objective of this study, it is believed that the shelf-margin environment provides excellent conditions for the generation of hydrocarbons as well as a heterogeneous complex of carbonate rocks, many having primary porosities greater than 30 percent. The immediate postdepositional history, diagenesis, lithification, and structural deformation by basinal subsidence serve to complicate exploration for these hydrocarbons. The ability of modern geophysics to define the shelf-margin facies does not answer the problems created by these sedimentary processes, but it greatly reduces the risk and more narrowly defines the limits of the trend within which these accumulations can be found. By progressive integration of shelf-margin research with well control and geophysics, a further definition of the most prospective locations along the shelf margin can surely be anticipated.

Shelf Slope

Geophysically, the shelf-margin - upper shelf-slope interface occurs as a relatively sharp increase in seismic dip rate. The shelfal reflection events immediately above or below the reflections mapped as the Edwards or Sligo, which become weak or extinct through the shelf-margin portion of the record section, commonly reappear south of this interface. Down-to-the-basin faulting which is uncommon along the shelf margin also becomes pronounced along the upper shelf slope. This faulting, which for the most part is purely a matter of geophysical interpretation, is substantiated by a further examination of the model. The pronounced lithologic variation between these two facies, the basinward and upward progradation of the shelf-margin complex, the tremendous variation in sediment accumulation, and basinal subsidence all provide legitimate logic for the conclusion that this faulting is real. In one example, the steep facies relationship and/or faulting which occurs at the shelf-margin - shelf-slope interface is exemplified by well control, a seismic record, and velocity surveys from two wells drilled on either side of the Sligo shelf-margin - shelf-slope interface. The record section in figure 10 is a composite of two adjacent dip lines joined by a strike line. Velocity control establishes reliable picks for the tops of the Edwards and Sligo Formations. These two sections were obtained and mapped before acquisition of the velocity data, but because the Edwards shelf-slope - shelf-margin interface is not involved, there is no difficulty in tying all the Edwards tops along the composite line to the shelf velocity survey. The Sligo events how-

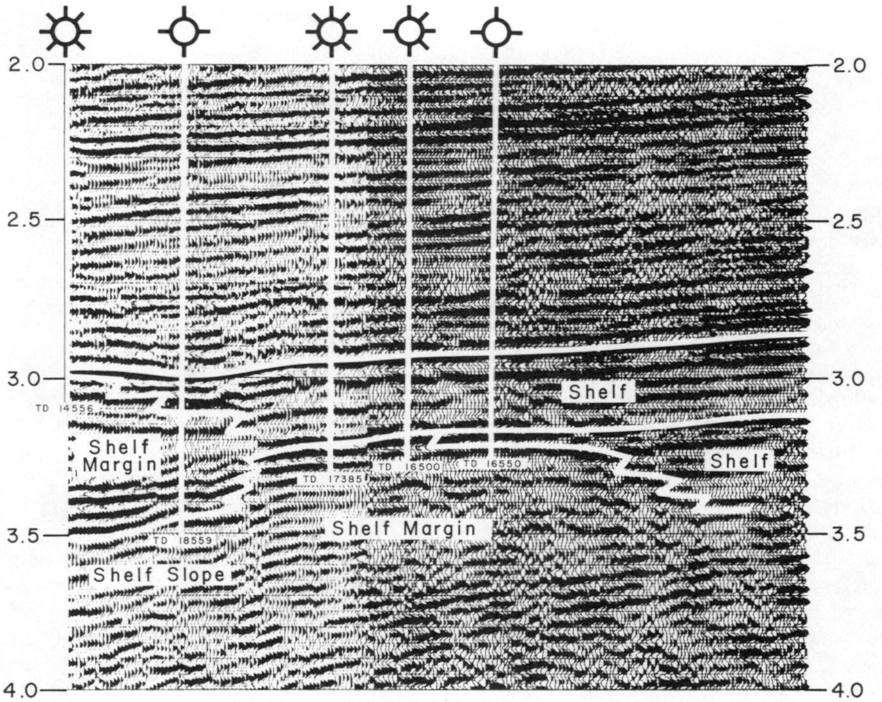

Figure 10a. Composite record section.

ever reflect the shelf-margin - shelf-slope interface. Well logs and samples from the Hunt #1 Osterloh and the Hunt #1 Schuenemann are definitely representative of shelf-margin and shelf-slope lithologies, respectively. The record section therefore is interpreted to reflect the facies relationship known to exist. A steep facies change and a possible fault are interpreted. Two hundred and eight feet of structural relief occurs on the top of the Edwards between the Osterloh and Schuenemann wells; on the other hand, 1,700 feet of structural relief occurs on the Sligo. A problem arises in determining which of the reflections south of the interface represents the Sligo. Without well or velocity information, jump correlation across the interface indicates a good tie reflecting approximately 100 milliseconds of displacement. However, velocity information from the Schuenemann well indicates that the Sligo reflection event is approximately 50 milliseconds deeper, and the apparent event correla-

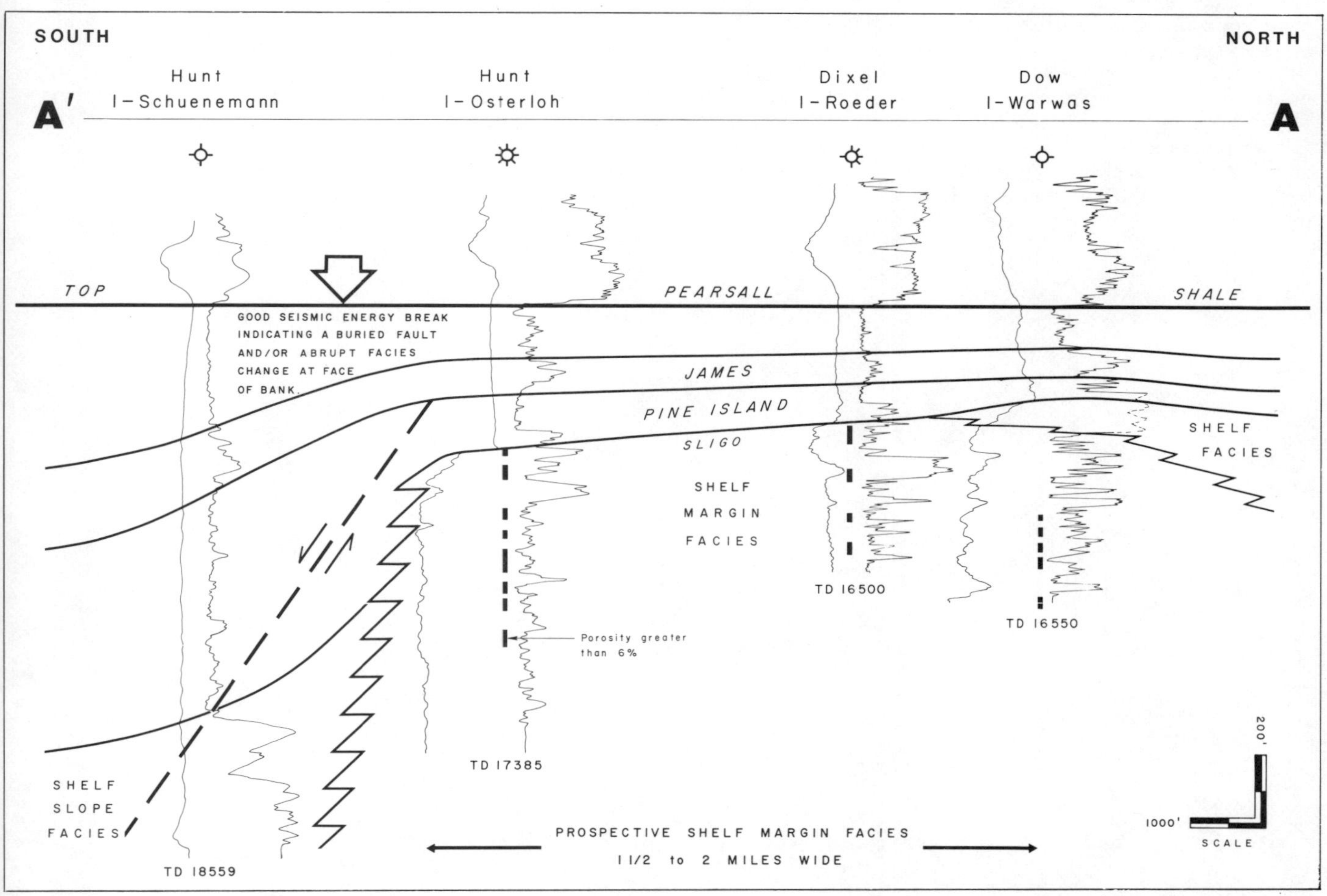

Figure 10b. Cross section A-A′.

tion across the interface is coincidental and false. This conclusion requires further consideration and possible reinterpretation of the shelf-margin - shelf-slope interface on all the dip lines within the study area. All the lines were previously interpreted to reflect a facies change and/or fault of minor magnitude, although the actual break or apparent displacement varies greatly from one line to another. The reevaluation of this portion of each line, in light of the data obtained from the relationship between the Schuenemann and Osterloh wells, leads to the following conclusion: without sufficient velocity and well information, determining which of the reflections south of the interface represents the Sligo is impossible. This condition probably exists all along the Edwards and Sligo shelf-margin - shelf-slope interfaces. As long as the geophysical conditions established between these two wells are found on other dip lines, the true Sligo reflection event south of the interface is not germane to shelf-margin hydrocarbon exploration. However, shelf margin talus and/or extreme fracturing in the shelf-slope facies along the hinge line which exists at the interface may be prospective. Hydrocarbons were generated in this environment and any major fracture system can easily contain accumulations.

The majority of the hydrocarbons generated from the abundance of flora and fauna which thrived within the shelf-margin facies remains there today, waiting for the innovative and persistent explorationist who, through the understanding, integration, and application of every exploratory tool at his disposal, may come to understand the environmental, diagenetic, and structural processes which complicate his efforts.

SUMMARY

The Lower Cretaceous shelf margin, composed almost entirely of sedimentary and biogenic carbonate strata, is one of the few sedimentary features common to the entire Gulf of Mexico. The rate of sediment accumulation along the shelf margin maintained a balance with basinal subsidence so that more than 2,000 feet of Edwards and 5,000 feet of Sligo carbonates can be identified. The tremendous hydrocarbon potential along these trends cannot be overemphasized.

Every explorationist has access to an abundance of good-quality seismic data and an ever increasing quantity of shelf-margin research and subsurface information. The combined strength of these exploratory tools can serve to define the major structural components and facies relationships of the shelf-margin complex. This approach can aid the industry in finding tomorrow's reserves today.

ACKNOWLEDGMENTS

We should like to acknowledge D.G. Bebout and R.G. Loucks for their invitation and encouragement to present this study and Larry L. Jones for his permission to do so. Sincere appreciation is expressed to Margaret D. Anderson for her preparation of the illustrations and to A.N. Francis for his rewarding conversations concerning depositional processes in carbonate environments.

REFERENCES

Bebout D. G., and Loucks, R. G., 1974, Stuart City Trend, Lower Cretaceous, South Texas: Univ. Texas, Austin, Bureau of Economic Geology Report of Investigations 78, 80 p.

Coogan, A. H., Bebout, D. G., and Maggio, Carlos, 1972, Depositional environments and geologic history of Golden Land and Poza Rica Trend, Mexico, an alternative view: AAPG Bull., v. 56, no. 8, p. 1419-1447.

Enos, Paul, 1974, Reefs, platforms, and basins of Middle Cretaceous in northeast Mexico: AAPG Bull., v. 58, no. 5, p. 800-809.

Perkins, Bob F., 1974, Paleoecology of a rudist reef complex in the Comanche Cretaceous Glen Rose Limestone of Central Texas: Geoscience and Man, v. VIII, p. 131-173.

SLIGO AND HOSSTON DEPOSITIONAL PATTERNS, SUBSURFACE OF SOUTH TEXAS

D. G. Bebout[1]

ABSTRACT

The Sligo/Hosston depositional wedge occurs over an area of thousands of square miles of South Texas. The wedge thickens from its pinchout updip to greater than 1,000 feet thick downdip at the shelf edge. The wedge can be subdivided into three major time-equivalent parts based on quartz-sandstone and carbonate facies and interpreted depositional environments: Hosston, lower Sligo, and upper Sligo. The Hosston dolomitic quartz sandstone, dolomite, and anhydrite were deposited on an arid tidal flat; the sandstone decreases and the dolomite increases in abundance upward in the Hosston. The lower Sligo is characterized by a number of cycles which consist of burrowed subtidal dolomite at the base and laminated supratidal dolomite and anhydrite at the top. The upper Sligo consists predominantly of subtidal skeletal wackestones to grainstones and thin oolite grainstones.

The vertical sequence described here indicates an overall transgressive trend to the Sligo/Hosston depositional wedge; however, detailed sections show that this transgression was accomplished by means of a number of progradational cycles each successive one of which was shifted landward of the previous one.

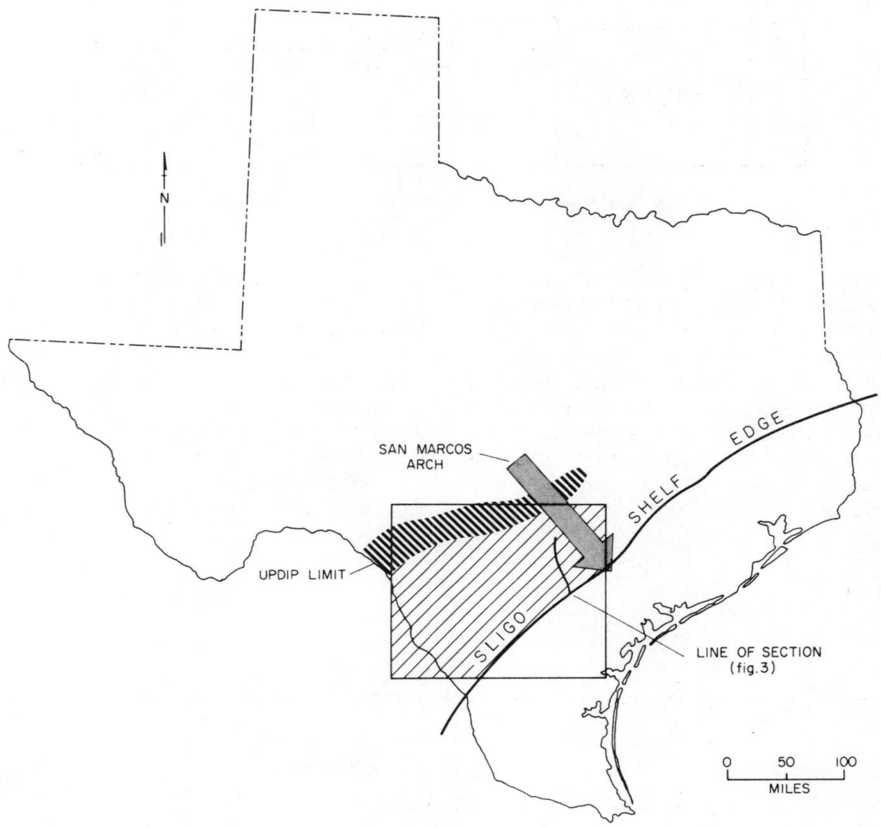

Figure 1. Study area bounded updip by the Sligo pinchout, downdip by the shelf edge, to the northeast by the San Marcos arch, and to the southwest by the international boundary with Mexico. The line of section is shown on figure 3.

INTRODUCTION

Regional Setting

The Sligo Formation occurs in the subsurface of South Texas over an area of thousands of square miles (fig. 1). Throughout this area the Sligo carbonates were deposited on a broad shelf under conditions ranging from shallow subtidal to intertidal and supratidal. The carbonate section terminates downdip at the shelf edge in a rudist reef and grainstone trend; updip the Sligo grades into a quartz sandstone and siltstone—the Hosston Formation (Sycamore Sandstone on the outcrop). The Sligo carbonates do not outcrop in Texas. The Sligo/Hosston Formations comprise a transgressive wedge of carbonates and quartz sandstones, respectively, which are Aptian in age (fig. 2) and represent lowermost Cretaceous rocks of Texas. Overlying the Sligo/Hosston is the Pearsall Formation and then the Glen Rose and Edwards limestones.

During the Lower Cretaceous the Texas Gulf Coast was characterized by continual subsidence. This Cretaceous subsidence is evidenced by thickening of the Sligo from its pinchout near the outcrop to greater than 1,000 feet downdip at the shelf edge (fig. 3). Continued subsidence during the Tertiary is indicated on the cross section by the depths to the top of the Edwards Limestone—350 feet updip to 13,485 feet downdip, a difference of more than 13,000 feet in elevation in less than 100 miles.

Approach

More than 200 wells penetrate at least part of the Sligo Formation in South Texas (fig. 4). Most of these wells penetrated only the top 50 to 100 feet of the Sligo and less than a quarter of the wells penetrate a significant section of the formation. Varying lengths of core are available from 47 wells, but most cores come from the top 50 to 100 feet. Consequently, the cross section on which this paper is based must be viewed in light of the existing limitations of control. The wells with continuous core throughout the Sligo Formation or a large portion of it are widely spaced and have been

[1]Bureau of Economic Geology, The University of Texas at Austin.

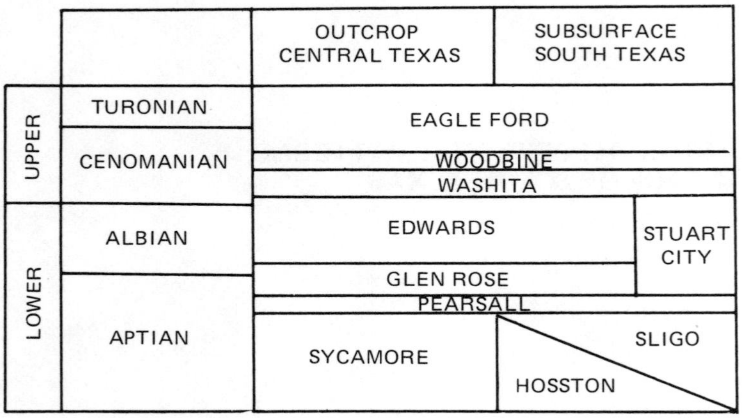

Figure 2. Cretaceous formations, Central and South Texas.

projected up to 30 to 40 miles into the section (fig. 4). Most of the remaining wells with limited core control are clustered in Atascosa, Frio, Medina, and Zavala Counties and are the result of an extensive exploration and drilling program of a single major oil company. Therefore, interpretations based on this study must be considered to be only very generally applicable to any specific area, and considerable variation should be expected. It is believed, however, that the broad depositional patterns developed here are meaningful and applicable over the wide area of study. A comprehensive report describing the facies distribution and diagenesis of the Sligo is now in preparation at the Bureau of Economic Geology.

Amsbury (1974) reported on the facies and depositional environments of the Sycamore Sandstone on the outcrop and the Hosston and Sligo Formations of the shallow subsurface of south-central Texas. His paper uses the Stanolind No. 1 Schmidt well as the downdip-most control; in contrast, the Schmidt well represents the most updip control for this study, thus forming an excellent overlapping point of correlation between Amsbury's study and this report.

SLIGO/HOSSTON LITHOLOGIES AND FACIES

In the Sligo/Hosston depositional wedge, four rock types are abundant: quartz sandstone and siltstone, anhydrite, dolomite, and limestone (fig. 5). Updip, the Hosston Formation lies with erosional unconformity on basement metamorphics; coarse conglomerates (fig. 6) mark this contact in several wells. Downdip, the Hosston and Sligo are underlain by Jurassic sediments, but few wells penetrate this contact and no cores across it have been located.

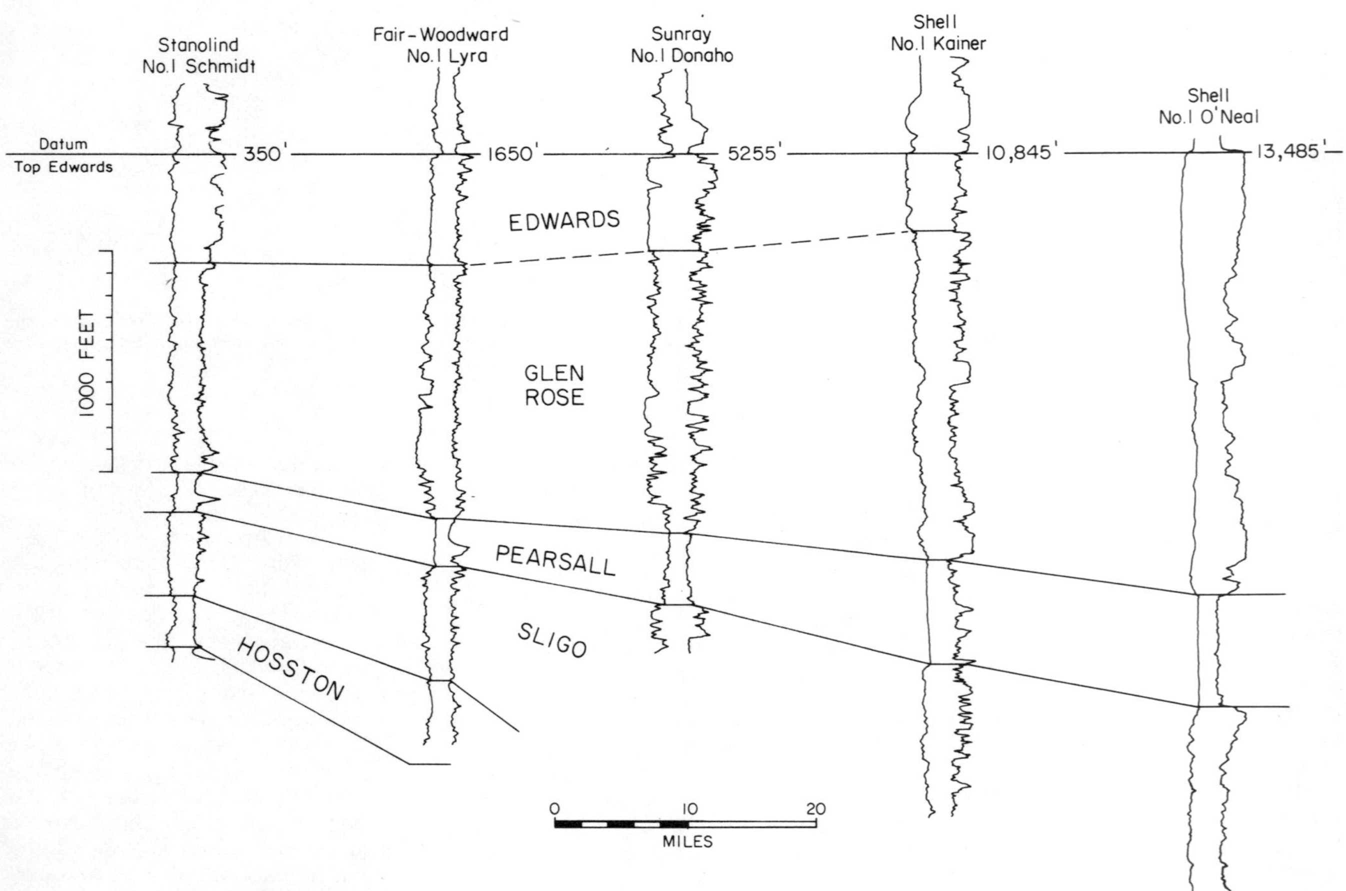

Figure 3. Electrical-log section of the Lower Cretaceous, South Texas.

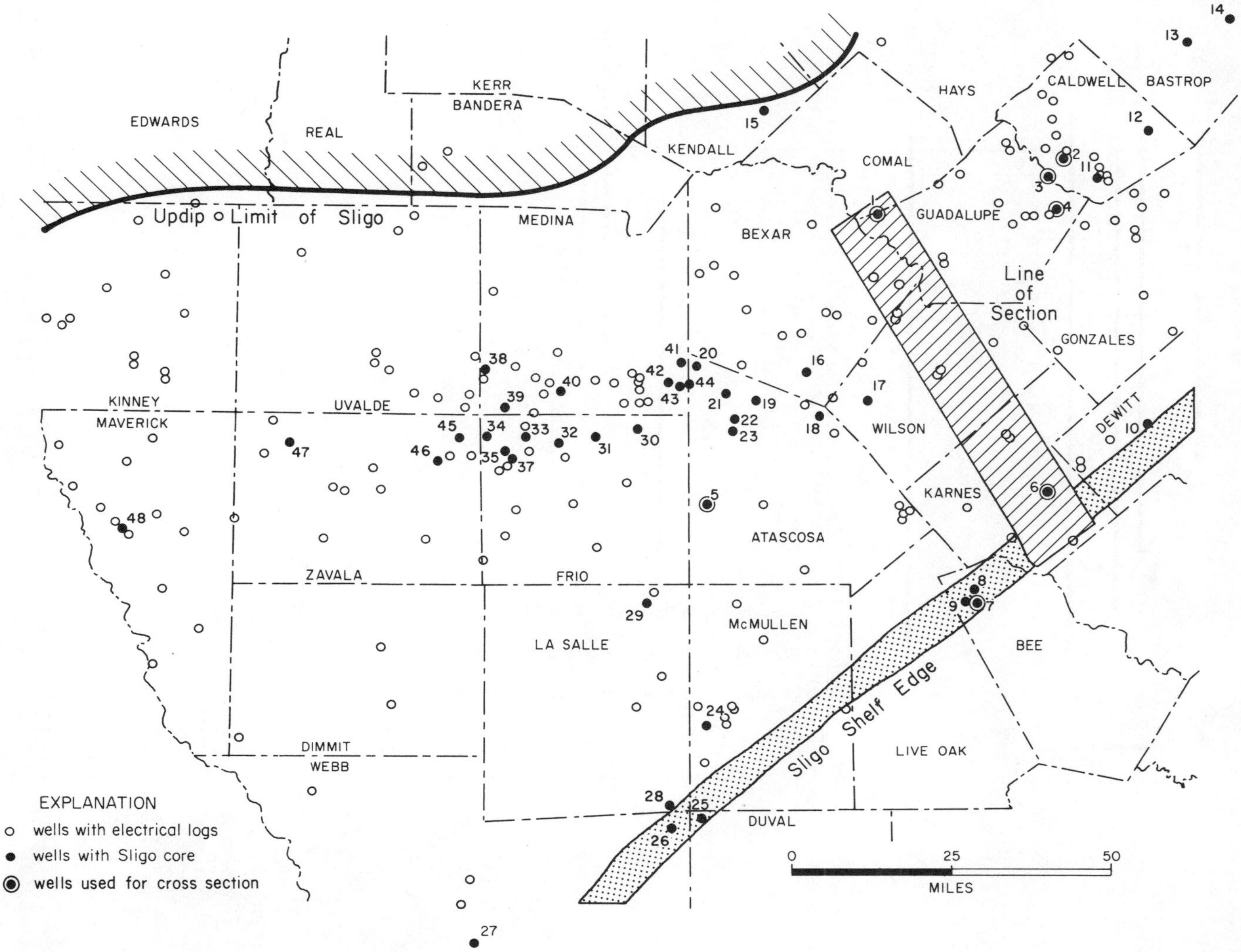

Figure 4. Sligo well and core control and location of detailed section (figs. 5, 20, 21, 23, 25).

The lower part of the Hosston Formation consists predominantly of dolomitic quartz sandstone and siltstone, dolomite, and anhydrite (fig. 5). The sandstone and siltstone facies are white to red and are laminated, cross laminated, rippled, and burrowed (figs. 7, 8, 9, 10). The dolomite is buff to light brown, dominantly mudstone, and alternately burrowed and laminated. Thin beds of nodular to mosaic anhydrite are associated with both the sandstone and siltstone and the dolomite sections. In the Magnolia No. 1 Baker core (fig. 7), the Hosston consists predominantly of sandstone and siltstone in the lower part and grades upward into dominantly anhydrite and dolomite. However, a 20-foot-thick sandstone unit also marks the top of the formation.

Updip the Sligo is readily subdivided into two parts: the lower part which consists largely of dolomite with minor amounts of limestone and anhydrite and the upper part which is dominantly limestone with lesser amounts of dolomite and scarce anhydrite (fig. 5). The lower Sligo consists of many cycles (fig. 11) which range from 30 to 65 feet in thickness. A number of carbonate facies are recognizable, but the cycles are characterized by highly burrowed dolomite and limestone in the lower part (fig. 12); toward the top of each cycle, algal-laminated and mud-cracked dolomite and thin beds of nodular to mosaic anhydrite become more common (figs. 13, 14, 15). The limited mulluscan fauna occurs sporadically throughout the cycles. Mollusk wackestone is the most common carbonate rock fabric, but pellet and intraclast wackestone and packstone occur near the top of the cycles.

The dominantly limestone upper Sligo is less than 100 feet thick at the updip end of the cross section (fig. 5) but thickens to more than 800 feet at the downdip end. Dolomite occurs in the lower half of the unit. In the Magnolia No. 1 Baker well (fig. 16), the carbonate fabrics represented are diverse and include burrowed skeletal and pelletal wackestones to grainstones (fig. 17). In other wells to the southwest, thin laminated and burrowed oolite grainstones are common (fig. 18). To the southeast, at the shelf edge, rudist grainstones and bafflestones occur along with burrowed wackestones and packstones (fig. 19).

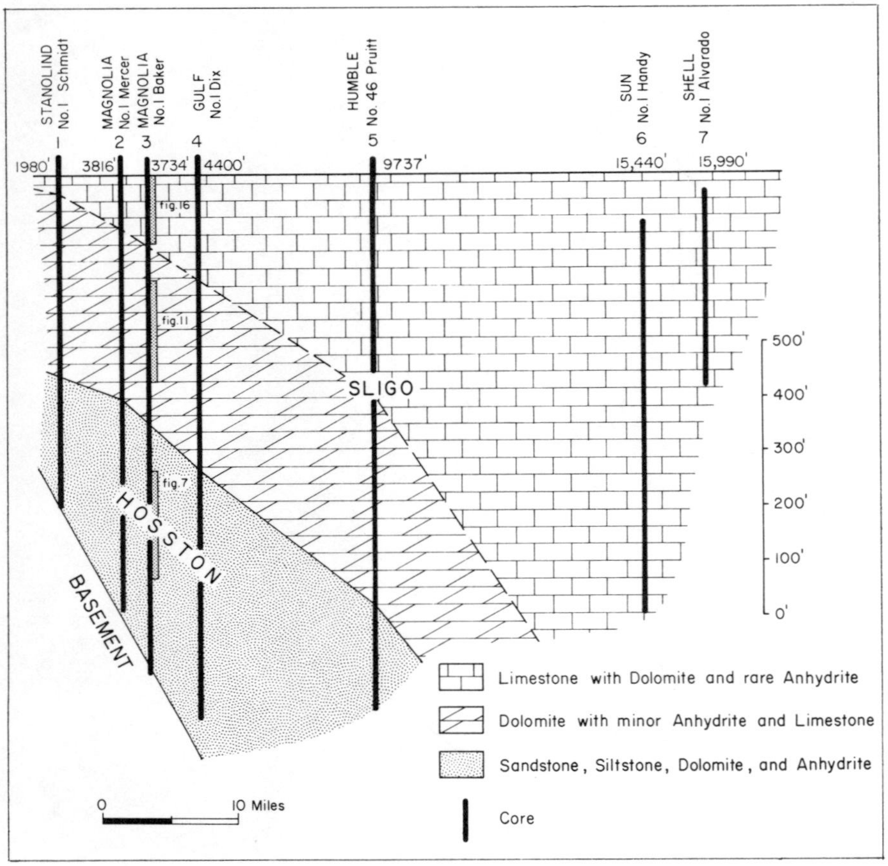

Figure 5

Figure 5. Lithologies of the Sligo/Hosston depositional wedge.

Figure 6. Red conglomeratic sandstone which marks the base of the Hosston Formation. From the Magnolia No. 1 Baker, Guadalupe County. a—4,660 ft; b—4,653 ft.

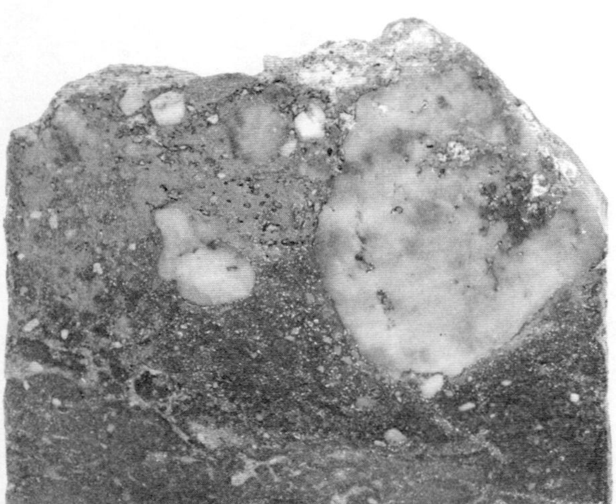

X I a.

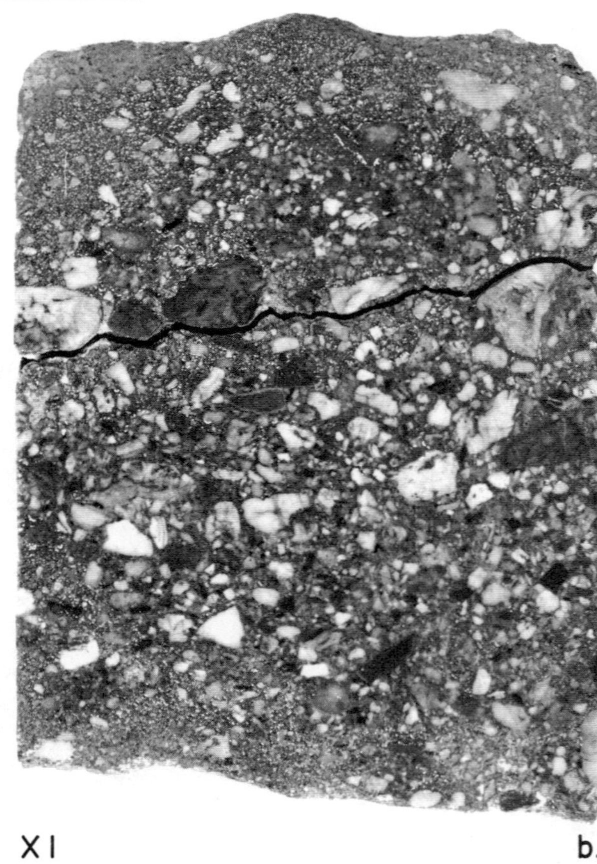

X I b.

Figure 6

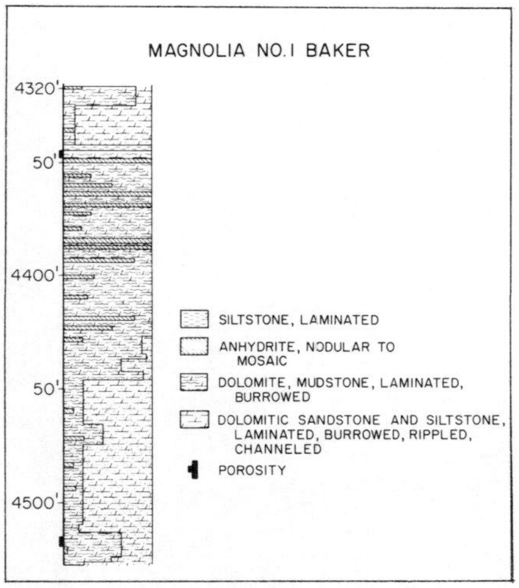

Figure 7. Hosston facies, South Texas.

Updip, mollusks, miliolids, and green algae are common to abundant in the upper Sligo unit. Downdip, the fossil diversity increases until, at the shelf edge, rudists (caprinids and toucasids), stromatoporoids, encrusting red algae, and corals are abundant.

DEPOSITIONAL HISTORY

Increase in fossil abundance and diversity can be readily demonstrated on the cross section (fig. 20) to occur in both upward and gulfward directions. For example, in Wells 1 and 2 (fig. 20) the Hosston and lower part of the Sligo contain mollusks as the only recognizable faunal constituent; toward the top of the Sligo, miliolids and green algae, along with the mollusks, are abundant and are important contributors to the sediment. Likewise, in a dip direction, the limited molluscan fauna of the updip Hosston and Sligo (Well 1) change to common mollusks, miliolids, and green algae in the middle of the platform (Well 4), to a diverse rudist, coral, mollusk, and miliolid fauna of the shelf edge (Wells 5 and 6). This change is accompanied by a decrease in quartz sandstone and siltstone, dolomite, and anhydrite (fig. 5), and a general increase in energy level is reflected by increase in the grain content of the carbonates. These vertical and lateral trends demonstrate the occurrence of more normal marine conditions toward the top of and gulfward from the Sligo/

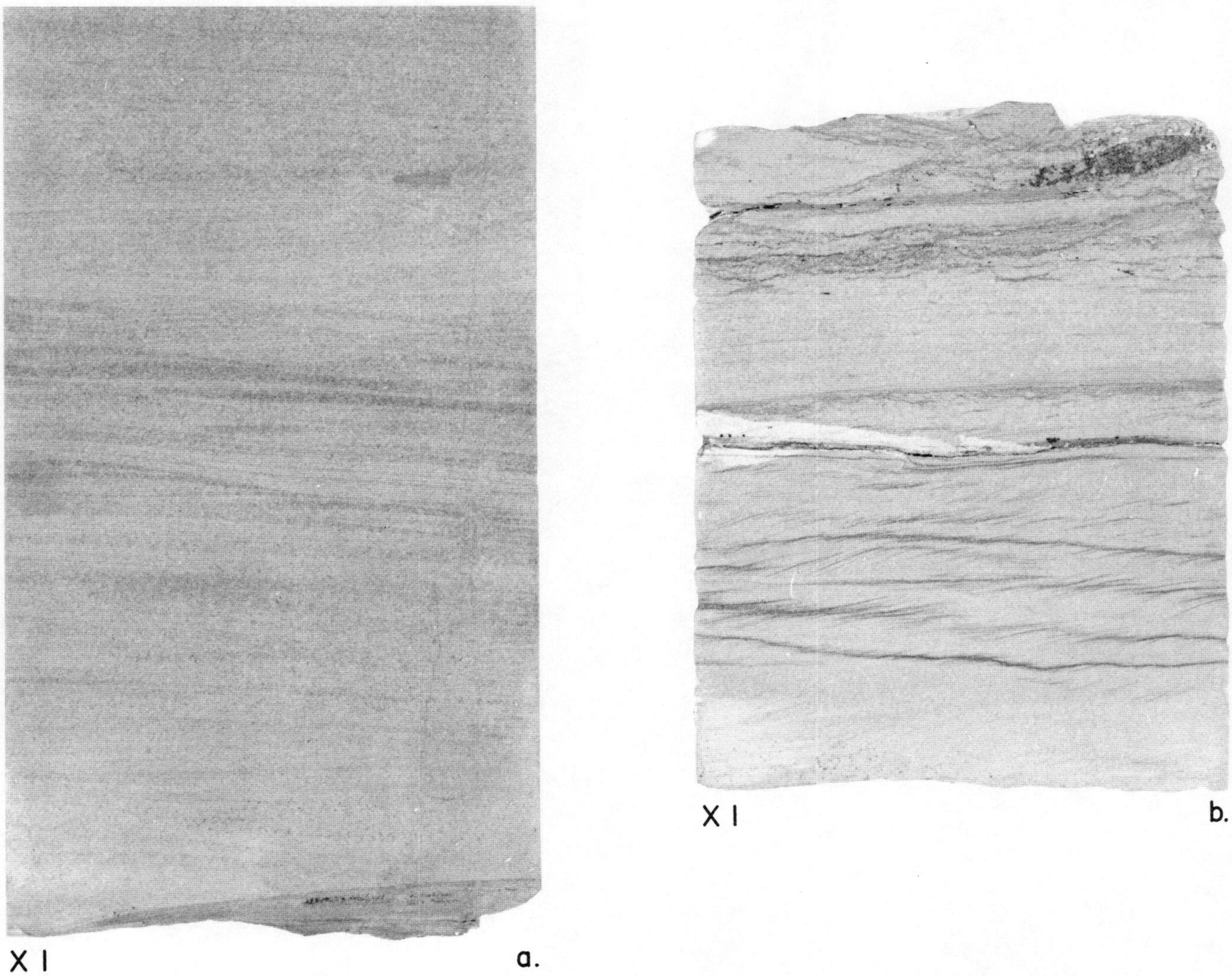

Figure 8. Laminated, cross-laminated, and rippled sandstone from the Hosston Formation. From the Magnolia No. 1 Mercer, Caldwell County, a–4,432 ft; b–4,411 ft.

Figure 9. Burrowed laminated sandstone from the Hosston Formation. a—from the Magnolia No. 1 Mercer, Caldwell County, 4,334 ft; b—from the Magnolia No. 1 Baker, Guadalupe County, 4,661 ft.

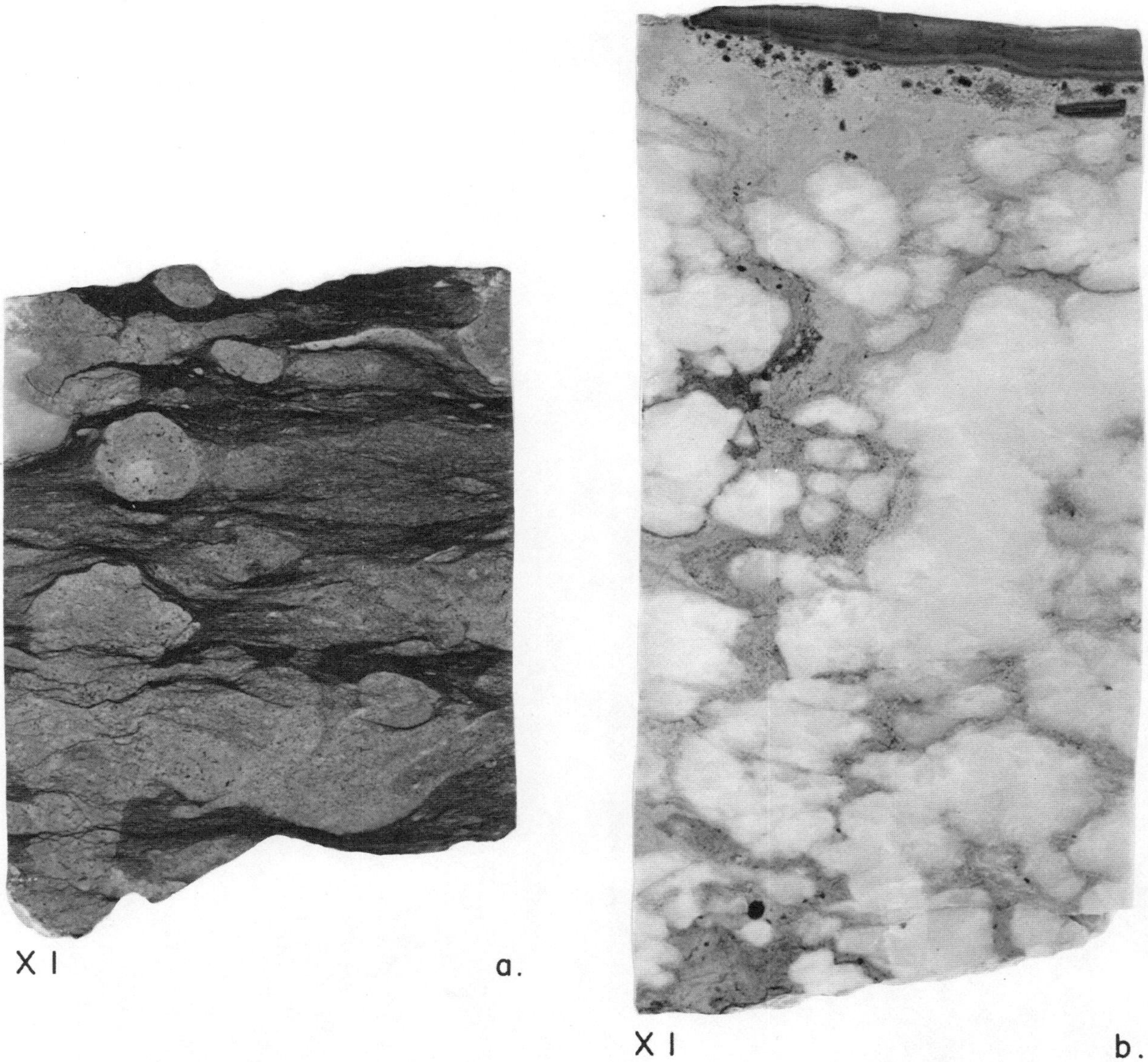

Figure 10. a—burrowed dolomite, Hosston Formation, from the Gulf No. 20 Dix, Guadalupe County, 5,046 ft; b—nodular-mosaic anhydrite and dolomite, Hosston Formation, from the Gulf No. 20 Dix, Guadalupe County, 5,063 ft.

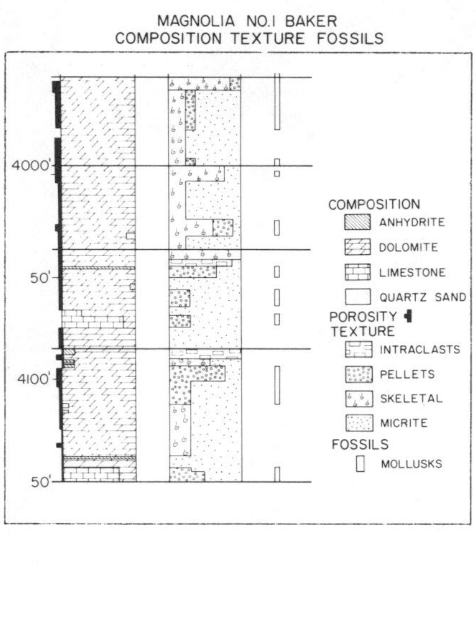

Figure 11. Lower Sligo facies and cycles, South Texas.

X I a. X I b.

Figure 12. Highly burrowed mollusk wackestone from the lower Sligo Formation. a—from the Magnolia No. 1 Baker, Guadalupe County, 4,162 ft; b—from the Magnolia No. 1 Mercer, Caldwell County, 4,041 ft.

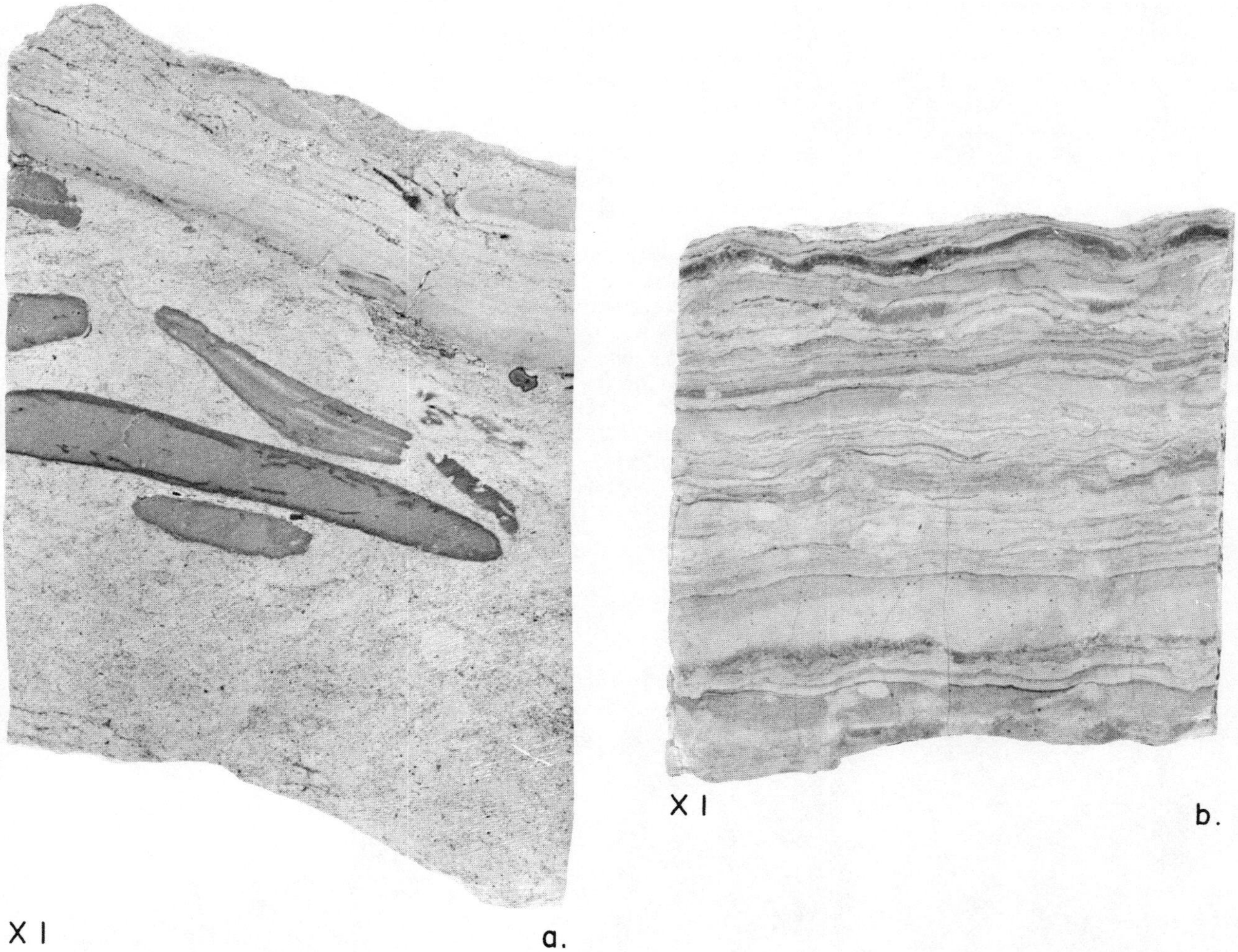

Figure 13. a—algal-laminated and intraclast dolomite; lower Sligo Formation from the Stanolind No. 1 Schmidt, Guadalupe County, 2,219 ft; b—algal-laminated dolomite, lower Sligo Formation from the Magnolia No. 1 Mercer, Caldwell County, 3,908 ft.

Hosston depositional wedge. Amsbury (1974) reported this same vertical change in the thinner updip Sligo.

Cores from wells at the updip end of the cross section contain a coarse conglomerate at the base of the Hosston Sandstone; this conglomerate probably represents alluvial-fan or braided-stream deposits. Updip, the Sycamore Sandstone was interpreted by Amsbury (1974) to have been deposited as point bars and alluvial fans. Amsbury also reported abundant caliche soils in the Sycamore and postulated from this a fairly dry climate at the time of deposition.

The Hosston laminated, rippled, and burrowed quartz sandstone and siltstone as well as the associated algal-laminated, mud-cracked, and burrowed dolomite and nodular to nodular-mosaic anhydrite are interpreted as having been deposited on a tidal flat (fig. 21). Much of the sandstone and siltstone was probably transported by wind to the tidal flat. An abundant but low-diversity fauna of mollusks and crustaceans inhabited the flats, as evidenced by the abundance of burrows in both the quartz sandstones and dolomites.

The lower Sligo cycles (fig. 11) represent numerous alternations from subtidal conditions (burrowed facies) in the lower part of each cycle to intertidal and supratidal conditions (algal-laminated, mud-cracked, and crossbedded facies) in the upper part. Interruptions in this progradational cycle occurred many times, as evidenced by the numerous incomplete cycles. The limited fauna and narrow range of carbonate fabrics suggest that this shoreline to very shallow-marine setting was typically of very low energy and restricted marine circulation.

The upper Sligo dominantly limestone section with more diverse faunal content and varied carbonate fabrics suggests a change to more normal-marine conditions (fig. 21). Along the line of section, subtidal conditions prevailed in the upper Sligo, as interpreted by diverse marine fauna and dominant wackestone facies. However, higher energy subtidal to intertidal conditions did exist on the platform; deposition of oolite and skeletal grainstones (fig. 18), probably as linear bars offshore from the tidal flat, resulted. Downdip, at the shelf edge, abundant grainstone and bafflestone facies along this narrow band indicate that high-energy conditions prevailed.

Most cores from downdip wells show a gradational contact between the top of the Sligo and the overlying Pine Island Shale of the Pearsall Formation. This gradational contact is indicated by the increase in terrigenous shale in the top 10 to 20 feet of the Sligo and the lack of a significant erosional surface at the top of the formation; numerous hardgrounds, indicated by erosional and bored surfaces (fig. 22), occur throughout the Sligo and are considered to be of minor local importance. Updip, on the other hand, Amsbury (1974) found widespread evidence of subaerial exposure of the top of the Sligo prior to deposition of the overlying shale. Loucks (1976, 1977) also found evidence of exposure of the top of the Sligo in some of the updip wells used both in his study of the overlying Pearsall Formation and in this Sligo study.

The overall vertical and lateral facies changes within the Hosston/Sligo Formations describe one huge transgressive depositional wedge (fig. 23). However, close examination of vertical changes within cored intervals suggests the existence of numerous progradational cycles. The effects of these cycles are most recognizable in the lower Sligo subtidal to supratidal part of the section because environmental conditions changed most drastically here in shoreline to nearshore positions. Each successive progradational wedge was displaced landward of the previous one as a result of sporadic sea-level rise or regional subsidence, thus resulting in an overall transgressive trend. Throughout this transgression, vertical growth of banks along the shelf edge resulted in their remaining in the same position; thus, the marine platform of shallow-shelf area expanded throughout deposition of the Sligo as the shoreline moved landward. The transgressive nature of the Sligo/Hosston depositional wedge was previously recognized by Amsbury (1974) in his study of the updip part of the section, by Stricklin and others (1971) from their study of these sediments on the outcrop of Central Texas, and by Bebout (1976) from a preliminary study of downdip Sligo.

A general model (fig. 24) depicting the environment of deposition

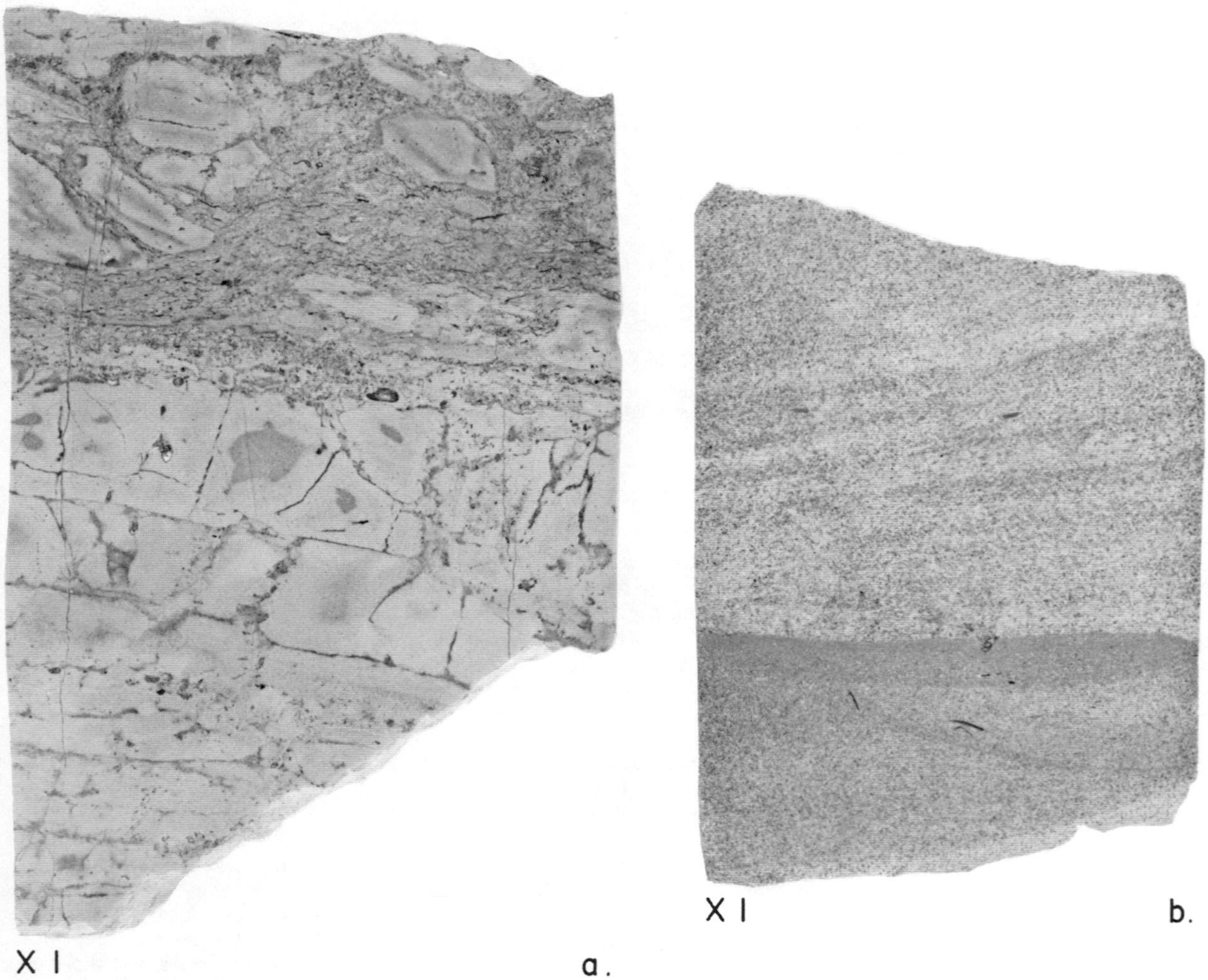

Figure 14. a—laminated and mud-cracked dolomitic mudstone, lower Sligo Formation, from the Magnolia No. 1 Baker, Guadalupe County, 3,880 ft; cross-laminated mollusk grainstone, lower Sligo Formation, from the Gulf No. 20 Dix, Guadalupe County, 4,803 ft.

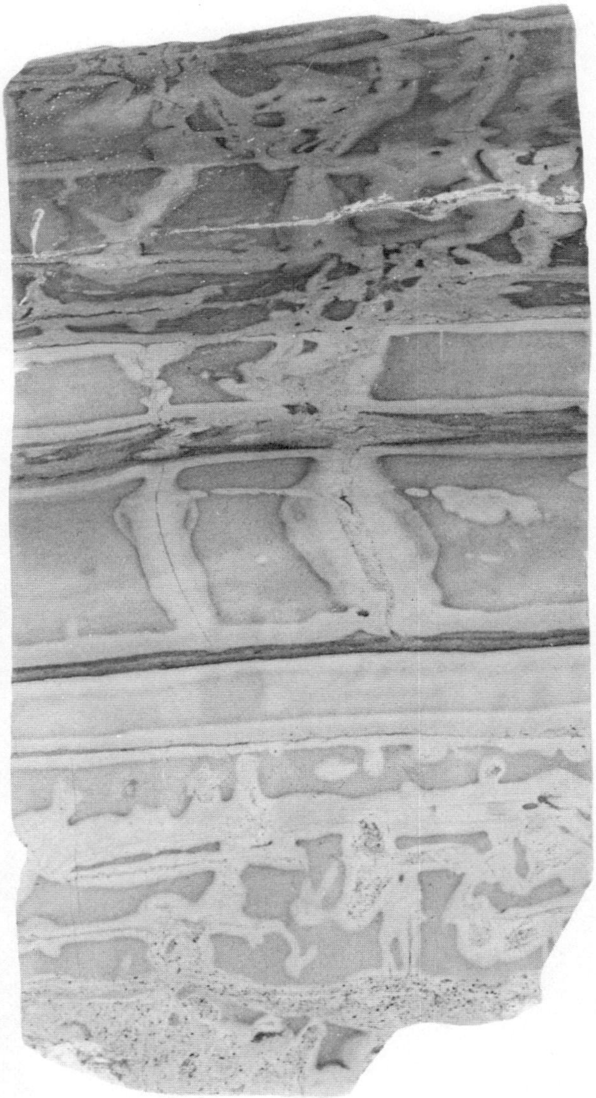

X I a.

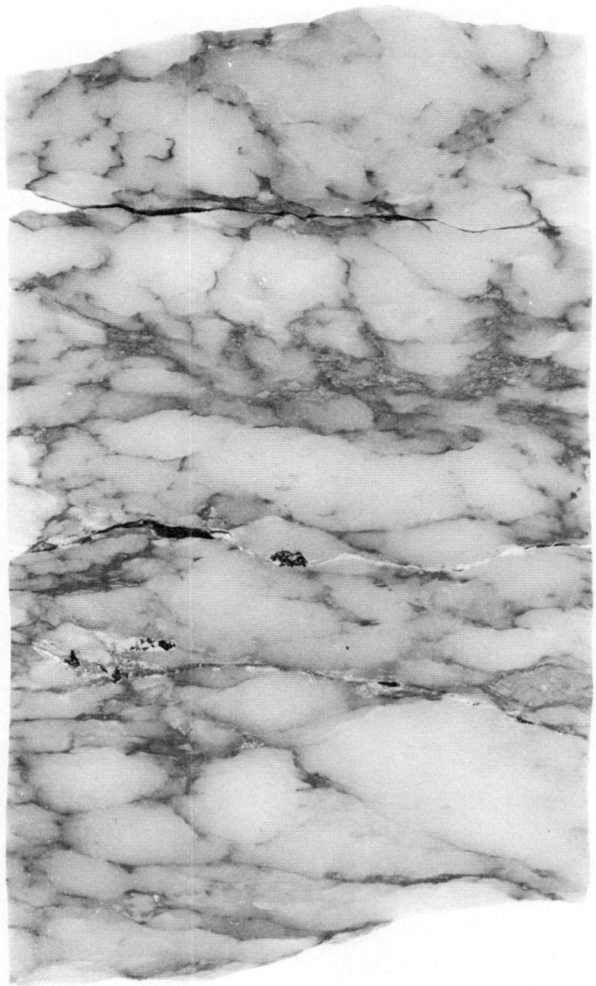

X I b.

Figure 15. a—burrowed laminated dolomitic mudstone, lower Sligo Formation, from the Magnolia No. 1 Mercer, Caldwell County, 3,994 ft. Fluid movement along permeable burrows and bedding planes caused oxidation of the marginal sediment. b—mosaic anhydrite, lower Sligo Formation, from the Magnolia No. 1 Baker, Guadalupe County, 4,418 ft.

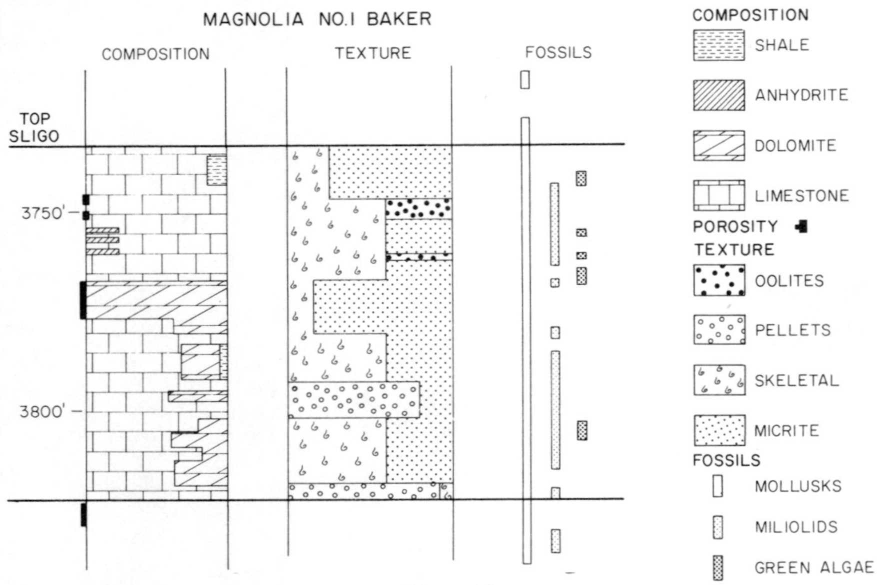

Figure 16

Figure 16. Upper Sligo facies, South Texas.

Figure 17. Burrowed mollusk-miliolid wackestone, upper Sligo Formation. a—from the Magnolia No. 1 Mercer, Caldwell County, 3,854 ft; b—from the Stanolind No. 1 Schmidt, Guadalupe County, 1,964 ft.

Figure 17

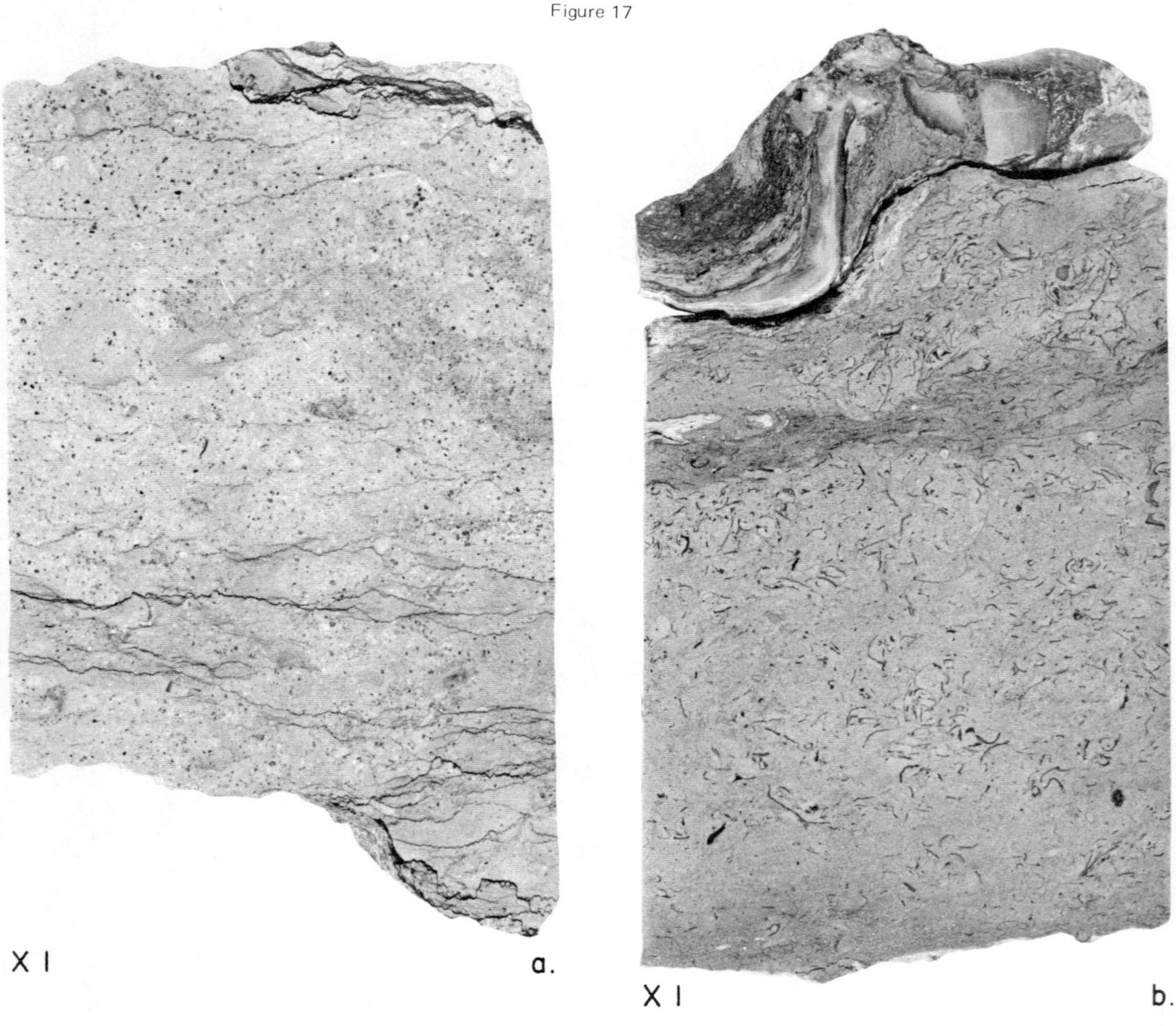

X1 a. X1 b.

Figure 18. Burrowed oolite grainstone, upper Sligo Formation. a—from the Tenneco No. 1 Roberts, Frio County, 6,937 ft; b—from the Magnolia No. 1 Baker, Guadalupe County, 3,750 ft.

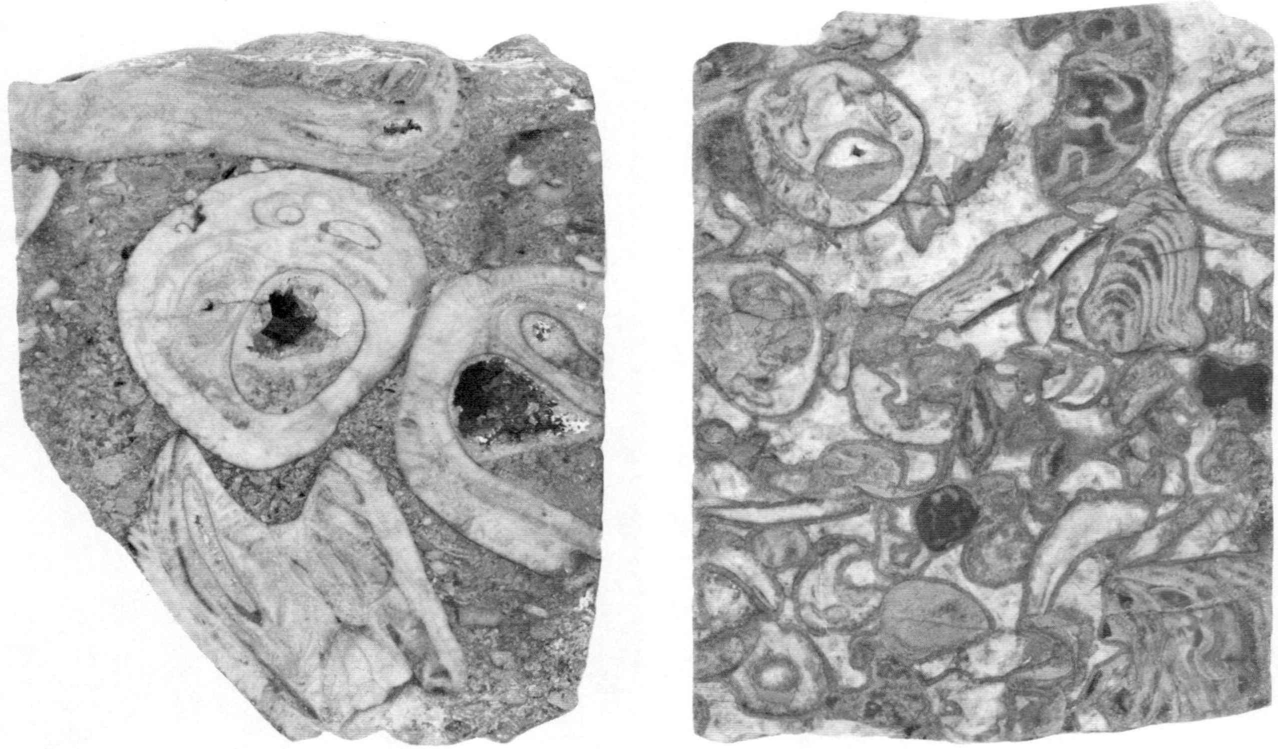

X I a. X I b.

Figure 19. Rudist grainstone, upper Sligo Formation, from the Gulf No. 1 Friedricks, Duval County. a—15,712 ft; b—15,647 ft.

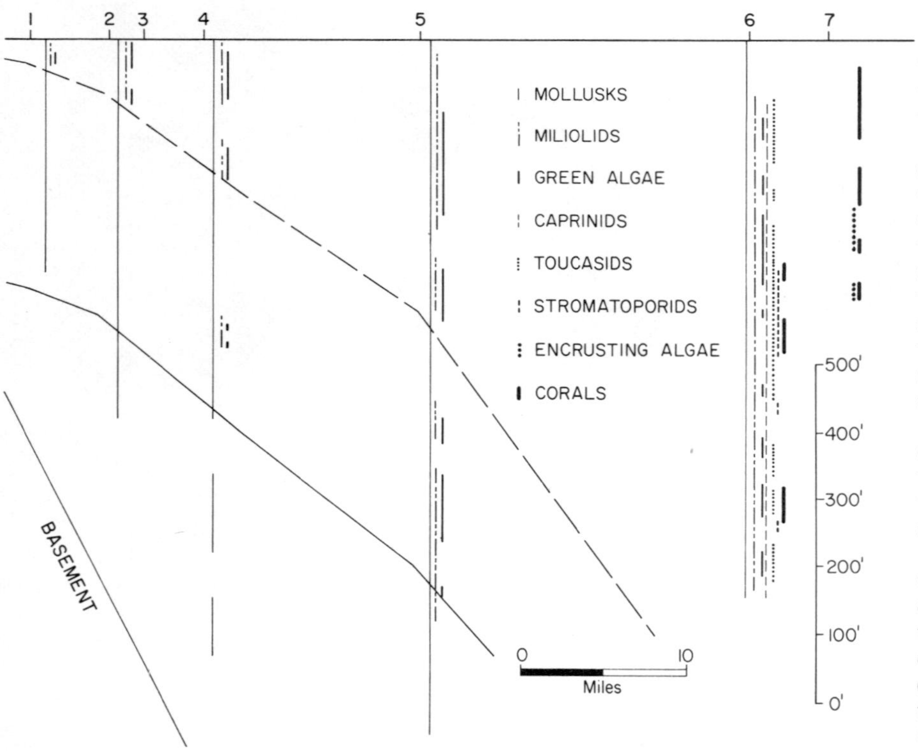

Figure 20. Fossil distribution, Sligo/Hosston, South Texas.

which prevailed during any period of time within the Sligo/Hosston depositional wedge consists of a very low-lying landmass and gently deepening shallow-water shelf. The shoreline was characterized by low energy and was extremely irregular and ill defined from the broad tidal-flat and intertidal algal-flat zones. Gypsum and anhydrite were deposited within the algal mats and carbonate and quartz-sandstone tidal flats, respectively. The presence of evaporites as well as abundant caliche, reported by Amsbury (1974), indicates a dry climate. The model described here for the Sligo/Hosston tidal-flat setting very closely resembles that of the Trucial Coast on the south side of the Persian Gulf (Butler, 1969; Kendall and Skipwith, 1969). The widespread occurrence of thin beds of oolites in the upper Sligo suggests that the oolites were transported as a series of sand waves probably oriented at an oblique angle to the shoreline, similar to those reported by Harms and others (1974) from Yucatan, Mexico. In Yucatan, oolite sand waves occur between Isla Mujeres and the Yucatan mainland as a result of steady

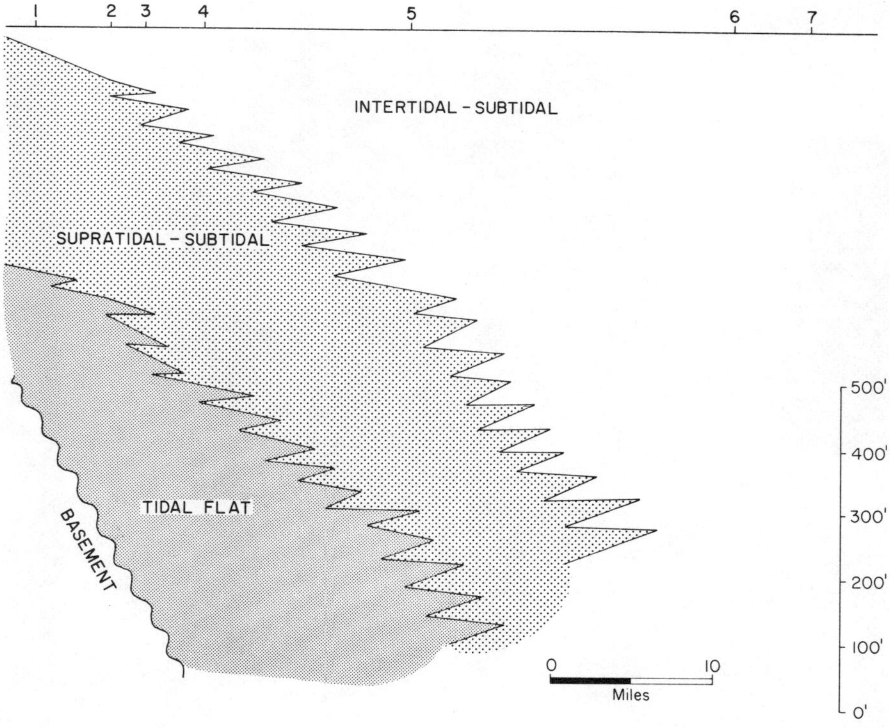

Figure 21. Depositional environments, Sligo/Hosston, South Texas.

northward-flowing currents parallel to the coast. At the Sligo shelf edge, discontinuous rudist and coral banks flourished and provided carbonate debris for the accumulation of extensive washover fans behind the banks a short distance landward.

POROSITY

Porosity (fig. 25) is best developed updip in the lower Sligo dolomite where it is mainly of the intercrystalline dolomite type. Visual estimation indicates that porosity in the dolomite ranges from 10 to 20 percent. In the upper Sligo limestones, intraparticle and vug porosity occur, particularly downdip in the shelf-edge facies. Here, however, porosity rarely reaches 5 to 10 percent.

ACKNOWLEDGMENTS

I should like to express my appreciation to D. E. Feray for first bringing to my attention the existence of large quantities of core from the Sligo and for aiding me in obtaining them. Special thanks are also due C. R. Burnett and Tenneco Oil Company for providing most of the subsurface data used in this report.

R. G. Loucks, Bureau of Economic Geology, contributed suggestions throughout the study; M. W. Longman, F. C. Lucas, and R. A. Schatzinger aided in logging the cores while employed as Research Assistants, Bureau of Economic Geology.

SELECTED REFERENCES

Amsbury, D. L., 1974, Stratigraphic petrology of Lower and Middle Trinity rocks on the San Marcos Platform, South Central Texas, *in* Aspects of Trinity Division Geology: Baton Rouge, Louisiana State University, Geoscience and Man, v. VIII, p. 1-36.

Bebout, D. G., 1976, Transgressive sandstone/carbonate sequence—The Hosston/Sligo Formations of South Texas (abstract): AAPG Bull., v. 60, p. 648.

Butler, G. P., 1969, Modern evaporite deposition and geochemistry of co-existing brines, the sabkha, Trucial Coast, Arabian Gulf: Journal of Sedimentary Petrology, v. 39, p. 70-89.

Harms, J. C., Choquette, P. W., and Brady, M. J., 1974, Carbonate sand waves, Isla Mujeres, Yucatan, *in* Field seminar on water and carbonate rocks of the Yucatan Peninsula, Mexico: New Orleans Geological Society, p. 123-147.

Kendall, C. G. St. C., and Skipwith, P. A. D'E., 1969, Holocene shallow-water carbonate and evaporite sediments of Khor al Bazam, Abu Dhabi, southwest Persian Gulf: AAPG Bull., v. 53, p. 841-869.

Loucks, R. G., 1976, Pearsall Formation, Lower Cretaceous, South Texas—depositional facies and carbonate diagenesis and their relationship to porosity: Univ. Texas, Austin, Ph. D. dissertation, 362 p.

______ 1977, Porosity development and distribution in shoal-water carbonate complexes—subsurface Pearsall Formation (Lower Cretaceous) South Texas: this volume.

Strickland, F. L., Smith, C. I., and Lozo, F. E., 1971, Stratigraphy of Lower Cretaceous Trinity deposits of Central Texas: Univ. Texas, Austin, Bureau of Economic Geology, Report of Investigations 71, 63 p.

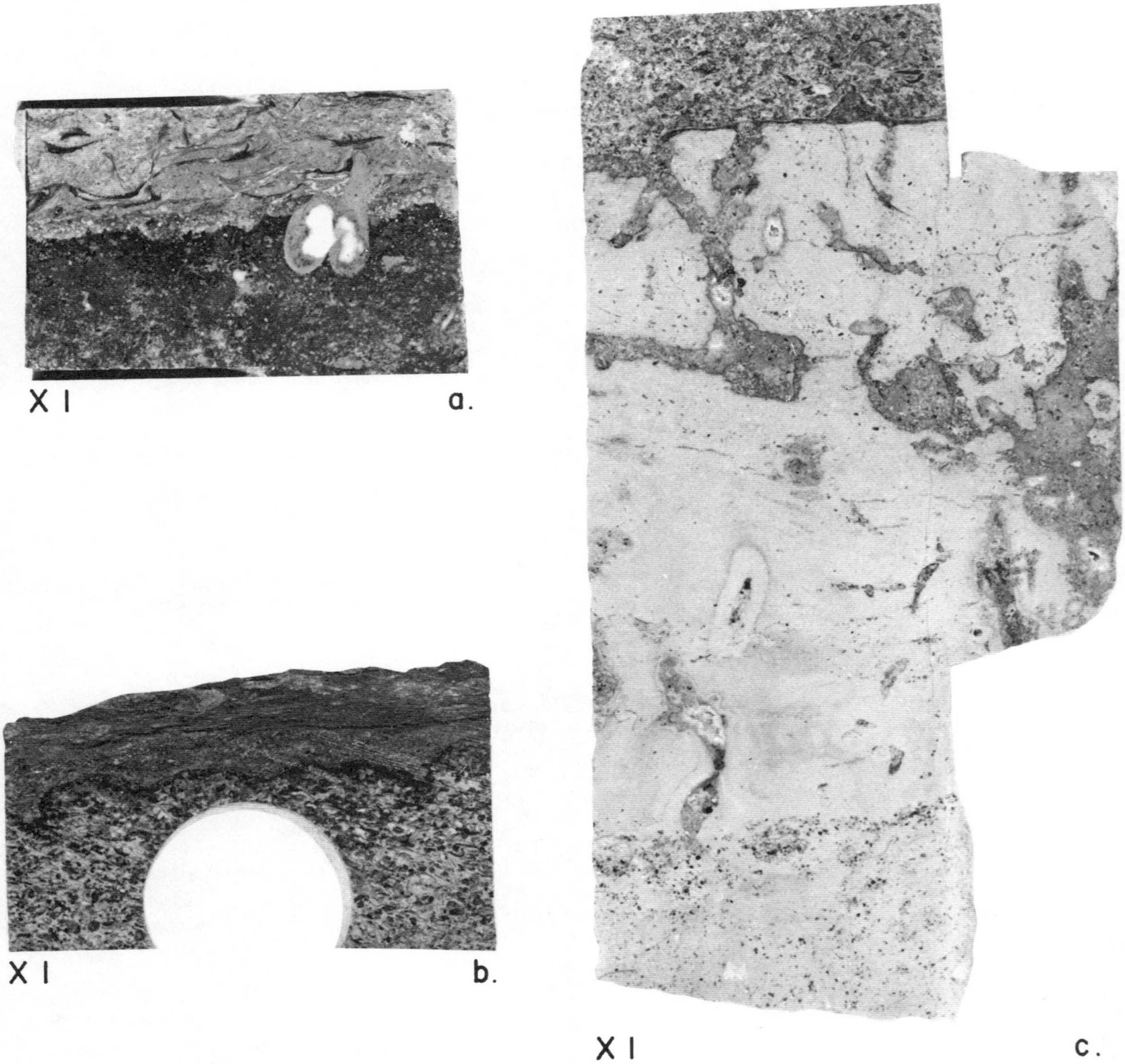

Figure 22. Hardened surfaces indicated by truncated (a, b) and bored (a, c?) surfaces from the upper Sligo Formation. a—from the Magnolia No. 1 Mercer, Caldwell County, 3,839 ft; b—from the Gulf No. 20 Dix, Guadalupe County, 4,844 ft; c—from the Tenneco No. 1 Ney, Medina County, 3,648 ft.

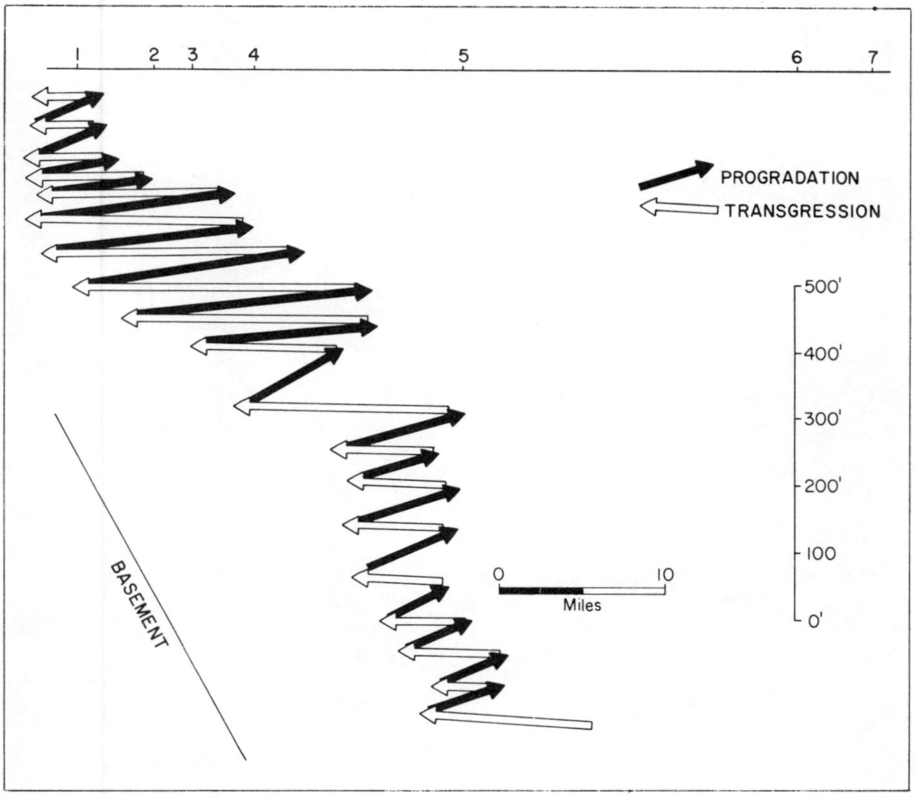

Figure 23. Diagrammatic relationship of the numerous progradational cycles to the overall transgressive character of the Sligo/Hosston depositional wedge.

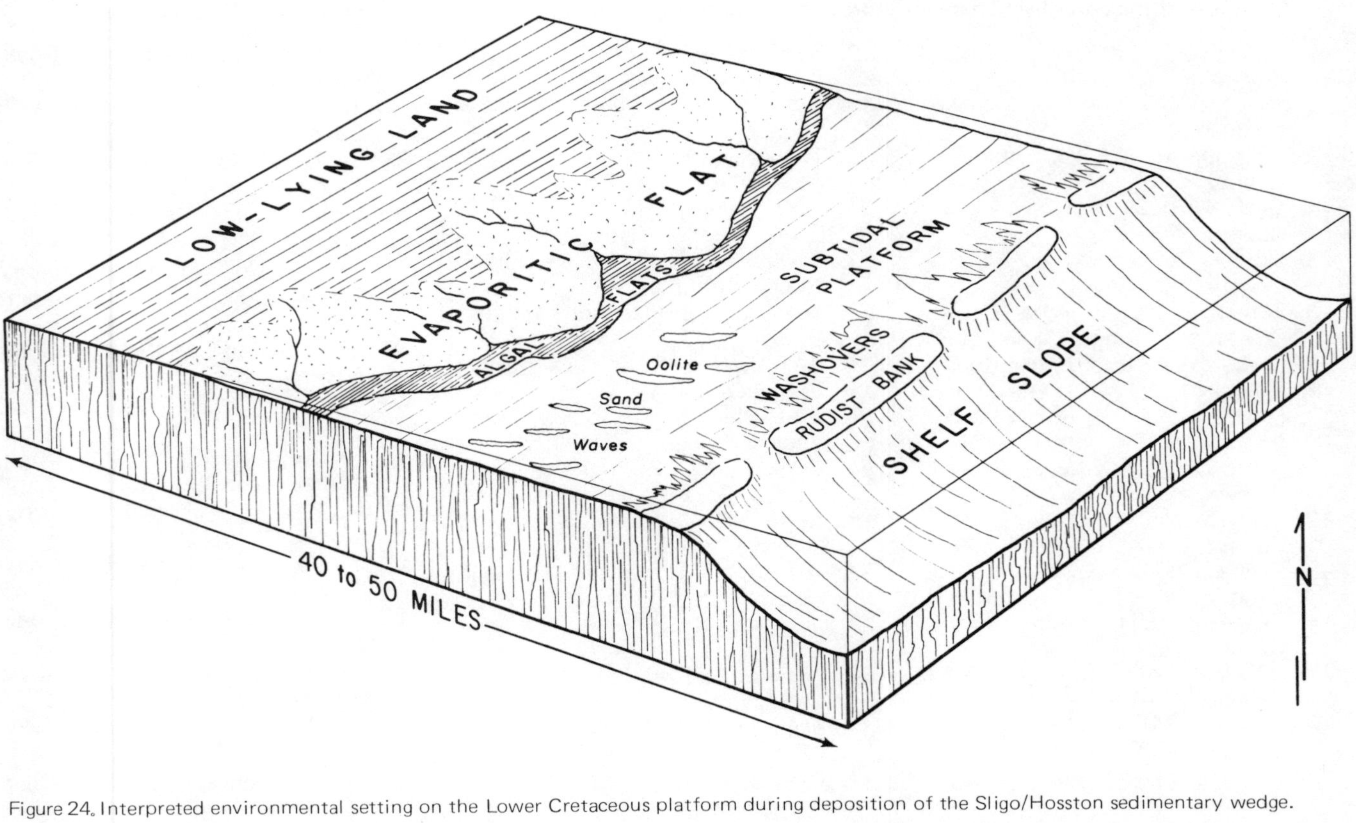

Figure 24. Interpreted environmental setting on the Lower Cretaceous platform during deposition of the Sligo/Hosston sedimentary wedge.

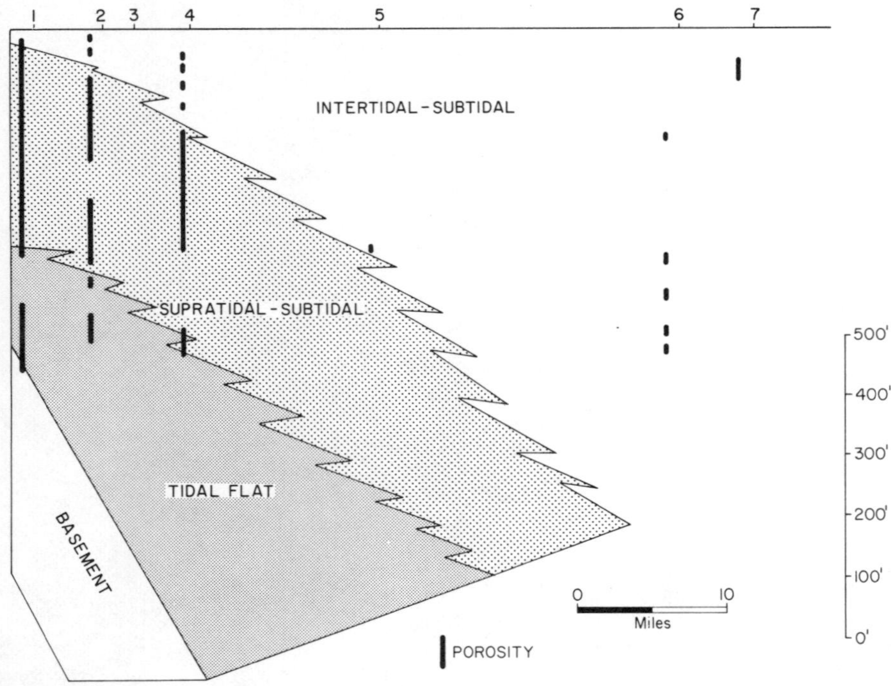

Figure 25. Porosity distribution in the Sligo/Hosston depositional wedge.

APPENDIX—WELLS WITH CORE AVAILABLE

(Most of the cores listed here are stored at the Bureau of Economic Geology, Well Sample and Core Library.)

1. Stanolind No. 1 Schmidt, Guadalupe County
2. Magnolia No. 1 Mercer, Caldwell County
3. Magnolia No. 1 Baker, Guadalupe County
4. Gulf No. 20 Dix, Guadalupe County
5. Humble No. 46 Pruitt, Atascosa County
6. Sun No. 1 Handy, Karnes County
7. Shell No. 1 Alvarado, Bee County
8. Shell No. 1 Roessler, Bee County
9. Shell No. 1 O'Neale, Bee County
10. Shell No. 1, Brown, De Witt County
11. Smith and Starr No. 1 Crowell, Caldwell County
12. Tenneco No. 1 Dickson, Caldwell County
13. Tenneco No. 1 Kauffman, Bastrop County
14. Tenneco No. 1 Sawicki, Bastrop County
15. G13 River Authority, Dam Site 7, Kendall County
16. Tenneco No. 1 Herrera, Bexar County
17. Tenneco No. 1 McKenzie, Wilson County
18. Tenneco-Pennzoil No. 1 Suggs, Atascosa County
19. Tenneco No. 1 Rogers, Atascosa County
20. Tenneco No. 1 P. R. Smith, Atascosa County
21. Tenneco-Pennzoil No. 1 J. J. Smith, Atascosa County
22. Tenneco-Pennzoil No. 1 Finch, Atascosa County
23. Tenneco No. 1 Climer, Atascosa County
24. Phillips No. 1 Washburn, McMullen County
25. Gulf No. 1 Friedricks, Duval County
26. Texaco No. 1 Watkins, Webb County
27. Hunt No. 1 Reuthinger, Webb County
28. Texaco No. 1 Canales, La Salle County
29. Tidewater No. 1 Wilson, La Salle County
30. Tenneco-Pennzoil No. 1 Edgar, Frio County
31. Tenneco No. 1 Sirianni, Frio County
32. Tenneco-Pennzoil No. 1 Wilbeck, Frio County
33. Tenneco-Pennzoil No. 1 Goad, Frio County
34. Tenneco No. 1 Machen, Frio County
35. Tenneco No. 1 Roberts, Frio County
36. Moncrief No. 1 Reiner, Frio County
37. Tenneco No. 1 Mack, Frio County
38. Tenneco No. 1 Ney, Medina County
39. Tenneco-Pennzoil No. 1 Hardie, Medina County
40. Tenneco No. 1 Wilson, Medina County
41. Tenneco-Pennzoil No. 1 Carroll, Medina County
42. Hughes and Hughes No. 1 Plachy, Medina County
43. Tenneco No. 1 Powell, Medina County
44. Moncrief No. 1 Collins, Medina County
45. Rowe No. 1 Kincaid, Zavala County
46. Tenneco No. 1 Nixon, Zavala County
47. Tenneco-Pennzoil No. 1 K, B, and M, Zavala County

POROSITY DEVELOPMENT AND DISTRIBUTION IN SHOAL-WATER CARBONATE COMPLEXES—SUBSURFACE PEARSALL FORMATION (LOWER CRETACEOUS) SOUTH TEXAS

R. G. Loucks[1]

ABSTRACT

The Lower Cretaceous Pearsall Formation in the subsurface of South Texas consists of contemporaneous carbonate and terrigenous clastic facies deposited in two major depositional systems: shoal-water carbonate complex and open, shallow-water shelf. Carbonate facies that were deposited in the high-energy environments are porous grainstones and boundstones. These are surrounded by a halo of nonporous lower energy packstones and wackestones. The open-shelf facies contains more terrigenous material than the carbonate shoal facies. Dominant facies on the shelf are oncolite packstone, terrigenous mudstone and shale, and mottled to interbedded carbonate wackestone and terrigenous mudstone.

Four stages of diagenesis are recognizable in Pearsall carbonate grainstones. Micrite envelopes, former aragonite cement, and broken grains are the dominant features in the first diagenetic stage which occurred in the marine environment. The next stage of diagenesis took place in a series of local meteoric-phreatic environments produced by partial subaerial exposure. The cements, fine-crystalline equant to bladed rim, medium-crystalline equant, and syntaxial calcite, indicate an oxidizing water chemistry varying in Mg and low in Fe. Leaching of aragonite shells created moldic porosity. Mg-calcite and aragonite grains stabilized to calcite in this stage.

Later with initial burial, medium- to coarse-crystalline equant calcite cement, low in Fe and Mg, precipitated from a regional meteoric ground-water system. Finally, at depths over 2,000 feet, quartz overgrowths, anhydrite, Fe-zoned baroque dolomite, and Fe-zoned coarse-crystalline equant calcite were precipitated. Hydrocarbons created a reducing environment. Dewatering of juxtaposed shale released Si and possibly Mg; along with stylolitization, it produced a hydraulic pump moving water through the rocks.

Approximately 95 percent of the high porosity and permeability is contained in the grainstone and boundstone facies; therefore distribution of porosity in the Pearsall Formation can be predicted from mapping depositional facies. Two major forms of porosity are primary interparticle and secondary moldic porosity.

INTRODUCTION

General

For many years the updip, shallow Cretaceous carbonates in South Texas have been explored for hydrocarbons. Most discoveries have been in the Edwards, Buda, and, in recent years, the Austin Chalk Formations. Unsuccessful targets have been the Sligo, Pearsall, and Glen Rose Formations even though they contain porous reservoirs that exhibit hydrocarbon shows.

A core study of the Pearsall Formation (fig. 1) in South Texas (fig. 2 and table 1) was initiated to aid exploration for hydrocarbons. Depositional facies and diagenetic history of the Pearsall Formation is important in understanding reservoir development and distribution.

Two major conclusions that will aid in future exploration in the Pearsall Formation are as follows:

(1) approximately 95 percent of the porosity occurs in grainstone and boundstone facies of three separate but superposed shoal-water carbonate complexes; consequently, mapping of these facies complexes outlines available reservoirs;

			NORTHEAST MEXICO	CENTRAL TEXAS	SOUTH TEXAS SUBSURFACE			EAST TEXAS SUBSURFACE		
LOWER CRETACEOUS	TRINITY DIVISION	UPPER	TAMAULIPAS FM.	GLEN ROSE FM. / HENSEL FM.	GLEN ROSE FM.		STUART CITY FM.(LOWER)	GLEN ROSE FM.		STUART CITY FM.(LOWER)
		MIDDLE	LA PEÑA FM.	HENSEL FM.	PEARSALL FORMATION	BEXAR MEMBER		PEARSALL FORMATION	BEXAR MEMBER	
				COW CREEK FM.		COW CREEK MEMBER			JAMES MEMBER	
				HAMMETT FM.		PINE ISLAND MEMBER			PINE ISLAND MEMBER	
		LOWER	CUPIDO FM.	SYCAMORE FM.	SLIGO FM. / HOSSTON FM.			SLIGO FM. / HOSSTON FM.		

Figure 1. Trinity Group Correlation chart.

[1]Bureau of Economic Geology, The University of Texas at Austin.

(2) reservoirs formed early and would have accommodated and retained any migrating petroleum if suitable traps had existed.

Paleogeographic Setting—Central and South Texas

A broad, shallow-water shelf up to 120 miles wide existed in South Texas during deposition of the Pearsall Formation. This broad shelf is divided into several paleogeographic provinces (fig. 3).

A complex of highland areas that formed the landward boundary of the shallow-water shelf included the Llano uplift, Devils River uplift, and Pandale anticline. The Pearsall Formation and several other Lower Cretaceous formations pinch out depositionally and erosionally against these source areas (fig. 4).

The San Marcos arch is a structural nose trending southeastward from the Llano uplift. The arch was rarely positive but was an area that subsided slower than the East Texas basin to the northeast and the Maverick basin to the southwest. This slower rate of subsidence resulted in deposition of a much thinner Pearsall section in the area of the arch.

Two other positive areas that produced thinning of strata are the Pearsall arch that projects into the Maverick basin from the northeast and the Tamaulipas platform in northeast Mexico, southwest of the Maverick basin.

Another geographic province that affected deposition of the Pearsall Formation was the complex composed of the Maverick basin, Atascosa trough, and Karnes trough. These negative areas were sites of thicker accumulations of sediments.

No recognizable shelf-edge barrier of either a reef or bank type in South Texas existed in the middle part of the Trinity Group. Presumably, the Pearsall shelf edge was in a position similar to the Sligo shelf edge (fig. 3). Seaward of the Pearsall shelf edge was the deep, ancestral Gulf of Mexico basin.

General Geologic History and Stratigraphy of the Trinity Group in Central and South Texas

The Trinity Group is divided into three transgressive-regressive cycles of

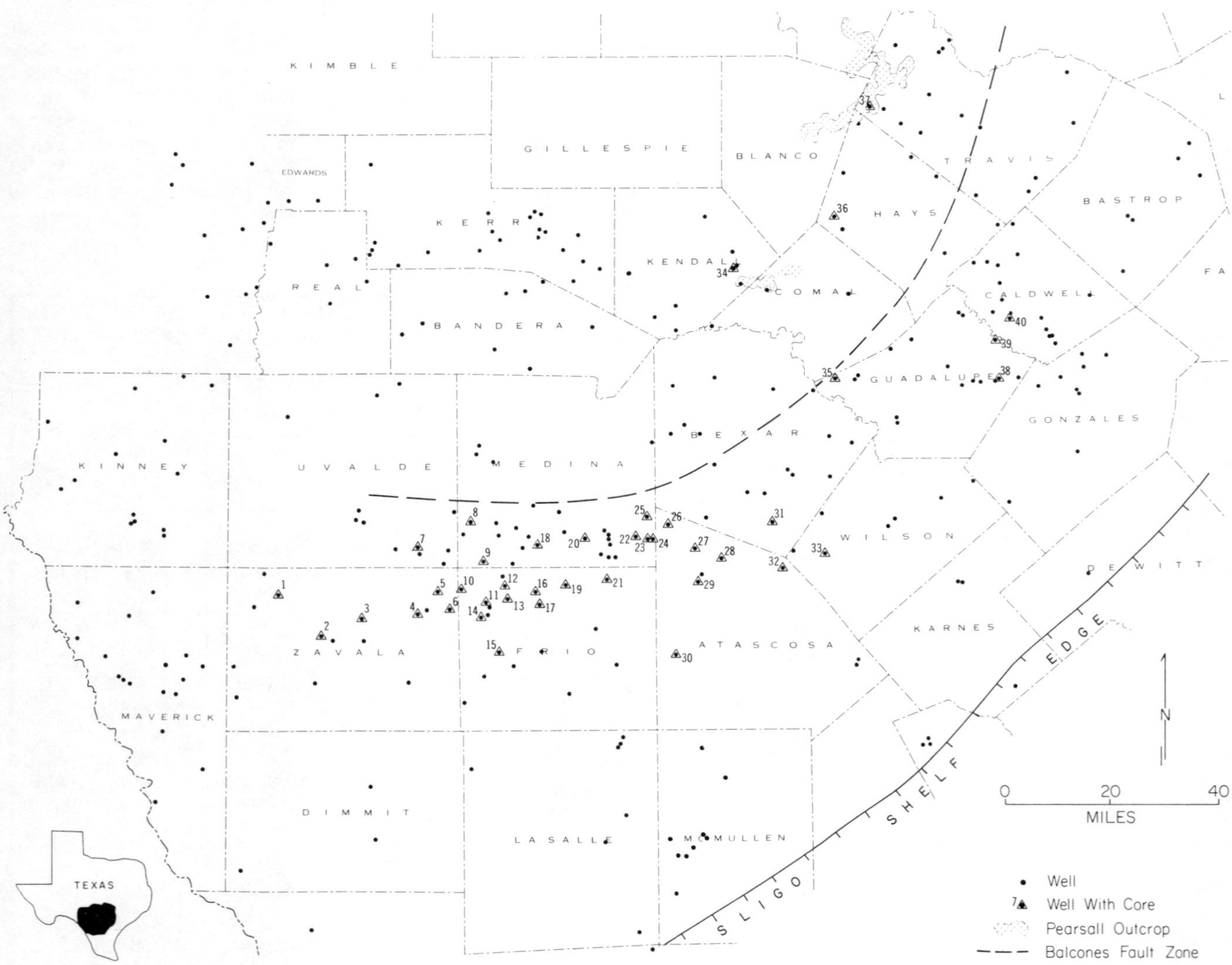

Figure 2. Well location and core control. Names of wells with core are identified on table 1.

Table 1. Names of wells with core that were used in this study.

LOCATION #	WELL NAME	COUNTY
1	Tenneco-Pennzoil Un., Inc. #1K.B.&M.	Zavala
2	Continental Oil Co. #1 Ike T. Pryor Jr. and others	Zavala
3	Tenneco #1 Kiefer	Zavala
4	Tenneco #1 Nixon	Zavala
5	Rowe #1 Kincaid	Zavala
6	Zavala Prop. #1 Murphy	Zavala
7	Tenneco-Pennzoil Un., Inc. #1 Kincaid	Uvalde
8	Tenneco #1 W.J. Ney Jr. Trustee	Medina
9	Tenneco-Pennzoil Un., Inc. #1 K.E. Hardie	Medina
10	Tenneco-Pennzoil Un., Inc. #1 Machen	Frio
11	Tenneco-Pennzoil Un., Inc. #1 Roberts	Frio
12	Tenneco-Pennzoil Un., Inc. #1 Goad	Frio
13	Tenneco-Pennzoil Un., Inc. #2 Goad	Frio
14	Tenneco #1 Mack	Frio
15	Tenneco-Pennzoil Un., Inc. #1 H.A. Holff	Frio
16	Tenneco-Pennzoil Un., Inc. #1 W.A. Wilbeck	Frio
17	Tenneco #1 Stoker	Frio
18	Tenneco #1 Roy Wilson	Medina
19	Tenneco #1 Sirianni	Frio
20	Ralph A. Johnston #1A Howard	Medina
21	Tenneco-Pennzoil Un., Inc. #1 E.E. Edgar	Frio
22	Hughes and Hughes #1 Placky	Medina
23	Tenneco #1 Powell	Medina
24	W.A. Moncriel #1 Joe E. Collins	Medina
25	Tenneco-Pennzoil Un., Inc. #1 J.W. Carroll	Medina
26	Tenneco #1 P.R. Smith	Atascosa
27	Tenneco-Pennzoil Un., Inc. #1 J.J. Smith	Atascosa
28	Tenneco #1 Rogers	Atascosa
29	Tenneco #1 Climmer	Atascosa
30	Humble #1 Pruitt	Atascosa
31	Tenneco #1 Herrera	Bexar
32	Tenneco-Pennzoil Un., Inc. #1 Suggs	Atascosa
33	Tenneco #1 McKenzie	Wilson
34	Guadalupe-Blanco River Authority Dam Site 7	Kendall
35	Stanolind #1 Schmidt	Guadalupe
36	Shell Development Co. #2 Craft Ranch	Hays
37	Shell Development Co. #1 Hamilton Pool	Travis
38	Gulf #1-Dix	Guadalupe
39	Magnolia Petroleum Co. #1 Baker	Guadalupe
40	Magnolia Petroleum Co. #1 Mercer	Caldwell

sedimentation (Lozo and Stricklin, 1956); the middle cycle is the Pearsall Formation (figs. 1,4). In the Central Texas outcrop area (fig. 2), each cycle is a terrigenous clastic-carbonate couplet with an upper erosional surface separating it from the next cycle above.

The lower Trinity transgressive-regressive cycle is divided into the Sycamore Sandstone (Hosston Formation of subsurface) and the Sligo Limestone, which in Texas occurs only in the subsurface (fig. 4). The Sycamore, Hosston, and part of the Sligo Formations are transgressive parts of the lower Trinity cycle (Bebout, 1976, 1977). On outcrop the Sycamore Sandstone lies unconformably on folded Paleozoic strata (Lozo and Stricklin, 1956), as does the updip part of the Hosston Formation; however, downdip the Hosston Formation overlies marine Jurassic carbonates (Bebout and Loucks, 1974). The Sligo Limestone, in transitional contact with the Hosston Sandstone below, is the upper transgressive and final regressive phase of the lower part of the Trinity Group (Bebout, 1976, 1977). The regressive shoreline only exposed the updip area of the Sligo Formation and created an erosional unconformity on which the Hammett Shale was deposited (fig. 4).

According to Lozo and Stricklin (1956), the middle part of the Trinity Group in outcrop consists of the Hammett Shale below (transgressive phase) and the Cow Creek Limestone above (regressive phase) and is equivalent to the subsurface Pearsall Formation that is composed of the Pine Island Shale, Cow Creek Limestone, and Bexar Shale members. The Bexar Shale member was not incorporated by Lozo and Stricklin into their transgressive-regressive couplet concept. Bexar Shale is inferred to have been eroded in the updip outcrop area (Forgotson, 1957; Tucker, 1962a, 1962b; Stricklin and others, 1971), but I propose, as Imlay (1945) did, that the subsurface Bexar Shale member is the time equivalent of part of the lower section of the Hensel Sandstone in Central Texas. Forgotson, Tucker, and Stricklin and others, who worked along the San Marcos arch and in East Texas, believe that these units are not equivalent and are separated by an unconformity. Loucks (1976) discusses this problem.

Lozo and Stricklin (1956) divided the upper part of the Trinity Group into the Hensel Sandstone (transgressive phase) and Glen Rose Limestone (regressive phase). In this paper, however, the lower part of the Hensel Sandstone is included with the middle part of the Trinity Group where the Hensel Formation is equivalent to the Bexar Shale member (fig. 4).

Even though the Trinity Group is composed of three minor transgressive-regressive cycles, it represents transgressive deposition. Each subdivision of the Trinity Group was deposited progressively farther landward onto the Llano uplift. The upper part of the Glen Rose Formation was deposited higher on the Llano uplift than any earlier Trinity deposits (fig. 4).

Along the Pearsall shelf edge, only two major transgressive-regressive cycles can be recognized—the Sligo and Stuart City cycles. The latter cycle extends into the Fredericksburg Group (Bebout and Loucks, 1974). Thinner cycles such as the Pearsall cycle can be recognized in updip areas but not downdip near the Lower Cretaceous shelf edge.

PEARSALL ENVIRONMENTS OF DEPOSITION

General

Two major depositional systems are recognized in the area of investigation: shoal-water carbonate complex and open, shallow-water shelf. Several other major depositional systems border the area, such as the basinal system to the south and fluvial-deltaic-strandplain systems to the north.

The block diagram (fig. 5), illustrating inferred distribution of depositional environments, does not

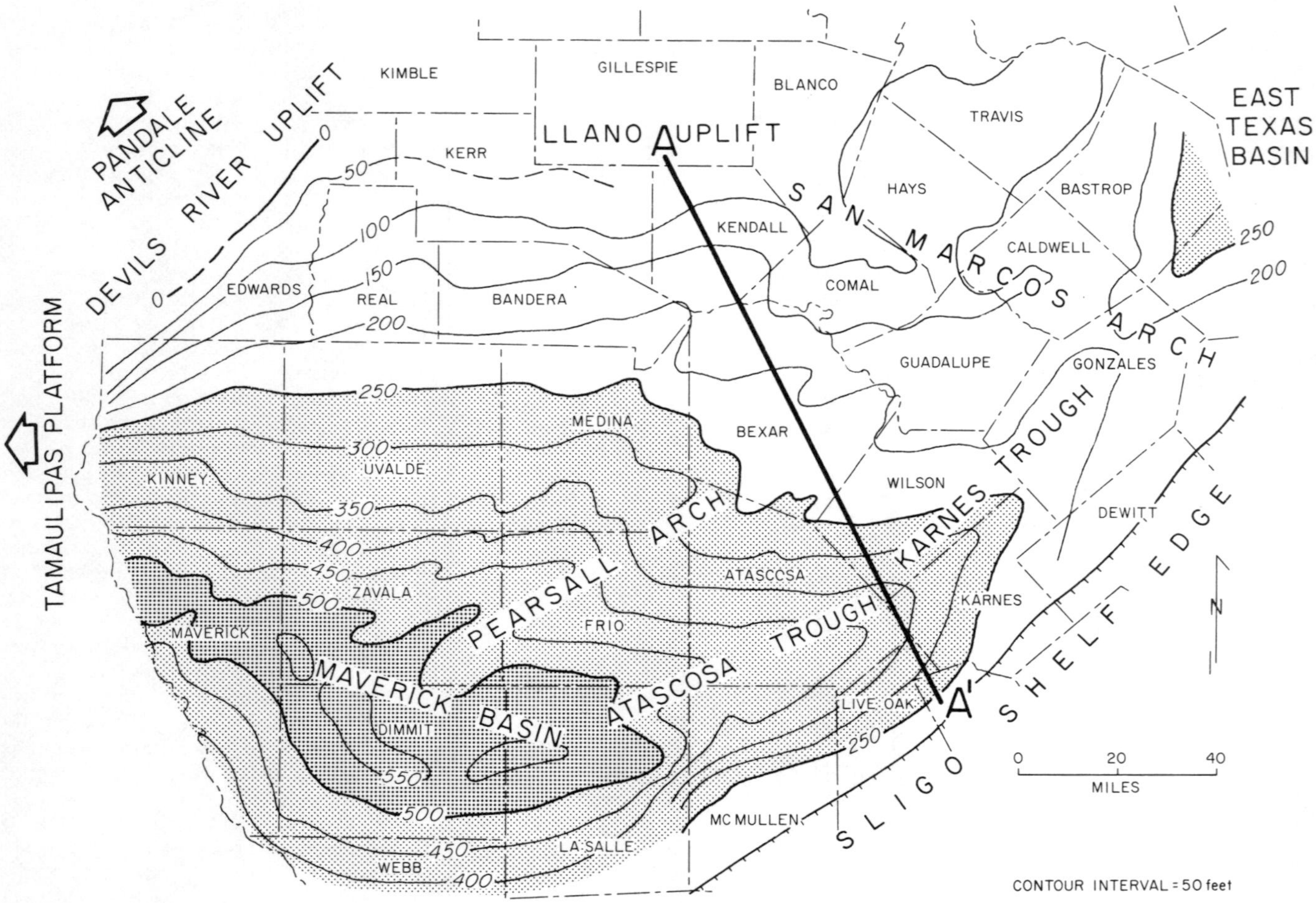

Figure 3. Isopachous map of Pearsall Formation with major paleogeographic features.

depict a specific period of time or precise geographic distribution of inferred environments. The facies, inferred depositional environments, and depth of water for the Pearsall environments are described in table 2. The facies elements of the Pearsall depositional systems are described in table 3 and examples are illustrated in figures 6, 7, 8, and 9. Facies relationships of the Pearsall Formation are well displayed by three cross sections (figs. 10, 11, 12) constructed from core and electrical-log data.

Shoal-Water Carbonate Complexes

The Pearsall Formation contains three shoal-water carbonate complexes that consist of high-energy grainstones and boundstones surrounded by a halo of lower energy packstones and wackestones. One carbonate assemblage comprises the upper part of the Cow Creek Limestone member and two assemblages occur in the Bexar Shale member (fig. 13). The oldest complex is in the Cow Creek member and has the widest areal distribution, whereas the two younger complexes were restricted west of the San Marcos arch by lateral terrigenous clastic deposition (fig. 14).

Upper Part of the Cow Creek Limestone Member

Carbonate facies of this member are the most areally extensive in the Pearsall Formation and range in thickness from a few feet to 136 feet. This member thins onto the Devils River and Llano uplifts and thins towards the Pearsall shelf edge (fig. 15). The San Marcos arch exerted little effect on sedimentation rates. The upper part of the Cow Creek Limestone member thickens in the Maverick basin, Atascosa trough, and Karnes trough and thins in the Uvalde embayment.

An isopachous map of argillaceous-poor carbonates of the Cow Creek shoal-water carbonate complex (fig. 16) was constructed from core data and the high spontaneous-potential curve on electrical logs (fig. 13). This isopachous map shows the thickness of the shoal-water carbonate association and its position along the flanks of the Uvalde embayment (fig. 16).

The upper part of the Cow Creek Limestone member exhibits the most areally extensive shoal-water carbonate complex in the Pearsall Formation (fig. 16). The complex comprises sheet deposits where it prograded in shallow water, but in deeper water it occurs as linear buildups or banks.

One of the shoal-water assemblages trends northeastward from Frio County through Travis County and is parallel to both the updip limit of the upper part of the Cow Creek Limestone member and to the Pearsall shelf edge (fig. 16). The depositional pattern was not affected by the San Marcos arch because isopachous contour lines do not bend around the arch (figs. 15, 16). Thickest section of the carbonate trend is at the southwest termination of the buildup where it directly overlies a depositional thick in the lower part of the Cow Creek Limestone member (Loucks, 1976).

Another shoal-water carbonate assemblage in northern Maverick

County and Kinney County is parallel to the Devils River uplift and the Tamaulipas platform (fig. 16). Water depth decreased in direction of these positive areas and produced high-energy environments.

The initiation of deposition of these high-energy carbonate sediments is well documented by Stricklin and Smith (1973) in the Travis-Burnet-Blanco area. They report that the initial beach sands (dashed outline in the grainstone facies in fig. 17) formed in a shoreline embayment where they were able to prograde southward protected from southwesterly flowing longshore currents. The carbonate beach-shoreface association continued to prograde seaward and also south-westward towards the deeper Maverick basin (fig. 16). It was able to extend into this basin of generally deeper water because a relic high in the lower part of the Cow Creek Limestone member produced a local shallow platform. Westward of this shallow platform in the Maverick basin, water became too deep for southwesterly progradation to continue. Deposition

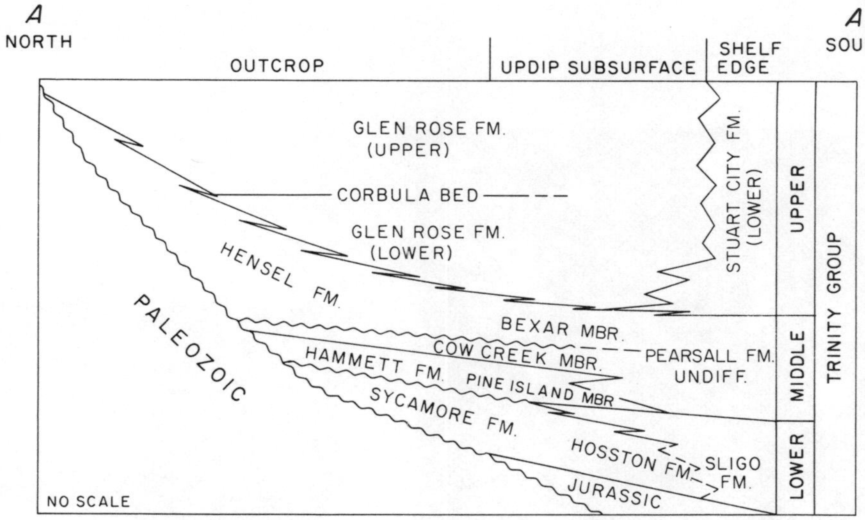

Figure 4

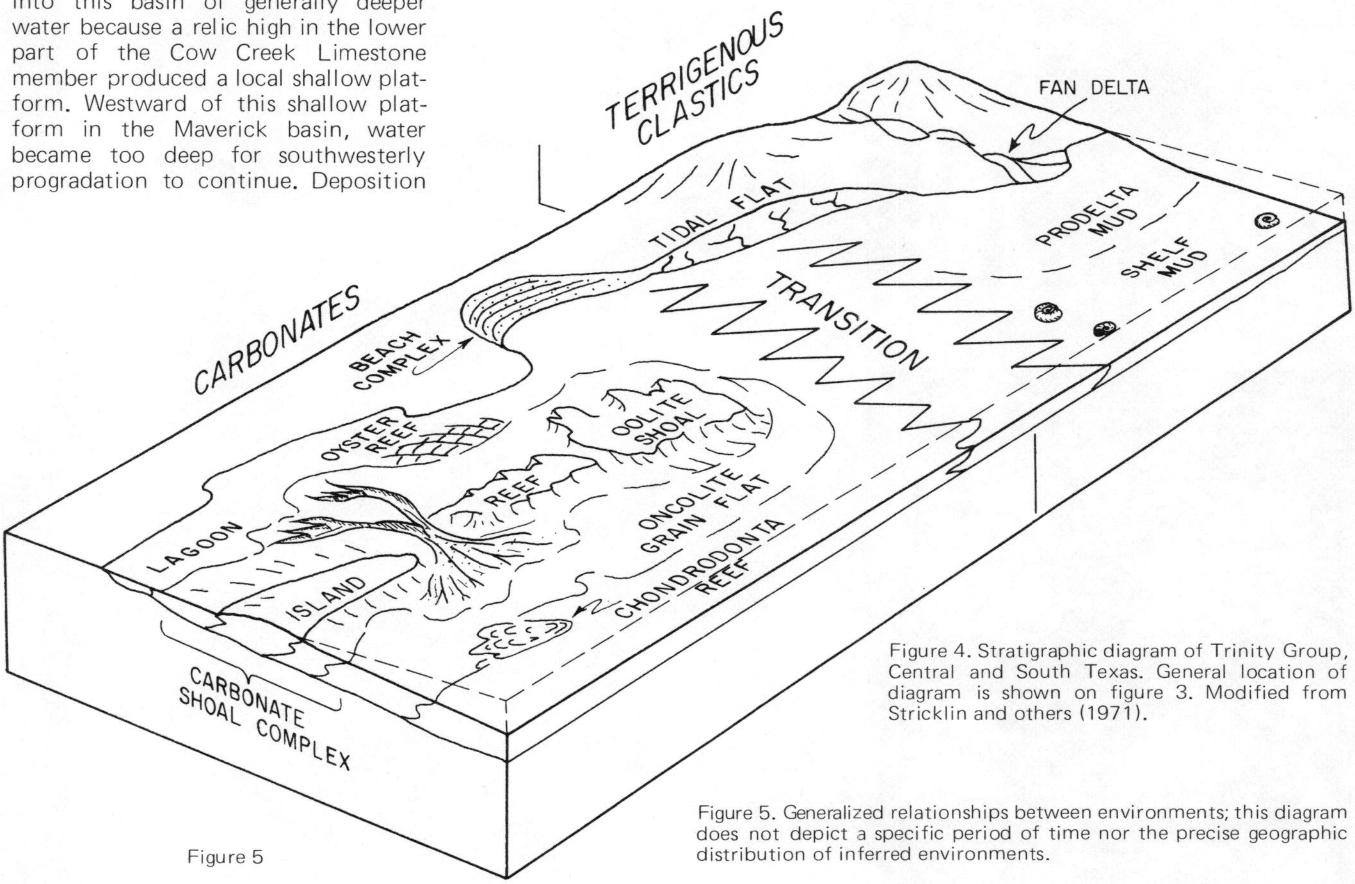

Figure 4. Stratigraphic diagram of Trinity Group, Central and South Texas. General location of diagram is shown on figure 3. Modified from Stricklin and others (1971).

Figure 5

Figure 5. Generalized relationships between environments; this diagram does not depict a specific period of time nor the precise geographic distribution of inferred environments.

(page 102)

Figure 6. (a) Well-sorted and rounded medium-grained mollusk grainstone. Well-developed micrite envelopes (1) and some grain micritization (2). "Ring" megaquartz (3) and length-slow chalcedony (4) replacing grains. Middle part of the Bexar Shale member, Tenneco #1 Ney (3,410 ft). Thin section. Crossed nicols. Bar = 0.5 mm.
(b) Oolitic coated grainstone. Grains have up to several oolitic layers. Upper part of the Cow Creek Limestone member, Tenneco #1 Stoker (7,011 ft). Thin section. Bar = 0.5 mm.
(c) Echinoid-mollusk grainstone. Coarser grainstone at top is separated from finer grainstone at bottom by a well-developed stylolitic contact. Within coarser grainstone is a sharp erosional contact that cuts both grains and submarine (beachrock?) cement. Lower part of the Bexar Shale member, Tenneco #1 Ney (3,422 ft). Core slab. Bar = 1.0 cm.
(d) Poorly sorted echinoid-mollusk packstone. Contains large fragments of oysters (1) and serpulid worm tubes (2). Lower part of the Bexar Shale member, Tenneco #1 Ney (3,491 ft). Core slab. Bar = 1.0 cm.
(e) Echinoid-mollusk wackestone. Some oyster shells are bored and several irregular clay laminae are at top of slab. A whole ***Chondrodonta*** is present at (1). Lower part of the Bexar shale member, Tenneco #1 Roberts (6,603 ft). Core slab. Bar = 1.0 cm.

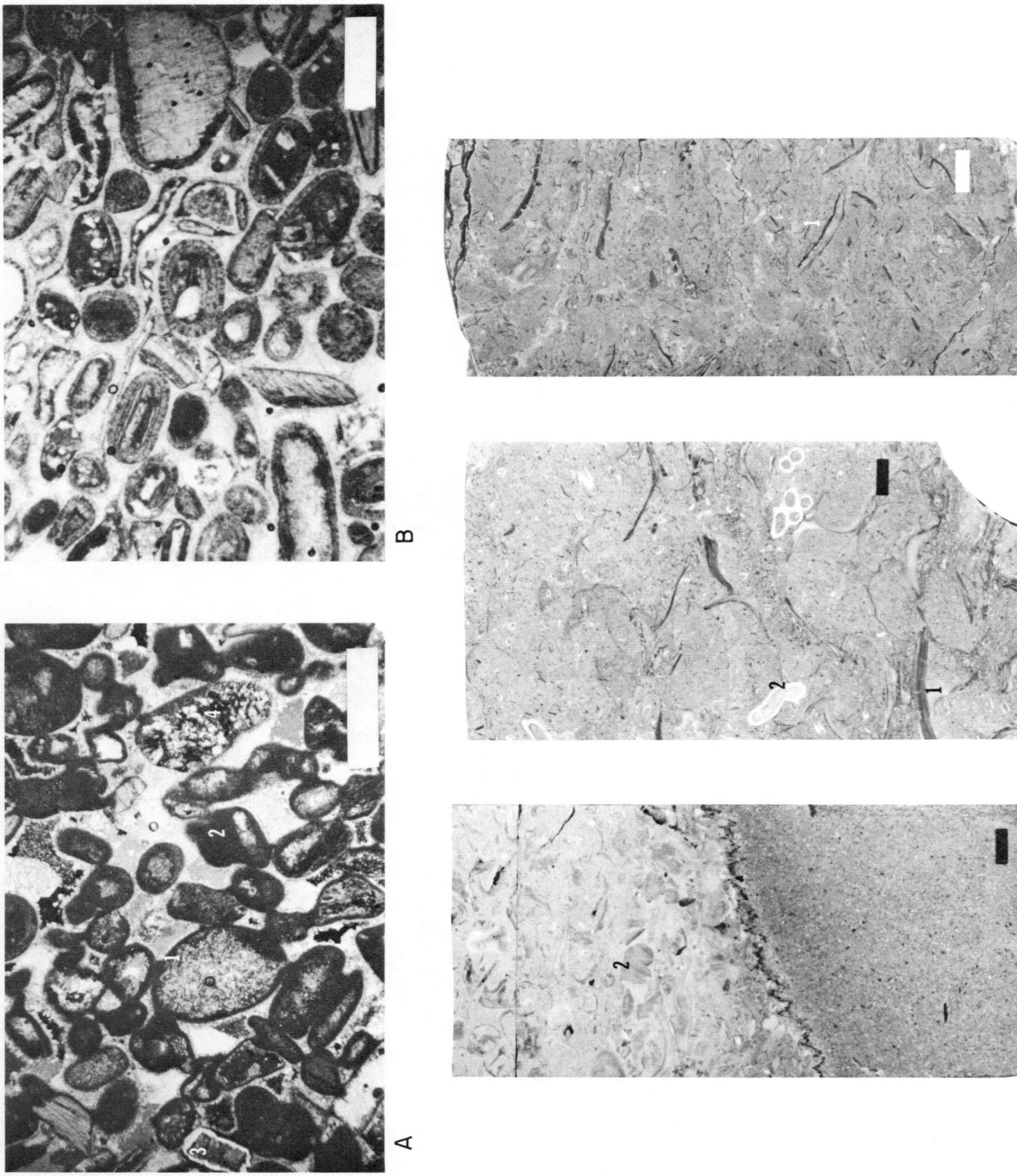

Figure 6

of the carbonate complex continued upward, and the complex formed the eastern boundary of the Uvalde embayment. The southwest strike orientation of the carbonate bank in the eastern half of the investigation area agrees with the southwesterly longshore current and wave approach direction from the southeast that is proposed by Stricklin and Smith (1973).

The carbonate environments prograded shelfward from shallower waters bordering the Devils River and the Llano uplifts. This extension of the carbonate facies complex forms the northern and western borders of the Uvalde embayment, which suggests that the embayment is not structural in origin but was created by different rates of sedimentation.

Grainstones are the most prominent facies in the carbonate bank assemblage. They were deposited in several high-energy environments discussed earlier (fig. 5, table 2). Beachrock cement and caliche within grainstone facies indicate local subaerial exposures on islands. Scattered islands were probably associated with the carbonate environments, and carbonate sand environments (grainstone facies) may have been in part connected to the mainland updip.

Coralgal-stromatoporoid-rudist boundstone facies developed on a salient on the shelf in Frio-Medina County area. The boundstone facies is approximately perpendicular to calculated wave approach from the southeast (fig. 17). Its development in this area is evidence for clear water and good circulation. Faunal and floral diversity in this facies is in marked contrast to more restricted echinoid-mollusk fauna of other facies (table 3).

An echinoid-mollusk packstone facies developed seaward of the grainstone facies, indicating a lower energy, deeper water environment. Seaward of the packstone facies in the Uvalde embayment is an echinoid-mollusk wackestone. These wackestones may have accumulated in deeper water where currents were weak and allochems were concentrated only by storms, as evidenced by graded beds and layers of concentrated shell.

Seaward of the packstone facies and east of the Uvalde embayment is a widespread oncolite facies which developed in an oncolite grainflat on an open, shallow-water shelf. The grainstones did not prograde over the oncolite facies; oncolites were not mixed with the grainstones, which indicates that this facies did not begin to develop until final stages of the deposition of this carbonate assemblage (figs. 10, 11, 12).

In the Uvalde embayment, a muddy echinoid-mollusk wackestone facies grades seaward into a mixed terrigenous mudstone - lime wackestone. Many of the carbonates contain quartz silt and sand which may indicate that a terrigenous facies occurs updip. Bushaw (1968) interprets similar facies found in East Texas to have been deposited on a continental alluvial plain.

A recent analog of the upper Cow Creek shoal-water carbonate complex is Joulters Cay in the Great Bahama Bank. In the area of the cay, reefs fringe a large oolite shoal with tidal

Table 2. Depositional environments of facies in the Pearsall Formation and their associations that define "carbonate complexes." Water depths are inferred.

FACIES	ENVIRONMENTS OF DEPOSITION AND PROBABLE WATER DEPTH
Terrigenous mudstone-siltstone	Tidal mudflat. Water depth 0 - 15 feet.
Echinoid-mollusk grainstone Oolitic coated grainstone	Beaches, tidal channels, spits, intertidal shoals, tidal bars, shallow subtidal sandflats. Water depth 0 - 30 feet.
Coralgal-stromatoporoid-rudist boundstone Coralgal-stromatoporoid-rudist packstone-grainstone	Patch reefs and associated flanking debris. Water depth 1 - 30 feet.
Echinoid-mollusk packstone Echinoid-mollusk wackestone Oolitic packstone Oyster packstone-boundstone Toucasiid wackestone	Shallow water protected lagoon and lee side of islands and intertidal shoals. Water depth less than 30 feet.
Echinoid-mollusk packstone Echinoid-mollusk wackestone Muddy echinoid-mollusk wackestone Oyster packstone-boundstone	Open, shallow-water shelf seaward to shoal-water carbonate complex. Water depth 20 - 30 feet.
Oncolite wackestone-packstone	Oncolite grainflat. Water depth 5 - 15 feet.
Terrigenous mudstone-siltstone Mixed terrigenous mudstone - lime wackestone Muddy echinoid-mollusk wackestone Terrigenous mud-shale and ammonite-bearing Oyster packestone-boundstone mud-shale	Open, shallow-water shelf and embayment. Water depth 0 - 60 feet.
Planktonic foraminifer wackestone	Open marine. Water depth greater than 60 feet.

(page 104)

Figure 7. (a) Oolitic coated grainstone with a shale layer on top, representing the close proximity of sediment types and probable mixing by storm processes. Middle part of the Bexar Shale member, Tenneco #1 Mack (7,434 ft). Bar = 1.0 cm.
(b) Calichified oolitic coated grainstone. Caliche-related features are irregular laminae (1), root burrows (2), desiccation cracks (3) (some filled with irregular laminae) and pisolitic grains. This zone of caliche represents an exposed surface at the top of the upper part of the Cow Creek Limestone member, Tenneco #1 Sirianni (6,143 ft). Core slab. Bar = 1.0 cm.
(c) Oolitic packstone at bottom grading up into a "storm" deposit consisting of oolitic sediment, mud clasts, and argillaceous material. Sequence is topped by an argillaceous mud drape. Lower part of the Bexar Shale member, Tenneco #1 Sirianni (6,134 ft). Core slab. Bar = 1.0 cm.
(d) Coralgal-stromatoporoid-rudist boundstone. The ***Radiolites*** is bored by pholad pelecypods (1). Surrounding fill is packstone. Upper part of the Cow Creek Limestone member, Tenneco #1 Siranni (6,187 ft). Core slab. Bar = 1.0 cm.
(e) Coralgal-stromatoporoid-rudist boundstone. A wackestone-packstone matrix fills in between a branching coral. Upper part of the Cow Creek Limestone member, Tenneco #1 Sirianni (6,180 ft). Core slab. Bar = 1.0 cm.

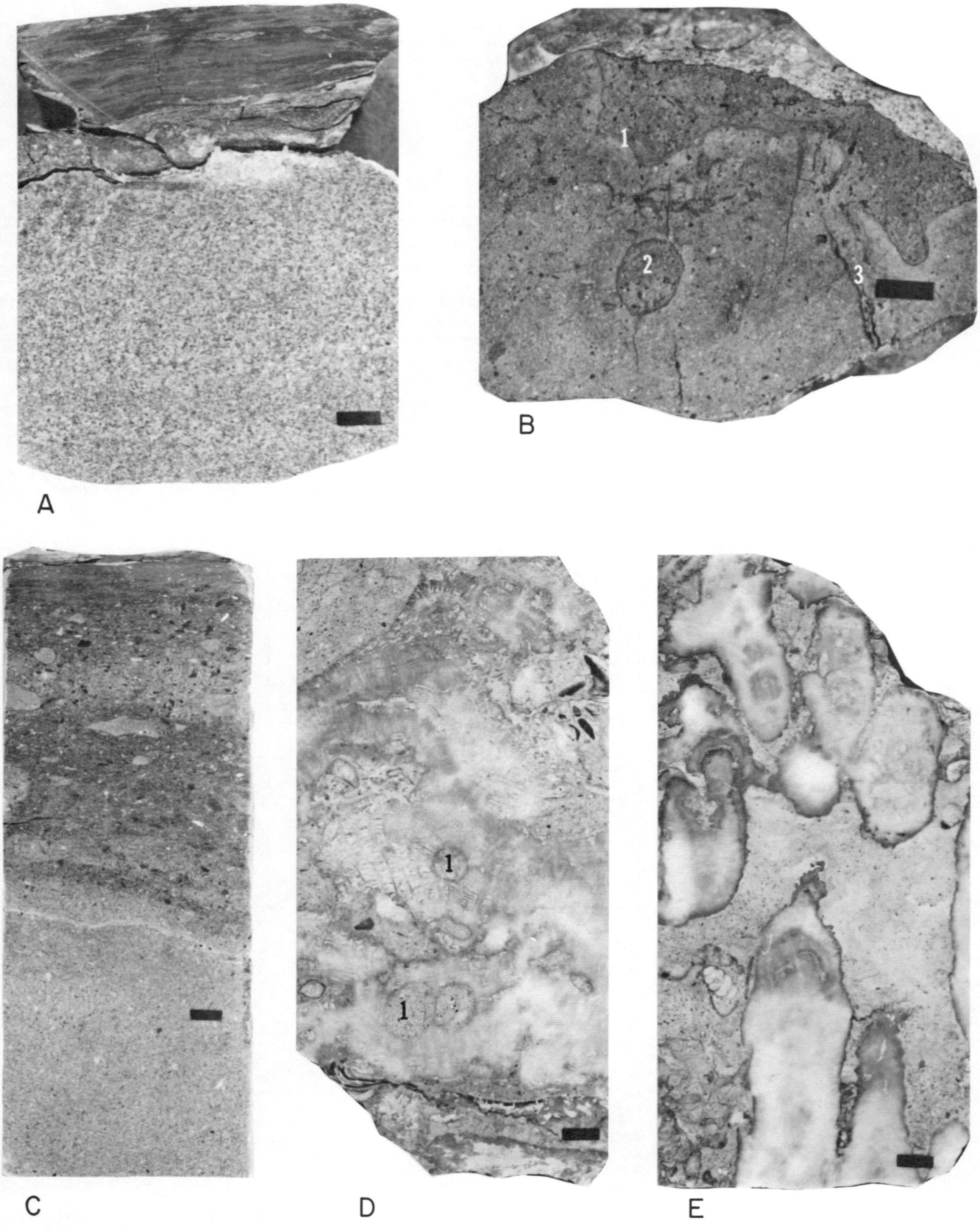

Figure 7

Table 3. Facies descriptions.

FACIES	DOMINANT ALLOCHEMS	STRUCTURE	THICKNESS RANGE (FT)	POROSITY CHOQUETTE AND PRAY, (1970)*
Echinoid-mollusk grainstone (figs. 6a, 6c)	Mollusks, echinoids, oysters, worm tubes	Horizontal laminae, cross laminae, graded beds, scoured surfaces, bored marine hardgrounds, burrows, stylolites	I to 65	0 to 30% BP & MO
Echinoid-mollusk packstone (fig. 6d)	Mollusks, echinoids, oysters, worm tubes	Distinct burrows, scoured surfaces, graded fining-upward sequences, irregular clay laminae, stylolites	1 to 37	0 to 5% minor BP
Echinoid-mollusk wackestone (fig. 6e)	Mollusks, echinoids, oysters, worm tubes	Distinct burrows, scoured surfaces, irregular clay laminae, stylolites, commonly homogenous	1 to 40	None observed
Oolitic coated grainstone (figs. 6b, 7a, 7b)	Oolitic coated grains, mollusks, echinoids, oysters	Horizontal laminae, cross laminae, graded beds, scoured surfaces, burrows, stylolites	1 to 29	0 to 25% BP & MO
Oolitic packstone (fig. 7c)	Oolitic coated grains, mollusks, echinoids, oysters	Burrows, scoured surfaces, irregular clay laminae, stylolites	1 to 8	Rare
Coralgal-stromatoporoid-rudist boundstone (figs. 7d, 7e)	Corals, stromatoporoids, ***Radiolites, Spongiomorphida, Solenopora,*** boring clams, mollusks, echinoids	Geopetal pellet fill, shelter formed voids, stylolites	3 to 32	0 to 20% MO Minor BP & WP
Coralgal-stromatoporoid-rudist packstone-grainstone	Corals, stromatoporoids, ***Radiolites, Spongiomorphida, Solenopora,*** mollusks, echinoids	Burrows, stylolites	1 to 10	5 to 10% BP & MO
Oyster packstone-boundstone (fig. 8e)	Oysters, ***Chondrodonia***, gryphaeids, echinoids, worm tubes	Irregular clay laminae, current concentrated and aligned shell laminae	1 to 20	None observed
Toucasiid wackestone (fig. 8d)	Toucasiids, miliolids, ostracods, mollusks	Irregular clay laminae	3 to 4	Rare
Oncolite wackestone-packstone (figs. 8a, 8b)	Oncolites, cerithiid gastropods, encrusting foraminifers, mollusks, echinoids, oysters, worm tubes, oolites, intraclasts	Burrows, irregular clay laminae, scoured surfaces, fining-upward sequences, stylolites	2 to 48	Rare
Muddy echinoid-mollusk wackestone (fig. 8c)	Oysters, mollusks, echinoids (some unbroken), worm tubes, pellets	Burrows, irregular clay laminae, scoured surfaces, fining-upward sequences, "horsetail stylolites" (Roehl, 1967)	1 to 57	None observed
Calcareous terrigenous mudstone (figs. 9a, 9b)	Oysters, mollusks, crustacean parts, worm tubes, echinoids (some unbroken)	Burrows, crude to well laminated ripples, shell laminae, mottled silt and mud, homogenous chert nodules (replaced anhydrite)	1 to 69	None observed
Mud-shale and ammonite-bearing mud-shale (figs. 9d, 9e)	Crustacean parts, thin-shelled mollusks (swimming type), oysters, ***Chondrodonta,*** ammonites	Fissile, rare burrows, crude laminations	1 to 75	None observed
Mixed terrigenous mudstone - lime wackestone (fig. 9c)	Oysters, mollusks, echinoids (some unbroken), worm tubes, intraclasts, pellets	Interbedded, mottled, burrowed, and swirled terrigenous mudstone and lime wackestone, also shell laminae in wackestone and soft sediment fractures	1 to 54	None observed

*BP = Interparticle porosity
MP = Moldic porosity
WP = Intraparticle porosity

channels, islands, and a shelf lagoon behind it (Malek-Aslani, 1973). The modern example contains many depositional facies similar to the upper part of the Cow Creek Limestone member.

Lower Part of the Bexar Shale Member

The lower part of the Bexar Shale member is separated from the upper part of the Cow Creek Limestone member by a muddy limestone interval that thickens to the west and south (figs. 10, 11). This unit ranges in thickness from 0 to 190 feet (fig. 18). It thins towards the Llano and Devils River uplifts to the north and northwest and thickens southwestward into the Maverick basin and Atascosa trough. The Karnes trough is not well defined by the isopachous patterns of the lower part of the Bexar Shale member. The lower part of the Bexar Shale is very thin over the carbonate buildup of the upper part of the Cow Creek Limestone member in the northern San Marcos arch area (fig. 18). The northeastward-trending isopachous thin in the San Marcos arch area coincides with the thick isopachous trend in the upper part of the Cow Creek Limestone member (figs. 15, 16).

The presence of the Uvalde embayment is indicated by thick Bexar deposits. The embayment appears to have been filled with terrigenous mudstones near the end of the late lower Bexar deposition, thus creating the thick. Thicker sections in Frio and Medina Counties comprise the lower shoal-water carbonate complex in the lower part of the Bexar Shale member similar to the carbonate complex in the underlying Cow Creek Limestone member.

The lower Bexar shoal-water carbonate buildup was restricted to the periphery of the Uvalde embayment; however, it did prograde farther into the embayment than the Cow Creek carbonate complex (fig. 14). A terrigenous system to the east prohibited carbonate bank development in that region and the terrigenous clastic environments prograded southwestward and westward from the San Marcos arch area over the carbonate buildup in the lower part of the Bexar Shale member (fig. 10). Updip, the Bexar carbonate sediments exhibit a sheet geometry, but downdip in deeper water, they aggraded to form banks (fig. 19).

The lower Bexar Shale carbonate facies are similar to the upper Cow Creek facies except that the coralgal-stromatoporoid-rudist boundstone was developed only in Medina County (fig. 20). Laterally echinoid-mollusk packstone facies surrounds echinoid-mollusk grainstone and oolitic grainstone. Between the packstone and the predominantly terrigenous facies are muddy echinoid-mollusk facies.

Terrigenous Bexar Shale facies east of the carbonate complex are the fine-grained distal portion of a terrigenous depositional system that prograded from the Llano uplift. The fluvial system for this terrigenous association is preserved in the coarse-grained Hensel Sandstone. Loucks (1976) has shown that the feeder system is one or more fan deltas similar to the Gascoyne Delta (Johnson, 1974) in western Australia. The arid climate (Inden, 1974), facies (Campbell, 1962; Inden, 1974), and development of contemporaneous carbonates and terrigenous clastics on the shelf indicate a fan-delta feeder system that was sporadically activated during deposition of the Hensel. This sporadic influx of terrigenous sediment onto a carbonate-producing shelf temporarily inhibited carbonate deposition and produced facies such as the mixed terrigenous mudstone - lime wackestone facies.

The mixed terrigenous mudstone - lime wackestone facies and terrigenous mudstone-siltstone facies in the Uvalde embayment may indicate a terrigenous sediment source from the area of the Tamaulipas platform. Again, these facies appear to be the fine-grained distal part of a deltaic system. Apparently, the sediments were reworked and transported into the embayment by storm, tidal, and longshore currents. The seaward transition from mudstone and mud-shale facies into an ammonite-bearing mud-shale facies coincided with more open and deeper water deposition.

Middle Part of the Bexar Shale Member

The middle part of the Bexar Shale member thins rapidly towards the Devils River and Llano uplifts and pinches out farther seaward on to these highlands than the lower part of the Bexar Shale member (fig. 21). It also thins onto the San Marcos arch and towards the Pearsall shelf edge. Presence of the Uvalde embayment at this time is indicated by a depositional thin on the isopachous map (fig. 21). The Maverick basin was not well developed because shoal-water carbonates prograded into and filled part of the basin (figs. 21, 22). The Atascosa trough is expressed on the isopachous map as a depositional thick; as in the lower part of the Bexar Shale member, the Karnes trough was not well developed.

The shoal-water carbonate complex of the middle part of the Bexar Shale member was restricted to the western half of the investigation area as was the carbonate system in the lower part of the Bexar Shale member (figs. 14, 22). This circumstance again indicates that, in the San Marcos arch area, terrigenous sediments were being deposited at this time. The carbonate

(page 107)

Figure 8. Oncolite grain. Algal-coated pelecypod shell (dark spots in shell are pyrite) with encrusting foraminifers (1) in algal coating. Upper part of the Cow Creek Limestone member, Tenneco #1 Halff (9,856 ft). Thin section. Bar = 0.5 mm.
(b) Oncolite packstone with poorly developed stylolites along irregular clay laminae. Upper part of the Cow Creek Limestone member, Moncrief #1 Collins (4,774 ft). Core slab. Bar = 1.0 cm.
(c) Muddy echinoid-mollusk wackestone with irregular clay laminae and large echinoid (?) burrows. White grains are echinoid plates. Lower part of the Bexar Shale member, Tenneco #1 Wilbeck (6,384 ft). Core slab. Bar = 1.0 cm.
(d) Toucasiid wackestone. Whole and broken shells in mud matrix cut by "horsetail" stylolites (1). Lower part of the Bexar Shale member, Tenneco #1 Ney (3,414 ft). Core slab. Bar = 1.0 cm.
(e) Silty dolomitic oyster packstone with gryphaeids and also containing a scoured surface (1) and current-oriented shells. Lower part of the Cow Creek Limestone member, Tenneco #1 P.R. Smith (4,337 ft). Core slab. Bar = 1.0 cm.

(page 108)

Figure 9. (a) Sequence of echinoid-mollusk wackestone with silicified evaporite nodule (1) overlain by a thin shale bed and laminated siltstone. Upper part of the Bexar Shale member, Tenneco #1 Wilson (4,205 ft). Core slab. Bar = 1.0 cm.
(b) Calcareous mudstone with whole echinoids (1). Lower part of the Cow Creek Limestone member, Humble #1 Pruitt (9,648 ft). Core slab. Bar = 1.0 cm.
(c) Mixed terrigenous mudstone - lime wackestone. Sequence of burrowed muddy echinoid-mollusk wackestone (1), fossiliferous terrigenous mudstone (2), muddy echinoid-mollusk wackestone (3), and mud-shale (4). Upper part of the Cow Creek Limestone member, Tenneco #1 Kiefer (7,727 ft). Core slab. Bar = 1.0 cm.
(d) Fissile clay-shale with thin siltstone layer. Lower part of the Bexar Shale member, Tenneco #1 Edgar (5,892 ft). Core slab. Bar = 1.0 cm.
(e) Ammonite (***Acanthoplites,*** sp.) in mud-shale. Lower part of the Bexar Shale member, Tenneco #1 Climer (6,537 ft). Core slab. Bar = 1.0 cm.

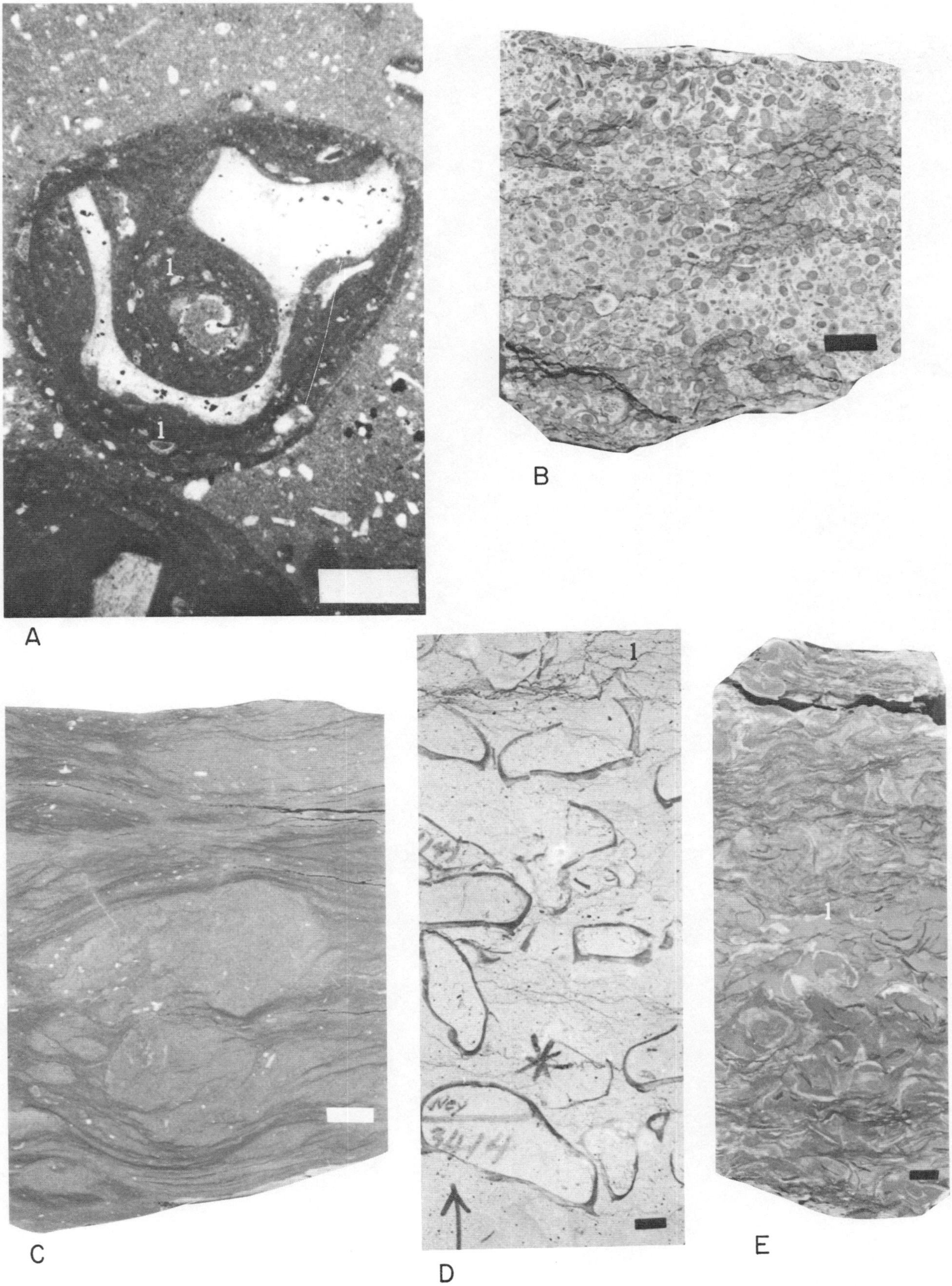

Figure 8

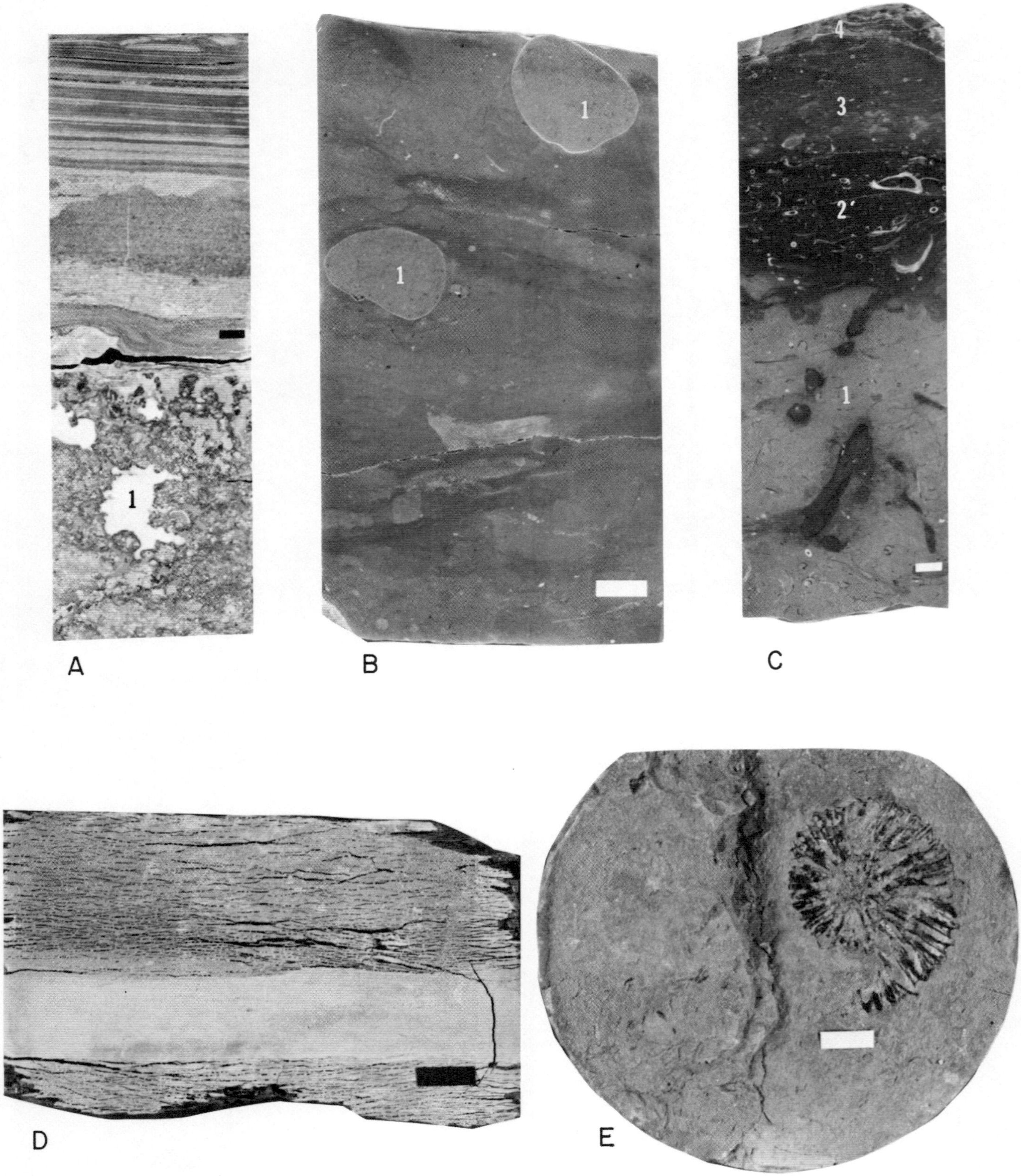

Figure 9

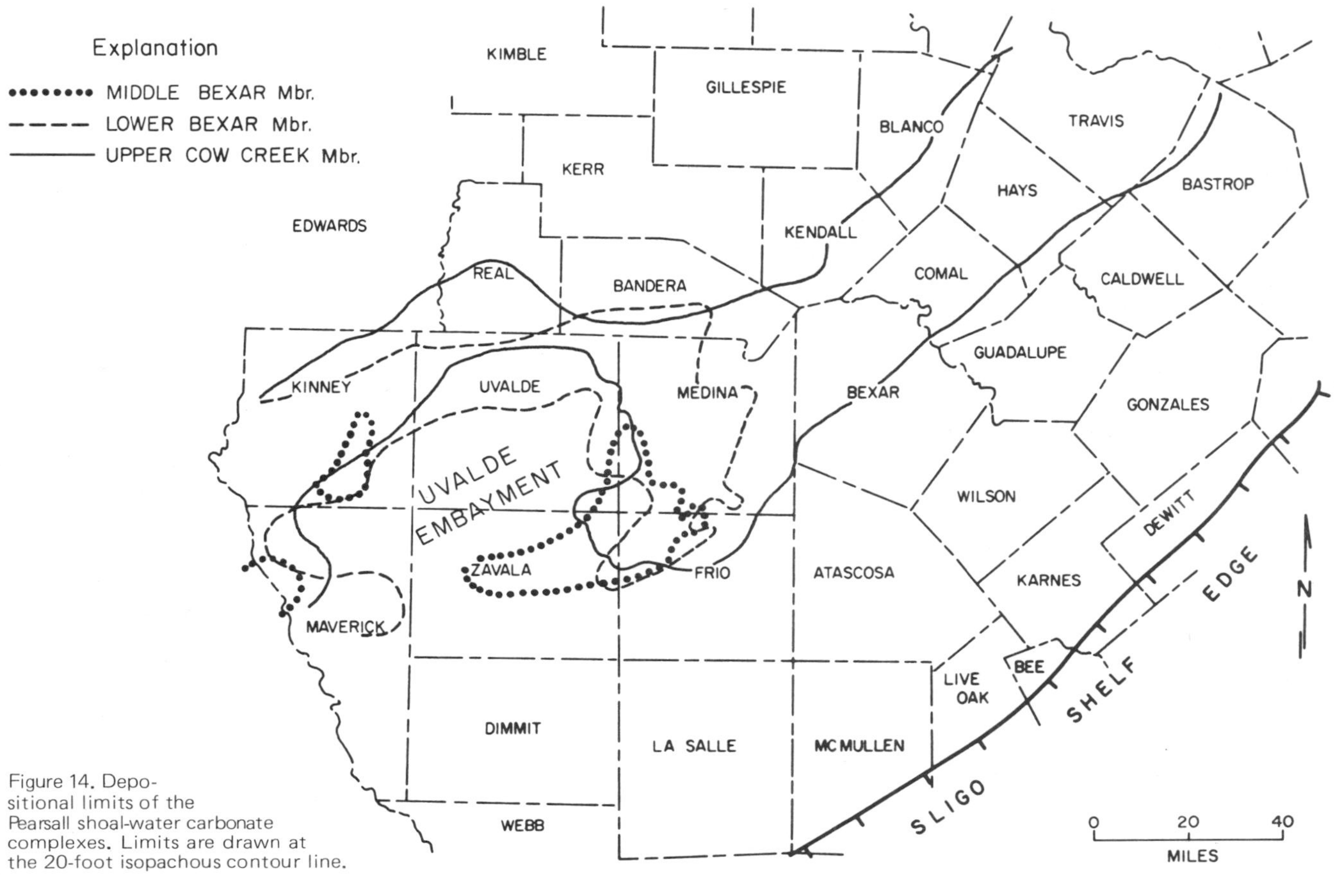

Figure 14. Depositional limits of the Pearsall shoal-water carbonate complexes. Limits are drawn at the 20-foot isopachous contour line.

development of a mixed terrigenous mudstone - lime wackestone facies indicated a waning of terrigenous sediment input from the Llano area. The mixed-sediment facies were produced by the sporadic influxes of the terrigenous sediments, probably by fan deltas, into a carbonate-producing area.

DIAGENESIS OF THE PEARSALL GRAINSTONES

General

The petrography of the Pearsall carbonate grainstones indicates four diagenetic stages (figs. 24, 25) as shown by Loucks (1976b). Diagenesis began in the marine environment where carbonate allochems were initially deposited. Later, some sediments were subaerially exposed, allowing a head of meteoric water to develop in carbonate sands. The most pronounced diagenesis occurred in the meteoric and mixing-zone stage. Following this local meteoric stage, a regional meteoric ground-water system developed that produced the third stage of diagenesis. Finally, after deeper burial, a late stage of diagenesis is taking place.

Stages of Diagenesis

Marine diagenetic stage

The original carbonate sands were composed of a mixed mineralogy of aragonite, Mg-calcite, and calcite. Each mineralogy was relatively stable in seawater. Two dominant features that formed during marine diagenesis were micrite envelopes and grain breakage, both of which had a profound effect upon porosity.

Micrite envelopes (fig. 26a-26d) were formed by boring algae and fungi (Bathurst, 1971). Davis (1970) and Perkins and Halsey (1971) observed that micrite envelopes in carbonate sediments are concentrated in the sands of the shallow, nearshore, high-energy environments where sunlight is sufficient for algal growth and little mud matrix exists to block pores. Inden (1972) noted a similar distribution of micritized and enveloped grains in outcropping Cow Creek facies. The envelopes may have been aragonite (Bathurst, 1966) or Mg-calcite (Winland, 1968).

Grainstones rarely contain whole or articulated shells and, therefore, contain rare intraparticle pore space (fig. 6a). Shells were broken by biological and mechanical processes. It was in the marine environment that most intraparticle porosity was destroyed.

Beachrock or submarine cement of the type described by Shinn (1969) and by many authors in Bricker (1971) was observed in a few intraclasts (fig. 27d) and in hardgrounds (fig. 17c). The intraclasts are all from the oolitic coated grainstone facies and are composed of grains cemented by an isopachous rim of blunt-tip crystals of what may have been fibrous aragonite cement followed by a micrite cement (fig. 27d). The clasts are interpreted to have formed as beachrock and were later eroded and deposited with the oolite sands. Many examples of beachrock cements have been described in Holocene and Pleistocene sediments (Bricker, 1971). An example of in situ beachrock was observed associated with an erosional surface in an echinoid-mollusk grainstone which is interpreted to have been deposited in a beach environment (figs. 6c, 27d).

Meteoric and mixing-zone stage

Second and most pronounced stage of diagenesis began when some

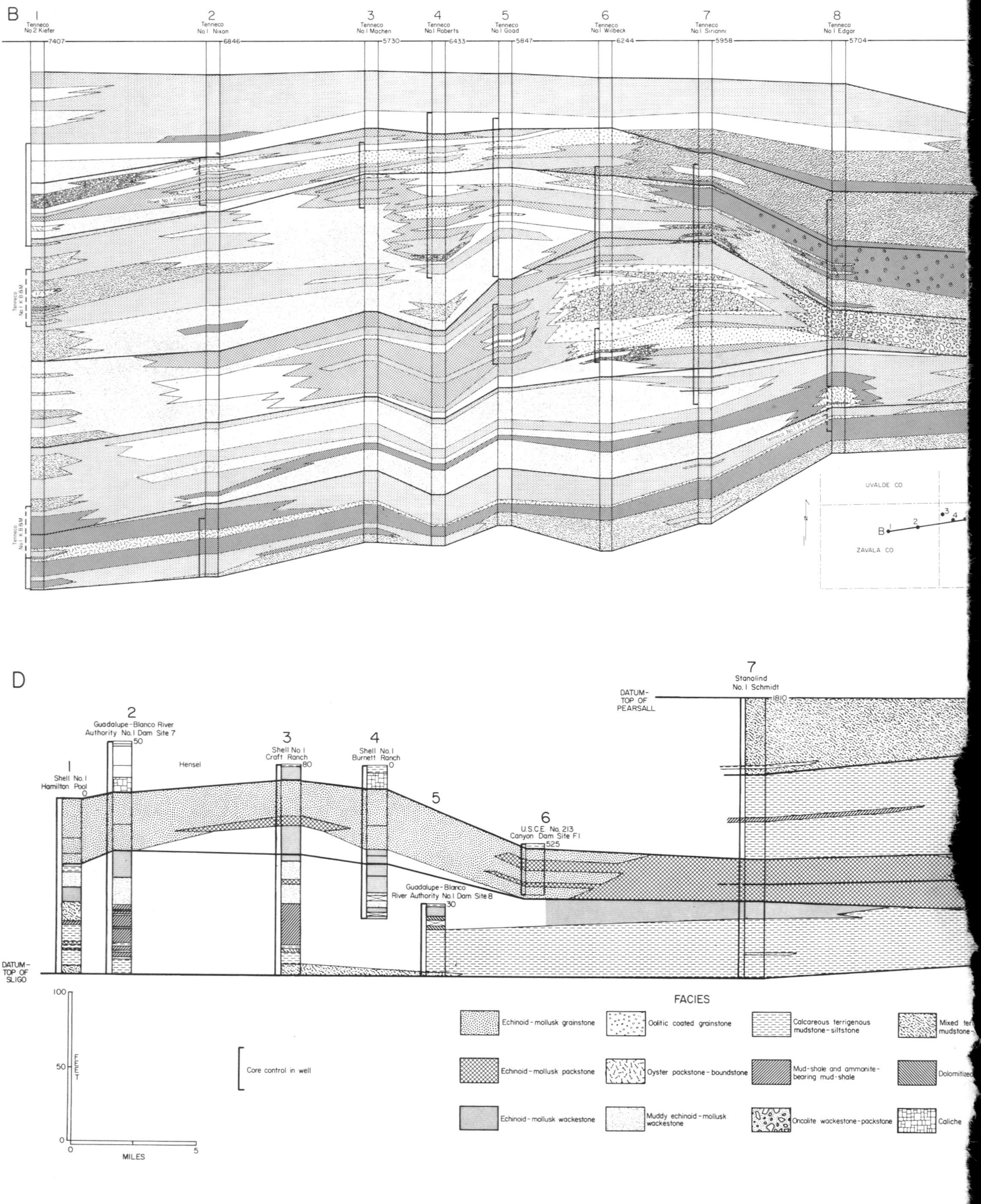

B
1 Tenneco No.2 Kiefer
2 Tenneco No.1 Nixon
3 Tenneco No.1 Machen
4 Tenneco No.1 Roberts
5 Tenneco No.1 Goad
6 Tenneco No.1 Wilbeck
7 Tenneco No.1 Sirianni
8 Tenneco No.1 Edgar
7407
6846
5730
6433
5847
6244
5958
5704
Tenneco No.1 K.B.&M.
Tenneco No.1 K.B.&M.
UVALDE CO
ZAVALA CO
D
1 Shell No.1 Hamilton Pool
2 Guadalupe-Blanco River Authority No.1 Dam Site 7
3 Shell No.1 Craft Ranch
4 Shell No.1 Burnett Ranch
5
6 U.S.C.E. No. 213 Canyon Dam Site F1
7 Stanolind No.1 Schmidt
Guadalupe-Blanco River Authority No.1 Dam Site 8
Hensel
DATUM-TOP OF PEARSALL
DATUM-TOP OF SLIGO
1810
50
80
0
0
525
30
100
50
0
FEET
0
5
MILES
Core control in well
FACIES
Echinoid-mollusk grainstone
Oolitic coated grainstone
Calcareous terrigenous mudstone-siltstone
Mixed ter mudstone-
Echinoid-mollusk packstone
Oyster packstone-boundstone
Mud-shale and ammonite-bearing mud-shale
Dolomitized
Echinoid-mollusk wackestone
Muddy echinoid-mollusk wackestone
Oncolite wackestone-packstone
Caliche

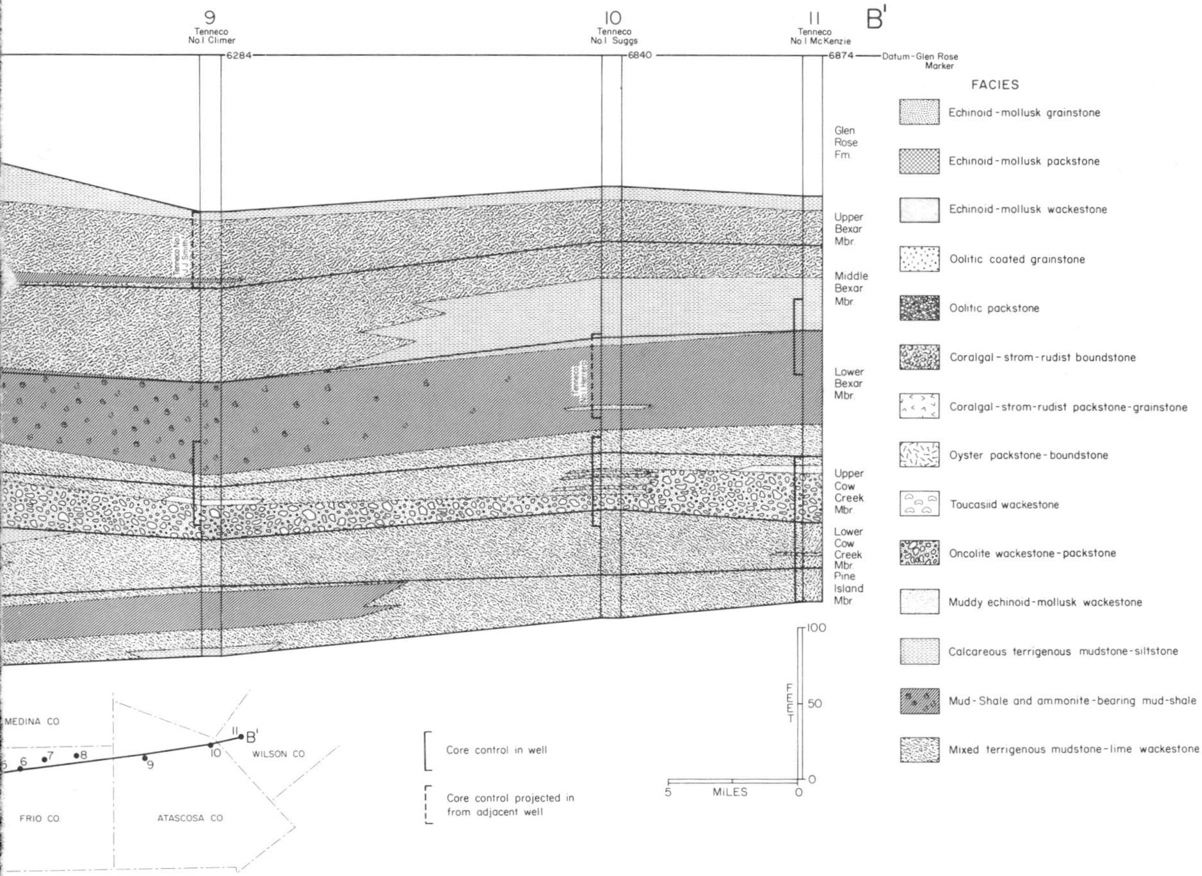

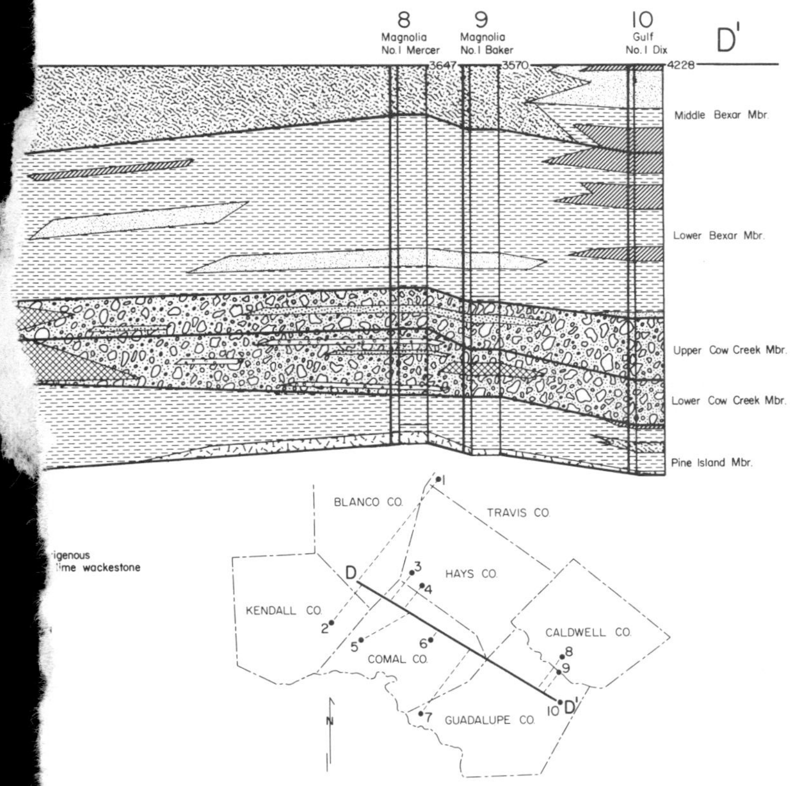

Figure 10. Pearsall Formation stratigraphic facies cross-section B-B'.

Figure 11 *(page 110a)*

Figure 12. Pearsall Formation stratigraphic facies cross-section D-D'.

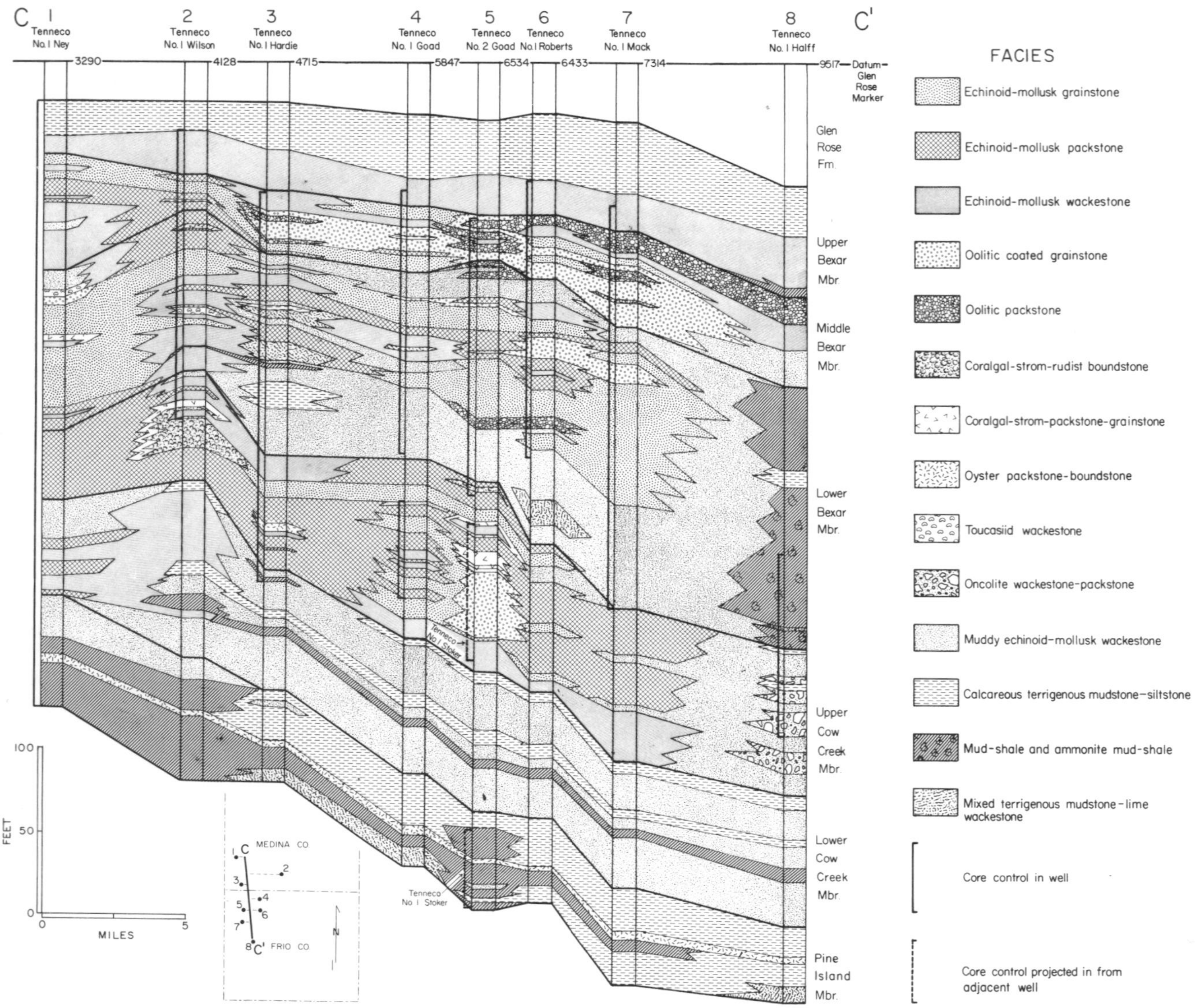

Figure 11. Pearsall Formation stratigraphic facies cross-section C-C'.

bank in this member did not completely surround the Uvalde embayment. In Zavala County, the bank prograded farther into the Uvalde embayment and Maverick basin than the two earlier carbonate systems (fig. 14).

The volume of terrigenous clastic sediments entering the area decreased during the middle part of the Bexar deposition, but the shoal-water carbonate assemblage was still restricted to the west and east flanks of the Uvalde embayment where facies define a sheet deposit (fig. 22). A mixed terrigenous clastic-carbonate facies association was deposited over a widespread area east of the embayment (fig. 23).

The progradation of the shoal-water carbonate environments into the Uvalde embayment, an area of previously deeper water, may indicate that a decrease in water depth occurred on the shelf at this time. The central area of the carbonate complex is composed of a crossbedded oolitic grainstone facies and is laterally surrounded by echinoid-mollusk packstone and wackestone facies. The abundance of oolites indicates that this area was dominated by shoaling water.

In the Uvalde embayment and seaward of the middle Bexar carbonate buildup, a muddy echinoid-mollusk wackestone was deposited. The

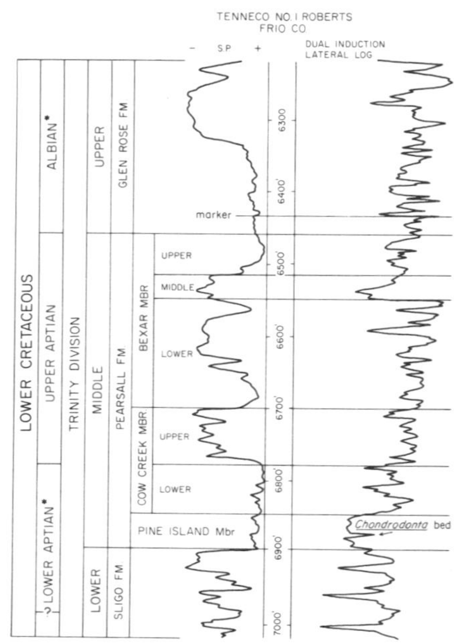

Figure 13. *(right)* Electrical log through Pearsall Formation in South Texas showing three shoal-water carbonate complexes. Marker bed at 6,400 ft is used as the datum for cross-sections B-B', C-C', and D-D' (figs. 10, 11, 12). (*) Ages are from Young (1974).

of carbonate sediments were exposed above sea level. A head of meteoric water developed within carbonate sands exposed on islands and/or prograding shorelines. Tolman (1937) explains that a lens of fresh ground water can develop over a body of saline ground water. If the fresh-water table is 1 foot above sea level, then the fresh-water lens can project 40 feet into the salt-water body. Periodic development of fresh-water lenses of that magnitude could have easily affected all grainstones. An analogous modern fresh-water lens in a salt-water body is shown by Steinen and Matthews (1973).

Evidence for several subaerial exposures has been observed in the Pearsall Formation. Exposures are indicated by an erosional beachrock surface (fig. 6c) and caliche (fig. 7b). Subaerial surfaces indicate that the underlying carbonate sediments were periodically exposed to meteoric waters. All sediments were not necessarily exposed at one time. As an island developed, sediments in that area underwent meteoric diagenesis while equivalent sediments outside the influence of the meteoric lens remained stable. At the top of the Cow Creek Limestone section, however, a prominent erosional surface may have developed over much of the area in Frio and Medina Counties, allowing extensive meteoric diagenesis during the period of exposure. At the outcrop, Stricklin and others (1971) also found an erosional surface at the top of the Cow Creek Limestone member.

Three main diagenetic environments developed in association with subaerial exposure: meteoric vadose zone, meteoric phreatic zone, and mixing zone between the meteoric phreatic and marine phreatic (Folk, 1973a; fig. 24). The meteoric vadose zone formed above the water table and was in contact with the atmosphere. Water percolates through it to the phreatic zone, and the only water retained in the vadose zone was held by surface tension and capillary forces at grain contacts. Characteristic features of vadose diagenesis are pendulus and meniscus cements (Dunham, 1969; Land, 1970). These cements were not observed in this investigation although Inden (1972) identified a few examples from Cow Creek outcrops.

Most meteoric diagenesis occurred in the meteoric phreatic zone (Land, 1970; Steinen and Matthews, 1973; Steinen, 1974) which was located below the water table where pore spaces were filled with water. Water in

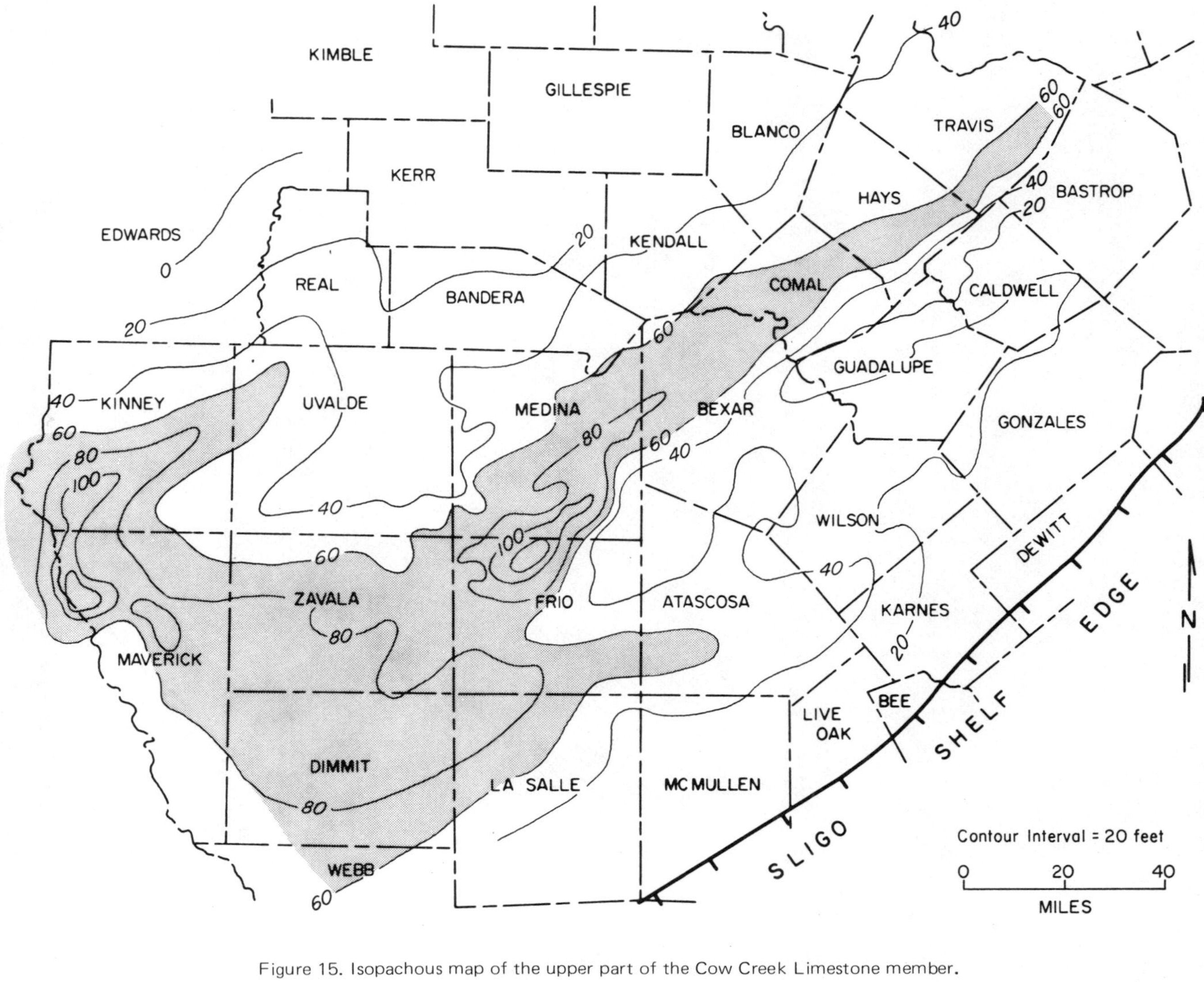

Figure 15. Isopachous map of the upper part of the Cow Creek Limestone member.

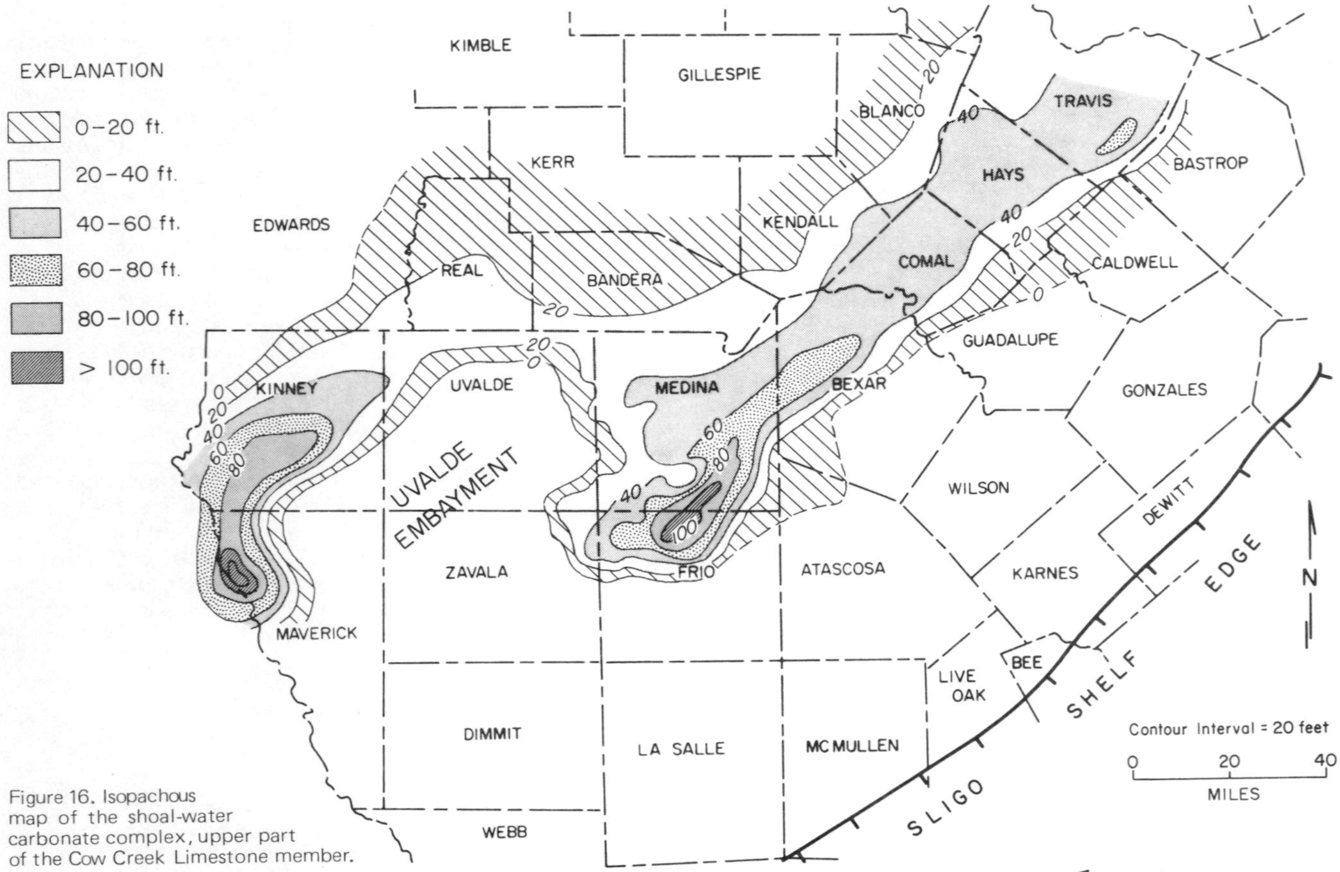

Figure 16. Isopachous map of the shoal-water carbonate complex, upper part of the Cow Creek Limestone member.

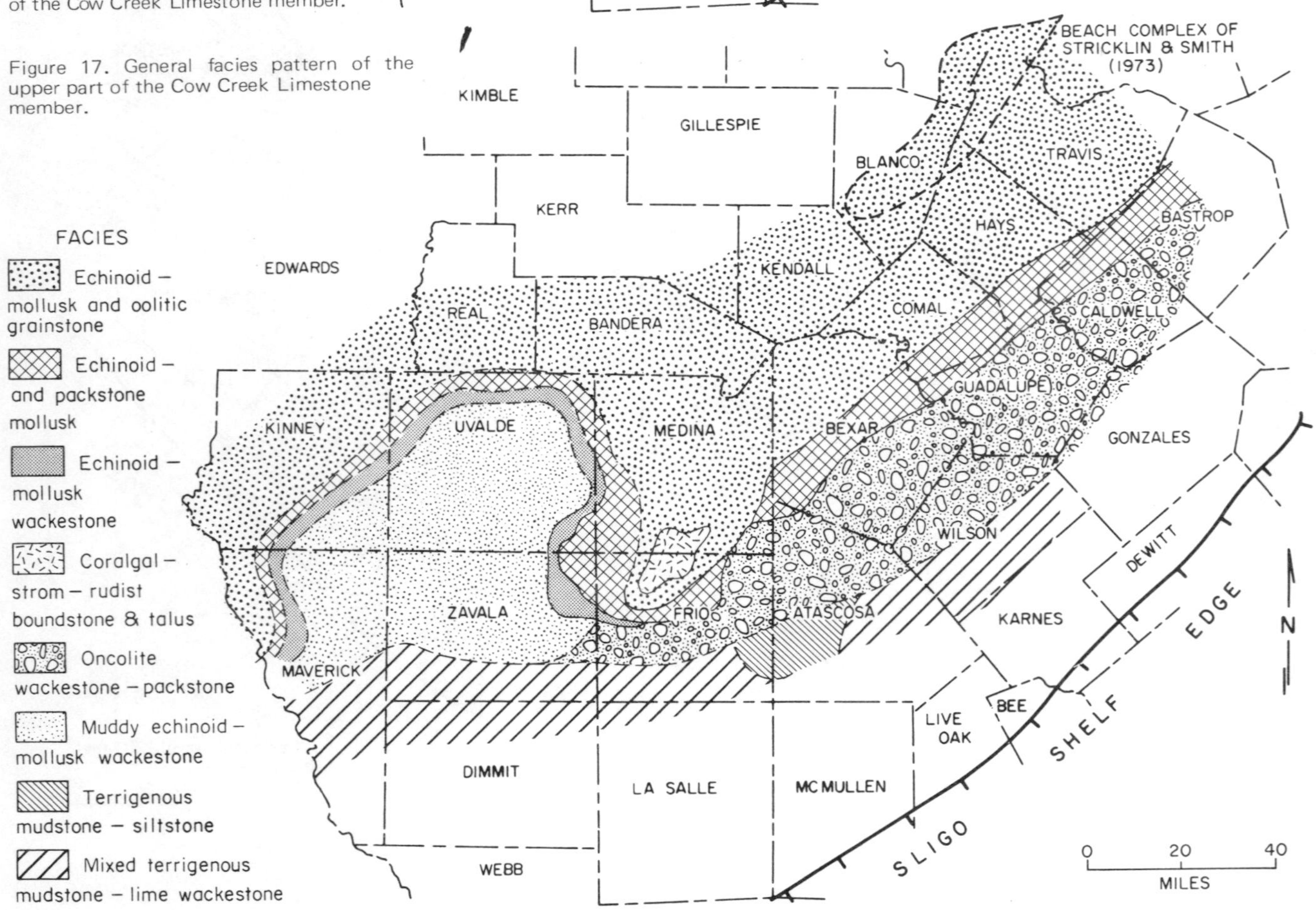

Figure 17. General facies pattern of the upper part of the Cow Creek Limestone member.

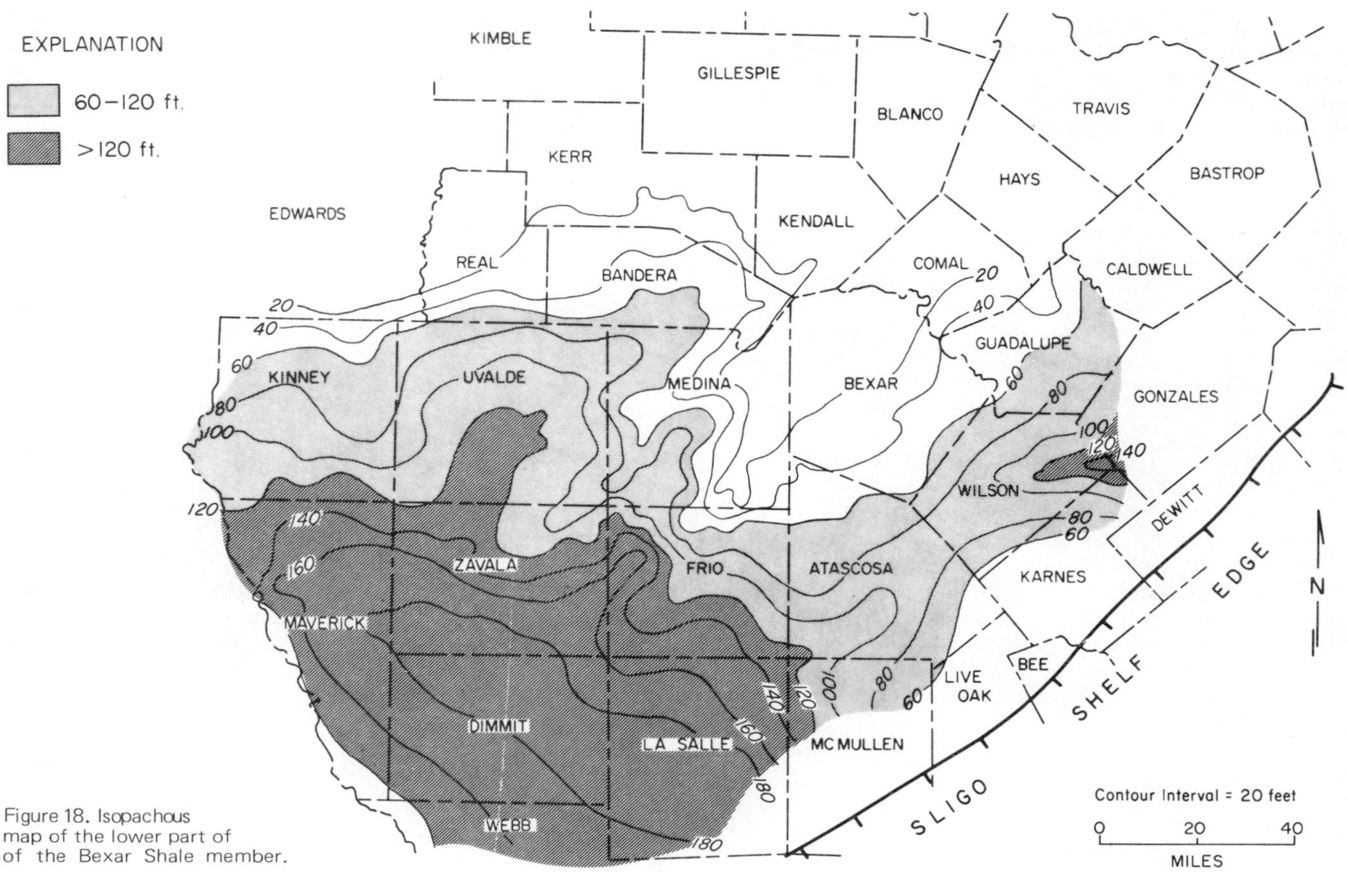

Figure 18. Isopachous map of the lower part of of the Bexar Shale member.

Figure 19. Isopachous map of the shoal-water carbonate complex, lower part of the Bexar Shale member.

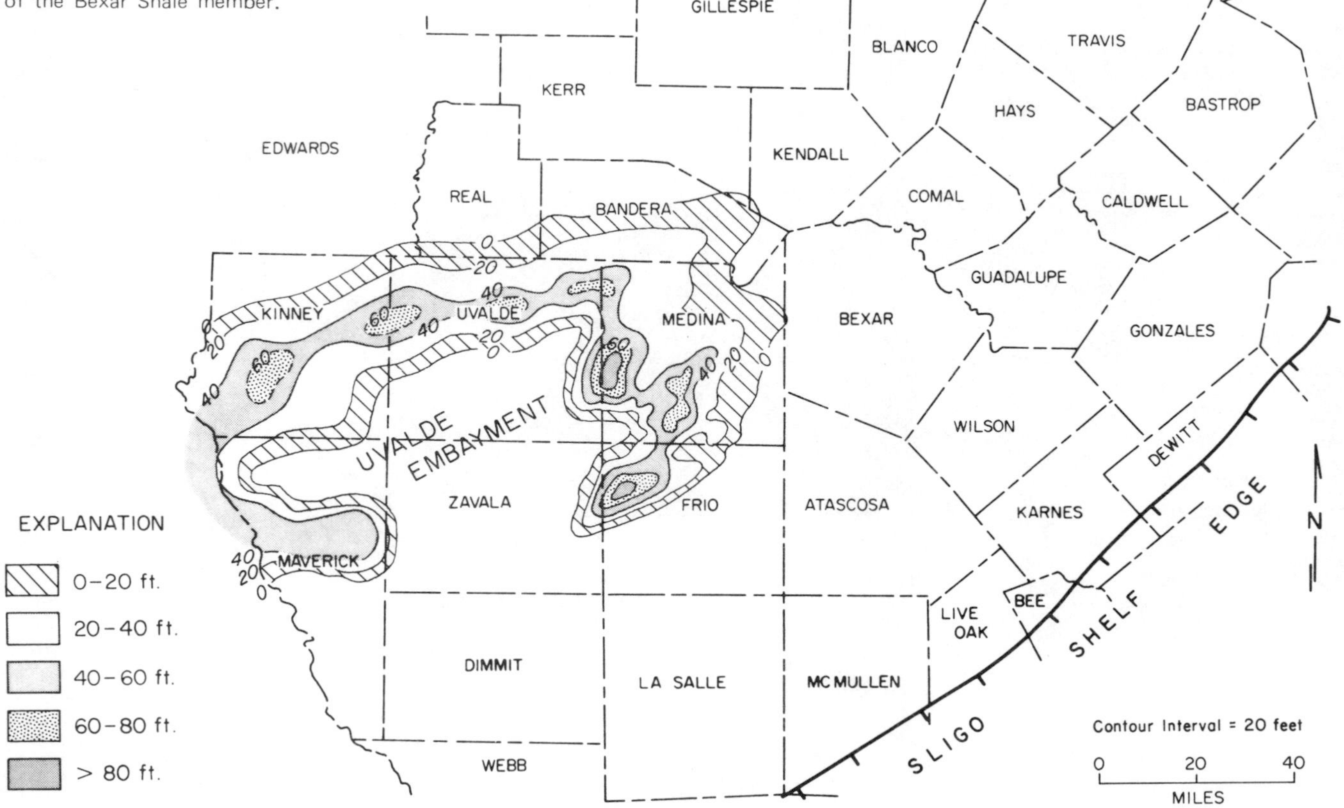

this zone was low in Mg and was undersaturated with respect to aragonite; therefore, some polymineralic sediments that were relatively stable in the marine environment were unstable in the meteoric environments (Stehli and Hower, 1961). A mass stabilization process began, in which aragonite was either neomorphosed to calcite or was leached, which resulted in the formation of cavities. The high-Mg calcite could have stabilized either by expelling excess Mg by exsolution or by absorbing more Mg and forming dolomite (Land, 1967). The high Mg/Ca ratio of 5/1 in seawater was not present in the meteoric zone; consequently, the Mg-poisoning effect which inhibited the precipitation of calcite in seawater was absent (Folk, 1973a). Aragonite was more soluble than calcite and dissolved, leaving the water saturated with calcite (Stanley, 1966). Thus, low-Mg calcite precipitated as a cement in water saturated with calcite that had a Mg/Ca ratio less than 2/1 (Folk and Land, 1972; Folk, 1973a).

Stabilization of the high-Mg calcite allochems (echinoids, red algae, micrite envelopes) occurred predominantly by the exsolution of Mg. This loss of Mg is shown by trace element analysis which indicates that these allochems are now calcite (Loucks, 1976). Trace element analysis also shows that a few dolomite crystals several microns in size formed in originally high-Mg calcite echinoid plates. Similar occurrences were observed by Land and Epstein (1970) in the meteoric stabilization of red algae and by MacQueen and others (1974) in crinoid plates.

Dissolution of aragonite allochems was very common in the meteoric environment in the Pearsall Formation, and some grains would have totally disappeared if it were not for Mg-calcite micrite envelopes. Dissolved shells were probably the principal source of $CaCO_3$ for the early cements present in this zone (Friedman, 1964), although some $CaCO_3$ may have been derived from seawater in the mixing zone. Examples of neomorphism of aragonite to calcite in which the original shell structure is retained (fig. 27b) are rare in the Pearsall grainstones.

The diagenetic features of stabilization of the meteoric phreatic environment in the Pearsall grainstones are very similar to those described by Land (1970) in the Pleistocene of Bermuda (fig. 27e, 27f) and to those described by Steinen and Matthews (1973) and Steinen (1974) in the Pleistocene of Barbados.

Cementation of the sediment occurred concurrently with the stabilization of the polymineralic allochems assemblage in the meteoric phreatic zone. The predominant cement is a fine-crystalline equant to bladed calcite cement that rims both outside of grains and inside of leached shells (figs. 26a-26d, 27a, 28d). Similar cements occur in the Pleistocene of Bermuda (fig. 27b, 27f). Rim cement

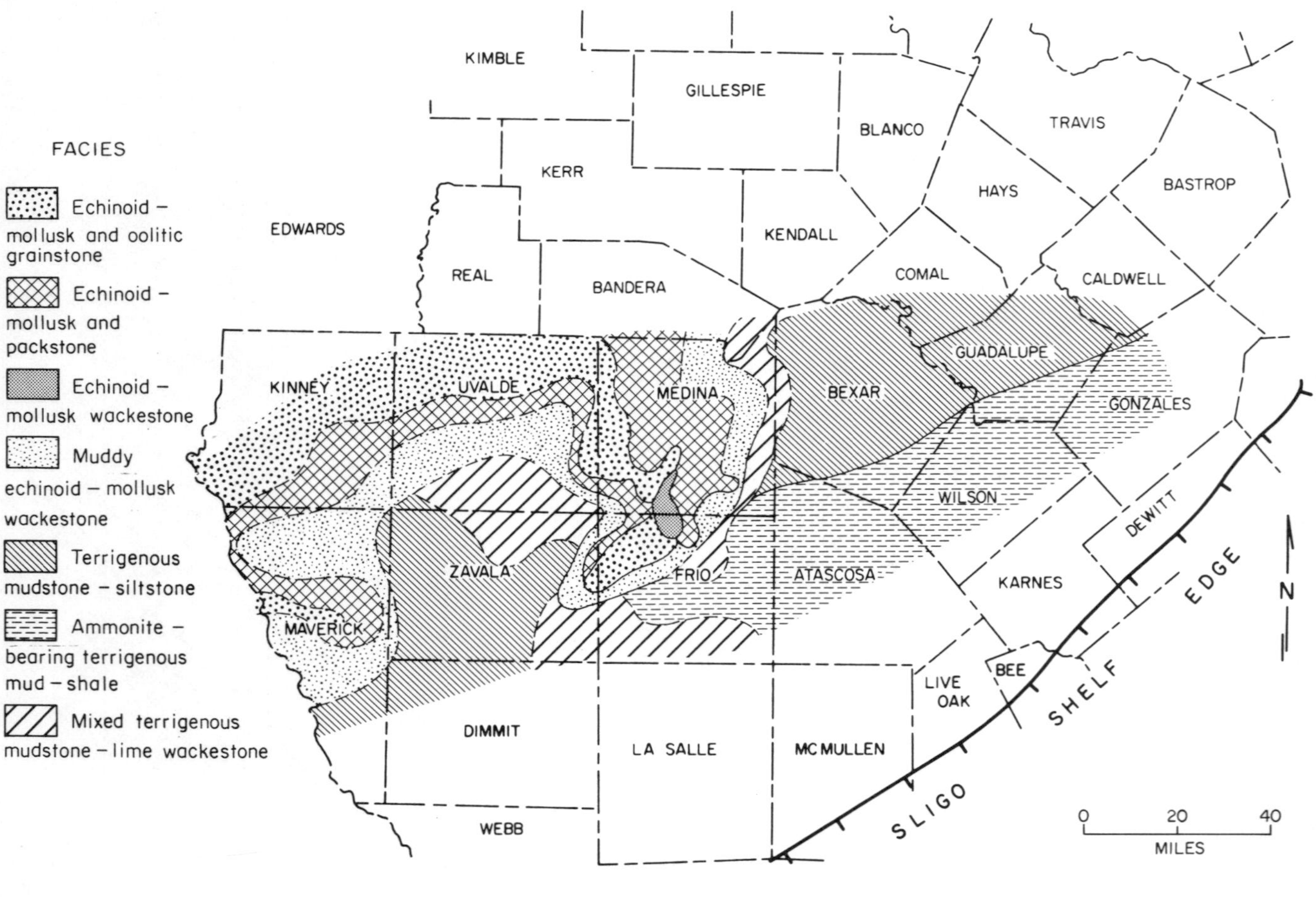

Figure 20. General facies pattern of the lower part of the Bexar Shale member.

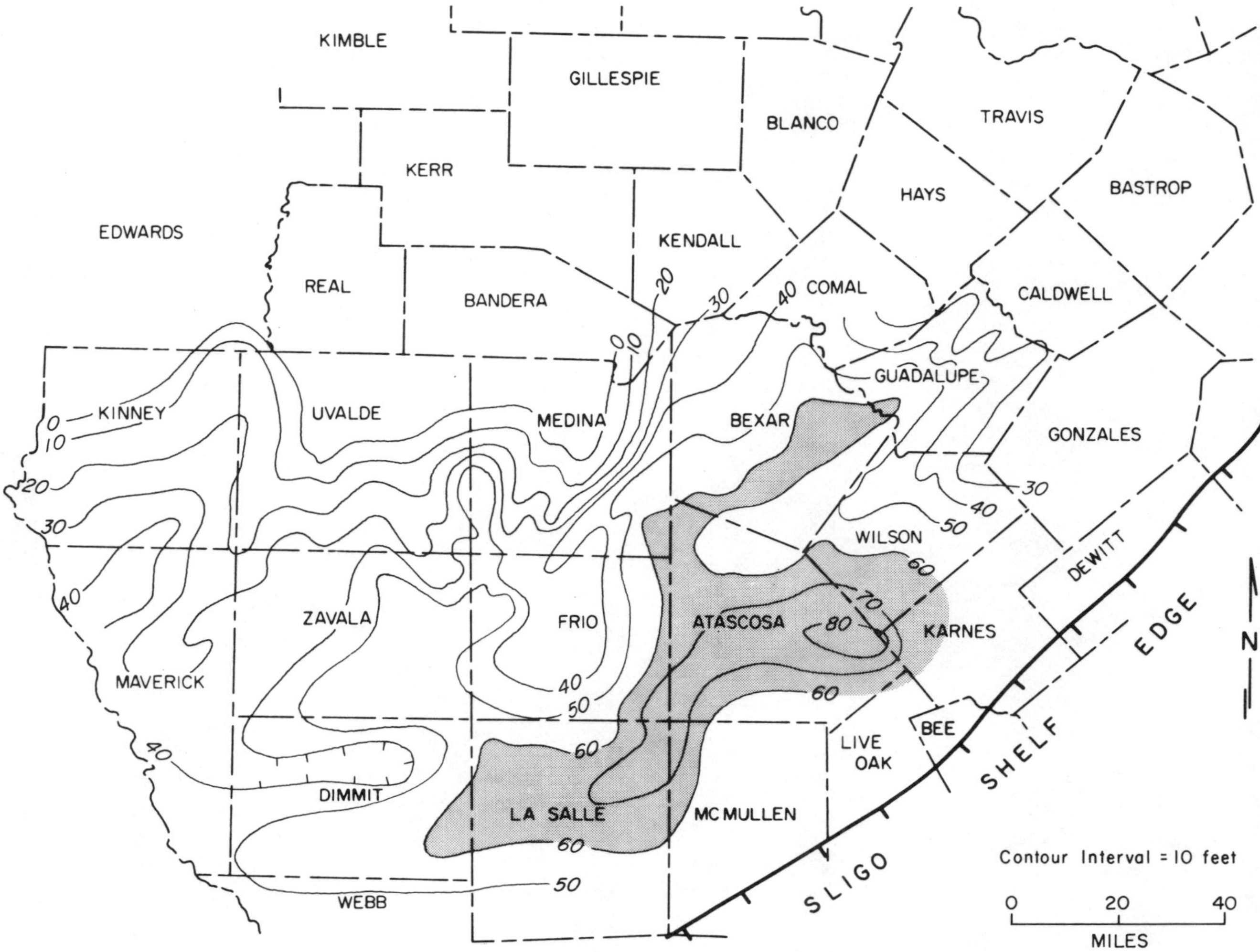

Figure 21. Isopachous map of the middle part of the Bexar Shale member.

in molds occurs only within grainstones; in rocks with a mud matrix, molds are generally filled only with coarse-crystalline equant spar (fig. 26f).

In echinoid fragments, syntaxial cement filled pore spaces and developed large overgrowths around the grains (fig. 27a). Syntaxial cement grew in competition with, but much faster than, rim cement, indicating that the cements formed simultaneously (fig. 27a). The syntaxial cement around echinoid grains is therefore meteoric in origin. Thin sections from the Pleistocene of Bermuda show that syntaxial cements also formed in competition with rim cements (fig. 27f).

Solution cavities of leached shells are empty or were partly to totally filled with calcite cement. One type of fill begins with fine-crystalline calcite spar in the center (fig. 26c, 26d). This cavity fill is a common feature in Pleistocene rocks that have undergone diagenesis in the meteoric environment (Friedman, 1964; Land, 1970). Another type of fill observed in the Pearsall leached grains started with a fine-crystalline rim cement followed in sharp contact by coarse-crystalline equant calcite (fig. 26a, 26d). This sharp change in crystal size means that the two cement types are not transitional and that they formed in different environments. Coarse-crystalline spar that is in sharp contact with rim cement probably formed in a later regional ground-water flow system or in the deeper subsurface environment. Medium-crystalline spar that is transitional with rim cement in leached grains is closely related to the rim cement and formed in nearly the same meteoric diagenetic environment.

Commonly the medium- to coarse-crystalline, equant calcite cement is difficult to classify according to time of formation because this type of cement can be closely associated with either early or late diagenetic features. Folk and Land (1972) and Folk (1973a) propose that sparry equant calcite should form in any calcite-precipitating environment low in Mg; therefore, it would be common in early to late stages of diagenesis.

I have observed some tentative qualitative features of early and late equant calcite cements that may be used to differentiate them. In the Pearsall Formation earlier equant cements are normally finer than later equant calcite cements [medium crystalline (0.1 mm) rather than coarse crystalline (0.25 to 1.5 mm)] and are more gradational in size with first-generation rim cement (fig. 26a, 26c, 26d). Also, early equant calcite contacts are commonly not as straight as contacts between the later crystals (fig. 26a, 26c, 26d, 26e). Late equant calcite commonly forms one or several large crystals in a pore space (fig. 26a, 26c).

Trace element analysis by electron microprobe shows coarser crystalline equant calcites to have lower Mg contents than any of the other cements

Figure 22.
Isopachous map of the shoal-water carbonate complex, middle part of the Bexar Shale member.

Figure 23. General facies pattern of the middle part of the Bexar Shale member.

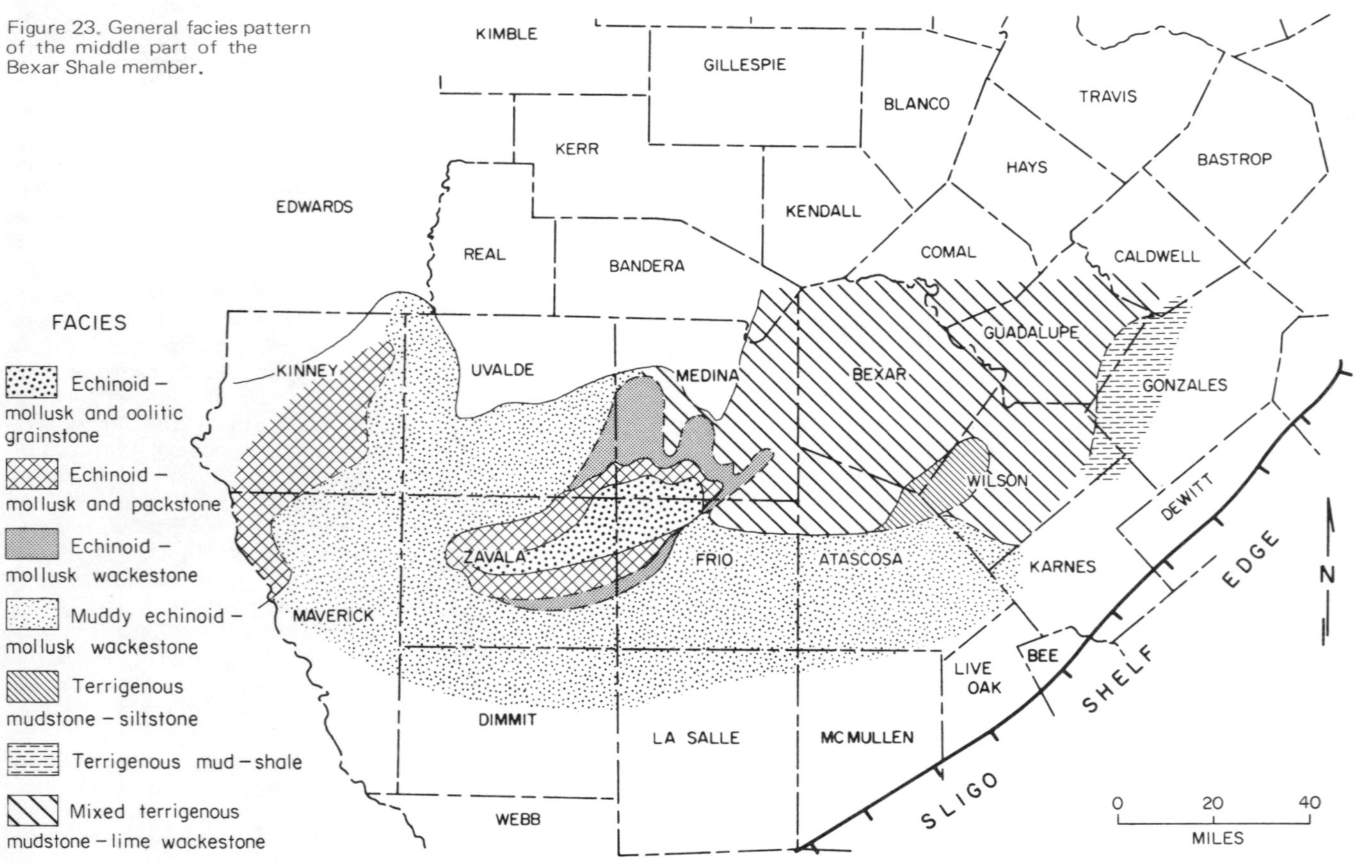

(average value 1.1± 0.4 mole percent $MgCO_3$). Some equant calcites which show higher Mg values similar to those of early rim cements may have formed earlier than crystals with significantly lower Mg values.

Staining for iron (Lindholm and Finkelman, 1972) indicates that equant calcite is both Fe-poor and Fe-rich (fig. 26a, 26c, 26d, 26e). Trace element analysis shows the average Fe content of all equant calcite to be 2,450±2,400 ppm (high standard deviation is due to zoning). Some equant calcite is well zoned with iron ranging from 1,850 ppm to 11,200 ppm. Some nonzoned Fe-poor equant calcite therefore probably formed in an Fe-poor diagenetic environment similar to the early nonferroan rim calcite, whereas zoned ferroan calcite is the result of later diagenesis.

Examples of early equant calcite in Pleistocene grainstones are described by Land (1970), Steinen and Matthews (1973), and Steinen (1974). Land (1970) shows that contacts between equant calcite crystals are somewhat irregular. The largest equant calcite crystals observed by Land (1970) were 0.25 mm with an average of 0.09 mm. These crystals are similar in size to those interpreted as early equant calcite crystals in the Pearsall

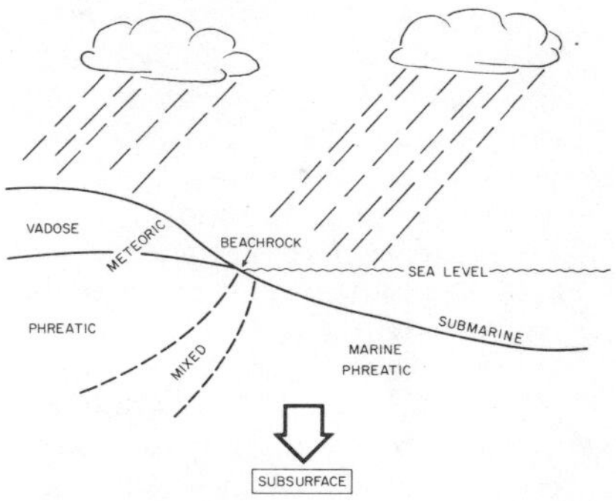

Figure 24. Major diagenetic environments. Modified from Folk (1973a).

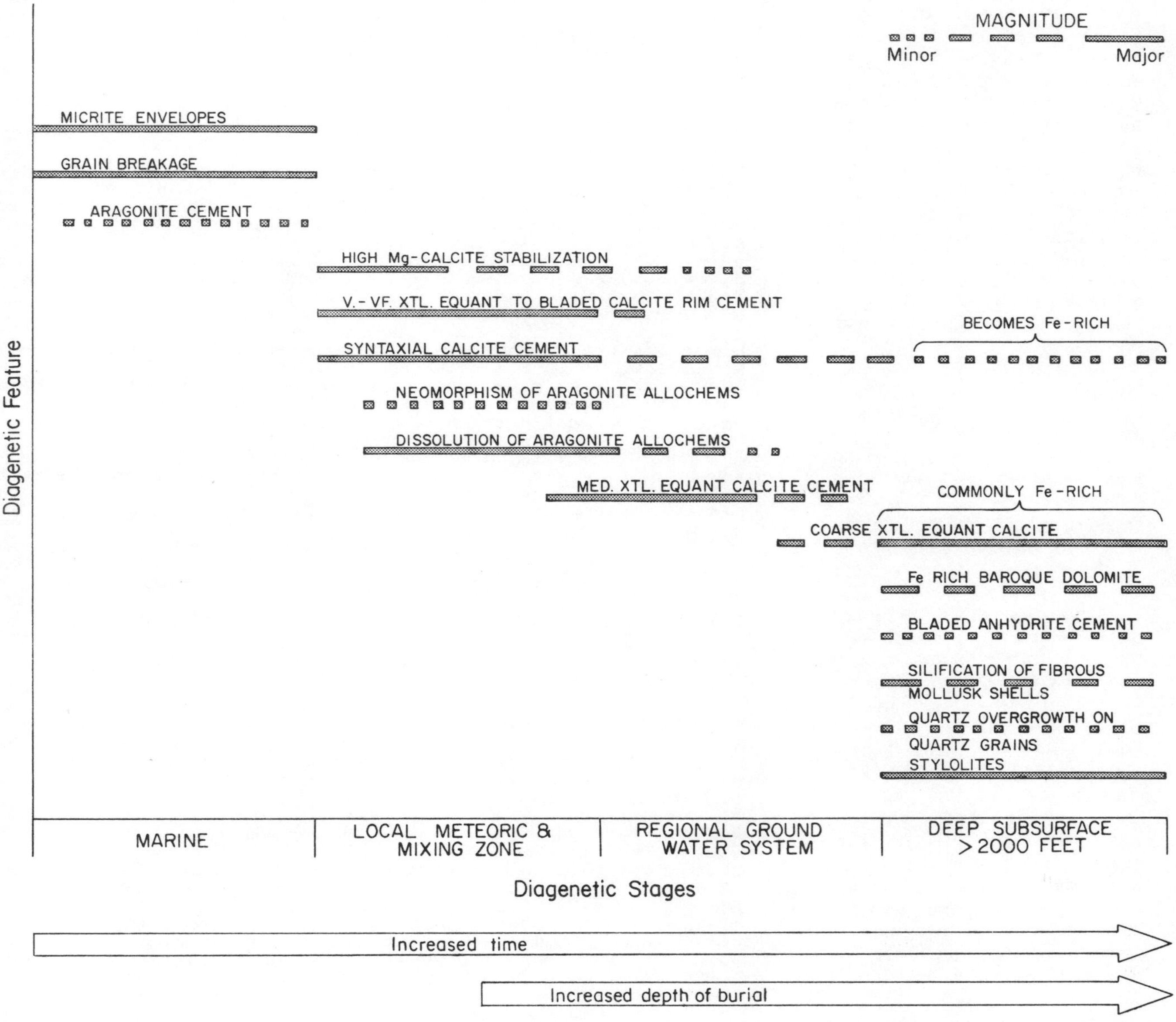

Figure 25. Relation of diagenetic features to diagenetic stages.

grainstones but much smaller than coarse-crystalline equant calcite (1 mm and greater) interpreted to have formed late. Equant calcite shown in studies of meteoric cements by Friedman (1964) and Gavish and Friedman (1969) is rarely coarser than 0.15 mm. Although some crystals have straight contacts, the contacts are commonly irregular.

Colley and Davis (1969) report an example of Pleistocene ferroan-calcite cement from the New Hebrides but state that it is the only known example at the time of their publication. This unusual occurrence of ferroan calcite was found interbedded with Fe-rich volcanics. As indicated by its almost total absence in Pleistocene carbonates, ferroan calcite is not a typical early near-surface cement.

In the mixing zone between the meteoric-phreatic and marine-phreatic environments, the Mg/Ca ratio is higher than in the meteoric zone and aragonite allochems are not as unstable. From the Cow Creek outcrop, Inden (1972) concludes that the number of neomorphosed grains of former aragonite shells retaining original shell structure increases downward through the foreshore facies. He attributed this increase of neomorphism to the presence of the high-Mg mixing zone below the meteoric-phreatic zone.

Another feature that may be a product of the mixing zone is the occurrence of large (up to 0.2 mm), well-developed bladed rim cements (figs. 26b, 28d). Mg concentration is high in this zone, and some Mg poisoning (Folk, 1973a) of calcite crystals probably takes place, resulting in a more bladed crystal than formed in the poorer Mg meteoric-phreatic environment.

Another possible interpretation of this well-developed bladed cement is that it is former submarine aragonite or Mg-calcite cement; however, it formed on both sides of micrite rims (figs. 26b, 28d), indicating that it filled a leached shell. The former aragonite shell was stable in the marine environment and was not dissolved until it was in contact with meteoric water (Bathurst, 1971). Thus, the bladed cement formed after leaching of the shell and after the sediment was out of the marine environment. Meteoric phreatic or mixing zone seems to be the most likely place for this bladed cement to have formed.

Regional ground-water flow stage

Early stage meteoric diagenesis occurred in local areas where fresh-

(page 119)

Figure 26. (a) Former aragonite allochem preserved by micrite envelope (1). Envelope is coated by fine-crystalline equant to bladed Fe-free calcite rim cement (2). In the allochem cavity the Fe-free rim is in sharp contact with coarse-crystalline equant ferroan calcite (3). The remaining pore space inside and outside the allochem is filled by anhydrite (4). Middle part of the Bexar Shale member, Tenneco #1 Goad (5,955 ft). Thin section stained with alizarin red-S and potassium ferrocyanide. Bar = 0.25 mm.
(b) Medium-crystalline bladed calcite rim cement (1) on both sides of micrite envelope (2). Rim cement on one side projects into primary interparticle porosity (3) while on the other side it projects into secondary moldic porosity (4). Middle part of the Bexar Shale member, Tenneco #1 Mack (7,539 ft). Thin section. Bar = 0.25 mm.
(c) Well-developed micrite envelope around former aragonite grain (1). Medium-crystalline equant Fe-free calcite formed in the fossil cavity (2) whereas coarse-crystalline equant ferroan calcite precipitated in the primary interparticle pore space (3). Middle part of the Bexar Shale member, Tenneco #1 Goad (5,955 ft). Thin section stained with alizarin red-S and potassium ferrocyanide. Bar = 0.25 mm.
(d) Outline of allochems preserved by micrite envelopes (1). Allochem molds are filled with two sizes of equant calcite; one has medium-crystalline equant calcite (2) and the other has a gradation from fine- to coarse-crystalline equant calcite (3). Middle part of the Bexar Shale member, Tenneco #1 Roberts (6,538 ft). Thin section stained with alizarin red-S and potassium ferrocyanide. Bar = 0.25 mm.
(e) Coarse-crystalline well-zoned ferroan calcite. Upper part of the Cow Creek Limestone member, Tenneco #1 Sirianni (6,149 ft). Thin section stained with potassium ferrocyanide and alizarin red-S. Bar = 0.5 mm.
(f) Coarse-crystalline equant calcite mold-fill in an echinoid-mollusk wackestone. Coarse-crystalline calcite is the first-generation cement in this void (no early fine-crystalline equant or bladed cement). Upper part of the Bexar Shale member, Tenneco #1 Stoker (7,011 ft). Thin section. Bar = 0.25 mm.

(page 120)

Figure 27. (a) Competitive cementation between syntaxial overgrowth (1) on echinoid fragment (2) and bladed rim cement (3) on mollusk grains (4). The rim cement is thickest on the side of the mollusk that is away from the echinoid fragment. Syntaxial cement grew faster than the rim cement and completely enclosed it. Upper part of the Cow Creek Limestone member, Tenneco #1 Stoker (7,234 ft). Thin section. Bar = 0.25 mm.
(b) Preserved growth lines in a neomorphosed mollusk shell. The aragonite shell neomorphosed to bladed calcite crystals that cut across growth lines. Crossed nicols. Lower part of the Bexar Shale member, Tenneco #1 Ney (3,413 ft). Thin section. Bar = 0.5 mm.
(c) Grainstone cemented by isopachous neomorphosed-calcite cement (marine) (1) followed by a micritic cement or fill (2). The cement is associated with an erosional surface that has cut both grains and cements (see fig. 6c). Lower part of the Bexar Shale member, Tenneco #1 New (3,422 ft). Thin section. Bar = 0.25 mm.
(d) Intraclast contains grains cemented with isopachous neomorphosed-calcite submarine cement (1). Grains are coated with an isopachous rim followed by a micrite cement. Upper part of the Cow Creek Limestone member, Tenneco #1 Stoker (7,205 ft). Thin section. Bar = 0.5 mm.
(e) Bermuda Pleistocene example of a grainstone showing a fine-crystalline bladed rim cement followed by a medium-crystalline equant cement in the interparticle pore space. Mollusk shells preserved by micrite envelopes. Solution-cavity fill is a gradation from fine-crystalline bladed rim to medium-crystalline equant cement. Thin section loaned by L.S. Land. Bar = 0.25 mm.
(f) Bermuda Pleistocene example of a grainstone showing a competitive cementation between syntaxial cement (1) around an echinoid fragment (2) and fine-crystalline bladed rim cement (3) around other allochems. Thin section loaned by L.S. Land. Bar = 0.25 mm.

(page 121)

Figure 28. (a) Euhedral quartz overgrowth on oolitic coated quartz grain replaces oolitic coating (1), and incorporates early meteoric rim cement (2). Other grains (3) show dissolution of grains with well-developed micrite envelopes that were later filled with a fine-crystalline equant calcite cement. Upper part of the Cow Creek Limestone member, Tenneco #1 Sirianni (6,222 ft). Thin section. Bar = 0.5 mm.
(b) Euhedral megaquartz replacing fibrous mollusk shell. Some fibrous structure of shell (1) is enclosed in the megaquartz. Middle part of the Bexar Shale member, Tenneco #1 Roberts (6,538 ft). Thin section. Bar = 0.25 mm.
(c) Length-slow chalcedony (1) and "ring" megaquartz (2) replacing mollusk shell. The two silica minerals grade into each other. Crossed nicols. Middle part of the Bexar Shale member, Tenneco #1 Roberts (6,538 ft). Thin section. Bar = 0.25 mm.
(d) Bladed anhydrite cement (1) filling large pore in an echinoid-mollusk grainstone. The micrite envelope (2) is all that remains of the allochem in the upper right. A well-developed bladed rim cement (3) occurs on both sides of the micrite envelope. Upper part of the Cow Creek Limestone member, Tenneco #1 Stoker (7,212 ft). Thin section. Bar = 0.5 mm.
(e) Baroque dolomite (1) and anhydrite (2) that competed for space in a cavity indicating that they are contemporaneous. Crossed nicols. Upper part of the Cow Creek Limestone member, Tenneco #1 Stoker (7,212 ft). Thin section. Bar = 0.5 mm.
(f) Well zoned ferroan baroque dolomite with curved crystal faces. Upper part of the Cow Creek Limestone member, Tenneco #1 Sirianni (6,221 ft). Thin section. Bar = 0.5 mm.

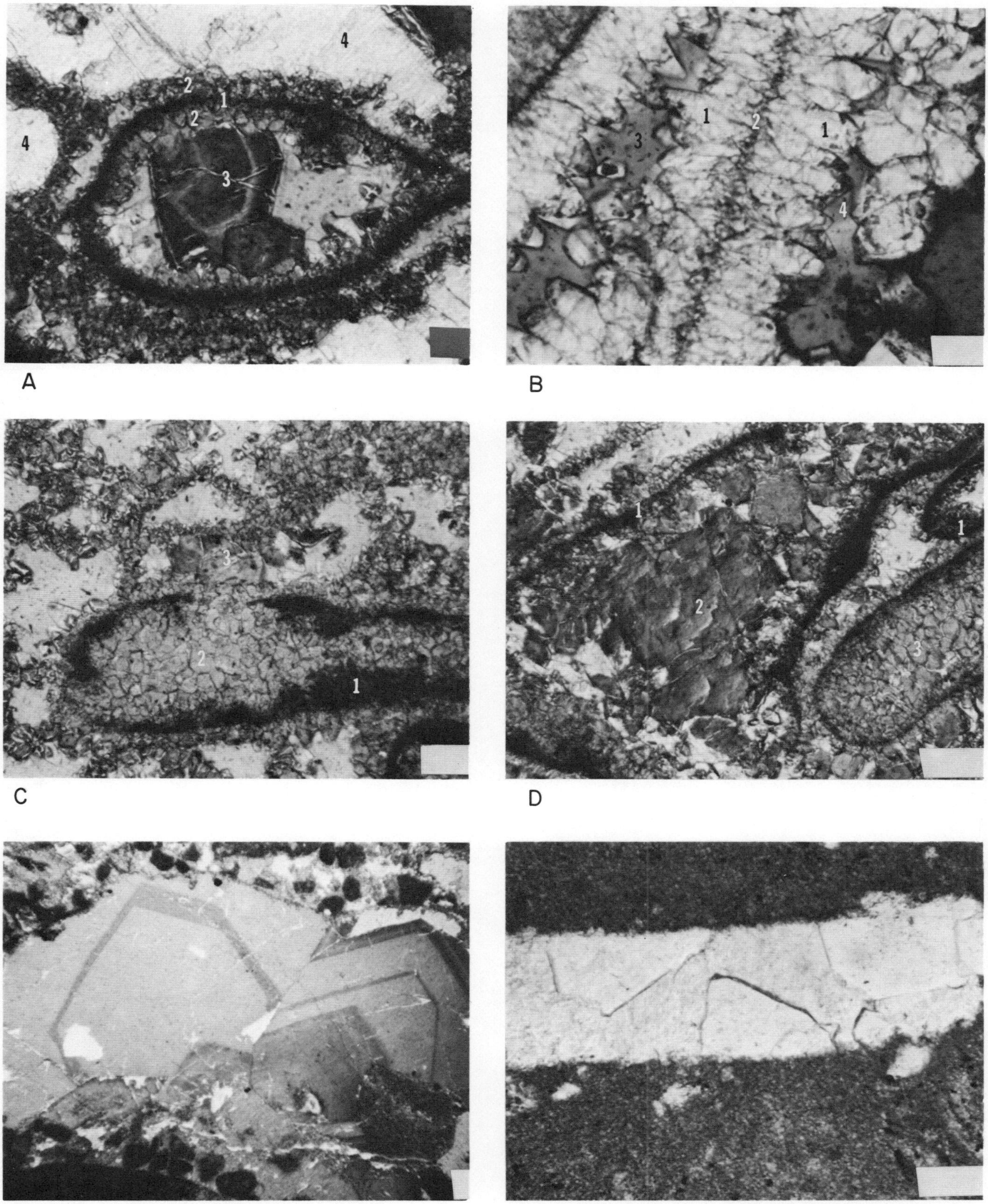

Figure 26

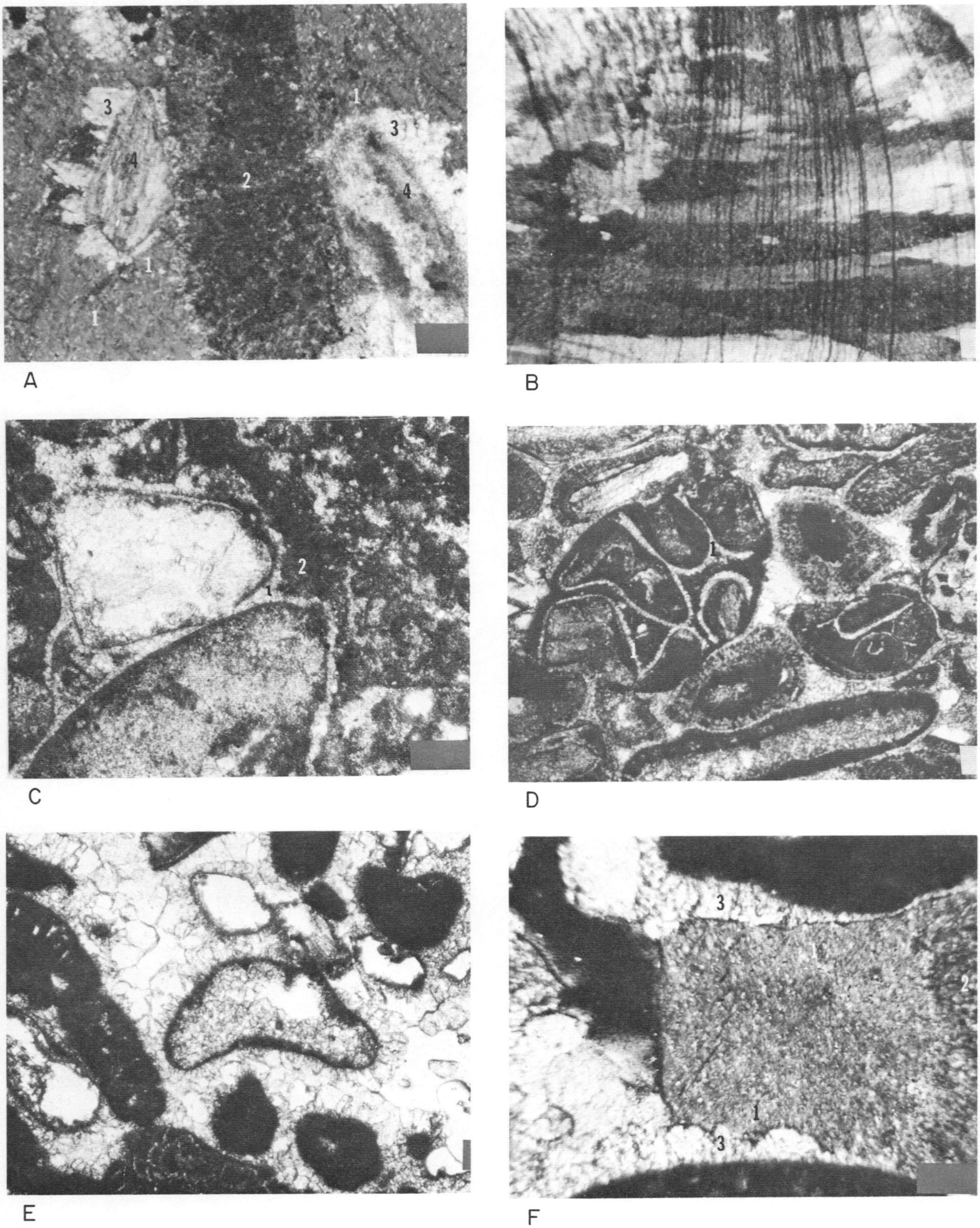

Figure 27

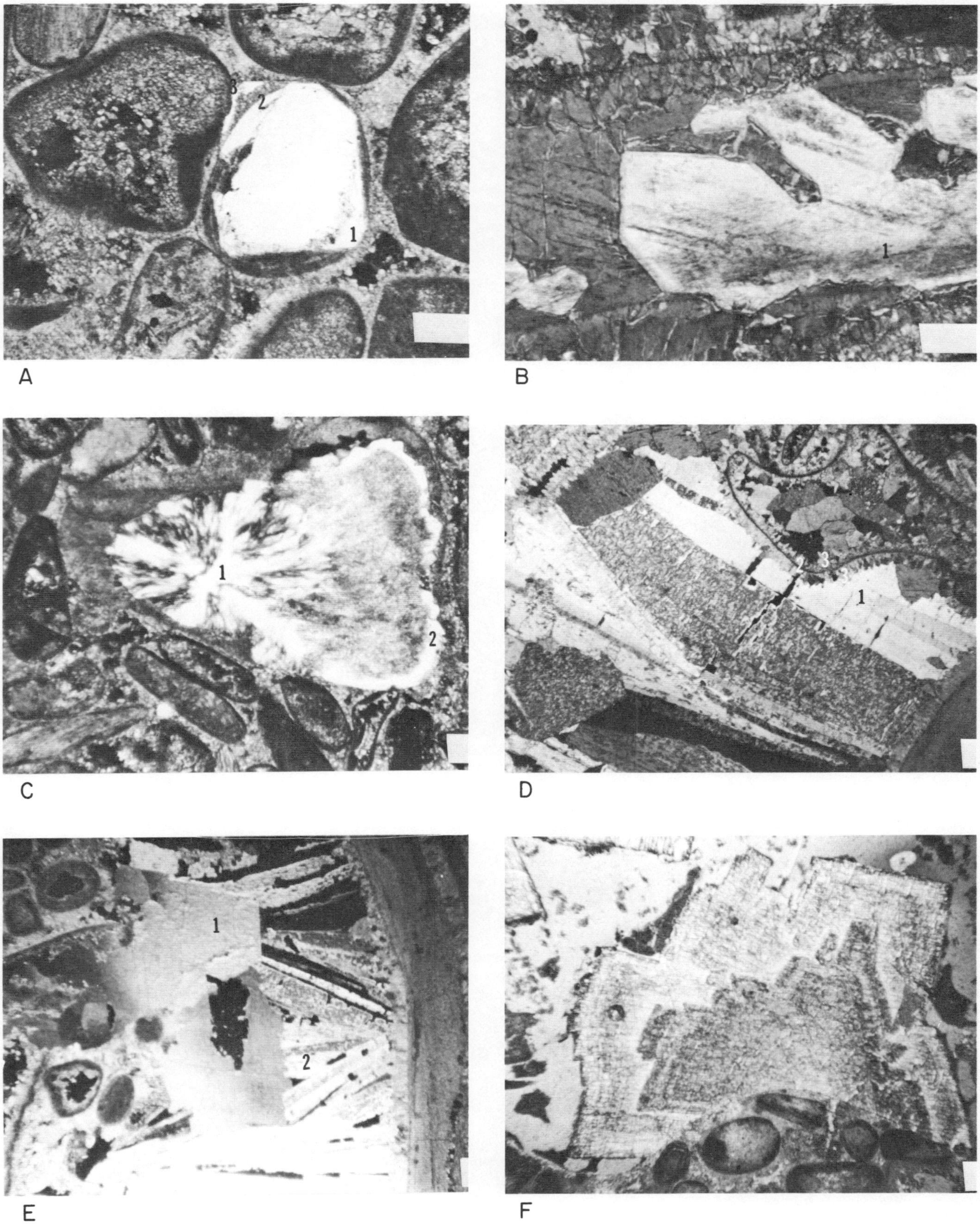

Figure 28

water lenses developed in the sediment. This stage was followed by a later, more regional meteoric ground-water flow system that had a limited effect on grainstone diagenesis.

Inden (1972) interpreted the coarse-crystalline calcite as a product of a regional ground-water flow diagenetic stage. This ferroan calcite could only be precipitated in an environment of negative Eh if the pH of the water was in equilibrium with carbonate rocks (Garrels and Christ, 1965). If Inden (1972) was correct in his assumption that the paleo-ground-water flow system was in equilibrium with carbonate and had a negative Eh, then ferroan calcite is a logical product of the regional ground-water system. The present shallow ground-water systems in Central Texas, however, are oxidizing or have a positive Eh (Rightmire and others, 1974). Iron-rich coarse-crystalline equant cement may have formed much later and may have been associated with a deep subsurface stage of diagenesis. On the other hand, Fe-poor medium-crystalline calcite may have started precipitating in the regional flow system and later added the ferroan calcite during the deep subsurface stage. Ferroan calcite will be discussed further in the section on the deep subsurface diagenetic environment.

Cementation during the regional ground-water flow stage left much of the porosity unaffected. Diagenesis was probably minor because ground-water circulation was limited. An arid climate may have limited the amount of water available to the system, and lack of downdip porosity in the mud-rich facies may have limited the flow of ground water. Absence of an open-flow system would tend to favor less cementation because there would be no recharge of $CaCO_3$ for cementation. If an open-flow system had sufficient water and dissolved $CaCO_3$ available to it, porosity might have been totally occluded. This stage of diagenesis was probably the most critical in preserving the porosity observed in the Pearsall grainstones.

Deep subsurface diagenetic stage

The last stage of diagenesis occurred in the deep subsurface (approximately 2,000 to 8,000 feet). The diagenetic features of this environment only occur where the Pearsall Formation was deeply buried. In the area where the Pearsall crops out, the rocks were never buried more than 1,500 feet before they were exhumed (Keith Young, personal communication). In the Pearsall outcrop area, Inden (1972) did not describe features attributed to deeper subsurface diagenesis noted in this investigation. It is, therefore, assumed that deep subsurface diagenetic features required a depth of burial greater than 2,000 feet.

This stage of diagenesis is interpreted to have started before the Miocene when the Balcones faulting occurred. Waters associated with the faulting locally leached some baroque dolomite that formed during the deep subsurface stage of diagenesis (Loucks, 1976).

Fluids that caused part of the subsurface diagenesis probably originated by dewatering of shales at depth. Montmorillonite alters to illite at about 8,000 feet at 100°C in the Gulf Coast area (Powers, 1969, 1967; Burst, 1969). During this inversion, water and silica that are bound to montmorillonite surfaces are released (Pollard, 1971). This water is forced into the adjacent porous grainstones. Also, hydrocarbons (indicated by oil and gas shows) would have produced a negative Eh environment. Temperature of the subsurface environment, as recorded on electrical-log runs, ranges from 50° to 100°C.

Silica from shale-derived water was deposited as euhedral quartz overgrowths around quartz sand grains (fig. 28a). These overgrowths incorporated the early meteoric phreatic calcite rim cement (fig. 28a), indicating that the quartz overgrowths are later than the meteoric diagenesis stage. The overgrowths also replaced carbonate grains adjacent to quartz sand grains.

Silica replaced evaporite nodules in adjacent facies which, in turn, released sulfate into the ground-water system. Other sulfate could have been contributed by the Glen Rose Limestone above or from the Sligo Limestone below the Pearsall Formation. Both of these formations contain evaporites in the area of this investigation (D. G. Bebout, personal communication; Bebout, 1977). The sulfate was reprecipitated as long bladed crystals of anhydrite cement in pore spaces (fig. 28d, 28e). Anhydrite fills fractures that cut both grains and meteoric phreatic cements that coat grains, indicating that anhydrite formed later than the meteoric stage of diagenesis.

Sulfate-rich water may have also influenced precipitation of coarse-crystalline, white to clear baroque dolomite (Folk, 1973a) with strongly curved crystal faces (figs. 28e, 28f) and replacement of original fibrous calcite mollusk shells with length-slow chalcedony (fig. 28c) and megaquartz (fig. 28d), as suggested by Folk and Pittman (1971) and Siedlecka (1972). Iron-rich (up to 18 mole percent $MgCO_3$) baroque dolomite and anhydrite (fig. 28) were precipitated contemporaneously. The high-Fe content of the dolomite indicates that it was formed in a strongly reducing environment. Baroque dolomite, as well as anhydrite, filled fractures that cut both grains and meteoric cements, indicating a later origin.

Precipitation of baroque dolomite may have been caused by mixing of subsurface brines with relatively low-salinity diagenetic water expelled from the shales. This is similar to the mixing model for dolomite formation proposed by Land (1973) and Folk and Land (1975). However, Folk and Land's source of low-salinity water was meteoric, whereas the source of low-salinity water in the Pearsall originated in the subsurface. Dorfman and Kehle (1974) state that water released from Gulf Coast shales during deep diagenesis has a salinity of approximately 5,000 ppm, much lower than the salinity of the brine in the porous Pearsall grainstone (20,000 to 200,000 ppm). Also, in the initial brine the Mg/Ca or Mg + Fe/Ca ratios may have been high because of calcium removed during precipitation of coarse-crystalline, equant calcite. When the two waters mixed, the Mg/Ca or Mg + Fe/Ca remained high but salinity was reduced to favor precipitation of dolomite.

Length-slow chalcedony and megaquartz replaced only shell material that was originally calcite. Allochems that were originally aragonite were not replaced by silica. The replacement of mollusk shells with silica, therefore, occurred after the meteoric-phreatic diagenetic stage, or some of the original aragonite shells also would have been replaced (Jacka, 1974).

As noted in the previous discussion of the regional ground-water flow system environment, much of the coarsely crystalline equant calcite is high in iron; however, many ground-water systems today have positive Eh values which could not produce ferroan calcite. The ferroan calcite was deposited, therefore, during an initial stage of diagenesis in the deep surface environment where the presence of organic material would have created the necessary negative Eh environment (Collins, 1975; L.S. Land, personal communication). Much of the ferroan equant calcite would have to have formed early during this stage because it commonly appears to have been precipitated before anhydrite and baroque dolomite, as indicated by their physical relationships in thin sections; however, some ferroan equant calcite is later. In addition, early precipitation of calcite in this environment would have raised the Mg/Ca or Mg + Fe/Ca ratios, thus favoring formation of ferroan dolomite.

A common feature in Pearsall grainstones is the occurrence of well-developed stylolites which formed by pressure solution at depth (Wagner, 1913). Formation of stylolites may have exerted several different influences on the deep subsurface environment. The rock that was dissolved would be an important source of $CaCO_3$ for cementation (Bathurst, 1958). Loucks (1976) found that δC^{13} values of late cements (+2.0°/₀₀ PDB) are quite similar to predicted values for late cements (+1.0°/₀₀ PDB) that would be precipitated from carbonate associated with fluids from stylolitization. Reduction of rock volume by the dissolution of rock along stylolites may have acted as a hydraulic pump that created a driving force which, along with the driving force from shale dewatering, moved fluids through the grainstones. Fluid motion would be necessary for continued cementation in the subsurface because a totally closed system would not be able to recharge itself with $CaCO_3$ and SiO_2 needed to produce the observed diagenetic features. Stylolitization, a pressure-solution process, could dissolve calcite in the presence of fluids in equilibrium with calcite in the manner expressed by Riecke's Principle (Johnston and Niggli, 1913). A calcite grain under stress at its contacts will dissolve and add more $CaCO_3$ to a solution already saturated with respect to calcite. From the supersaturated solution, calcite will precipitate in pore spaces not under stress. Consequently, stylolitization is a mechanism for supplying $CaCO_3$ for cementation, transporting $CaCO_3$-bearing fluids, and allowing $CaCO_3$ to be dissolved in the presence of $CaCO_3$-saturated fluids.

A summary of porosity development and preservation as related to the stages of diagenesis is outlined in figure 29.

POROSITY DISTRIBUTION

Knowledge of porosity distribution within a formation is important in petroleum exploration. If the effect of diagenesis on porosity in various facies is understood, then facies maps can also be used as porosity maps. Approximately 95 percent of the porosity in the Pearsall Formation occurs in grainstone and boundstone facies, with porosity best developed and preserved in the former. Porosity in the boundstone facies is less predictable and less abundant. A core from the Tenneco #1 Sirianni well demonstrates the relationship of porosity and permeability to different textural facies (fig. 30). All permeability values and some of the porosity values shown are from core analysis, whereas the remaining porosity values were visually estimated from core.

A porosity cross section (fig. 31), similar to stratigraphic facies cross-section B-B′ (fig. 12), shows that porosity is closely related to grainstone facies over a wide area (50 miles). Thus, by mapping distribution of the facies and by understanding diagenetic effects on the facies, the porosity distribution in the Pearsall Formation can be predicted.

PETROLEUM POTENTIAL

Certain facies within the subsurface Pearsall Formation in South Texas are prospective petroleum reservoirs which produce in several areas in Texas. In East Texas oil is produced from porosity in a boundstone facies (Bushaw, 1968). Gas production in Maverick County and southern La Salle County is from fracture porosity. In the Frio-Medina County area where Tenneco Oil Company conducted an extensive exploration effort for stratigraphic traps, no commercial petroleum has been discovered in the Pearsall Formation (Burnett, 1974).

Several ideas about petroleum potential in Frio and Medina Counties can be proposed. First, grainstone facies formed an excellent reservoir with high porosity and permeability. The reservoir formed early and would have accommodated any petroleum migration.

Second, grainstones are in close proximity to highly organic terrigenous mudstones and shales that may have been good source rocks; core analyses indicate oil and gas shows in grainstone facies. Nevertheless, these shows may indicate that only a limited amount of hydrocarbon was generated and that the source rocks are still thermally immature.

Probably the major limiting factor for hydrocarbon accumulation in the Pearsall grainstones is absence of a good trapping mechanism at time of hydrocarbon migration. Hydrocarbons may have accumulated in local stratigraphic and paleotopographic traps which will probably be discovered with additional drilling. The large regional faults that form more extensive traps, however, did not form until the Tertiary period (Murray, 1961). These faults, the Balcones and Luling fault zones, would have provided

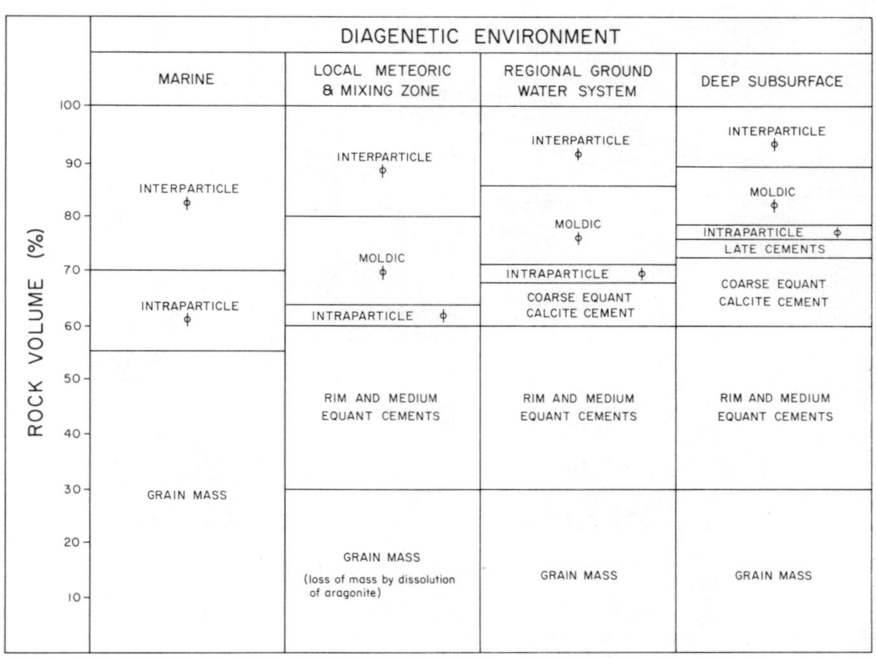

Figure 29. Porosity summary as related to idealized diagenetic history of a grainstone.

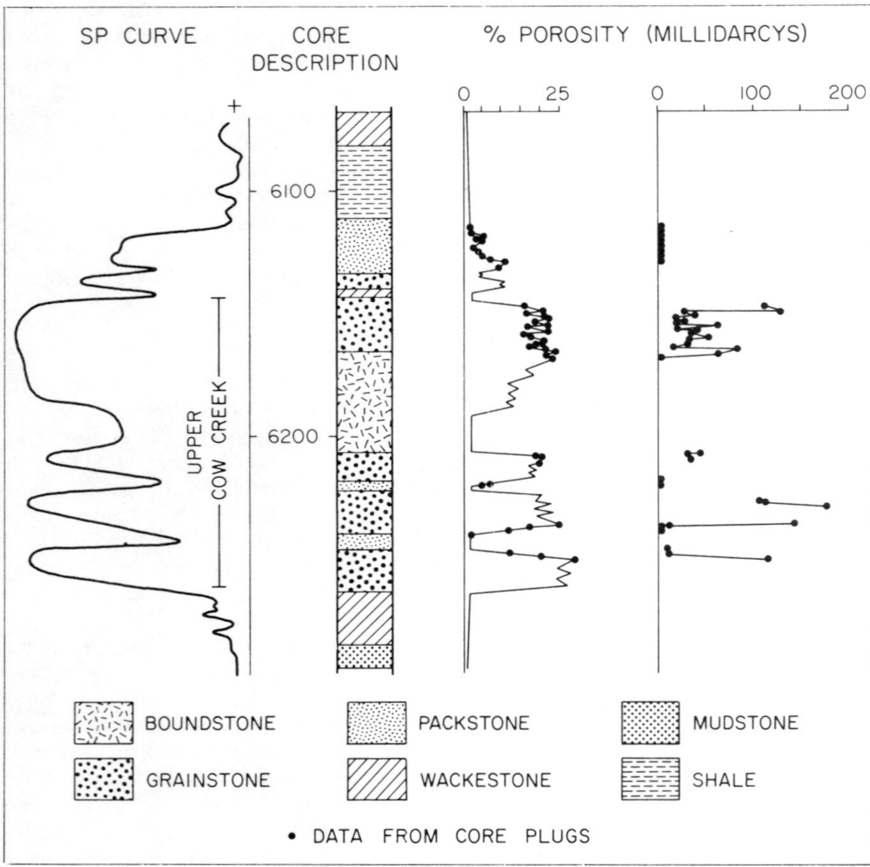

Figure 30. Relationships of porosity and permeability to lithic composition in a core from the Tenneco #1 Sirianni well in Frio County.

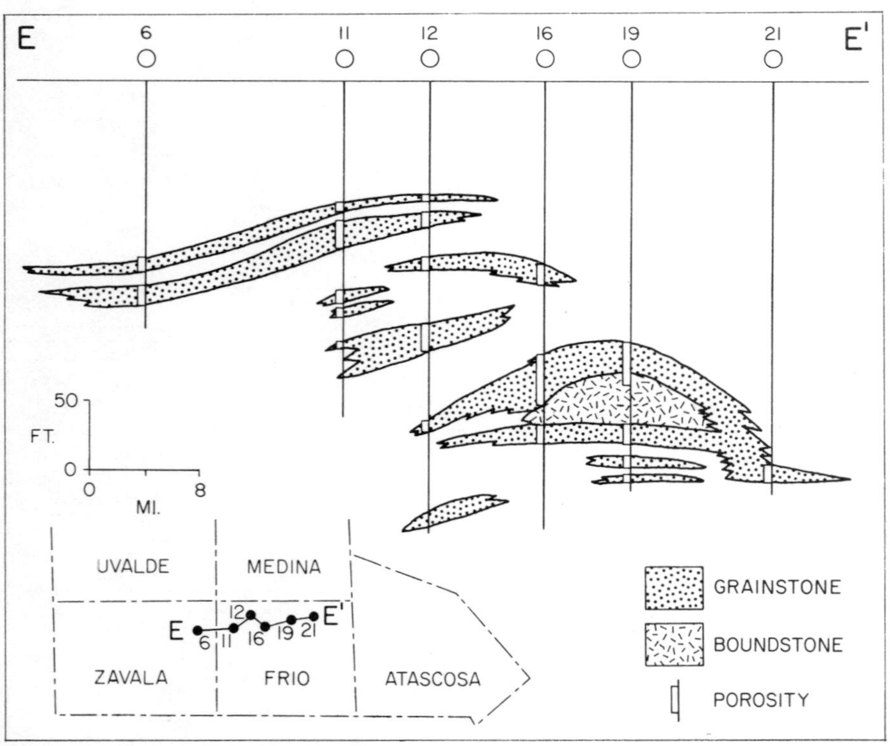

Figure 31. Simplified stratigraphic cross-section B-B' (fig. 10) showing major grainstone and boundstone facies and occurrence of porosity.

excellent traps had they developed earlier in the fluid history of the Pearsall. Most hydrocarbons that migrated prior to the Tertiary period were lost at the surface or may have trapped in other formations updip from this area or higher in the section.

The Pearsall Formation is still a good petroleum prospect considering its high porosity and permeability, relatively shallow depth, and possible stratigraphic and paleotopographic traps. If not drilled as a primary target, it certainly should be tested as a secondary objective.

CONCLUSIONS

The Lower Cretaceous Pearsall Formation in South Texas contains three vertically superposed shoal-water carbonate complexes which form prospective petroleum reservoirs. The carbonate complexes of the upper part of the Cow Creek Limestone are the most areally extensive, whereas the complexes in the lower and middle parts of the Bexar Shale member have a smaller areal distribution because of the effect of contemporaneous terrigenous clastic deposition in the San Marcos arch area.

Approximately 95 percent of the porosity in the Pearsall Formation is in grainstone and boundstone facies of the shoal-water carbonate complexes. Consequently, facies distribution maps outline potential reservoirs (figs. 16, 19, 22). Furthermore, diagenetic analysis of the grainstone indicates that porosity has been present during all four stages of diagenesis (fig. 29) and would have accommodated and retained any petroleum if contemporaneous traps existed. Hopefully, further exploration will locate areas of potential commercial production in the porous shoal-water carbonate facies complexes of the Pearsall Formation in South Texas.

ACKNOWLEDGMENTS

I would like to thank the following people who contributed to this study: D.G. Bebout and J.H. McGowen of the Bureau of Economic Geology and L.S. Land, A.J. Scott, and Keith Young of the Department of Geological Sciences, The University of Texas at Austin. I would also like to thank L.F. Brown, Jr., and P.A. Mench of the Bureau of Economic Geology for reviewing this paper.

I am grateful to the Geology Foundation and the Owen Coates Fund of the Department of Geological Sciences at The University of Texas at Austin for financial assistance for part of this study.

A special thanks is extended to Tenneco Oil Company and particularly C.R. Burnette for providing cores and thin sections that formed the basis for my study. Also appreciated is the logistical support provided by D.C. Ratcliff and the staff at the Bureau of Economic Geology Well Sample and Core Library.

REFERENCES

Amsbury, D.L., 1974, Stratigraphic petrology of lower and middle Trinity rocks on the San Marcos platform, south-central Texas, *in* Perkins, B.F., ed., Geoscience and man: Vol. VIII, Aspects of Trinity Division geology: Louisiana State Univ., School of Geoscience, p. 1-35.

Bathurst, R.G.C., 1958, Diagenetic fabrics in some British Dinantian limestones: Liverpool Manchester Geological Journal, v. 2, p. 11-36.

_______ 1966, Boring algae, micrite envelopes and lithification of molluscan biosparites: Journal of Geology, v. 5, p. 15-32.

_______ 1971, Carbonate sediments and their diagenesis: Amsterdam, Elsevier, 620 p.

Bebout, D.G., 1976, Transgressive sandstone/carbonate sequence—The Hosston/Sligo Formations of South Texas (abs.): AAPG Bull., v. 60, p. 648.

_______ 1977, Sligo and Hosston depositional patterns, subsurface of South Texas: this volume.

_______ , and Loucks, R.G., 1974, Stuart City Trend, Lower Cretaceous, South Texas: A carbonate shelf-margin model for hydrocarbon exploration: Univ. Texas, Austin, Bureau of Economic Geology Report of Investigations 78, 80 p.

Bricker, O.P., ed., 1971, Carbonate cements: Baltimore, Johns Hopkins Press, 376 p.

Burnette, R.C., 1974, Search for stratigraphically trapped hydrocarbons in Lower Cretaceous of fifteen counties in South Texas (abstract): AAPG and SEPM Abstracts with Programs, v. 1, p. 14.

Burst, J.F., 1969, Diagenesis of Gulf Coast clayey sediments and its possible relation to petroleum migration: AAPG Bull., v. 53, p. 73-93.

Bushaw, D., 1968, Environmental synthesis of East Texas Lower Cretaceous: GCAGS Transactions, v. XVIII, p. 416-438.

Byrne, J.V., LeRay, D.O., and Riley, C.M., 1959, The chenier plain and its stratigraphy, southwestern Louisiana: GCAGS Transactions v. IX, p. 237-260.

Campbell, D.H., 1962, Petrography of the Cretaceous Hensel Sandstone, Central Texas: Univ. Texas, Austin, Master's thesis, 181 p.

Choquette, P., and Pray, L., 1970, Geologic nomenclature and classification of porosity in sedimentary carbonates: AAPG Bull., v. 54, p. 207-250.

Colley, H., and Davies, P.J., 1969, Ferroan and non-ferroan calcite cements in Pleistocene - Recent carbonates from the New Hebrides: Journal of Sedimentary Petrology, v. 39, p. 554-558.

Collins, A.G., 1975, Geochemistry of oilfield waters: New York, Elsevier Scientific Publishing Co., 496 p.

Cooper, J.D., 1964, Geology of the Spring Branch area, Comal and Kendall Counties, Texas: Univ. Texas, Austin, Master's thesis.

Davies, G.R., 1970, Carbonate bank sedimentation, eastern Shark Bay, western Australia: AAPG Memoir 13, p. 85-168.

Dorfman, M., and Kehle, R.O., 1974, Potential geothermal resources of Texas: Univ. Texas, Austin, Bureau of Economic Geology Geological Circ. 74-4, 33 p.

Dunham, R.J., 1969, Vadose pisolite in the Capitan Reef (Permian), New Mexico and Texas, *in* Friedman, G.M., ed., Depositional environments in carbonate rocks: SEPM Special Pub. 14, p. 182-191.

Folk, R.L., 1973a, Carbonate petrography in the post-Sorbian age, *in* Ginsburg, R.N., ed., Evolving concepts in sedimentology: Johns Hopkins Univ., Studies in Geology no. 21, p. 118-158.

_______ 1973b, Geochemical history of some aragonite and dolomites: Penrose Conference, Vail Colorado, Tx, lecture.

_______ 1975, Mg/Ca ratio and salinity: Two controls over crystallization of dolomite: AAPG Bull., v. 59, p. 60-68.

_______, and Land, L.S., 1972, Mg/Ca vs. salinity; a frame of reference for crystallization of calcite, aragonite and dolomite (abs.): GSA Abstracts with Programs, v. 4, p. 508.

_______, and Pittman, J.S., 1971, Length-slow chalcedony: A new testament for vanished evaporites: Journal of Sedimentary Petrology, v. 41, p. 1045-1058.

Forgotson, J.M., Jr., 1957, Stratigraphy of Comanche Cretaceous Trinity group: AAPG Bull., v. 41, p. 2328-2363.

Friedman, G.M., 1964, Early diagenesis and lithification in carbonate sediments: Journal of Sedimentary Petrology, v. 34, p. 777-813.

Garrels, R.M., and Christ, C.L., 1965, Solutions, minerals, and equilibria: New York, Harper and Row, 450 p.

Gavish, E., and Friedman, G.M., 1969, Progressive diagenesis in Quaternary to late Tertiary carbonate sediments: Sequence and time scale: Journal of Sedimentary Petrology, v. 39, p. 980-1006.

Imlay, R.W., 1945, Subsurface Lower Cretaceous formations of South Texas: AAPG Bull., v. 29, p. 1416-1469.

Inden, R.F., 1972, Paleogeography, diagenesis, and paleohydrology of a Trinity Cretaceous carbonate beach sequence, Central Texas: Louisiana State Univ., Ph.D. dissertation, 264 p.

_______ 1974, Lithofacies and depositional model for a Trinity Cretaceous sequence, Central Texas, *in* Perkins, B.F., ed., Geoscience and man, Vol. VIII, Aspects of Trinity Division geology: Louisiana State Univ., School of Geoscience, p. 35-52.

Jacka, A.A., 1974, Replacement of fossils by length-slow chalcedony and associated dolomitization: Journal of Sedimentary Petrology, v. 44, p. 421-427.

Johnson, D., 1974, Sedimentation in the Gascoyne River delta, western Australia: Univ. Western Australia, Ph.D. dissertation, 144 p.

Johnston, J., and Niggli, P., 1913, The general principles underlying metamorphic processes: Journal of Geology, v. 21, p. 610-612.

Land, L.S., 1967, Diagenesis of skeletal carbonates: Journal of Sedimentary Petrology, v. 37, p. 914-920.

_______ 1970, Phreatic versus vadose meteoric diagenesis of limestones: Evidence from a fossil water table: Sedimentology, v. 14, p. 175-185.

_______ 1973, Contemporaneous dolomitization of middle Pleistocene reefs by meteoric water, north Jamaica: Bull. of Marine Science, v. 23, p. 64-92.

_______ , and Epstein, S., 1970, Late Pleistocene diagenesis and dolomitization, north Jamaica: Sedimentology, v. 14, p. 187-200.

_______ , MacKenzie, F.T., and Gould, S.J., 1967, Pleistocene history of Bermuda: GSA Bull., v. 78, p. 993-1006.

Lindholm, R.C., and Finkelman, R.B., 1972, Calcite straining: Semiquantitative determination of ferrous iron: Journal of Sedimentary Petrology, v. 42, p. 239-242.

Loucks, R.G., 1975a, Depositional environments in a carbonate-terrigenous shelf system: Subsurface Lower Cretaceous Pearsall Formation of South Texas (abstract): GSA Abstracts with Programs, v. 7, p. 211.

_______ 1975b, Porosity development in carbonate grainstones in subsurface Lower Cretaceous Pearsall Formation of South Texas: AAPG and SEPM Abstracts with Programs, v. 2, p. 45.

_______ 1976, Pearsall Formation, Lower Cretaceous, South Texas—Depositional facies and carbonate diagenesis and their relationship to porosity: Univ. Texas, Austin, Ph.D. dissertation, 362 p.

Lozo, F.E., and Stricklin, F.L., Jr., 1967, Stratigraphic notes on the outcrop basal Cretaceous, Central Texas: GCAGS Transactions, v. VI, p. 67-78.

MacQueen, R.W., Ghent, E.D., and Davies, G.R., 1974, Magnesium distribution in living and fossil specimens of the echinoid ***Peronella lesueuri*** Agassiz, Shark Bay, western Australia: Journal of Sedimentary Petrology, v. 44, p. 60-69.

Malek-Aslani, M., 1973, Environmental modeling: A useful exploration tool in carbonates: GCAGS Transactions, v. XXIII, p. 239-244.

Murray, G.E., 1961, Geology of the Atlantic and Gulf Coastal Province of North America: New York, Harper and Bros., 692 p.

Perkins, R.D., and Halsey, S.D., 1971, Geological significance of microboring fungi and algae in Carolina shelf sediments: Journal of Sedimentary Petrology, v. 41, p. 843-853.

Pollard, C.O., Jr., 1971, Semidisplacive mechanism for diagenetic alteration of montmorillonite layers to illite layers: GSA Special Paper 134, Appendix, p. 79-93.

Powers, M.C., 1959, Adjustment of clays to chemical change and the concept of the equivalence level, ***in*** Clays and clay minerals, Vol. 2: New York, Pergamon Press, p. 309-326.

_______ 1967, Fluid-release mechanisms in compacting marine mudrocks and their importance in oil exploration: AAPG Bull., v. 51, p. 1240-1254.

Rightmire, C.T., Pearson, F. J., Jr., Back, W., Rye, R.O., and Hanshaw, B.B., 1974, Isotope techniques in ground-water hydrology: Vienna, Austria, Proceedings of a Symposium, v. 2, p. 91-207.

Shinn, E.A., 1969, Submarine lithification of Holocene carbonate sediments in the Persian Gulf: Sedimentology, v. 12, p. 109-144.

Siedlecka, A., 1972, Length-slow chalcedony and relics of sulfates—Evidences of evaporitic environments in the Upper Carboniferous and permian beds of Bear Island, Svalbard: Journal of Sedimentary Petrology, v. 42, p. 812-816.

Stanley, S.M., 1966, Paleoecology and diagenesis of Key Largo Limestone, Florida: AAPG Bull., v. 50, p. 1927-1947.

Stehli, F.G., and Hower, J., 1961, Mineralogy and early diagenesis of carbonate sediments: Journal of Sedimentary Petrology, v. 31, p. 358-371.

Steinen, R.P., 1974, Phreatic and vadose diagenetic modification of Pleistocene limestone: Petrographic observations from subsurface of Barbados, West Indies: AAPG Bull., v. 58, p. 1008-1024.

Steinen, R.P., and Matthews, R.K., 1973, Phreatic vs. vadose diagenesis: Stratigraphy and mineralogy of a cored borehole on Barbados, West Indies: Journal of Sedimentary Petrology, v. 43, p. 1012-1020.

Stricklin, F.L., Jr., and Smith, C.I., 1973, Environmental reconstruction of a carbonate beach complex, Cow Creek (Lower Cretaceous) Formation of Central Texas: GSA Bull., v. 84, p. 1349-1368.

_______ , Smith, C.I., and Lozo, F.E., 1971, Stratigraphy of Lower Cretaceous Trinity deposits of Central Texas: Univ. Texas, Austin, Bureau of Economic Geology Report of Investigations 71, 63 p.

Tolman, C.F., 1937, Ground water: New York, McGraw-Hill, 593 p.

Tucker, D.R., 1962a, Subsurface Lower Cretaceous stratigraphy, Central Texas: Univ. Texas, Austin, Ph.D. dissertation, 137 p.

_______ 1962b, Subsurface Lower Cretaceous stratigraphy, Central Texas, contributions to the geology of South Texas: South Texas Geological Society, p. 177-216.

Wagner, G., 1913, Stylolithen and Drucksuturen: Geol. Paleontol. Abhandl. (N.F.), v. 11, p. 101-128.

Young, Keith, 1974, Lower Albian and Aptian (Cretaceous) ammonites of Texas, ***in*** Perkins, B.F., ed., Geoscience and man, Vol. VIII, Aspects of Trinity Division geology: Louisiana State Univ., School of Geoscience, p. 175-228.

CONTRASTS IN CEMENTATION, DISSOLUTION, AND POROSITY DEVELOPMENT BETWEEN TWO LOWER CRETACEOUS REEFS OF TEXAS

C. A. Achauer[1]

ABSTRACT

The influence of cementation and dissolution on porosity is investigated by comparing two reefs with significantly different diagenetic histories. Reefs of the subsurface Sligo Formation in South Texas are buried to depths of 15,000 to 20,000 feet in a narrow belt along a shelf edge of regional extent. Cores from five wells distributed along 225 miles of the shelf edge show that porosity in the reef and backreef facies is persistently occluded by (1) radiaxial fibrous calcite, a cement whose origin appears to be related to the replacement of a synsedimentary marine cement, and/or (2) coarse calcite mosaic, a cement introduced later in the diagenetic sequence during a time of basin subsidence. Calcite cements of this type and sequence have also been reported to be detrimental to porosity in other ancient shelf-margin reefs. Therefore, the creation of porosity in shelf-margin reefs may depend on processes (dolomitization, fracturing, dissolution) which can offset the influence of cementation.

In contrast to the Sligo, large patch reefs in the outcropping lower Glen Rose Formation in south-central Texas record a simpler diagenetic history which preserved porosity. One reef, selected for detailed study, has experienced only one phase of diagenesis—namely, early subaerial exposure which created much moldic porosity and precipitated a single generation of calcite cement. After this exposure, no additional cements were introduced, despite burial beneath a few thousand feet of younger Cretaceous marine sediments and reexposure of the reef to subaerial weathering from Miocene to Recent time. Thus, the key to understanding porosity preservation in the Glen Rose reef lies in the role of paleohydrologic and geochemical conditions during burial and reexposure of the reef. Whether this can be predicted is a matter for future research.

[1]Atlantic Richfield Company, Dallas, Texas.

INTRODUCTION

In Lower Cretaceous limestone reefs of Texas, diagenesis of the reef rock reduced or preserved porosity. For example, Bebout (1974) and Bebout and Loucks (1974a, 1974b) have shown that porosity in the Stuart City reef trend of South Texas was significantly reduced by cementation. On the other hand, cementation within reefs of the James Limestone of East Texas was a relatively minor event and, consequently, porosity was preserved (Achauer, 1974). These opposing cases have inspired an investigation of factors which lead to reduction or preservation of porosity in two Lower Cretaceous reefs of Texas.

In order to define and appraise the factors which influence porosity, limestone reefs in the Sligo and Glen Rose Formations were selected for study (fig. 1). These particular reefs were chosen because preliminary studies demonstrated that porosity in Sligo reefs appeared to be controlled by cementation, whereas porosity in the Glen Rose appeared to be little affected by cementation. In fact, much of the primary and secondary porosity in the Glen Rose reef is still preserved—a fact that becomes immediately clear at the outcrop.

The primary objectives of this study were twofold: (1) to make petrographic examinations of core and outcrop samples for data pertaining to the diagenetic sequence and to the control of diagenesis on porosity, and (2) to obtain information on the distribution of pore-filling cements so that some idea of their areal control on porosity could be achieved. The results of these investigations are incorporated in this paper.

Two critical problems confronted in this investigation were the origin of pore-filling cements and the explanation of the conditions which preserve porosity. Neither of these problems was fully resolved. However, this paper does attempt to discuss the nature of the problems and to formulate a plan for further investigation.

SLIGO REEFS

General Geological Setting

Subsurface reefs of the Sligo Formation in South Texas are found at depths of 15,000 to 20,000 feet in a narrow belt along a shelf edge of regional extent. The location of the shelf edge is determined by well control (fig. 2). Near the shelf edge, up to 2,000 feet of reef and backreef facies have been penetrated (fig. 3). At present, however, well control is not sufficient for accurate delineation of reef facies versus backreef facies in the vicinity of the shelf edge.

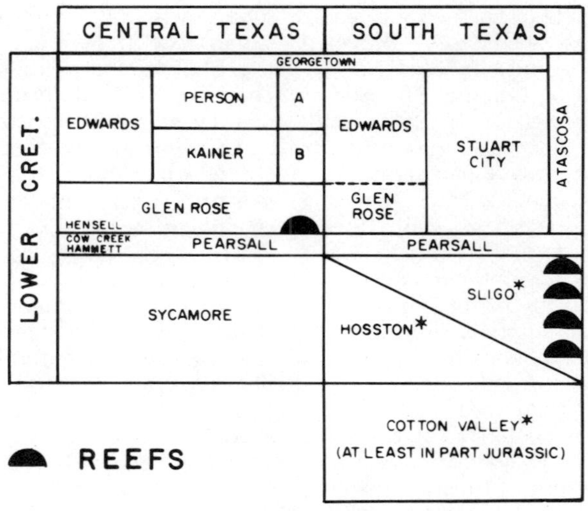

Figure 1. Diagram showing stratigraphic position of Sligo and Glen Rose reefs compared in this paper. Adapted from Bebout and Loucks (1974b).

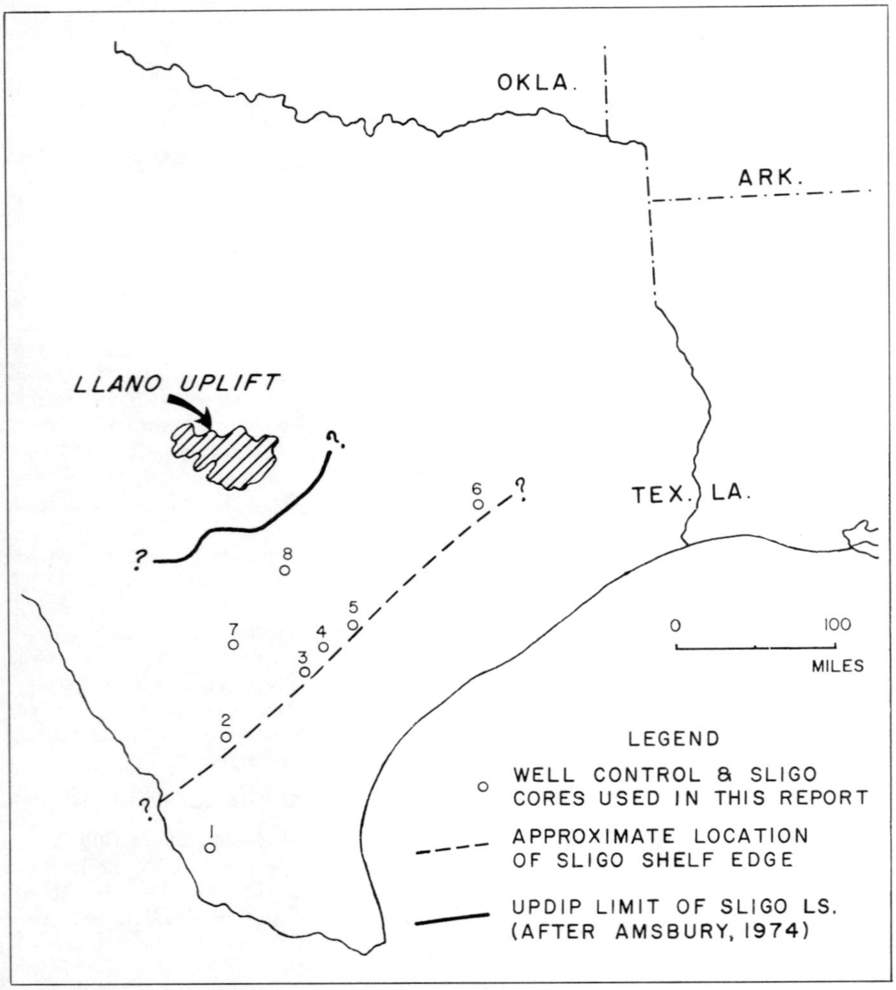

Figure 2. Map of Sligo shelf edge and core control in South Texas. Well numbers designate the following control points: (1) Gulf No. 1 Volpe, Zapata County, (2) Gulf No. 1 Friedrichs, Duval County; (3) Shell No. 1 O'Neal, Bee County; (4) Sun No. 1 Handy, Karnes County; (5) Shell No. 1 Brown, De Witt County; (6) Shell No. 1 Chapman, Waller County; (7) Humble No. 46 Pruitt, Atascosa County; (8) Pan Am No. 1 Schmidt, Guadalupe County.

Facies

The optimum development of reef facies along the Sligo shelf edge is seen in cores of the Shell No. 1 Brown, De Witt County and the Gulf No. 1 Friedrichs, Duval County. The Shell No. 1 Brown well is of particular importance because intermittent coring demonstrates a persistence of reef facies throughout 2,000 feet of the Sligo. In the Brown well, three principal reef facies are recognized (fig. 4): (a) boundstone comprised of caprinids, corals, hydrozoa, algae, and encrusting foraminifers(?); (b) rudistid and/or caprinid boundstone; and (c) reef-derived skeletal grainstone.

Backreef facies near the Sligo shelf edge are characterized by skeletal wackestone- and grainstone-bearing miliolid foraminifers, micrite pellets and lumps, and algal fragments. These particular facies are judged to be in close proximity to the main reefs because they are interbedded with subordinate units of reef-derived grainstone, packstone, and wackestone (fig. 5). The repetitious, but not necessarily cyclic, interbedding between reef facies and the subordinate backreef facies is shown on figure 6.

North of the Sligo shelf edge, only a general picture of facies and facies changes is available (fig. 3). Amsbury (1974) recorded nearshore facies changes from Sligo carbonates to sandstones and conglomerates, that is, a change from marine to terrestrial environment, in outcrops along the flanks of the Llano uplift. Between this transition zone and the Sligo shelf edge, lithologic data from a single well (Humble No. 46 Pruitt, Atascosa County) show the Sligo to consist of thinly interbedded lime mudstone, miliolid-bearing wackestone, microcrystalline dolomite, and oolitic grainstone. In addition, several very thin beds of algal-laminated dolomite, intraformational breccia, and nodular anhydrite of probable tidal-flat origin are present. Thus, in a general way, it appears that a broad shelf, fluctuating between subtidal and tidal-flat environments, extends from the Sligo shelf edge to the outcrop area of the Llano uplift. Many details of facies changes and facies distribution on this shelf need to be more fully documented.

Cementation Sequence

Four generations of cements have been observed in Sligo reefs, but by far the major reduction of pore space is accomplished by two generations: radiaxial fibrous calcite and coarse calcite mosaic. Radiaxial fibrous calcite and coarse calcite mosaic clearly exceed other cement generations in terms of relative abundance (fig. 6). In addition, thin sections (fig. 7) repeatedly show that most, if not all, primary pore space in the reef rock is filled by one or a combination of these two cements. Hence there is little doubt that radiaxial calcite and coarse calcite mosaic control porosity, and consequently, the origin and distribution of these cements become of fundamental importance in the attempt to evaluate reservoir potential.

Origin of Radiaxial Fibrous Calcite

Radiaxial fibrous calcite is a very distinctive fabric whose morphologic and optical properties have been described in detail by Bathurst (1971) and Kendall and Tucker (1973). The characteristics which easily distinguish the radiaxial fabric from other calcite fabrics include the fibrous nature of the crystals, the wavy extinction of the crystals under cross polarization, and the cloudy or dirty appearance of the crystals, a condition caused apparently by inclusions (fig. 7).

Opinion is divided over the origin of radiaxial fibrous calcite in ancient carbonates. Bebout and Loucks (1974a, 1974b) point out that radiaxial fibrous calcite fills skeletal moldic pores in the Stuart City reef. In view of this relationship, they contend that radiaxial fibrous calcite may represent a cement precipitated from meteoric waters in the phreatic zone following subaerial exposure of the reef. However, different petrographic and geochemical data have led others to the conclusion that radiaxial fibrous calcite *replaces* a synsedimentary marine cement (Kendall and Tucker, 1973; Davies and Krouse, 1975; Davies, 1977) or that it is capable of replacing metastable carbonate matrix and skeletal material (Ross, 1975).

In Sligo reefs, probably the most significant observation about radiaxial fibrous calcite is that it is restricted to the reef grainstone and boundstone at the shelf margin. This restricted occurrence poses these questions:

(1) Does the restriction mean that only the reefs along the shelf margin were subjected to subaerial exposure which led to precipitation of radiaxial fibrous calcite in the vadose or phreatic zones?

(2) Does the restriction mean that radiaxial fibrous calcite replaced a metastable, synsedimentary, carbonate cement which formerly was restricted to the reef facies at the shelf margin?

In regard to subaerial exposure, the radiaxial fibrous calcite was not precipitated in the vadose zone mainly because this calcite does not have the morphological characteristics of a vadose cement. Besides, the radiaxial fibrous calcite postdates the few occurrences of authentic vadose calcite in Sligo reefs (fig. 7). Alternatively, the radiaxial fibrous calcite may have been precipitated in the fresh-water phreatic zone. However, radiaxial fibrous calcite is not described or referred to as a meteoric water cement by several workers who have examined cements of the subaerial environment (Friedman and Kolesar, 1971; Matthews, 1971; Land, 1971; Multer, 1971; Nelson, 1971; and Supko, 1971).

Three reasons exist for considering that the radiaxial cement represents replacement of a synsedimentary marine cement. First, radiaxial fibrous calcite may represent a replacement fabric after an earlier or synsedimentary marine cement, as suggested by Kendall and Tucker (1973), Davies and Krouse (1975), and Davies (1977). Second, earlier cement may have been restricted to the shelf margin, in view of the fact that the radiaxial fibrous calcite is restricted to the shelf margin. Third, because a generation of coarse calcite mosaic postdates radiaxial fibrous calcite, an episode of ground-water movement may have resulted in the creation of two cementation fabrics:

(1) Initially the ground-water movement led to replacement of an earlier marine cement.

(2) After replacement, continuation of ground-water movement led to the direct precipitation of a coarse calcite mosaic.

These rationalizations set the stage for further petrographic and geochemical investigations (similar to those conducted by Davies, 1977) aimed at seeking evidence for the replacement origin and evidence that the ground water causing precipitation of the coarse calcite mosaic also caused replacement of an earlier marine cement. These further investigations are required before making final conclusions regarding the origin of radiaxial fibrous calcite in Sligo reefs.

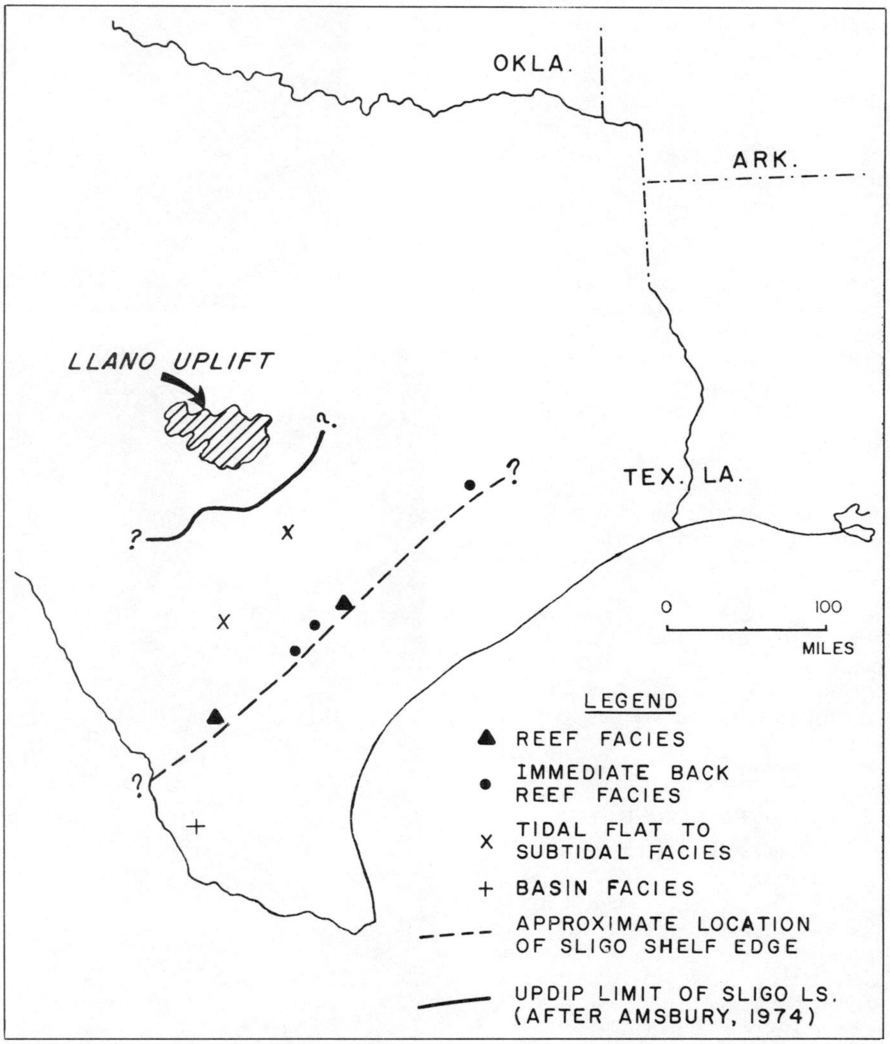

Figure 3. Generalized map showing location of facies control, Sligo limestone, South Texas.

Origin of Coarse Calcite Mosaic

Coarse calcite mosaic postdates the radiaxial fibrous calcite and is characterized by an aggregate of fine to coarse, anhedral to subhedral, calcite crystals. Staining of this cement by alizarin red S and potassium ferricyanide reveals that it is consistently poor in ferrous iron. Thus, the combination of texture, position in the cementation sequence, and paucity of ferrous iron greatly aids in the recognition of this cement generation everywhere along the Sligo reef trend.

The coarse calcite mosaic was precipitated following the formation of fractures in the reefs and prior to hydrocarbon migration into the reefs (fig. 8). This timing is revealed by coarse calcite mosaic filling fractures and by bitumen postdating coarse calcite mosaic in a fracture.

Distribution of Cements and Their Regional Influence on Porosity

The distribution of cements in five wells along 225 miles of the Sligo shelf edge (fig. 9) shows that both radiaxial fibrous calcite and coarse calcite mosaic are found in the reef facies, whereas only coarse calcite mosaic is found in the backreef facies. The net effect of these cements is that they substantially reduce porosity in each of the wells. Comparing petrographic data on cements with available core analyses of porosity and permeability of reef and backreef facies confirms this effect. Therefore, the implication is strong that radiaxial fibrous calcite and coarse calcite mosaic represent regional cementation events and may exert a regional control on porosity.

The reduction of porosity at ancient shelf margins by radiaxial fibrous calcite or a combination of radiaxial fibrous calcite and coarse calcite mosaic has also been reported by Bebout and Loucks (1974b), Walls (1976), and Davies (1977). These cases add further testimony that shelf margins can serve as preferred sites for extensive cementation.

For good reservoir conditions to occur in the Sligo reefs, the apparent regional effect of cementation must be offset by processes capable of creating porosity. These processes might include dolomitization, dissolution (leaching), and fracturing. Whether any of these processes have been imposed on Sligo reefs is a matter for further study. Nevertheless, it would be best to examine the possibilities of porosity creation in the Sligo reefs with these processes in mind.

GLEN ROSE REEF

General Geological Setting

In the Hill Country of south-central Texas, a northeast-trending belt of outcropping reefs occurs in the lower part of the Lower Cretaceous Glen Rose Formation (fig. 1). One of these reefs, which is spectacularly exposed along Blanco River in Hays County, has been selected for study. This reef is of particular interest and significance because it is one of the few reefs of Lower Cretaceous age composed dominantly of coral to be exposed on the North American Continent. Previous studies of this reef have dealt mainly with its paleontology (Wells, 1932) and its morphology and stratigraphic setting (Lozo and Stricklin, 1956).

The lower Glen Rose reef appears to be situated on a shallow-water shelf somewhat intermediate between the strandline and the shelf edge (fig. 10). As defined by Perkins (1974), the reef belt is separated from the strandline to the north by 10 to 40 miles of very shallow-water, molluscan limestones. To the south (or seaward), the reef belt is displaced into the subsurface by the Balcones fault. However, the southern edge of the reef belt may roughly correspond with the Balcones fault zone because several miles south-east of the fault zone, in Caldwell County, the equivalent of the reef interval is marked by normal Glen

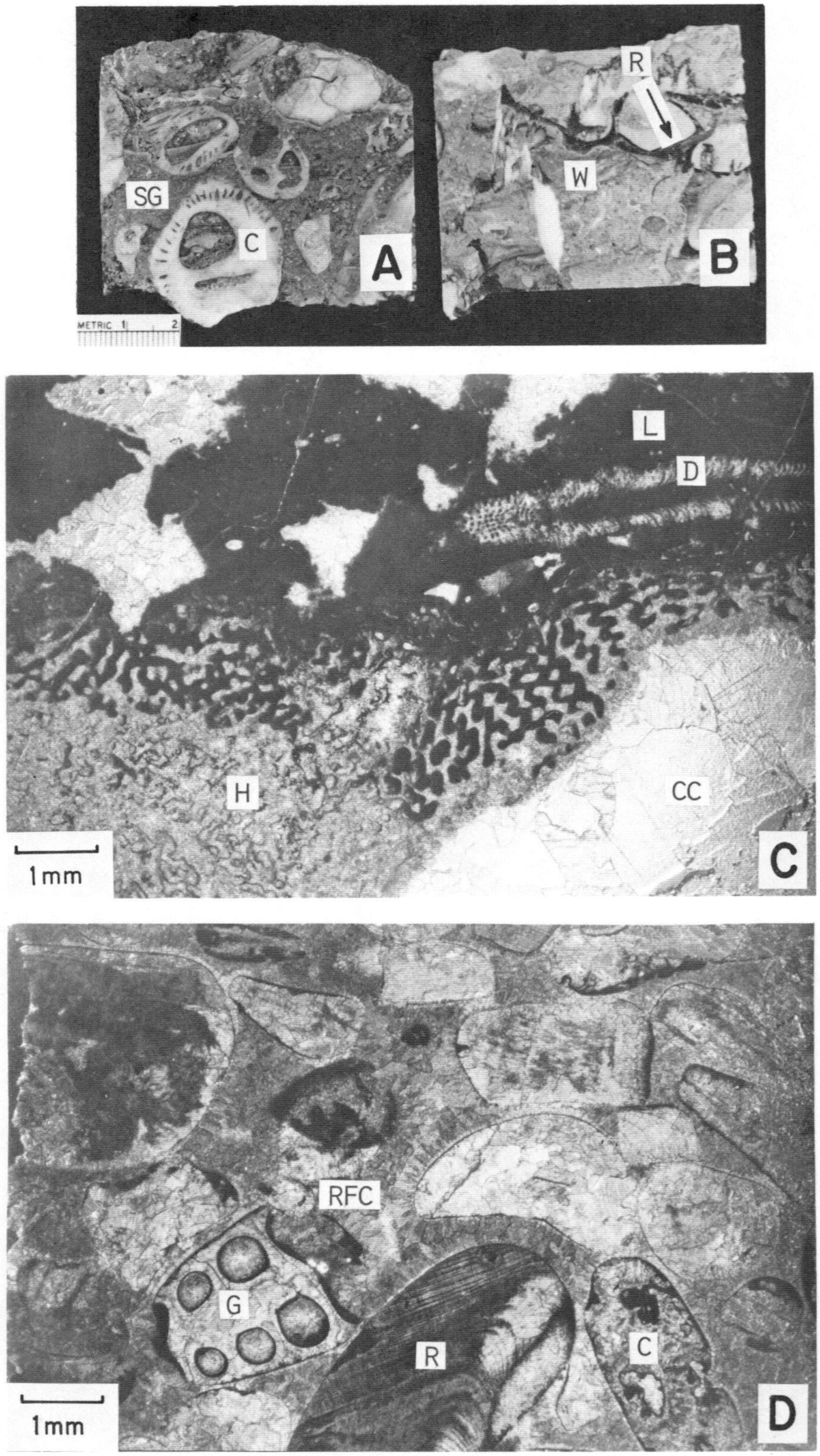

Figure 4. Photographs and photomicrographs of typical Sligo reef facies.

A. Caprinid boundstone—closely crowded, whole caprinids (C) in a matrix of skeletal grainstone (SG); Shell, Brown No. 1, 16,024 feet.

B. Rudistid boundstone—intergrown rudistid shells (R) in a wackestone matrix (W); Shell, Brown No. 1, 17,459 feet.

C. Hydrozoa-coral boundstone—partly recrystallized hydrozoa colony (H), a dasycladacean (algal) stem segment (D), lime mud (L), and coarse calcite mosaic (CC); plane light; Shell, Brown No. 1, 17,200 feet.

D. Reef-derived skeletal grainstone—rounded and abraided skeletal debris includes gastropod (G), rudistid (R), and coral (C) cemented by radiaxial fibrous calcite (RFC); plane light; Gulf, Friedrichs No. 1, 15,527 feet.

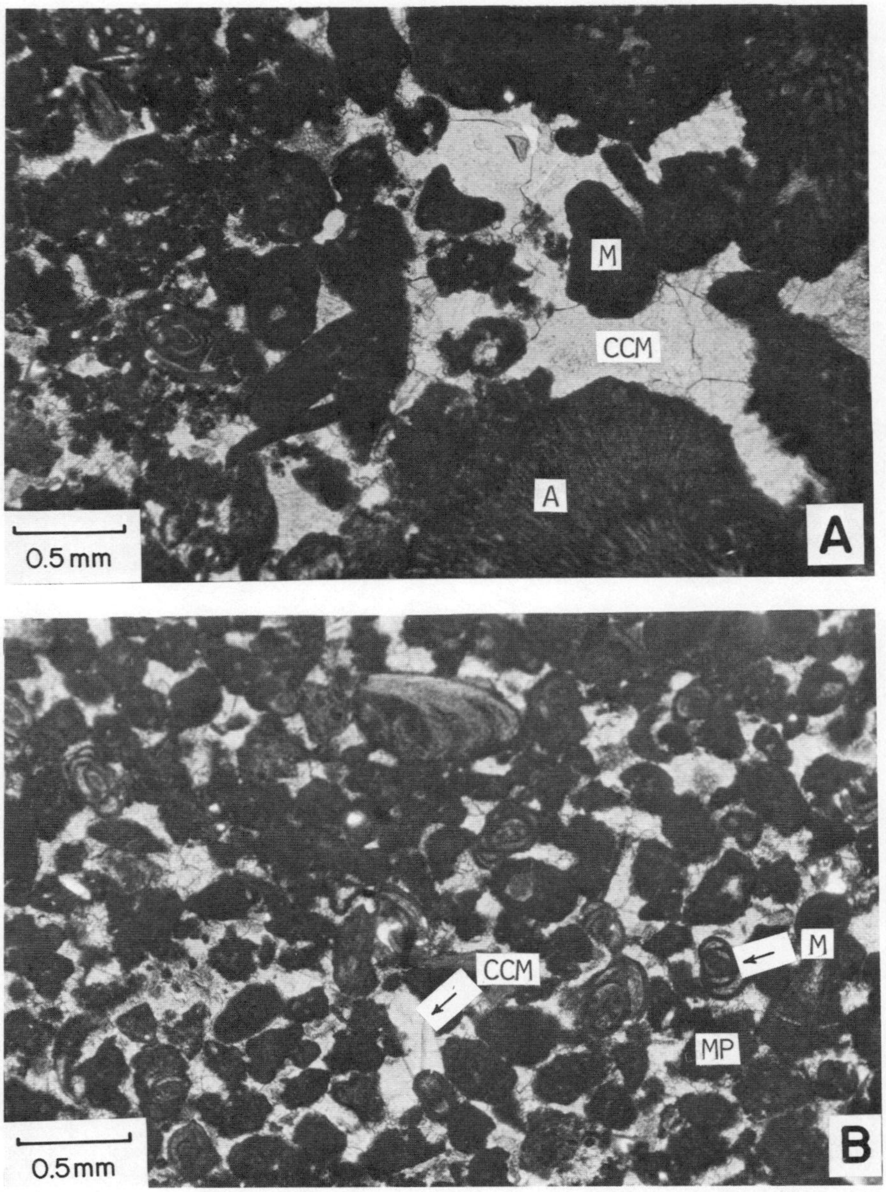

Figure 5. Photomicrographs of typical Sligo backreef facies.

A. Pelletal grainstone—with micritic pellets (M), and algal fragments (A) cemented by coarse calcite mosaic (CCM); plane light; Sun, Handy No. 1, 16,156 feet.

B. Miliolid pellet grainstone—with miliolids (M) and micritic pellets (MP) cemented by coarse calcite mosaic (CCM); plane light; Sun, Handy No. 1, 16,156 feet.

Rose shelf facies of wackestones bearing large clams, oysters, and the large foram, ***Orbitolina***. Thus, the Glen Rose shelf extends several miles to the southeast of the Balcones fault belt, and consequently, the lower Glen Rose reefs occupy an interior shelf position.

Facies

The vertical facies sequence (fig. 11) of the Glen Rose reef at "The Narrows" on Craft Ranch consists, from bottom to top, of the following facies: (1) massive coral boundstone with coral colonies preserved in growth position, (2) very thin rudistid boundstone in which the rudistids are attached to each other in growth position, (3) transitional zone of interbedded coral boundstone and reef-derived skeletal grainstone with whole shells of ***Chondrodonta*** (a pelecypod) and ***Nerinea*** (a gastropod), and (4) bedded, reef-derived skeletal grainstone whose bedding planes consistently dip at a low angle to the southeast. Evidently, the dips in facies (4) are depositional because dips on bedding planes in the overlying limestones are nearly horizontal. The top of facies (4) is marked by a calcrete zone (a form of caliche). This zone, about 1 foot thick, is not readily recognizable at the outcrop. Its position and significance were recognized from thin section studies of very closely spaced samples through the upper part of facies (4). The calcrete zone is overlain sharply by a partly dolomitized lime mudstone and wackestone of marine origin belonging to the Glen Rose Formation. Therefore, the caliche zone at the top of facies (4) is a paleocaliche of Lower Cretaceous age.

The vertical facies sequence of the Glen Rose reef definitely implies a gradual emergence of the reef from subtidal conditions to subaerial exposure through at least three stages. During the first stage, the coral boundstone, the rudistid boundstone, and the transitional zone were deposited under shallow-water, high-energy conditions conducive to the development of a reef. During the second stage, it appears that the reef was gradually buried by reef-derived grainstones. There is a strong possibility that these grainstones were deposited in a beach setting, judging primarily from the preservation of a low-angle depositional dip in the seaward direction. The third stage, marked by the paleocaliche (calcrete) zone, signifies the exposure of the reef to subaerial weathering. This period of subaerial weathering led to the development of much moldic porosity, a porosity which is still preserved.

Subaerial Diagenesis and the Development of Moldic Porosity

Subaerial diagenesis of the Glen Rose reef has left two significant diagenetic imprints: the formation of a calcrete zone at the top of the reef and the creation, by dissolution, of much skeletal moldic porosity, especially in reef-derived skeletal grainstone. Both of these diagenetic imprints have been described from Pleistocene and Quaternary carbonates whose history of subaerial exposure and fresh-water diagenesis is firmly documented.

The calcrete zone at the top of the Glen Rose reef bears a strong resemblance to calcretes formed during subaerial exposure of Quaternary carbonates. The Quaternary calcretes, as described by Seisser (1973), were formed at the top of eolanites by the precipitation of micrite as coatings around skeletal grains. Sometimes the micrite envelopes two or more grains, thereby producing a composite grain which looks like grapestone. According to Seisser, the micrite coatings or envelopes were precipitated in a zone at the top of the eolanites as a result of a combination of processes including dissolution, evaporation, and soil suction. Practically the same type of coatings and envelopes seen in the Quaternary calcrete are also found in the calcrete at the top of the Glen Rose reef (fig. 12).

Additionally, one of the very common results of fresh-water movement through metastable carbonate sands is the extensive dissolution of aragonitic skeletal fragments or oolitic grains, leaving moldic pore space. This solution phenomenon has been noted, especially in Pleistocene reefs where the record of fresh-water diagenesis is well documented (Matthews, 1967; Steinen, 1974; Friedman, 1975). The same moldic fabric, so prevalent in Pleistocene reefs, is also well developed in the Glen Rose reef (fig. 12). Thus, the moldic porosity in the reef together with the calcrete zone at the top of the reef present direct evidence that the Glen Rose reef was exposed to subaerial weathering and fresh-water diagenesis not long after deposition.

Post-Depositional Controls on the Preservation of Porosity

Following subaerial weathering, the Glen Rose reef was buried about 3,200 feet by Lower Cretaceous shelf carbonates and Upper Cretaceous marl, chalk, and shale. In Miocene time, the Glen Rose reef commenced to be uplifted as a result of movement along the Balcones fault system (Young, 1972). This set the stage for post-Miocene removal by erosion of much of the Lower and Upper Cretaceous strata above the reef. Eventually the Glen Rose reef was reexposed to subaerial weathering, which has continued to the present time. Despite this post-depositional history of burial and reexposure through nearly 108 million years of geological time, no additional cements were introduced into the reef. Thus the moldic porosity, created largely in Lower Cretaceous time, remains preserved.

The conditions leading to porosity preservation in the Glen Rose reef are largely speculative, but they must relate to the post-depositional history. For instance, water capable of precipitating carbonates might be expected to have moved through the buried reef either as a result of the compaction of surrounding sediments or by downdip movement of meteoric waters. How-

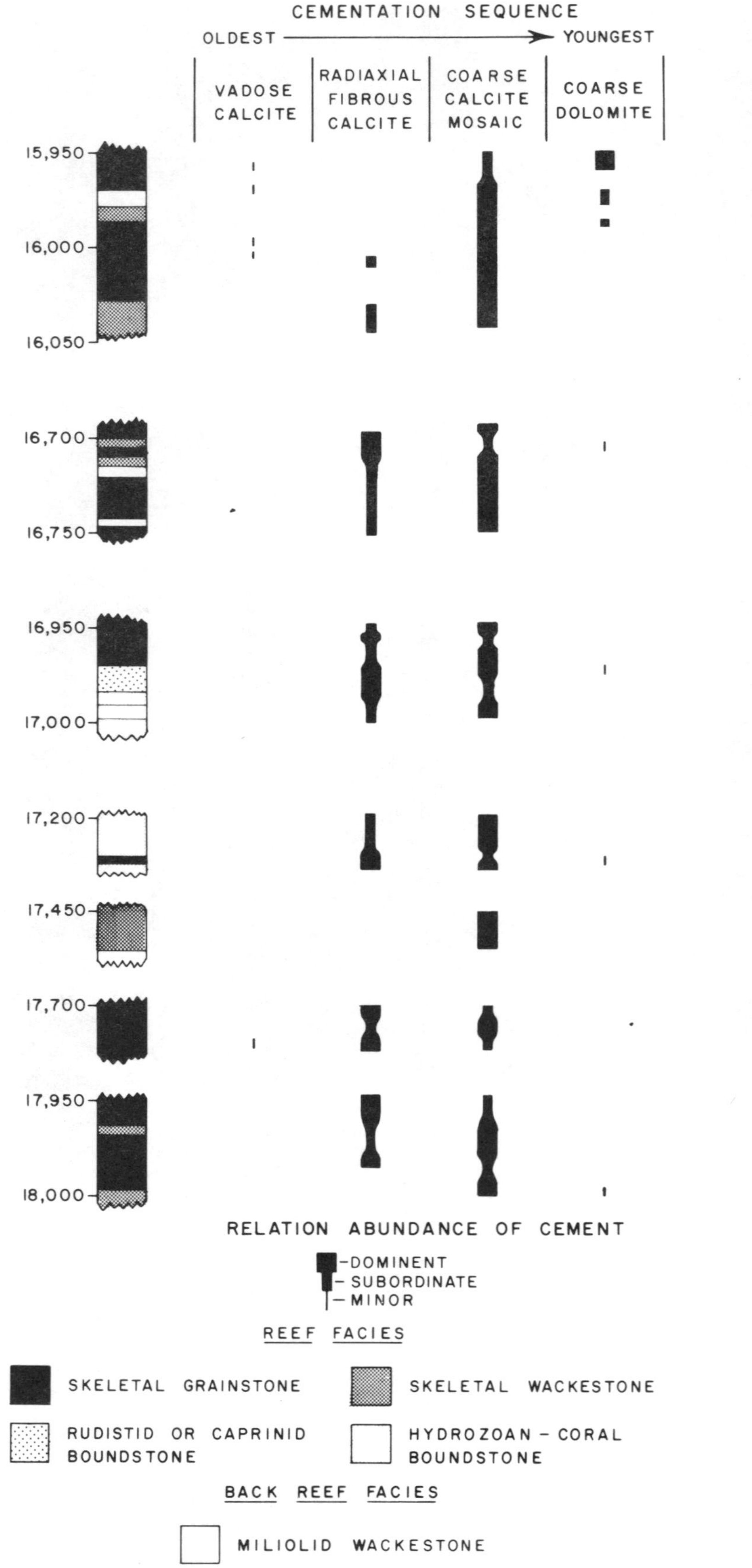

Figure 6. Graphic representation of facies and cementation in Sligo cores of Shell, Brown No. 1, De Witt County.

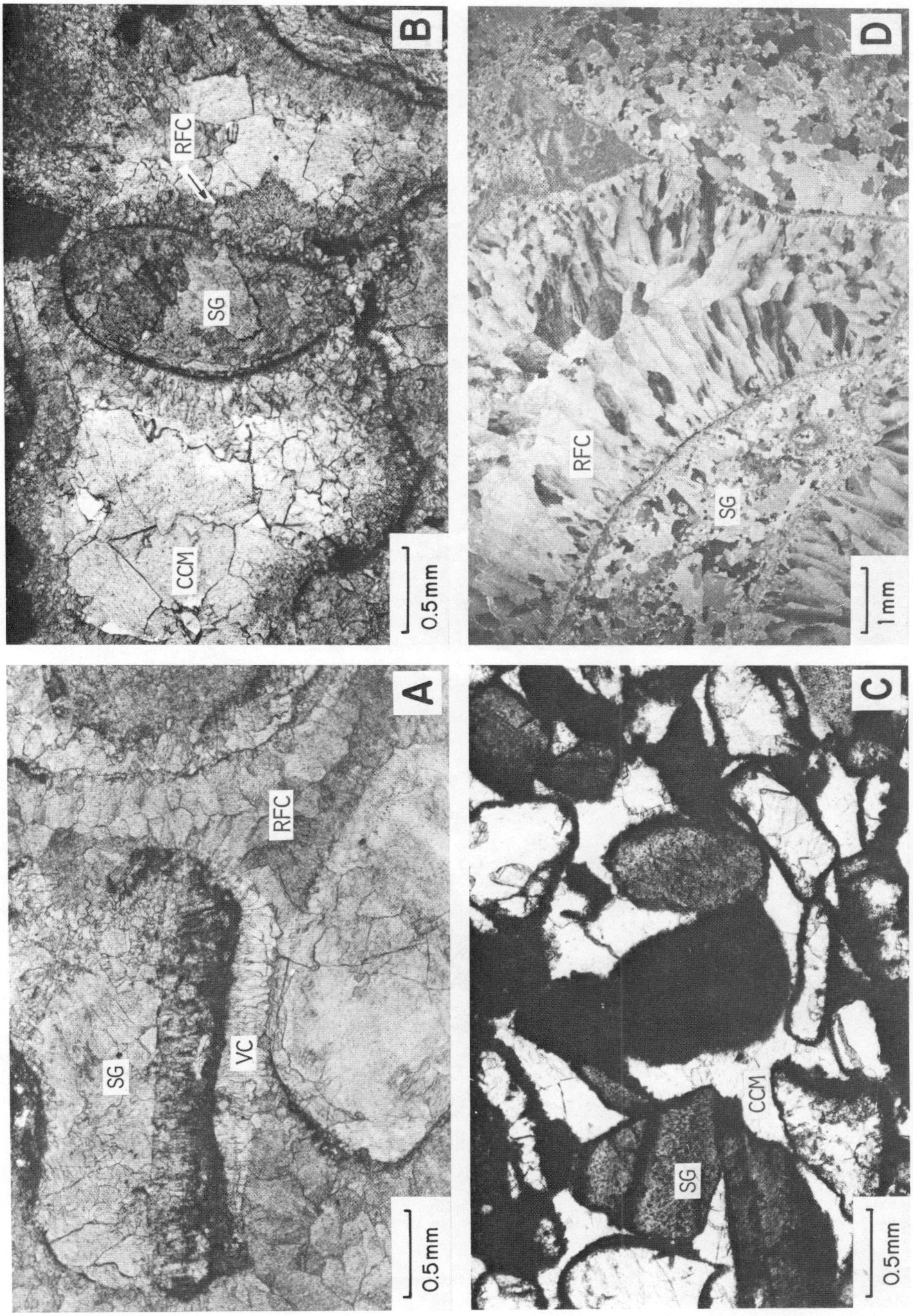

Figure 7

Figure 7. Photomicrographs of cementation sequence and fabrics in Sligo reef facies. (page 133)
A. Recrystallized skeletal grains (SG). The oldest cement is clear, vadose calcite (VC) in asymmetric form on the underside of a skeletal grain. The youngest cement is a pervasive, cloudy, radiaxial fibrous calcite (RFC); plane light; Shell, Brown No. 1, 17,720 feet.
B. Recrystallized skeletal grains (SG). The oldest cement is cloudy, radiaxial, fibrous calcite (RFC). The youngest cement is coarse calcite mosaic (CCM); plane light; Shell, Brown No. 1, 16,952 feet.
C. Skeletal grains (SG) exclusively cemented by coarse calcite mosaic (CCM); Shell, Brown No. 1, 16,720.5 feet.
D. Recrystallized skeletal grains (SG) exclusively cemented by radiaxial fibrous calcite (RFC). The alternating dark and light crystals reflect the undulose extinction under crossed Nicols; Shell, Brown No. 1, 16,979 feet.

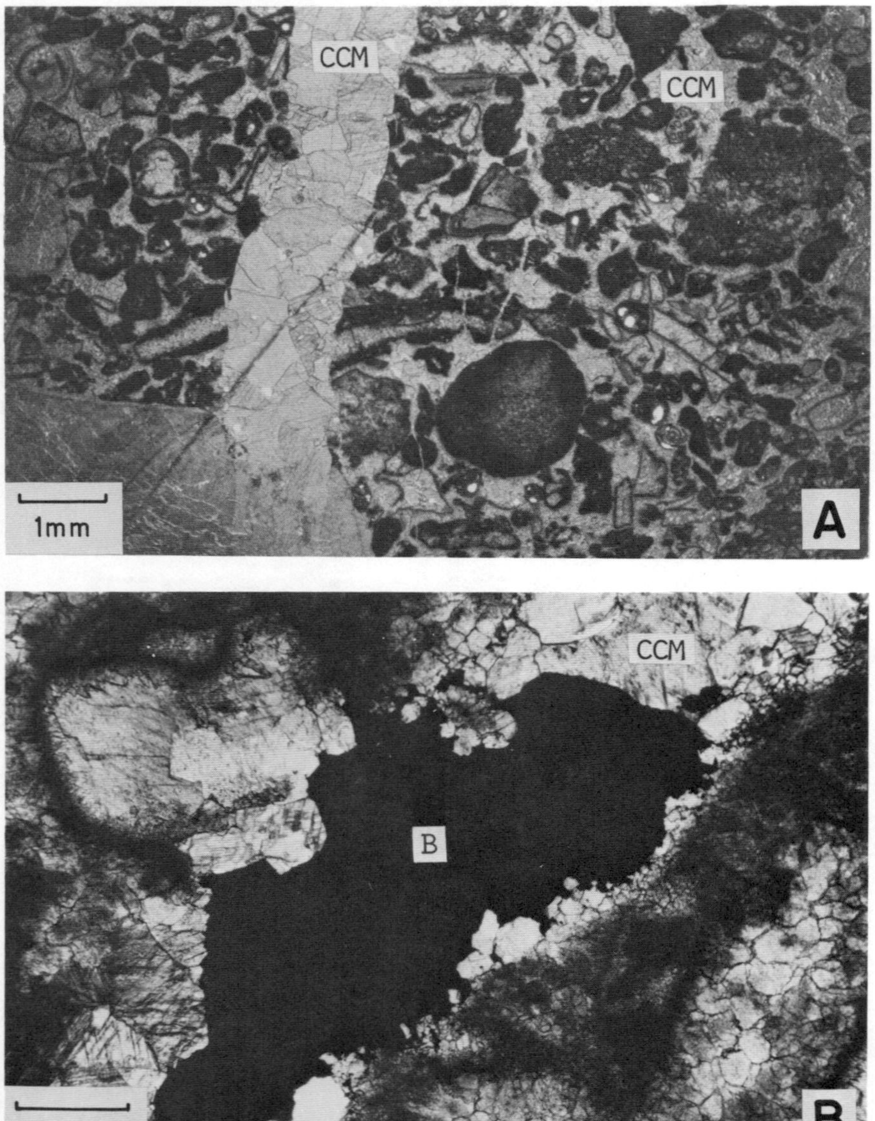

Figure 8. Photomicrographs of cementation fabrics and sequence in fractures of Sligo reef facies.
A. Skeletal grainstone. Note filling of fracture and cementation of skeletal grains by coarse calcite mosaic (CCM); plane light; Shell, Brown No. 1, 17,975 feet.
B. Enlarged view of fracture filling. The oldest fill in the fracture is coarse calcite mosaic (CCM) and the youngest fill is bitumen (B); plane light; Shell, Brown No. 1, 16,018 feet.

ever, neither of these conditions may have occurred to any significant degree if early lithification prevented compaction in the surrounding rocks and if the surrounding sediments were too impermeable for downdip water movement. Further examination and mapping of the compactibility and permeability of the surrounding rocks might show that a lack of ground-water movement during the burial phase may be the reason that additional carbonate cements were not precipitated in the reef. On the other hand, water may have moved through the reef in its burial history, but the chemical conditions were simply not conducive to the precipitation of a cement. It follows, therefore, that the key to a better understanding of porosity preservation lies in the investigation of paleohydrologic and geochemical conditions of the Glen Rose reef in its post-depositional history. Hopefully, in future studies, a greater emphasis will be given to the post-depositional history of carbonates so that porosity preservation can be more fully appraised and evaluated.

CONCLUSIONS

1. The reduction of porosity in Sligo reefs along the Sligo shelf edge is caused by cementation. Radiaxial fibrous calcite and coarse calcite mosaic are the principal cements involved in this porosity reduction.
2. The origin of radiaxial fibrous calcite cannot be resolved with certainty. However, in the case of the Sligo, the restriction of radiaxial fibrous calcite at the shelf margin may have a special meaning: radiaxial fibrous calcite may have replaced a syndepositional marine cement which was formerly restricted to the shelf margin. Future studies should test this hypothesis.
3. In a broad time framework, precipitation of the coarse calcite mosaic can be placed between fracturing of lithified reef rock and the introduction of hydrocarbon into the reefs. Work is needed to more closely determine the timing of the fracture fillings.
4. The processes causing cementation in Sligo reefs appear to be regional in scope. Consequently, other processes capable of ***creating*** porosity must be considered. These processes might include fracturing, dolomitization, and leaching.
5. The preservation of porosity, as exemplified by the Glen Rose reef, appears to be related to paleohydrologic and geochemical conditions operating on the reef during its post-depositional history. The understanding of these conditions is a formidable challenge which must be met if

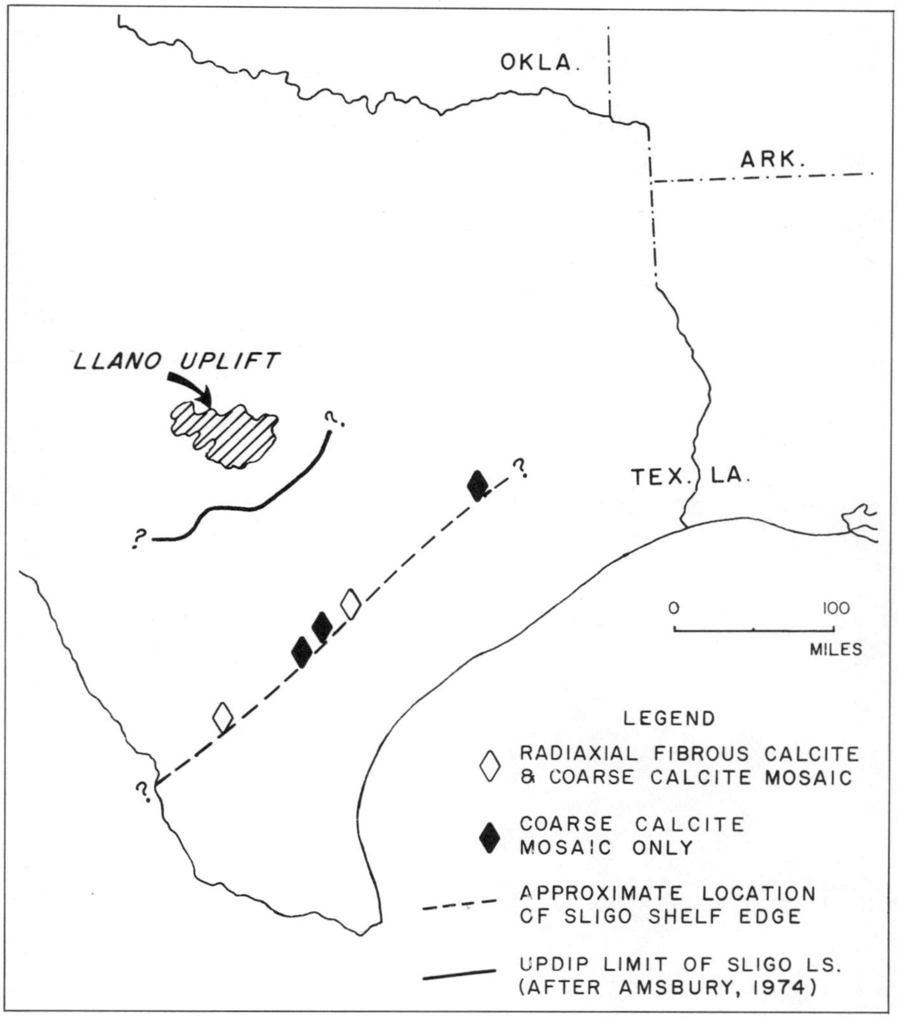

Figure 9. *(left)* Map showing distribution of elements in wells along the Sligo shelf edge, South Texas.

meaningful criteria for predicting the preservation of porosity are to be established.

ACKNOWLEDGMENTS

I would like to thank Alan Craft for permission to conduct field studies on "The Narrows" Ranch. I also appreciate the helpful discussions on various aspects of this study with Gene Martin, Otis Coulson, and R. S. Agatston of Atlantic Richfield Company. The Well Sample and Core Library of the Texas Bureau of Economic Geology was very helpful in locating cores and in providing facilities for sample examination.

REFERENCES

Achauer, C. W., 1974, Deposition and diagenesis of the James limestone (Lower Cretaceous) in the East Texas Basin (abs.): GCAGS Transactions, v. XXIV, p. 210.

Amsbury, D. L., 1974, Stratigraphic petrology of lower and middle Trinity rocks on the San Marcos Platform: Geoscience and Man, v. VIII, p. 1-35.

Bathurst, R. G. C., 1971, Carbonate sediments and their diagenesis: New York, Elsevier Publishing Company.

Bebout, D. G., 1974, Lower Cretaceous Stuart City shelf margin of South Texas: Its depositional and diagenetic environments and their relationship to porosity: GCAGS Transactions, v. XXIV, p. 138-159.

_______, and Loucks, R. G., 1974a, Cementation of carbonate grainstone body—Lower Cretaceous Stuart City Trend, South Texas: AAPG and SEPM Abstracts with Programs, v. 1, p. 4-5.

_______, and Loucks, R. G., 1974b, Stuart City Trend, Lower Cretaceous, South Texas: Univ. Texas, Austin, Bureau of Economic Geology Report of Investigations 78, p. 1-80.

Davies, G. R., 1977, Former magnesium calcite and aragonite submarine cements in Upper Paleozoic reefs of the Canadian Arctic: A summary: Geology, v. 5, p. 11-15.

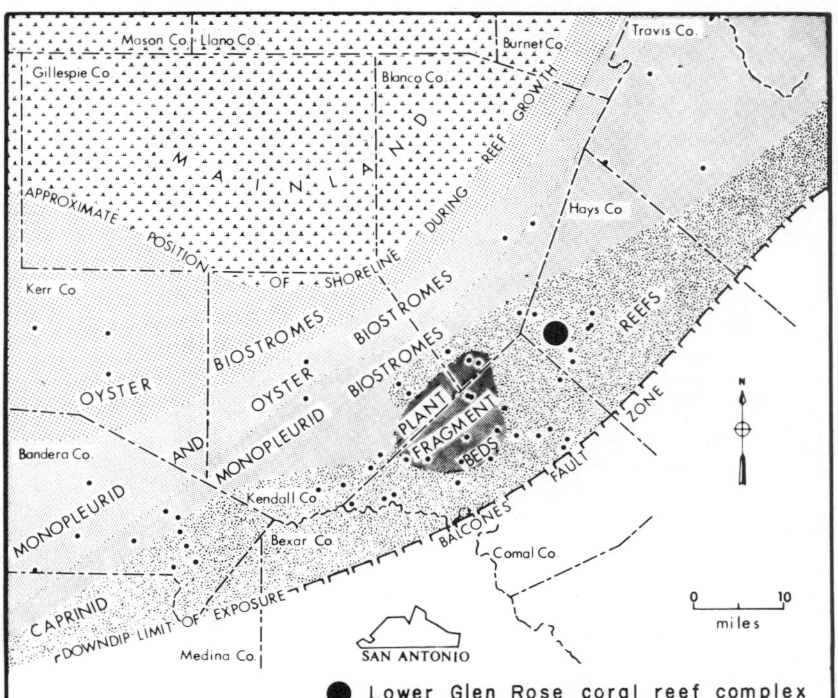

Figure 10. *(left)* Principal facies and paleogeography of the lower Glen Rose reef interval in Central Texas. Adapted from Perkins (1974).

______ , and Krouse, H. R., 1975, Carbon and oxygen isotopic composition of Late Paleozoic calcite cements, Canadian Arctic Archipelago—Preliminary results and interpretation: Canada Geological Survey Report on Activities, pt. B, Paper 75-1B, p. 215-220.

Friedman, G. M., 1975, Address of the retiring president, Society of Economic Paleontologists and Mineralogists: The making and unmaking of limestones or the downs and ups of porosity: Journal of Sedimentary Petrology, v. 45, no. 2, p. 379-398.

______ , and Kolesar, P. T., Jr., 1971, Fresh water carbonate cements, *in* Bricker, O. P., ed., Carbonate cements: Baltimore and London, Johns Hopkins Press, p. 122-126.

Ginsburg, R. N., and James, N. P., 1976, Submarine botryoidal aragonite in Holocene reef limestones, Belize: Geology, v. 4, p. 431-436.

Kendall, A. C., and Tucker, M. E., 1973, Radiaxial fibrous calcite: A replacement after acicular carbonate: Sedimentology, v. 20, p. 365-389.

Land, L. S., 1971, Phreatic versus meteoric diagenesis of limestones: Evidence from a fossil water table in Bermuda, Bricker, O. P., ed., *in* Carbonate cements: Baltimore and London, Johns Hopkins Press, p. 133-136.

Lozo, F. E., and Stricklin, F. L., Jr., 1956, Stratigraphic notes on the outcrop basal Cretaceous, Central Texas: GCAGS Transactions, v. VI, p. 67-78.

Matthews, R. K., 1967, Diagenetic fabrics in biosparites from the Pleistocene of Barbados, West Indies: Journal of Sedimentary Petrology, v. 36, no. 4, p. 1147-1153.

______ 1971, Diagenetic environments of possible importance to the explanation of cementation fabric in subaerially exposed carbonate sediments, *in* Bricker, O. P., ed., Carbonate cements: Baltimore and London, Johns Hopkins Press, p. 127-132.

Multer, H. G., 1971, Drusy mosaic growth in the Miami limestone, *in* Bricker, O. P., ed., Carbonate cements: Baltimore and London, Johns Hopkins Press, p. 137-138.

Nelson, H. F., 1971, Cementation in a Holocene chenier sand, *in* Bricker, O. P., ed., Carbonate cements: Baltimore and London, Johns Hopkins Press, p. 141-142.

Perkins, B. F., 1974, Paleoecology of a rudist reef complex in the Comanche Cretaceous Glen Rose limestone of Central Texas: Geoscience and Man, v. VIII, p. 131-173.

Ross, R. J., Jaanusson, V., and Friedman, I., 1975, Lithology and origin of Middle Ordovician calcareous mudmound at Meiklejohn Peak, southern Nevada: USGS Professional Paper 871, p. 1-48.

Seisser, W. G., 1973, Diagenetically formed ooids and intraclasts in South African calcretes: Sedimentology, v. 20, p. 539-551.

Steinen, R. P., 1974, Phreatic and vadose diagenetic modification of Pleistocene limestone: Petrographic observations from subsurface of Barbados, West Indies: AAPG Bull., v. 58, no. 6, p. 1008-1024.

Supko, P. R., 1971, "Whisker" crystal cement in a Bahamian rock, *in* Bricker, O. P., ed., Carbonate cements: Baltimore and London, Johns Hopkins Press, p. 143-146.

Figure 11. Photographs showing the vertical facies sequence of the Glen Rose reef at "The Narrows," Craft Ranch, Hays County, Texas.

A. General view of the sequence shows the coral reef (CR) overlain with sharp contact by bedded, reef-derived skeletal grainstone (SG).

B. Details of reef zonation—coral zone (C), rudistid zone (R), and transitional zone (T). The zone of reef-derived skeletal grainstone is out of view.

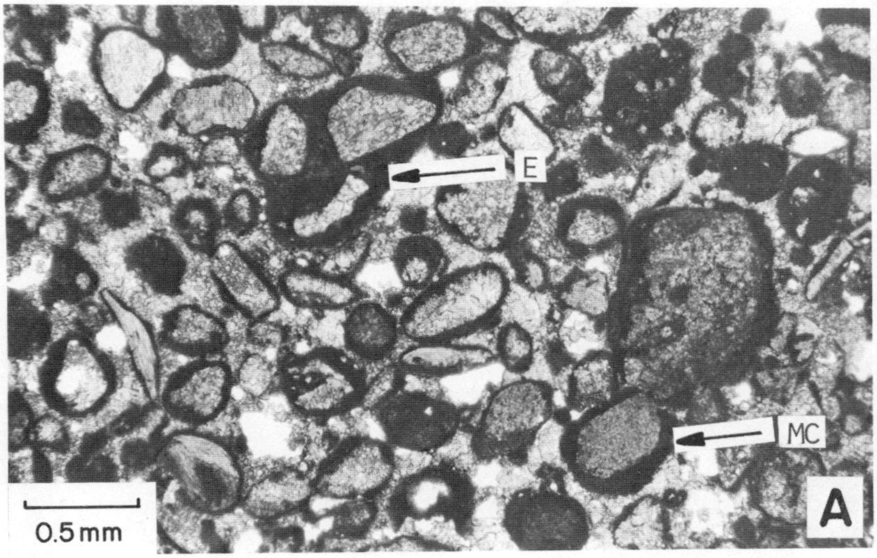

Figure 12. Photomicrographs of subaerial diagenetic features in the Glen Rose reef.

A. Photomicrograph of calcite zone at the top of the reef. Note micritic coatings (MC) on skeletal grains and the envelopment of several grains by micrite (E). These micritic coatings are believed to have been precipitated during subaerial exposure following deposition; plane light.

B. Skeletal grainstone with well-developed skeletal moldic porosity (SMP). This porosity results from the selective leaching of grains by movement of fresh water during subaerial exposure of the reef. At 7.5 feet below the top of the reef. Plane light.

Walls, R. A., 1976, Porosity reduction in reef margins by early cementation, Golden Spike Field, Alberta (abs.): AAPG Bull., v. 60, no. 4, p. 730.

Wells, J. W., 1932, Corals of the Trinity Group of the Comanchean of Central Texas: Journal of Paleontology, v. 6, no. 3, p. 225-256.

Young, K., 1972, Mesozoic history, Llano region, *in* Geology of Llano Region and Austin area: Univ. Texas, Austin, Bureau of Economic Geology Guidebook 13.

DIAGENESIS AND GEOCHEMISTRY OF A GLEN ROSE PATCH REEF COMPLEX, BANDERA COUNTY, TEXAS

Timothy J. Petta[1]

ABSTRACT

Rigid reef framework for the Pipe Creek patch reef complex (lower Glen Rose Formation) was produced by syngenetic rudist accretion, internal sedimentation, and submarine cementation of the reef frame and internal sediment. Clionid sponges bored the reef framework during accretion and produced cavities, calcareous silt, and peloids that can be easily mistaken for vadose features. Local pholad-bored surfaces developed on the reef crest(s) when vertical framework accretion exceeded local subsidence so that truncation and extensive bioerosion of the reef crest(s) occurred in the littoral zone. Small forereef beaches contain littoral cementation features. Submarine diagenesis of backreef beds included peloid induration and grain micritization.

Epigenetic diagenesis that affected the sequence at Pipe Creek is divided into three distinct phases: phase I—marine connate—closed, phase II—early fresh water—open, and phase III—late fresh water—open. During phase I, partial incongruent dissolution of magnesian calcite submarine cements and internal sediments in the caprinid reefs effectively raised the Mg/Ca ratio of the interstitial marine water. This water composition change stimulated dolomitization of clay-rich backreef lime muds by cloudy, 8 to 10 μm, anhedral to subhedral dolomite.

Fresh water began to displace marine connate water either during late Glen Rose or latest Fredericksburg time (phase II). The change from a closed marine to an open fresh-water system caused the final incongruent dissolution of magnesian calcite, partial dolomitization of the sediments by clear, 50 to 60 μm, euhedral dolomite, inversion of some aragonitic mollusks to calcite, and conversion of lime mud to lime mudstone (micrite). As the water became progressively enriched in CO_2, megascale dissolution of aragonitic allochems occurred. Moldic porosity developed during this phase has been preserved by the precipitation of intergranular equant sparry calcite. Clay-rich beds have recrystallized, indicating that clay materials have acted as nuclei for microspar and pseudospar. At the end of this phase, the rocks had been converted from predominately metastable (aragonite and magnesian calcite) to stable (calcite and dolomite) minerals.

Phase III is characterized by changes in rock fabric rather than mineralogy. Fractures and vugs that have developed through the Holocene epoch have been partially filled by bladed to micritic calcite cements that were precipitated in the meteoric phreatic and vadose zones.

Present-day values of some elements, notably Sr, in calcite cements and micrite are relatively low but do reflect original mineralogy. Higher Sr values within internal sediments (micrite), backreef micrite, and recrystallized mollusk fragments indicate an original high-Sr aragonite mineralogy. In addition, lower permeabilities of micrite prevented effective removal or flushing of Sr from the rocks by the modern ground-water system.

All the early fresh-water diagenetic features at Pipe Creek are thought to have evolved during burial. The small amount of diagenesis attributed to subaerial exposure during deposition only affected perireef lime grainstones. Because of pervasive submarine diagenesis, reef beds appear to have much lower permeability than adjacent grainstones, although vuggy porosity is well-developed in the reefs. Many features in Cretaceous rudist reefs that have been attributed to syndepositional meteoric water diagenesis may have been developed during burial. Waters responsible for diagenesis may have flowed downdip from emergent land areas, or in an updip direction preceding or in conjunction with hydrocarbon migration.

INTRODUCTION

This study is concerned with the construction of a diagenetic model for rudist reefs. In the western hemisphere, Cretaceous rudist reefs occur from the West Indies to Baja, California, and from Texas to Venezuela; in the eastern hemisphere, they extend from the western Mediterranean into the western Himalayas (Perkins, 1974). Subsurface rudist reefs and associated sediments that comprise the Golden Lane and Poza Rica Trends in Mexico are important hydrocarbon reservoirs (Coogan and others, 1972). Shelf-margin reefs across South Texas have been the target of extensive locally successful subsurface petroleum exploration (Bebout, 1974). Outcropping Cretaceous reefs in Mexico, Central and West Texas, the Caribbean, and the eastern Mediterranean are similar to those reported from the subsurface (Rose, 1963; Perkins, 1974; Bein, 1976).

In order to properly reconstruct the diagenetic events that have affected any reef sequence, a good understanding of the tectonic-stratigraphic setting, depositional framework, and paleoecology is essential. The rocks that comprise the Glen Rose Formation (Albian) on the Edwards Plateau are essentially flat lying and undisturbed by tectonism and contain numerous widely occurring stratigraphic marker beds which have enabled geologists to reconstruct a detailed stratigraphic framework. Perkins (1974) conducted a comprehensive paleoecological study of a caprinid reef interval that occurs within the lower member of the Glen Rose. He showed that rudist reefs within this interval accreted in a manner analogous to shelf-margin reefs associated with the subsurface Stuart City Trend in South Texas and the outcropping El Abra and subsurface Golden Lane - Poza Rica reef trends of Mexico.

One reef complex, exhumed by Red Bluff and Pipe Creeks in southeastern Bandera County, Texas, is probably the best exposed rudist reef in the world. Besides the outcrops, samples from 18 cores drilled adjacent to this reef were used. The combination of available material provides excellent stratigraphic control for the study within a 2 km^2 area.

[1]Shell Oil Company, Houston, Texas

Excellent exposure, lateral traceability into the adjacent subsurface, and location within a well-documented stratigraphic framework make the Pipe Creek reef especially appropriate for a diagenetic study. Diagenetic investigations into most ancient carbonate sequences must separate textural features imposed by late Cenozoic and Holocene hydrology from those related to syngenetic and earlier epigenetic hydrologic regimes. It is important to recognize which portions of a reef complex are most affected by the processes of syngenetic cementation, bioerosion, and internal sedimentation because these processes effectively destroy reef framework, reduce primary porosity, and may restrict epigenetic diagenesis to certain lithofacies or portions of lithofacies.

Regional Geologic Setting

The study area occupies a small portion of the eastern part of the Edwards Plateau (fig. 1). Faulting and subsequent uplift of the entire area began sometime between the Late Cretaceous period and the early Miocene epoch (Murray, 1961). Cretaceous rocks crop out on the plateau in a belt bounded on the south and east by the Balcones fault zone (fig. 1). This belt generally parallels the Early Cretaceous shoreline (Stricklin and others, 1971). The Cretaceous strata strike generally northeast to southwest and dip southeast at 9 to 16 m per km. Resistant dolostones of the upper Glen Rose and the lower Edwards Formations form caprock throughout the area, whereas the lower Glen Rose Formation is exposed along the many streams that dissect the plateau.

General Lower Cretaceous Tectonic: Stratigraphic Setting

Lower Cretaceous rocks in Texas are divided into the Trinity, Fredericksburg, and Washita divisions in ascending stratigraphic order (fig. 2). Each division of the Trinity is thought to represent a transgressive-regressive couplet separated by regional unconformities (Lozo and Stricklin, 1956) but more recent work by Inden (1972) on the Cow Creek - Hammet - Hensel relationships weakens the couplet hypothesis.

The thickness of the Glen Rose Formation ranges between 14 and 225 m on the Edwards Plateau (Young, 1967). Generally, the formation thins from the southeast to the northwest between the Balcones fault zone and the Llano uplift. In the South Texas subsurface near the shelf edge, the Glen Rose is equivalent to the lower portion of the Stuart City reef trend (Bebout, 1974). The Edwards - Stuart City is stratigraphically equivalent to the El Abra - Tamalipas - Tamabra limestones of northern Mexico (fig. 2).

Glen Rose Formation: Stratigraphy and Depositional Environments

The Glen Rose Formation was deposited in a variety of shallow-marine environments on the Comanche shelf behind the shelf-margin Edwards - Stuart City reef complex. Thin layers of lime mudstone to packstone that contain casts and steinkerns of the burrowing pelecypod ***Corbula*** sp. separate the Glen Rose into upper and lower members. Most of the lower member (fig. 3) is composed of normal shallow marine shelf sediments that contain a diverse mollusk assemblage (Young, 1967). During deposition of the lower member, normal shelf sedimentation was locally interrupted by caprinid reef and mound development. Following the reef phase, extensive intertidal flats characterized by stromatolites and ripple-marked calcareous sands accumulated (Stricklin and Amsbury, 1974). During this time the

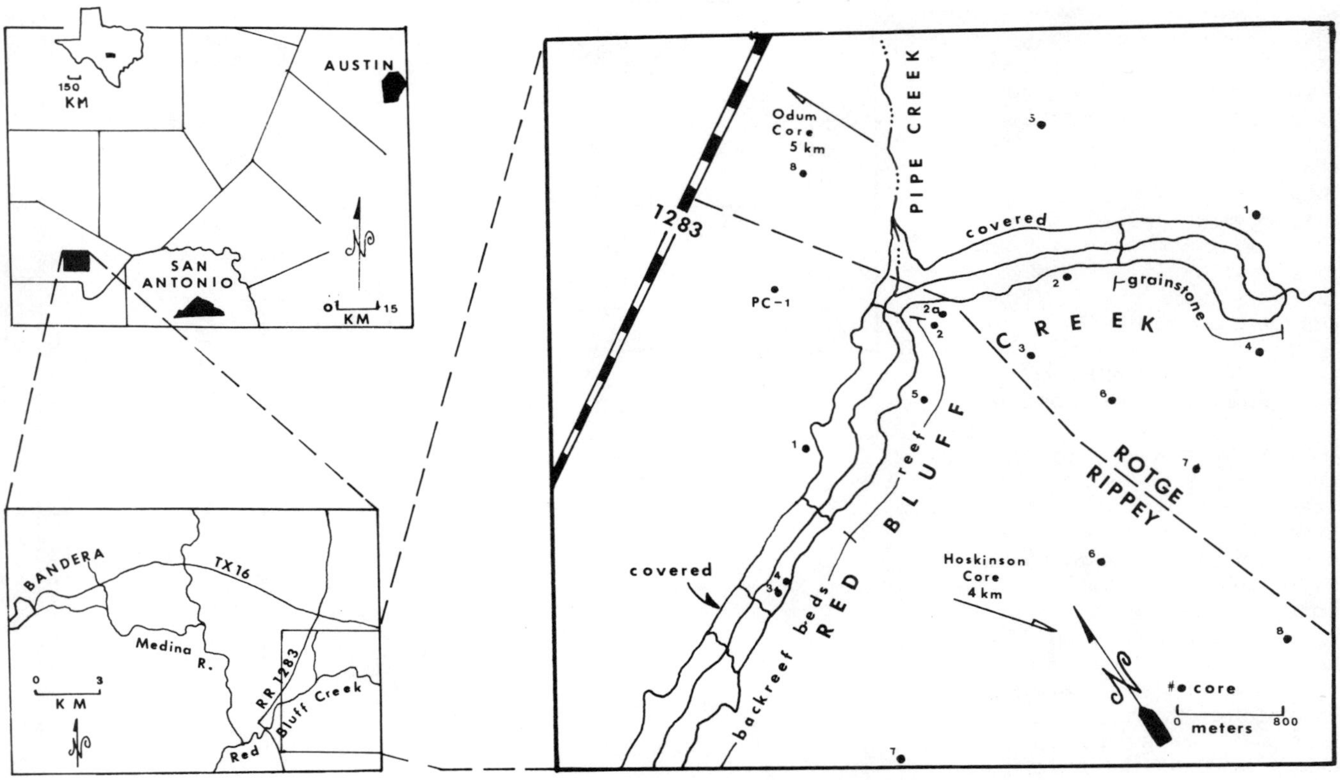

Figure 1. Location map of the study area.

deeper portions of the shelf southwest of the study area were filled with marine sediments that include beds which contain abundant fossil remains of the stirodont echinoid *Salenia texana* Credner (Perkins, 1974).

The *Corbula* bed that separates the lower and upper members of the Glen Rose Formation overlies the *Salenia* or stromatolite beds. This bed or series of beds contains remains of the pelecypod *Corbula harveyi* Hill (Perkins, 1974). Although in some localities the zone is represented by several resistant *Corbula*-bearing layers interbedded with calcareous shale, gypsum, or argillaceous, fossiliferous lime wackestone, nowhere do these layers exceed a total thickness of 1 m (Stricklin and Amsbury, 1974). The *Corbula* bed, persistent for over 8,000 km^2 in south-central Texas, is an excellent regional stratigraphic marker (Perkins, 1974).

The upper member of the Glen Rose is composed of dolostone, lime mudstone, and fine- to medium-grained fossiliferous lime wackestones that alternate with softer beds of calcareous clay and gypsum (Perkins, 1974). No reefs or mounds are reported from the upper member updip from the Cretaceous shelf margin.

Regional stratigraphic control provided by the *Corbula* and *Salenia* marker beds has allowed the position of the Glen Rose reef interval to be precisely located within the rock column (fig. 3). Within the study area, a bed of limestone 1 m thick that contains abundant pelecypod steinkerns directly overlies the reef-interval rocks and is used as a local stratigraphic marker. Reef-interval beds are underlain by a 10-m-thick lime wackestone bed that contains numerous tests of the foraminifer *Orbitolina texana* (fig. 3).

Reef-Interval Facies: Stratigraphic Relationship

The Glen Rose reef interval (Perkins, 1974) consists of four separate lithofacies: oyster biostrome, monopleurid biostrome, plant fragment, and caprinid reef (fig. 4). The oyster biostrome facies, nearest the paleoshoreline adjacent to the Llano uplift, is transitional with the monopleurid biostrome facies. Caprinid mounds occur in the area between the monopleurid biostrome and the caprinid reef facies. Tabular caprinid reefs and flanking grainstones completely surround the plant fragment facies which occur locally northeast of the study area. Individual reefs are separated by interreef lime grainstones or shallow-marine shelf lime wackestones.

Tridactyl dinosaur tracks atop several monopleurid biostromes have established that the water depth during deposition of that facies was between 1 and 4 m. Maximum water depths around caprinid reefs slightly downdip probably never exceeded 5 m (Perkins, 1974). Several pholad-bored zones within the reef facies have been cited as evidence for the subaerial exposure of some reefs as low-relief islands (Perkins, 1970). None of these surfaces can be traced landward with certainty. Similar surfaces occur at the top of several monopleurid biostromes, but in every case these surfaces are only locally developed.

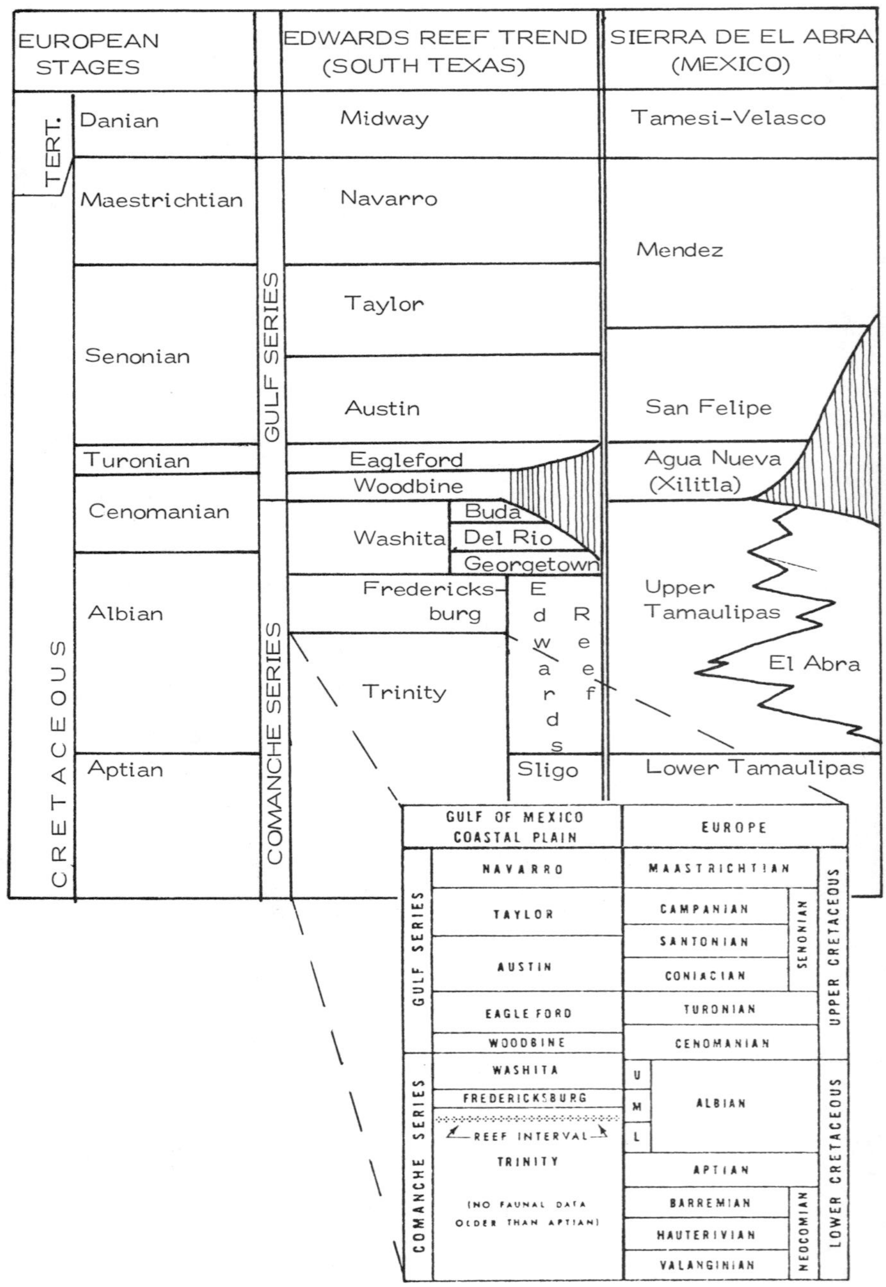

Figure 2. Subsurface correlation chart and Lower Cretaceous stratigraphic nomenclature for South Texas and northeastern Mexico (adapted from Rose, 1963; Perkins, 1974).

Caprinid Reef Facies Deposition

Cretaceous shelf-margin reefs are typified by a caprinid-coral framework that was bound by matlike hydrozoans, coralline algae, and encrusting foraminifers (Coogan and others, 1972; Bebout, 1974). Conversely, caprinid reefs on the shelf commonly lack organic binders but did form wave-resistant structures by accretion of a mutually supportive skeletal framework (Stricklin and others, 1971; Perkins, 1974). Shelf-dwelling rudists probably grew in water characterized by higher turbidity and lower energy than shelf-margin rudists. Fine-grained calcareous sediment that filtered into framework voids and submarine cementation probably enhanced the ability of shelf reefs to resist destruction by waves.

Rocks that underlie the caprinid reefs at Pipe Creek are predominantly foraminifer peloid lime mudstones and wackestones characterized by massive bedding, pervasive burrows, and abundant remains of ***Orbitolina texana*** and mollusks. The proportion of fine-grained carbonate material in these rocks generally decreases upward, suggesting a slight increase in local energy levels during deposition.

Most of the caprinid reefs examined have a thin (50 to 80 cm) basal layer of lime wackestone that contains recumbent caprinid skeletons. Many of the larger rudists are attached to other caprinids, oysters, and other large mollusks. Caprinid spat, perhaps transported by currents from the northeast, evidently attached to any available substrate, i.e., other mollusks (Perkins, 1974, personal communication). Upward growth on the relatively unstable muddy substrate resulted in "toppling" of young rudists. These rudists acted as substrate for later generations until caprinids completely colonized the area. Caprinids eventually constructed a mutually supportive, wave-resistant framework that was strengthened by contemporaneous internal sedimentation and cementation. Encrusting foraminifers and rhodophytes were minor contributors to framework stabilization.

Gastropods, thick-shelled slightly frilled oysters, requieniid rudists, and other less abundant bivalve mollusks also occur within the reefs. Echinoids, agglutinated and miliolid foraminifers, ostracods, and dasycladacean algae are locally abundant epifauna and shallow infauna that did not contribute to reef framework construction.

The main reef mass, approximately 250 m wide and extending laterally 4 to 5 km in the subsurface (fig. 5), is composed of a series of overlapping, juxtaposed tabular reefs that are 2 to 4 m thick. Total thickness of the reef mass is 16 m. Individual tabular reefs rarely exceed 150 m wide and 300 m long. Although the primary texture is lime boundstone, portions of the reef are composed of lenses of caprinid lime packstone that represent material eroded from adjacent reef framework and deposited between accreting reefs.

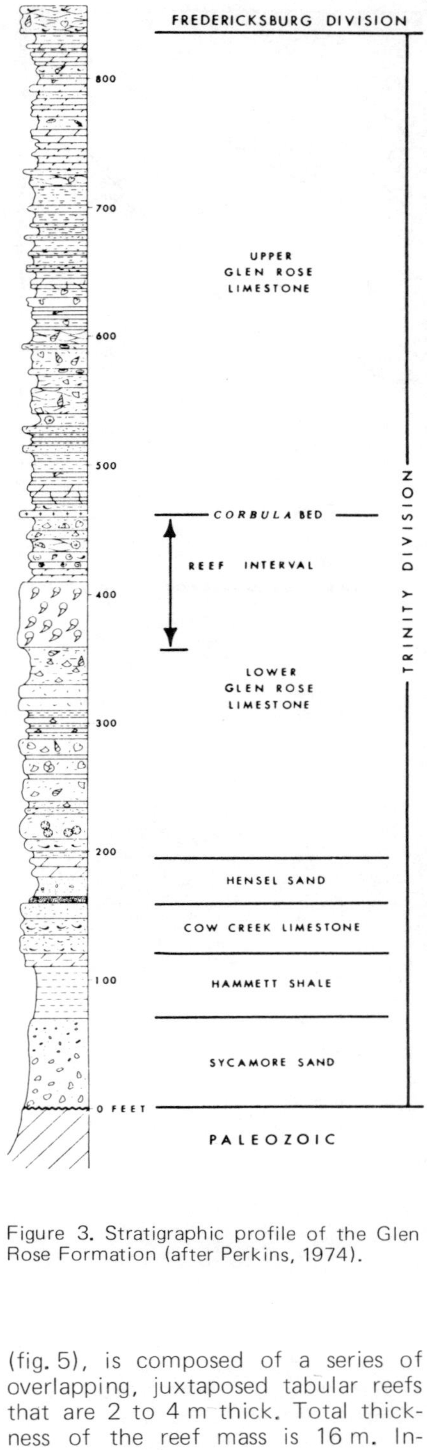

Figure 3. Stratigraphic profile of the Glen Rose Formation (after Perkins, 1974).

Lateral and vertical accretion of the reefs produced local low-energy conditions in the landward backreef area. Miliolid peloid lime wackestones and caprinid mounds that were deposited synchronously with the reefs are characteristic backreef rocks. The ubiquitous occurrence of miliolids in these sediments may indicate the presence of prolific quantities of codiacean algae and/or marine grass. Lime mud was probably produced by the disintegration of rhodophyte epibionts on marine grasses or aragonitic algae or both. The inverted conical cross sectional shape of the circular caprinid mounds suggests that the rudists in the backreef area were gradually smothered by lime mud. The mounds (fig. 6) are predominantly lime wackestone that contain recumbent rudists and lack submarine cement, organic binders, and bored surfaces.

Pholad borings are present in one core and within the outcropping reef mass at Pipe Creek. These borings are recognized by their smooth margins, semicircular to ovoid shape, and size (fig. 7). The outcropping bored surface (fig. 8) locally encrusted by oysters, overlies a thin layer of mollusk lime grainstone or caprinid lime boundstone. In one core through the reef 240 m southeast of the outcrop, a similar surface is developed in finely laminated miliolid lime wackestone. The wackestone has been brecciated and reworked into overlying coarse-grained intraclastic mollusk lime packstone. Some of the intraclasts are bored and coated by an isopachous micrite rim.

It is not known whether the bored surfaces are continuous or represent surfaces that developed at different times. Perkins (1974) suggested that they represent periods of reef emergence caused by eustatic fluctuations of sea level. This interpretation implies that regional tectonic uplift completely exposed the reef to subaerial processes that produced vadose cementation of the sediments and reef framework. During resubmergence of the reef mass, the surface was bored by pholad mollusks and encrusted with oysters. To support this hypothesis, evidence of an early fresh-water vadose and, presumably, phreatic diagenetic phase should be present.

An alternative explanation is that the bored surfaces represent submarine or intertidal hardgrounds that developed on the upper surfaces of reefs when upward accretion resulted in the uppermost portions of the reef(s) reaching mean sea level. Similar surfaces are developed on the crests of all modern coralgal reefs that have reached the littoral zone by vertical accretion.

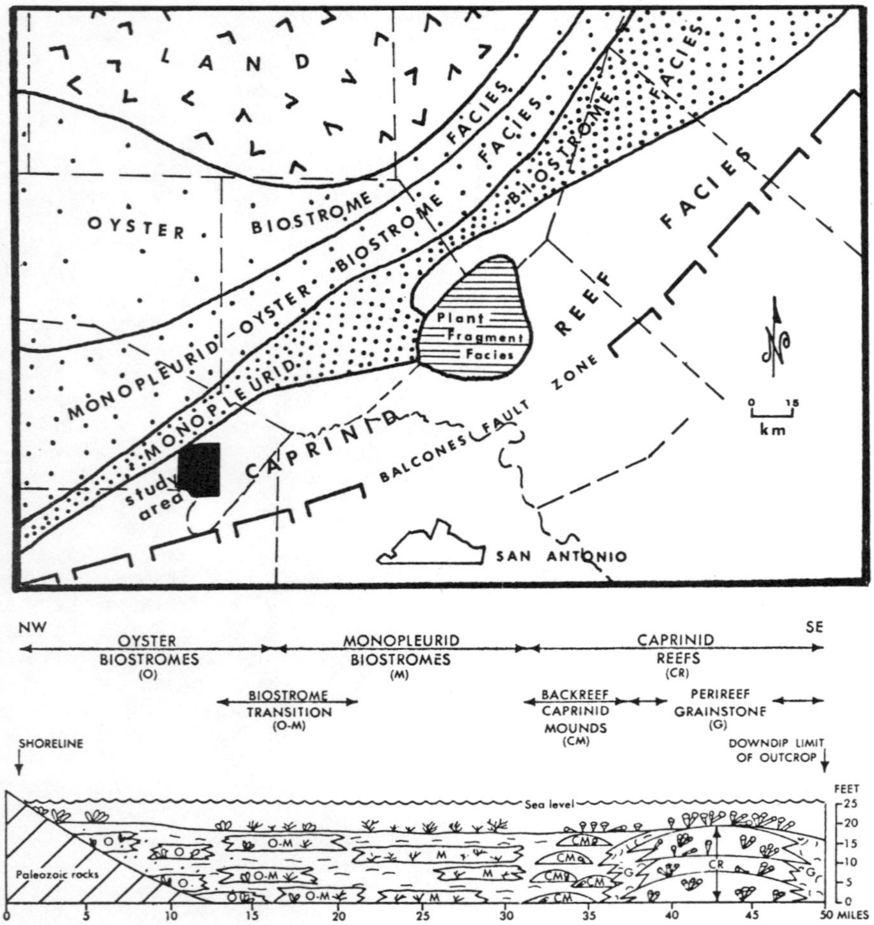

Figure 4. Map and cross section showing the distribution of the Glen Rose reef interval.

Rose (1970) interpreted a bored surface in the Edwards Formation as a series of locally developed submarine hardgrounds. Superimposed borings of a ***Lithophaga***-type bivalve at the top of these surfaces suggest that submarine cementation was followed by boring, sedimentation, cementation, boring, etc. Shinn (1969) described a similar diagenetic sequence from Holocene shallow-marine carbonate sediments in the Persian Gulf. The question of how the bored surfaces at Pipe Creek developed will be discussed in detail under the heading of diagenesis.

During the development of the bored surfaces, physical erosion of the caprinid reef crest(s) produced a supply of coarse mollusk sand that was reworked into small beaches on the seaward side of reefs or into a large submarine bar or beach complex that prograded northward. Migration of this sand body created more widespread low-energy conditions which enabled requieniid rudists to colonize the lagoon. Sedimentary structures within the rocks that comprise the large sand body are analogous to those normally developed within modern submerged bars and beaches (figs. 9, 10; table 1). Petrographic criteria must be used to determine if these grainstones underwent subaerial exposure contemporaneous with deposition.

Continued shelf subsidence allowed rudists to recolonize portions of the reefs where bored surfaces had developed. New reefs became a sediment source for the sand body which continued its migration. Cessation of reef growth discontinued the supply of sediment. Lime wackestones that represent the final depositional phase are often shaley and dolomitic. The dolostones do not, however, contain any sedimentary or diagenetic structures diagnostic of either Cretaceous or Holocene supratidal dolomite sequences. Therefore, any explanation of dolomitization must account for the seemingly sporadic distribution of dolomite in the section. The relationship of the reefs and contemporaneous sediments through time is shown by figure 11.

CAPRINID REEF FACIES DIAGENESIS

Introduction

Modern coralgal reefs are sites of simultaneous framework construction and destruction. Some organisms construct (coral, hydrozoans) and bind (coralline algae) the reef framework together; other organisms (pelecypods, echinoids, barnacles, sponges, and endolithic algae) bore into or rasp away the reef frame (Goreau and Hartman, 1963; Schroeder and Zankl, 1974). Many of these organisms are capable of producing sediment that fills primary and secondary framework voids or is removed by intraframework circulation, an effect caused by local tides and currents. Acicular and micritic aragonite and magnesian calcite cements are often precipitated in intraskeletal and interskeletal voids, lithify internal sediment, and further strengthen the reef framework (Schroeder, 1972).

Laminated sediment that resembles internal sediments in modern reefs fills interstices in rudist reefs that are a part of the Edwards - Stuart City reef trend in South Texas (Bebout, 1974). Shinn and others (1974) described similar features from subsurface Cretaceous rudist reefs in South Texas and from outcropping reefs in northern Mexico. In both cases, the internal sediments were associated with contiguous acicular sparry calcite cements that line primary framework cavities and intercalate the internal sediment. Bein (1976) and Aguayo

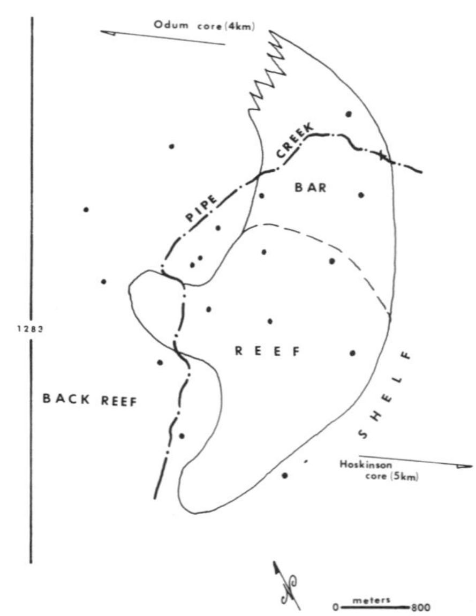

Figure 5. Map of the facies distribution associated with the Pipe Creek rudist reef complex. Black dots represent core holes studied.

Figure 6

Figure 8

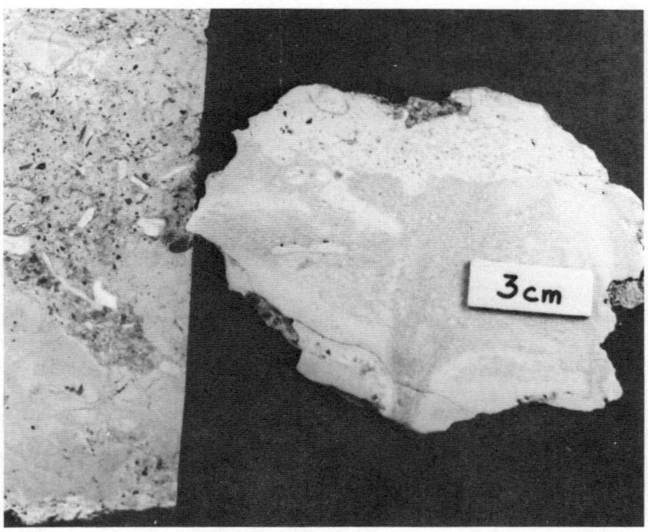

Figure 7

Figure 9

Figure 6. Caprinid mound and intermound sediment contact at Pipe Creek. Note that the intermound sediments drape over the mound. The mound is composed of caprinid lime wackestone. The caprinids are mostly recumbent, although they are often articulate.

Figure 7. Pholad-bored surfaces in slabbed surfaces of the Rotge 6 core (left) and the outcrop at Pipe Creek (right). Both surfaces are overlain by poorly sorted lime packstone.

Figure 8. Pholad-bored surface (arrow) at the top of the lower reef mass at Pipe Creek. Notice that the surface is relatively planar but does contain minor irregularities.

Figure 9. Accretion beds of the foreshore portion of the carbonate sand body (unit IIb) that prograded over marine shelf lime wackestone. Accretion beds (a) dip southeast at 5 to 8 degrees.

Figure 10. Festoon crossbedded lime grainstones of the upper offshore zone (unit IIc) at Pipe Creek. Note the small ripples in the festoon trough. The trough dips northeast at 3 to 4 degrees.

Figure 10

(1976) concluded that submarine cementation, organic boring, and internal sedimentation were contemporaneous with rudist framework accretion in Cretaceous fringing reefs from Israel and the El Abra rudist reef complex of Mexico respectively. Although similar diagenetic features have been noted in the Pipe Creek reef complex, their origin has been attributed to vadose diagenesis that was the result of reef emergence (Perkins, 1974). Cursory inspection of the material from Pipe Creek revealed that clionid sponge borings, peloidal internal sediments, and fibrous sparry calcites are present. Spurr (1975) noted that peloids and calcareous silt produced by endolithic clionid sponges may be mistaken for pellet and crystal silts that are thought to represent meteoric vadose diagenesis.

Although the reefs are now composed of calcite and the nonreef rocks are composed of calcite and protodolomite, both rock types contain organisms that were originally aragonite (caprinid), calcite (oyster), and magnesian calcite (rhodophytes and miliolids) (table 2). Mineralogic changes through time were undoubtedly accomplished by pore waters with chemical characteristics that were different from the original marine interstitial water. Table 3 is a summary of calcareous cement morphology and trace element composition related to water chemistry. Generally, submarine cements are composed of aragonite or magnesian calcite, contain high amounts of Sr, Mg, and Na, and have a fibrous or equant habit. Conversely, fresh-water cements are calcite, contain rather low Sr, Mg, and Na, and are equant. More than one generation of calcareous cement may occupy a pore space, evidence that the chemical composition of the interstitial water has changed through time. Careful use of carbonate skeletal and cement textures, cement morphology, and calcite trace element chemistry is crucial to the proper interpretation of the paragenesis of the rocks at Pipe Creek.

Syngenetic Diagenesis: Reef

Framework Boring

The development of rudist reef framework has been discussed in the preceding section. Primary framework voids were formed between living caprinids. Framework degradation by organic degradation was contemporaneous with framework accretion. Discontinuous micrite rims that surround caprinid skeletons and line pallial canals are overlain by sparry calcite cement (fig. 12). Such micrite rims are common in ancient carbonate rocks (Bathurst, 1966). Aragonitic and magnesian calcite rims are common in modern carbonate sediments (Winland, 1968).

Bathurst (1966) attributes the origin of these rims to organic boring and skeletal degradation by endolithic algae and fungi. Intensive boring produces an anastomosing network of tubules that penetrate the outer margin of the grain. Following the death of the borer, the tubules are filled by lime mud, producing a micrite rim. The exact process by

Table 1. Comparison of sand units at Pipe Creek.

UNIT	DESCRIPTION	DOMINANT ALLOCHEMS	CONTACT WITH BEDS: UNDERLYING	OVERLYING	INTERPRETATION
I	Accretion and festoon crossbedded lime grainstone. Graded bedding within accretion sets. Sorting moderate to good. 1.8 to 2.5 m thick. Pinches out against a caprinid reef (SW) and unit IIa (NE). Not present in cores.	C, R, P, DA, CA, E, and I (base of cosets)	Sharp	Sharp	Perireef beach
IIa	Massive-bedded lime wackestone to packstone. Wackestones are pervasively burrowed. 2.7 to 3.2 m thick. Pinches out into lenticular toucasid lime wackestone or unit IIb. Poor sorting and rounding. Present in cores.	P and G in packstones; E, F, CA, PE, R in wackestones	Sharp	Gradational	Backbeach/bar lagoon with spillovers
IIb	Accretion-bedded lime grainstone. 4.0 m thick. Sequence fines upward. Graded beds within accretion sets. Very good sorting and rounding. Grades into unit IIc (SE) and unit IIa (NW). Present in cores.	C, R, P, PE, I	Sharp	Sharp	Foreshore
IIc	Festoon crossbedded lime packstone. Small-scale ripples in troughs trend NE. 4.0 m thick. Grades into unit IIb (NW) and mollusk peloid lime wackestone (SE). Moderate sorting and rounding. Present in cores.	C, R, P, PE, I	Sharp (scour)	Sharp	Upper offshore

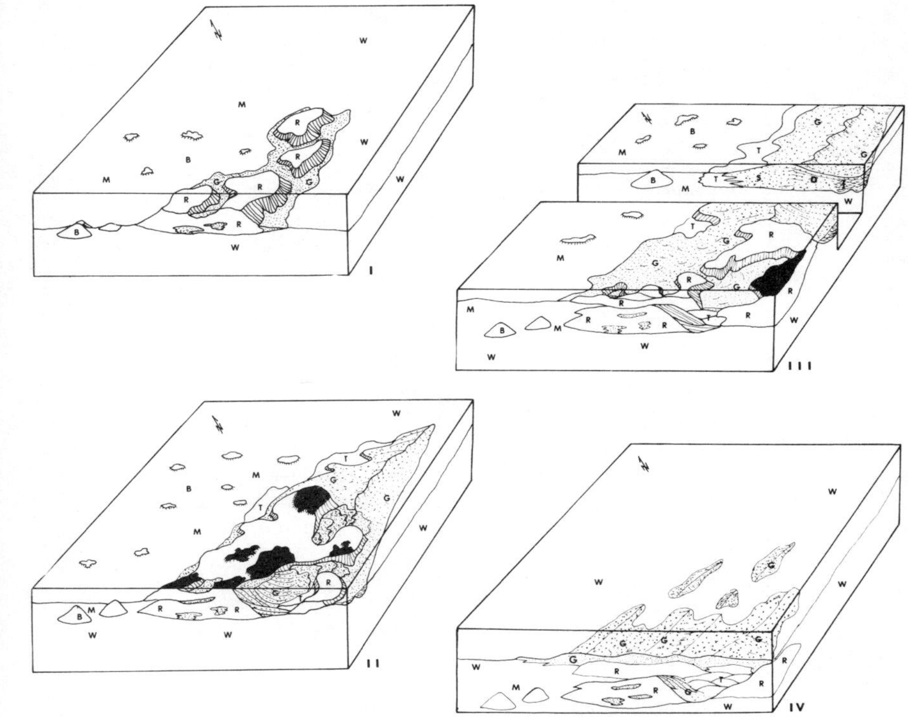

Figure 11. Idealized block diagram (I) showing the early development of the caprinid reef facies at Pipe Creek. This is a generalized diagram based on outcrop and core data. Reefs (R) shelter the backreef area where caprinid mounds (B) and miliolid peloid lime wackestones (M) accumulate. Some lime packstone and grainstone deposits (G) have resulted from physical erosion of the reefs. The surrounding shelf is the site of mollusk peloid lime wackestone to packstone deposition (W). The top of the block represents mean sea level. (II) Idealized block diagram showing the development of bored surfaces. Portions of the reef complex (R) have reached the littoral zone by vertical accretion. These areas are indicated by black. Reworking of the reef crests has provided sediment that is incorporated within a submerged bar (G) that prograded northward. This sand body shelters more of the shelf. In response to the lowered energy levels behind the bar, caprinid mounds (B), miliolid lime wackestones (M), and requieniid lime wackestones (T) accumulate. (III) Idealized block diagram showing further development of the carbonate sand body as the reefs (R) are continually reworked into coarse mollusk sediment. Note the relationship between upper offshore (f), foreshore accretion (a), and backshore washover (s) beds. (IV) Final phase of caprinid reefs facies development. Shelf subsidence has outstripped caprinid growth rates and has resulted in the crests of the reefs being reworked into blanket sands. Since the reefs no longer acted as energy baffles, the sand blankets are laterally contiguous to normal shelf mollusk peloid lime wackestone and packstones (W).

which the lime mud is introduced into the vacated borings is unknown (Bathurst, 1971). Lloyd (1969), however, showed that the carbon isotope values of Holocene micrite rims in the Bahamas closely approximates those common to most genera of chlorophytes. Therefore, algal photosynthesis is probably an important factor in the development of micrite rims in shallow-water carbonate sequences.

Besides the indirect evidence for skeletal framework boring by algae and fungi, other more recognizable borings are present that show that boring and reef accretion were contemporaneous. The size and shape of these borings are variable but the features are generally subspherical to ovoid (fig. 13). Boundaries between the borings and the framework are sharp and well defined, generally having a characteristic serrate or scalloped morphology (fig. 14). These borings occur not only within invertebrate framework skeletons but also within internal sediments where the scalloped margins of the borings are partly obscured (fig. 15). In some samples, sediment-filled borings are transected by other borings. In the case of these superimposed borings, the first boring was filled with sediment, the sediment was cemented in the marine environment, and the sediment-filled boring was rebored (fig. 16). The occurrence of borings that transect internal sediments and other borings is an indication that framework accretion, boring, cementation, and internal sedimentation were contemporaneous.

The morphology of the excavations formed by Cretaceous organisms, including sponges, was described by Perkins (1971) (fig. 17). Examination of Holocene coralgal reef framework from Jamaica reveals that modern clionid sponge galleries are identical to those within the caprinid reef framework at Pipe Creek (fig. 18). Many sponges evolved during the Paleozoic (Reiswig, 1973). Although Reiswig did not specifically mention clionid sponges, Fenton and Fenton (1932) described clionid-type borings in mollusk fragments from the Devonian of Iowa. The earliest positive evidence of these sponges is from the Mississippian of Oklahoma where sponge borings and silt are associated with algal mounds (Bonen, 1975).

Investigations by Goreau and Hartman (1963) and Spurr (1975) indicate that clionid sponges alter a significant volume of the modern coralgal reef framework in Jamaica and produce peloids and calcareous silt during boring. Rutzler and Rieger (1973), Futterer (1974), and Moore and others (1976) have described calcareous silt produced by these sponges. The silt, 20 to 80 μm in diameter, has a characteristic semi-triangular shape. One face of the sponge chip is convex outward whereas the other faces are slightly concave. This morphology is the result of the manner in which the sponge excavates the substrate (Rutzler and Rieger, 1973). Clionid silt and peloids have been recognized as important constituents of internal sediments of Jamaican reefs (Spurr, 1975).

A comparison of peloid size from Cretaceous caprinid and modern scleratinian reefs (table 4) seems to suggest that most of the Cretaceous peloids may have been produced by clionids. Note that most of the sediments within the framework are produced internally and do not require transport from outside the reef framework into the interior. However, the coarser layers of mollusk lime packstone were probably swept into some primary cavities by marine currents, possibly tidal.

Internal Sedimentation

Primary caprinid reef framework cavities that formed during reef accretion are filled with laminated to

Table 2. Inferred original mineralogy of Cretaceous organisms. Information from Chave (1954), Johnson (1971), Majewske (1969), and Milliman (1974).

GRAIN TYPE	INFERRED ORIGINAL MINERALOGY	ORIGINAL SHELL STRUCTURE	PRESENT SHELL STRUCTURE
Caprinidae	Aragonite	Lamellar	Recrystallized
Requieniidae	Aragonite + Calcite	Lamellar (A) and prismatic (C)	Lamellar layer recrystallized, prismatic layer unaltered
Ostrea sp.	Aragonite + Calcite	Compound foliose/ lamellar	Foliose/lamellar
Inoceramus sp.	Aragonite + Calcite	Lamellar (A) and prismatic (C)	Recrystallized or leached lamellar, unaltered prismatic
Gastropoda	Calcite or Aragonite or both	Complex, prismatic, lamellar, cross-lamellar, or foliose	Calcite layers unaltered, aragonite layers recrystallized or replaced by sparry calcite cement
Chlorophyta	Aragonite	Micron-sized needles	Sparry calcite
Rhodophyta	Magnesian calcite (10-21 mole percent $MgCO_3$)	Complex network of micron-sized crystals	Micron-sized calcite crystals
Ecinodermata	Magnesian calcite (10-30 mole percent $MgCO_3$)	Monocrystalline	Monocrystalline
Ostracoda	Calcite	Prismatic	Prismatic
Foraminifers	Magnesian calcite (0-24 mole percent $MgCO_3$)	Micron-sized equant to prismatic crystals	Micron-sized equant or prismatic crystals

Table 3. Ion strength, environment, and resultant carbonate cement morphology (adapted from Folk, 1974).

TRACE ELEMENT			ENVIRONMENT	CRYSTAL HABIT
Mg	Na	Sr		
High	High	Low (Mg calcite) High (aragonite)	Hypersaline to normal marine, beachrock, sebkha, submerged reefs, etc.	Steep rhombs of Mg calcite with vertically oriented flutings. Fibers of Mg calcite and aragonite grow rapidly in the c direction, slower lateral growth attributed to selective Mg poisoning. Crystals limited in width to a few microns
Low	High	Low	Mainly connate subsurface waters	Complex polyhedra and anhedra of calcite. Lack of Mg allows unhampered growth and equant habit
Low	Low-Moderate	Low	Meteoric phreatic, deep subsurface, meteoric-connate mixed waters	Complex polyhedra and anhedra of calcite. Lack of Mg results in equant crystals (often) that may be very coarse crystalline
Low	Low	Low	Meteoric vadose, caliche, streams, and lakes	Simple unit rhombohedra of calcite
Low	Low	Low	Streams, lakes, caliche	Calcite micrite. Also, calcite sheets or hexagonal crystals with basal pinacoids. Sheet structure on edges visible because of rapid lateral growth

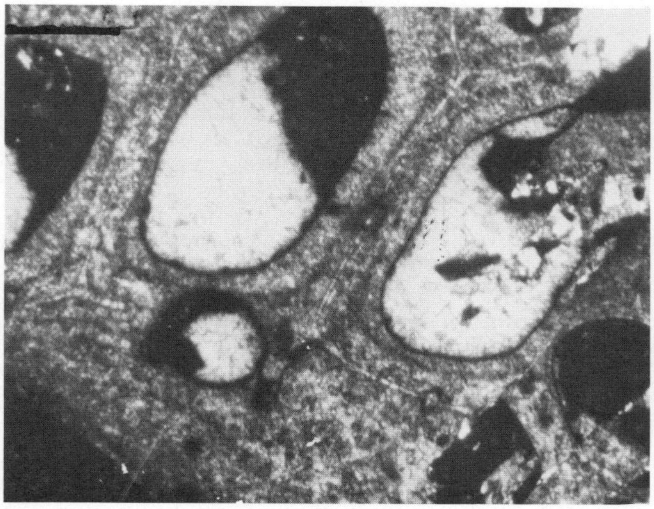

Figure 12. Photomicrograph showing a well-developed micrite rim around a caprinid pallial canal. Note the geopetal filling of the canal. Peloidal micrite is overlain by equant sparry calcite. Bar scale is 200 microns.

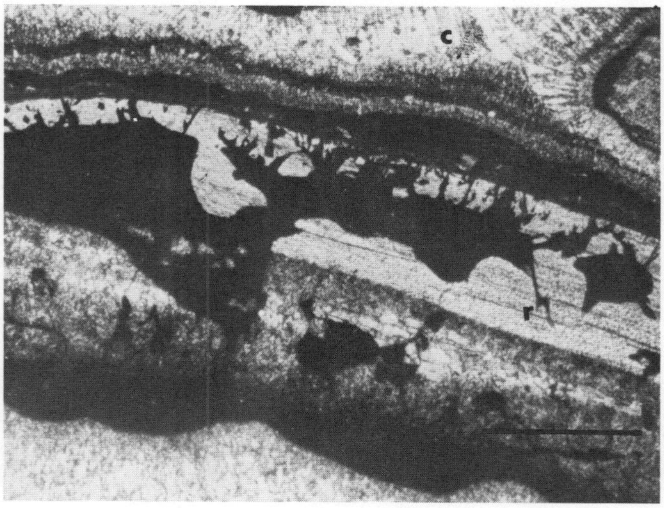

Figure 13. Photomicrograph showing a series of clionid borings that transect a caprinid shell in the reef framework at Pipe Creek. Layers of micritic and fibrous cement (c) occupy the framework cavity next to the caprinid (R). Bar scale is 1 cm.

massive peloidal and/or coarse- to fine-grained fossiliferous lime wackestone to packstone. Caprinid, requieniid, and gastropod mantle cavities and clionid borings are also filled with these sediments. Allochems in these cavity-filling sediments include peloids, angular calcite silt, mollusk fragments and molds, articulate and disarticulate ostracodes, and foraminifers. Subspherical to oval pores are also common (fig. 19). The predominant allochems in the internal sediment are peloids that have a bimodal size distribution: 18 to 52 μm. and 70 to 300 μm. When peloid layers are laminated, peloid size generally decreases upward, producing a graded appearance to the laminae. Most peloids are mutually supportive and generally encrusted with equant (P.E.$_2$C) or fibrous (P.F.$_3$C) sparry calcite cement (fig. 20).

Peloid layers are often truncated by an overlying peloid layer (fig. 21), fossiliferous lime wackestone to packstone, or irregular layers of micrite and fibrous (P.F.$_3$C) sparry calcite cement (fig. 22). These relationships imply that the deposition of peloids in framework cavities was often interrupted by influxes of fossiliferous sediment derived from adjacent cavities or from outside the caprinid framework. Fibrous cement layers that intercalate with layers of cavity-filling sediment and fibrous cement that coat peloids and skeletal debris show that periods of cementation in the marine environment alternated with periods of sedimentation. These observations are similar to those reported by studies of cavities within modern coralgal reef framework.

Sediments within the framework of Bermuda algal cup reefs are described as fine- to coarse-grained, peloidal to fossiliferous deposits that are often sorted within individual cavities (Ginsburg and others, 1971). Jamaican reefs commonly contain primary and secondary cavities that are infilled by laminated to massive, lithified magnesian calcite-rich pelleted micrite that contains a minor amount of aragonite (Land and Goreau, 1970; Spurr, 1975) (fig. 22). Other layers that contain calcareous silt, mollusk fragments, ostracodes, and foraminifers are associated with peloidal layers in Jamaica (fig. 23) and Bermuda (Ginsburg and others, 1971).

During laboratory experiments using clionid sponges, Spurr (1975) discovered that, in addition to calcareous silt, these organisms also produced loosely bound aggregates of pelleted micrite. The shape of individual peloids varied from irregular agglomerates to discrete oval subspherical particles. Because the size of these particles is identical to the size of peloids in the internal sediment of the Jamaica reefs, Spurr concluded that clionids were a major contributor of peloids to the internal sediment. A short-term experiment using four species of Pacific clionid sponges was conducted by C. H. Moore, Jr., and the author at Eniwetok Atoll. This experiment demonstrated that clionid sponges produce peloids similar to those described by Spurr. A tropical storm prematurely interrupted the experiment and prevented photographic documentation.

Calcareous silt is common to Cretaceous internal sediments. These sedimentary particles have a shape that resembles silt chips produced by sponges (fig. 24). The diameter of the convex side of these silt chips is comparable to the diameter of individual cerrae on the margin of clionid galleries, confirming the sponge origin of the silt. Recognizable clionid silt is not a major constituent of Cretaceous internal sediments, but small pores that have comparable size and shape are common (fig. 25). Most of the clionid borings occur within reef-framework caprinids, originally aragonite (table 2). Introduction of fresh water during epigenetic diagenesis no doubt dissolved the sponge chips and produced the pores. Clionid silt is most common in samples from caprinid mounds. Because oysters and requieniids, originally calcite (table 2), are more prevalent in the mounds, one must suspect that more clionid silt in the mounds was calcite and was less susceptible to fresh-water dissolution. A similar occurrence was reported by Spurr (1975) for Jamaican Pleistocene reef sediments. In this case aragonite silt was dissolved whereas the calcite silt was not.

Syngenetic Cementation

Multiple generations of isopachous fibrous (P.F.$_4$C) sparry calcite cement, separated by 2- to 5-μm-thick layers of micrite, line most framework voids. The cement nucleated on the micrite rims that coat skeletal framework and other allochems, internal sediment layers, and peloids (fig. 26). Individual crystals, 10 to 15 μm wide and 100 to 150 μm long, are normal to the nucleation surface (fig. 11). The thin micrite layer that separates generations of fibrous calcite may represent boring of the cement by endolithic algae and fungi during periods of noncementation, a common occurrence in modern reefs (Schroeder and Zankl, 1974). At the base of framework cavities, the cement grades into a thinner layer of fibrous sparry calcite cement (P.F.$_3$C). This cement is clouded with inclusions, forms layers 20 to 80 μm thick, and has planar terminations. When traced to the side of voids, it is obvious that individual layers of fibrous cement on the floor of voids are equivalent to one layer of fibrous cement on the side of the cavity (fig. 27). Multiple layers of P.F.$_3$C are separated by internal sediment or by a layer of micrite that contains marine fossils. Lack of internal sediments on the sides of cavities allowed cement nucleated on these surfaces to grow considerably thicker than its counterparts on the cavity floor. Since these cement layers nucleated on sediment that is often truncated by clionid borings, the internal sediments were probably cemented in the marine environment.

Shinn (1969) found that acicular aragonite and magnesian calcite cements precipitated in the bottom of large intergranular pores and burrows were overlain by lime mud washed into the pores. This mud effectively stopped further cement growth by sealing off the cemented layer from circulating interstitial marine water. Growth of cement on the top and sides of the pores continued uninterrupted by the influx of lime mud. Similar processes can be used to explain the diagenetic sequence within reef cavities at Pipe Creek.

The distribution and mineralogy of modern submarine cements are apparently largely controlled by the interstitial water chemistry which may vary between cavities (Ginsburg and others, 1971). Schroeder (1972) suggested that the introduction of algal filaments into cavities may be a mechanism that effects changes in the microenvironmental chemistry. Decaying organic tissue and bacteria may also promote the precipitation of calcium carbonate (Berner, 1968; Mitterrer, 1971).

In the Cretaceous reefs, the bottom portions of cavities were the sites of cement growth and deposition of peloidal, fossiliferous sediment. The peloids were probably largely pro-

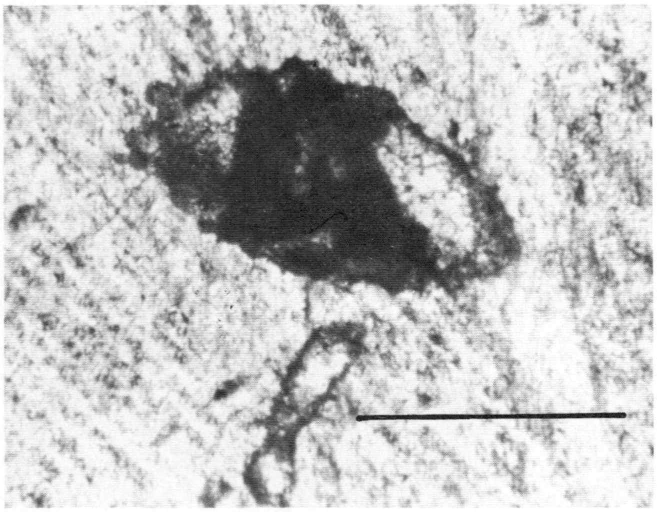

Figure 14

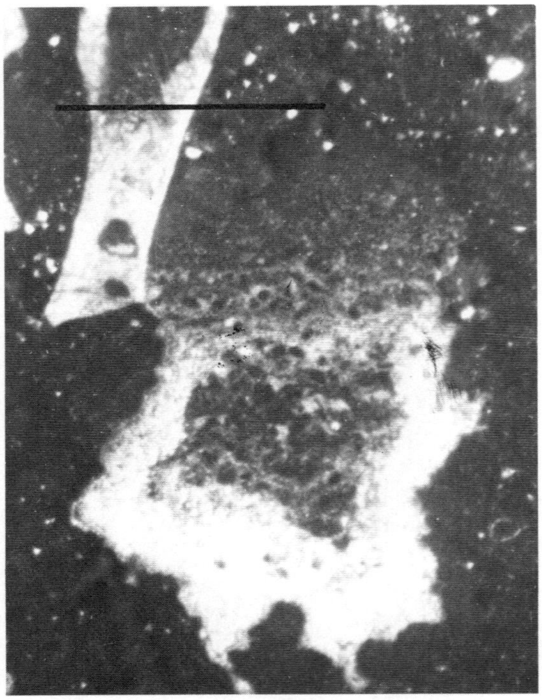

Figure 15

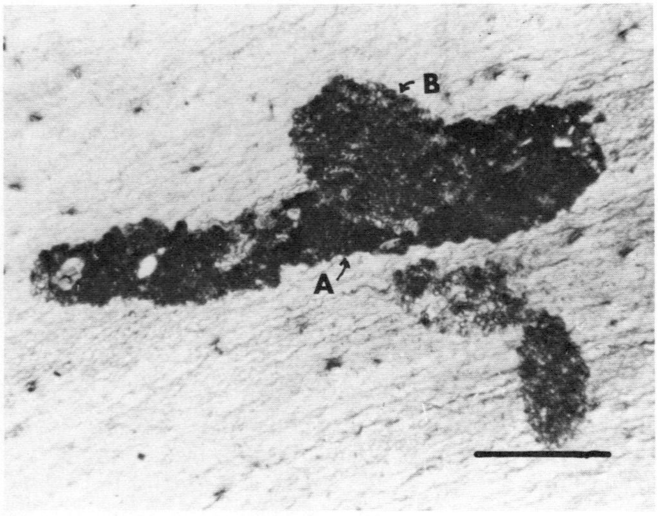

Figure 16

Figure 14. Photomicrograph of a clionid boring in the caprinid reef frame showing the characteristic cerrate or scalloped margin of the boring. Bar scale is 500 microns.

Figure 15. Photomicrograph of a clionid boring that transects the internal sediment of a primary framework cavity at Pipe Creek. Gallery is filled with internal sediment, peloids, and sparry calcite cement. Bar scale is 500 microns.

Figure 16. Photomicrograph of superimposed clionid borings. Boring A was filled with sediment that was cemented, then rebored. Boring B obviously transects boring A and the sediment. Note the scalloped margins of each boring. Bar scale is 250 microns. Cretaceous, Pipe Creek.

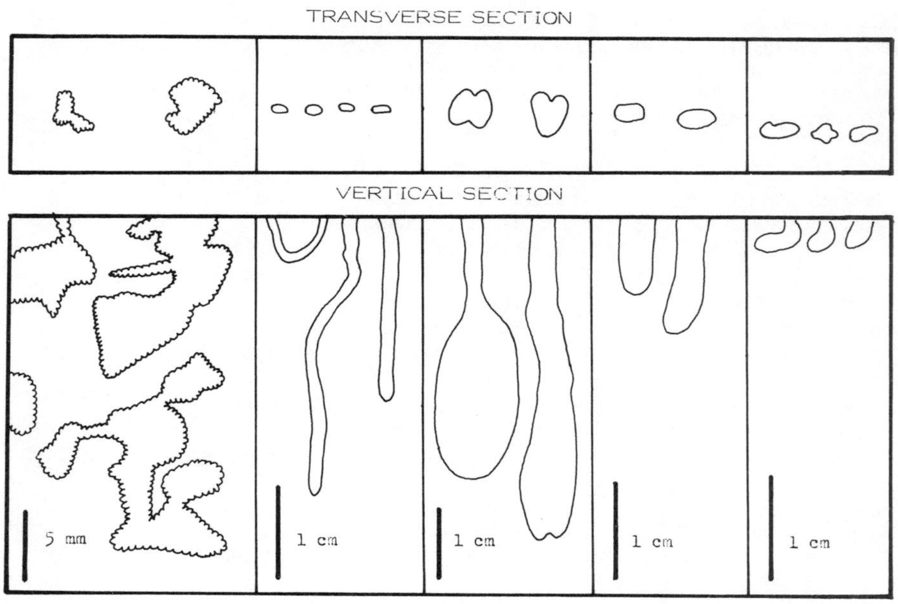

Figure 17. Diagrammatic comparison of the morphology of excavations made by commonly occurring rock-boring organisms (modified from Perkins, 1971).

duced by clionid sponges. Submarine fibrous cement indurated the Pipe Creek reef internal sediment as in modern reefs. Intercalation of internal sediment layers with fibrous calcite implies that the sediment was lithified shortly after deposition. Clionid borings that penetrate this sediment are indisputable evidence of its early marine origin.

The isopachous fibrous sparry calcite cements in the caprinid reefs are strikingly similar to acicular magnesian calcite cement crusts that are common in Holocene reefs in Jamaica (Land, 1969) and Bermuda (Ginsburg and others, 1971; Schroeder, 1972) and in submarine hardgrounds in the Persian Gulf (Shinn, 1971). Isotopic composition and mineralogy of these Holocene cements confirm their submarine origin (Shinn, 1971). Aragonitic fibrous crusts are also common in modern reefs (Ginsburg and others, 1971; Davies and Kinsey, 1973; Friedman and others, 1974) but are less common than magnesian calcite cement. Schroeder (1972) found that magnesian calcite "palisade" cements (fibers 20 to 80 μm long and up to 5 μm wide) could be distinguished from contemporaneous aragonite cements in Bermuda reefs by their more orderly crystal orientation and smooth, regular surface outline. Because the fibrous crusts in the Pipe Creek reefs closely resemble "palisade" cements, they may have originally been composed of magnesian calcite.

Data indicate that clionid degradation of reef framework in the Cretaceous was perhaps not as effective as during the Holocene (table 5). Lower volume infestation was possibly coupled with lower calcium carbonate generation rates resulting in reduced internal sediment production in the Cretaceous reefs. In turn, less sediment production resulted in more pore space available for cementation. Higher rates of lime mud production and turbidity on the Comanche shelf may have had detrimental effects on clionid life processes.

The implication of the data obtained from the study is that initial lithification of reef beds occurred in the marine environment and was concurrent with reef-framework accretion. This conclusion contrasts with the previous interpretation which proposed that the reefs were lithified within the meteoric vadose zone as a result of several periods of reef emergence. Pholad-bored surfaces that truncate portions of the reef mass are now thought to represent submarine hardgrounds rather than exposure surfaces. Growth of the semi-indurated reef framework into the littoral zone was accomplished when caprinid growth rate exceeded the shelf subsidence rate.

Syngenetic Diagenesis: Nonreef

Grain Diagenesis

Organic boring and micritization of allochems, induration of peloids, and formation of intraclasts occurred in nonreef beds at the water-sediment interface. Peloidal micrite fills clionid borings in large oyster, requieniid, and caprinid fragments. Small amounts of clionid silt occur in backreef sediments. Interstices in caprinid mounds are often floored with 20- to 80-μm-diameter calcite crystals that are thought to be clionid chips (fig. 28). Requieniids and oysters, originally calcite, are more common in the mounds than the reefs. Therefore, a large proportion of clionid silt in the bioherms was calcite and thus more

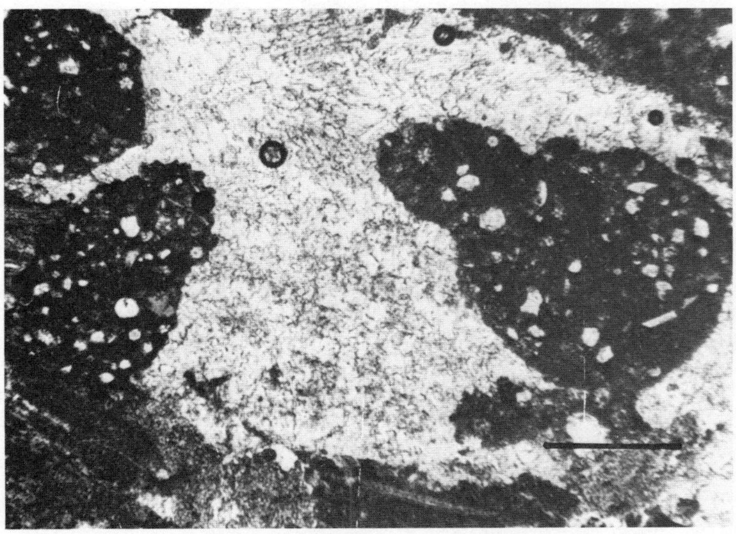

Figure 18. Photomicrograph of clionid borings in Holocene scleractinian reef framework, Discovery Bay, Jamaica. Borings are filled with internal sediment. Bar scale is 250 microns. Thin section courtesy of Clyde H. Moore, Jr.

Table 4. Comparison of physical characteristics and the occurrence of recent peloids within scleractinian reef frameworks and possible clionid sponge peloids indigenous to rudist frameworks.

PELOIDS		MINERALOGY OR INFERRED MINERALOGY	OCCURRENCE	REFERENCE
Shape	Size			
Irregular agglomerates, ellipsoidal to spherical	10-45 μm	Magnesian calcite plus aragonite	Laboratory experiment, produced by Caribbean clionid sponges	Spurr (1975)
Irregular agglomerates, ellipsoidal to spherical	14-45 μm	Magnesian calcite plus aragonite	Within the coralgal reef framework as internal sediment (Jamaica)	Spurr (1975)
Spherical to ellipsoidal	N.R.	Magnesian calcite (18.5 mole percent $MgCO_3$)	Within the coralgal reef framework as internal sediment (Jamaica)	Land and Goreau (1970)
Spherical to ellipsoidal	50 μm	Magnesian calcite plus aragonite	Laboratory experiment, produced by Pacific clionid sponges	This study
Spherical to oval	N.R.	N.R.	Internal sediments of algal cup reefs (Bermuda)	Ginsburg and others (1967)
Spherical to ellipsoidal	18-52 μm	Magnesian calcite plus aragonite	Internal sediment of Pipe Creek rudist reef framework	This study

stable than aragonite during freshwater diagenesis. Thus, the silt is more common in the mounds, whereas small pores in the internal sediment of the caprinid reefs represent dissolved sponge chips.

Micrite rims, well developed in lime packstones and grainstones, often contain endolithic fungal borings (1 to 2 μm in diameter) and larger algal borings (50 to 60 μm in diameter) (figs. 29, 30). Rims are not particularly well developed in the mud-supported rock types common to backreef sediments at Pipe Creek.

Rounded, well-sorted, spherical to oval peloids comprise 12 to 49 percent of the allochems in the lime wackestones of the shelf and the backreef beds. Three sizes of peloids were noted. Approximately 10 percent of the peloids are 100 to 200 μm in diameter and appear to be miliolid foraminifers whose tests have been filled with micrite. Ovoid peloids with long diameters up to 600 μm are micritized mollusks. Grain micritization by centripetal replacement (Bathurst, 1971) was probably responsible for the formation of these peloids (fig. 31).

Most of the peloids in nonreef beds average 40 to 50 μm in diameter. These peloids probably represent fecal pellets because they are well rounded, well sorted, and occur within beds that have undergone extensive bioturbation. The preponderance of burrows in the backreef miliolid peloid lime wackestones suggests that marine mollusks (Kornicker, 1962; Folk, 1967) and/or polychaete worms (Moore, 1939; Schaefer, 1972), capable of ingesting large quantities of lime mud, actively pelletized these rocks. Peloids in nonreef beds have not been subjected to extensive compaction, a phenomenon characteristic of most ancient pelleted carbonate sequences (Beales, 1965). Obviously these peloids were cemented during or immediately after deposition. Fecal pellets in the Bahamas are often indurated at the water-sediment interface (Illing, 1954; Bathurst, 1971).

Syngenetic Cementation

Fibrous and micritic submarine cements (fig. 32) have been reported from marine/brackish zones within the Holocene intertidal and subtidal beachrock zones in the Persian Gulf (Taylor and Illing, 1969). The micrite cement is composed of micron-sized aragonite, occurs at grain contacts, and commonly contains peloids. Similar cements composed of magnesian calcite have been reported from beachrock sequences in the Mediterranean and the Caribbean (Alexandersson, 1969; Moore, 1973; Sibley and Murray, 1973). Evidence that algae probably play an important role in the formation of these cements was presented by Moore (1973).

Allochems within the uppermost portions of the accretion-bedded grainstone unit are coated by isopachous fibrous (P.F.$_3$C) sparry calcite cement. Individual fibers are clouded with inclusions and were probably precipitated in the submarine phreatic zone. Microstalactitic and meniscus micrite cements overlie isopachous fibrous cements in the Pipe Creek rocks (fig. 32). These micrite cements are asymmetrical, thickening on the undersides of grains and at grain contacts. This growth habit suggests a vadose origin for the cement. However, crusts were also observed to thicken on the upper surfaces of allochems. This seemingly contradictory cement distribution was explained by Moore (1971, 1973). He found that micrite cements in subtidal beachrock sequences from Grand Cayman were generally associated with anastomosing networks of algal tubules. The algae most often attach at grain contacts and may induce the precipitation of micritic magnesian calcite during algal photosynthesis. The resultant micrite cements mimic meniscus cements and defy gravity. Moore and others (1972) described similar cements from a carbonate beach sequence within the Edwards

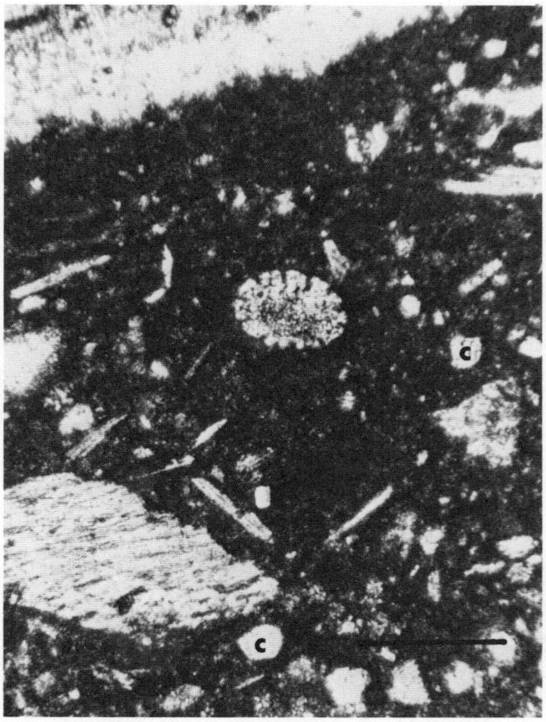

Figure 19

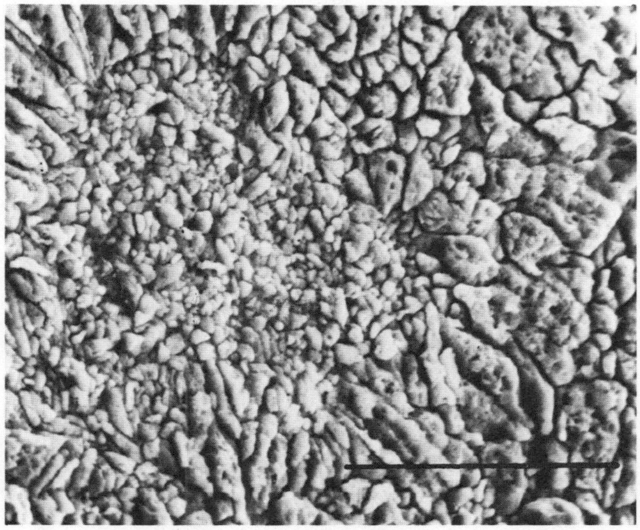

Figure 20

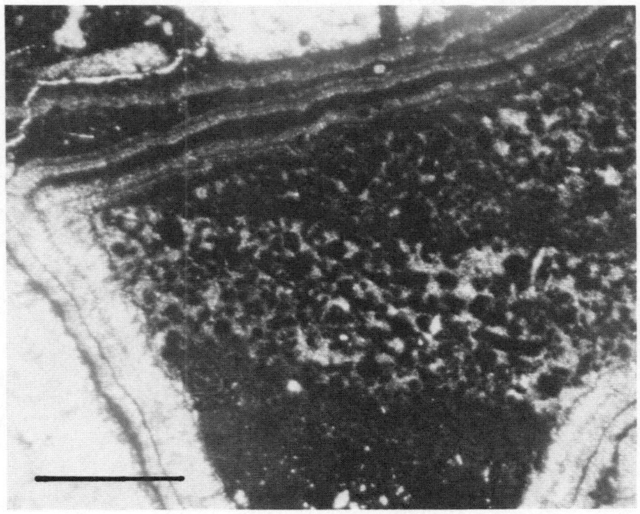

Figure 21

Figure 19. Photomicrograph of internal sediment that contains mollusk and echinoid debris and sponge chips (c). Cretaceous, Pipe Creek tabular reef. Bar scale is 250 microns.

Figure 20. SEM photomicrograph showing a peloid within the internal sediment of the Pipe Creek reef surrounded by a layer of fibrous sparry calcite. Bar scale is 50 microns.

Figure 21. Photomicrograph of layered peloidal internal sediment within a cement-lined primary framework cavity of the Pipe Creek reef. Bar scale is 250 microns.

Limestone at the Round Mountain quarry in Texas. Therefore, the similarity between these modern and Cretaceous cements and the cements at Pipe Creek in the grainstones of unit I implies that the latter cements may have originally been magnesian calcite and were precipitated in the shallow-marine phreatic zone.

Keystone vugs, indicative of the upper intertidal zone (Dunham, 1971), are associated with the uppermost beds of unit I (fig. 32). These vugs are equivalent in size to six or eight adjacent grains. They form in modern sediments in the upper part of the swash zone as air escapes from the sediments during wave backwash. Carbonate beach sands are most conducive to the preservation of these features because of rapid rates of littoral zone cementation. Keystone vugs have been reported from Cretaceous beach sequences at Round Mountain (Moore and others, 1972) and Cow Creek (Inden, 1972). Keystone vugs at the top of the unit I grainstone at Pipe Creek suggest that the unit was deposited as a beach and that the upper portion of the adjacent reef mass was probably also subaerially exposed.

Distribution of inversion textures within the lime grainstones that compose the carbonate sand body (unit IIa-c) suggests that these sediments may have been affected by fresh water during deposition. Point-count data from five vertical transects through the outcropping portion of the sand body indicate that inversion textures are most common in upper offshore and backshore sediments and that mollusk molds are more common in foreshore beds (fig. 33). Meteoric water undersaturated with respect to $CaCO_3$ percolated downward through the sediments and effected dissolution of aragonite mollusk fragments in the foreshore beds. Emergence of the sand body may have been the result of storms which effectively piled carbonate sand above mean sea level, similar to modern carbonate sand bodies on the Yucatan platform (Folk, 1967). Mollusk molds are outlined by noncollapsed micrite rims (fig. 34), evidence that the precipitation of equant intergranular sparry calcite cement prevented the collapse of molds. The $CaCO_3$ required for cementation was probably provided from aragonite dissolution as proposed by Bathurst (1958). As aragonite was dissolved in the upper portion of the foreshore, marine water in the backshore and upper offshore beds became increasingly saturated with respect to calcite. When supersaturation

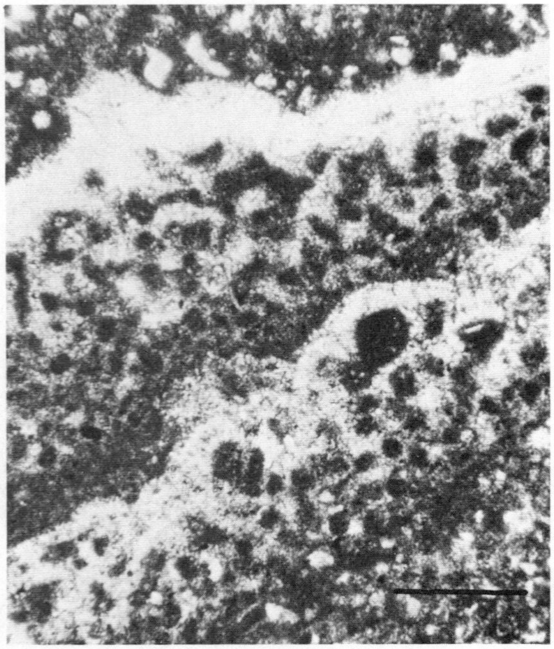

Figure 22

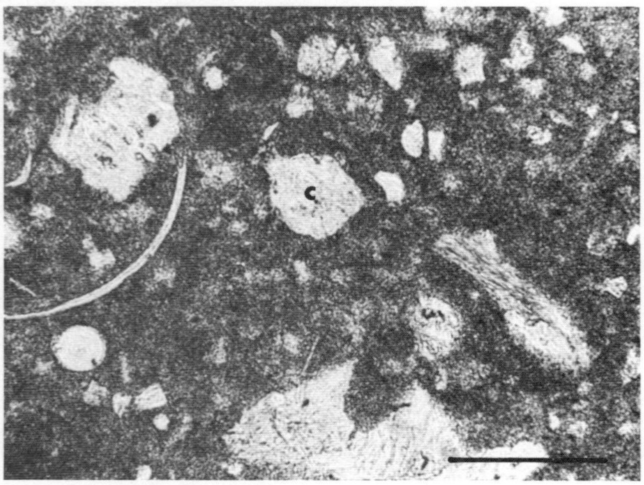

Figure 23

Figure 22. Photomicrograph of layered peloidal sediment and fibrous cement (magnesian calcite?) within a primary framework cavity of the Holocene reef at Discovery Bay, Jamaica. Compare these layers and peloids to those shown by figure 21. Bar scale is 250 microns. Thin section courtesy of Clyde H. Moore, Jr.

Figure 23. Photomicrograph of internal sediment containing a clionid sponge chip (c). Holocene scleractinian reef, Discovery Bay, Jamaica. Bar scale is 100 microns. Thin section courtesy of Clyde H. Moore, Jr.

Figure 24. Photomicrograph of a clionid sponge chip in internal sediment. Cretaceous, Pipe Creek tabular caprinid reef. Bar scale is 100 microns.

Figure 25. Photomicrograph of vugs (v) in internal sediment within the caprinid reef framework that may represent dissolved sponge chips. Bar scale is 100 microns.

Figure 26. Photomicrograph showing the relationship of submarine fibrous cement layers within a primary framework cavity at Pipe Creek. Lowermost fibrous cements overlie micrite that was cemented prior to fibrous cementation. Internal sedimentation retarded the cement growth in the bottom of the cavities. Note that the cement fibers at the sides of the cavity are not as dirty as those on the floor of the cavity. Bar scale is 500 microns.

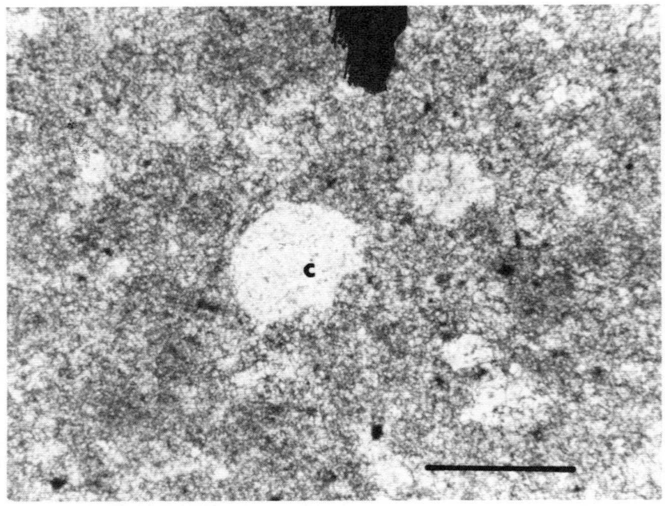

Figure 24

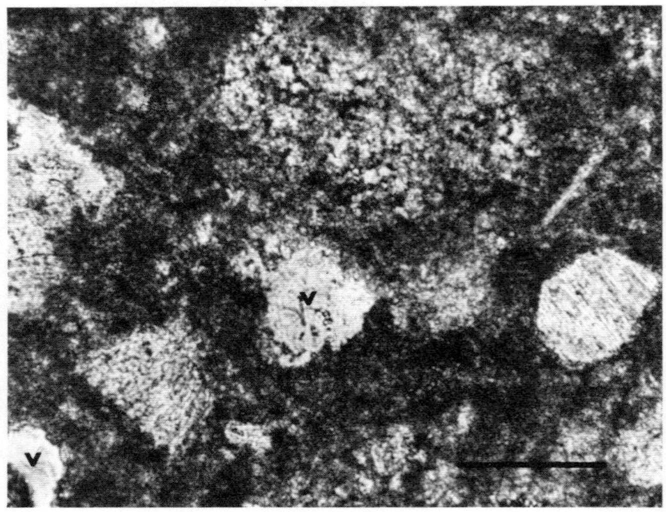

Figure 25

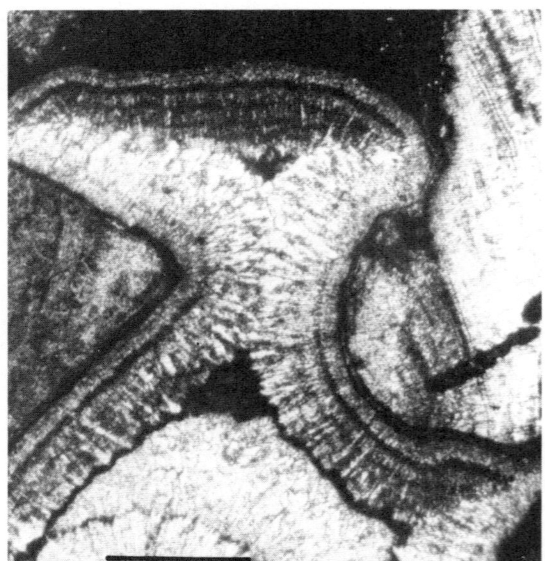

Figure 26

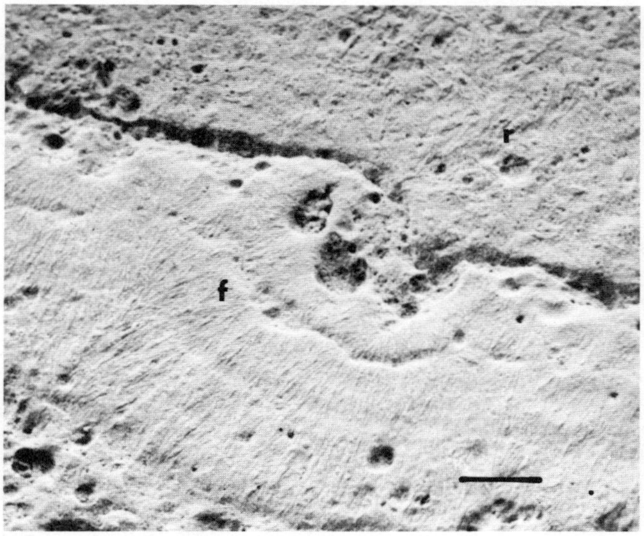

Figure 27. SEM photomicrograph showing three generations of fibrous sparry calcite (f) that coat a rudist skeleton (r) in the reef framework at Pipe Creek. Bar scale is 50 microns.

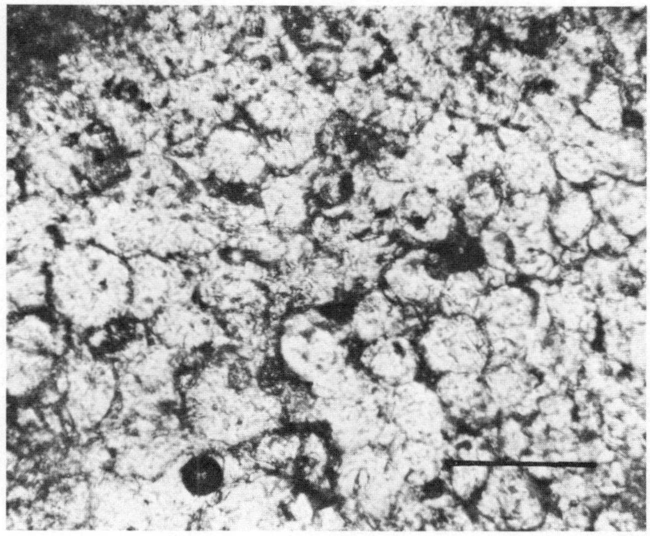

Figure 28. Photomicrograph of clionid silt within a cavity or burrow in a caprinid mound at Pipe Creek. Note the characteristic size and shape of the sponge chips. They are easily mistaken for microspar or vadose silt. Bar scale is 100 microns.

Table 5. Comparisons of primary textures within a rudist reef (Pipe Creek) and a Holocene coral reef (Discovery Bay, Jamaica). Holocene data courtesy of Clyde H. Moore, Jr.

THIN SECTION(S)	PIPE CREEK CAPRINID REEF			
	FRAMEWORK	VOID	INTERNAL SEDIMENT	CEMENT
1	42.5	3.5	14.0	40.0
1	40.0	3.5	25.5	31.0
1	24.0	5.0	61.5	9.5
1	25.5	4.5	54.0	16.0
1	38.0	4.5	36.0	21.5
1	40.5	5.0	39.0	15.5
1	32.0	3.5	44.0	20.5
1	11.0	7.5	75.0	6.5
1	50.5	7.0	29.5	13.0
1	35.5	2.5	58.5	3.5
MEAN=	33.9	4.7	43.7	17.7
	DISCOVERY BAY CORAL REEF			
2	30.0	5.0	53.0	12.0
3	8.0	5.0	77.0	10.0
4	13.0	8.0	71.0	4.0
6	19.0	7.0	68.0	6.0
MEAN=	17.0	7.0	69.0	7.0

occurred, a thin rind of isopachous equant sparry calcite was precipitated in pore spaces in the backshore and upper offshore beds in the fresh- or brackish-water phreatic zone.

A similar distribution of moldic as opposed to inversion textures was reported for Cretaceous carbonate beach sequences from the Edwards Limestone (Moore and others, 1972) and the Cow Creek Limestone (Inden, 1972). These sequences are characterized by beachrock lithoclasts, keystone vugs, intertidal cements, and an overlying caliche and/or supratidal dolomite. With the exception of one or two possible beachrock lithoclasts in the Odum core, none of these features is present in the Pipe Creek sand body. Because the distance between the outcrop and the Odum core is 4 km (fig. 5), it is possible that portions of the sand body now in the subsurface were emergent as beaches that acted as intake areas for meteoric water. The trend of moldic as opposed to inversion textures within the sand body certainly suggests that a freshwater lens was present during deposition. Shelf subsidence and ensuing burial resulted in dispersion of the lens and a return to marine interstitial water conditions.

Intraclasts are common to all of the grainstone units but are most abundant in the lime packstones of unit IIa (fig. 35). Because these intraclasts are composed of peloid lime wackestone to packstone, typical marine shelf sediment, syngenetic cementation of portions of the marine shelf is suggested. Semi-indurated shelf sediment was reworked as intraclasts and washed over the crest of the sand body into the backshore zone. Well-rounded lithoclasts of mollusk lime packstone occur at the top of unit IIb in the Odum core and provide further evidence of submarine cementation. These lithoclasts are composed of extensively bored mollusk fragments cemented by isopachous rinds of fibrous calcite spar and may represent reworked lime grainstone that was deposited and cemented in the littoral or shallow sublittoral zone as beachrock.

Unit IIa, a series of coarse-grained washover deposits interbedded with requieniid lime wackestones, contains large caprinid fragments whose pallial canals are filled with fibrous (P.F.$_3$C) sparry calcite cement overlying peloidal micrite. Not all of these geopetal structures are aligned with one another (fig. 36). Orientation of these geopetal structures and their association with superimposed borings show that sedimentation and cementation within the pallial canals and borings occurred before the fragments were deposited as part of the grainstone unit.

Other grainstones contain more convincing evidence of syngenetic cementation. Lenticular deposits of mollusk lime packstone to grainstone comprise portions of the reef. Well-rounded, moderately sorted mollusk fragments in these grainstones are coated with continuous, isopachous micrite rims that are associated with algal and fungal borings. Isopachous fibrous sparry calcite surrounds grains. The individual fibers are clouded with inclusions and often contain polygonal sutures (fig. 37) characteristic of the phreatic zone (Shinn, 1969). Where fibrous cement is absent, peloidal micrite forms isopachous and meniscus crusts around grains. Evidently these grainstones were cemented in the littoral and shallow sublittoral zones during deposition (fig. 38).

EPIGENETIC DIAGENESIS

Epigenetic diagenesis in this paper includes all processes that have effectively converted metastable carbonate sediments into stable limestone and/or dolomite. These processes (cementation, dissolution, inversion, and dolomitization) occurred in the presence of interstitial waters that, for the most part, had chemical compositions and salinities markedly different from the original marine water. Nonmarine calcareous cements are composed of clear, equant, well-formed crystals that are easily distinguished from typically cloudy, fibrous to bladed marine cements (Folk, 1974). Because shallow-marine carbonate sediments are usually subjected to one or more fresh-water stages, it is not uncommon to observe that two or more cementation episodes have affected a single pore.

Aragonite and magnesian calcite, although metastable, are the most common minerals in Holocene shallow-marine carbonate sediments, whereas calcite is the most stable calcium carbonate phase in fresh water (Stehli and Hower, 1961; Chave, 1962). Magnesian calcite having more than 12.0 mole percent $MgCO_3$ is more soluble than aragonite in marine

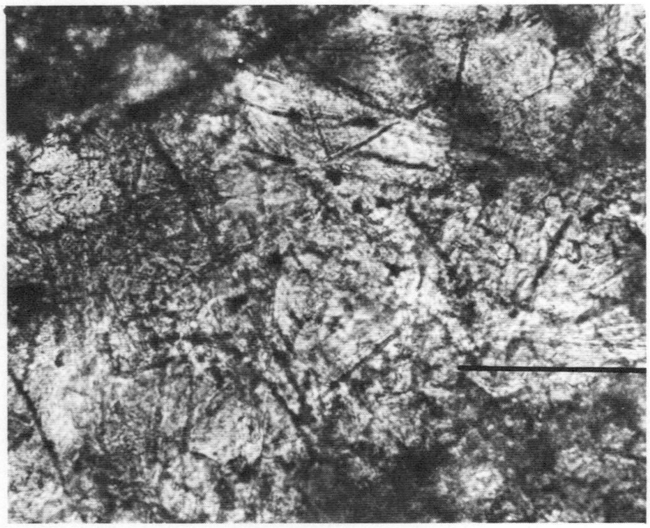

Figure 29

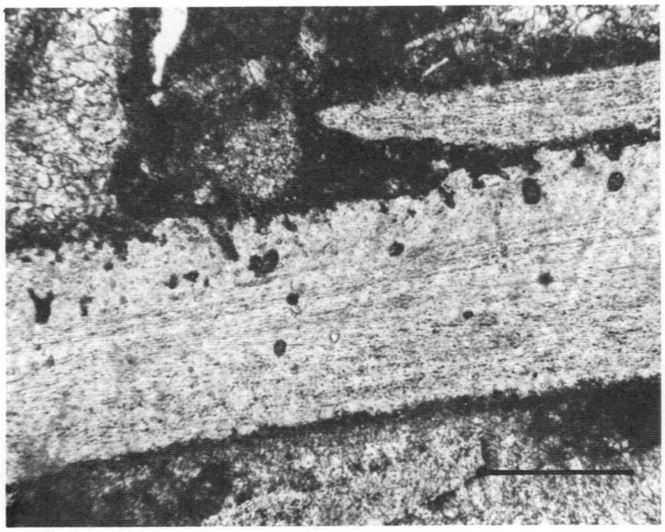

Figure 30

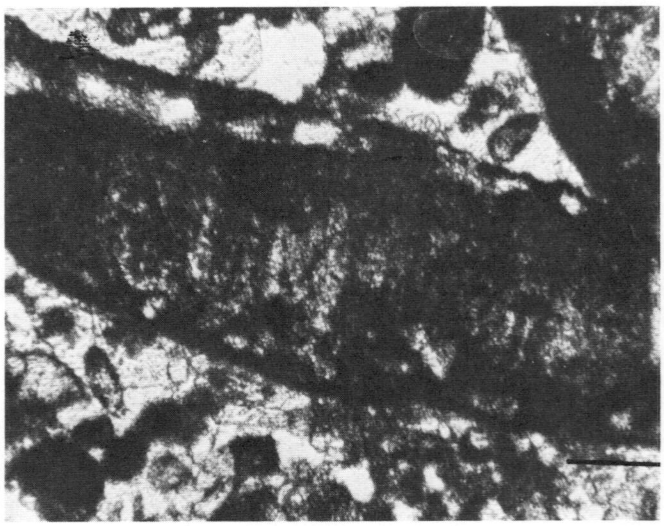

Figure 31

Figure 29. Photomicrograph of fungal borings in a mollusk grain, Pipe Creek. These microborings are typically straight and have diameters between 1 and 3 microns. Bar scale is 50 microns.

Figure 30. Photomicrograph showing algal borings that penetrate the outer margin of a mollusk grain, Pipe Creek. Coalesced borings of this type are thought to be responsible for the production of micrite rims. Bar scale is 250 microns.

Figure 31. Photomicrograph of a large peloid produced by centripetal replacement of a mollusk fragment by endolithic borers. Bar scale is 250 microns.

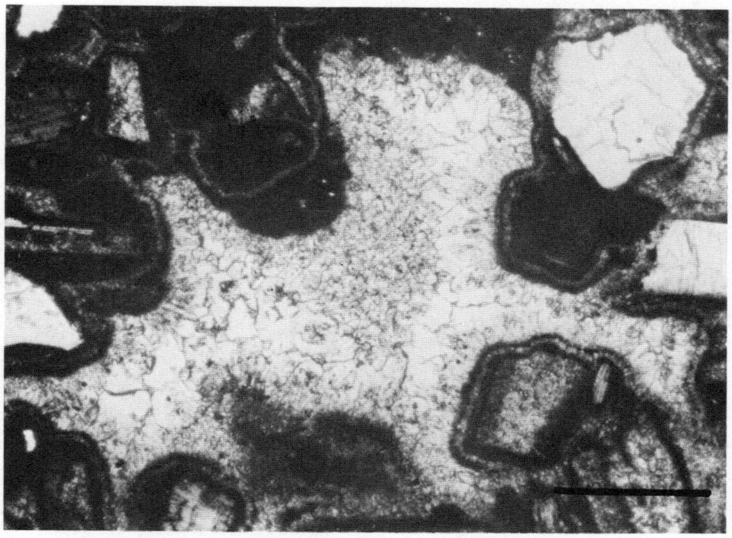

Figure 32. Photomicrograph showing a keystone vug at the top of a perireef grainstone (unit I). Fibrous isopachous cement (submarine phreatic?) is overlain by micritic meniscus cement (marine vadose?). The vug was filled later by equant sparry calcite (fresh-water phreatic). Bar scale is 250 microns.

water (Weyl, 1967). Berner (1975) demonstrated that, theoretically, magnesian calcites which contain greater than 7.0 mole percent $MgCO_3$ are less stable than coexisting aragonite in normal sea water. According to Berner, magnesian calcite will attain equilibrium with marine water through time by partial incongruent dissolution to intermediate magnesian calcite having between 2.0 and 7.0 mole percent $MgCO_3$ within its lattice. Therefore, the substitution of Ca^{+2} in a closed chemical system containing marine water could effectively raise the Mg/Ca ratio of the interstitial water and perhaps induce local dolomitization.

Both magnesian calcite and aragonite are unstable in fresh water but tend to equilibrate by different methods (Land, 1967; Winland, 1971; Berner, 1975). Fresh waters that may be undersaturated with respect to calcium carbonate remove Mg^{+2} from magnesian calcite by exsolution or incongruent dissolution and congruently dissolve aragonite (Land, 1967). Widespread dissolution may eventually result in the supersaturation of the pore water with respect to the least soluble phase, calcite, which is precipitated as intragranular or intergranular sparry calcite cement (Purdy, 1968). If cementation is not contemporaneous with dissolution, the relatively fragile micrite rims which often surround allochem molds are easily crushed during compaction (Dodd, 1966). Noncollapsed skeletal molds may be filled by calcite cements that are either syntaxial or epitaxial, depending on the timing of cementation.

In contrast to mineral stabilization by dissolution-precipitation, inversion textures are produced when the aragonite-calcite conversion (fluids present) occurs without an intervening mold stage (Folk, 1965). Inversion may involve dissolution-precipitation on a microscale as documented by Bathurst (1958) who noticed a distinct boundary or reaction front in partially inverted aragonite skeletons. The most diagnostic feature of inversion is the retention of organic laminae from the original shell. These laminae commonly transect calcite crystals (pseudospar of Folk, 1965) and usually continue across crystal boundaries (Bathurst, 1971). Also pseudospar crystals often penetrate the surrounding matrix along a jagged recrystallization front. The conversion of predominately aragonitic lime mud into lime mudstone (microcrystalline calcite or micrite of Folk, 1965) is probably accomplished by microscale dissolution-precipitation and cementation in the presence of fresh water. Holocene lime muds in Florida Bay and Discovery Bay, Jamaica, contain marine interstitial water and are not undergoing any mineralogical changes. Berner (1967) showed that Florida Bay muds near the northwest portion of the bay were losing Mg^{+2} by incongruent dissolution, but these sediments are annually affected by brackish waters derived from the adjacent Everglades.

Crickmay (1945) reported that some aragonite skeletons from the subsurface of Lau and Guam were recrystallized but that others from the same stratigraphic horizon were not. Evidently the recrystallization of aragonitic skeletons is controlled somewhat by shell micro-architecture or other factors as proposed by Banner and Woods (1964).

Schmaltz (1956) and Schlanger (1964) observed that calcitization of aragonitic shell material in Pleistocene and Pliocene epoch sediments from Funafuti and Guam occurred in the fresh-water phreatic zones immediately below exposure surfaces which developed during lowered stands of sea level. Middle Pleistocene epoch mollusks from Bermuda have inverted in the fresh-water phreatic zone in less than 10^5 years (Land, 1967, 1970).

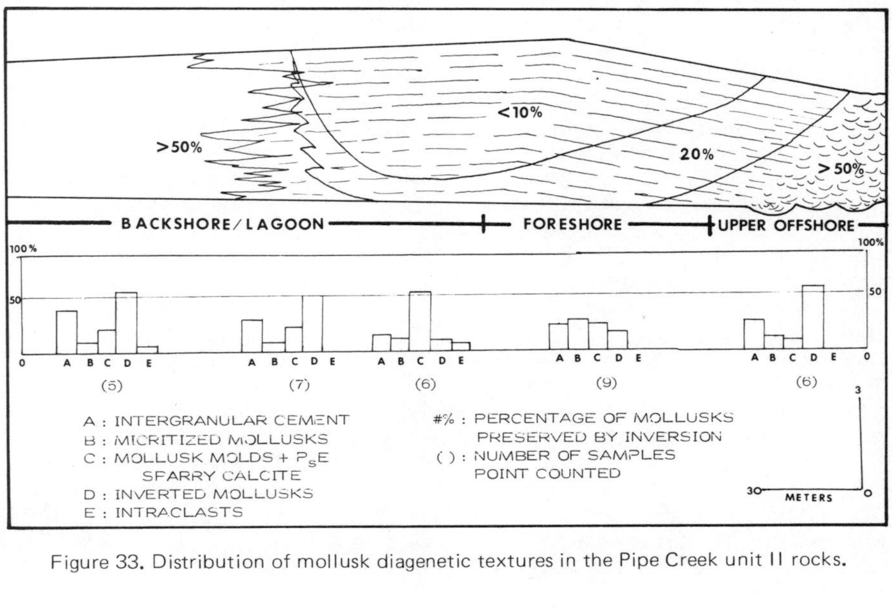

Figure 33. Distribution of mollusk diagenetic textures in the Pipe Creek unit II rocks.

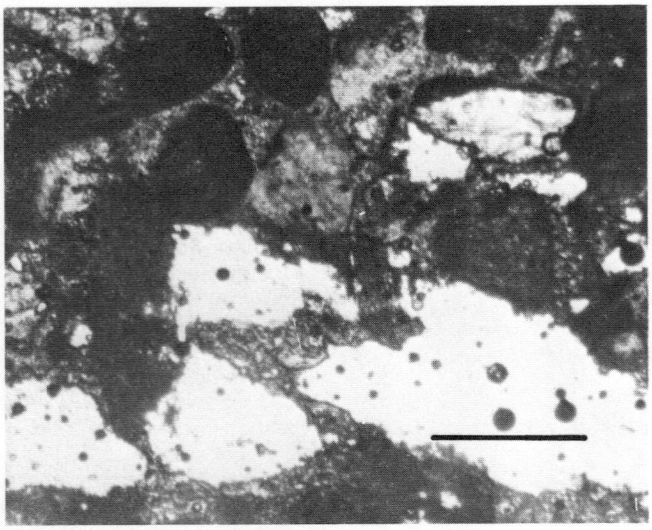

Figure 34

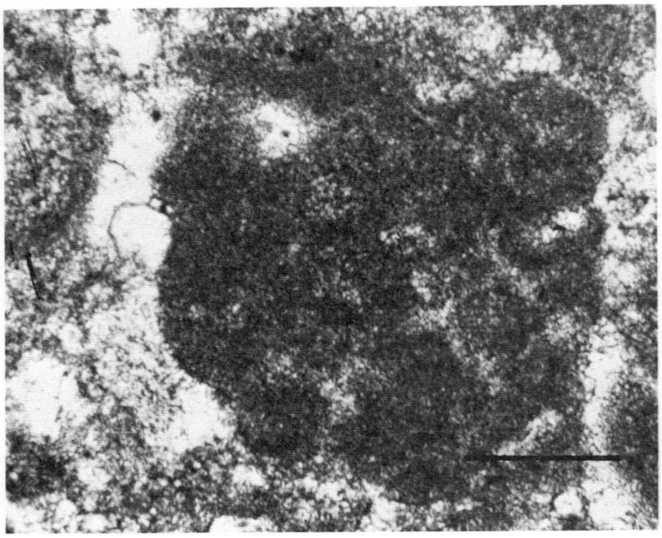

Figure 35

Figure 36

Figure 34. Photomicrograph showing moldic porosity developed in a foreshore lime grainstone-packstone at Pipe Creek. Noncollapsed mollusk mc!ds have been slightly enlarged during most recent dissolution phase. Bar scale is 250 microns.

Figure 35. Photomicrograph of an intraclast composed of peloid lime wackestone from the base of the festoon crossbedded lime grainstone (unit IIc). Bar scale is 250 microns.

Figure 36. Photomicrograph of misaligned geopetal structures in a mollusk fragment within the washover lime packstones of unit IIa. Cementation of the internal sediment occurred when the grain was in two different orientations, indicating submarine cementation. Bar scale is 250 microns.

Gavish and Friedman (1969) reported that inversion took place within 2×10^5 years in Pleistocene shallow-marine carbonate sediments from Israel. The geological evidence for fresh-water phreatic inversion of aragonite to calcite is well documented. Although, theoretically, aragonite should transform to calcite in sea water, this process of marine inversion must be dependent upon time, relative permeability of the sediment, temperature, and pressure. Most shallow-marine carbonate sediments may be exposed to fresh water shortly after deposition and burial. Presumably, once fresh water enters the sediments rapid inversion can take place.

Although the aragonite-calcite transformation, either by megascale dissolution-precipitation or inversion (microscale dissolution-precipitation?), can occur in fresh water with concomitant textural change, many fossils have apparently not been affected by either phenomenon. Land (1967) reported that magnesian calcite allochems now in the fresh-water phreatic zone of Bermuda have converted to calcite without attendant textural change. Land proposed that Ca^{+2} liberated from aragonite during congruent dissolution has substituted for Mg^{+2} in the calcite lattices of coralline algae, echinoderms, and miliolid foraminifers. This exsolution or incongruent dissolution (Land, 1967) was reproduced in laboratory experiments (Schroeder, 1969). Winland (1968) suggested that micrite rims and submarine-cement textures are faithfully preserved in many ancient carbonate rocks because their original magnesian calcite mineralogy was converted to calcite by incongruent dissolution.

Using all these observations, the paragenesis of the caprinid reef facies at Pipe Creek was reconstructed. Epigenetic diagenesis of the Cretaceous sediments that were originally composed of predominantly aragonite and magnesian calcite to rocks composed of calcite and dolomite followed a logical and predictable sequence.

Phase I: Marine Connate—Closed System

Dolomite within the caprinid reef facies at Pipe Creek is restricted to portions of the backreef and marine shelf lime mudstones and wackestones. Sedimentary structures normally associated with supratidal sequences are absent from these sporadically distributed dolomite zones.

More than 80 percent of the dolomite crystals average 8 to 10 μm in diameter. These subhedral crystals (fig. 39) are cloudy, probably because of fluid and/or solid inclusions within the lattice. Approximately 15 to 20 percent of the dolomite is composed of clear, euhedral, 50-μm-diameter rhombs. The more diminutive dolomite has replaced matrix only, whereas the larger rhombs partially replace matrix and allochems (fig. 40). Dolomitization is often incomplete; the percentage of dolomite ranges between 0.1 and 100.00 volume percent of dolostones and dolomitic limestones. The relative proportion of dolomite per sample varies with the amount of terrigenous clay (fig. 41). X-ray diffraction analysis indicates that the carbonate phase is disordered protodolomite that contains 43.0 to 48.0 mole percent $MgCO_3$ (table 6).

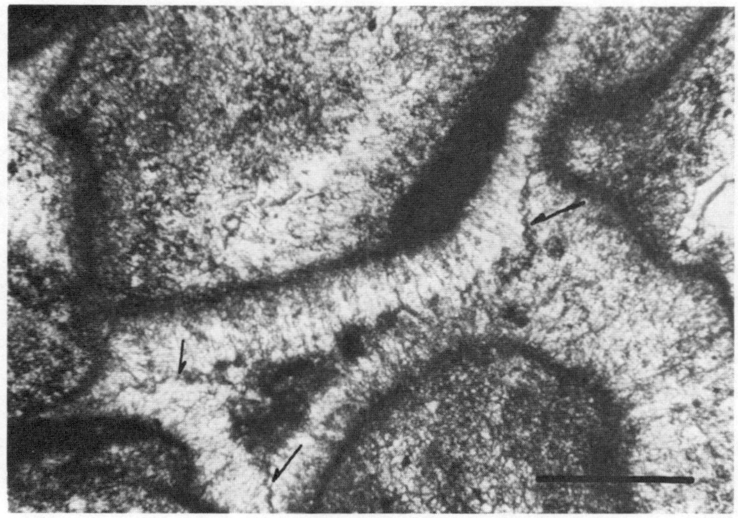

Figure 37. Photomicrograph showing polygonal cement sutures (arrows) that indicate that this cement was precipitated in the phreatic zone. The cement is fibrous and clouded with inclusions. This possible submarine cement is from a lens of lime packstone within the uppermost portion of the lower tabular reef mass at Pipe Creek. Bar scale is 250 microns.

Several studies have reported the association of dolomite and clay minerals (Payton, 1966; Schmidt, 1969; Mossler, 1971). These studies conclude that the release of absorbed Mg^{+2} from phyllosilicate clay minerals can promote dolomitization. Other researchers have proposed that clay minerals may react with calcite to form dolomite (Zen, 1959) or act as nucleation sites or "seeds" for dolomite crystals (Kahle, 1965). I propose that the clay minerals at Pipe Creek acted as nucleation sites for dolomitization during initial burial in marine connate water.

Magnesian calcites should have equilibrated through time through partial incongruent dissolution to magnesian calcite having 2.0 to 7.0 mole percent $MgCO_3$ (Berner, 1975). Volumetric estimates of various lithofacies at Pipe Creek were made from isopachous maps and cross sections in order to estimate the amount of Mg^{+2} originally in the sediments. Data derived from mass-balance calculations indicate that the major sources of Mg^{+2} were submarine cements and internal sediments associated with the reef framework (table 7). The rearrangement of Mg^{+2} to form dolomite is not similar to the magnesian calcite to dolomite transformation of coralline algae from Eniwetak described by Schlanger (1963). Mass-balance calculations suggest that at least 30 times the amount of Mg^{+2} than is present in the Pipe Creek dolomite was available for dolomitization if the magnesian calcite comprising submarine cements and internal sediments converted from an average 15 mole percent $MgCO_3$ to 7 mole percent $MgCO_3$. This incongruent-dissolution phase released Mg^{+2} and depleted the connate water in Ca^{+2}, effectively raising the Mg/Ca ratio of the interstitial water. Folk and Land (1975) have proposed that an increased Mg/Ca ratio of water either with or without an attendant salinity increase would stimulate dolomitization. Dolomitization that affected

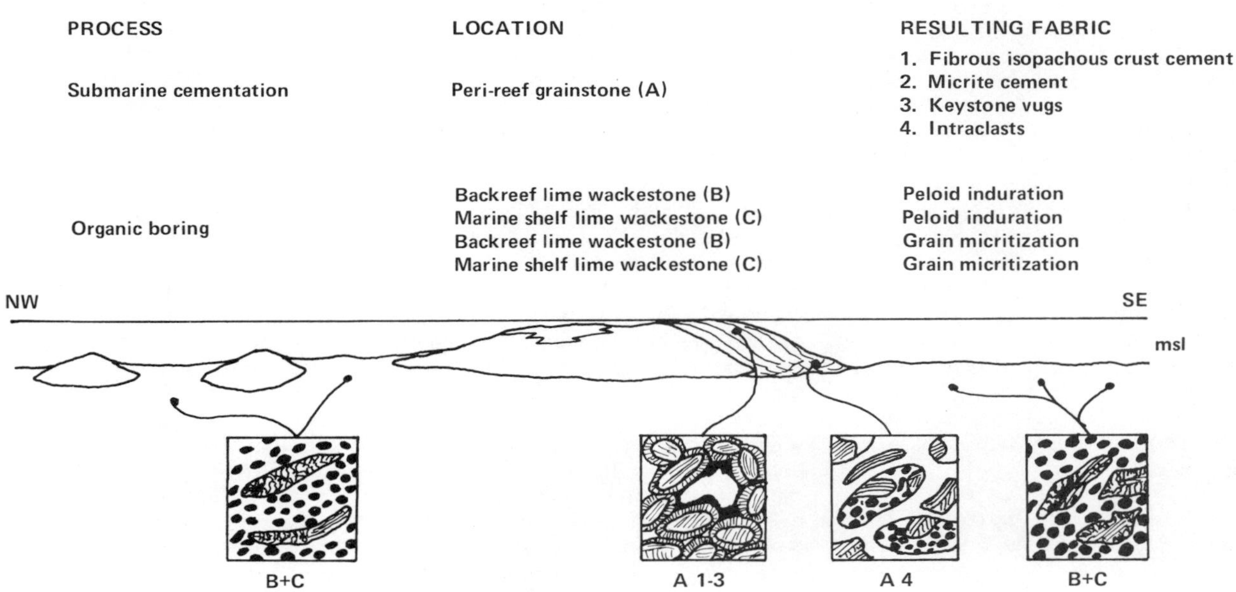

Figure 38. Distribution of syngenetic diagenesis and resultant fabrics in nonreef beds.

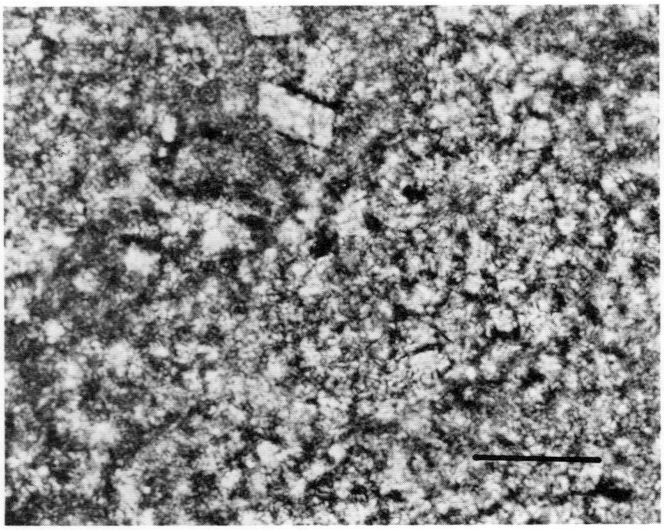

Figure 39

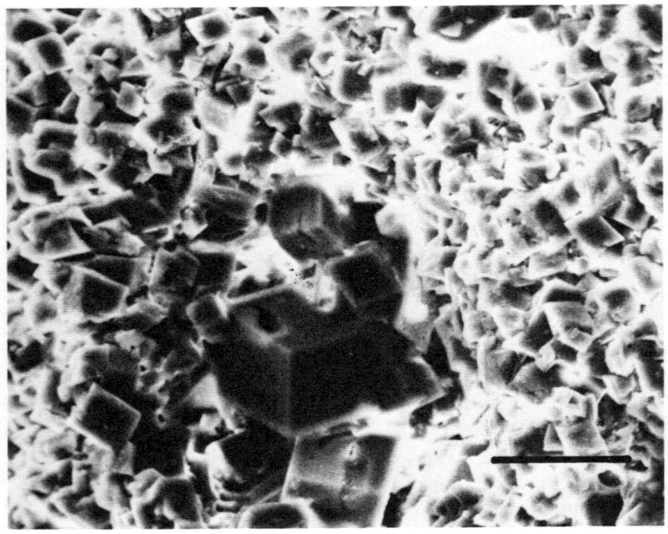

Figure 40

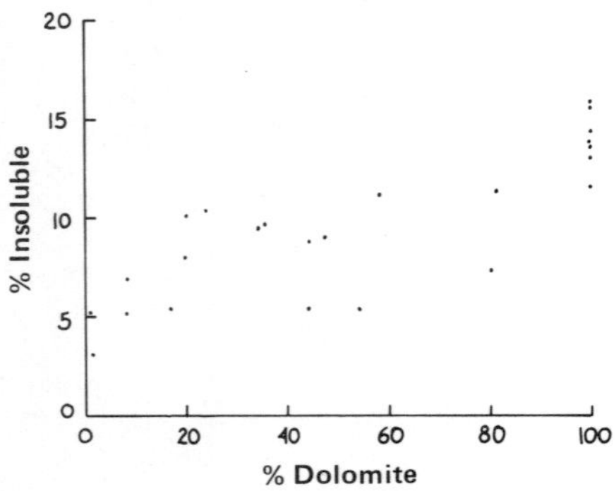

Figure 41

Figure 39. Small subhedral to euhedral primary protodolomite with larger euhedral secondary protodolomite rhombs. Dolomitic portion of a mollusk lime wackestone, Pipe Creek. Bar scale is 100 microns.

Figure 40. SEM photomicrograph of primary and secondary protodolomite from the sample shown in figure 39. Bar scale is 50 microns.

Figure 41. Graph of dolomite percentage vs. insoluble percentage for Pipe Creek dolostones. The dolomite is protodolomite containing an average of 43.0 mole percent $MgCO_3$. The clay fraction is composed of quartz, kaolinite, illite, and montmorillonite.

clay-rich beds was apparently interrupted by fresher water that flushed excess Mg^{+2} from the study area.

Atomic absorption analysis indicates that the 8- to 10-μm dolomite crystals at Pipe Creek are characterized by low Sr^{+2} and high Na^{+1} values. The mean value of strontium (73 ppm) compares favorably with values published for this element in other ancient dolomites (fig. 42). Holocene supratidal and subtidal dolomites average between 600 and 900 ppm Sr^{+2} (Behrens and Land, 1972; Land and Hoops, 1973). Ancient dolomites commonly have much lower values, presumably because of physical contact with fresh water which flushed or diluted trace elements (Land and others, 1975). Therefore, the Sr^{+2} values of the Cretaceous dolomite suggest a post-dolomitization fresh-water phase of diagenesis.

The sodium content of the Pipe Creek dolomites is extremely high when compared to other ancient dolomites but is comparable with Holocene dolomite (table 6). Chloride analysis by titration indicated that the anomalously high sodium content cannot be explained by either halite inclusions (none noted by X-ray) or fluid inclusions of marine water (table 6). Therefore, it is assumed that the sodium values reflect Na^{+1} released by clay during acid digestion of the samples for wet chemical analysis.

Most of the lime mud within the study area was probably produced by the disintegration of codiacean algae (aragonite), rhodophyte epibionets (magnesian calcite) from marine grass blades, or both. Folk (1965) noted that between 10 and 1,000 aragonite needles would be required to produce one 3-μm polyhedron of calcite (micrite). The difference in relative surface area between aragonite needles and micrite polyhedra would enhance the reactivity of the needles with interstitial water. This reasoning suggests that dolomitization occurred before the conversion of lime mud to mudstone (micrite). At the end of the marine connate phase of diagenesis, the sediments and rocks at Pipe Creek were composed of aragonite, intermediate magnesian calcite (2 to 7 mole percent $MgCO_3$), and dolomite.

Phase II: Early Fresh Water—Open System

The Glen Rose reef interval rocks are overlain by upper Comanchean through Gulfian (middle Albian through Maastrichtian) rocks that have a thickness ranging between 500 and 750 m (Young, 1967). Uplift of portions of the Comanche shelf may have begun as early as the middle of the Cretaceous period. However, reworked Upper Cretaceous limestone pebbles and fossils that occur on the downthrown (Gulf) side of the Balcones fault in Miocene rocks are evidence of fault movement during the early Miocene epoch (Murray, 1961).

Prior to extensive regional uplift associated with faulting, meteoric water intake areas had probably developed on the Comanche shelf. The upper portion of the lower member and the entire upper member of the Glen Rose contain a large proportion of supratidal-dolomite and tidal-flat sequences. These extensive carbonate mudflats may have acted as intake areas for meteoric water that flushed downdip through the underlying sediments. In addition, the top of the

Table 6. Sodium and strontium values for dolostones from the published literature and dolostones at Pipe Creek.

MINERALOGY	DOLOMITE ORIGIN	AGE	Sr (ppm)	Na (ppm)		REFERENCE
Dolostone	Secondary	-	174-181	251-391		Weber (1964)
Dolomite	Secondary	-	84.5	209-264		"
Dolomite	Supratidal	-	-	200-900		Fritz and Katz (1972)
Dolomite	Primary	-	-	70-200		"
Dolomite	Secondary	-	-	150		"
Dolomite	Hydrothermal	-	-	100		"
Dolomite	Secondary	Ordovician	37	190		Badiozamani (1973)
Limestone	Primary	Ordovician	50-180	-		"
Dolostone	Supratidal	Eocene	90	300		Land and others (1975)
Dolostone	Supratidal	Ordovician	50-180	-		"
Dolomite	Subtidal	Holocene	840-970	1010-2500		Behrens and Land (1972)
Dolomite	Supratidal	Holocene	620-640	1780-3050		Land and Hoops (1973)
					Cl (ppm)	
Dolostone	Secondary	Cretaceous	87	N.D.	N.D.	
Dolostone	Secondary	"	77	1858	1168	
Dolostone	Secondary	"	11	2180	3390	
Dolostone	Secondary	"	100	1979	1337	
Dolostone	Secondary	"	91	2628	990	
Dolostone	Secondary	"	123	3987	941	
Dolostone	Secondary	"	33	630	1350	
Mean			73	2210		

Table 7. Mass balance calculations that indicate that the total magnesium required to produce the dolomite at Pipe Creek is much less than the amount available. The large surplus indicates that the geochemical system became open prior to more complete dolomitization. The open system effectively flushed excess Mg^{+2} downdip.

Magnesium Source	Available Mg^{+2} (Moles)
Miliolid lime wackestones	4.5×10^7
Caprinid reef cements and internal sediments	7.0×10^8
Interstitial water	9.0×10^7
Total Mg^{+2} available	8.3×10^8
Mg^{+2} in dolomite	4.9×10^7
Surplus magnesium	7.8×10^8

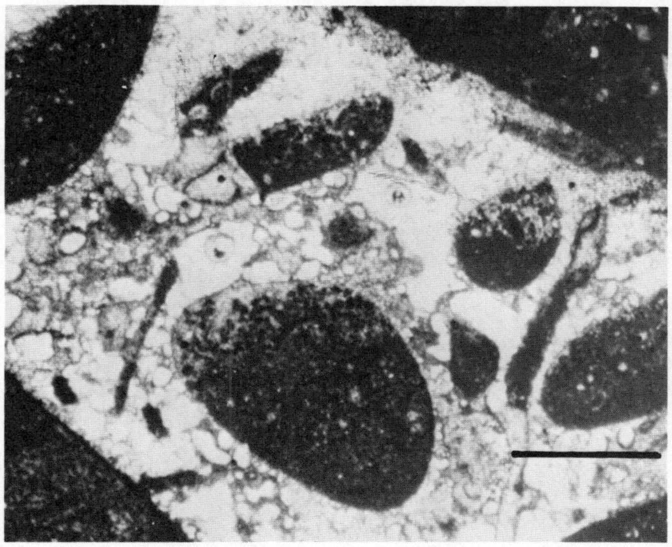

Figure 42

Figure 42. Photomicrograph that shows a caprinid replaced by sparry calcite euhedra. The aragonitic shell wall was dissolved and the resulting mold filled by cement. The cement has been subsequently partly leached. Mold of the rudist is outlined by a micrite rim. Pipe Creek caprinid mount. Bar scale is 250 microns.

Fredericksburg Group is marked by a pronounced, iron-stained unconformity that is thought to have developed during a marine regression around the Llano region of Central and south-central Texas (Moore, 1964). Therefore, meteoric water intake areas could have developed during latest Fredericksburg time.

Some caprinids, requieniids, gastropods, and other mollusk fragments are preserved by inversion textures within the caprinid reef facies. The resulting pseudospar has the following characteristics: slightly curved crystal boundaries that do not form enfacial junctions, linear arrangement of organic inclusions that transect calcite mosaic and continue across crystal boundaries (fig. 12), faint brown color, and slightly undulose extinction. Approximately 22 percent of the caprinids within tabular reefs and 5 percent of the caprinids within backreef mounds were subjected to inversion. The remaining caprinids have undergone dissolution-precipitation (fig. 42). The microscale phenomenon was interrupted by megascale dissolution as the interstitial water became progressively undersaturated with respect to $CaCO_3$.

During the fresh-water phase, equant sparry calcite was precipitated in fossil molds. This calcite (figs. 43 and 44) is distinguished from pseudospar by its clarity, lack of inclusions, and planar crystal boundaries. Incomplete cementation within some molds or later dissolution of calcite by modern ground water has resulted in the retention of some secondary porosity (fig. 42).

Final incongruent dissolution of magnesian calcite removed the remaining Mg^{+2} from coralline algae, echinoderms, miliolids, submarine cements, and pelleted internal sediments. The Mg/Ca ratio of the interstitial water probably varied considerably if the system were open and if the hydrostatic head that developed updip were of great enough magnitude to keep interstitial water flowing through the sediments (Benson and others, 1972). As the Mg/Ca ratio of the pore waters decreased, a second very minor stage of dolomitization occurred, which resulted in the formation of dolomite which averages 50 μm in diameter, is composed of well-formed euhedra, and is optically clear. Presumably this dolomite formed in fresh water when the Mg/Ca ratio was approximately 1 (Hanshaw and others, 1971; Folk and Land, 1975).

Nonreef lime wackestones within the caprinid reef facies have converted to micrite, microspar, and pseudospar. Bathurst (1971) noted that fresh water seems to be necessary to promote the conversion of lime mud to mudstone, although internal sediments within modern reefs composed to a large extent of lime mud are often thoroughly cemented under marine conditions. Evidently the interstitial water chemistry of reefs, perhaps influenced by organic decay and/or photosynthesis, is often conducive to the precipitation of calcium carbonate cements (Schroeder and Zankl, 1974). However, the thick sequences of Holocene lime mud (aragonite and magnesian calcite) in Discovery Bay (Jamaica) and Florida Bay do not contain any evidence of lithification in the presence of marine pore water (Berner, 1966; Moore, 1974, personal communication).

The term "grain growth" was used by Bathurst (1958) to describe the mud to micrite conversion. Bathurst felt that micrite formed by solution and precipitation of overgrowth cement on preexisting nuclei. Folk (1965) argued in favor of porphyroid neomorphism where grains, starting from nuclei spaced 1 to 5 μm apart, grow in the presence of fluids until they coalesce into micrite (microcrystalline calcite). This micrite is composed of anhedral calcite crystals that have diameters of between 3 and 5 μm. The original needles or plates and the voids between them are thereby replaced during this microscale dissolution-precipitation.

In some Pipe Creek samples the micrite has recrystallized to microspar and pseudospar. The microspar often contains numerous "floating" peloids that are not mutually supportive. Many irregular patches of micrite occur within microspar crystals and outline crystal boundaries. The overall microspar distribution is random and

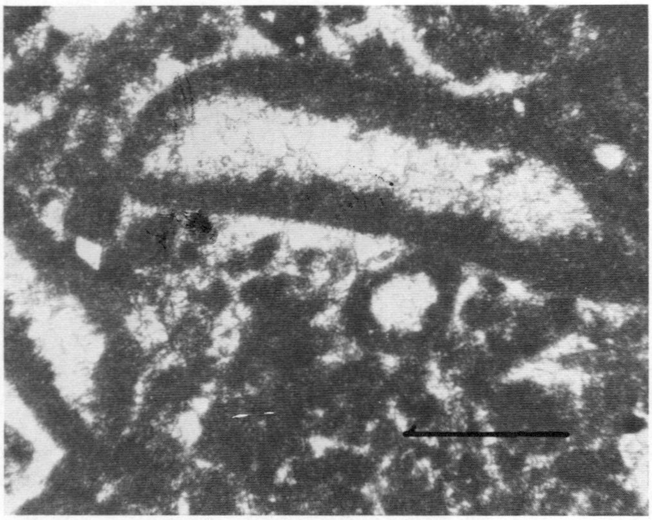

Figure 43

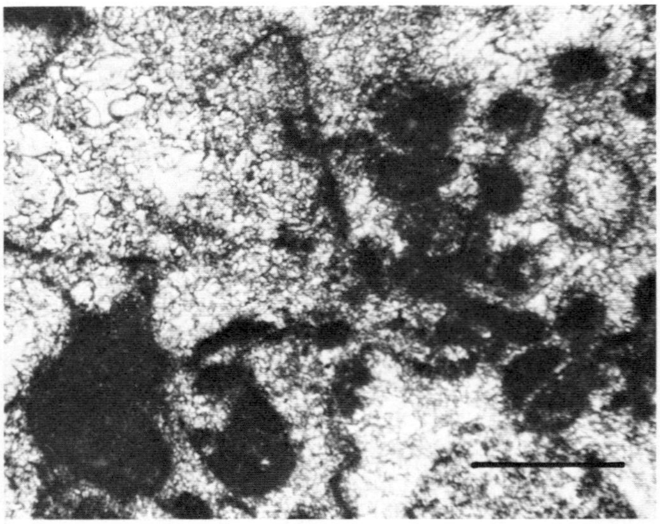

Figure 44

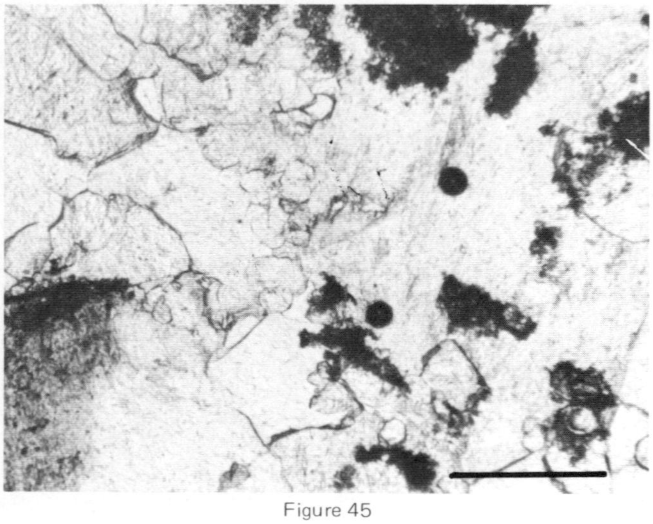

Figure 45

Figure 43. Photomicrograph of a mollusk mold filled with PsE sparry calcite. Noncollapsed micrite rim shows that dissolution of aragonite shell material was accompanied by intergranular cementation that prevented rim collapse. Bar scale is 250 microns.

Figure 44. Photomicrograph of dasycladacean algae and a mollusk mold. Both the mollusk shell and the thalli of the green alga have been dissolved and replaced by sparry calcite (PsE variety). Algal utricles are filled with micrite. Pipe Creek. Bar scale is 250 microns.

Figure 45. Photomicrograph of pseudospar replacing micrite of lime wackestone in the backreef facies at Pipe Creek. Note the micrite inclusions within crystals and along crystal boundaries. Bar scale is 250 microns.

Table 8. Trace element values for various carbonates from the published literature and from Pipe Creek samples. All Pipe Creek data are by atomic absorption spectrophotometry. A = aragonite, C = calcite, MC = magnesian calcite.

TYPE OF SAMPLE	OCCURRENCE	INFERRED ORIGINAL MINERALOGY	Mole % $MgCO_3$	Wt. % Fe	(ppm) Na	(ppm) Sr
P. FC cement	Submarine cement in reef frame	MC or A	0.02	0.42	784	71
"	"	"	0.01	0.00	38	37
"	"	"	0.03	0.09	398	34
"	"	"	0.02	0.00	23	15
			Mean:		311	39
Peloid packstone	Internal sediment, reef frame	MC or A	0.08	0.04	170	125
"	"	"	0.05	0.08	343	206
"	"	"	0.06	0.04	186	138
"	"	"	0.08	0.12	195	139
Micrite	Filling in caprinid pallial canal	"	0.01	0.14	323	33
			Mean:		243	128
P. E sparry calcite	Geopetal filling, reef frame	C	0.08	0.03	198	267
"	"	C	0.14	0.04	263	246
"	Vug filling, reef frame	C	0.02	0.11	16	32
"	Geopetal filling, unit I g.s.	C	0.06	0.27	34	2
"	Intergranular cement, unit I g.s.	C	0.07	5.85	35	17
P. BC Scalenohedral	Fracture filling, reef frame	C	0.02	1.51	402	74
Toucasia sp.	Backreef lime wackestone	C	0.01	0.00	1151	125
"	Unit IIa spillover lens p.s.	C	0.01	0.10	1202	1058
Inverted mollusk	Unit IIb foreshore g.s.	A	0.01	0.08	2798	812
"	Unit IIc upper offshore g.s.	A	0.01	0.09	1733	967
"	"	A	0.01	0.00	211	108
"	"	A	0.01	0.00	1719	137
"	Unit IIc upper offshore p.s.	A	0.02	0.03	1158	374
			Mean:		1425	512

Cement Morphology or Allochem	Mineralogy	Age	Mole % $MgCO_3$	Wt. %Fe	Sr (ppm)	Na (ppm)	Reference
Equant	C*	-	-	-	1200	-	Kinsman (1969)
Equant-fibrous	C**	-	-	-	700-10000	-	"
Equant-fibrous	C***	-	-	-	350	-	"
Fibrous	A	Holocene	0.14- .31	-	7542-9249	-	Moore (1973)
Bladed	MC	Holocene	-	-	1025	-	"
Micritic	MC	Holocene	18.73	-	1182-2736	-	"
Micritic	A	Holocene	-	-	-	4580	Moore (1970)
Micritic	MC	Holocene	13.0-15.0	-	-	-	Alexandersson (1972)
Equant	C	Cretaceous	0.32-2.72	0.06- .26	-	-	Benson and others (1972)
Equant	C	Pennsylvanian	2.43	0.13	-	-	"
Equant	C	Mississippian	0.16	0.12	-	-	"
Equant	C	Devonian	0.69	0.05	-	-	"
Oolite	A	Holocene	0.18- .28	-	9800	-	Kinsman (1969)
Oolite	A	Holocene	0.19- .82	-	9590	-	"
Coral	A	Holocene	0.33- .64	-	8500	-	"
Chlorophyte algae	A	Holocene	0.14- .22	-	8740	-	"
Mollusks	A, A + C	Holocene	-	-	800-4000	-	Lowenstam (1963)
Lime mud (Persian Gulf)	A	Holocene	-	-	9390	-	Kinsman (1969)
Average Ancient Limestone	C	-	-	-	250-700	-	"

*Theoretical calcite in equilibrium with sea water
**Closed-system calcite after aragonite
***Open-system calcite after aragonite

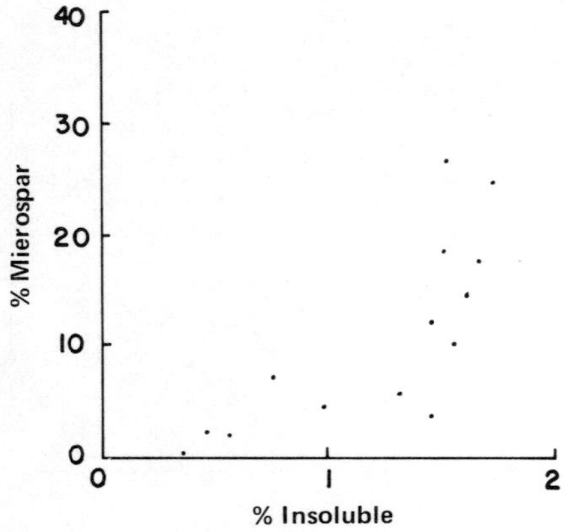

Figure 46. Graph of insoluble residue percentage as opposed to recrystallized matrix percentage for backreef lime wackestones at Pipe Creek.

in many thin sections is absent. Often the microspar has grown into pseudospar, a product of porphyroid neomorphism (fig. 45). Terrigenous clay minerals in quantities exceeding 2.0 percent are thought to enhance the recrystallization of micrite by providing nucleation sites for calcite (Folk, 1965; Bausch, 1968). Fresh water also promotes micrite recrystallization by removing absorbed Mg^{+2} (Berner, 1966; Folk, 1974). Insoluble-residue analysis shows that there is an apparent correlation between clay content and the occurrence of microspar and pseudospar within nondolomitic Pipe Creek lime mudstone and wackestones (fig. 46). Samples that contain 1.5 to 2.0 percent insoluble residue contain an average of 15 percent microspar and pseudospar.

Elements such as magnesium, strontium, iron, and sodium—generally present in trace amounts in limestone—are particularly useful when combined with cement morphology in determining the original pore-water chemistry and evolution of Holocene rocks and sediments (Folk, 1974). Unfortunately, extrapolation of similar techniques to ancient rocks is difficult because of mineralogical changes in response to changing interstitial-water composition through time. For example, Benson and others (1972) found that the Mg and Fe content of Mesozoic and Paleozoic calcite cements is highly variable. In some cases the trace element composition changes at a petrographic boundary between cement generations. In other cases this relationship is not observed.

In general, Sr^{+2} is a large cation that is easily accommodated within the aragonite lattice, whereas Mg^{+2} and Fe^{+2} are preferred by calcite and dolomite because of their smaller ionic radii (Deer and others, 1966). However, small amounts of Sr^{+2} are present in most calcite and minor amounts of Mg^{+2} are reported from aragonite. The presence of Na^{+1} in Holocene micritic intertidal cements was reported by Moore (1973). The ionic radius of Na^{+1} is similar to the radius of Ca^{+2} (Deer and others, 1966). Evidently the abundance of this monovalent cation in sea water allows it to be incorporated into $CaCO_3$ lattices that prefer divalent cations.

The distribution of Sr^{+2} and Na^{+1} within cements and internal sediment is an indicator of the paleohydrology at Pipe Creek. Fibrous submarine cements within the reef framework contain moderate Na^{+1} and low Sr^{+2} (table 8). Petrographic observations suggest that these cements were originally magnesian calcite which would have contained low Sr^{+2} because the relatively large ionic radius of Sr^{+2} prohibits the incorporation of large amounts of Sr within the calcite lattice. Peloidal internal sediment adjacent to the fibrous cement has slightly higher values of Sr^{+2} (table 8). The slightly higher average Sr value of the internal sediments suggests that their origin was magnesian calcite plus aragonite. Fresh ground waters have flushed most of the Sr from the cement and the internal sediment but some memory of original mineralogy is apparently retained.

Equant sparry calcite that fills geopetal structures within the reef framework is relatively enriched in Na and Sr (table 8). This equant calcite most often overlies and is in optical continuity with a single generation of cloudy fibrous sparry calcite that may have originally been aragonite. The petrographic relationship implies that inversion of the fibrous aragonite was accomplished during precipitation of pore-filling equant calcite. Sr^{+2} and Na^{+1} from the aragonite were reincorporated into the equant calcite. Lowered porosity and permeability of the reef framework, because of earlier submarine diagenesis, enhanced the reincorporation of these trace elements into the equant cements by reducing the flow rate of subsurface water through the reef framework. Conversely, equant cements in nonreef beds have very low Sr^{+2} and Na^{+1} values (table 8). These data indicate that the nonreef equant cements were precipitated in a relatively open geochemical system where Sr and Na were flushed downdip.

Recrystallized (inverted) mollusk fragments and backreef micrite contain the highest Sr and Na values of any of the samples analyzed (table 8). Because Holocene mollusks contain 800 to 5,000 ppm Sr^{+2} (Lowenstam, 1963), it seems reasonable to assume that aragonitic Cretaceous mollusks were composed of high-Sr and perhaps high-Na aragonite. Calcite after aragonite in an open system should theoretically contain less than 350 ppm Sr (Kinsman, 1969). The probability of retention of rather high trace element values by mollusk shells at Pipe Creek may have been enhanced by the intergranular equant sparry calcite that coats the mollusk shells. Perhaps the equant calcite has acted as a shield against further removal of Sr and Na from the shells by fresh water. Similar reasoning can be applied to the backreef micrite. Holocene algal aragonitic lime mud contains about 9,000 ppm Sr^{+2} (Kinsman, 1969). During inversion of the Cretaceous lime mud (aragonite?), Sr was reincorporated into the micrite because of relatively lower water flow rates through the muddy textured nonreef units. Subsequent fresh-water flushing has failed to effectively remove Sr and Na from the micrite because of reduced flow rates in the open system.

Values of Mg and Fe for the Pipe Creek rocks are very low (table 8). Magnesium was removed from submarine magnesian calcite cements, internal sediments, and allochems by incongruent dissolution. Fresh-water equant sparry calcite cements are also low Mg, indicating that the hydrologic system that removed Mg^{+2} was open and effectively flushed Mg^{+2} from the study area. Only one of the calcite cements rendered a positive reaction to staining for iron. Calcite cements that contain ferrous iron are thought to represent fresh-water phreatic precipitation from waters that have low Eh values (Evamy, 1969). The absence of ferrous iron does not imply that the cement was precipitated in the vadose (oxidizing) zone, but only that phreatic waters did not have a low Eh or that little dissolved iron was available. Hensington (1962) reported that modern ground waters in the Llano region and the Edwards Plateau are mostly iron deficient. The ground waters responsible for the precipitation of equant sparry calcite at Pipe Creek were probably similar to modern ground waters in the region.

During the early fresh-water phase, the mineralogy of the caprinid

Table 9. Diagenetic phases for the caprinid reef interval at Pipe Creek. The sequence was established by petrographic and geochemical data.

	0	I	I	III
GRAIN DIAGENESIS	Grain micritization, micrite rims formed, clionid boring of reef framework	Incongruent dissolution of MC	Incongruent dissolution of MC, megascale dissolution of A, inversion of A to C	Some dissolution of allochems (c)
INTERGRANULAR/ MATRIX DIAGENESIS	Cementation of reef frame by MC, A, or both. Cementation of internal sediment. Perireef sands cemented in the littoral zone. Micritic and fibrous cements	No textural change. Dolomitization of clay-rich backreef beds with 8 to 10 μm dolomite	Lime mud to lime mudstone (micrite), dolomitization (50 μm rhombs), equant calcite cements. Micrite to microspar and pseudospar	Fracturing and some dissolution, bladed to micritic calcite cements. ?Micrite to microspar and pseudospar?
INFERRED ORIGINAL MINERALOGY	A, MC (10-20 mole percent $MgCO_3$), and C	C, A, and MC (10-20 mole percent $MgCO_3$)	C, A, MC (2-7 mole percent $MgCO_3$), and D	C + D
FINAL MINERALOGY	Same	C, A, MC (2-7 mole percent $MgCO_3$), and D	C + D	C + D
INFERRED PORE WATER COMPOSITION	Marine, high Na, low Sr, Mg/Ca=3	Marine, high Na, low Sr, Mg/Ca=10+	Brackish to fresh, low to moderate Na and Sr, Mg/Ca=1	Fresh, low Na, low Sr, Mg/Ca=1
POROSITY TYPE	Minor porosity interframework, intergranular porosity in sands	Intercrystalline porosity in dolomitic sands	Melodic porosity in reefs, mounds, and sands. Intergranular porosity partly occluded by cementation	Development of vuggy and fracture porosity—not fabric dependent. Minor, local cementation of vugs and fractures
POSSIBLE TIME SEQUENCE	Syndepositional	Initial burial to upper Trinity or upper Fredericksburg	Upper Trinity or upper Fredericksburg	Cenozoic

reef facies was altered from aragonite, dolomite, and intermediate (2 to 7 mole percent $MgCO_3$) magnesian calcite to calcite and dolomite. Mineral stabilization was accomplished entirely in the subsurface and was not dependent on subaerial exposure of the reef-interval units.

Phase III: Later Fresh-Water Epigenetic Diagenesis—Open System

Diagenesis of the rocks has continued through the present (table 9). Fractures are developed in caprinid reefs and mounds. Because they transect allochems, matrix, and cement, these features are thought to represent a response of the rocks to regional tectonic stress. Scalenohedral bladed sparry calcite often lines these fractures. A reddish brown micrite associated with these cements is often draped over individual bladed crystals (fig. 47). The micrite contains small flecks of limonite which impart a red color to the micrite. This calcite is probably the product of recent subaerial weathering. As undersaturated meteoric water percolates downward through limestones, it becomes more saturated with respect to $CaCO_3$. Water ponded in fractures and vugs may precipitate micrite during evaporation. Large solution vugs, common in surface samples from Pipe Creek, are often filled with microstalactitic equant to bladed sparry calcite cement. This cement represents precipitation in the modern vadose zone.

Vugs and fractures developed after the entire rock sequence had been converted to the relatively stable calcite and protodolomite. Undoubtedly enormous quantities of fresh water have passed through the rocks since the Pleistocene epoch when the modern regional drainage pattern developed. Autocthonous sparry calcite (Bathurst, 1971) was precipitated in vugs and fractures from waters that reached carbonate saturation levels as they flowed through the limestones.

CONCLUSIONS

The caprinid reef facies of the Glen Rose reef interval may be subdivided into reef and nonreef beds. Rigid reef framework for the Pipe Creek reef complex, composed of juxtaposed tabular caprinid reefs, was produced by mutually supportive rudists, submarine cementation, and internal sedimentation. Clionid sponges bored the reef framework, cement, and internal sediment, producing calcareous silt, peloids, and cavities that can be easily mistaken for diagenetic vadose crystal silt, pellet silt, and solution cavities. Superimposed borings that transect contemporaneous rudist framework, fibrous cements, and internal sediments are common and support the idea that framework accretion, organic boring, submarine cementation, and internal sedimentation were syngenetic.

Locally developed pholad-bored surfaces that truncate reef masses probably represent submarine hard-

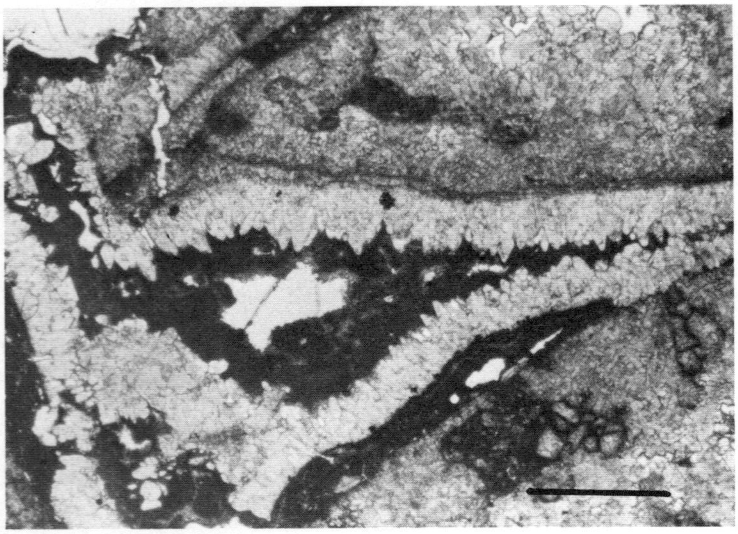

Figure 47. Photomicrograph showing bladed scalenohedral sparry calcite that fills a fracture. Reddish brown micrite overlies this cement. Bar scale is 250 microns.

grounds. These surfaces developed when vertical-reef accretion rates exceeded basin subsidence, resulting in the truncation and extensive bioerosion of portions of the cemented reef crest(s) in the littoral and/or shallow sublittoral zones.

Submarine diagenesis also affected nonreef beds. Perireef grainstones that probably represent small forereef beaches contain fibrous cements, micritic cements, and keystone vugs that indicate cementation in the shallow sublittoral and littoral zones. A large elongate sand body, thought to have developed as a submarine bar-beach complex, began to accumulate when the reef crest(s) approached sea level. The distribution of moldic and inversion textures in mollusks suggests that portions of the sand body acted as meteoric water intake areas.

Induration of peloids in the backreef and marine shelf lime wackestones was accomplished at or immediately below the water-sediment interface. Large peloids were produced by centripetal replacement of mollusk fragments by endolithic borers (algae and fungi).

During initial burial of the sediments, partial incongruent dissolution of magnesian calcite in the closed marine interstitial water system effectively increased the Mg/Ca ratio of the water. Clay-rich beds were selectively dolomitized, probably because the clay particles acted as nuclei for 5- to 10-μm dolomite crystals. This dolomite is low Sr, reflecting a later freshwater diagenetic phase, and high Na, perhaps because of sodium associated with the clay minerals.

Fresh water began to enter the caprinid reef facies during upper Glen Rose and/or latest Fredericksburg time(s). Final incongruent dissolution converted any remaining magnesian calcites to stable calcites that faithfully retain primary textures and fabrics. When the Mg/Ca ratio of the pore water approached unity and the salinity of the water was significantly lowered, a fresh-water dolomite, characterized by clear, euhedral rhombs 50 μm in diameter, partially replaced matrix and, rarely, allochems. The chemical system became progressively open because of the hydrostatic head developed in updip meteoric water intake areas. Conversion of lime mud to micrite and inversion of some aragonitic allochems occurred during the initial fresh-water phase.

Water that was undersaturated with respect to $CaCO_3$ dissolved most aragonitic allochems and created moldic porosity. Simultaneous precipitation of equant sparry calcite around these pores enabled the molds to be preserved. Later, autocthonous calcium carbonate was precipitated as intergranular and intragranular calcite cement which may partly or wholly occlude porosity.

Analysis of certain elements present in trace amounts in Cretaceous cements, internal sediments, and inverted mollusk shells was useful to the paleohydrologic reconstruction. Relatively low Mg values indicate that fresh water has effectively removed this element from nondolomitic portions of the caprinid reef facies. Low Fe values within calcite cements suggest that the fresh waters were either oxidizing, iron deficient, or both. No discernable present-day compositional trends of Mg or Fe were observed. However, the Sr and Na distribution of calcites analyzed does indicate some trends.

Fibrous submarine cements within the reef frame are now composed of very low Sr calcite probably because their original magnesian calcite lattice contained only small amounts of this element. Peloidal internal sediments yield slightly higher Sr values because their original composition was magnesian calcite and aragonite. Equant sparry calcite cements in the reef framework are relatively enriched in Sr and Na. Pervasive submarine diagenesis evidently resulted in lower porosity and permeability of the reef framework and produced a semiclosed chemical and hydrologic system that allowed Sr and Na, liberated from aragonite and magnesian calcite, to be reincorporated into the fresh-water equant sparry calcite. Backreef lime wackestones are also enriched in Sr and Na, again reflecting slower water-flow rates and a semiclosed system that prevented effective removal of these elements and allowed their reincorporation into micrite. Mollusk fragments that inverted to calcite retain the highest Na and Sr values of the samples analyzed. The original content of these elements in the mollusk shells was probably quite high. Present-day values reflect the original trace element content of the shells and suggest that equant sparry calcite that surrounds these grains has acted as a buffer between the allochems and later pore waters so that trace element dilution has not been totally effective. At the end of the initial fresh-water stage of diagenesis, the sediments and rocks had been converted to rocks composed of the relatively stable mineralogies calcite and dolomite. Moldic porosity created within the reef framework was probably only locally well connected because most of the submarine cement and internal sediment (originally magnesian calcite) was converted to calcite without attendant textural change. Some additional moldic porosity was developed within the grainstones derived from the reefs, but intergranular cementation partly reduced the effective porosity.

Diagenesis has continued through the Holocene. Fractures and vugs that transect primary rock fabric probably developed during the Cenozoic era. Vuggy and fracture porosity is partly filled with micritic and bladed to equant sparry calcite cements that have been precipitated in the meteoric vadose and phreatic zones.

Results of this study may be useful to paragenetic reconstructions of other rudist reef complexes in the subsurface where less control is available. Contemporaneous reef-framework accretion, organic boring, cementation, and internal sedimentation, common in Holocene scleractinian reefs, also affected Cretaceous rudist reefs. Careful petrographic analysis is the key to the proper environmental and diagenetic reconstruction of this, or any, reef sequence. Once the petrographic and diagenetic framework is established, subtle chemical trends within cements and allochems may be applied to the paleohydrologic reconstructions of other rudist reef complexes.

All the early fresh-water diagenetic features at Pipe Creek were developed during burial, presumably by subsurface waters that entered sediments younger than the caprinid reef facies. A minimum amount of diagenesis is attributed to subaerial exposure during deposition and only affected perireef lime grainstones. Because of pervasive submarine diagenesis that affected reef beds, the greatest porosity is preserved in the perireef grainstones.

ACKNOWLEDGMENTS

The author wishes to thank Clyde H. Moore, Jr., Bob F. Perkins, and Jeffrey S. Hanor for their comments and suggestions during this study. This paper is, in part, a condensation of the author's Ph. D. research conducted at Louisiana State University. My thanks to Shell Oil Company for providing typing assistance.

REFERENCES

Aguayo, J.E.C., 1976, Sedimentary environments and diagenesis of El Abra Limestone at its type locality, eastern Mexico (abstract): AAPG Annual Meeting, p. 644.

Alexandersson, T., 1969, Recent littoral and sub-littoral high magnesian calcite lithification in the Mediterranean: Sedimentology, v. 12, p. 47-62.

Badiozamani, K., 1973, The Dorag dolomitization model—Application to the Middle Ordovician of Wisconsin; Journal of Sedimentary Petrology, v. 43, p. 965-984.

Banner, F. T., and Woods, G. V., 1964, Recrystallization in microfossiliferous limestones: Geological Journal, v. 4, p. 21-34.

Bathurst, R.G.C., 1958, Diagenetic fabrics in some British Dinantian limestones: Liverpool and Manchester Geological Journal, v. 2, p. 11-36.

________ 1966, Boring algae and micrite envelopes, and lithification of molluscan biosparites: Geological Journal, v. 5, p. 15-32.

______ 1971, Carbonate sediments and their diagenesis: Developments in sedimentology 12: Elsevier, 620 p.

Bausch, W. M., 1968, Clay content and calcite crystal size of limestones: Sedimentology, v. 10, p. 71-75.

Beales, F. W., 1965, Diagenesis in pelleted limestones, *in* Pray, L. C., and Murray, R. C., eds., Dolomitization and limestone diagenesis—A symposium: SEPM Special Pub. 13, p. 49-70.

Bebout, D. G., 1974, Lower Cretaceous Stuart City shelf margin of South Texas: Its depositional and diagenetic environments and their relationship to porosity: GCAGS Transactions, v. 24, p. 56-84.

Behrens, E. W., and Land, L. S., 1972, Subtidal Holocene dolomite, Baffin Bay, Texas: Journal of Sedimentary Petrology, v. 42, p. 151-161.

Bein, A., 1976, Rudistid fringing reefs of Cretaceous shallow carbonate platform of Israel: AAPG Bull., v. 60, p. 258-272.

Benson, L. V., Achauer, C. W., and Matthews, R. K., 1972, Electron microprobe analyses of magnesium and iron distribution in carbonate cements and recrystallized sediment grains from ancient carbonate rocks: Journal of Sedimentary Petrology, v. 42, p. 803-811.

Berner, R. A., 1966, Chemical diagenesis of some modern carbonate sediments: American Journal of Science, v. 264, p. 1-36.

______ 1967, Comparative dissolution characteristics of carbonate minerals in the presence and absence of aqueous magnesium ion: American Journal of Science, v. 265, p. 45-70.

______ 1968, Calcium carbonate concretions formed by the decomposition of organic matter: Science, v. 159, p. 195-197.

______ 1975, The role of magnesium in the crystal growth of calcite and aragonite from sea water: Geochimica et Cosmochimica Acta, v. 39, p. 489-504.

Bonen, R. M., 1975, Comparison of ecology and sedimentation in Pennsylvanian (Morrowan) bioherms of northeast Oklahoma with modern patch reefs in Jamaica and the Florida Keys: Univ. Oklahoma, Ph. D. dissertation, 194 p.

Chave, K. E., 1954, Aspects of the biogeochemistry of magnesium, 1. Calcareous marine organisms: Journal of Geology, v. 62, p. 266-283.

______ 1962, Factors influencing the mineralogy of carbonate sediments: Limnology and Oceanography, v. 7, p. 218-223.

Coogan, A. H., Bebout, D. G., and Maggio, C., 1972, Depositional environments and geologic history of the Golden Lane and Poza Rica trends, Mexico—An alternative view: AAPG Bull., v. 56, p. 1419-1447.

Crickmay, G. W., 1945, Petrography of limestones, *in* Ladd, L. S., and Hoffmeister, J. E., eds., Geology of Lau, Fiji: B. P. Bishop Museum Bull., v. 181, p. 211-250.

Davies, G. R., and Kinsey, D. W., 1973, Organic and inorganic factors in Recent beachrock formations, Heron Island, Great Barrier Reef: Journal of Sedimentary Petrology, v. 43, p. 59-60.

Deer, W. A., Howie, R. A., and Zussman, J., 1966, An introduction to the rock-forming minerals: London, Longman Press, p. 473-503.

Dodd, J. R., 1966, Processes of conversion of aragonite to calcite with examples from the Cretaceous of Texas: Journal of Sedimentary Petrology, v. 36, p. 733-741.

Dunham, R. J., 1962, Classification of carbonate rocks according to depositional texture, *in* Classification of carbonate rocks—A symposium: AAPG Memorial, p. 108-121.

______ 1971, Meniscus cement, *in* Bricker, O. P., ed., Carbonate cements: Baltimore, Johns Hopkins Univ. Press, p. 297-300.

Evamy, B. D., 1969, The precipitational environment and correlation of calcite cements deduced from artificial staining: Journal of Sedimentary Petrology, v. 39, p. 787-793.

Fenton, C. L., and Fenton, M. A., 1932, Boring sponges in the Devonian of Iowa: American Midland Naturalist, v. 13, p. 42-54.

Folk, R. L., 1965, Some aspects of recrystallization in ancient limestones, *in* Dolomitization and limestone diagenesis—A symposium: SEPM Special Pub. 13, p. 14-48.

______ 1967, Sand cays of the Alacran reef, Yucatan, Mexico—Morphology: Journal of Geology, v. 75, p. 412-437.

______ 1974, The natural history of crystalline calcium carbonate; Effect of Mg-content and salinity: Journal of Sedimentary Petrology, v. 44, p. 40-53.

______ , and Land, L. S., 1975, Mg/Ca ratio and salinity: Two controls over crystallization of dolomite: AAPG Bull., v. 59, p. 60-68.

Friedman, G. M., Amiel, A. J., and Schneidermann, N., 1974, Submarine cementation in reefs; Examples from the Red Sea: Journal of Sedimentary Petrology, v. 44, p. 816-825.

Fritz, P., and Katz, A., 1972, The sodium distribution of dolomite crystals: Chemical Geology, v. 10, p. 237-244.

Futterer, D. K., 1974, Significance of the boring sponge Cliona for the origin of fine-grained material of carbonate sediments: Journal of Sedimentary Petrology, v. 44, p. 79-84.

Gavish, E., and Friedman, G. M., 1969, Progressive diagenesis in Quaternary to Late Tertiary carbonate sediments; Sequence and time scale: Journal of Sedimentary Petrology, v. 39, p. 980-1006.

Ginsburg, R. N., Shinn, E. A., and Schroeder, J., 1967, Submarine cementation and internal sedimentation within Bermuda reefs: GSA Abstracts, p. 78.

______ , Marszalek, D. S., and Schneidermann, N., 1971, Ultrastructure of carbonate cements in a Holocene algal reef of Bermuda: Journal of Sedimentary Petrology, v. 41, p. 472-482.

Goreau, T. F., and Hartman, W. D., 1963, Boring sponges as controlling factors in the formation and maintenance of coral reefs: American Association for the Advancement of Science Pub., v. 75, p. 25-54.

Hanshaw, B. B., Back, W., and Deike, R. G., 1971, A geochemical hypothesis for dolomitization by groundwater: Economic Geology, v. 66, p. 710-724.

Hensington, E. R., 1962, Water diagenesis in the Lower Cretaceous Trinity aquifers of Central Texas: Baylor Geological Studies Bull. 3, 37 p.

Illing, L. V., 1954, Bahaman calcareous sands: AAPG Bull., v. 38, p. 1-95.

Inden, R. F., 1972, Paleogeography, diagenesis, and paleohydrology of a Trinity Cretaceous beach sequence, Central Texas: Louisiana State Univ., Ph. D. dissertation, 264 p.

Johnson, J. H., 1971, An introduction to the study of inorganic limestones: Colorado School of Mines Quarterly, v. 66, p. 195.

Kahle, C. F., 1965, Possible roles of clay minerals in formation of dolomite: Journal of Sedimentary Petrology, v. 35, p. 448-453.

Kinsman, D.J.J., 1969, Interpretation of Sr^{+2} concentrations in carbonate minerals and rocks: Journal of Sedimentary Petrology, v. 39, p. 486-508.

Kornicker, L. S., 1962, Evolutionary trends among mollusk fecal pellets: Journal of Paleontology, v. 36, p. 829-839.

Land, L. S., 1967, Diagenesis of skeletal carbonates: Journal of Sedimentary Petrology, v. 37, p. 914-930.

______ 1969, Submarine lithification of Jamaican reefs, *in* Bricker, O. P., ed., Carbonate cements: Baltimore, Johns Hopkins Univ. Press, p. 59-62.

______ 1970, Carbonate mud: Production by epibiont growth on ***Thalassia testinudum*** grass: Journal of Sedimentary Petrology, v. 40, p. 1361-63.

______ , and Goreau, T. F., 1970, Submarine lithification of Jamaican reefs: Journal of Sedimentary Petrology, v. 40, p. 457-462.

______ , and Hoops, G. K., 1973, Sodium in carbonate sediments and rocks; A possible index to the salinity of diagenetic solutions: Journal of Sedimentary Petrology, v. 43, p. 614-617.

______ , Salem, M.R.I., and Morrow, D. W., 1975, Paleohydrology of ancient dolomites: Geochemical evidence: AAPG Bull., v. 59, p. 1602-1625.

Lloyd, R. M., 1971, Some observations on recent sediment alteration ("micritization") and the possible role of algae in submarine cementation, *in* Bricker, O. P., ed., Carbonate cements: Baltimore, Johns Hopkins Univ. Press.

Lowenstam, H. A., 1963, Biologic problems relating to the composition and diagenesis of sediments, *in* Donnelly, T. W., ed., The earth sciences—Problems and progress in current research: Univ. Chicago Press, p. 137-195.

Lozo, F. E., and Stricklin, F. L., Jr., 1956, Stratigraphic notes on the outcrop basal Cretaceous, Central Texas: GCAGS Transactions, v. 6, p. 67-78.

Majewske, O. P., 1969, Recognition in invertebrate fossil fragments in rocks and thin sections: Brill, Leiden, Netherlands, 101 p.

Milliman, J. D., 1974, Recent marine carbonates: Springer-Verlag, 328 p.

Mitterer, R. M., 1971, Comparative amino acid composition of calcified and non-calcified polychaete worm tubes: Comp. Biochem. Physiol., v. 38B, p. 405-409.

Moore, C. H., Jr., 1964, Stratigraphy of the Fredericksburg division, south-central Texas: Univ. Texas, Austin, Bureau of Economic Geology Report of Investigations 52, p. 1-48.

______ 1971, Recent intertidal cements—Their mineralogy, texture and significance, Grand Cayman, B.W.I. (abs.): AAPG Bull., v. 54, p. 861.

______ 1973, Intertidal carbonate cementation, Grand Cayman, B.W.I.: Journal of Sedimentary Petrology, v. 43, p. 591-602.

______ , Smitherman, J. N., and Allen, S. H., 1972, Pore system evolution in a Cretaceous carbonate beach sequence: Proceedings, 24th International Geological Congress, Montreal, Section 6, p. 124-136.

______ , Graham, E. A., and Land, L. S., 1976, Sediment transport and dispersal across the deep fore-reef and island slope (-55 m to -305 m), Discovery Bay, Jamaica: Journal of Sedimentary Petrology, v. 46, p. 174-187.

Moore, H. B., 1939, Faecal pellets in relation to recent marine sediments—A symposium: AAPG Pub., p. 516-524.

Mossler, J. H., 1971, Diagenesis and dolomitization of the Swope Formation (upper Pennsylvanian), southeast Kansas: Journal of Sedimentary Petrology, v. 41, p. 962-970.

Murray, G. E., 1961, Geology of the Atlantic and Gulf Coastal province of North America: New York, Harper Bros., p. 175-177.

Payton, C. E., 1966, Petrology of the carbonate members of the Swope and Dennis Formations (Pennsylvanian), Missouri and Iowa: Journal of Sedimentary Petrology, v. 36, p. 576-601.

Perkins, B. F., 1970, Genetic implications of rudist reef architecture (abs.): AAPG Bull., v. 54, p. 863-64.

______ 1971, Traces of rock-boring organisms in the Comanche Cretaceous of Texas, *in* Trace fossils; A field guide: Louisiana State Univ., School of Geoscience Miscellaneous Pub. 71-1, p. 137-47.

______ 1974, Paleoecology of a rudist reef complex in the Comanche Cretaceous, Glen Rose Limestone, central Texas, *in* Perkins, B. F., ed., Aspects of Trinity Division geology—A symposium: Louisiana State Univ., Geoscience and Man, v. 8, p. 131-173.

Purdy, E. G., 1968, Carbonate diagenesis, an environmental survey: Geol. Romana 7, p. 184-228.

Reiswig, H. M., 1973, Population dynamics of three Jamaican demospongiae: Bull. Marine Science, v. 23, p. 177-189.

Rose, P. R., 1963, Comparison of type El Abra of Mexico with the "Edwards reef trend" of south central Texas: Corpus Christi Geological Society Annual Meeting Field Trip, Perigrina Canyon and Sierra del El Abra, p. 57-64.

______ 1970, Stratigraphic interpretation of submarine vs. subaerial discontinuity surfaces: An example from the Cretaceous of Texas: GSA Bull., v. 81, p. 2787-97.

Rutzler, K., and Rieger, G., 1973, Sponge burrowing; Fine structure of ***Cliona lampa*** penetrating calcareous substrata: Marine Biology, v. 21, p. 144-62.

Schaefer, W., 1972, Ecology and paleoecology of marine environments, *in* Craig, C. Y., ed., Univ. Chicago Press, p. 415-430.

Schlanger, S. O., 1963, Subsurface geology of Eniwetok Atoll: USGS Professional Paper 260-BB, p. 991-1066.

______ 1964, Petrology of the limestones of Guam: USGS Professional Paper 403-D, p. 1-52.

Schmaltz, R. F., 1956, The mineralogy of the Funafuti drill cores and its bearing on the dolomite problem (abs.): Journal of Sedimentary Petrology, v. 26, p. 185-186.

Schmidt, V., 1965, Facies diagenesis and related reservoir properties in the Gigas Beds (upper Jurassic), northwestern Germany, *in* Pray L. C., and Murray, R. C., eds., 1969, Dolomitization and limestone diagenesis: SEPM .Special Pub. 13, p. 124-168.

Schroeder, J. H., 1972, Fabrics and sequences of submarine carbonate cements in Holocene Bermuda cup reefs: Geologische Rundschau, v. 61, p. 708-30.

______ , and Zankl, H., 1974, Dynamic reef formation: A sedimentological concept based on studies of Recent Bermuda and Bahama reefs, *in* Proceedings of the 2nd International coral reef symposium, v. 2. Great Barrier reef committee: Brisbane, p. 413-28.

Shinn, E. A., 1969, Submarine lithification of Holocene carbonate sediments in the Persian Gulf: Sedimentology, v. 12, p. 109-44.

______ 1971, Aspects of diagenesis of algal cup reefs in Bermuda: GCAGS Transactions, v. 21, p. 387-394.

______ , Bloxsom, W. E., and Lloyd, R. M., 1974, Recognition of submarine cements in Cretaceous reef limestones from Texas and Mexico: AAPG Annual Meeting Abstracts, v. 1, p. 82-83.

Sibley, D. F., and Murray, R. C., 1973, Marine diagenesis of carbonate sediment, Bonaire, Netherlands Antilles: Journal of Sedimentary Petrology, v. 42, p. 168-178.

Spurr, M. R., 1975, Reef bioerosion and sediment production by excavating sponges; Families Clionidae and Adociidae: Louisiana State Univ., Master's thesis, 132 p.

Stehli, F. G., and Hower, J., 1961, Mineralogy and early diagenesis of carbonate sediments: Journal of Sedimentary Petrology, v. 31, p. 358-371.

Stricklin, F. L., Jr., Smith, C. I., and Lozo, F. E., 1971, Stratigraphy of lower Cretaceous Trinity deposits of central Texas: Univ. Texas, Austin, Bureau of Economic Geology Report of Investigations 71, 63 p.

______ , and Amsbury, D. L., 1974, A low-relief carbonate shelf, middle Glen Rose deposits, central Texas: Louisiana State Univ., Geoscience and Man, v. 8, p. 53-66.

Taylor, J.M.C., and Illing, L. V., 1969, Holocene intertidal calcium carbonate cementation, Qatar, Persian Gulf: Sedimentology, v. 12, p. 69-107.

Weber, J. N., 1964, Trace element composition of dolostone and dolomite and its bearing on the dolomite problem: Geochimica et Cosmochimica Acta, v. 28, p. 1817-68.

Weyl, P. K., 1967, The solution behavior of carbonate materials in sea water: Univ. Miami, Studies in Tropical Oceanography, v. 5, p. 178-228.

Winland, H. D., 1968, The role of high-Mg calcite in preservation of micrite envelopes and textural features of aragonitic sediments: Journal of Sedimentary Petrology, v. 38, p. 1320-25.

______ 1971, Non-skeletal deposition of high-Mg calcite in the marine environment and its role in the retention of textures, *in* Bricker, O. P., ed., Carbonate cements: Baltimore, Johns Hopkins Univ. Press, p. 278-284.

Young, K. E., 1967, Comanche Series (Cretaceous), south-central Texas, *in* Hendricks, L., ed., Comanchean (lower Cretaceous) stratigraphy and paleontology of Texas: SEPM Permian Basin Section Pub., 67-8, p. 9-29.

Zen, E-An, 1959, Clay mineral—Carbonate relations in sedimentary rocks: American Journal of Science, v. 257, p. 29-43.

MIDDLE GLEN ROSE (CRETACEOUS) FACIES MOSAIC, BLANCO AND HAYS COUNTIES, TEXAS

Arthur W. Cleaves[1]

ABSTRACT

An 80-foot interval from the middle part of the Glen Rose Limestone has been sampled at 35 surface localities in Central Texas for the purpose of reconstructing vertical and lateral changes of depositional environment on the northwestern part of the San Marcos platform. A marker unit, the *Corbula* interval, crops out in the center of the stratigraphic interval and serves as a datum for interpreting carbonate facies changes above and below the marker.

The middle Glen Rose was deposited as a mosaic of shoal-water lithotopes in a broad, shallow lagoon behind the Gulf Coast reef trend. Over much of the platform, the sea was sufficiently shallow to allow for the formation of local tidal-flat offlap sequences. Hence, intertidal and supratidal units make up a significant proportion of the individual local facies successions.

In Blanco and Hays Counties, there are two distinct patterns of vertical facies change. Closer to the Llano uplift (Blanco County), three to five offlap cycles are present within the 80-foot interval. In all sections, the highest of the cycles contains the *Corbula* interval. These cycles involve a gradational trend from shallow subtidal through supratidal facies. Each cycle is commonly bracketed by sharply defined bedding planes. The cycles represent local marine offlap sequences and result from the progradation of carbonate mudflats into the shelf sea. Further to the east and more distant from the Llano uplift (Hays County), the facies mosaic lacks the superposed succession of local regressional cycles. Subtidal mudstones and fossiliferous marls comprise the bulk of the section both above and below the *Corbula* interval.

The difference in the vertical facies pattern for the two areas may result from their position in relation to the ancient shoreline. Because the Llano uplift was emergent during the accumulation of middle Glen Rose lithotopes, the outcrops closer to the uplift could be expected to contain abundant evidence of tidal-flat sedimentation. Mud mounds and small islands adjacent to land may have served as nuclei for the development of local offlap sequences. Florida Bay is an excellent Holocene analogue for these deposits. Further to the southeast, the shelf sea was slightly deeper and lacked the nuclei necessary to initiate offlap sequences. One complete cycle, the *Corbula* interval cycle, crops out in both areas and may record a brief period of emergence for part of the San Marcos platform. Above the *Corbula* interval, no complete cycles are present for approximately 150 feet, with subtidal marl and mudstone being the dominant lithologies for both areas.

[1]Department of Geology and Geological Engineering, University of Mississippi

LOWER CRETACEOUS DEPOSITIONAL SYSTEMS, WEST TEXAS

R. W. Scott[1] and E. J. Kidson[1]

ABSTRACT

Two surface stratigraphic cross sections of Lower Cretaceous rocks in West Texas contain four depositional systems: coastal plain, carbonate shelf, platform shelf margin, and shelf basin. These systems consist of lithofacies, megafossil paleocommunities, and palynomorph assemblages that represent specific environments. The vertical succession of deeper and shallower, or low- and high-energy facies indicates depocenter cycles of subsidence and progradation. The diversity and abundance of palynomorph morphotypes seem to be reliable environmental tools because they vary with other environmental indicators.

These depositional systems comprise depositional sequences which are transgressive-regressive stratigraphic units bounded at least in part by unconformities: the Trinity, Fredericksburg, lower Washita sequences of the Lower Cretaceous, and the upper Washita sequences of the Upper Cretaceous. Some of the sequence boundaries are synchronous within the limits of a paleontologic zone, but others are clearly time-transgressive. These sequences can be correlated in cross sections from a detailed measured section at Fort Stockton to Big Bend National Park and to the southern Quitman Mountains.

INTRODUCTION

General

The Lower Cretaceous section in West Texas and parts of adjacent Mexico is well suited to applying the concept of depositional systems and to testing the synchroneity of potential seismic reflecting surfaces. The section can readily be divided into sequences that have virtually synchronous boundaries as well as diachronous boundaries. The sequences consist of depositional systems comprised of environmentally significant lithofacies and paleocommunities, which indicate depocenter cycles of subsidence and progradation. The resulting depositional model provides a test of the environmental significance of palynomorph assemblages in this part of the Albian-Cenomanian section.

The concept of depositional systems has become a key to the interpretation of seismic-stratigraphic sequences (Brown and Fisher, 1976; Mitchum and others, 1976). As process-related assemblages of lithofacies (Fisher and others, 1969), ***depositional systems*** are three-dimensional packages of lithosomes that represent a physiographic environmental complex such as alluvial, deltaic, or shelf systems. Superjacent and lateral depositional systems commonly comprise transgressive-regressive stratigraphic packages bounded by regional unconformities or their synchronous conformities; these stratigraphic units are called ***depositional sequences*** (Mitchum and others, 1974). The boundary surfaces of depositional sequences commonly are seismic reflecting surfaces. The basic assumption is that these regional surfaces are approximately synchronous (Brown and Fisher, 1976); this assumption is based on the hypothesis that depositional sequences are a response to or a product of eustatic sea level changes (Mitchum and others, 1974). A depositional sequence, then, represents the time interval of one transgressive-regressive sedimentary cycle. Depositional sequences represent different scales of time, some spanning 1 to 2 million years, others up to tens of million years. The inferred cycles of sea-level change then become correlative tools in themselves. By attribution, seismic sequences take on the connotation of time, although, admittedly, they are not formal lithostratigraphic or chronostratigraphic units.

Mitchum's concept of a depositional sequence is very similar to Lozo's concept of ***division*** (Lozo and Stricklin, 1956; Lozo, 1959). A division is a set of strata that are related both environmentally and tectonically and are separated from adjacent divisions by widespread, abrupt conformable or unconformable surfaces. A division is the product of one sedimentary cycle and consists of a basal siliciclastic system overlain by a carbonate system (Lozo and Stricklin, 1956). Divisions are inferred to reflect tectonic activity in the source area, and sequences are interpreted to be the response to eustatic sea-level changes. Conceptually, depositional sequence and division are chronostratigraphic units because they represent "rocks formed during a specific interval of geologic time" (Hedberg, 1971). As used in the Texas Lower Cretaceous, these units are of stage rank and have all the attributes, except paleontological, of a stage.

Paleocommunities of Lower Cretaceous deposits along with lithofacies are a basis for interpreting depositional environments (Scott, 1976). Paleocommunities are recurring suites of dominant or characteristic fossils that lived together and probably interacted to some degree (Kauffman and Scott, 1976). In north-central Texas, the facies and paleocommunities suggest a depositional model of subsidence and deposition of dark-gray mud followed by progradation either of a shallow-shelf carbonate from the south or of a shoreface-deltaic-estuarine system from the north (Scott and others, in press), so that within the depocenter, cycles of subsidence and progradation produce the succession of facies within the depositional system. The general relation of sequences, systems, cycles, and environments is shown on table 1.

The West Texas region is part of the Comanche shelf that lies landward of the Ancestral Gulf of Mexico (fig. 1). The shelf in this region consisted of the positive Coahuila platform which was between the Chihuahua trough on the west and the Sabinas basin on the east. Within these shelf basins, several transgressive events successively flooded the Coahuila platform. The lithostratigraphic framework is well known for the northern Chihuahua trough (DeFord and Haenggi, 1970) and parts of the platform (Hay-Roe, 1957; Brand and DeFord, 1958; Amsbury, 1958; Twiss, 1959; Underwood, 1963; Smith, 1970; Maxwell and others, 1967). However, stratigraphic terminology in the Fort Stockton area is in flux (fig. 2), and pending publication of work by C. I. Smith, F. E. Lozo, and J. B. Brown, temporary formational names are used here in quotes.

[1] Amoco Production Company, Tulsa, Oklahoma

Table 1. Hierarchy of Concepts

DEPOSITIONAL SEQUENCES
Transgressive-regressive stratigraphic units bounded in part by unconformities; products of tectonic or eustatic processes

DEPOSITIONAL SYSTEMS
3-D set of facies formed within physiographically related set of environments

DEPOCENTER CYCLES
Successional facies produced by intrabasinal differences in rates of subsidence and sedimentation

DEPOSITIONAL ENVIRONMENTS
Specific physical, chemical and biological conditions are inferred from sets of lithofacies and paleocommunities

The basal Cretaceous sandstone overlying Triassic and Paleozoic rocks can be called the Antlers Formation following Fisher and Rodda (1966). The next overlying limestone interval has some of the characteristics of Rose's (1972) Fort Terrett Formation. Above this is a 225-ft-thick (67-m) interval of dark-gray shale and thin argillaceous limestone beds that underlie the 125-ft (37-m) limestone caprock sequence. The formational terminology of North Texas applied to this interval by Adkins (1927) is inadequate, and Young's (1967) term for the lower part, "University Mesa Marl," has never been defined. The Boracho Formation (Brand and DeFord, 1958) is a limestone and shale sequence overlying the Finlay Limestone and underlying the Buda Limestone near Kent, Texas, 80 miles west. For the purpose of this report, the name "Boracho" is applied to the lower limestone and shale overlying the "Fort Terrett Formation," which is a lithic equivalent of the Finlay Formation. The upper caprock limestone interval is lithologically like the Fort Lancaster Formation of C. I. Smith (in Smith and others, 1974) near Sheffield, 70 miles east, where the Fort Lancaster overlies the Fort Terrett and underlies the Buda. Thus, the Boracho and Fort Lancaster have a lateral facies relation.

Paleontological Control

Biostratigraphy of the ammonites in West Texas is well known (Adkins and Winton, 1920; Young, 1967, 1974). Although detailed biostratigraphic studies of other taxa are not available, enough is known of ammonite and foraminifer ranges so that virtually synchronous biostratigraphic markers can be traced throughout the area (fig. 3). The Aptian-Albian boundary is well defined by the succession of ***Dufrenoyia justinae*** (Hill) and ***Douvilleiceras mammillatum*** (Schlotheim) in the middle member of the Quitman Formation (Jones and Reaser, 1970; Young, 1974) and in the basal 80 feet (27 m) of the Cuchillo Formation in the Sierra Pilares range in Chihuahua (DeFord and Haenggi, 1970). In Coahuila, the ***Dufrenoyia justinae*** zone is in the La Peña Formation, and the ***Douvilleiceras mammillatum*** zone is in the lower part of the Glen Rose (Smith, 1970). The ***D. mammillatum*** zone also occurs in the basal part of the upper member of the Bluff Formation near Presidio, Texas (Amsbury, 1958).

The top occurrence of ***Orbitolina texana*** (Roemer) is partly a function of environmental changes in a regressing sequence. However, it is a widespread biozone boundary that has a low level of chronostratigraphic precision. Its top can be picked in most sections (fig. 3), and it approximates a synchronous surface because its slope conforms generally with that of the Aptian-Albian boundary. Smith's (1970, p. 29) caution that ***O. texana*** is not a precise correlation tool should be heeded, however, because we have found ***Orbitolina*** sp. in the Fort Terrett Formation west of Sanderson, Texas, and ***Orbitolina*** spp. range into the Cenomanian in the Tethyian Realm.

Likewise, the top occurrence of ***Dictyoconus walnutensis*** (Carsey) is controlled partly by environment. Its local top can be used to approximate a very generalized biostratigraphic marker with equally low precision. ***D. walnutensis*** is in the upper part of the Finlay Limestone in the southern Quitman Mountains (Jones and

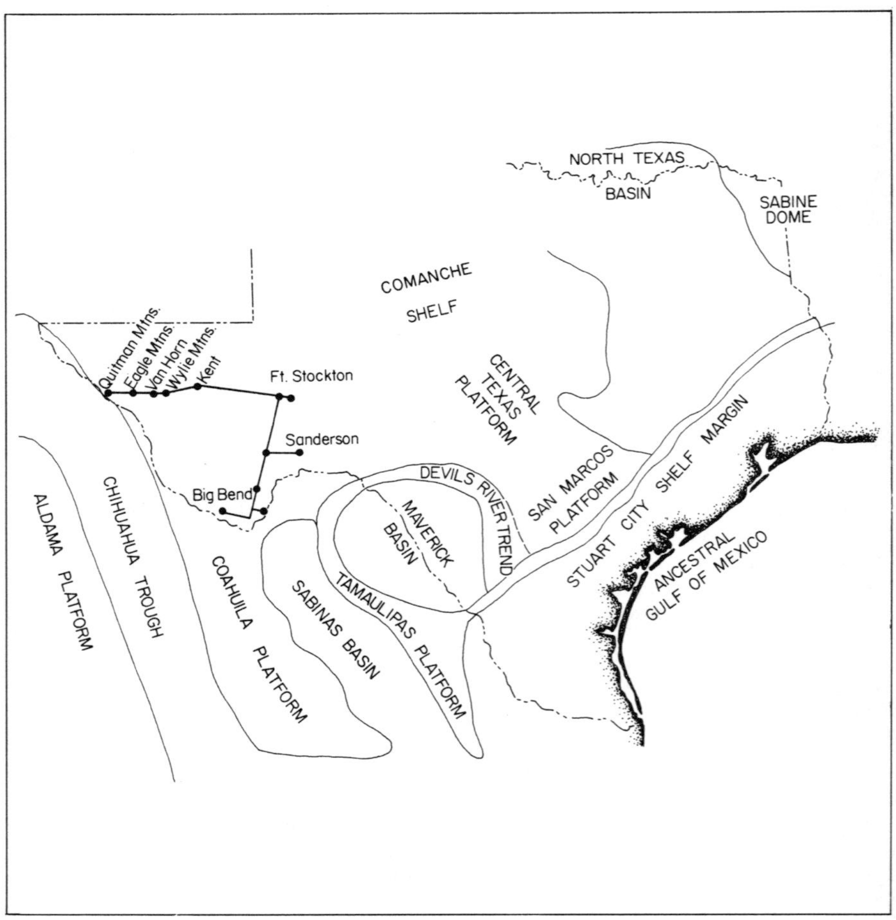

Figure 1. Major Early Cretaceous paleographic features.

Reaser, 1970) and in the Eagle Mountains (Underwood, 1963). In the El Cuervo area of Mexico, it is in the basal 90 to 135 feet (27 to 40 m) of the Finlay (DeFord and Haenggi, 1970), and in Big Bend it ranges to the top of the Del Carmen Limestone. Based on these occurrences, it would appear that the Finlay becomes somewhat younger from the Coahuila platform into the Chihuahua trough. If so, the Finlay would be a progradational shelf facies of the carbonate shelf depositional system. Therefore, its upper surface, a potential seismic reflector, would not be synchronous. This hypothesis obviously needs further testing.

The most reliable biozone is the ***Craginites serratescens*** (Cragin) range zone because the species is readily recognized, short ranging, and widespread. Furthermore, this is one in a sequence of ammonite zones (Young, 1967) in the shaly part of the section. This zone approaches a high level of chronostratigraphic precision.

The limestone interval of the "Fort Lancaster" and Santa Elena contains no ammonites; however, it has distinctive microfossils which define the ***Micritosphaera ovalis*** assemblage zone. The name fossil is a new indeterminant taxon (Scott, in press). This zone extends from the top of the ***Dictyoconus walnutensis*** zone to the base of the ***Kimbleia albrittoni*** zone and includes the local top of ***Coskinolinoides texanus, Barkerina barkerensis***, and ***Pseudocyclammina hedbergi***. As defined, the ***Micritosphaera ovalis*** zone overlaps several ammonite zones in the Sue Peaks and "Boracho."

The contact in West Texas between the upper limestone unit in the section (Loma Plata, Espy, Boracho, Santa Elena, or "Fort Lancaster") and the overlying shale (Del Rio) or sandstone (Eagle Mountains) is abruptly gradational in many places, suggesting that it is a facies change. It would be a distinct seismic reflector in the subsurface. The biostratigraphy adjacent to this boundary is not precise; however, the basal Cenomanian ***Plesioturrilites brazoensis*** (Roemer) zone (Young, 1967) is in the upper 100 ft (30 m) of the Espy Limestone in the Eagle Mountains, and typical Del Rio oysters are in the basal Eagle Mountains Sandstone (Underwood, 1963). The same succession spans the contact of the Espy and Eagle Mountains in the southern Quitman Mountains (Jones and Reaser, 1970). ***Plesioturrilites brazoensis*** also occurs at the top of the Boracho Limestone at Kent, Texas (Brand and DeFord, 1958). At Fort Stockton, the caprinid ***Kimbleia albrittoni*** (Perkins) zone marks the top of the "Fort Lancaster Formation." This species is widespread at the top of the limestone sequence below the Del

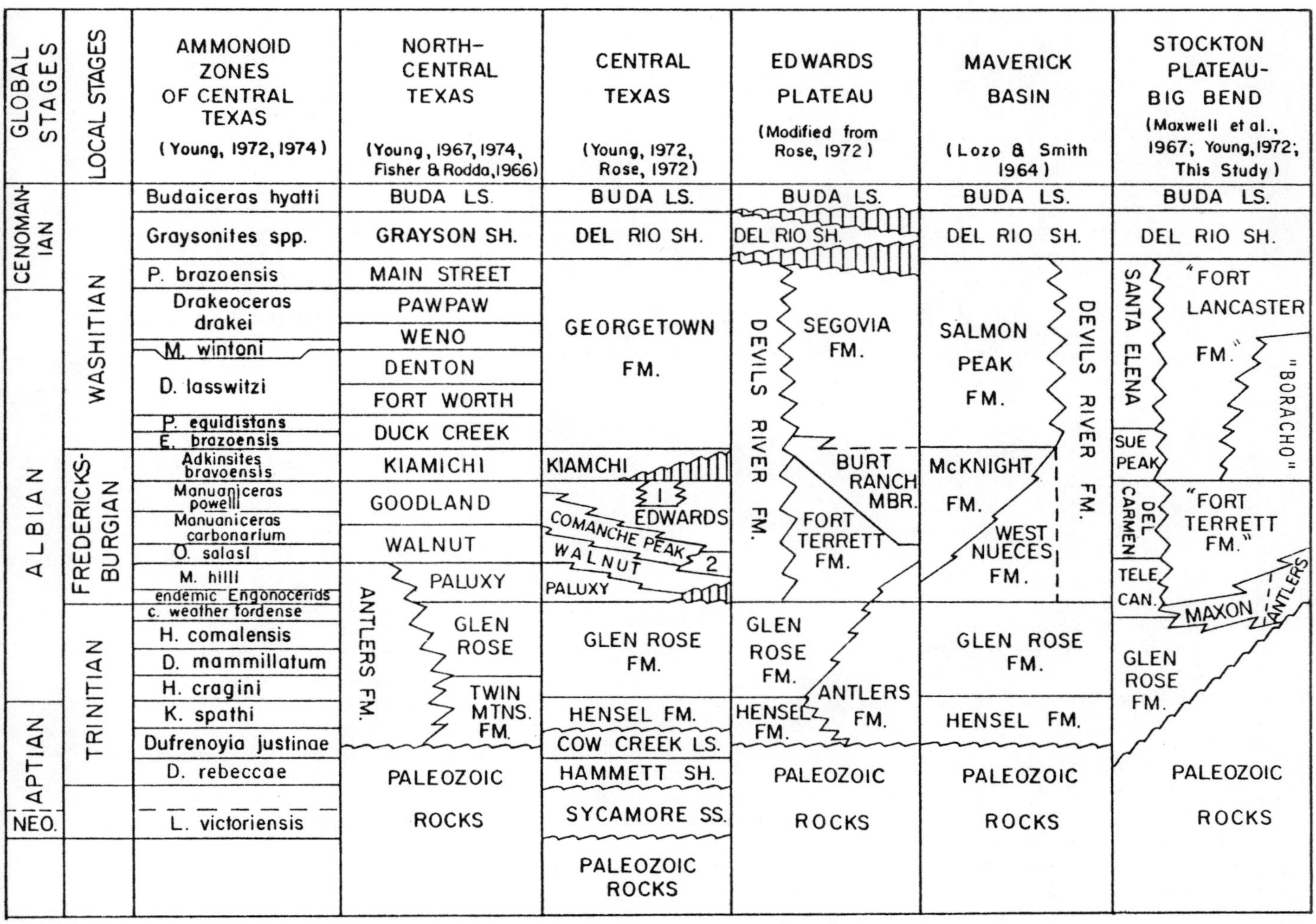

Figure 2. Stratigraphic correlation chart. 1 = moffatt lentil; 2 = whitestone lentil; Tele. Can. = Telephone Canyon Fm.

Rio, and Coogan (1973) considered it to be uppermost Albian. However, its apparent lateral continuity with the *P. brazoensis* zone supports a basal Cenomanian age. At any rate, this limestone-shale boundary is synchronous within the level of biostratigraphic precision now available, even though in many places it indicates a gradual change from a carbonate to a clay depositional regime.

Depositional Sequences

Four depositional sequences can be recognized in the east-west cross section from Fort Stockton to the southern Quitman Mountains (fig. 3) and in the north-south cross section from Fort Stockton to Big Bend Park (fig. 4). Each sequence consists of a basal siliciclastic unit and an upper carbonate unit. These "couplets" (Smith, 1970) or "divisions" (Lozo and Stricklin, 1956) consist of pairs of formations: the Trinity sequence is La Peña - Glen Rose; the Fredericksburg sequence is Telephone Canyon - Del Carmen and Antlers - "Fort Terrett"; the lower Washita sequence is Sue Peaks - Santa Elena and "Boracho" - "Fort Lancaster"; the upper Washita sequence is Del Rio - Buda. The basal siliciclastic unit is either sandstone or shale, and the upper carbonate unit is either massive-bedded limestone or thin-bedded limestone and marl. Because these sequences record transgressive-regressive events, their boundaries must be diachronous; however, this diachroneity may not be recognizable within the limits of precision of standard biostratigraphic tools. Each of the 16 ammonoid zones of the Albian in Texas (Young, 1972) represents an average duration of 500,000 to 875,000 years, assuming the Albian duration of either 8 million years (Van Hinte, 1976) or 14 million years (Kauffman, 1970). Unfortunately, such precision is not achieved in practice because ammonoids are restricted to certain facies, which are then correlated to nonammonoid facies by other fossils, by superposition, or by lithologic continuity. So testing the synchroneity or diachroneity of depositional sequence boundaries awaits a more integrated approach. Within our present range of error, the top of the Cupido Limestone (Smith, 1970) and the top of the Santa Elena Limestone appear to approximate separate synchronous surfaces. The top of the Glen Rose grades into the Telephone Canyon Formation in Big Bend Park and is correlative with the Bluff Formation in the southern Quitman Mountains, which likewise grades into the overlying Cox Formation. Although this boundary would appear as a planar surface on seismic records, it is time-transgressive on a regional scale. Similarly, the tops of the "Fort Terrett" and Finlay Formations appear to form a planar, synchronous reflector; however, the "Fort Terrett" top apparently is older at Fort Stockton where it underlies the *Manuaniceras powelli* zone than is the top of the Finlay westward in the Eagle Mountains which underlies the next younger *Adkinsites bravoensis* zone (Twiss, 1959). So the sequence boundaries are diachronous where closely spaced time-stratigraphic control is available, but the boundaries are "synchronous" on a broader time scale.

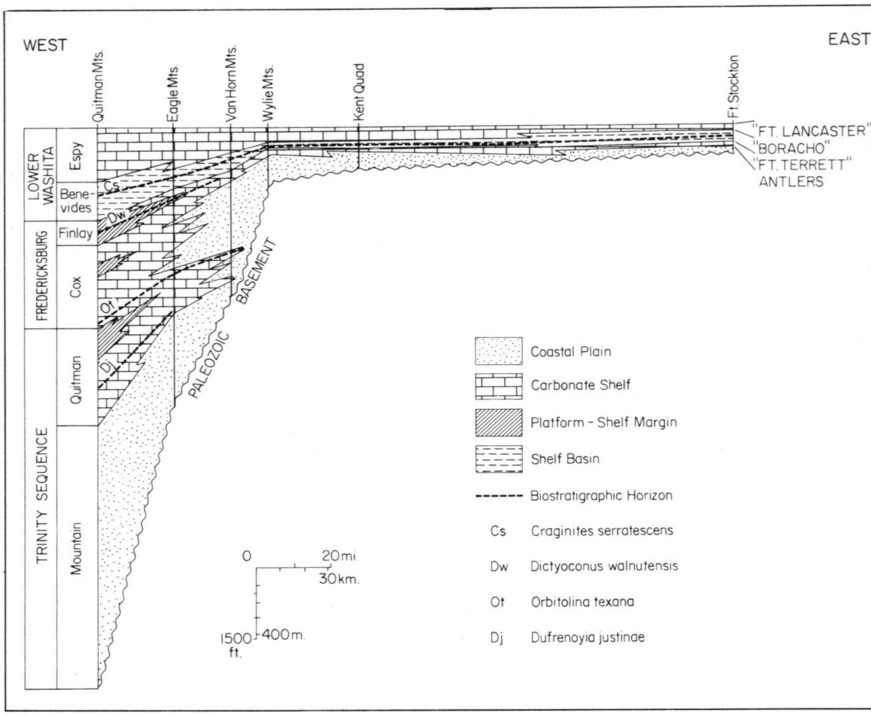

Figure 3. Depositional systems of Chihuahua trough to Coahuila platform. Exaggeration x70.

DEPOSITIONAL SYSTEMS

Four depositional systems are characterized by sets of lithofacies, paleocommunities, palynomorph assemblages, and stratigraphic relations (figs. 5, 6). The *coastal-plain* system consists of sandstone and mudstone that overstep older basement rocks. The *carbonate shelf* system consists of thin-bedded limestone updip to the *platform/shelf-margin* system, which is characterized by massive-bedded limestone. The *shelf-basin* system is comprised of dark-gray shale set within the carbonate shelf and platform/shelf-margin systems. A fifth depositional system, the *deep basin* system, was not encountered in our sections, but its facies are described by Rose (1972), Enos (1974), and Bebout and Loucks (1974). Both slope and basin-floor deposits are included in this system.

Coastal Plain System

Thin-bedded and trough-crossbedded sandstone interbedded with yellow-tan to red-gray siltstone, mudstone, and shale are characteristic. The sandstone ranges from fine to coarse grained with local graded sets. Quartz is the dominant mineral and rounded chert nodules occur locally. On the east and west sides of the Marathon basin, this basal sandstone unit is mapped as the Maxon Sandstone (King, 1937). It grades above into the overlying "Fort Terrett Formation" and grades south into the Telephone Canyon Formation and possibly into the Glen Rose (Maxwell and others, 1967). In the Fort Stockton area, the Antlers Formation is 140 ft (42 m) thick (Adkins, 1927). This facies is generally unfossiliferous and oversteps older Mesozoic and Paleozoic rocks. It represents the transgressive clastic wedge that accumulated in the coastal plain environmental complex; it is analogous to the Hensel and Paluxy sandstones in

Central Texas (Moore and Martin, 1966; Inden, 1974).

Carbonate Shelf System

This complex system consists of many limestone facies and paleocommunities representative of several distinct environments. The facies comprise laterally persistent, tabular lithosomes or interbedded successions representative of tidal-flat, restricted- and nearshore-shelf, open-lagoon, open-shelf, and shelf-to-basin transition environments.

The tidal-flat environment is represented by laminated to thin-bedded dolomite mudstone, stromatolitic dolomite mudstone, collapse breccia, shale, fossiliferous lime wackestone, and interbedded lime mudstone, oyster lime packstone, and pellet wackestone. Dolomite rhombs are secondary. Fossils tend to be concentrated into transported layers interbedded with fine-grained wackestone and mudstone. The faunal composition is highly variable from sample to sample: oysters, snails, and ostracodes at one place, miliolids, sponge spicules, and ostracodes elsewhere. Dinosaur footprints upon stromatolites are preserved in the upper part of the "Fort Terrett Formation" northeast of Fort Stockton. In the basal part of the Fort Lancaster Formation east of Sanderson, a 6-ft-thick (2-m), ripple-laminated, dolomitic calcisiltite contains a 10- to 20-cm-thick crossbedded channel fill. This unit overlies a surface bored by bivalves in a lime wackestone containing transported snails and articulated bivalves. In the Fort

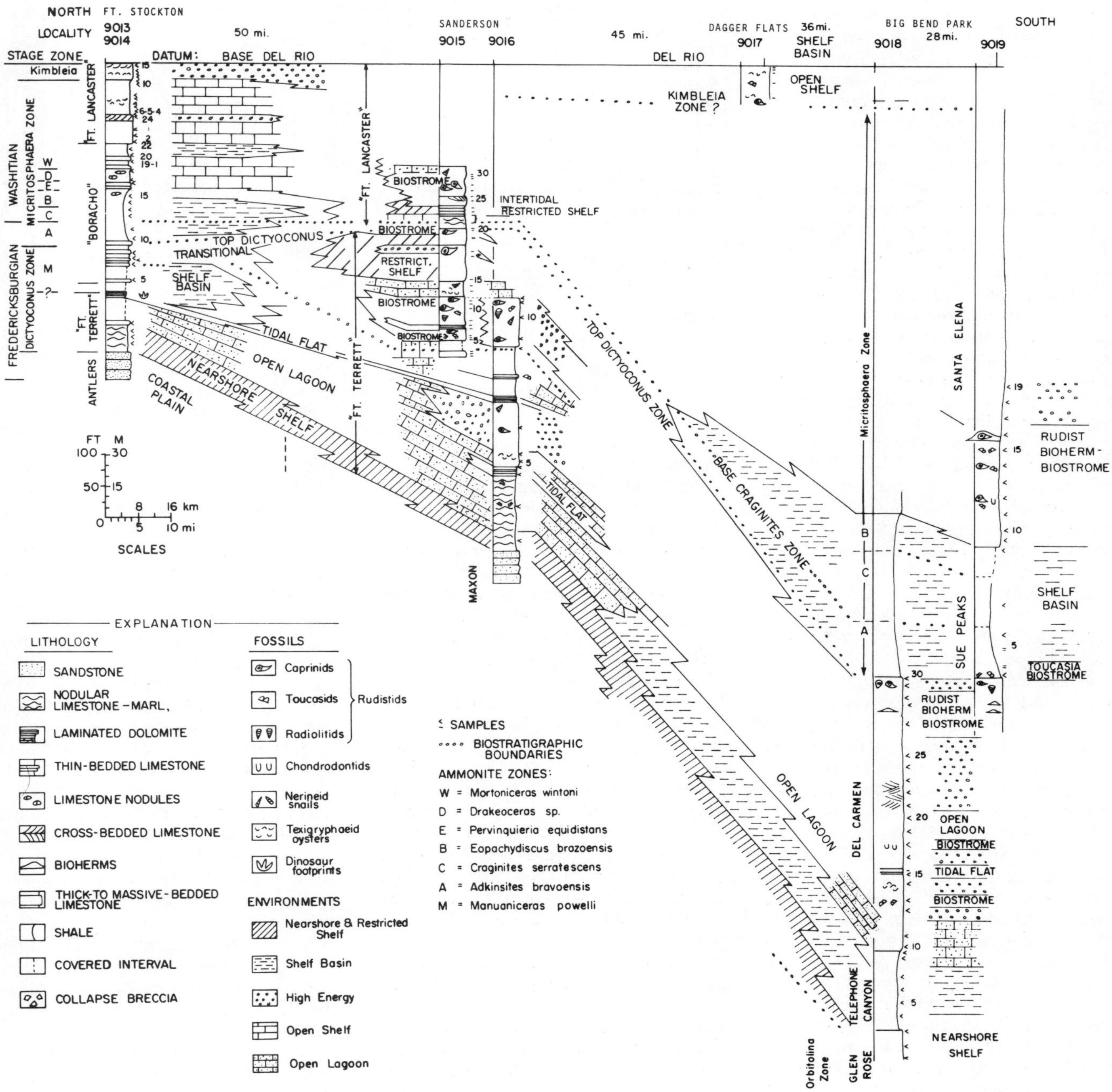

Figure 4. Stratigraphic cross section from Pecos County to Brewster County, Texas, of Comanchean strata, Fredericksburgian and Washitian stages.

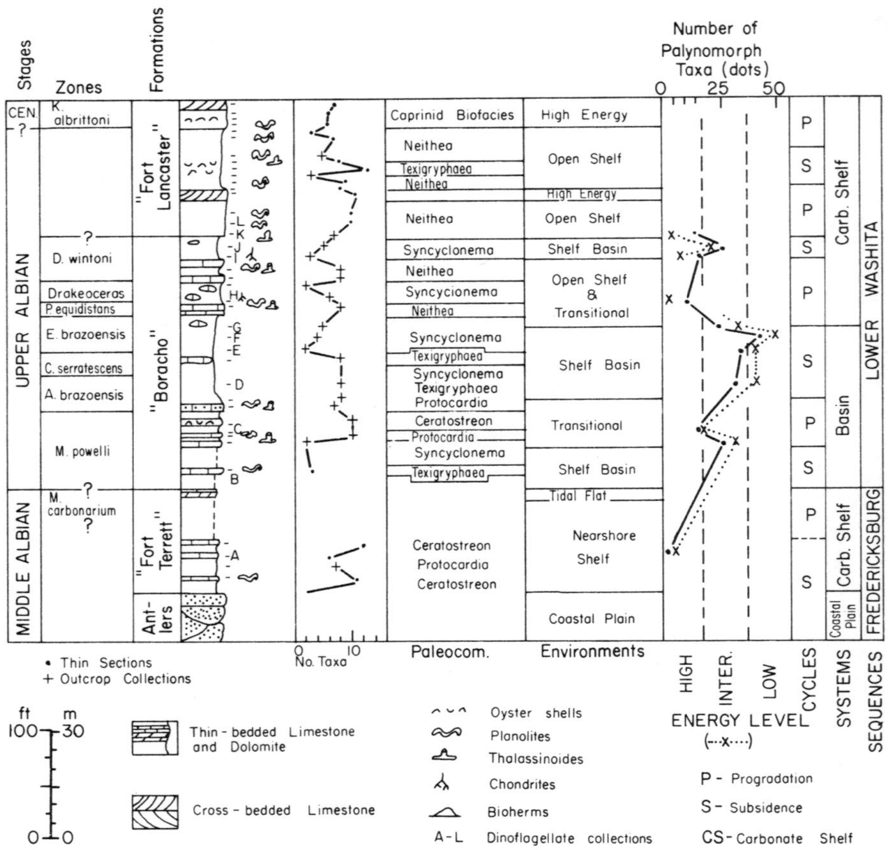

Figure 5. Fort Stockton composite section measured at east and Big Mesas northeast of Fort Stockton. Lettered samples indicate both megafossil and palynomorph samples.

Terrett, Segovia, and part of the Devils River, limestone and dolomite collapse breccias indicate the former presence of evaporites that were deposited in the tidal-flat complex (Rose, 1972). Lithologies of the tidal-flat system are sandwiched within biostromal, shelf or high-energy facies in the "Fort Terrett," Del Carmen, and "Fort Lancaster" Formations. They are like supratidal and intertidal facies in the Edwards Group (Rose, 1972, p. 51), thin-bedded dolomite in the Hensel Sandstone (Inden, 1974), and stromatolitic beds in the Glen Rose (Stricklin and Amsbury, 1974).

The nearshore- and restricted-shelf environments are represented by lime mudstone, wackestone, and packstone containing a low-diversity assemblage of ostracodes, ***Permocalculus*** (red algae), oysters, thin- and thick-walled miliolids, ***Acicularia*** (green algae), echinoids, gastropods, and in a few samples, various other foraminifers in low abundance. Generally, fossils make up no more than 20 percent of the rock, and many grains have micrite rims or even algal coatings. Locally, ***Thalassinoides*** burrows are the only evident organic remains in the very light-gray, massive lime mudstones of the "Fort Terrett Formation." Where the facies is transitional with the underlying sandstone unit, it is sandy, and agglutinate arenaceous foraminifers are locally common. At Fort Stockton the basal "Fort Terrett" contains diverse assemblages of infaunal bivalves in nodular marl. This facies is interbedded with the mollusk-skeletal algal lime packstone, suggesting an alternation of restricted and open conditions.

Facies of the nearshore and restricted shelf were deposited either nearshore where environmental changes may have been unpredictable or behind island, bar, or reef barriers. The echinoids and diverse mollusks indicate normal marine salinities and oxygen content for at least part of the time. The algae indicate shallow, nearshore depths; transported lime packstone intraclasts indicate occasional storms so that the restrictive conditions probably varied from time to time and were not constant. This facies is comparable with the restricted shallow-marine environment in the Edwards Group (Rose, 1972) and the restricted lagoon in the Glen Rose (Stricklin and Amsbury, 1974); however, it differs in that evaporites were not developed at our localities.

Open-lagoon miliolid lime wackestone formed on the shallow shelf landward of the shelf edge and was characterized by miliolid lime wackestone and lime packstone. Some characteristic fossils are ***Toucasia*** and the foraminifers, ***Barkerina, Cuneolina, Pseudocyclammina, Nummoloculina***, and ***Orbitolina***, along with various dasyclad algae, ***Acicularia*** and ***Nemerus***. This facies generally is even, thin to medium bedded, and bioturbated; it comprises part of the "Fort Terrett" and "Fort Lancaster" Formations. It is equivalent to the open shallow-marine, low-wave energy environment of Rose (1972), which occurs in the West Nueces, Fort Terrett, and Segovia Formations. In Coahuila, the platform interior (Smith and Bloxsom, 1974) is represented in part by the Glen Rose and Telephone Canyon Formations. In Texas the Glen Rose open-shelf environment of Stricklin and Amsbury (1974) is similar.

Thin rudistid lime wackestone and grainstone beds are interbedded with other shelf facies. The rudistid biostrome lime wackestone facies in the "Fort Terrett" and "Fort Lancaster" Formations is characterized by common to abundant toucasids, caprinids, and caprotinids in a lime mud matrix. Locally corals, foraminifers, and skeletal algae are present. The massive, tabular beds are 30 cm to 1 m thick. The rudistids generally are in various attitudes and range from complete specimens to fragments. Few are in living position but have not been transported or winnowed; the deposits accumulated in the neighborhood of the living organisms. Generally, fossils are 20 to 35 percent of the rock, pellets are not commonly preserved, and rarely do fossils make a textural framework. The biotic composition and relative abundance patterns can be used to distinguish paleocommunities. Locally, such as the "caprock" parts of the "Fort Lancaster Formation" around Fort Stockton, caprinid grainstone beds are intercalated within the shelf facies. A high-energy barrier bar or beach environment is suggested by large-scale tabular planar bedding and small-scale, wedge planar crossbedding. Comparable facies in a shelf system is the Cow Creek Limestone, which has accretionary beaches represented by planar- and trough-crossbedded calcarenite (Stricklin and Smith, 1973; Inden, 1974), and the Whitestone Limestone member of the Walnut

Formation, which is an offshore bar-channel grainstone complex (Moore and Martin, 1966).

The open-shelf and shelf-to-basin transitional mollusk-echinoid lime wackestone is characterized by lime wackestone with diverse bivalves and gastropods, echinoids, foraminifers, calcareous algae, and calcispheres. Generally, fossils comprise 15 to 50 percent of the rock, pellets are no more than 10 percent, and some samples are partly dolomitized or recrystallized to spar. This facies is present as thin, wavy limestone beds in the "Boracho Formation" at Fort Stockton and in the lower Segovia, "Fort Lancaster," Telephone Canyon, upper Del Rio, and Buda Formations. It represents quiet-water deposition below normal wave base (Rose, 1972). Infaunal bivalves are commonly diverse. Where the interbedded marl-calcareous shale becomes thicker and where discontinuous nodular limestone zones predominate, the open shelf becomes transitional with the shelf basin. Near Fort Stockton the transitional zone is represented by a sandy, brown-gray, nodular wackestone bed recognized as marker bed "D" by Adkins (1927). This bioturbated bed has diverse bivalves and some important ammonites.

Texigryphaea spp. biostromes, 0.5 to 12 m thick, are persistent in the middle of the Segovia and Fort Lancaster Formations. Because *Texigryphaea* biostromes mark the shelf to shelf-basin transition in north-central Texas (Scott and others, in press), they are included as part of the open-shelf facies. Oyster biostromes in the Segovia developed in the shallow, restricted shelf.

Platform/Shelf-Margin System

This system consists of facies deposited mainly in high-energy environments; the important lithofacies are: pellet-mollusk-echinoid packstone, pellet-miliolid grainstone/packstone, caprinid grainstone/packstone, echinoid-caprinid-algal packstone, and pellet-oyster packstone. Fossils comprise from 10 to 50 percent and pellets from 2 to 30 percent of the rock. These transported deposits of angular, poorly to well-sorted shell fragments are medium to thick bedded; internally the strata are homogeneous, parallel laminated, or cross laminated, with generally tabular planar accretion beds. Parts of the Del Carmen and the Santa Elena Limestones in Big Bend consist of this facies.

Comparable facies are widespread in Lower Cretaceous rocks in Texas. Grainstone to packstone bars and channels are developed on the shelf margin in the Glen Rose (Smith and Bloxom, 1974) and in the Stuart City (Bebout and Loucks, 1974). Rose (1972) interpreted these bioclastic calcarenites to have accumulated in an open, shallow-marine, moderate to high wave-energy environment.

Rudistid bioherms and biostromes comprise an important part of the platform/shelf-margin system in the Del Carmen and Santa Elena Limestones. Caprinid bioherms up to 30 ft (10 m) thick or more are lenticular buildups on the east face of Mesa de Anguila in Big Bend Park. The bio-

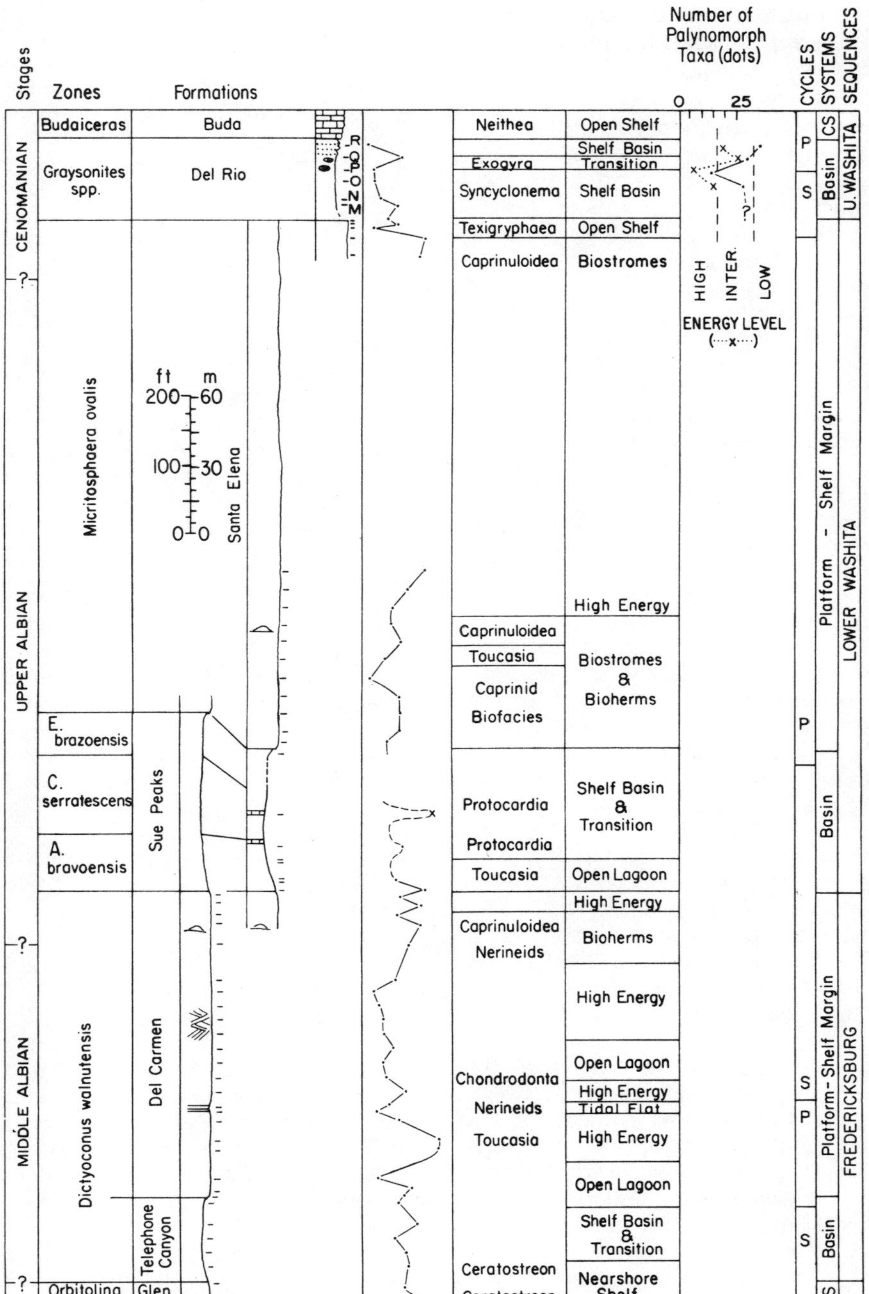

Figure 6. Big Bend Park composite section measured at Marufo Vega Trail, Santa Elena Canyon, and Dagger Flats. Lettered samples indicate both megafossils and palynomorph samples.

stromes of caprinids, radiolitids, chondrodontids, and toucasids are tabular units 30 cm to 1 m thick and are comprised of toucasid lime wackestone/packstone, caprinid lime wackestone/packstone, and chondrodontid lime wackestone. Few specimens are in living position, but generally are not transported.

The caprinid biostromes and bioherms in our measured sections have most of the features of backreef caprinid mounds (Perkins, 1974), although careful study of the mounds in the Del Carmen and Santa Elena in Big Bend National Park may show that caprinid reefs are also present. A rudistid bioherm-biostrome lime wackestone facies occurs also in shelf-edge deposits of the Glen Rose Formation (Smith and Bloxom, 1974) and the Stuart City Limestone (Bebout and Loucks, 1974), but none were found in our sections. In the Devils River Limestone, rudistid biostromes and small caprinid bioherms are interbedded with lime grainstones, forming a complex facies. In the Edwards Group of Central Texas, Fisher and Rodda (1969) and Roberson (1972) described a comparable rudistid biohermal-biostromal facies which consists of a core of whole rudistids in a lime mud matrix surrounded by flank deposits of rudistid-fragment packstone.

A lesser part of the platform/shelf-margin system consists of miliolid lime wackestone and tidal-flat facies in the Del Carmen and Santa Elena Formations. They are very similar to those in the carbonate shelf system.

Shelf-Basin System

This system consists mainly of dark-gray calcareous shale with thin interbeds of oyster lime mudstone, calcisphere-spicule lime packstone, and sparse mollusk-echinoid wackestone. The shale has sparse to common bivalve molds, pyrite, and lime-mudstone nodules. The oyster shell fragments are concentrated into layers at the base and within the lime mudstone indicating transport by density currents (Scott and others, 1975). The mollusk-echinoid lime wackestone in the Big Bend sections contains a diverse bivalve fauna and probably represents tongues of the shelf-margin lithotope prograding into the basin. The shale facies makes up most of the Telephone Canyon and Sue Peaks Formations and near Fort Stockton comprises most of the "Boracho Formation"; it also makes up most of the McKnight Formation.

This basinal environment developed upon the Comanche shelf and was below normal wave base; the fauna indicates that bottom waters were oxygenated. This basin lay landward of the Coahuila carbonate platform and is like the Lower Cretaceous North Texas - Tyler basin but unlike the deep basinal lime mudstones seaward of the Stuart City shelf edge (Bebout and Loucks, 1974).

PALEOCOMMUNITIES AND PALYNOMORPH ASSEMBLAGES

Most of the fossil deposits of paleocommunities consist of specimens preserved where they lived but disturbed by burrowers, scavengers, or gentle currents (disturbed-neighborhood deposits, Scott, 1976). In this study, the paleocommunities are based on general outcrop observations and thin section studies rather than on uniform quadrant samples.

Three rudistid paleocommunities, well developed at Sanderson, are the caprinid, caprinid-toucasid, and toucasid communities. These are preserved as disturbed-neighborhood biostromes. Only in the Del Carmen and Santa Elena Limestones in the Big Bend and in the Devils River Limestone are bioherms developed. The paleocommunities are characterized by the relative-abundance patterns of several taxa.

The caprinid biostrome community is dominated by various species of *Caprinuloidea* ranging in abundance on the outcrop from 20 to 30 percent; other megafossils are generally less than 1 to 2 percent: caprotinids, toucasids, chondrodontids, and colonial corals. Pellets and fragments of various mollusks and echinoids are common, but foraminifers are scarce or absent. The rock fabric is wackestone to packstone.

The caprinid-toucasid community consists of subequal proportions of *Caprinuloidea* and *Toucasia*, each ranging in abundance from 5 to 10 percent on the outcrop and thin section. Caprotinids are in most of the samples in the same abundances. Less commonly occurring and less abundant are ***Eoradiolites, Chondrodonta***, and ***Cladophyllia***. Various miliolids and skeletal algae are locally common in the wackestone fabric.

The ***Toucasia*** community is dominated by indeterminant species of this genus, which range from 5 to 20 percent in outcrop and thin section abundance. Caprinids are rarely over 2 percent, and colonial corals, ***Chondrodonta***, or ***Monopleura*** are present locally. The wackestone also has ***Acicularia***, miliolids, and echinoids.

Each of these three communities lived upon a shallow, low-energy shelf. At Sanderson they are interbedded with facies of the open lagoon and restricted shelf, and at Big Bend they are interbedded with open-lagoon and higher energy deposits. The low diversity and moderate abundances in given samples suggest that environmental conditions were restrictive in some way, possibly by higher salinity conditions on the restricted shelf. The dominance of epifaunal suspension feeders indicates that the sea water contained adequate food resources but that little organic matter accumulated in the sediment. The substrate was stable and somewhat firm to support the reclining and cementing rudistids.

A number of different paleocommunities characterize other facies. The nerineid-snail open-lagoon community and the ***Chondrodonta*** biostrome community indicated a backreef shelf in the Mural Formation of Arizona (Scott and Brenckle, 1977); both are also found in the Del Carmen Limestone in Big Bend Park. The nearshore oyster community of the Mural is represented here by ***Ceratostreon texana*** in the "Fort Terrett Formation." Several north-central Texas communities (Scott and others, 1973; Scott and Root, 1975; Scott, 1976) indicate the succession from open shelf to transitional shelf to shelf basin: ***Neithea-Planolites, Plicatula-Neithea-Lima, Texigryphaea***, and ***Nuculana-Corbulid-Syncyclonema*** communities. These communities indicate similar environments in the "Boracho Formation" and the "Fort Lancaster Formation" at Fort Stockton, in the "Fort Terrett" at Sanderson, and in the Sue Peaks and Del Rio Formations at Big Bend.

The palynomorph assemblages of this study are composed of both marine and nonmarine elements. The nonmarine component is made up of land-derived spores and pollen (column 1 of the histograms, fig. 7). The marine component is composed of dinoflagellates or allied acritarchous organic-walled members of the phytoplankton association, which is found on the same microscope slide. Spores and pollen tend to concentrate in the nearshore sediments where phytoplankton numbers are typically lowest, and they become decreasingly important in a more marine direction as marine phytoplankton become more diverse. Fossil dinoflagellates are known to be a resting spore or cyst of motile thecate dinoflagellates. Upon excystment, the cell content leaves the cyst to secrete a new biflagellated thecae leaving the empty cyst to settle to the bottom where it may be preserved in the fossil record. All fossil dinoflagellates are considered to be photosynthetic and so it is necessary that the cyst between encystment, and excystment remain within the light-penetrating zone of the water column. If the water is shallow, the excystment will be within the euphotic zone, but if the water is deep then the cyst must evolve an elaborate flotation capacity to retard its settling velocity.

Downie and others (1971) noted that distinct dinoflagellate associations could be related to lithologies in early Tertiary sediments, and they inferred ecologic control over this distribution. Many biostratigraphers have observed similar floral discrimination between different lithologies but have lacked adequate controlled sections to go beyond these empirical observations. The environmental interpretations applied here are based on the following assumptions: (1) specific palynomorph morphotypes are adapted to particular environments; (2) the most consistent environmental parameter that controls dinoflagellate distribution in the euphotic zone is wave action or water energy; (3) morphologic end members that have adapted to very high-energy or very low-energy environments can be readily identified; (4) displacement of dinoflagellate cysts from their adapted environment to a different environment before they are incorporated into the sediments is an important factor but one that does not typically blend the resulting associations beyond our ability to resolve environmental boundaries; (5) each stage throughout the geologic column has its own particular morphotypes adapted to the environmental spectrum, and the species of this study are only applicable to the Albian-Cenomanian.

The typical high-energy, usually nearshore, dinoflagellate assemblage consists of species with distinctly separated wall layers (cavate construction), thick, unornamented cyst periphragms (walls), and very often peridinioid (approaching fusiform) shape. The typical low-energy (open-marine) dinoflagellate assemblage consists of species which are highly ornamented and have thin single-walled construction with long, usually tubiform processes radiating from a small spherical central body which encapsulates the protoplasm of the cyst. Although the interspecific variation of this morphotype is consid-

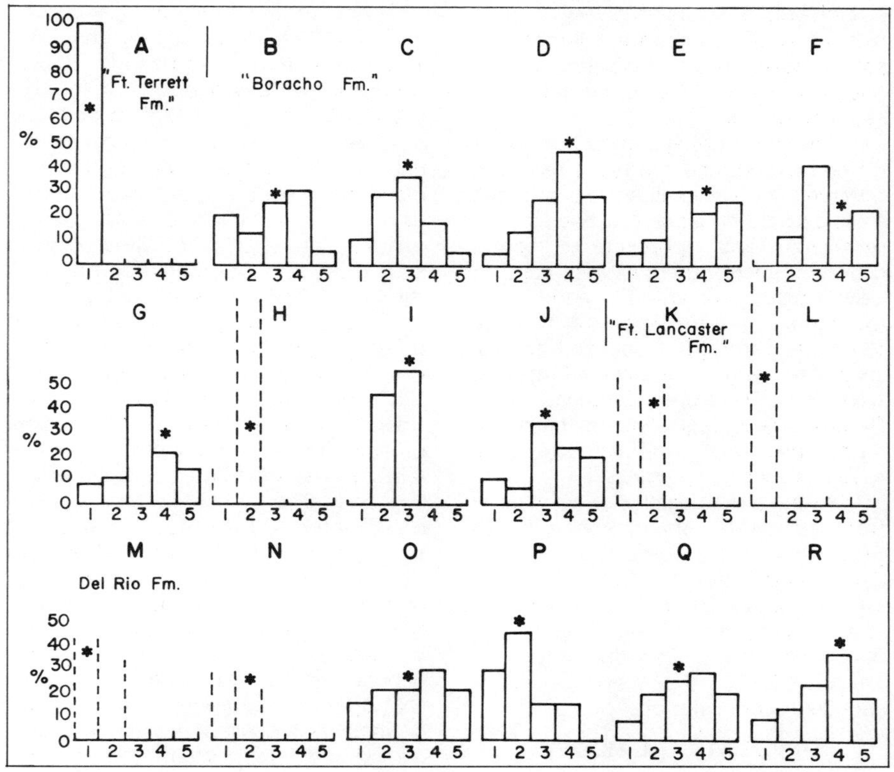

Figure 7. Percent taxa of each palynomorph morphotype from Fort Stockton (A-L) and from Dagger Flats (M-R). 1 = land-derived spores and pollen. 2 = heavy-walled cavate dinoflagellate cyst assemblage. 3 = thin-walled cavate and nonparatabulate proximate dinoflagellate cyst assemblage. 4 = paratabulate proximate and chorate dinoflagellate cysts with distally closed processes. 5 = chorate dinoflagellate cysts with distally open (tubiform) processes. Samples H, K, L, M and N did not contain sufficient numbers of palynomorphs to be diagnostic. * = The most abundant morphotype in an entire population.

erable, it may be noted that the general patterns are consistent. This elaborately evolved morphology may be explained as an adaptation for increasing frictional resistance to sinking by increasing surface area. The selective pressure in this environment is very efficient—if the cyst sinks below the light compensation point before excystment, then its genetic component is lost from the gene pool.

The foregoing discussion establishes high-energy, nearshore and low-energy, open-marine assemblages that may be end members of an environmentally controlled continuum of dinoflagellate occurrences. A less easily circumscribed variety of morphologically distinct cysts can be demonstrated to typify the "in-between" conditions. Many species may be grouped into a morphological lineage intermediate to the above end members. Primarily because of mixing that occurs prior to settling on the bottom, clear-cut examples are hard to find, but critical floral comparisons can segregate assemblages dominated by species of ***Cyclonephelum, Palaeoperidinium, Cleistosphaeridium***, and species of acritarchs (column 3 of fig. 7). Those more delicate species which appear to be better adapted for flotation and inhabit quieter waters are represented by species of ***Spiniferites, Gonyaulacysta***, and long-processed species of ***Cleistosphaeridium*** (column 4 of fig. 7). These two assemblages are representative of the intermediate environments for the Albian-Cenomanian samples of this study. Each time interval throughout the column has a slightly different assemblage makeup that responds to this environmental belt.

The basic dinoflagellate data from this study are contained in the series of five-division histograms of figure 7. Note that the morphotype having the highest number of species does not always have the greatest number of specimens.

A summary of all of the information contained on figure 7 is represented by the environmental curves of figures 5 and 6. These curves indicate high-energy, generally nearshore to nonmarine samples at the left and low-energy marine assemblages at the right. After the environmental curve was plotted, the species diversity curve was plotted using total number of palynomorph taxa. The parallelism of the two independent curves may only be coincidental.

The morphologic distribution of different dinoflagellate types has been observed for several years, but the cross-discipline control needed to segregate large populations across different environmental belts has not been readily available. In addition, a complex indexing technique (computer storage) is needed to discriminate large numbers of species through different geologic ages. Because these requirements have not always been met, studies of this type have been slow in coming.

The dinoflagellates recovered from this study, when compared with European reference material, may be dated latest Albian for the oldest samples and earliest Cenomanian for the youngest samples with the boundary within the unfossiliferous interval between the "Fort Lancaster" and Del Rio Formations.

DEPOSITIONAL MODEL

A synthetic depositional model (fig. 8) combines elements from different stratigraphic units and consists of at least 14 major environments. This model is not intended to depict conditions for any one zone. It follows the model of Rose (1972) as much as possible in order to stabilize terminology. However, modifications are made to permit a wider applicability of this model. The major geographic features strongly controlled depths and, thus, deposition environments of the Lower Cretaceous sequence.

Different sets of environments reflect depocenter cycles of basinal deepening or subsidence and basinal filling or progradation (figs. 5, 6). Subsidence is represented by a succession of environments culminating in (overlain by) the shelf-basin deposits or by the coastal plain to nearshore shelf succession near Fort Stockton (Antlers to middle "Fort Terrett"). Progradation is represented by a succession of environments culminating in tidal-flat, shelf, or high-energy deposits. The boundaries of these cycles within the depocenter do not necessarily correspond with the system or sequence boundaries because the latter generally correspond with formational boundaries and represent a dominant environmental complex rather than a vertical succession from shallow to deep or deep to shallow facies. Depocenter cycles are probably most directly a response to local sedimentation rates rather than to eustatic or tectonic changes.

Thin to massive homogeneous beds or graded beds indicate that deep-basin sediments (not included in this study) were lime muds deposited either by slow setting or by density currents (Rose, 1972; Enos, 1974). Burrows, pyrite, and chert are present locally. The rock is generally dark gray but is light yellow gray in the Salmon Peak Limestone and part of the Devils River Limestone. Megafossils are rare, but the microfossil assemblage is diverse: planktonic foraminifers, tintinnids, calcispheres, nannoconids, and radiolarians. These rocks were deposited below wave base in a pelagic environment. This facies is present in the Maverick basin as the Salmon Peak Limestone, which accumulated after open communication with the Ancestral Gulf was established. The Salmon Peak is not typical of deep-basin deposits such as the upper Tamaulipas Formation in Mexico.

Slope deposits were quite variable, consisting of both breccias and muddy fossiliferous sediments (Coogan and others, 1972, p. 1435; Enos, 1974, p. 804-806). In the Stuart City Limestone slope facies are intraclast grainstone, echinoid packstone, and echinoid-mollusk wackestone (Bebout and Loucks, 1974). No slope deposits were recognized in the outcrops measured.

The platform/shelf-margin depositional system consisted of a complex set of environments: various types of reefs and bioherms and high-energy deposits in bars, banks, channels, washover fans, and tidal deltas. The presence of ecologically complex reefs analogous to the Mural reefs (Scott and Brenckle, 1977) is suggested by facies such as coral-stromatoporoid boundstone, coral-caprinid boundstone, and caprinid-coral wackestone in the Stuart City (Bebout and Loucks, 1974) and by the caprinid-micritic facies of Coogan and others (1972, p. 1431-1435) in the Golden Lane - Poza Rica area of Mexico. These reefs grew in shallow, open-marine environments at or just below wave base. The caprinid bioherms are smaller, lenticular bodies that lie in a grainstone complex like the Del Carmen, Santa Elena, and in the Devils River. Their lower species diversity suggests a high degree of specialization, perhaps because of restrictive

A. ENVIRONMENTS AND FACIES DURING FREDERICKSBURGIAN STAGE

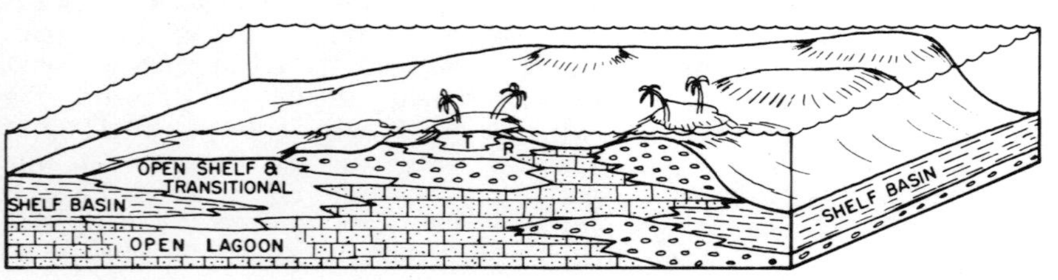

B. ENVIRONMENTS AND FACIES DURING EARLY WASHITIAN STAGE.
T = TIDAL FLAT AND R = RESTRICTED SHELF ENVIRONMENTS.

Figure 8. Interpretive diagrams of facies and environments from the margin of the Coahuila platform northward.

A. Environments and facies during Fredericksburg stage.

B. Environments and facies during early Washita stage. T = tidal flat and R = restricted shelf environments.

conditions. These bioherms need to be examined in detail.

The high-energy pellet-bioclast lime grainstone of the shelf margin consists of various microfacies besides those described in this study: the coarse and fine bioclastic facies in Golden Lane - Poza Rica area (Coogan and others, 1972, p. 1435) and the rudistid grainstone and algae-encrusted miliolid-coral-caprinid packstone in the Stuart City (Bebout and Loucks, 1974). Facies of this type were called "open shallow marine, moderate to high-wave energy" by Rose (1972). These facies are interbedded with the reefal facies and lay both seaward and landward of the shelf edge. Thinner accumulations are also found on the shallow shelf between the restricted shelf and the open shelf, where wave energy intersects the sloping sea floor. These facies comprise much of the Del Carmen, Santa Elena, and Devils River Limestones and are beds or lenses within the "Fort Lancaster," "Fort Terrett," and Segovia Formations.

An open lagoon of the carbonate shelf system generally lay landward of the shelf margin (fig. 8A). It was a wide, shallow, virtually flat substrate protected from breaking waves, but wind-generated waves winnowed the muds, concentrating skeletal fragments into poorly bedded units. Locally, tides and waves deposited some muddy or clean carbonate sands. However, the most common sedimentary facies was thin-bedded, bioturbated miliolid wackestone containing some pellets. Environmental conditions were somewhat uniform across this broad area, and consequently, some species were widespread and others were clustered into dense and distinct colonies. Examples of this are monopleurid bouquets in the Glen Rose (Perkins, 1974) and in the Mural (Scott and Brenckle, 1975) or toucasid biostromes in the Del Carmen, Santa Elena, and Fort Lancaster Formations. This environment is comparable to the shallow shelf or lagoonal miliolid-calcarenite, miliolid-micritic, and molluscan-micritic facies of Coogan and others (1972, p. 143) and the platform facies of Enos (1974). Bebout and Loucks (1974) described a shelf lagoon in which miliolid, mollusk, and toucasid wackestones were deposited; their mollusk-miliolid grainstone represented locally higher energy deposits in the lagoon.

The nearshore and restricted-shelf environments of the carbonate shelf system differed by the relative input of terrigenous sediment and dolomitic to evaporitic minerals. The nearshore shelf (fig. 8A) lay seaward of the coastal plain, and local offshore bars and channels formed; generally, however, gently sloping plane merged landward with the paralic environments. The restricted shelf (fig. 8B), on the other hand, lay between tidal-flat areas upon positive features and the open lagoon. Conditions fluctuated on this restricted shelf, producing cyclic sets of fossiliferous muds and bioturbated muds (Rose, 1972). These environments are represented by facies in the "Fort Terrett," Segovia, "Fort Lancaster," and Del Carmen Formations.

The tidal-flat facies complex accumulated on the Central Texas platform in the Kirschberg Lagoon (Fisher and Rodda, 1969). The facies of this area are distinctive, whether the environment was a true lagoon or a sabkha as suggested by Rose (1972). The tidal flat includes the intertidal zone, supratidal flats, tidal channels, ponds, or even larger isolated playas. The resulting sediment complex consists of collapse breccias, stomatolitic laminated muds, bioturbate pelletal muds, flat-pebble conglomerates, and dolomites. Only a few of the open-lagoon taxa are preserved: miliolids, ostracodes, and some oysters. This environment is represented in the "Fort Terrett," Segovia, "Fort Lancaster," Del Carmen, and Devils River Formations.

The coastal plain system consisted of the shoreface, deltaic, and fluviatile environments which are well defined by their sedimentary structures, texture, and biota. Sandstones and mudstones in the Maxon and Antlers sandstones represent this complex. The common trough crossbedding and absence of marine or brackish faunas suggest that alluvial plains and deltas were widespread and shoreface slopes were uncommon.

The shallow open-shelf environment was a broad, level bottom below normal wave base (fig. 8B) which sloped gently into the shelf basin. A transitional zone of sediments marks this relation. The mollusk-echinoid wackestone of the "Boracho Formation" near Fort Stockton was deposited in the shallow open-shelf environment and, in many ways, is similar to parts of the Georgetown and associated formations in north-central Texas. This substrate was populated by several different paleocommunities: ***Neithea-Planolites, Plicatula-Thalassinoides***, and ***Kingena*** biostromes. The transitional zone consists of interbedded shale and limestone and supported the ***Plicatula-Thalassinoides*** and ***Texigryphaea*** communities. This facies represents an eastward extension of the Chihuahua trough in which correlative basinal units were deposited. The "Boracho" is also correlative with the Kiamichi-Duck Creek and Tucumcari-Purgatoire Formations northward. Southward the transitional environment is represented by parts of the Sue Peaks and Telephone Canyon Formations.

The shelf-basin depositional system of terrigenous muds accumulated in an epicontinental depression (fig. 8B) and are represented by part of the "Boracho" near Fort Stockton and by part of the Telephone Canyon and Sue Peaks units in the Big Bend. These basins were deeper than wave base, but probably not deeper than about 200 ft (60 m), judging from their sediment thicknesses. A common benthic fauna of ***Syncyclonema*** and juvenile oysters, among others, suggests that bottom waters were partly oxygenated. Locally, in the Maverick basin, restricted conditions lead to the deposition of euxinic black laminated muds and evaporites (Rose, 1972).

REFERENCES

Adkins, W. S., 1927, The geology and mineral resources of the Fort Stockton Quadrangle: Univ. Texas Bull. 2738, 166 p.

_______ , and Winton, W. M., 1920, Paleontological correlation of the Fredericksburg and Washita Formations in north Texas: Univ. Texas Bull. 1945, 128 p.

Amsbury, D. L., 1958, Geology of the Pinto Canyon area, Presidio County, Texas: Univ. Texas, Austin, Bureau of Economic Geology Geologic Quadrangle Map 22.

Bebout, D. G., and Loucks, R. G., 1974, Stuart City trend, Lower Cretaceous, South Texas: Univ. Texas, Austin, Bureau of Economic Geology Report of Investigations 78, 80 p.

Brand, J. P., and DeFord, R. K., 1958, Comanchean stratigraphy of Kent Quadrangle, Trans-Pecos, Texas: AAPG Bull., v. 42, p. 371-386.

Brown, L. F., Jr., and Fisher, W. L., 1976, Seismic facies reflection patterns: Examples from Brazilian rift and pull-apart basins: AAPG and Society of Exploration Geophysicists, Notes for Stratigraphic Interpretation of Seismic Data School, 48 p.

Coogan, A. H., 1973, New rudists from the Albian and Cenomanian of Mexico and South Texas:

Revista del Instituto Mexicano del Petroleo, v. 5, p. 51-76.

________, Bebout, D. G., and Maggio, C., 1972, Depositional environments and geologic history of Golden Lane and Poza Rica Trend, Mexico, an alternative view: AAPG Bull., v. 56, p. 1419-1447.

DeFord, R. K., and Haenggi, W. T., 1970, Stratigraphic nomenclature of Cretaceous rocks in northeastern Chihuahua, *in* Seewald, Ken, and Sundeen, Dan, eds., The geologic framework of the Chihuahua tectonic belt: West Texas Geological Society Pub. 71-59, p. 175-196.

Downie, C., Hussain, M. A., and Williams, G. L., 1971, Dinoflagellate cyst and acritarch associations in the Paleogene of southeast England: Louisiana State Univ., Proceedings, American Association of Stratigraphic Palynologists, Geoscience and Man, v. 3, p. 29-35.

Enos, Paul, 1974, Reefs, platforms, and basins of Middle Cretaceous in northeast Mexico: AAPG Bull., v. 58, p. 800-809.

Fisher, W. L., Brown, L. F., Jr., Scott, A. J., and McGowen, J. H., 1969, Delta systems in the exploration for oil and gas: Univ. Texas, Austin, Bureau of Economic Geology Special Pub., 212 p.

________, and Rodda, P. U., 1966, Nomenclature revision of basal Cretaceous rocks between the Colorado and Red Rivers, Texas: Univ. Texas, Austin, Bureau of Economic Geology Report of Investigations 58, 20 p.

________, and Rodda, P. U., 1969, Edwards Formation (Lower Cretaceous), Texas: Dolomitization in a carbonate platform system: AAPG Bull., 53:55-72.

Hay-Roe, Hugh, 1957, Geology of Wylie Mountains and vicinity, Culberson and Jeff Davis Counties, Texas: Univ. Texas, Austin, Bureau of Economic Geology Geologic Quadrangle Map 21.

Hedberg, H. D., ed., 1971, Preliminary report on chronostratigraphic units: Montreal, 24th International Geological Congress, International Subcommittee on Stratigraphic Classification, Report 6, 39 p.

Inden, R. F., 1974, Lithofacies and depositional model for a Trinity Cretaceous sequence, Central Texas: Louisiana State Univ., Geoscience and Man, v. 8, p. 37-52.

Jones, B. R., and Reaser, D. F., 1970, Geology of Southern Quitman Mountains, Hudspeth County, Texas: Univ. Texas, Austin, Bureau of Economic Geology Geologic Quadrangle Map 39.

Kauffman, E. G., 1970, Population systematics, radiometrics and zonation—A new biostratigraphy: Proceedings, North American Paleontological Convention, Part F, p. 621-666.

________, and Scott, R.W., 1976, Basic concepts of community paleoecology, *in* Scott, R. W., and West, R. R., eds., Structure and classification of paleocommunities: Stroudsburg, Pennsylvania, Dowden, Hutchinson, and Ross, p. 1-28.

King, P. B., 1937, Geology of the Marathon Region, Texas: USGS Professional Paper 187, 148 p.

Lozo, F. E., 1959, Stratigraphic relations of the Edwards Limestone and associated formations in north-central Texas: Univ. Texas, Austin, Bureau of Economic Geology, Pub. 5905, p. 1-20.

________, and Smith, C. I., 1964, Revision of Comanche Cretaceous stratigraphic nomenclature southern Edwards Plateau, southwest Texas: GCAGS Transactions, v. 14, p. 285-307.

________, and Stricklin, F. L., Jr., 1956, Stratigraphic notes on the outcrop basal Cretaceous, Central Texas: GCAGS Transactions, v. 6, p. 67-78.

Maxwell, R. A., Lonsdale, J. T., Hazzard, R. T., and Wilson, J. A., 1967, Geology of Big Bend National Park, Brewster County, Texas: Univ. Texas, Austin, Bureau of Economic Geology, Pub. 6711, 320 p.

Mitchum, R. M., Jr., Vail, P. R., and Sangree, J. B., 1974, Regional stratigraphic framework from seismic sequences (abs.): GSA Abstracts with Programs, v. 6, no. 7, p. 873.

________, Vail, P. R., Sangree, J. B., and Thompson, S., III, 1976, Stratigraphic interpretation of seismic reflection patterns in depositional sequences: AAPG and Society of Exploration Geophysicists, Notes for Stratigraphic Interpretation of Seismic Data School, 24 p.

Moore, C. H., Jr., and Martin, K. G., 1966, Comparison of quartz and carbonate shallow marine sandstones, Fredericksburg Cretaceous, Central Texas: AAPG Bull., v. 50, p. 981-1000.

Perkins, Bob F., 1974, Paleoecology of a rudist reef complex in the Comanche Cretaceous Glen Rose Limestone of Central Texas: Louisiana State Univ., Geoscience and Man, v. 8, p. 131-174.

Roberson, D. S., 1972, The paleoecology, distribution and significance of circular bioherms in the Edwards Limestone of Central Texas: Baylor Geological Studies, Bull. 23, 35 p.

Rose, P. R., 1972, Edwards Group, surface and subsurface, Central Texas: Univ. Texas, Austin, Bureau of Economic Geology Report of Investigations 74, 198 p.

Scott, R. W., 1976, Trophic classification of benthic communities, *in* Scott, R. W., and West, R. R., eds., Structure and classification of paleocommunities: Stroudsburg, Pennsylvania, Dowden, Hutchinson, and Ross, p. 29-66.

________ in press, ***Micritosphaera***, a new Cretaceous microfossil: Journal of Paleontology.

________, and Brenckle, P. L., 1975, A model of Lower Cretaceous coral-algal-rudist reefs (Mural Formation, southeast Arizona): Amoco Production Company, Research Department Report F75-G-16, 68 p.

________, and Brenckle, P. L., 1977, Biotic zonation of Lower Cretaceous coral-algal-rudist reef, Arizona: 3d International Symposium on Coral Reefs.

________, Laali, H., and Fee, D. W., 1975, Density-current strata in Lower Cretaceous Washita Group, north-central Texas: Journal of Sedimentary Petrology, v. 45, p. 562-575.

________, Laali, H., Fee, D. W., and Magee, R., in press, Epeiric depositional models for the Lower Cretaceous Washita Group.

________, and Root, S. A., 1975, Community structure and depositional environments of Lower Cretaceous Washita strata, north-central Texas: AAPG and SEPM Annual Meeting Abstracts, v. 2, p. 94.

________, and others, 1973, Benthic community succession in a Cretaceous carbonate to shale transition: GSA Abstracts with Programs, v. 5, no. 7, p. 801.

Smith, C. I., 1970, Lower Cretaceous stratigraphy, northern Coahuila, Mexico: Univ. Texas, Austin, Bureau of Economic Geology Report of Investigations 65, 101 p.

______ , and Bloxsom, W. E., 1974, The Trinity Division and equivalents of northern Coahuila, Mexico: Louisiana State Univ., Geoscience and Man, v. 8, p. 67-76.

______ , Charleston, S., and Brown, J. B., 1974, Lower Cretaceous shelf, platform, reef and basinal deposits, southwest Texas and northern Coahuila: West Texas Geological Society and SEPM, Permian Basin Section, Pub. 74-64, 33 p.

Stricklin, F. L., Jr., and Smith, C. I., 1973, Environmental reconstruction of a carbonate beach complex, Cow Creek (Lower Cretaceous) Formation of Central Texas: GSA Bull., v. 84, p. 1349-1368.

______ , and Amsbury, D. L., 1974, Depositional environments on a low-relief carbonate shelf, middle Glen Rose Limestone, Central Texas: Louisiana State Univ., Geoscience and Man, v. 8, p. 53-66.

Twiss, P. C., 1959, Geology of Van Horn Mountains, Texas: Univ. Texas, Austin, Bureau of Economic Geology Geologic Quadrangle Map 23.

Underwood, J. R., Jr., 1963, Geology of Eagle Mountains and vicinity, Hudspeth County, Texas: Univ. Texas, Austin, Bureau of Economic Geology Geologic Quadrangle Map 26.

Van Hinte, J. E., 1976, A Cretaceous time scale: AAPG Bull., v. 60, p. 498-516.

Young, K., 1967, Ammonite zonation, Texas Comanchean (Lower Cretaceous). SEPM, Permian Basin Section, Pub. 67-8, p. 65-70.

______ 1972, Cretaceous paleogeography: Implications of endemic ammonite faunas: Univ. Texas, Austin, Bureau of Economic Geology Geological Circular 72-2, 13 p.

______ 1974, Lower Albian and Aptian (Cretaceous) ammonites of Texas: Louisiana State Univ., Geoscience and Man, v. 8, p. 175-228.

DEPOSITION AND DIAGENESIS OF THE FORT TERRETT FORMATION (EDWARDS GROUP) IN THE VICINITY OF JUNCTION, TEXAS

by Alonzo D. Jacka[1]

ABSTRACT

Excellent exposures of Fort Terrett Formation (Edwards Group) carbonates occur along new roadcuts of Interstate 10 in the vicinity of Junction, Texas. The Fort Terrett contains well-developed depositional cycles, most of which consist of supratidal, intertidal, and subtidal facies. The Fort Terrett also exhibits complex diagenetic cycles that include calcitization, dolomitization, sulfate emplacement and dissolution, silicification, and, in some cases, dedolomitization. Depositional and diagenetic patterns reflect superimposition of the following processes: (1) eustatic changes in sea level and associated climatic fluctuations, (2) seaward progradations of supratidal, intertidal, and subtidal facies, and (3) subsidence.

Limestone intervals exhibit abundant evidence of having stabilized in fresh-water diagenetic environments. Supratidal deposits represent sabkhas and were penecontemporaneously dolomitized according to the Persian Gulf model. In many cycles, portions of subtidal facies have also been dolomitized, possibly in zones of mixing at the bases of fresh-water lenses.

Primary porosity was formed in grainstones but very little was preserved. Much secondary porosity was formed in limestone intervals through selective dissolution of crystalline aragonitic shells and ooids. Most secondary porosity in limestones has been occluded by calcite cements, but some has been selectively preserved within hollow micrite envelopes, on micritized foundations, and in large micrite-walled solution vugs. Secondary intercrystalline and moldic porosity were formed and extensively preserved in dolostone intervals.

Much porosity was created within collapse breccias which were formed through dissolution of sulfates. In thin collapse breccias most porosity has been occluded by deposition of internal sediment and meteoric cements. Considerable porosity still exists in a thick collapse zone in the upper Fort Terrett.

Tertiary porosity is very well developed and extensively preserved in most dolostone intervals and in many limestone units. Tertiary porosity was created through dissolution of sparry anhydrite which extensively replaced limestones and dolostones after secondary voids had been formed and filled by cements. Pulverulent limestones and dolostones represent soft, powdery materials which once contained exceedingly high concentrations of sparry replacement anhydrite.

INTRODUCTION

Excellent exposures of Edwards Group carbonates occur along new roadcuts of Interstate 10 in the vicinity of Junction, Texas (fig. 1).

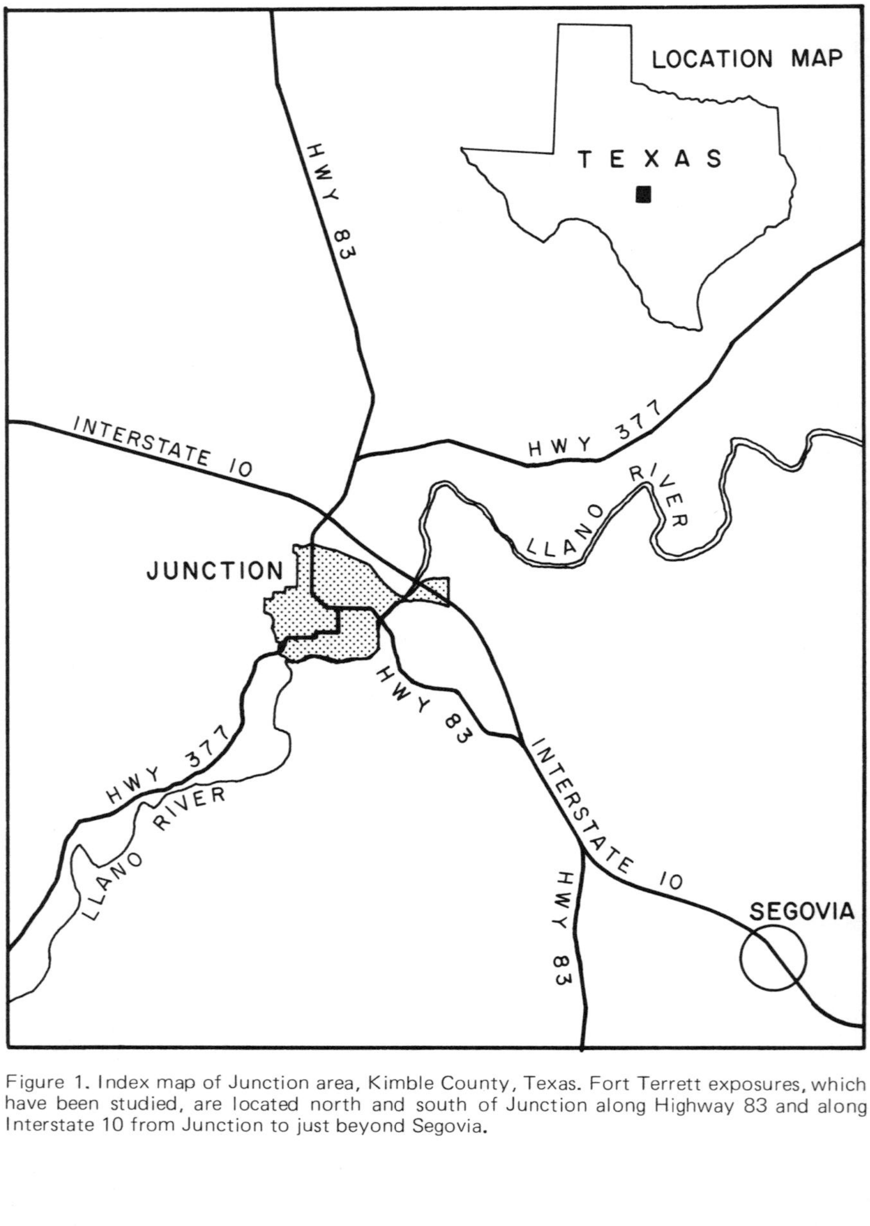

Figure 1. Index map of Junction area, Kimble County, Texas. Fort Terrett exposures, which have been studied, are located north and south of Junction along Highway 83 and along Interstate 10 from Junction to just beyond Segovia.

[1]Texas Tech University, Lubbock, Texas

Detailed studies of the Fort Terrett have been initiated in conjunction with the Texas Tech carbonate course which is presently being conducted from the Texas Tech University Center at Junction.

Rose (1972) established the stratigraphy and generalized environmental-lithostratigraphic framework of the Edwards Group, surface and subsurface of Central Texas. In the Junction area, the Edwards Group consists, in ascending order, of the Fort Terrett and Segovia Formations (Rose, 1972). No detailed studies of deposition or diagenesis of Edwards carbonates in the Junction area have been published.

The Fort Terrett Formation exhibits some well-developed depositional cycles and some very interesting diagenetic patterns. The following categories of diagenesis are included: (1) calcitization, (2) dolomitization, (3) emplacement and dissolution of sulfates, (4) dedolomitization, and (5) silicification.

Fort Terrett exposures included within this study occur along Highway 83 north and south of Junction and along Interstate 10 from Junction to beyond Segovia (fig. 1).

DEPOSITIONAL CYCLES

The Fort Terrett Formation is approximately 55 m thick in the Junction area and contains well-developed depositional cycles. Depositional cycles range in thickness from less than 1 m to approximately 10 m, and as many as 17 disconformities are represented in the Fort Terrett interval. Many cycles record the following sequence of events: (1) rapid eustatic rise in sea level and transgression; (2) stillstand of sea level and seaward progradation of subtidal, intertidal, and supratidal facies; (3) rapid transgression, stillstand, and progradation (fig. 2B, 2C). This type of sequence was not accompanied by significant erosion; however, the uppermost surfaces of supratidal (sabkha) deposits constitute minor disconformities and are abruptly overlain by subtidal deposits that contain thin, reworked basal intervals.

Some Fort Terrett cycles reflect the following events: (1) eustatic rise in sea level and rapid transgression; (2) stillstand of sea level and progradation, forming a vertical sequence of subtidal, intertidal, and supratidal deposits; (3) eustatic lowering of sea level and erosion of supratidal or supratidal, intertidal, and possibly some of the subtidal facies; (4) eustatic rise in sea level and rapid transgression. In a case such as this, a disconformity and thin reworked zone may separate two vertically juxtaposed subtidal deposits (fig. 3A). Sequences in which overlying subtidal facies are separated by disconformities from underlying subtidal or intertidal facies could also be explained by the following succession of events: (1) eustatic rise and transgression; (2) stillstand and progradation, but supratidal or supratidal and intertidal facies did not reach the Junction area; (3) a minor lowering of sea level (stadial) that exposed subtidal or intertidal facies, but little erosion occurred.

A major collapse breccia zone, approximately 9 m thick, occurs in the upper portion of the Fort Terrett Formation in the Junction area (fig. 4B). This zone was formed by dissolution of underlying sulfates of the Kirschberg evaporite unit (see Fisher and Rodda, 1969; Rose, 1972). The collapse breccia reflects a major eustatic lowering of sea level, prolonged subaerial exposure, and a change in climate from arid or semiarid to one of increased rainfall (interpluvial to pluvial). This climatic cycle may record a full glacial interval. The only relict of subjacent sulfates, which were dissolved to form the collapse breccia, consists of silica pseudomorphs of anhydrite nodules ("cauliflower" cherts, see fig. 5A). A climatic fluctuation superimposed upon a low stand of sea level suggests glacial control of such a cycle. Most Fort Terrett cycles probably reflect stadials (minor oscillations within glacial cycles).

CRITERIA FOR RECOGNITION OF DEPOSITIONAL ENVIRONMENTS

Subtidal lithofacies include biowackestones and packstones, biopelwackestones, packstones and grainstones, and oobiopel-wackestones, packstones, and grainstones. Subtidal facies exhibit intensive bioturbation by random, fodinichnial burrows (fig. 3D). Several examples of large subtidal algal domes occur in the upper third of the Fort Terrett (fig. 2D). Small and relatively large current ripple structures occur in some grainstones.

Intertidal deposits include most of the same lithofacies found in subtidal deposits. Vertically oriented domichnial burrows are predominant in intertidal facies. Intertidal grainstones contain eroded wave ripples, consisting of flat- to round-crested forms. Intraclasts are much more abundant in intertidal than in subtidal facies. Most intraclasts seem to have been derived from mud polygons or desiccated stromatolites of the supratidal zone. Many lenticular layers probably represent tidal channels. Algal stromatolites consisting of relatively small domes and flat laminar morphologies are abundantly represented in intertidal facies.

Supratidal (sabkha) deposits originally contained anhydrite nodules (figs. 3, 4A) and were penecontemporaneously dolomitized according to the Persian Gulf model of Butler (1969) and Kinsman (1969). They also contain mud-crack casts, flat-pebble conglomerates, and fenestral cavities (bird's-eye vugs). Supratidal stromatolites commonly are crinkled like those illustrated by Ginsburg (1957). Former anhydrite nodules in supratidal deposits are now represented by molds, or they have been replaced by silica (figs. 3, 4A, 5A).

The supratidal facies ranges up to 1 m in thickness. The 1-m thicknesses of Fort Terrett supratidal facies relate very closely with those of modern sabkha deposits described by Kinsman (1969) and Butler (1969). Kinsman (1969) has defined a coastal sabkha with reference to a salt-water table lying 1 m or less from the surface.

EARLY DIAGENESIS IN MARINE ENVIRONMENT

Micritization of crystalline shells and ooids is abundantly evidenced in all stratigraphic intervals of the Fort Terrett. Small, thin-shelled invertebrates and bioclasts were most susceptible to micritization. High proportions of ooids in oolitic grainstones exhibit intensive micritization (fig. 6A, 6B). In many, the concentric structure

(page 184)

Figure 2. Characteristics of Fort Terrett exposures along Interstate 10 between Junction and Segovia (fig. 1).
A. A disconformable contact separates Glen Rose (GR) and Fort Terrett (FT). Fissures in the upper Glen Rose clastic interval contain material from the overlying Fort Terrett. A thin reworked zone, containing chert pebbles and other material reworked from the Glen Rose, immediately overlies the disconformity. The basal Fort Terrett consists of nodular, calcareous shale which grades upward into limestone.
B. Several relatively thin depositional cycles consisting, in descending order, of supratidal (SP), intertidal, and subtidal (S) deposits are illustrated.
C. Two depositional cycles consisting of supratidal (SP), intertidal (i), and subtidal (S) are shown.
D. Large algal domes (probably subtidal) occur immediately above the highway and in other intervals in the middle and upper portions of the exposure.

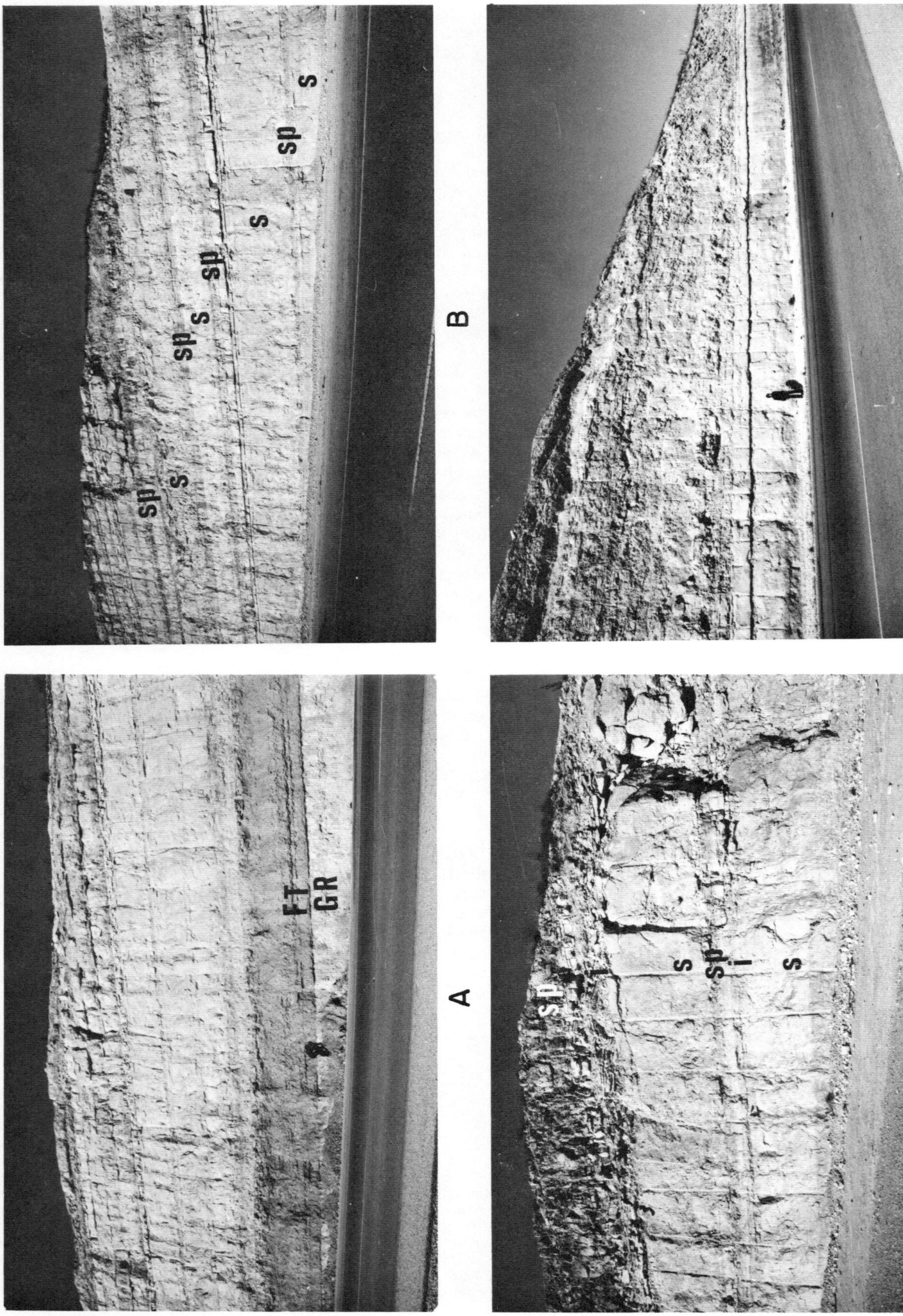

Figure 2

is barely discernible, or it has been completely obliterated and ooids have been converted to peloids (fig. 6A, 6B). Kendall and Skipwith (1969) have shown that intensive micritization commonly results from activity of endolithic blue-green algae. Winland and Matthews (1974) have postulated that micritization on the Bahama platform represents degrading recrystallization.

Micrite envelopes are abundantly represented throughout the Fort Terrett. Many have sharp contacts with grain boundaries and could represent submarine or intertidal, micrite cement films. Many others, especially those which have formed on relatively large, thick shells or bioclasts, appear to have been formed by endolithic algae.

Submarine and beachrock (marine vadose) cements were precipitated within several Fort Terrett grainstones (fig. 6A, 6B). Bladed to fibrous films probably represent original magnesian calcite or aragonite cements. Commonly, grains have bilaminar cement films: an inner micrite envelope and an outer bladed to fibrous film (James and others, 1976). Jacka (1974a) has described examples of bilaminar cement films from meteoric and marine vadose environments. In larger voids, bladed to fibrous films can be seen to thicken downward as gravitational (microstalactitic) cement in beachrock. Müller (1971), Taylor and Illing (1971), and Jacka (1974a) have described similar examples of gravitational cements. The only difference between submarine and beachrock cements is the presence of gravitational thickening of films in beachrock.

DIAGENESIS OF FORT TERRETT DOLOSTONES

Distribution of Limestones and Dolostones in Fort Terrett Cycles

The Fort Terrett contains limestones, dolomitic limestones, and dolostones. Within well-developed depositional cycles, uppermost supratidal deposits are intensively to completely dolomitized. In most cycles, supratidal dolostones are underlain by intertidal and subtidal limestones. In three cycles, supratidal dolostones are successively underlain by limestone, which includes intertidal and upper subtidal facies, and dolostone (lower subtidal zone). In two cycles, dolomitization extends beneath the supratidal facies and includes intertidal and upper subtidal facies, but the lower portion of the subtidal zone is now limestone.

Petrography and Diagenetic Mechanisms

Supratidal dolostones represent sabkha deposits which were penecontemporaneously dolomitized according to the Persian Gulf model of Butler (1969) and Kinsman (1969). Anhydrite nodules, which had formed in sabkha deposits, were subsequently dissolved or replaced by silica. In supratidal dolostones, dolomite crystals range in diameter from 1 to 20 microns. In cycles wherein dolomitization of intertidal or subtidal facies took place, dolomite crystals are generally larger and range in diameter from 5 to 80 microns. In several intervals, burrows have been selectively dolomitized in subtidal limestones. In cases wherein subtidal facies have been intensively dolomitized, dolomite crystals in burrows are larger than in the adjacent matrix.

The mechanism by which subtidal or intertidal facies became dolomitized in some cycles but not in others is not understood. Conceivably, dolomitization of subtidal or intertidal facies could have occurred at the base of a coastal fresh-water lens in the zone of mixing. Dolomitization associated with a fresh-water lens has been postulated by Hanshaw and others (1971), Land (1973), and Badiozamani (1973). Badiozamani (1973) termed this mechanism the dorag ("mixed blood") model. Just how the aforementioned patterns of dolomitization of Fort Terrett subtidal and, in rare cases, intertidal facies could be explained in terms of the dorag model is uncertain.

Steinen (1974) has provided the only documented case history study of a coastal fresh-water lens which has been superimposed upon metastable carbonates on the island of Barbados. Steinen analyzed borehole data that successively penetrated a meteoric vadose, meteoric phreatic, and marine phreatic zone below the lens. Analysis of carbonate mineralogy revealed that no dolomitization has taken place at the base of or below the lens. At this point in time, the efficacy of the dorag dolomitization mechanism has not been established. No dolomitization mechanism, other than the sabkha mechanism (Persian Gulf model), has been demonstrated.

From study of middle Permian surface and subsurface carbonates of the Permian Basin, Jacka (1975) has inferred that dolomitization of aragonitic lime mud or shells results in formation of neomorphic, rhombic-crystal fabrics. In cases wherein dolomite directly replaced calcite, calcite textures are paramorphically replicated. Replacement of dolomite by calcite (dedolomitization) is also paramorphic and the rhombic crystal integrity is preserved. All dolostone intervals in the Fort Terrett are neomorphic, which indicates replacement of predominantly aragonitic lime muds. Within many dolostones and dolomitic limestones, numerous examples of partial paramorphic dolomitization of calcitic shells were noted. In such cases the calcitic shell fabrics exhibit perfect optical continuity through dolomitized areas (fig. 7D). Jacka (1975) has inferred that only neomorphic dolomitization of aragonite produces the classical intercrystalline porosity of sucrosic dolostone (see Murray, 1960).

Intercrystalline porosity was formed and preserved in some Fort Terrett dolostone intervals (especially in subtidal facies) when rhombic dolomite crystals replaced original aragonitic lime muds to form neomorphic fabrics (fig. 7B). Much of this intercrystalline porosity may have been filled by anhydrite which was subsequently dissolved.

During dolomitization many crystalline calcitic and aragonitic(?) shells and ooids were dissolved, as first noted by Murray (1960). Fossil molds and oomolds with relatively smooth borders are abundantly recorded in Fort Terrett dolostones. It is also possible that most or many of these

(page 186)

Figure 3. Detailed characteristics of Fort Terrett exposures along Interstate 10 between Junction and Segovia and along Highway 83 north of Junction (fig. 1).
A. A disconformity (D) separates two subtidal facies (S). Cobbles and pebbles of material reworked from the subjacent subtidal deposit lie immediately above the disconformity. A flat cobble lies just to the left of the scale and between the two S's. Above the disconformity, a subtidal deposit (S) grades upward into intertidal (I) and supratidal (SP) facies. The supratidal facies contains molds of anhydrite nodules.
B. A depositional cycle, consisting of supratidal (SP), intertidal (I), and subtidal (S) facies, is shown. Within the intertidal zone, laminoid-nodular patterns of dark chert (C) record replacement of algal stromatolites. In the subtidal facies, silicified fodinichnial burrows (B) are represented by dark chert. Molds of anhydrite nodules occur in the supratidal zone.
C. A portion of a cycle, consisting of intertidal (I) and supratidal (SP) facies, is shown. Most anhydrite nodules in the basal portion of the supratidal zone, to the left and right of the pocket knife, have been replaced by light chert (C).
D. Intensive, random fodinichnial burrows within a subtidal deposit have been etched into bold relief by weathering.

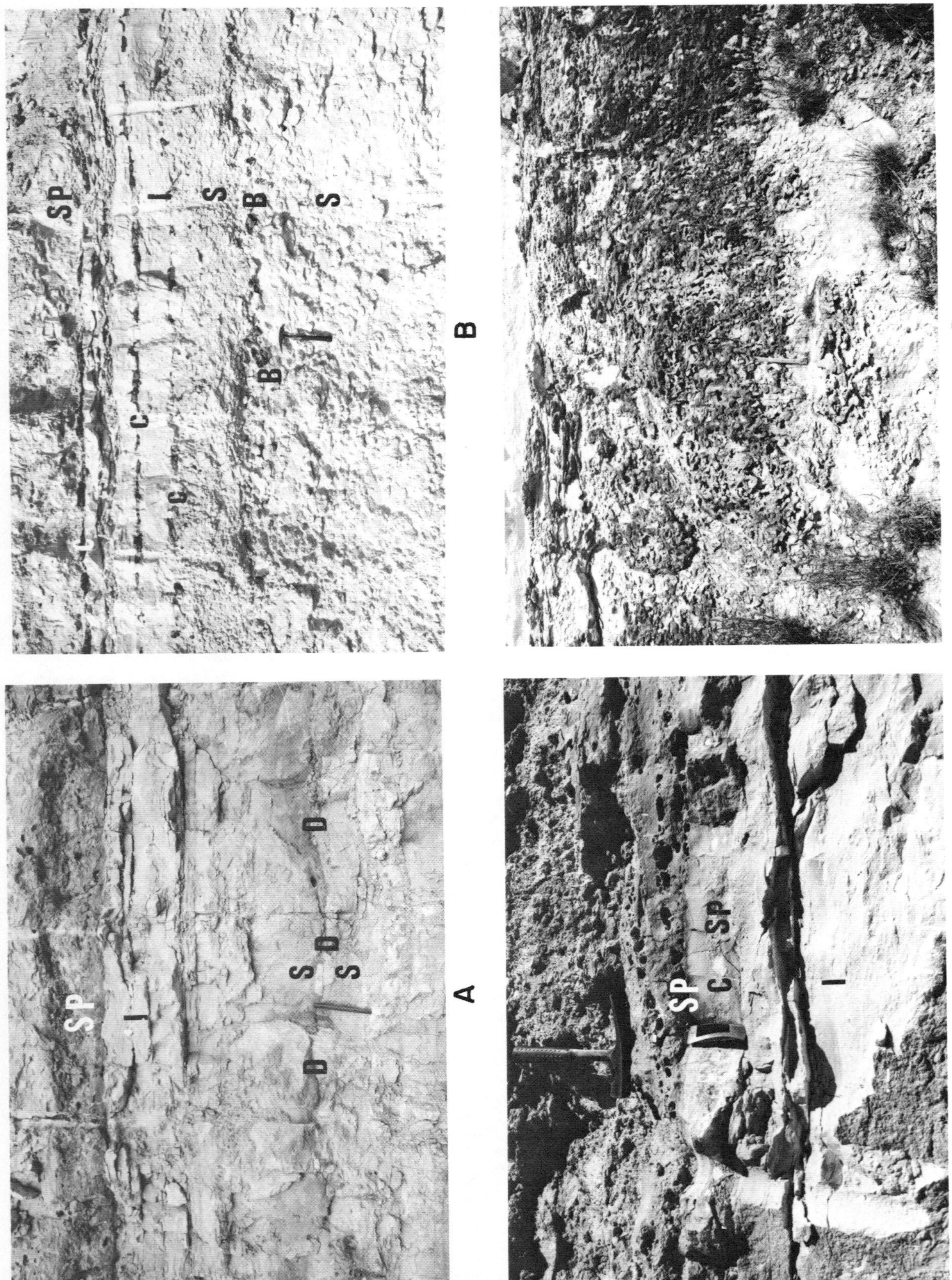

Figure 3

molds represent cases wherein anhydrite replaced allochems, but not the adjacent matrix, and subsequently became dissolved. Many examples of anhydrite replacing shells and not invading the matrix have been noted in Permian dolostones of the Permian Basin.

Many dolostones still contain calcite shells or bioclasts. In most cases peloids are partially to completely dolomitized. No ooids have been noted in Fort Terrett dolostones; they apparently either dissolved or were thoroughly micritized to form structureless peloids.

Sulfate Replacement

After, and also possibly during, dolomitization, much replacement of carbonate by large crystals or crystal aggregates of sparry anhydrite occurred. Sulfates nucleated within shells and ooids and then began to replace the matrix as blocky, rectangular (in cross section) extensions of sparry, replacement anhydrite crystals (fig. 8A, 8B). Borders of replacement anhydrite crystals commonly appear cloudy in thin section (fig. 8A), owing to inclusions of unreplaced carbonate. The phenomenon of sparry replacement anhydrite with rectangular terminations was first described by Murray (1960) who termed it replacement anhydrite. After Murray's description, no additional reports of this phenomenon have appeared in the literature.

My experience indicates that replacement anhydrite is exceedingly abundant in subsurface Permian dolostones of the Permian Basin, in Minnelusa carbonates of the Powder River Basin, and in Edwards equivalents of the J.F.S. Field, Dimmit County, Texas (Maverick Basin). Examples of sparry replacement anhydrite from the Permian Basin and Maverick Basin are depicted in figure 8A and 8B.

The former presence of abundant sparry replacement anhydrite in the Fort Terrett is of more than passing interest. All of the replacement anhydrite has been dissolved from Fort Terrett exposures, leaving only subtle indications of its former presence—the stairstep molds (figs. 8, 9, 10). Stairstep molds are also abundantly represented in both outcrop and subsurface of middle Permian dolostones of the Permian Basin and in Edwards equivalents in the subsurface of the Maverick Basin.

Silicification of Dolostones

Silica is abundantly represented in Fort Terrett dolostones. It occurs as replacements of carbonates and sulfates and as void-filling cements.

Silica cements have been found only in dolostones. Some small molds are completely filled by length-fast chalcedony (chalcedonite). Other voids are filled by chalcedonite and drusy quartz. Many larger molds are filled successively by: (1) length-fast chalcedony, (2) length-slow chalcedony (quartzine), and (3) length-fast chalcedony (fig. 5D). In dolostones silica cements most commonly occur in molds formed by dissolution of shells, ooids, and sulfates(?).

In many supratidal deposits at least some nodular anhydrite has been replaced by silica as laminoid-nodular zones (fig. 3B, 3C). In some cases, dense anhydrite nodular mosaics (chicken wire anhydrite) have been replaced to form "cauliflower" cherts (fig. 5A). Analysis of thin sections reveals that silica, which has replaced anhydrite, consists of spherulitic and polygonal length-slow chalcedony, megaquartz, and "blossoms" of flamboyant megaquartz, all of which contain inclusions of anhydrite (fig. 5B).

Most silica in dolostones occurs as replacements of organic-rich materials, such as shells, peloids, ooids, burrows, and stromatolites (Jacka, 1974b). In many cases silica replacement is restricted to allochems, but it commonly invades the matrix to form nodules. Peloids and finely crystalline matrix have been replaced by microquartz. Calcite and magnesian calcite shells have been replaced by length-slow chalcedony, megaquartz, or by both length-slow chalcedony and megaquartz (fig. 5C). The megaquartz has highly undulose extinction (Jacka, 1974b). Silica replacement of stromatolites has produced laminoid-nodular zones, similar to those formed by replacement of nodular anhydrite (fig. 3B). Algal stromatolites have been replaced by microquartz. Bird's-eye vugs within stromatolites have been filled by length-fast chalcedony.

No instances of sparry replacement anhydrite having been replaced by silica were noted within dolostones, suggesting that silicification preceded emplacement of second-generation sparry anhydrite. As previously mentioned, some silicification of nodular anhydrite, which formed penecontemporaneously within the sabkhas, took place.

(page 188)

Figure 4. Detailed characteristics of Fort Terrett exposures along Interstate 10 from Junction to beyond Segovia.

A. Supratidal dolostone (SP) contains molds of anhydrite nodules. The supratidal zone grades downward into intertidal (I) and subtidal (S) facies.

B. Major collapse breccia zone at the top of the Fort Terrett is shown. This zone records dissolution of Kirschberg sulfates and probably reflects a major eustatic-climatic fluctuation.

C. A pulverulent limestone interval occurs alongside the highway.

D. A pulverulent dolostone layer (P) is exposed in the middle of the photo.

(page 189)

Figure 5. Petrographic aspects of silica in Fort Terrett Formation. Photos B, C, and D are micrographs.

A. Outcrop samples of "cauliflower" chert which has replaced nodular mosaics of anhydrite which formed in supratidal environments.

B. Nodular anhydrite has been replaced by length-slow chalcedony and flamboyant megaquartz (Q). The silica contains abundant inclusions of anhydrite (A). Each smallest division equals 31 microns. Crossed polarizers.

C. Pelecypod shell in limestone has been partially replaced by length-slow chalcedony (C) and megaquartz (Q) with highly undulose extinction. Each smallest division equals 38 microns. Crossed polarizers.

D. Mold in dolostone interval has been filled successively by: length-fast chalcedony (F), length-slow chalcedony (S), and length-fast chalcedony (F). Each smallest division equals 90 microns. Crossed polarizers.

(page 190)

Figure 6. Photomicrographs illustrating depositional and diagenetic features of the Fort Terrett. All were photographed in plane light.

A. Oolitic-peloid grainstone. Particles are coated by bladed to fibrous submarine cement films which originally consisted of high magnesium calcite. Most ooids have been micritized and concentric laminae are barely discernible. Each smallest division equals 0.038 mm (38 microns).

B. Oolitic-peloid grainstone. Unmicritized ooids consisted of crystalline aragonite and were coated by micrite envelopes (dark rims) and bladed films of beachrock cement; subsequently, crystalline aragonitic ooids dissolved in a fresh-water diagenetic environment, leaving hollow micrite envelopes. Porosity has been selectively preserved in hollow micrite envelopes and on micritized bioclasts and peloids, because these constitute the most hostile substrates for nucleation of cements. Each smallest division equals 38 microns.

C. Intraclast wackestone. Matrix has been recrystallized to pseudospar in fresh-water diagenetic environment, and sparry anhydrite was dissolved leaving a mold (right side of micrograph) which later became partially filled by scalenohedral calcite cement. Intraclasts (dark) were not recrystallized. Each smallest division equals 38 microns.

D. Biopeloid-grapestone packstone. Micrite matrix has been intensively recrystallized in a fresh-water diagenetic environment, and hollow micrite envelopes were formed and subsequently filled by calcite cement. Calcite shells, peloids, micrite envelopes, grapestone grains (G), and bioclasts were not recrystallized. Each smallest division equals 90 microns.

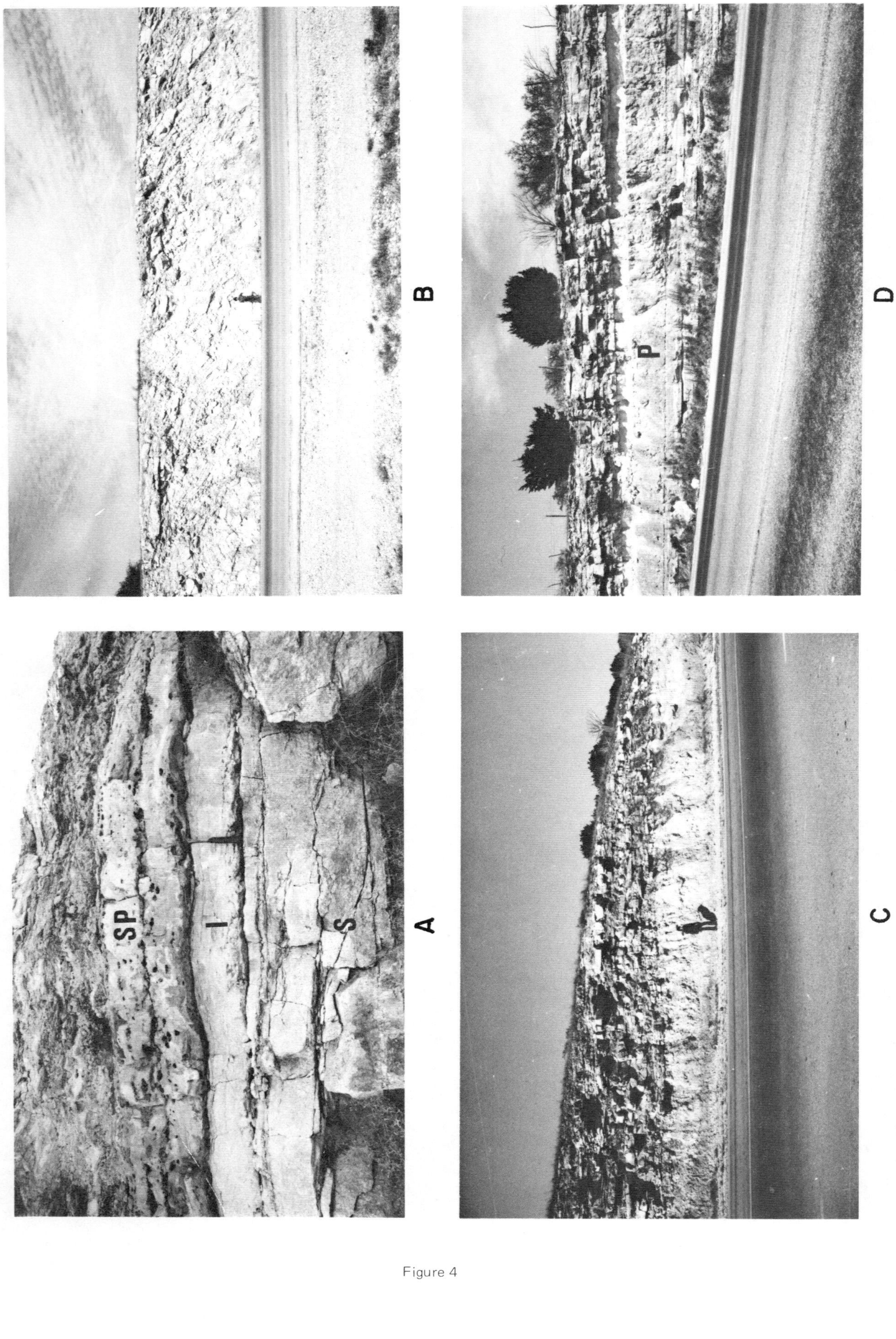

Figure 4

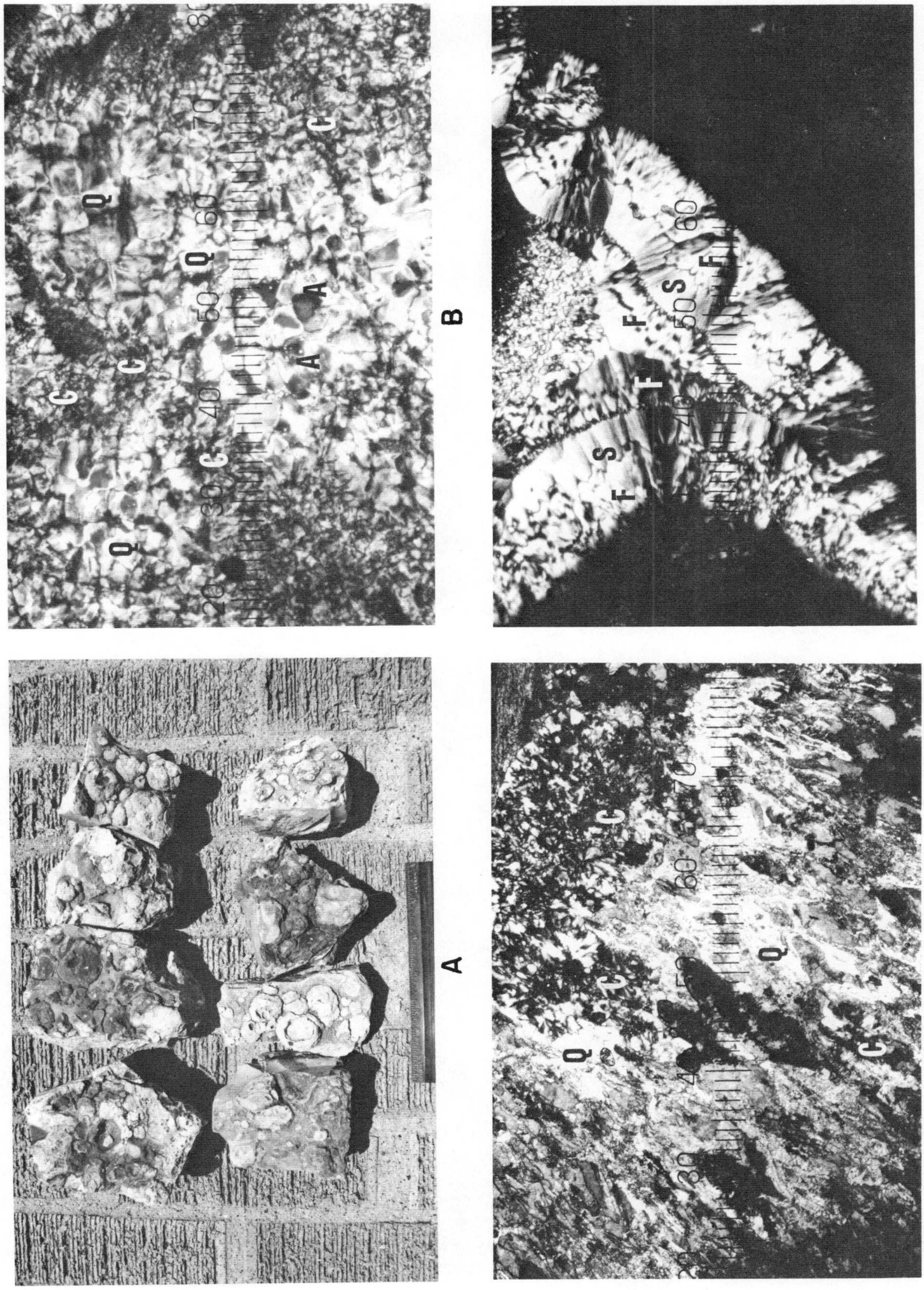

Figure 5

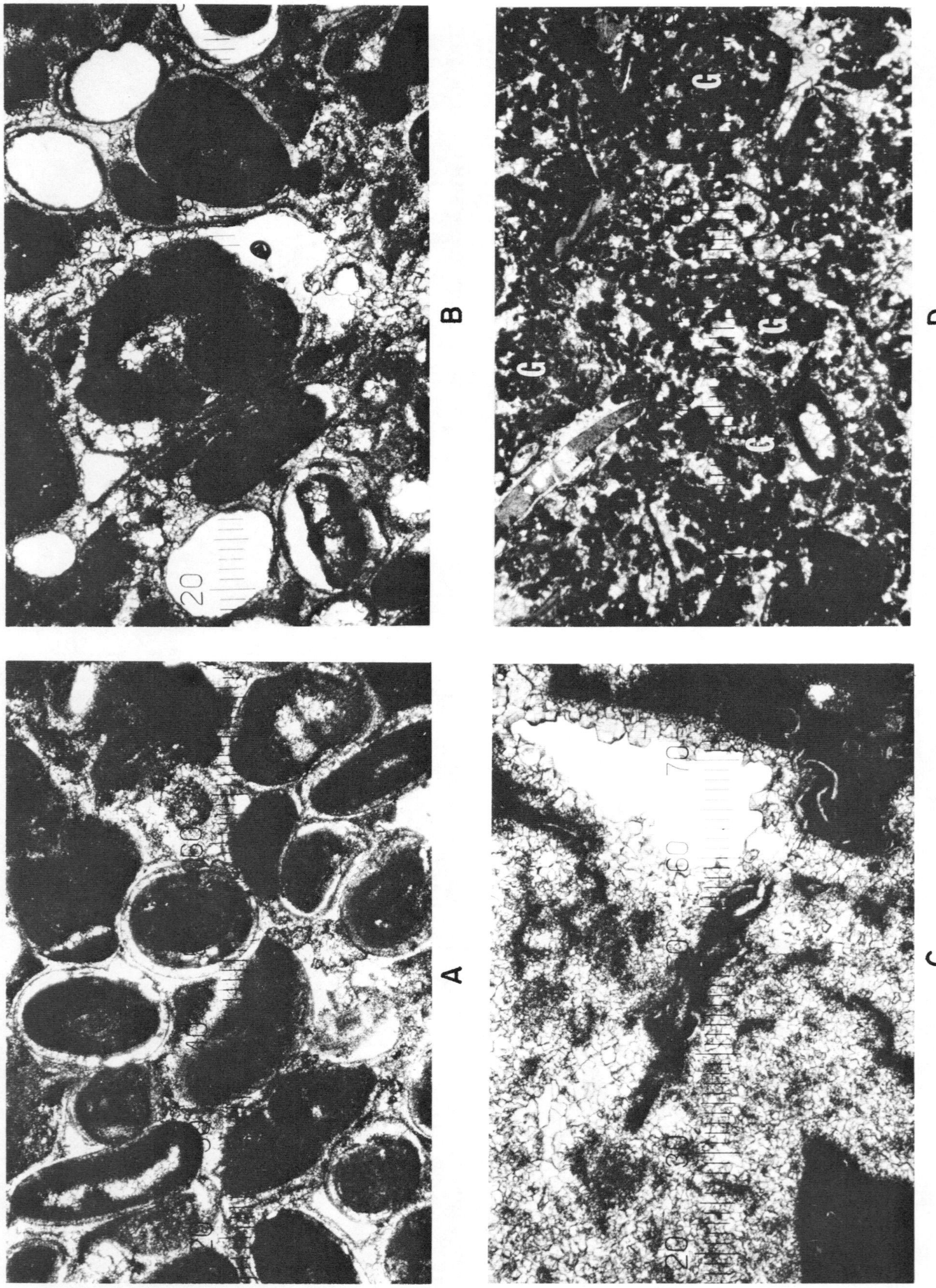

Figure 6

Dedolomitization

Dedolomitization is abundantly represented in the Fort Terrett. An example of complete dedolomitization of an interval that is at least 0.6 m thick was noted (fig. 7D). This dedolomitized zone occurs just below the major collapse breccia. Dissolution of sulfates, which caused formation of collapse breccia, probably produced ground waters with a high Ca/Mg ratio; these high-calcium waters probably caused calcite to replace dolomite, according to the mechanism of DeGroot (1967). Many examples of small-scale dedolomitization have been observed. In most of these cases, large calcite cement crystals have engulfed small dolomite cement crystals which had been precipitated around margins of vugs. Many of these engulfed dolomite crystals have been partially to completely dedolomitized.

In all examples noted, rhombic dolomite crystals were paramorphically replaced by calcite.

DIAGENESIS OF FORT TERRETT LIMESTONES

Petrography and Diagenetic Mechanisms

All Fort Terrett limestone intervals bear abundant evidence of having stabilized in fresh-water diagenetic environments, according to the format established by Matthews (1968). Nearly all crystalline aragonite shells and ooids were selectively dissolved in meteoric diagenetic environments. Those aragonitic shells and ooids which became micritized soon after deposition in the marine environment were not dissolved. Many examples of intensive recrystallization of original lime mud matrix occur in limestone intervals (fig. 6C, 6D). Jacka and Brand (1977) have shown that such coarse aggrading recrystallization may reflect a meteoric phreatic overprint. Many limestone intervals contain dolomite and quartz phorphyroblasts. The dolomite crystals are rhombic and the quartz consists of small doubly terminated crystals.

in several limestone intervals aragonitic shells were dissolved before lithification of matrix had been completed, and molds partially collapsed to form crumbly fractures. Subsequently, molds and fractures became enlarged by solution to form irregular vugs and channels, respectively. Most of these secondary solution cavities became partially to completely filled by dolomite and calcite cements, with calcite predominating. Calcite cements consist of scalenohedra and blocky, equant crystals.

In intervals which contain large solution cavities, voids are commonly floored by vadose internal sediment (both carbonate and clastic materials), and they also contain gravitational (microstalactitic) cements. These usually consist of elongated crusts of drusy calcite scalenohedra. Commonly, dripstone occurs in largest voids, and the laminar structure has recrystallized to radial fibrous calcite (fig. 11C). It is possible that the original dripstone consisted of aragonite which was replaced by radial fibrous calcite.

Silicification

Much silicification is evidenced in limestone intervals. Occurrences of doubly terminated quartz porphyroblasts have been previously cited. Most replacement silica nucleated within organic-rich components, such as shells, ooids, peloids, burrows, and stromatolites (figs. 3B, 5D). It commonly invaded the surrounding matrix to form nodules. Replacement of stromatolites usually forms laminoid-nodular zones (fig. 3B). It is inferred that silica replacement occurred after aragonitic shells had been selectively dissolved, because only magnesian calcite and calcite shells have been replaced. Molds of aragonitic shells or bioclasts contain calcite cements (fig. 11A). Both magnesian calcite and calcite shells have been replaced by either length-slow chalcedony or megaquartz or by both varieties of silica (fig. 5C). Micrite matrix, peloids, and stromatolites have been replaced by microquartz. No examples of silica cements were noted in limestones.

It is inferred that silica, for replacements, was derived from meteoric ground waters; silicification thus constitutes part of the fresh-water diagenetic syndrome.

Sulfate Replacement in Limestone

Most limestone intervals contain stairstep molds (figs. 8D, 9, 11A) like dolostone intervals. As in dolostones, anhydrite nucleated within crystalline shells and ooids. No example of silica replacement of sulfates was noted in Fort Terrett limestones, and accordingly, it is inferred that anhydrite replacement occurred after silicification.

PULVERULENT LIMESTONES AND DOLOSTONES

Several stratigraphic intervals within the Fort Terrett consist of soft

(page 192)

Figure 7. Photomicrographs illustrating some petrographic aspects of Fort Terrett dolostone intervals. All micrographs were exposed in plane polarized light.
A. Light area in center of micrograph represents neomorphic dolomitization of a burrow within a limestone interval. Thin section was stained with Alizarin Red S. Dolomitized burrow has excellent intercrystalline and moldic porosity, while the surrounding limestone is very dense. Each smallest division equals 90 microns.
B. Neomorphically dolomitized subtidal facies has excellent intercrystalline and moldic porosity (light). Each smallest division equals 38 microns.
C. Fibrous calcitic pelecypod shell has been partially dolomitized in two areas (D). Note optical continuity of calcitic fibers through dolomite, and the calcitic fabric has been preserved. The thin section was stained with Alizarin Red S. Each smallest division equals 38 microns.
D. Excellent example of dedolomite. Calcite has paramorphically replaced dolomite, and the rhombic morphology of original neomorphic dolomite crystals has been preserved. Each smallest division equals 12.5 microns.

(page 193)

Figure 8. Photomicrographs illustrating morphology of sparry replacement anhydrite and molds which result from its solution.
A. Sparry replacement anhydrite in subsurface sample from middle Permian Yates Formation, Lea County, New Mexico. Note rectangular extensions and terminations and cloudy borders, representing inclusions of unreplaced carbonate. Plane light. Each small division equals 90 microns.
B. Large unit crystal of sparry anhydrite is replacing a peloid grainstone in subsurface sample of Edwards-McKnight interval, J.F.S. Field, Dimmit County, Texas. Note blocky rectangular extensions around margin of crystal. Crossed polarizers. Each small division equals 90 microns.
C. Stairstep mold records dissolution of sparry replacement anhydrite from Fort Terrett dolostone unit. Plane light. Each smallest division equals 38 microns.
D. Stairstep molds in Fort Terrett limestone interval record dissolution of replacement anhydrite. Plane light. Each smallest division equals 90 microns.

(page 194)

Figure 9. Photomicrographs illustrating petrographic characteristics of pulverulent limestones in the Fort Terrett.
A. Close spacing of stairstep molds, which reformed through dissolution of sparry replacement anhydrite, is depicted. Plane light. Each smallest division equals 90 microns.
B. "Shotgun" pattern of porosity represents molds formed by dissolution of relatively large and small areas of replacement anhydrite. Some smaller molds record dissolution of dolomite porphyroblasts. Plane light. Each smallest division equals 90 microns.
C. Highly porous pulverulent limestone with relatively small stairstep molds and dolomolds (rhombic outlines) is illustrated. Plane light. Each smallest division equals 38 microns.
D. Two dolomite porphyroblasts are "caught in the act" of selectively dissolving from limestone matrix. Dissolved portion appears black under crossed polarizers. Each smallest division equals 10 microns.

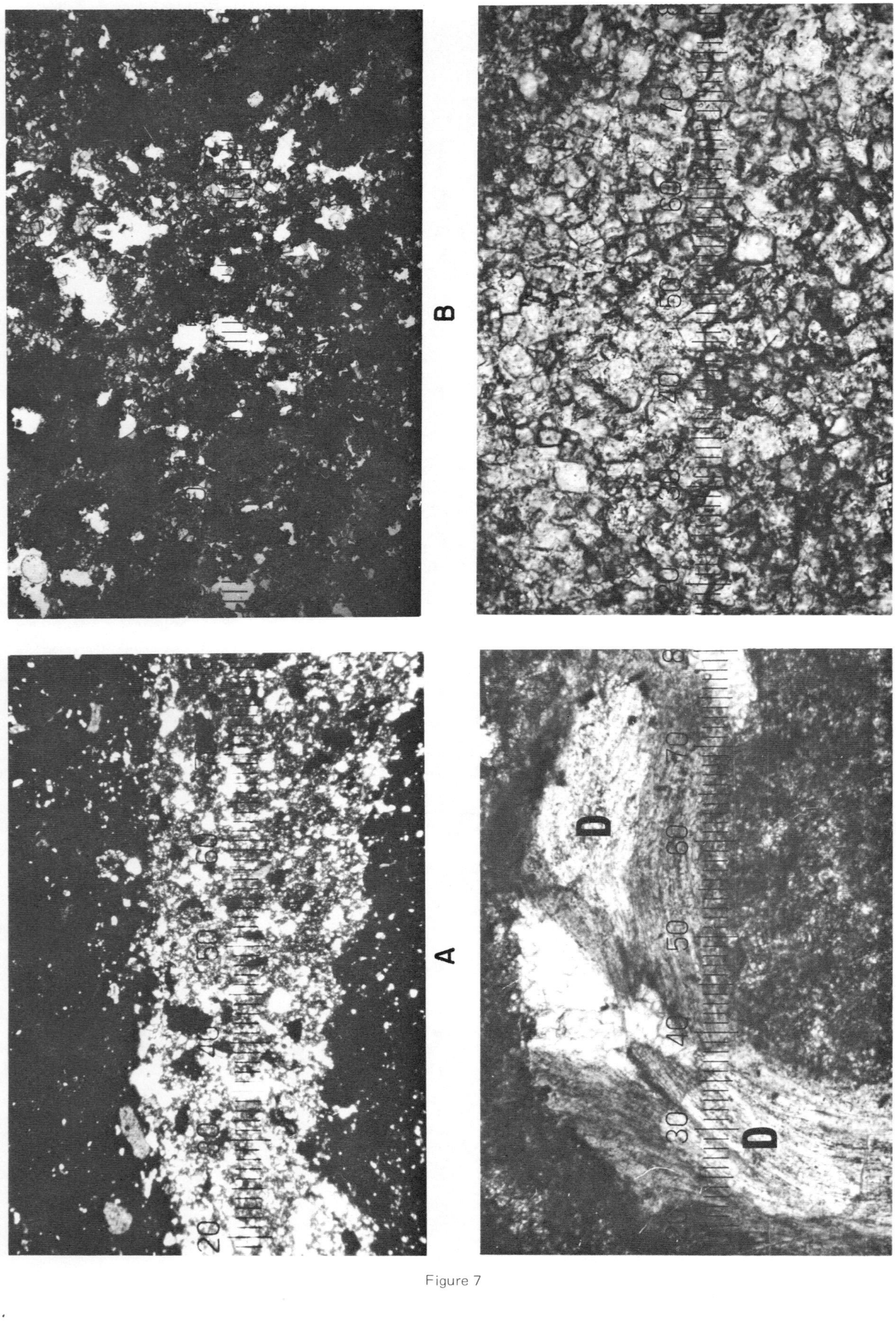

Figure 7

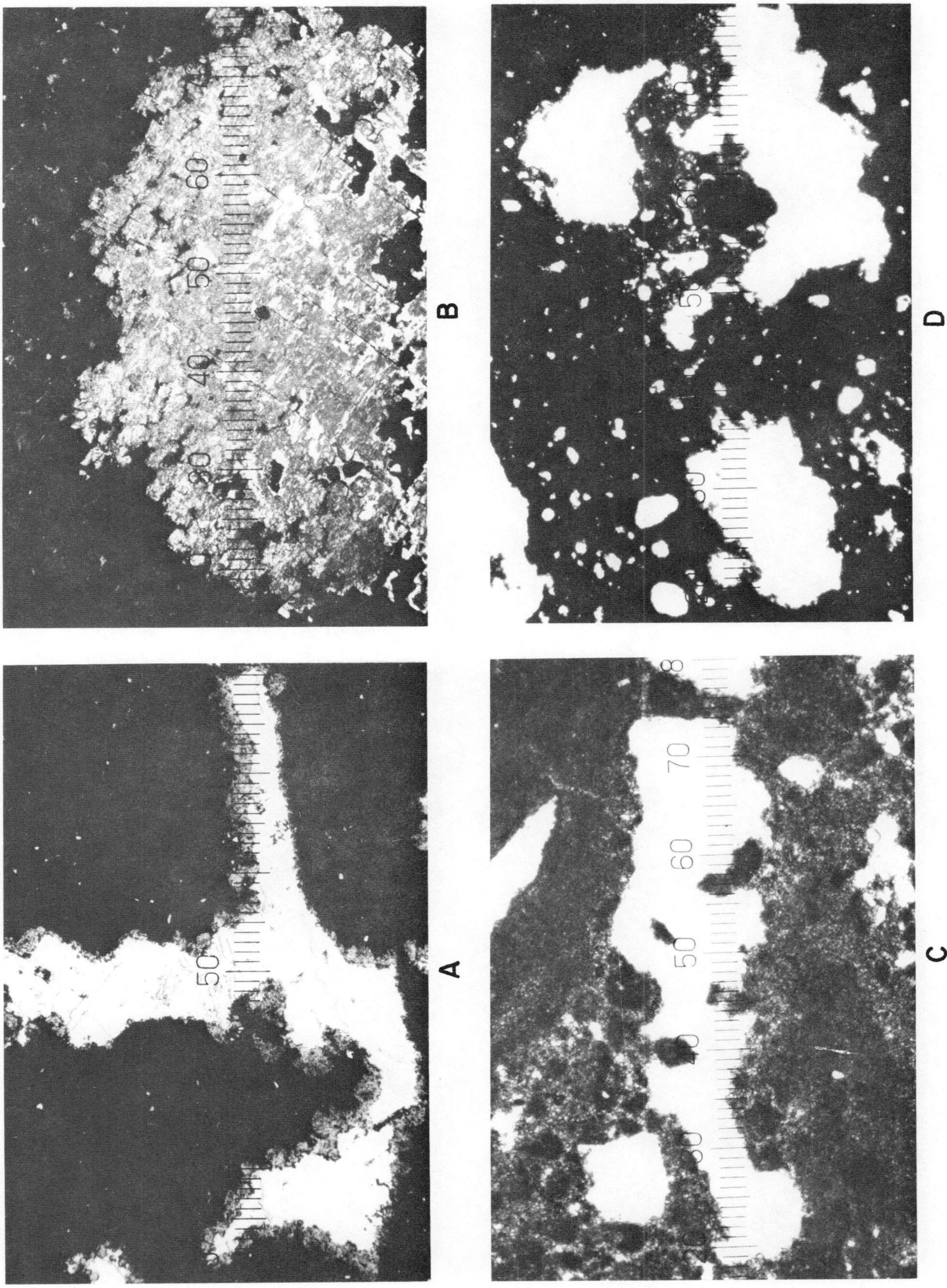

Figure 8

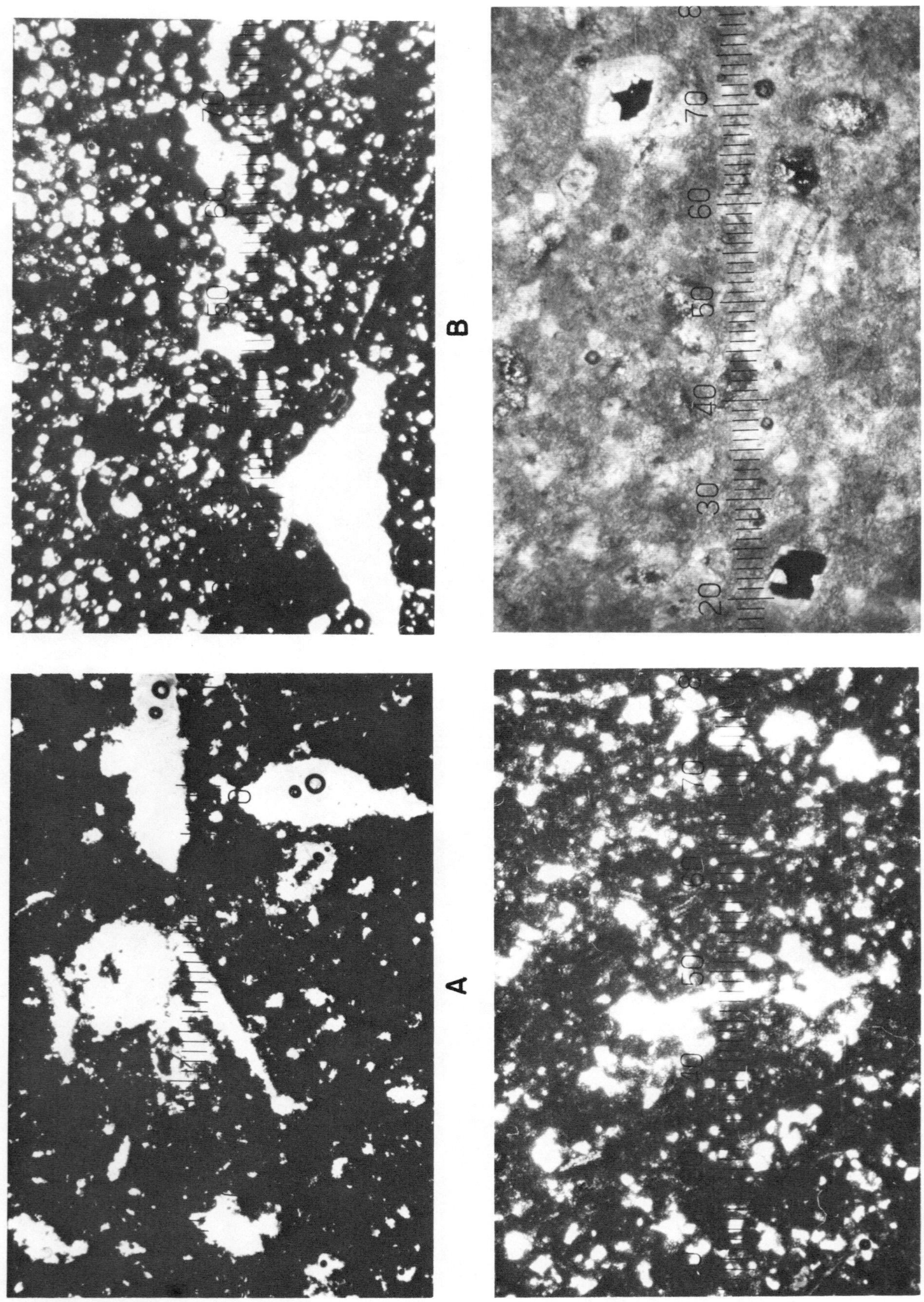

Figure 9

powdery limestone and dolostone. The materials crumble into powder when pressed tightly between thumb and forefinger. Rose (1972) describes such material as pulverulent limestone and dolostone, or pulverulite. Pulverulite layers are exposed in new roadcuts along Interstate 10 (fig. 4C, 4D).

Twelve samples of pulverulent limestone and dolostone were impregnated with epoxy and thin sections were prepared and studied. Pulverulent dolostones contain "excessive" development of intercrystalline and moldic porosity (fig. 10). Most molds are of the stairstep variety that records dissolution of sparry replacement anhydrite (fig. 10). It is also possible that "excessive" intercrystalline porosity was once occupied by anhydrite. Many examples of this phenomenon have been noted in Permian carbonates of the Permian Basin and in the Minnelusa of the Powder River Basin.

Pulverulent limestones also contain a hyperabundance of stairstep molds and dolomolds (fig. 9). In two thin sections of pulverulent limestone, all dolomite porphyroblasts have completely dissolved. In several thin sections (not just from pulverulent limestones) examples of partially corroded dolomite porphyroblasts were noted (fig. 9D). To my knowledge, dolomoldic porosity has not previously been noted or described in the literature. In one thin section of pulverulent limestone, a completely dolomitized area (except for calcitic bioclasts) representing a burrow was observed. Within this dolomitized burrow, many hollow dolomite crystals are represented by outer rims, the centers having been dissolved (fig. 10C). In addition to hollow dolomite crystals, many fragile, irregular remnants of dolomite crystals occur. Thus, solution of dolomite was not restricted to dolomite porphyroblasts.

The question of when anhydrite was dissolved to form pulverulent limestones and dolostones is not easily answered. In most pulverulent dolostone intervals, stairstep molds contain at least a small amount of dolomite cement. In two pulverulitic dolostone intervals, stairstep molds contain: (1) dolomite and (2) calcite cements (fig. 10D). The calcite cement crystals are much larger than the dolomite crystals that rim the stairstep molds. Many dolomite crystals, which have been engulfed by calcite cement crystals, have been dedolomitized. In cases where stairstep molds contain cements, it is possible that dissolution of sulfates may have occurred in the Cretaceous. In cases where stairstep molds are completely open, anhydrite solution may be a relatively recent event. No examples of sulfates were noted in any of the samples which were analyzed. Pulverulent limestones and dolostones both represent "testaments to vanished sulfates."

POROSITY RELATIONSHIPS

Primary Porosity

Primary porosity was formed in grainstones. In some grainstone intervals, much primary porosity was occluded by precipitation of subtidal and beachrock cements, which have been previously described. In such cases remaining voids were largely to completely filled by scalenohedral calcite cements which were precipitated within meteoric diagenetic environments. Very little primary porosity was preserved within grainstone lithofacies. In fact, much more secondary porosity has been preserved in grainstones than primary porosity. This secondary porosity was created by selective dissolution of crystalline aragonitic shells, bioclasts, and ooids (fig. 6B).

Secondary Porosity in Collapse Breccias

A thick (approximately 9 m) collapse breccia zone occurs in the upper interval of the Fort Terrett. As previously indicated, this occurrence reflects dissolution of underlying sulfates of the Kirschberg evaporite unit during a major eustatic lowering of sea level, which was associated with increased rainfall. Large voids were created among the collapsed and fractured blocks and many of these became enlarged by solution to form microcavernous porosity. Much of this porosity became occluded by deposition of internal sediment and dripstone cements. Much solution and precipitation of cements has been taking place during the latest episode of subaerial exposure. Excellent porosity has been preserved within the major collapse breccia zone.

Two thin collapse breccias occur below the thick collapse zone at the top of the Fort Terrett. These zones were formed within Fort Terrett time and also reflect eustatic-climatic fluctuations. Porosity in the smaller scale collapse zones has been largely occluded by internal sedimentation and precipitation of meteoric cements.

Secondary Porosity in Dolostones

Intercrystalline porosity was formed and has been preserved in several dolostone intervals (fig. 7B). Most dolostones also contain smooth-walled molds which appear to reflect selective dissolution of ooids, shells, and bioclasts during dolomitization. All dolostones contain stairstep molds which were formed by dissolution of anhydrite. It is possible that smooth-walled molds and intercrystalline voids also record solution of anhydrite.

Several examples of selective dolomitization within burrows were observed. In all cases dolomitized burrows had much greater porosity than surrounding limestones (fig. 7A). Dolomitized burrows contain intercrystalline and moldic porosity.

Secondary Porosity in Limestones

Much secondary porosity was formed in limestone intervals. It was initiated by selective dissolution of crystalline aragonitic allochems in fresh-water diagenetic environments, which were associated with low stands of sea level and increased rainfall.

Preservation of secondary porosity in limestones has been highly selective. Preservation of leached porosity has been most favored within hollow micrite envelopes and on micritized shell foundations and peloids. These constitute the most hostile substrate for precipitation of calcite cements. The second most hostile substrate for precipitation of calcite cements is micrite matrix-walled solution voids. Calcite cements are precipitated on crystalline calcite shell foundations much faster than on

(page 196)

Figure 10. Photomicrographs illustrating petrographic features of pulverulent dolostones in Fort Terrett. All were photographed in plane polarized light.

A. Extremely high porosity is depicted. Both stairstep molds and "excessive" intercrystalline voids record dissolution of anhydrite. Each smallest division equals 90 microns.

B. Notice extremely high porosity which includes some relatively small stairstep voids. Each smallest division equals 90 microns.

C. Micrograph depicts intensively dolomitized zone occurring within a pulverulitic limestone (probably a dolomitized burrow). The matrix consists of highly corroded remnants of once solid dolomite crystals. Several hollow dolomite crystals (H) are visible. The cores of these crystals have selectively dissolved to form "pillbox" structures. Each smallest division equals 10 microns.

D. Stairstep molds in dolostone interval are shown. All of these molds contain some dolomite cement crystals (rhombic). Two stairstep molds contain coarse sparry calcite cement (C) which was precipitated after dolomite cement. These stairstep molds may have formed by solution of anhydrite in Cretaceous time. Each smallest division equals 38 microns.

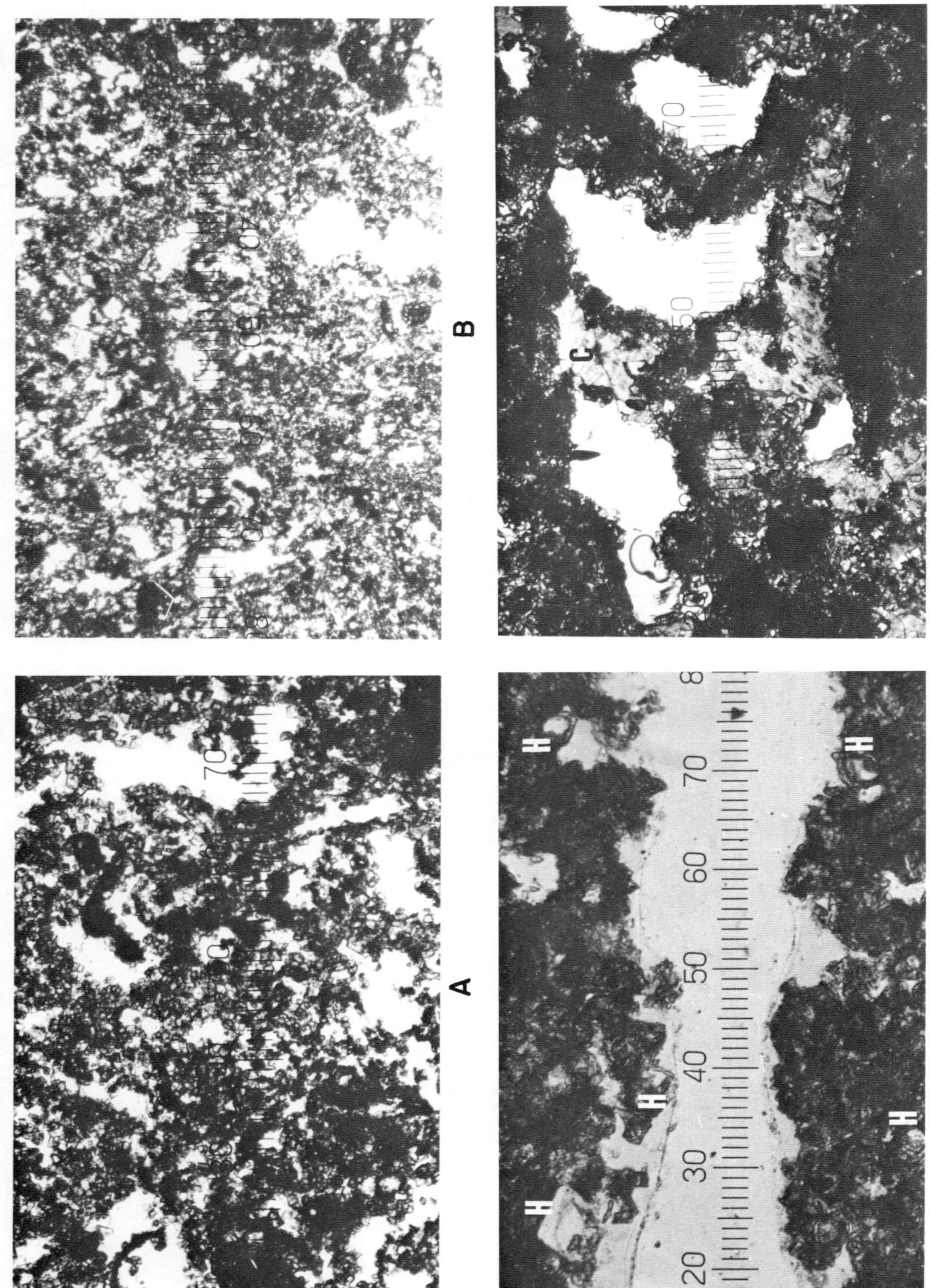

Figure 10

prismatic or fibrous foundations. These relationships to a large extent reflect degree of epitaxial response (see Jacka, 1974a). Precipitation of calcite cements in crystalline calcite foundations requires only a crystal growth step, involving extension of preexisting crystals as optically continuous overgrowths. Precipitation on micritized or micrite matrix-walled voids may require a nucleation step before crystal growth can occur.

Tertiary Porosity in Limestones

In both limestones and dolostones of the Fort Terrett, a third generation of voids was formed which constitutes tertiary porosity. Third-generation voids were formed after most secondary voids had been partially to completely filled by cements.

Most limestone units exhibit the following diagenetic sequence: (1) aragonitic shells were selectively dissolved during mineralogic stabilization by fresh ground waters, and the molds were subsequently filled to varying degrees by calcite cements; (2) sparry anhydrite was emplaced, first replacing shells and then invading the matrix; (3) sparry anhydrite was dissolved by a second incursion of fresh ground waters, and stairstep molds were formed (fig. 9). Commonly, open stairstep molds were noted adjacent to aragonitic shell molds which are filled by calcite cement (fig. 11A). In some limestone intervals, stairstep voids are filled by calcite cement (fig. 11D).

Tertiary Porosity in Dolostones

In most dolostone intervals, open stairstep molds occur adjacent to and transecting secondary molds which are filled by dolomite and calcite cements (fig. 11B, 11C). Implantation of sparry anhydrite in dolostones was similar to that recorded in limestones. Anhydrite nucleated in shells and then began to invade the matrix.

In one thin section from a dolostone interval, the mold of a bioclast contained some small dolomite cement crystals and was filled by gravitational calcite cement which was precipitated in a meteoric vadose environment (fig. 11C). The original microstalactitic cement now consists of radial fibrous calcite. Adjacent to this cement-filled mold are open stairstep molds which contain no cement. The following paragenesis is thus recorded.

(1) During dolomitization of a subtidal facies, some bioclasts dissolved.
(2) Sea level was lowered and the mold became filled by gravitational cement during subaerial exposure.
(3) Extensive replacement by sparry anhydrite took place. Whether anhydrite also was implanted during subaerial exposure or after the ensuing rapid submergence is not certain.
(4) Replacement anhydrite was dissolved in a fresh-water diagenetic environment. This step represents at least the second time the interval was subjected to fresh-water diagenesis.

Most supratidal dolostones were subjected to two episodes of sulfate emplacement: (1) gypsum and nodular anhydrite were formed penecontemporaneously during dolomitization and (2) sparry anhydrite was subsequently emplaced after secondary molds had formed and received cements. Such alterations of sulfate emplacement and fresh-water leaching reflect eustatic-climatic fluctuations.

I have analyzed porosity relationships in the Edwards-McKnight interval in J.F.S. Field, Dimmit County, Texas. It has been determined that production from the main gas pay zone is from tertiary stairstep molds in a dolostone. Anhydrite was dissolved from the dolostone in a fresh-water diagenetic environment after secondary intercrystalline voids had been filled by what is now a solid hydrocarbon. The stairstep molds contain no solid hydrocarbon and are clearly tertiary in origin (see Jacka and Stevenson, 1977).

CONCLUSIONS

1. In the vicinity of Junction, Texas, the Fort Terrett Formation contains well-developed depositional cycles, consisting of subtidal, intertidal, and supratidal (sabkha) facies. Cycles are separated by disconformities, which record degrees of erosion ranging from very small to moderate.

2. The Fort Terrett also exhibits well-developed and complex diagenetic cycles that include dolomitization, calcitization, sulfate emplacement dissolution, silicification, and, in some cases, dedolomitization.

3. Depositional and diagenetic cycles reflect interaction of the following processes: (a) glacially controlled eustatic sea level changes and associated climatic fluctuations, (b) progradation of subtidal, intertidal, and supratidal facies, and (c) subsidence.

4. A major collapse breccia constitutes the upper portion of the Fort Terrett and was formed by dissolution of the Kirschberg evaporite unit. This dissolution reflects a major eustatic-climatic fluctuation and may record a full glacial interval. Below this major collapse zone, at least two thin collapse zones were formed, and these may represent smaller scale fluctuations.

5. Early diagenesis in the marine environment is represented by extensive micritization of foraminifers, small bioclasts, and ooids, by formation of micrite envelopes, and by precipitation of subtidal and intertidal (beachrock) cements.

6. Supratidal deposits represent sabkhas; according to the Persian Gulf model, they were subjected to penecontemporaneous gypsum precipitation and conversion to anhydrite and dolomitization.

(page 198)

Figure 11. Photomicrographs illustrate porosity relationships of stairstep molds in limestones and dolostones. All micrographs were exposed in plane polarized light.
A. Limestone interval. An open stairstep mold (S) occurs adjacent to a cement-filled mold formed by solution of an aragonitic bioclast (A) within a fresh-water diagenetic environment. The stairstep mold formed through dissolution of sparry replacement anhydrite after the aragonitic bioclast mold had been filled by calcite cement; the anhydrite mold thus constitutes tertiary porosity. Each smallest division equals 25 microns.
B. Dolostone interval. An open tertiary stairstep mold (left center of micrograph) transects a secondary shell mold (S) which is partially filled by some small dolomite cement crystals and completely filled by large sparry calcite cement crystals. The secondary mold was probably formed during dolomitization. Replacement by and dissolution of sparry replacement anhydrite occurred after the secondary mold had been filled by cement. Thin section was stained with Alizarin Red S. Each smallest division equals 38 microns.
C. Dolostone interval. Bioclast on left side of micrograph dissolved during dolomitization to form a secondary mold. A few small dolomite cement crystals were precipitated in the mold; it subsequently became filled by gravitational (microstalactic) dripstone cement which was precipitated in a meteoric vadose diagenetic environment. The gravitational cement consists of radial-fibrous calcite which formed by recrystallization of dripstone or replacement of original aragonite. Open tertiary stairstep molds occur to the right of the cement-filled mold; these were created by dissolution of sparry replacement anhydrite after the secondary mold had been filled by calcite cement. Precipitation of meteoric gravitational cement and dissolution of anhydrite record two distinct episodes of fresh-water diagenesis. Each small division equals 90 microns.
D. Limestone interval. A stairstep mold has been completely filled by sparry calcite cement. Such stairstep molds may have been formed and filled by calcite cement in Fort Terrett time. Each smallest division equals 25 microns.

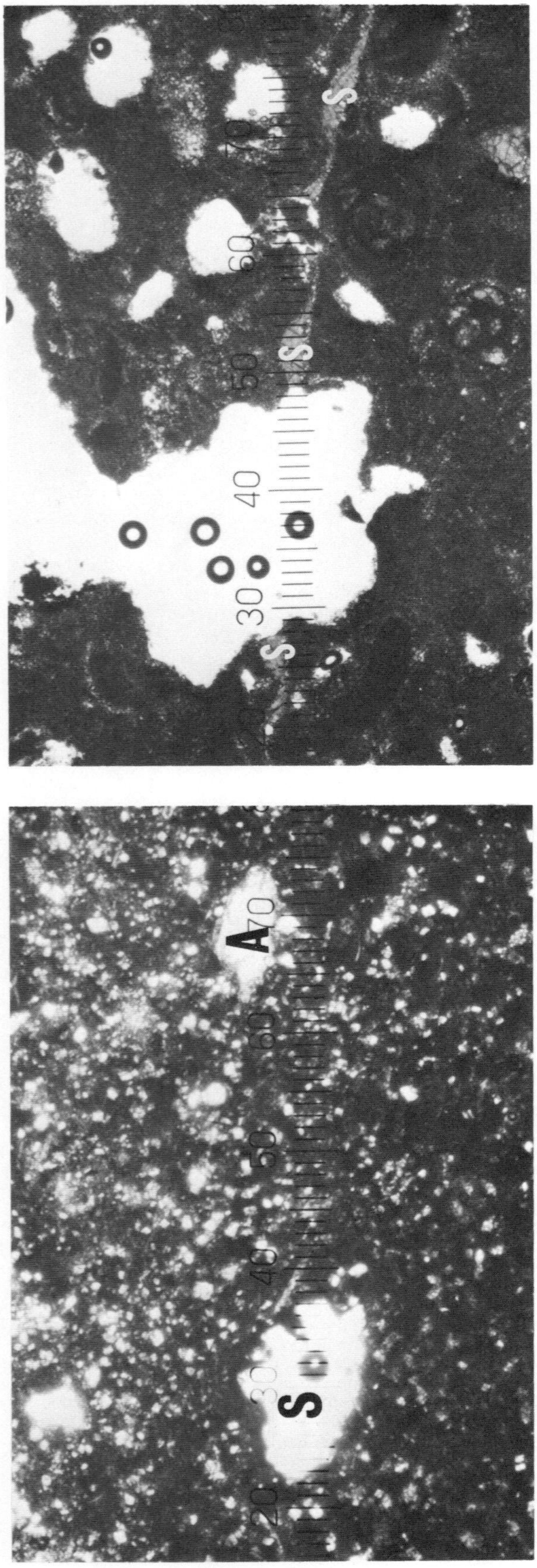

A B C D

Figure 11

7. Within most depositional cycles supratidal dolostones are underlain by intertidal and subtidal limestones. In some cycles supratidal dolostones are underlain successively by intertidal and subtidal limestones and dolostone (lower portion of subtidal facies).

8. Nearly all limestones bear abundant evidence of having stabilized within fresh-water diagenetic environments. The mechanism (or mechanisms?) of dolomitization of subtidal facies has not been clearly revealed; some intervals could have been dolomitized according to the dorag model—in the zone of mixing at the base of a coastal, fresh-water lens.

9. Silica exhibits a strong preference for nucleation within organic-rich carbonate materials and anhydrite nodules. In supratidal dolostones, silica occurs as replacement of anhydrite nodules and stromatolites to form laminoid-nodular zones. Laminoid-nodular zones of chert have also formed as replacements of intertidal and subtidal stromatolites. Anhydrite has been replaced by length-slow chalcedony, megaquartz, and flamboyant megaquartz. In dolostone intervals, including subtidal, intertidal, and supratidal facies, stromatolites, burrows, peloids, and lime mud matrix have been replaced by microquartz. Crystalline shells and ooids have been replaced by length-slow chalcedony and megaquartz. Length-fast chalcedony, length-slow chalcedony, and drusy quartz all occur as void-filling precipitates within dolostone intervals.

In limestones silica occurs as replacements of shells, ooids, peloids, burrows, stromatolites, and lime mud matrix. The petrographic varieties are the same as those found in dolostones for each counterpart. Silica was emplaced within limestones after aragonite shells had been dissolved and molds filled by cement.

10. Very little primary porosity has been preserved in grainstones. Most primary porosity has been occluded by precipitation of marine and meteoric cements. In some grainstones considerable secondary porosity has formed from dissolution of crystalline aragonitic ooids and bioclasts. Much of this porosity has been preserved within hollow micrite envelopes.

11. Much secondary porosity was formed in wackestone and packstone limestones; it was initiated through selective solution of crystalline aragonitic shells and ooids. Relatively little of this secondary porosity has been preserved. Preservation of porosity in limestones was heavily biased toward retention in hollow micrite envelopes and within cavities of micritized shells. These materials constitute the most hostile substrates for nucleation of calcite cements. Micrite matrix-walled voids constitute the second most favored substrate for preservation of porosity.

12. Considerable secondary intercrystalline and moldic porosity were formed in dolostones and have been extensively preserved.

13. Excellent porosity exists in the thick collapse breccia zone atop the Fort Terrett. Original porosity probably has been enhanced by relatively recent solution.

14. Throughout the Fort Terrett most preserved porosity is of tertiary origin. It was created by dissolution of sparry anhydrite which invaded both limestone and dolostone intervals after mineralogic stabilization. Tertiary porosity consists of open stairstep molds which occur adjacent to and transecting secondary voids which were filled to varying degrees by cements. Sparry replacement anhydrite constitutes a second generation of sulfate emplacement within many supratidal dolostones.

15. Pulverulent limestones and dolostones crumble into powder when pressed tightly between thumb and forefinger. Pulverulent dolostone contains "excessive" intercrystalline and moldic porosity. Most molds are of the stairstep variety. Pulverulitic limestones contain "excessive" stairstep molds and dolomolds. Both pulverulent limestones and dolostones constitute testaments to vanished sulfates.

ACKNOWLEDGMENTS

I am grateful to the Geosciences Department at Texas Tech University, especially to Donald R. Haragan, for encouragement and enthusiastic support which has enabled me to conduct a graduate course in carbonates from the Texas Tech Center at Junction. The excellent cooperation and support of Robert Packard, Academic Director of the Texas Tech Center at Junction, is gratefully acknowledged.

The Geosciences Department provided thin sections and photographic reproduction for this report. Mike Gower prepared the thin sections and Robert Suddarth did the photography.

Gratitude is expressed to Suburban Propane Gas Corporation, San Antonio, for granting me permission to comment on diagenetic and porosity relationships found within Edwards equivalents in J.F.S. Field, Dimmit County, Texas (Maverick Basin).

REFERENCES

Badiozamani, K., 1973, The Dorag dolomitization model-application to the Middle Ordovician of Wisconsin: Journal of Sedimentary Petrology, v. 43, p. 965-98.

Butler, G. P., 1969, Modern evaporite deposition and geochemistry of coexisting brines, the sabkha, Trucial Coast, Arabian Gulf: Journal of Sedimentary Petrology, v. 39, p. 70-89.

DeGroot, K., 1967, Experimental dedolomitization: Journal of Sedimentary Petrology, v. 37, p. 1216-20.

Fisher, W. L., and Rodda, P. U., 1969, Edwards Formation (Lower Cretaceous), Texas: Dolomitization in a carbonate platform system: AAPG Bull., v. 53, p. 55-72.

Ginsburg, R. N., 1957, Early diagenesis and lithification of shallow-water carbonate sediments in south Florida, *in* Regional aspects of carbonate deposition: SEPM Special Pub. No. 5, p. 80-100.

Hanshaw, B. B., Back, W., and Deike, R. G., 1971, A geochemical model for dolomitization by groundwater: Economic Geology, v. 66, p. 710-24.

Jacka, A. D., 1974a, Differential cementation of a Pleistocene carbonate fanglomerate, Guadalupe Mountains: Journal of Sedimentary Petrology, v. 44, p. 85-92.

______ 1974b, Replacement of fossils by length-slow chalcedony and associated dolomitization: Journal of Sedimentary Petrology, v. 44, p. 421-27.

______ 1975, Observations on paramorphic and neomorphic Permian dolostones and associated porosity relations (abs.): AAPG and SEPM Annual Meeting, Program Bull., p. 38-39.

______ , and Brand, J. P., 1977, Biofacies and development and differential occlusion of porosity in a lower Cretaceous (Edwards) reef: Journal of Sedimentary Petrology, v. 47, no. 1.

______ , and Stevenson, J. E., 1977, The J.F.S. Field, Dimmit County, Texas: Some unique aspects of Edwards-McKnight diagenesis: GCAGS Transactions, v. 27.

James, N.P., Ginsburg, R.N., Marszalek, D.S., and Choquette, P.W., 1976, Facies and fabric specificity of early subsea cements in shallow Belize (British Honduras) reefs: Journal of Sedimentary Petrology, v. 46, p. 523-545.

Kendall, C.G., and Skipwith, A.D., 1969, Holocene shallow-water carbonate and evaporite sediments of Khor al Bazam, Trucial Coast, Persian Gulf: AAPG Bull., v. 53, p. 841-69.

Kinsman, D.J.J., 1969, Modes of formation, sedimentary association and diagnostic features of shallow water and supratidal evaporites: AAPG Bull., v. 53, p. 830-40.

Land, L. S., 1973, Contemporaneous dolomitization of Middle Pleistocene reefs by meteoric water, north Jamaica: Bull. of Marine Science, v. 23, p. 64-92.

Matthews, R. K., 1968, Carbonate diagenesis: Equilibration of sedimentary mineralogy to the subaerial environment; Coral cap of Barbados, West Indies: Journal of Sedimentary Petrology, v. 38, p. 1110-1118.

Müller, G., 1971, Gravitational cement: An indicator of subaerial diagenetic environment, *in* Bricker, O. P., ed., Carbonate cements: Baltimore, Johns Hopkins Press, p. 310-20.

Murray, R. C., 1960, Origin of porosity in carbonate rocks: Journal of Sedimentary Petrology, v. 30, p. 59-89.

Rose, P. R., 1972, Edwards Group, surface and subsurface, central Texas: Univ. Texas, Austin, Bureau of Economic Geology, Report of Investigations 74, 198 p.

Steinen, R. P., 1974, Phreatic and vadose diagenetic modification of Pleistocene limestone: Petrographic observations from subsurface of Barbados, West Indies: AAPG Bull., v. 58, p. 1008-1024.

Taylor, J.C.M., and Illing, L. V., 1971, Variation in recent beachrock cements, Qatar, Persian Gulf, *in* Bricker, O. P., ed., Carbonate cements: Baltimore, Johns Hopkins Press, p. 32-35.

Winland, H. D., and Matthews, R. K., 1974, Origin and significance of grapestone, Bahama Islands: Journal of Sedimentary Petrology, v. 44, p. 921-27.

CARBON AND OXYGEN ISOTOPIC EVOLUTION OF WHOLE ROCK AND CEMENTS FROM THE STUART CITY TREND (LOWER CRETACEOUS, SOUTH-CENTRAL TEXAS)

D. R. Prezbindowski[1]

ABSTRACT

The Stuart City Trend consists of a shelf-margin buildup of Middle Cretaceous carbonates, now buried to depths of 10,000 to 18,000 feet in south-central Texas. Whole rock analyses of 92 samples from 16 wells along a 250-mile strike section show a δO^{18} range of -5.9°/oo to -2.7°/oo and a δC^{13} range of -.7°/oo to +5.1°/oo relative to PDB (figs. 1, 2). Oxygen isotopes become lighter toward the southwest (fig. 3). Whole rock values of δC^{13} indicate that vadose diagenesis was not volumetrically important.

Individual cements were also analyzed. The two predominant cement sequences are: (1) fibrous crust, (2) inclusion-rich radiaxial, and (3) clear spar, or (1) fibrous crust, (2) inclusion-rich spar, and (3) clear spar. Inlcusion-rich radiaxial cements show δO^{18} values closely grouped about a mean of -2.6°/oo PDB and δC^{13} values between -29.1°/oo and +3.2°/oo PDB. Inclusion-rich spar cements likewise show δO^{18} values closely grouped about the mean of -2.8°/oo PDB and δC^{13} values ranging from -7.4°/oo to +3.8°/oo PDB. In contrast, the clear blocky spars exhibit a wider range of δO^{18} values, from -6.6°/oo to -2.3°/oo with a mean of -5.2°/oo PDB; δC^{13} values range from -5.5°/oo to +4.5°/oo PDB. No significant isotopic differences were observed in the final generation of clear blocky spar cement between depths of 10,300 feet and 20,400 feet.

The whole rock and cements are not in oxygen isotopic equilibrium with sampled formation fluids. Individual cements maintain an isotopic memory of successive cementation events during burial.

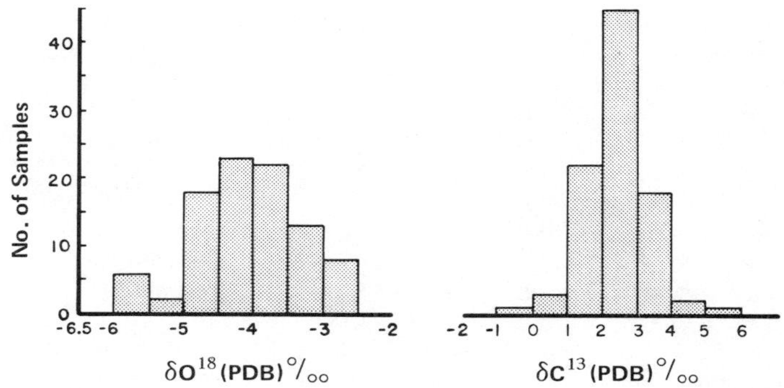

Figure 1. Histograms of oxygen and carbon isotopic composition (relative to PDB standard) of 92 whole rock samples, from cores taken from 16 wells along a 440-km strike section of the Lower Cretaceous, Stuart City Trend.

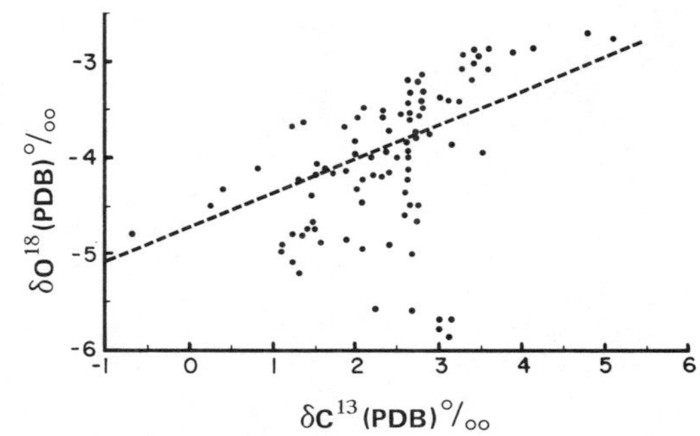

Figure 2. Scattergram plot of oxygen versus carbon isotopic composition of 92 whole rock samples, the same as illustrated in figure 1. The linear regression equation is $\delta O^{18} = .409(\delta C^{13})-4.98$. The Pearson product-moment correlation coefficient of this relationship is .49 and is significant at the .05 probability level.

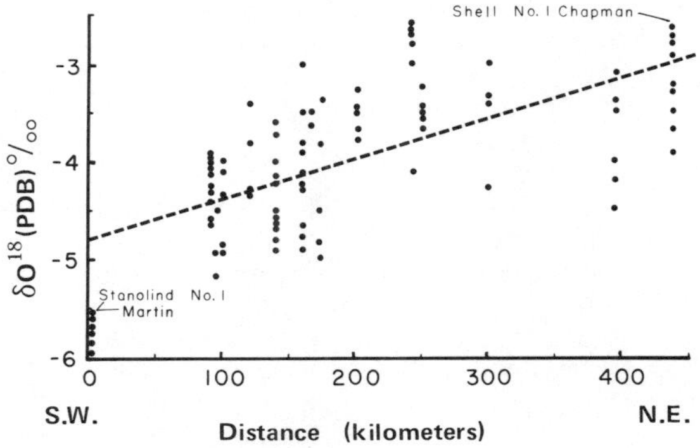

Figure 3. Scattergram plot of oxygen isotopic composition of 92 whole rock samples, the same as depicted in figures 1 and 2, versus distance from the initial southwesterly well (Stanolind No. 1 Martin, La Salle County, Texas) to the most northeasterly well (Shell No. 1 Chapman, Waller County, Texas). The linear regression equation is δO^{18} = .0044(distance)-4.81. The Pearson product-moment correlation coefficient of this relationship is .66 and is significant at the .05 probability level.

[1]Department of Geological Sciences, The University of Texas at Austin

HYDROGEN AND OXYGEN ISOTOPIC COMPOSITION OF CHERT FROM THE EDWARDS GROUP, LOWER CRETACEOUS, CENTRAL TEXAS

Lynton S. Land[1]

ABSTRACT

The oxygen isotopic composition of nodules and beds of replacement chert of the Edwards Group from outcrop and subsurface of Central Texas ranges between +30.2 and +33.9°/₀₀ relative to standard mean ocean water (SMOW). D/H ratios vary between -26 and -86°/₀₀ relative to SMOW. Although only 8 samples from the subsurface (maximum depth 3,400 m) and 25 from outcrop have been completely analyzed to date, burial to approximately 3.5 km (approximately 130°C) does not appear to have affected the δO^{18} but may have slightly affected the δD of these microquartz cherts. Other forms of silica from both outcrop and subsurface, including megaquartz filling molds resulting from the dissolution of evaporite nodules, replacement megaquartz, and replacement chalcedony, are depleted in O^{18} relative to the chert nodules, with values as light as +25.5°/₀₀.

Microquartz chert probably formed from opal C-T precursors and replaced carbonate sediments which formed and were altered in a variety of environments ranging from hypersaline through meteoric. The variation in the oxygen isotopic composition of waters responsible for the diagenesis of these rocks must have exceeded by at least a factor of 2 the observed 4°/₀₀ range in δO^{18} of these cherts. Possible contamination of chert by later generations of silica (depleted in O^{18}) cannot be excluded as contributing to the observed range and, if present, would further limit the possible range in isotopic composition of water forming the chert. In addition, the δO^{18} of Edwards chert is virtually identical to that of chert from European Cretaceous chalk which formed on open-marine shelves. On this basis, silicification under a wide variety of salinity-temperature conditions is ruled out. Contemporaneous silicification under evaporitic conditions is ruled out by calibration with all available silica-water temperature scales. Silicification must have occurred either contemporaneously from solutions having a δO^{18} close to SMOW or in the shallow subsurface at elevated temperature and from O^{18}-enriched water.

INTRODUCTION

As part of an on-going geochemical-petrographic study of Lower Cretaceous carbonates of Central Texas, nodular chert and other forms of quartz have been analyzed for δO^{18} and δD. Chert and other forms of silica were collected from outcrop and subsurface core. Samples are keyed to Rose's (1972) and Mueller's (1975) measured sections and Rose's (1972) and Bebout and Loucks's (1974) core logs. In the laboratory, chert nodules were broken and examined for inhomogeneities. Representative samples were taken from near the centers of nodules or beds (some rinds of nodules were sampled as well), and approximately 3 grams were pulverized in a tungsten-carbide ball mill to pass a 62 micrometer sieve. The powders were leached with concentrated HCl to remove carbonate, washed extensively with distilled water, and dried. All samples gave X-ray diffraction patterns of α-quartz.

Oxygen was liberated from 15 to 20 mg samples using BrF_5 (Clayton and Mayeda, 1963). Seven samples were replicated at random during the course of the study. The maximum difference found between samples was 0.6°/₀₀, with a mean of 0.25°/₀₀. NBS #28 quartz yielded a δO^{18} value of +9.7 (Friedman and O'Niel, 1977).

Water was extracted from chert following the procedure outlined by Knauth (1972). Approximately 0.5 gram of chert was evacuated overnight (15±1 hours) in a Pt crucible under high vacuum at room temperature. Liquid nitrogen was then placed in a high surface area, spiral trap adjacent to the sample, and the sample and trap were isolated from the vacuum pumps. Using a resistance furnace, the sample was heated cherry red (800±25°C) for four hours, at which time the furnace and trap were evacuated through the trap. The sample was heated an additional hour, and after a second evacuation, liquid nitrogen condensible gases were transferred to another trap and raised to dry ice temperature to liberate CO_2 and possibly some hydrocarbons. Water thus purified was converted to H_2 (Godfrey, 1962). Analysis was accomplished by bracketing the unknown between standard mean ocean water (SMOW) and NBS #1A (-182.2°/₀₀). Correction for H^3 by this method gives nearly identical results to those obtained by extrapolating ion beam intensities to zero pressure (Friedman, 1953). The values are reproducible ±5°/₀₀.

DATA

Data obtained to date (fig. 1, table 1) demonstrate a relatively narrow spread of both δD and δO^{18} values for the microquartz chert. Other samples of silica from both outcrop and subsurface are depleted in O^{18} with respect to chert. Some chert from the subsurface appears to be enriched in D relative to samples from outcrop. The δO^{18} values of chert presently buried at temperatures up to approximately 130°C and in contact with formation waters of variable isotopic composition (unpublished data) are identical to chert from outcrop, indicating little or no interaction between the chert and the formation fluid. Some δD values may, however, be somewhat changed owing to burial.

INTERPRETATION

If the oxygen isotopic composition of the chert has not been altered since its deposition, what information is preserved? Rather than try to interpret the significance of the isotope ratios based on isotopic geochemical arguments, I will try to interpret the significance of the isotope ratios on the basis of what we know about the origin and diagenetic history of these rocks.

The rocks were deposited cyclically in a variety of depositional environments and altered in a variety of diagenetic environments ranging from hypersaline to meteoric (Rose, 1972; Mueller, 1975; Loucks, 1976). Lowenstam (1961) demonstrated that

[1]Department of Geological Sciences, The University of Texas at Austin.

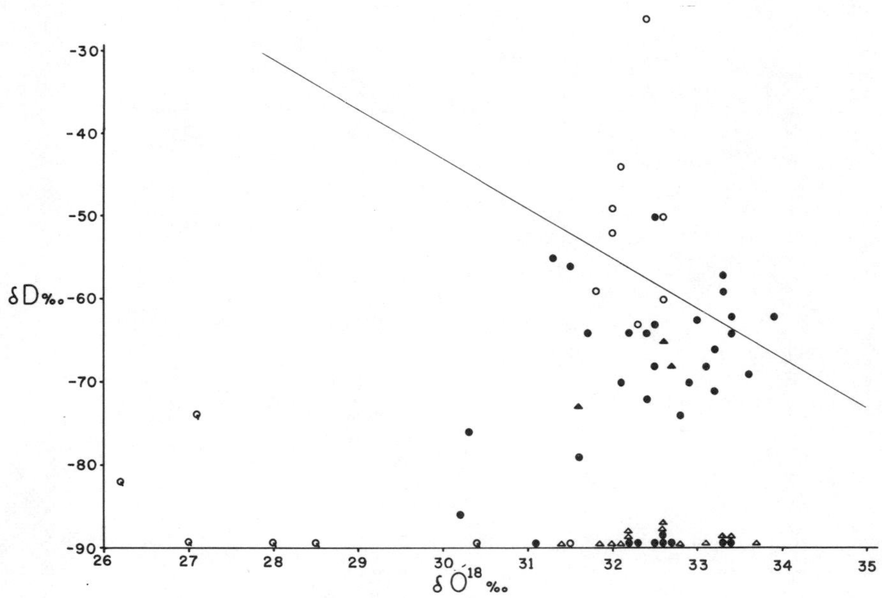

Figure 1. Scatter diagram of δD versus δO^{18}, both relative to SMOW. Solid circles are Edwards microquartz chert from outcrop; open circles are Edwards microquartz chert from the subsurface. Solid triangles are from Knauth and Epstein, 1976. Quartz samples are designated by Q's. Samples for which δD values are not available are plotted tangent to the abscissa. Open triangles are Cretaceous chert from European chalk (Scholle, 1977). Diagonal line is Knauth and Epstein's "locus of marine cherts."

Cretaceous temperatures and the oxygen isotopic composition of the Cretaceous ocean were very similar to the present. Therefore, we are constrained to a depositional temperature maximum of about 35°C (Kinsman, 1976) and perhaps a winter minimum of 15°C. The diagenetic temperature range would have been more restricted.

Regardless of the isotopic composition of the Cretaceous ocean on the SMOW scale, Cretaceous hypersaline water could not have been more enriched than about 6‰ (Lloyd, 1966) relative to the Cretaceous ocean. Contemporaneous meteoric waters could not have been more depleted than about -4‰, judging from the isotopic composition of modern meteoric water in coastal areas (Land and Epstein, 1970; Taylor, 1974). Therefore, we might expect that about 10‰ variation in the isotopic composition of water was associated with the genesis and early diagenesis of these rocks. Temperature variations in the depositional environment of at most 20°C (corresponding to about 5‰ variation in δO^{18}) would in part cancel changes in isotopic composition of the water because warmer periods would certainly correspond to periods of increased evaporation. Conservatively, we should expect that if Edwards chert formed in all possible depositional and diagenetic environments, it should display a range in δO^{18} of at least 10‰. Clearly it does not; therefore, silicification (or the transformation of opal C-T chert to microquartz chert?) must have occurred under relatively restricted conditions of temperature and δO^{18} H_2O.

In addition, the range in isotopic composition of the Edwards chert is almost identical to that of the chert which replaced Cretaceous chalk studied by Scholle (1977) from northwest Europe, Great Britain, and the North Sea (fig. 1). The averages of the two sets of data are within 0.1‰ of each other, with a "grand" average of 32.5‰. The chalk is interpreted as having accumulated on an open-marine shelf.

Three silica-water "paleotemperature" scales are currently available (Clayton and others, 1972; Knauth and Epstein, 1976; Labeyrie, 1974). Using an average δO^{18} of 32.5, table 2 presents calculated temperatures using a reasonable range of δO^{18} H_2O. Clearly all three relations yield unreasonable temperatures if contemporaneous evaporated waters are involved. Knauth and Epstein's (1972) temperature scale suggests slightly enriched marine water (which is typical of tropical surface waters) to be the primary agent of silicification, whereas other equations suggest that a water significantly meteoric in character is involved.

Three lines of evidence lead me to conclude that Knauth and Epstein's (1976) paleotemperature is more nearly correct and that silicification took place via a fluid whose oxygen isotope ratio was essentially that of normal sea water. First, Edwards chert replaces a number of carbonate rock types which formed in several depositional environments and were altered in a variety of hydrologic environments. The Edwards chert is isotopically similar to European chert found exclusively in chalk. Most authors interpret chert as having formed early in the history of the rocks (Pittman, 1959), one of the main lines of evidence being the compaction of carbonate sediment around nodules. Because chert in European chalk formed in a normal marine environment and because both sets of data are nearly identical, we must conclude that Edwards chert had the same origin. Knauth's (1974) contention that the isotopic composition of Cretaceous chert from north-central Texas reflects the environment of deposition is not confirmed by this study. The average isotopic composition of chert replacing supratidal beds is a few tenths per mil enriched relative to the average chert replacing subtidal beds, suggesting a possible environmental influence. But the relationship is too often reversed to be useful. Chert associated with preserved evaporites (Gypsum Quarry, table 1) is no different isotopically from chert found in other localities. The observed range of isotope ratios must reflect different conditions of silicification, but 86 percent of Edwards chert samples fall within a 2‰ range of O^{18}. This variation is considerably less than that which must have been present in waters associated with the origin of the carbonate rocks. In addition, no conclusive evidence shows that chert formed contemporaneously with the deposition of the sediments and from the same fluid which deposited the carbonate sediment. The chert may have formed over a period of time during early diagenesis when the pore fluid was very different from that present when the chertified bed was deposited.

Second, the sea was the reservoir of water in which the Edwards Limestone and European chalk were deposited. Even during times of periodic emergence and development of meteoric water lenses during Edwards deposition, deeper water would remain marine in origin. Marine formation water would "float away" meteoric water lenses during times of platform submergence. It is tempting to speculate that perhaps one reason why chert nodules in the Edwards Formation are commonly restricted to single horizons might be salinity strati-

Table 1. Part 1 - Samples from outcrop.

Locality	Bed Samples	Mineralogy*	δO^{18}	δD
Cherry Spring (Mueller, 1975)	cycle 3, burrowed subtidal	quartz	26.2	-82
	cycle 4, burrowed subtidal		30.2	-86
	base Kirshburg breccia		31.6	-79
	topmost cycle, laminated supratidal		32.2	-64
	middle Kirshburg breccia, nodule		32.4	-72
	middle Kirshburg breccia, disseminated		30.3	-76
	cycle 7, laminated supratidal		33.0	-62
Kerrville (Mueller, 1975)	base	quartz	27.0	
	cycle 1, laminated		32.6	
	cycle 2, base	quartz	27.1	-74
	cycle 2, laminated		33.4	
	cycle 5, laminated		32.5	-63
	cycle 2, burrowed		33.3	
	cycle 8, transition		32.7	
	cycle 11, base		33.4	-62
Gypsum Quarry (Rose, 1972)	3′ below gypsum, center of nodule		32.5	-50
	3′ below gypsum, rind of nodule		33.2	-66
	32′, above gypsum		33.3	-59
	39′, above gypsum		33.6	-69
	44′, above gypsum		32.4	-64
Medina Mountain (Rose, 1972)	165′, subtidal		32.5	-68
	170′		33.3	-57
	220′		31.7	-64
	226′		33.9	-62
	236′, center of nodule		32.1	-70
	236′, rind of nodule		32.8	-74
Vanderpool (Rose, 1972)	180′		31.1	
	190′		31.3	-55
	206′		32.6	
	210′		32.9	-70
	240′		32.3	
	246′		33.2	-71
	250′		32.5	
	263′, collapse breccia		31.5	-56
	280′		32.2	
	406′, subtidal, rind of nodule		33.4	-64
	406′, subtidal, core of nodule		33.1	-68

Table 1. Part 2 - Samples from the subsurface (Rose, 1972; Bebout and Loucks, 1974).

Well Designation and County	Depth	Mineralogy*	O^{18}	D
Mobil No. 40 Mercer, Caldwell County	2,116′		32.0	-48
	2,329′		31.8	-59
	2,350′		32.4	-26
	2,365′, rind of nodule		32.6	-60
	2,365′, core of nodule		32.0	-52
Lone Star (Sun) No. 1A Tom, Atascosa County	10,679′		32.6	-50
Tenneco No. 1 Wiatrek, Karnes County	11,049′		32.1	-44
	11,084′		32.3	-63
	11,197.5′	chalcedony	31.5	
Humble No. 1 Dillworth, McMullen County	11,332′	quartz	28.5	
	11,873′	quartz	25.5	
Tenneco No. 1 Alamo Lumber, Live Oak County	13,676′	quartz	28.0	
Tenneco No. 1 Schulz, Live Oak County	13,695′	quartz	30.4	

*Microquartz chert unless otherwise noted.

Table 2. Possible temperatures of silicification (°C).

Character and Isotopic Composition (SMOW) of Water	Clayton, and others, 1972[1]	Knauth and Epstein, 1976[2]	Labeyrie, 1974[3]
hypersaline, +4 ‰	54.4	40.6	52.2
normal marine, 0 ‰	35.9	22.8	35.8
meteoric, -4 ‰	20.3	7.8	19.4

1/ $10^3 1n\alpha = 3.38\ (10^6 T^{-2}) - 3.4$, T in °K, quartz
2/ $10^3 1n\alpha + 3.09\ (10^6 T^{-2}) - 3.29$, T in °K, microquartz chert
3/ $T = 5 - 4.1\ (\delta SiO_2 - \delta H_2O - 40)$, T in °C, biogenic opal

fication within the sediment pile. Any flow of water (transporting silica) within a salinity stratified system (in this case within the accumulating limestone pile) will be contained by density gradients. Light water cannot flow into dense water. Because of the horizontal extent of these deposits, salinity stratification would be nearly horizontal and essentially parallel to bedding on the outcrop scale. Circulating marine fluids beneath meteoric lenses (Kohout, 1960) might then preferentially transport silica along specific horizons, aided by porosity-permeability differences; preferential silicification results.

Finally, analysis of Miocene chert from north Jamaica (Land, in preparation), where the conditions of formation of the chert are even better known, concludes that only Knauth and Epstein's (1976) relation fits the observed data.

CONCLUSIONS

Chert in the Edwards Group, Central Texas, is best interpreted as having formed from a water of quite restricted isotopic composition near SMOW (which might, in cases, have had increased salinity owing to the dissolution of previously deposited evaporites), at temperatures near 25°C. Some chert has been buried to depths of at least 3.5 km (130°C) without alteration of its oxygen isotopic composition.

ACKNOWLEDGMENTS

I wish to thank Harry Mueller, who initiated collections on which this study is based, and Paul Knauth and Mike Lloyd for profitable discussions. Lonnie Martin, Harry Mueller, and Robert Killian assisted with laboratory preparation. This research was supported by the National Science Foundation, grant EAR76-17774, and the Geology Foundation of The University of Texas at Austin.

REFERENCES

Bebout, D. G., and Loucks, R. G., 1974, Stuart City Trend, Lower Cretaceous, South Texas: Univ. Texas, Austin, Bureau of Economic Geology, Report of Investigations No. 78, 80 p.

Clayton, R. N., and Mayeda, T. K., 1963, The use of bromine pentafluoride in the extraction of oxygen from oxides and silicates for isotopic analysis: Geochimica et Cosmochimica Acta, v. 27, p. 43-52.

______, O'Neil, J. R., and Mayeda, T. K., 1972, Oxygen isotope exchange between quartz and water: Journal of Geophysical Research, v. 77, p. 3057-3067.

Friedman, I., 1953, Deuterium content of natural waters and other substances: Geochimica et Cosmochimica Acta, v. 4, p. 89-103.

______, and O'Neil, J. R., 1977, Data of geochemistry, Chapter KK: Compilation of stable isotope fractionation factors of geochemical interest: USGS Professional Paper 440-KK, 12 p.

Godfrey, J. D., 1962, The deuterium content of hydrous minerals from the east-central Sierra Nevada and Yosemite National Park, Geochimica et Cosmochimica Acta, v. 26, p. 1215-1245.

Kinsman, D. J. J., 1976, Evaporites: Relative humidity control of primary mineral facies: Journal of Sedimentary Petrology, v. 46, p. 273-279.

Knauth, P. A., 1972, Oxygen and hydrogen isotope ratios in cherts and related rocks: California Institute of Technology, Ph. D. dissertation, 369 p.

______ 1974, Relationship of the isotopic composition of cherts to the depositional environment of carbonates (abs.): GSA Abstracts with Programs, v. 7, no. 7, p. 828.

______, and Epstein, S., 1976, Hydrogen and oxygen isotope ratios in nodular and bedded cherts: Geochimica et Cosmochimica Acta, v. 40, p. 1095-1108.

Kohout, F. A., 1960, Cyclic flow of salt water in the Biscayne Aquifer of southeastern Florida: Journal of Geophysical Research, v. 65. p. 2133-2141.

Labeyrie, L., 1974, New approach to surface seawater paleotemperature using O^{18}/O^{16} ratios in silica of diatom frustules: Nature, v. 248, p. 40-42.

Land, L. S., and Epstein, S., 1970, Late Pleistocene diagenesis and dolomiticization, North Jamaica: Sedimentology, v. 14, p. 187-200.

Lloyd, R. M., 1966, Oxygen isotope enrichment of sea water by evaporation: Geochimica et Cosmochimica Acta, v. 30, p. 801-814.

Loucks, R. G., 1976, Pearsall Formation, Lower Cretaceous, South Texas: Depositional facies and carbonate diagenesis and their relationship to porosity: Univ. Texas, Austin, Ph. D. dissertation, 362 p.

Lowenstram, H. A., 1961, Mineralogy, O^{18}/O^{16} ratios, and strontium and magnesium contents of Recent and fossil brachiopods and their bearing on the history of the oceans: Journal of Geology, v. 69, p. 241-260.

Mueller, H. W., III, 1975, Centrifugal progradation of carbonate banks: A model for deposition and early diagenesis, Ft. Terrett Formation, Edwards Group, Lower Cretaceous, Central Texas: Univ. Texas, Austin, Ph. D. dissertation, 300 p.

Pittman, J. S., 1959, Silica in Edwards limestone, Travis County, Texas, *in* Ireland, H. A., ed., Silica in sediments: SEPM Special Pub. No. 7.

Rose, P. R., 1972, Edwards Group, surface and subsurface, Central Texas: Univ. Texas, Austin, Bureau of Economic Geology, Report of Investigations No. 74, 198 p.

Scholle, P. A., 1977, Chalk diagenesis and its relation to petroleum exploration problems, or, oil from chalks: A modern miracle?: AAPG Bull.

Taylor, H. P., Jr., 1974, The application of oxygen and hydrogen isotope studies to problems of hydrothermal alteration and ore deposition: Economic Geology, v. 69, p. 843-883.

SIGNIFICANCE OF FRESH-WATER LIMESTONES IN MARINE CARBONATE SUCCESSIONS OF PLEISTOCENE AND CRETACEOUS AGE

Robert B. Halley[1] and Peter R. Rose[2]

ABSTRACT

Fresh-water lime sediments may be deposited over tens of thousands of square kilometers during subaerial exposure of marine carbonate platforms. Such deposits, only slightly above sea level, are presently found covering portions of the Florida-Bahamas carbonate platform. Analogous ancient fresh-water limestones can be identified in Pleistocene limestones of the south Florida platform and Cretaceous limestones of the Central Texas platform.

The co-occurrence of a variety of features provides a guide for the identification of fresh-water limestones in marine carbonate sequences. These include: (1) exceptional color (gray or dark gray); (2) lime mudstone lithology; (3) single, isolated 1 to 2 m thick homogeneous beds; (4) mottled or burrowed internal structures; (5) irregularly cracked and void-riddled fabric; (6) rare fossils, usually gastropods and ostracods, exceptionally rare marine fossils; (7) evidence of early lithification, and (8) position at disconformities in carbonate sequences as evidenced by subaerial exposure criteria (leached fossils, caliche, erosional surfaces and so forth).

Recognition of fresh-water limestones in carbonate sequences provides the stratigrapher with evidence of disconformities that might otherwise be overlooked. Occurrences of fresh-water limestones also imply paleo-fresh-water diagenesis, knowledge of which may help the stratigrapher understand or predict the occurrence of porosity related to subaerial exposure and stratigraphic-type hydrocarbon accumulations.

INTRODUCTION

Fresh-water limestones, or marlstones, occur throughout the geologic record in a variety of depositional settings. They occur within lacustrine rocks, within deltaic deposits and in shallow-marine carbonate sequences. This paper is concerned with the recognition and significance of fresh-water limestones in the last of these, the seemingly anomalous occurrences of fresh-water limestones within shallow-water marine carbonate rock successions. It is here that they gain particular stratigraphic and, by inference, diagenetic importance. Occurrences of fresh-water limestones within marine carbonate successions indicate subaerial exposure, at least local and perhaps regional, and they imply that the underlying marine sediments may have been altered by fresh water. If fresh-water limestones can be recognized, they can be a valuable criterion for identifying exposure surfaces and disconformities in ancient carbonate sequences and can be valuable in explaining the origin and distribution of some diagenetic porosity.
tribution of some diagenetic porosity.
aspects of Holocene fresh-water sediment accumulation follows. Also following are descriptions of occurrences in Pleistocene rocks of south Florida and Cretaceous rocks of Central Texas, as examples of the potential interpretive significance of fresh-water limestone in ancient carbonate platforms.

HOLOCENE FRESH-WATER MARLS ON CARBONATE PLATFORMS

Among the better known modern fresh-water carbonate accumulations are the marls of southern Florida, which attracted attention 90 years ago. Heilprin (1887) recognized the fresh-water limestones exposed along the Caloosahatchee River and their modern counterparts in the marls of the Okeechobee Wilderness. These deposits cover thousands of square kilometers of south Florida, and some have been given formal stratigraphic recognition, for example, the Lake Flirt Formation of Sellards (1919). Other deposits have been designated as soils—Flamingo Marl, Ochopee Marl, and Perrine Marl of Davis (1943). These fresh-water deposits are continuous over large areas covering almost 8,000 km^2 in south Florida, and isolated patches extend significantly beyond the major areas of accumulation.

A prerequisite for marl formation is persistent or at least seasonal ponding of fresh water. Taft and Harbaugh (1964) explained that during the rainy summer season of south Florida, water collects in poorly drained sawgrass marshes to a depth of 15 or 20 cm. This water contains calcium carbonate derived from underlying limestones, which is precipitated on floating algal masses (periphyton). Gleason (1972) and Gleason and Spackman (1974) described the chemical setting in which precipitation takes place on algal peryphyton. The precipitate is mud-sized (less than 10 μ) crystals or crystal aggregates of low magnesium calcite.

Calcite precipitates within algal mats floating on the water surface and around any object on which algae live. These precipitations are not lithified, but accumulate on the bottom as calcite mud. Stromatolitic structures may form, and individual algal filaments may become encrusted with calcite, but these features are rarely preserved a few centimeters below the surface. The addition of clumps of calcite-laden algae to the bottom may account for the common spongy texture of fresh-water limestones. This may also account for the pseudo-brecciated or clotted fabric described by Pettijohn (1957, p. 411).

When pure, fresh-water Florida muds are gray, often light gray (N7, on Geological Society of America Rock Color Chart, 1970), but they vary through shades of brownish gray (5YR4/1) and dark gray (N3). They are often associated with peats, but the transition is usually sharp. The muds contain few fossils, but fresh-water gastropod shells and ostracods may be locally abundant. Rare marine shells may be present, reworked from underlying deposits or "rafted" into the muds by storm tides and animals. Roots of grasses, herbaceous plants, and trees commonly riddle the sediments.

Fresh-water calcite muds accumulate over very porous and permeable Pleistocene limestones in south

[1]U.S. Geological Survey, Fisher Island Station, Miami Beach, Florida

[2]Energy Reserves Group, Houston, Texas

Florida. Much of this area is less than 2 m above sea level so that the water table is always close to the surface. South Florida receives much of its annual rainfall in summer, and the water table rises above ground level to form large ponded areas. This water-table rise increases surface drainage which, in turn, limits the depth of water (and the depth of fresh-water marl accumulation) to less than 1 m (and rarely over 2 m) over much of the Everglades. Although usually less than 1 m thick, a maximum thickness for the Lake Flirt Marl of 2.6 m has been recorded by Schroeder (1954). Marls can accumulate at higher elevations if the underlying soil or rock has poor drainage, but the bulk of south Florida marls appears to be accumulating within a few meters of sea level. Because marls are a near-sea-level deposit, they are most likely to be deposited at specific places during a transgressive-regressive cycle, just before marine invasion and just after withdrawal of the sea.

The Pleistocene rock underlying the marls shows evidences of fresh-water alteration. Commonly, there is a subaerial crust developed at the upper surface of the limestone (Multer and Hoffmeister, 1968). Aragonitic shell material is often dissolved, leaving molds of shells. Clasts of the underlying marine limestone may be present in the base of the marl.

Fresh-water lime muds are terrestrial deposits and should not be confused with marine supratidal deposits. Supratidal marine sediments are formed in shallow marine shelf environments and deposited above sea level during storms (Shinn and others, 1965; Shinn and others, 1969). Supratidal sediments occupy tracts between the sea and fresh or brackish water ponds and swamps and may also form mud islands as in Florida Bay (Taft and Harbaugh, 1964). The original mineralogy and fauna of fresh-water marls and marine supratidal sediments reflect their respective origins. The former is composed almost entirely of low-magnesium calcite and contains little shelly fauna, that, when present, is predominantly fresh-water gastropods and ostracods. The latter sediments are aragonite and high-magnesium calcite and may contain abundant pellets, marine foraminifers, and other marine fossils.

Holocene fresh-water marls are not frequently encountered in areas of marine carbonate sedimentation. Besides the south Florida deposits already mentioned, the only other well-documented occurrence is that of the Great Bahama Bank (Black, 1933; Monty, 1967, 1973; Hardie and Monty, 1976). According to this literature, fresh-water marls clearly do not occur in arid climates (south shore of the Persian Gulf or Shark Bay, western Australia) nor are they characteristic of small carbonate islands (Barbados, Bermuda, Bikini, Guam). It is concluded here that they are most likely to be found as carbonate platform deposits in areas of humid climate.

Although seemingly surficial deposits and thus more liable to erosion and removal, fresh-water marls are preserved in the geologic record. Marlstones are well known in the Pleistocene of the Florida platform, and they are described here for the first time from the Lower Cretaceous Central Texas platform.

PLEISTOCENE MARLSTONES OF SOUTH FLORIDA

Fresh-water limestones were recognized in south Florida as early as 1887 by Heilprin. Sellards (1919) included multiple marlstone beds outcropping along the Caloosahatchee River in the Fort Thompson Formation. Since that time, these rocks have been traced in the subsurface, and their distribution most recently mapped by DuBar (1974). Figure 1 is a reproduction of this distribution with some modification by the authors which shows that marlstones underlie more than 10,000 km^2 of south Florida, although the extent of individual beds may be much more limited. Several attributes distinguish marlstones from other Pleistocene marine limestones of south Florida. The following description of these features, together with the description of the Cretaceous examples in the next section, should provide a guide for recognizing marlstones in marine platform limestones.

Color

Pleistocene marlstones are distinctly darker than the enclosing marine limestones. The marlstones vary from light gray (N7) to medium dark gray (N4). Rarely they are olive gray (5Y4/1). Occasionally they contain dark clasts, sometimes even black clasts (N2 or N1), apparently a weathering product (Ward and others, 1970). Similar black clasts occur in marine limestones immediately overlying the marlstones, and they will be discussed in more detail in a subsequent section. The color agents of the marlstones are unknown but might be organic matter which occurs in the rocks at levels below 0.3-percent organic carbon (Palacas, personal communication, 1976).

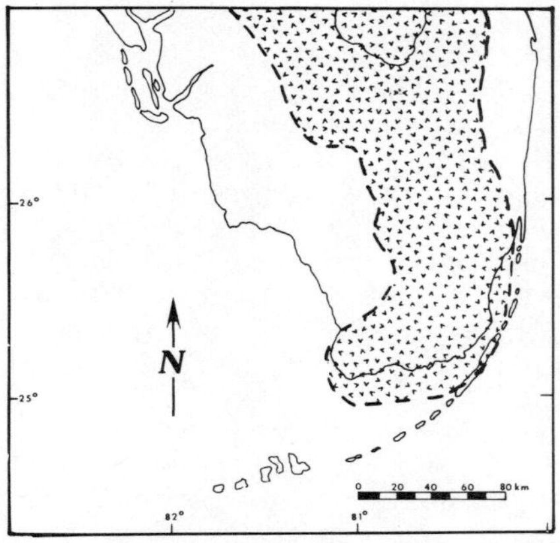

Figure 1. Approximate subsurface distribution of Pleistocene fresh-water limestones (patterned) in southern Florida.

Fauna

Although not common in marlstones, fossils can sometimes be locally abundant. Most common are thin-walled, unornamented gastropods, usually preserved as molds, and species of ***Pomacea, Helisoma***, and ***Ameria*** are most frequent. Next, ostracod valves, usually disarticulated or fractured, are most frequent. Other fossils are extremely rare, but undeniably marine forms occur in marlstones, presumably blown or washed into areas of marl formation. Some singular occurrences are thought to have been transported by animals, particularly birds.

Fabric

Pleistocene marlstones are lime mudstones, following the classification of Dunham (1962). Minor constituents include fine quartz sand and silt and clay, forming as much as 15 percent of the rock by weight. Marlstones exhibit few primary sedimentary structures. Laminations are absent. The rock typically has a churned, thoroughly burrowed or clotted appearance. Individual burrows usually cannot be distinguished, but occasional vertical tubes (5 mm by 20 cm, fig. 2a) are present. These may be enlarged by solution and might be either burrows or root tubes. Often marlstones contain myriads of small voids (50 μm by 3 mm), sometimes spar-filled, resembling cracks (but not clearly desiccation cracks). These voids, possibly left by the removal of organic matter, together with color variations, impart a distinctive clotted appearance to the rock. Although difficult to describe, this fabric is one of the most characteristic features of Pleistocene marlstones and is illustrated in figures 3a and 3b.

Lithification

Marlstones show a variety of features which suggest early lithification. Brecciation and cracking are common at the upper surfaces of marlstones, and overlying marine sediments fill these cracks, indicating lithification prior to the start of marine sedimentation. Rounded pebbles and cobbles of fresh-water limestone occur in marine limestones (fig. 4a) which overlie the marlstone beds. These clasts are bored, probably by sponges, giving further evidence of lithification prior to marine inundation. Large round holes (25 cm by 1 m) in marlstones filled from above with marine sediments are thought to be subaerial solution holes (pot holes).

Local Succession

As many as seven marlstones have been reported from a single borehole in the Pleistocene of Dade County, Florida (Schroeder and others, 1958). The thickest marlstone is about 1.5 m and most are less than 1 m thick. The local succession through one of these beds is varied and complex. The limestone immediately underlying the marlstone may contain fresh-water internal sediment which infills holes (burrows, solution holes, root tube holes?). These infillings have been observed to extend as much as 30 cm into the underlying bed. The upper surface of the marine limestone, directly underneath the fresh-water limestone, often shows evidence of subaerial exposure. Commonly this surface is solution pitted and discolored (gray, brown, or red) and may be coated with a caliche crust similar to that described from the present surface of exposed Florida limestones (Multer and Hoffmeister, 1968). A similar caliche may overlie the marlstone. Between caliche crusts lies the marlstone bed, 0.1 to 0.5 m thick, occasionally with a brecciated upper surface. Clasts of the underlying marine limestone may be present in the base of the fresh-water marl (fig. 5a). Occasionally a disconformity or diastem is indicated within the marlstone. An example is illustrated in figure 2a. The diastem is marked by a color change from gray to brown-gray and a sharp horizon breaking the mottled texture of the rock. Pebbles and clasts lie in the marlstone matrix near the horizon. The limestone overlying the marlstone is typically coarse and shelly and often contains clasts of marlstone and caliche near the contact.

Regional Distribution and Stratigraphic Significance

Schroeder and others (1958, fig.8) correlated seven fresh-water limestone beds between five core holes over a distance of about 500 m. They showed that some beds disappear laterally. They used this evidence of discontinuity in marlstones to refute earlier correlations by Parker and Cooke (1944), who believed marlstones to be the result of major exposure of the south Florida platform during Pleistocene glaciations. Perkins (1974) recognized that although marlstones were discontinuous because of erosion or nondeposition, they did lie at disconformities recognized by other criteria (laminated crusts, paleosols, root structures, vadose sediment, bored surfaces, and corrosion zones). He has been able to correlate these disconformities over large parts of south Florida.

LOWER CRETACEOUS FRESH-WATER LIMESTONES, CENTRAL TEXAS

Distinctive nodules and discontinuous layers of dark-gray lime mudstone, here interpreted to be fresh-water limestone (or marlstone),

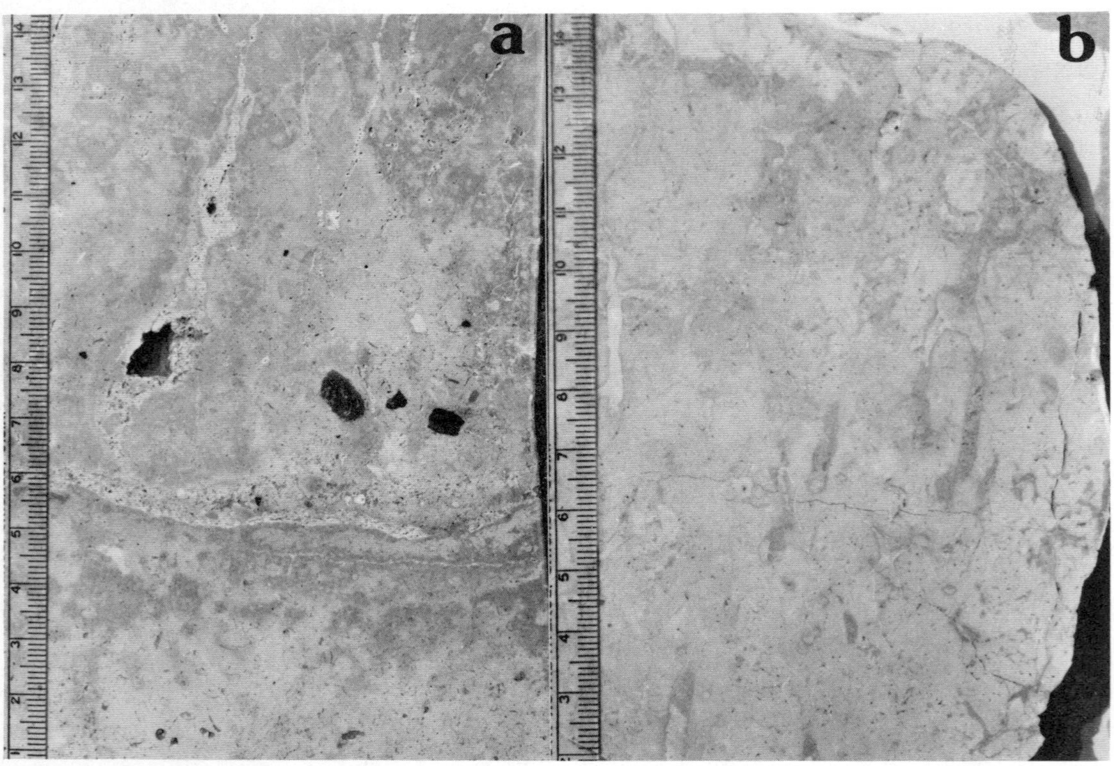

Figure 2. (a) Pleistocene fresh-water limestone illustrating clotted texture and vertical burrow or root-tube structures in upper half. Prominant disconformity passes through the middle of the rock slab with dark clasts above. (b) Cretaceous fresh-water limestone illustrating churned or clotted texture with vertical burrow or root structures.

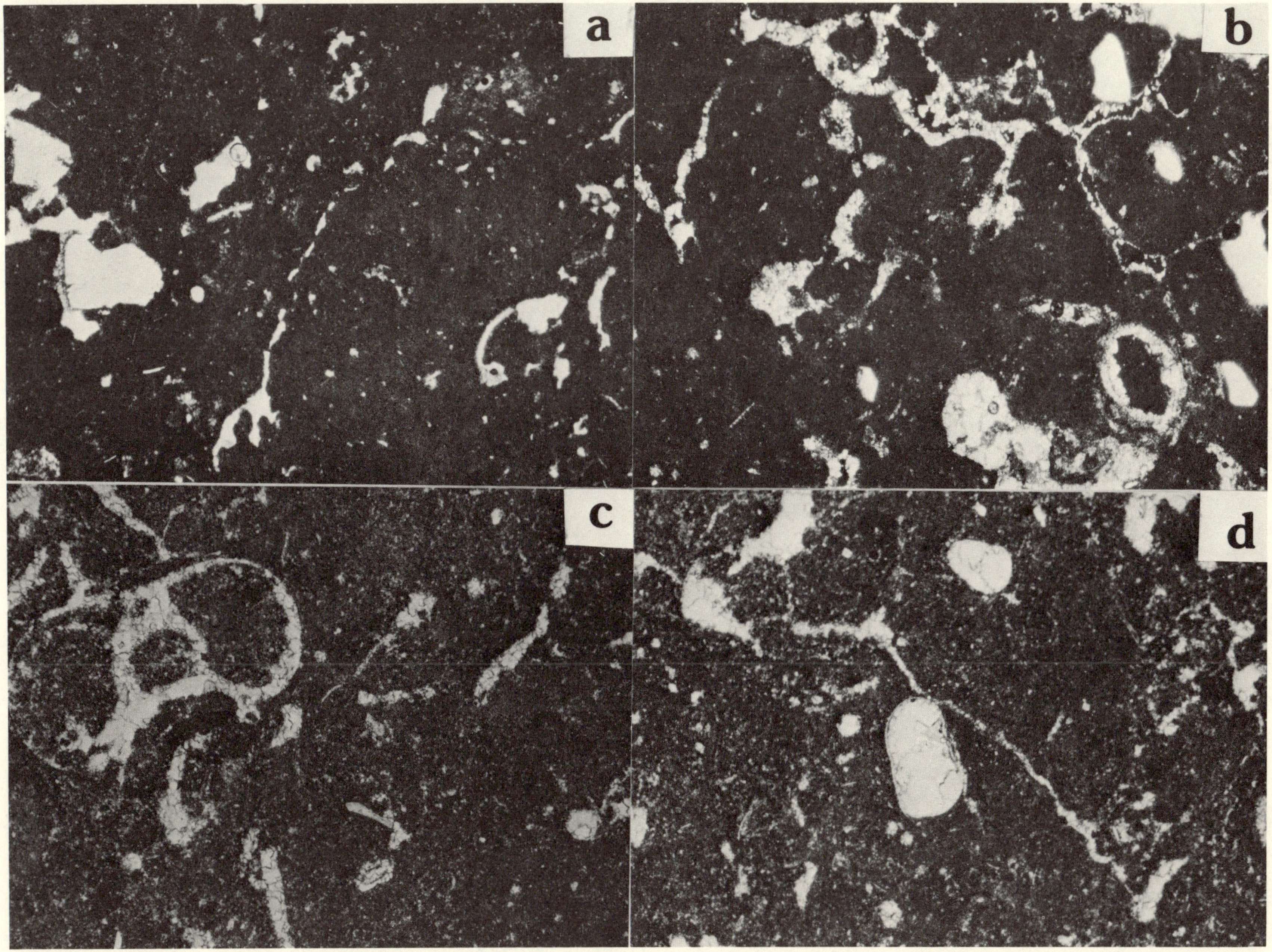

Figure 3. Photomicrographs of Pleistocene and Cretaceous fresh-water limestones. (a) and (b) Pleistocene fresh-water lime mudstones illustrating irregular voids and cracks, partially infilled with calcite cement in (b). (c) and (d) Cretaceous fresh-water lime mudstones illustrating spar-filled voids and cracks that give rock its "spongy" texture. Field of view in all photomicrographs is 2.6 mm across.

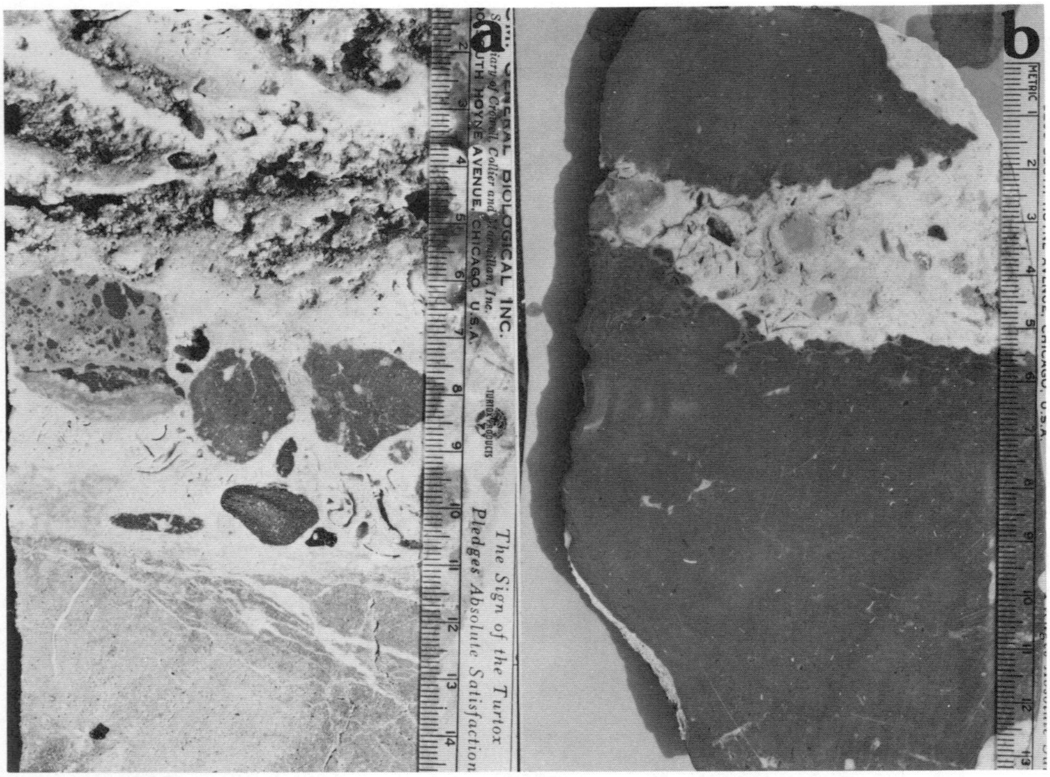

Figure 4. (a) Pleistocene fresh-water limestone overlain by marine limestone containing clasts of fresh-water limestone bored by marine organisms. Contact between marlstone and marine limestone is at 11.5 cm. Sample is a core slab. (b) Rock slab through upper surface of Cretaceous fresh-water limestone overlain by marine limestone. Contact is at 5.4 cm, and dark clast within marine limestone overlies the contact.

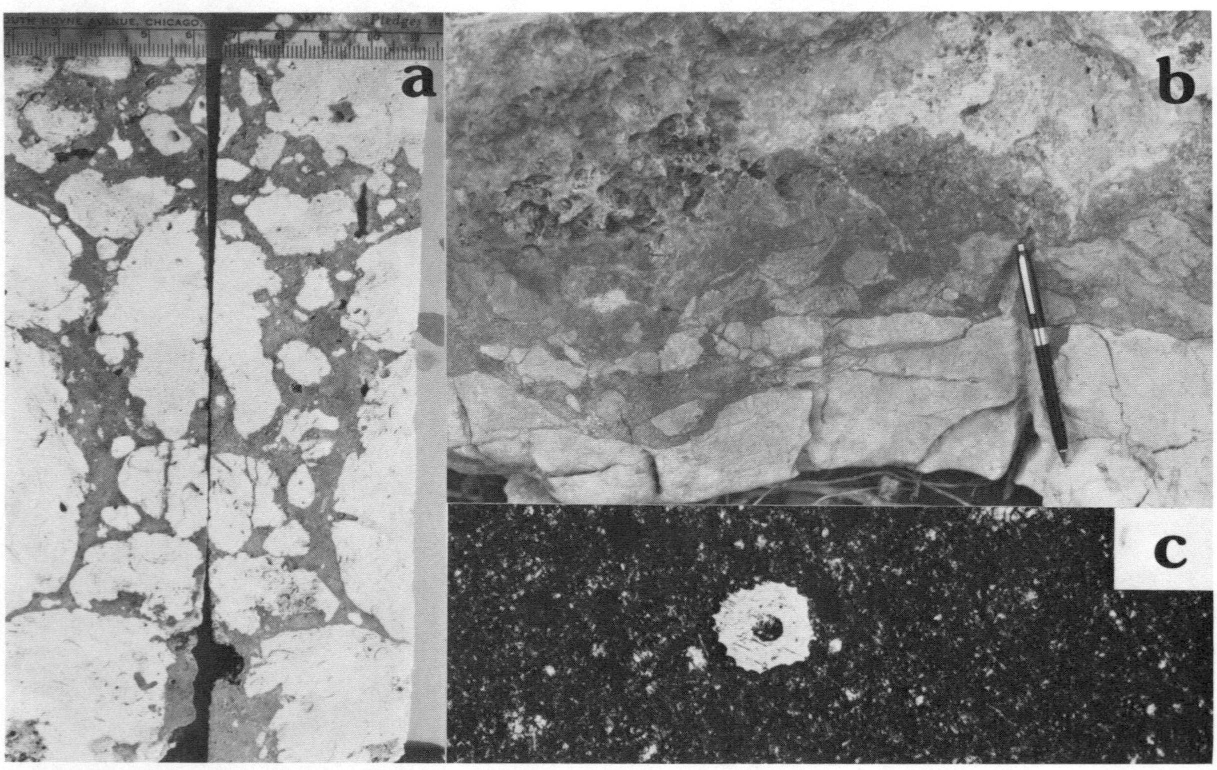

Figure 5. (a) Sawed core slab of south Florida Pleistocene limestone near base of fresh-water limestone. The rock is a conglomerate containing light-colored clasts of marine limestone set in a matrix of dark fresh-water limestone. (b) Outcrop photograph of Cretaceous fresh-water limestone with clasts of normal marine limestone near base. (c) Unidentified microfossil of Cretaceous marlstone, 150 μm in diameter, with ribbed outer surface and central void.

occur in association with a persistent and widespread key bed near the top of the Edwards Group (Lower Cretaceous) in the Edwards Plateau region of west-central Texas. The key bed was first mentioned by Calvert (1928), who identified the unique ledge-and-slope interval in Val Verde County near the top of the otherwise indivisible Devils River Formation, a lateral equivalent of the Edwards Group. This marker, called informally the "Calvert Slope," lent itself readily to surface structural mapping, and the subsequent discovery of distinctive black limestone nodules in the "Calvert Slope" by Humble Oil and Refining Company geologists during the late 1950's made it a reliable mapping horizon now known informally as the "Black Bed." Rose (1972) described features suggesting that the Black Bed sequence represented regional subaerial exposure. Subsequent investigations by Halley and Rose have now resulted in a more complete description and interpretation based on studies of rocks exposed at localities (listed in table 1) of the depositional and diagenetic attributes of the Black Bed succession.

Color

The characteristic gray (N3-N6) lime mudstones that occur in this succession are distinctly darker than the enclosing light-gray to cream-colored shallow-marine limestones or yellowish-gray caliche. Apparently the dark color is not caused by large concentrations of organic material, inasmuch as analyses indicate that less than 0.3-percent organic carbon is present, but very fine disseminated algal debris and minor amounts of ferrous iron and manganese throughout the sediment may account for the dark color.

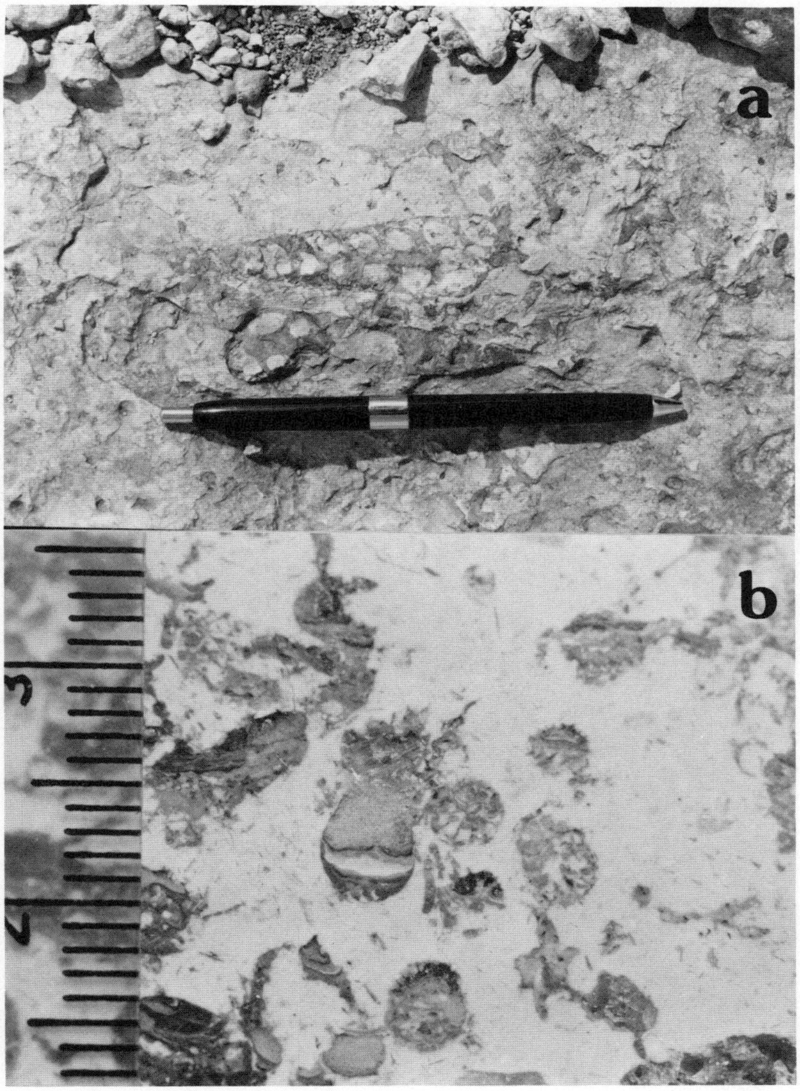

Figure 6. (a) Outcrop photograph of Cretaceous marine limestone underlying marlstone. Mold of large gastropod has been filled with dark fresh-water limestone. (b) Rock-slab photograph of same rock containing partially and completely leached coral fragments (septa still visible near arrow) now filled with laminated fresh-water limestone and calcite spar.

Fauna

Fossils are rare in the dark lime mudstones, but indeterminate microfossils (fig. 5c), fine molluscan fragments, whole gastropod shells, and ostracod carapaces are present in very small numbers.

Fabric

Fresh-water limestones of the Black Bed sequence are almost invariably lime mudstone, following the classification of Dunham (1962). Generally they are featureless and homogeneous, but many nodules show a clotted texture and many small spar-filled cracks (figs. 2c, 2d). Their fabric is identical to that of Pleistocene marlstone from Florida. The only known exception is at locality 15 in northeastern Edwards County, where a dark-gray calcisiltite limestone occurs as laminated internal sediment-filling solution holes and molds after aragonitic fossils dissolved away from within the underlying marine limestone (fig. 6). After emplacement of the dark laminated silt-sized lime sediment, incompletely filled coral molds were then filled by sparry mosaic calcite cement.

Lithification

Dark limestones of the Black Bed sequence contain (or are associated with) at least three features that suggest early lithification:

1. The dark limestone contains scattered spar-filled cracks, possibly desiccation or weathering cracks.
2. It characteristically occurs as smooth, rounded pebbles or cobbles weathering out of the "Calvert Slope" (fig. 7a), and in some localities pebbles are present in overlying marine limestone just above a disconformity (as clasts in a very-coarse-grained marine limestone that forms the resistant upper ledge in the Black Bed succession).
3. Some large dark clasts have been bored extensively by pholad-type clams prior to incorpora-

Table 1. Key localities of Cretaceous Black Bed outcrops, Central Texas

15. Quarry on east side of U. S. Highway 377 about 4.8 km north of its intersection with State Highway 41 east of Rocksprings, Edwards County: Black Bed exposed in base of quarry, with about 3 m of massive limestone above. Black Bed occurs both as nodules in middle caliche and as laminated fillings of solution holes developed in colonial corals contained in underlying limestone.
16. About 3.2 km east of intersection of U. S. Highway 377 and State Highway 41, Edwards County: short section measured beside ranch road just south of State Highway 41 on Morris ranch about 909 m east of ranch headquarters, from Black Bed to Buda (interval 13.6 m). About 0.3 m above Black Bed is limestone conglomerate resting upon distinct erosional surface.
68. Quarries about 0.4 km southwest of highway intersection of U. S. 83 and State Highway 41 (Garven Store) on Will Auld Ranch, extreme northeast Real County: distinct erosion surface developed above Black Bed, with cracks and fissures in the Black Bed filled by overlying tan calcarenite that contains pebbles of Black Bed; also root holes (or borings?) in the Black Bed; compare with locality 16.
69. Outcrops about 0.8 km east of U.S. Highway 83 beside paved ranch road leading to Will Auld Ranch headquarters, extreme southwestern Kerr County, about 3.2 km south of Garven Store: very large rounded blocks of Black Bed, some bored, crop out on either side of the road; Black Bed occurs as rounded cobbles in tan calcarenite of locality 68; measured interval from Black Bed to Del Rio very small in this vicinity, 0.6 to 1.5 m.
70. Roadcut on U.S. Highway 83, 22.6 km north of Garven Store, western Kerr County; bored surface at base of calichified middle recessive interval.

tion in coarse-grained shallow marine limestones, thus indicating that the dark sediment was indurated (fig. 7b).

Local Succession

Figures 8, 9, and 10 illustrate most of the features typically associated with the Black Bed found at locality 15. Main stratigraphic elements of the sequence are:

(a) an underlying resistant light-gray skeletal marine limestone that forms a prominent bench;
(b) an intervening soft, "calichified" interval 1.5 to 2.4 m thick that forms a recessive slope; and
(c) an overlying light-gray to cream-colored, coarse-grained, porous skeletal marine limestone 1.8 to 2.4 m thick that forms a resistant ledge.

As described, dark-gray laminated limestone fills solution holes in the upper part of the underlying skeletal limestone, the upper surface of which is locally bored by pholad-type clams, suggesting early lithification but not necessarily subaerial exposure (Purser, 1969; Shinn, 1969; Rose, 1970).

The most common occurrence of the dark-gray limestone is as smooth, rounded cobbles and pebbles weathering out of the middle, recessive, calichified slope. Fresh quarry exposures of this interval show a deeply altered, massive, punky, fine-crystalline limestone or "caliche" with one or two hard, thin sucrosic limestone beds near the top. Dark lime mudstone cobbles are scattered through the massive altered or "calichified" zone.

The dark lime mudstone also occurs as an intermittent bed up to 1 ft thick in the basal part of the upper resistant limestone. This bed thins to zero from the top by erosion. The rock becomes darker gray upward, suggesting that the darker color may be related to length of weathering. The upper surface of this dark limestone bed is locally irregular and bored by pholad-type clams (Purser, 1969; Shinn, 1969; Rose, 1970). Large dark-gray limestone fragments, some bored, are incorporated into the overlying light-gray coarse-grained marine limestone, which also contains lithoclasts composed of other limestone types different from the darker lime mudstone.

All the evidence indicates that two exposure surfaces are present in the Black Bed succession: one at the base of the calichified recessive middle slope and another at the base of the

Figure 7. (a) Rounded cobbles of fresh-water limestone weathering out of the "Calvert Slope." Weathering is thought to be Cretaceous, because unweathered limestone overlies the "Calvert Slope" at this quarry locality (locality 68). (b) Rock slab through contact between Black Bed and overlying marine grainstone. Upper surface of Black Bed is bored by marine organisms, probably pholads (marine boring clams).

overlying resistant limestone ledge. The middle recessive interval (the "Calvert Slope") is therefore interpreted to represent a Cretaceous weathering unit or "paleo-caliche" containing clasts and lenses of dark-gray, fresh-water limestone or marlstone that formed in lakes and ponds on an exposed carbonate terrane analogous to present-day south Florida.

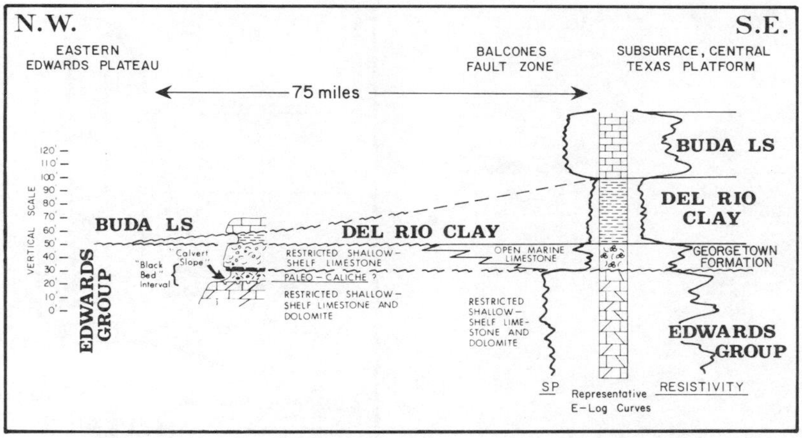

Figure 8. Regional correlation of "Black Bed" interval from Edwards Plateau outcrop to subsurface of Central Texas platform.

Regional Distribution and Stratigraphic Significance

Although strong evidence was lacking, Rose (1972, p. 48) interpreted the Black Bed as a disconformity and correlated it to the southeast with the widespread disconformity between the Edwards Group and overlying Georgetown Formation in the Balcones Fault Zone and in the subsurface beneath the Gulf Coastal Plain. Compelling evidence exists to document this disconformity at the outcrop and particularly in the subsurface, such as: invariably sharp, irregular contact; truncated shells and spar-filled fractures; small solution caves just below the top of the Edwards that are filled with Edwards clasts and Georgetown matrix; soil breccia at the top of the Edwards; abrupt upward change from very shallow restricted marine to open marine fauna at the boundary; up to 30 m of stratigraphic truncation at the boundary (Rose, 1972).

The known area covered by this well-documented disconformity between the Edwards and Georgetown includes approximately 13,000 km^2 comprising parts or all of 11 counties and covering the crest of the Central Texas platform, the dominant structural positive that affected Lower Cretaceous shelf deposits of the region (fig. 11). Known area in which the Black Bed occurs (Brown, personal communication, 1975) covers the crest and flank of the Central Texas platform, covering an area of over 10,000 km^2, comprising parts or all of eight counties.

In the Edwards Plateau no Georgetown Formation is present, and the Del Rio Clay rests directly upon the Edwards Group, whereas, to the southeast, Del Rio rests upon Georgetown. Therefore, the interval from the Black Bed to the top of the Edwards is interpreted to be equivalent to the Georgetown Formation of the Balcones fault zone and the subsurface. If this correlation is correct, the open-marine Georgetown must change facies northwestward to shallow-shelf Edwards-type sediments. In those Balcones fault zone outcrops and shallow subsurface areas that are closest to the Black Bed outcrop localities and on strike with them, the Georgetown Formation is about 6 m thick, approximately the same thickness as the counterpart interval between the Black Bed and top of Edwards (Rose, 1972). Moreover, this correlation is entirely compatible with thickness patterns and correlation of more conventionally identified underlying and overlying stratigraphic units associated with this interval.

The area that was subaerially exposed before Georgetown deposition was an elongate carbonate terrane along the crest of the Central Texas platform that terminated to the southeast at the Stuart City reef trend (fig. 11). This exposed land surface must have been an extremely flat lowland not unlike present southern Florida, covering a total area perhaps 11 km wide and 320 km long, or 36,000 km^2.

Figure 9. Locality 15 illustrating (a) patches of fresh-water limestone overlying marine limestone in foreground, (b) soft recessive weathering "Calvert Slope," and (c) overlying skeletal marine limestone that forms resistant ledge.

Diagenetic Significance

Subaerial exposure is one of the most important geologic events affecting diagenesis of carbonate rocks and in different situations may cause either preferential cementation (porosity occlusion) or solution (porosity enhancement). Accordingly, the recognition of previously unperceived subaerial exposure surfaces (disconformities) may allow the stratigrapher to understand existing patterns of porosity development and even to infer the existence of commercial reservoir rocks in undrilled areas. Thus, the identification of additional criteria indicating subaerial exposure, such as fresh-water limestone deposits that are contained within overall

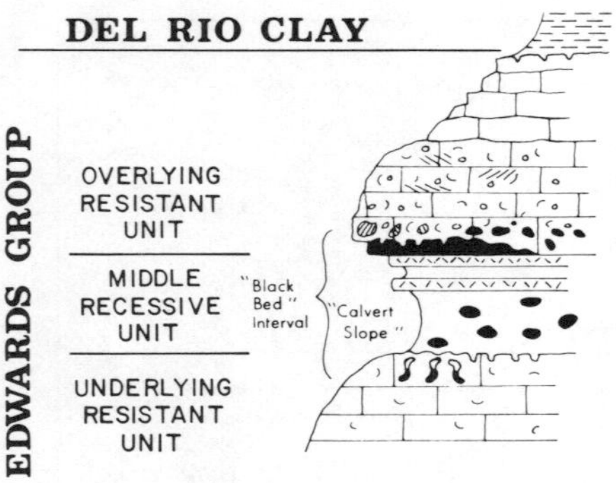

Figure 10. Detailed section of the "Black Bed" interval.

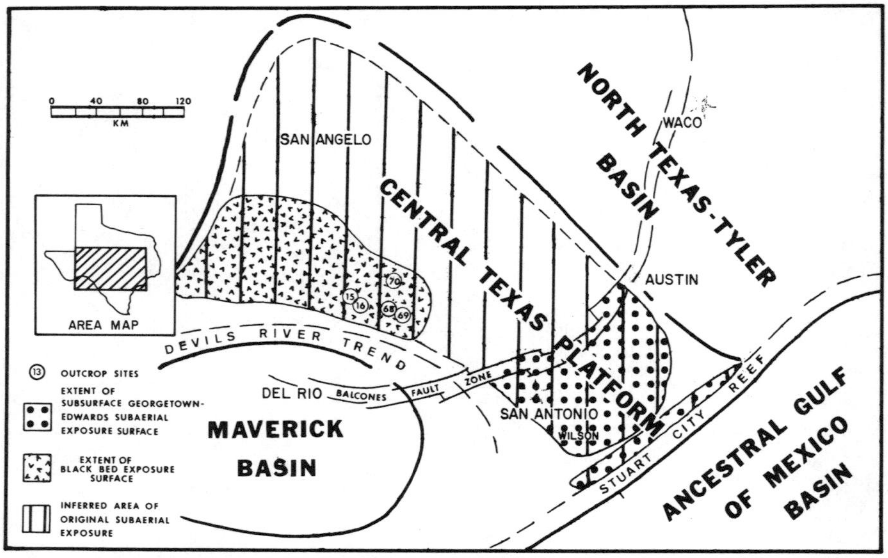

Figure 11. Major structural features of the Central Texas platform showing the distribution of the inferred total Black Bed subaerial exposure surface (vertical hatch), the known extent of the subsurface Georgetown-Edwards subaerial exposure surface (dots), and the known aerial extent of the Black Bed (pattern). Compare fresh-water limestone extent with that shown in figure 1.

marine carbonate successions, may lead indirectly to the discovery of new petroleum accumulations. Evidence at hand indicates that fresh-water limestones or marlstones are indeed distinctive, that they occur in characteristic lithic successions, and that their occurrence is compatible with known regional stratigraphic patterns.

In the example presented here, the Lower Cretaceous Edwards Group, the prevailing regional pattern of subsurface porosity development is that porosity and permeability are highest near the top of the Edwards and diminish downward progressively (if erratically) with depth, below the subaerial exposure surface at the top of the Edwards Group. In the Edwards Plateau, geophysical borehole measurements of porosity have not been carried out in numbers sufficient to make quantified generalizations. Moreover, the long Tertiary history of subaerial exposure and ground-water flow through exposed Edwards rocks in the Edwards Plateau area has doubtless severely altered previously developed porosity patterns. Accordingly, in the outcrop area where the recognizable fresh-water limestones are developed, we are unable to say whether or not the downward-diminishing porosity development is duplicated in the subsurface Edwards.

CONCLUSIONS

Fresh-water limestones share many features with normal marine limestones that allow them to be easily overlooked in a thick succession of carbonate rocks. However, these combined attributes (table 2) are distinctive and provide persuasive arguments by analogy for the existence of fresh-water limestones in marine carbonates. Other criteria not listed here may simplify identification of fresh-water limestones; the presence of a strictly terrestrial fauna or flora is an example.

Table 2. Attributes of Pleistocene and Cretaceous Fresh-water Limestones.

Attribute	Pleistocene	Cretaceous
Color	Gray - N4-N7 Olive gray - 5Y4/1	Gray - N3-N6 Brownish gray - 5YR4/1 Yellowish gray - 5Y7/2
Lithology	Lime mudstone	Lime mudstone
Bedding	Single homogeneous beds up to 1.5 m thick or as erosional remnants	Single homogeneous beds up to 1.0 m thick or as erosional remnants
Sedimentary structures	Homogeneous or mottled, no lamination, rare vertical solution holes	Homogeneous, churned or mottled, laminated only where infilling voids in underlying marine limestone, rare filled vertical burrows or root tubes
Fabric	Fine-grained, irregular cracks and holes	Fine-grained, irregular spar-filled cracks and holes
Fossils	Rare gastropods and ostracods, may be locally abundant, very rare marine fossils	Rare gastropods and ostracods, problematic mircofossil, very rare marine fossils
Lithification	Evidence of early lithification; upper surface brecciated, clasts in overlying marine limestone, clasts bored by marine organisms	Evidence of early lithification; upper surface brecciated and bored by marine organisms, clasts in overlying marine limestone
Sequence	Within marine limestones; overlain and underlain by caliche crust, underlying marine limestone shows evidence of leaching and early lithification with clasts in base of fresh-water limestone bed	Within marine limestones; underlain or overlain by caliche, as erosional cobbles within caliche profile, underlying marine limestone shows evidence of leaching
Areal distribution	About 13,000 km^2	About 10,000 km^2

However, many fresh-water limestones may not contain such diagnostic fossils. Perhaps some do not share all the attributes described, but the appearance of these features in Pleistocene and Cretaceous fresh-water limestones as well as in their Holocene counterparts suggests broad applicability of the attributes listed in table 2.

The significance of recognition of fresh-water limestones in carbonate sequences extends beyond the intrinsic interest in this type of limestone. The recognition of fresh-water limestones may allow the identification of previously unperceived disconformities and paleo-diagenetic environments. This, in turn, may help the stratigrapher to understand or even predict the occurrence and distribution of diagenetic porosity related to subaerial exposure, and may result in the discovery of stratigraphic-type oil and gas accumulations.

ACKNOWLEDGMENTS

The authors thank James Palacas of the U.S. Geological Survey, Denver, for organic analyses of dark fresh-water limestones and Johnnie B. Brown of Midland, Texas for unpublished information regarding the distribution of the Edwards Black Bed. Eugene A. Shinn and Barbara H. Lidz read and improved earlier versions of the manuscript.

REFERENCES

Black, M., 1933, The algal sediments of Andros Island, Bahamas: Philosophical Transactions, Royal Society of London, Series B, v. 222, p. 165-192.

Calvert, W.R., 1928, Geologic features of Val Verde County, Texas and adjacent area: Oil and Gas Journal, v. 26, no. 36, Jan. 26, p. 81-82, 85.

Davis, J.H., 1943, The natural features of southern Florida: Florida Geological Survey Bull. 25, 311 p.

DuBar, J.R., 1974, Summary of the Neogene stratigraphy of south Florida, *in* Oaks, R.Q., Jr., and DuBar, J.R., eds., Post Miocene stratigraphy, Central and Southern Atlantic coastal plain: Utah State Univ. Press, p. 206-231.

Dunham, R.J., 1962, Classification of carbonate rocks according to depositional texture, *in* Ham, W.E., ed., Classification of carbonate rocks: AAPG Memoir 1, p. 108-121.

Gleason, P.J., 1972, The origin, sedimentation and stratigraphy of a calcitic mud located in the southern freshwater Everglades: Pennsylvania State Univ., Ph.D. dissertation, 355 p.

_______, and Spackman, W., Jr., 1974, Calcareous periphyton and water chemistry in the Everglades, *in* Gleason, P.J., ed., Environments of South Florida: Present and past: Miami Geological Society Memoir 2, p. 146-181.

Heilprin, A., 1887, Explorations on the west coast of Florida and in the Okeechobee wilderness: Philadelphia, Pennsylvania, Transactions, Wagner Free Institute of Science, v. 1, p. 32-33.

Monty, C., 1967, Distribution and structure of Recent stromatolitic algal mats, eastern Andros Island, Bahamas: Ann. Soc. Geol. Belgium, Bull. 90, p. 55-100.

_______ 1973, Recent algal stromatolitic deposits, Andros Island, Bahamas: Preliminary report: Geologische Rundschau, Band 61, Heft no. 2, p. 742-783.

_______, and Hardie, L. A., 1976, The geological significance of the freshwater blue-green algal calcareous marsh, *in* Walter, M., ed., Stromatolites: New York, Developments in Sedimentology, Elsevier Scientific Publishing Co., v. 20, p. 447-477.

Multer, H. G., and Hoffmeister, J. E., 1968, Subaerial laminated crusts of the Florida Keys: GSA Bull., v. 79, p. 183-192.

Parker, G.G., and Cooke, C.W., 1944, Late Cenozoic geology of southern Florida with a discussion of the ground water: Florida Geological Survey Bull. 27, 119 p.

Perkins, R.D., 1974, Discontinuity surfaces as a stratigraphic tool: The Pleistocene of south Florida: GSA Abstracts with Programs, v. 6, no. 7, p. 908-909.

Pettijohn, F.J., 1957, Sedimentary Rocks: Harper's Geoscience Series, p. 411.

Purser, B.H., 1969, Syn-sedimentary marine lithification of Middle Jurassic limestones of the Paris Basin, France: Sedimentology, v. 12, no. 3-4, p. 205-230.

Rose, P.R., 1970, Stratigraphic interpretation of submarine ***versus*** subaerial discontinuity surfaces: An example from the Cretaceous of Texas: GSA Bull., v. 81, p. 2787-2798.

_______ 1972, Edwards Group, surface and subsurface, Central Texas: Univ. Texas, Austin, Bureau of Economic Geology, Report of Investigations 74, 198 p.

Schroeder, M.C., 1954, Stratigraphy of the outcropping formations in south Florida: Southeastern Geological Societies Guidebook, 8th Field Trip, p. 18-36.

_______, Klein, H., and Hoy, N.D., 1958, Biscayne aquifer of Dade and Broward Counties, Florida: Florida Geological Survey, Report of Investigations, no. 17, figure 5.

Sellards, E.H., 1919, Geologic section across the Everglades, Florida: Florida Geological Survey, 12th Annual Report, p. 67-76.

Shinn, E.A., 1969, Submarine lithification of Holocene carbonate sediments in the Persian Gulf: Sedimentology, v. 12, no. 1-2, p. 109-144.

_______, Ginsburg, R.N., and Lloyd, R.M., 1965, Recent supratidal dolomite from Andros Island, Bahamas, *in* Pray, L.C., and Murray, R.C., eds., Dolomitization and limestone diagenesis, a symposium: SEPM Special Pub. no. 13, p. 112-123.

Shinn, E.A., Lloyd, R.M., and Ginsburg, R.N., 1969, Anatomy of a modern carbonate tidal flat, Andros Island, Bahamas: Journal of Sedimentary Petrology, v.39, p. 1202-1228.

Taft, W.H., and Harbaugh, J.W., 1964, Modern carbonate sediments of southern Florida, Bahamas and Espiritu Santo Island, Baja California: A comparison of their mineralogy and geochemistry: Stanford Univ. Publications in the Geological Sciences, v. 8, p. 1-64.

Ward, W.C., Folk, R.L., and Wilson, J.L., 1970, Blackening of eolianite and caliche adjacent to saline lakes, Isla Mujeres, Quintana Roo, Mexico: Journal of Sedimentary Petrology, v. 40, p. 548-555.

FACIES, DIAGENESIS, AND POROSITY DEVELOPMENT IN A LOWER CRETACEOUS BANK COMPLEX, EDWARDS LIMESTONE, NORTH-CENTRAL TEXAS

Ralph S. Kerr[1]

ABSTRACT

An Edwards carbonate bank complex developed on the structurally positive Belton high in north-central Texas. Mobile carbonate sandbars, rudist reefs, and beaches of the shallow-water bank differ strikingly from time-equivalent deeper water muds of the Tyler basin to the north and from supratidal dolomites and evaporites of the Central Texas platform to the south.

Early stages of bank development are well exposed near Belton, Texas, and exhibit the following progradational vertical facies succession: open-marine basin, shallow grain shoal, rudist patch reef, exposed beach, restricted lagoon, and intertidal-supratidal mudflats. Paleocaliche, desiccation cracks, and algal boundstone suggest subaerial exposure of the uppermost unit.

The major diagenetic changes occurred early and predate regional exposure of the bank complex at the end of Edwards time. Most calcite cementation is early and associated with local meteoric water tables where fine-crystalline bladed crusts, syntaxial overgrowths, solution-cavity fill, and medium- to coarse-crystalline equant calcite were precipitated. Early Mg calcite bladed crusts also formed in the marine or mixing-zone environments. Late-stage calcite cementation was limited to coarse-crystalline equant calcite.

Most of the dolomite is of multi-stage fresh-water mixing origin. Dolomitization of lime muds and calcite cements occurred early, in association with local meteoric water tables, and late, as the result of a meteoric ground-water system. Fine-crystalline anhedral dolomite in lagoonal and tidal-flat facies is of hypersaline origin.

The shoreface and foreshore facies of the beach complex have up to 35-percent (thin section estimated) moldic and interparticle porosity as a result of (1) high original interparticle porosity, (2) early cementation to reduce compaction, (3) aragonite allochem dissolution to produce moldic porosity, (4) subaerial exposure to reduce further cementation, and (5) an early seal to retard fluid migration. The lagoonal and tidal-flat facies have up to 20-percent (estimated) moldic and intercrystalline porosity because of (1) dissolution of aragonitic allochems and (2) partial dolomitization and subsequent dissolution of the lime mud.

INTRODUCTION

A high-energy carbonate bank complex developed along the gently sloping northeastern margin of the Central Texas platform during deposition of the Lower Cretaceous Edwards Limestone in north-central Texas. The early stages of bank development are well exposed in a quarry located approximately 2 miles northeast of the town of Belton, Texas (fig. 1).

Because of the extent of the exposure (pl. 1), the Belton Quarry provides an unusually good opportunity to explore the three-dimensional relations of depositional and diagenetic facies in an open-marine to supratidal sequence.

Paleogeographic Setting

Early Cretaceous sedimentation was controlled by a combination of structural and regional depositional features (fig. 2). Fredericksburg-age rocks were deposited on the broad, flat Comanche shelf that covered most of present-day Texas except for the southeastern Gulf Coastal Plain. On its southeastern side, the shelf sloped steeply into the deeper waters of the Ancestral Gulf. The central portion of the shelf, known as the Central Texas platform, was a broad swell developed over the structurally positive San Marcos arch, which trends roughly perpendicular to the shelf edge (fig. 3).

The northeastern flank of the platform sloped gently towards the shallow North Texas - Tyler basin and was the site of rudist patch reef and carbonate sand accumulation. This report examines in detail one such accumulation, known as the Moffat Lentil (fig. 4). Shoaling in the area of the Moffat Lentil may have been accentuated by the presence of the Belton high (Tucker, 1962a, 1962b)—a broad positive feature located about 100 miles north of and trending

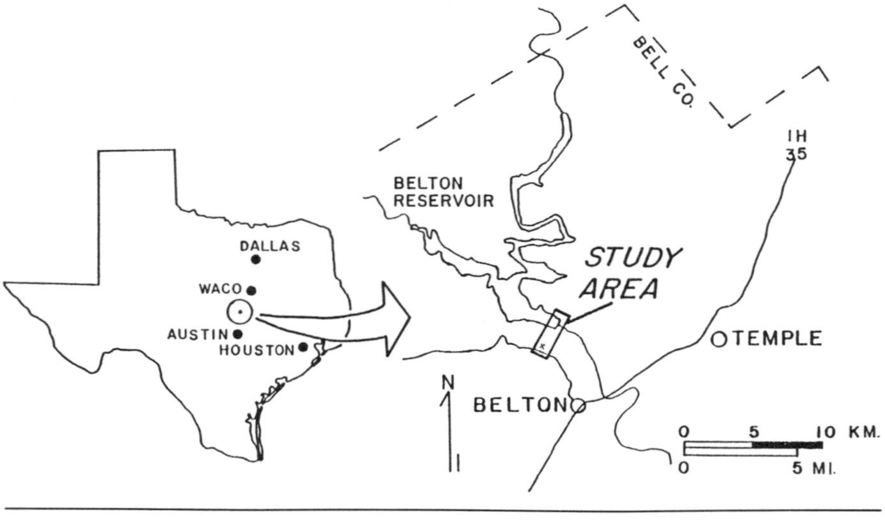

Figure 1. Location map showing the study area. The Belton Quarry is located at the southern boundary of the outline.

[1]Shell Oil Company, Houston, Texas

Belton Quarry

Plate 1

Plate 1. Photomosaic of the western half on Belton Quarry (measured sections 7 through 10). Exposed on the quarry face are the crossbedded grainstone lithofacies (beach complex), burrowed wackestone lithofacies (lagoon), and the burrowed and laminated mudstone lithofacies (intertidal and supratidal flats). Staff painted in 0.5-m increments.

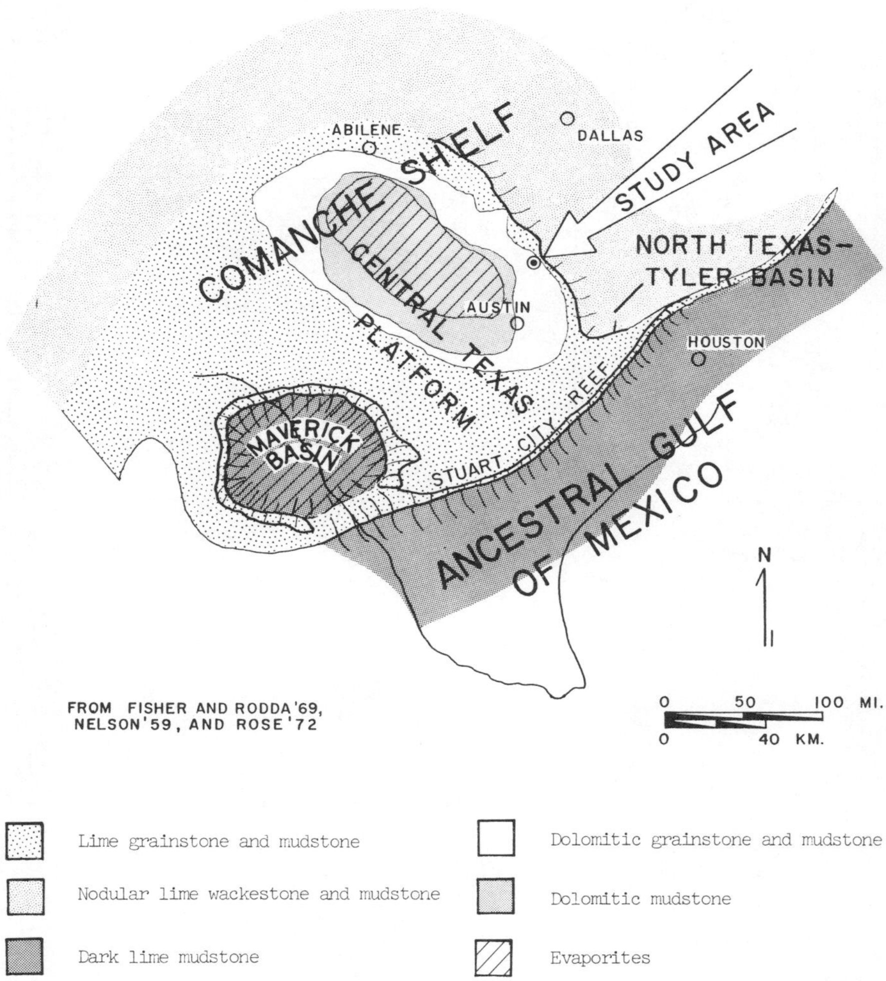

Figure 2. Paleogeography of Edwards Group in Texas. Major structural and regional depositional features and distribution of lithologic types are indicated.

parallel to the San Marcos arch (fig. 3) on the Central Texas platform.

Stratigraphy

In the study area, the Edwards Formation thickens from north to south by means of a facies change into the underlying Comanche Peak Formation (fig. 4). Thus, the Edwards grainstones of the Moffat Lentil in central Bell County are equivalent in time to the Comanche Peak wackestones of the North Texas - Tyler basin.

This regular southward thickening is interrupted in the central part of the county where the Edwards attains an unusual thickness of 125 ft. Moore (1961, 1964) described the grainstones of this thick Edwards feature and named it the Moffat Lentil. He interpreted the lentil as a buildup of the Fredericksburg surface; "this limestone body replaced 30 feet of the Comanche Peak below and extends some 60 feet above the general upper Fredericksburg surface" (Moore, 1964). Brown (1975) documented the northward progradation of these platform-margin facies into the North Texas - Tyler basin.

The lowermost Washita unit is the Kiamichi Shale, a transgressive terrigenous marl, that pinches out by onlap around the Belton high (fig. 5). It is probable that the area over the Belton high was exposed at the end of Fredericksburg time and that the initial transgression during Kiamichi (early Washita) time was not sufficient to inundate the entire Fredericksburg surface, particularly in the area of the Moffat Lentil. Nelson (1959) noted a direct regional relationship between the absence of the Kiamichi Shale and what he termed post-lithification dolomite.

BELTON QUARRY FACIES

Six major lithofacies have been identified in the Belton Quarry, on the basis of lithology, texture, primary sedimentary structures, degree and type of bioturbation, and fossil assemblages. The vertical succession of facies and their interpreted depositional environments are shown on figure 6, an idealized section through the quarry. The idealized section is a composite of 12 measured sections: 10 spaced along the quarry wall and 2 in a newly opened pit in the quarry floor (fig. 7).

The vertical sequence, from bottom to top, consists of (1) plant-rich skeletal wackestone of the open-marine Comanche Peak Formation, (2) alternating skeletal grainstone and burrowed wackestone deposited as mobile grain shoals, (3) rudist bafflestones associated with shallow-water patch reefs, (4) accretion-bedded skeletal grainstone of a prograding beach complex, (5) dolomitized wackestone deposited in a restricted low-energy lagoon, and (6) intertidal burrowed and supratidal laminated mudstone capping the sequence. Paleocaliche, desiccation cracks, marsh deposits, and caliche-algal boundstone suggest repeated subaerial exposure of this uppermost unit.

Figure 7 shows the distribution of the major facies along the quarry wall and in the central pit. All the facies are laterally persistent along the length of the wall, although they vary in thickness. The depositional slope as determined from the true dip of the accretion-bedded foreshore is to the northeast. The quarry face trends obliquely downslope between measured sections 1 through 9 but upslope between sections 9 and 10. The major characteristics of each facies are summarized graphically in figure 8 and illustrated in plates 2 through 14.

Based on the vertical facies succession, the following depositional model is proposed (fig. 9). At the base of the sequence, in the open or partially restricted marine waters of the North Texas - Tyler basin, lime muds and an admixture of skeletal debris were deposited. Communities of seagrasses caused muds to be trapped, and their rhizomes stabilized the sediment substrate forming broad, topographically elevated mudbanks. The major processes were sediment baffling by the seagrasses and bioturbation by organisms.

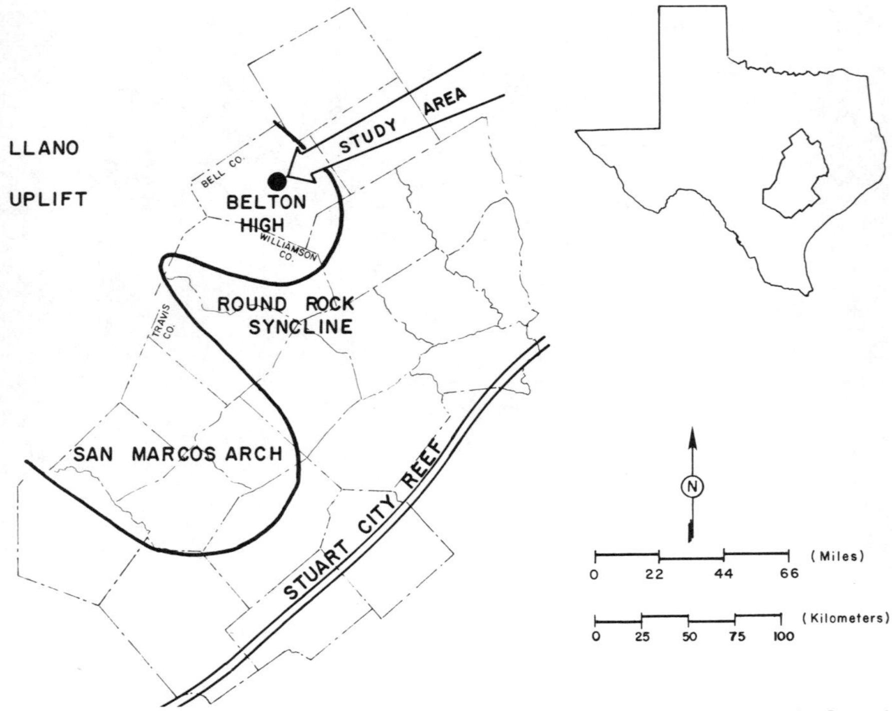

Figure 3. Major structural elements affecting deposition during Edwards time in Central Texas. From Tucker (1962b).

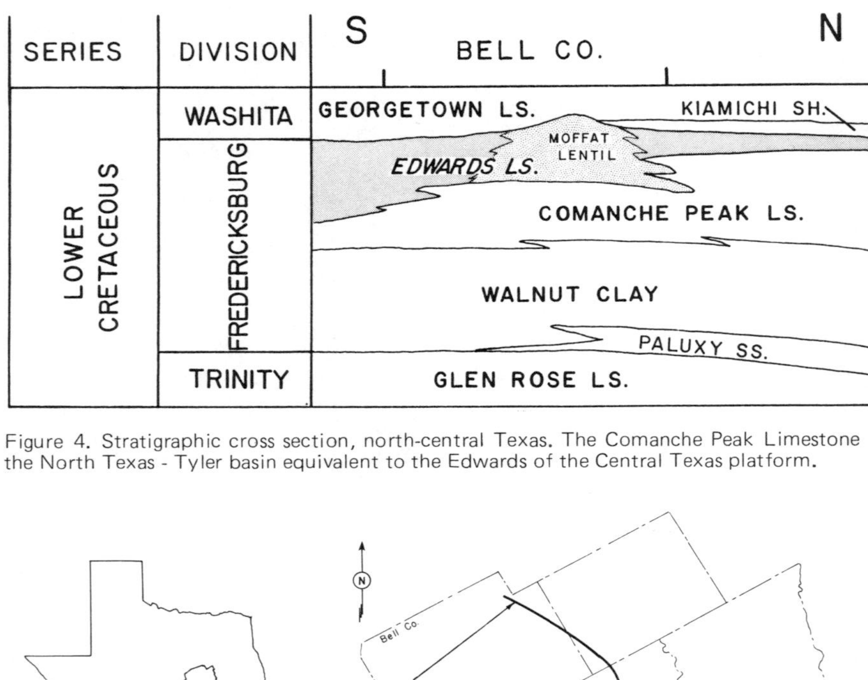

Figure 4. Stratigraphic cross section, north-central Texas. The Comanche Peak Limestone is the North Texas - Tyler basin equivalent to the Edwards of the Central Texas platform.

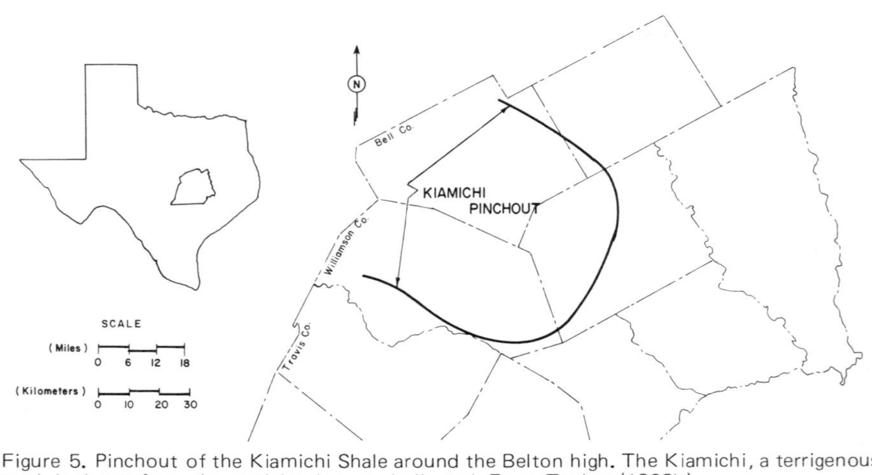

Figure 5. Pinchout of the Kiamichi Shale around the Belton high. The Kiamichi, a terrigenous marl, is absent from the semicircular area indicated. From Tucker (1962b).

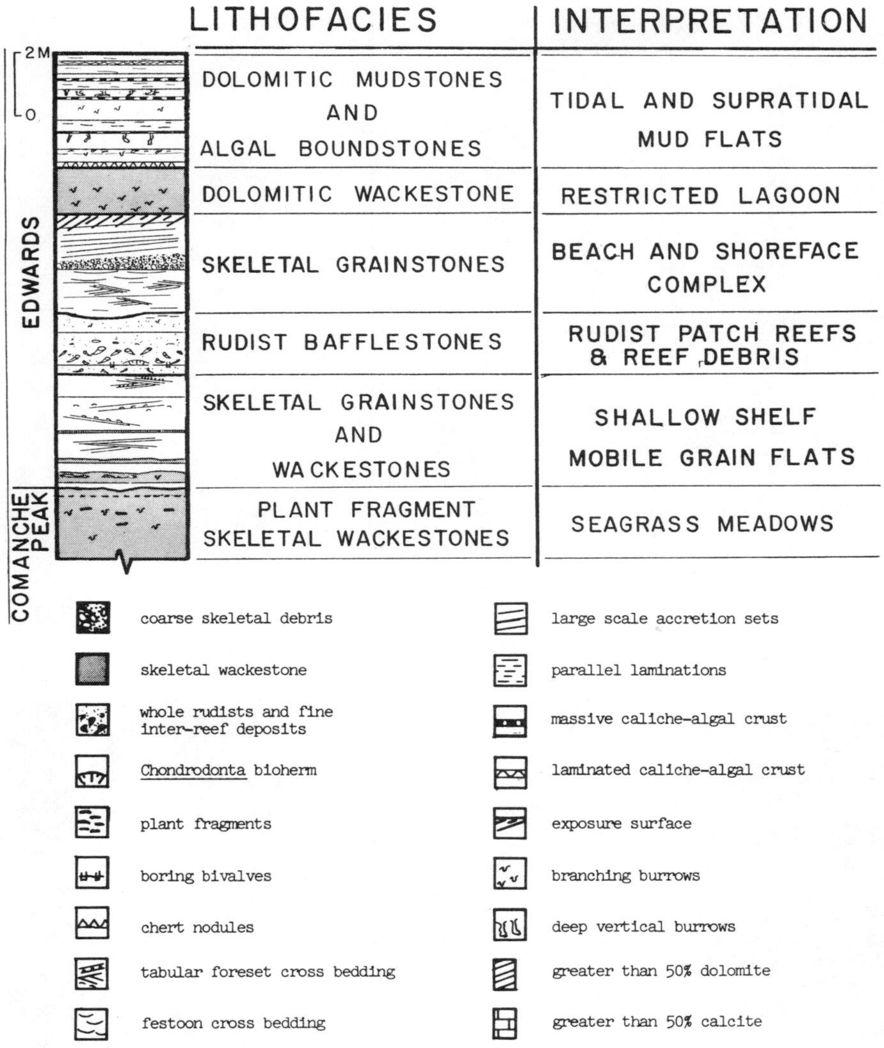

Figure 6. Composite vertical measured section, Belton Quarry. Symbols applicable to figures 6-13.

Patch reefs developed in the shallower water further landward on the shelf, perhaps associated with the topographically elevated and stabilized mudbanks. The large amount of skeletal debris produced by these reefs was reworked into mobile grainstone bars trending parallel to the shoreline. Storms moved sediment in a general landward direction, and finer sediments, trapped in the lee of the patch reefs, formed the muddier interreef deposits.

As sediments were reworked landward, shoals developed. The shoals became emergent and shoreline processes became dominant. Longshore currents transported coarse shell debris along strike to build spits, and storm washovers transported skeletal sands into the low-energy lagoon behind the shoals.

In the lagoon, bioturbation was the major process, and although the number of individual organisms was relatively large, the species diversity was small because of the restrictive physical conditions imposed by varying temperatures and salinities.

The sequence was capped by prograding intertidal and supratidal flats. Burrowing was common on the lower flats but decreased landward and upward in subaerially exposed units. At the end of Fredericksburg time, the Moffat Lentil was exposed to subaerial diagenesis (Nelson, 1959; Moore, 1964; Rose, 1972).

BELTON QUARRY DIAGENESIS

Early and late stages of diagenesis have been recognized in the carbonates of the Belton Quarry. "Early" diagene-

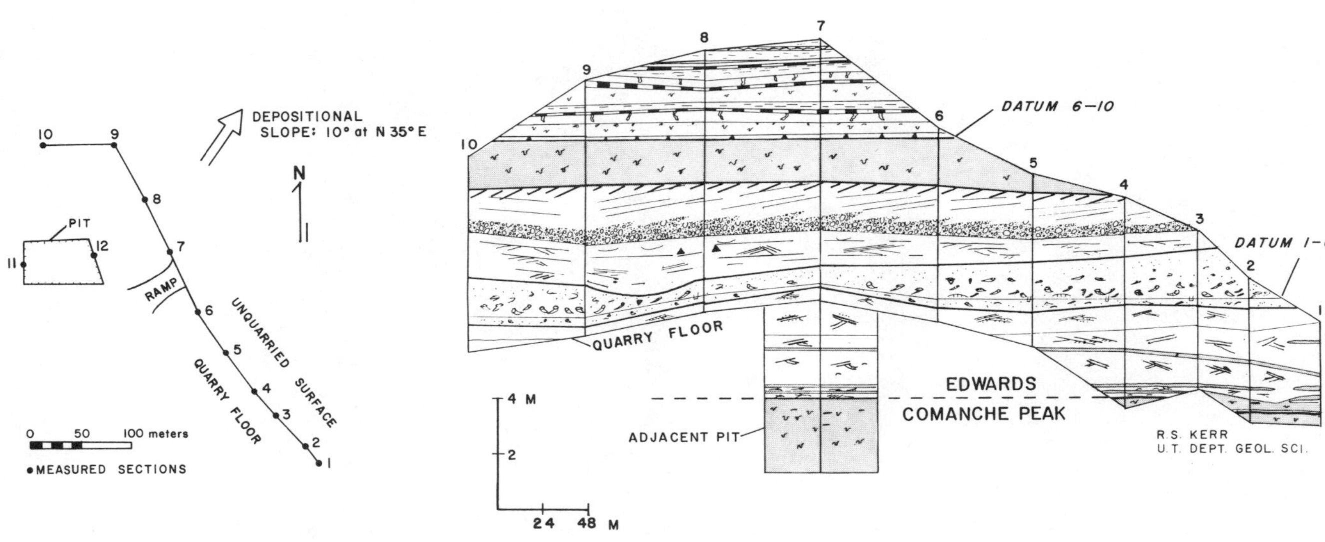

Figure 7. Location map of measured sections and facies cross section, Belton Quarry.

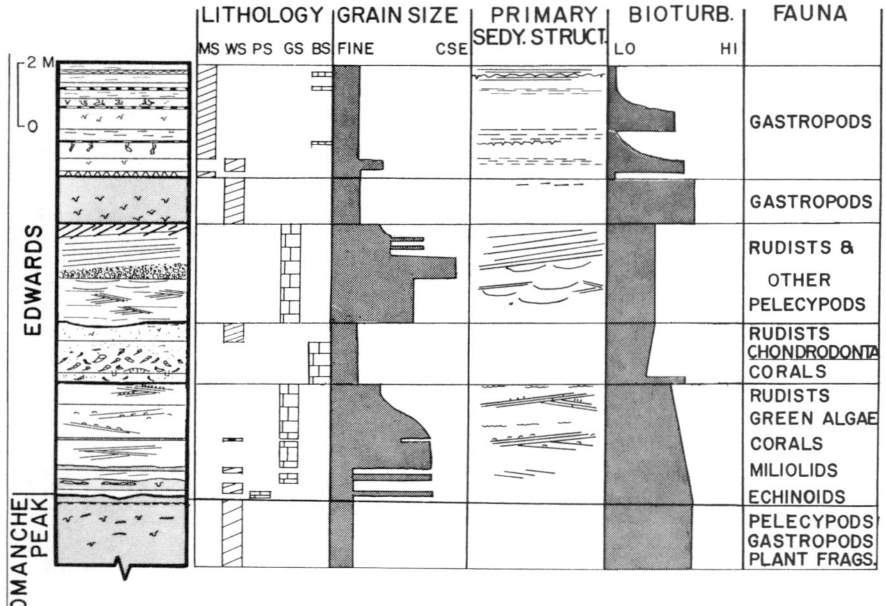

Figure 8. Major characteristics of the Belton Quarry lithofacies.

Plate 2

Plate 3

Plant-Rich Wackestone Facies
Plate 2. Intercalation of uppermost Comanche Peak (darker lenticular units) and Edwards grainstones of overlying facies. Sedimentation was continuous during Comanche Peak - Edwards time, the transition between these two lithologies indicating the early development of the high-energy Edwards Moffat Lentil. Measured section 1.

Plant-Rich Wackestone Facies
Plate 3. Three types of plant remains in the Comanche Peak: (1) fragment with a fibrous structure oriented parallel to the elongation of the blade, (2) fragment with concentric banding, and (3) fragment with central canal and fibers branching from the canal at angles of approximately 45°. Evidence suggests that these fragments are the remains of an Early Cretaceous seagrass community. Bar = 1 cm.

Alternating Skeletal Grainstone and Wackestone Facies
Plate 4. Outcrop of alternating grainstone and wackestone. Wackestone lithology is dominant at the base of the quarry, but the grainstone units become thicker upward in the section. Grainstones are lighter colored units. Increments on staff are 0.5 m. Measured section 2.

Plate 4

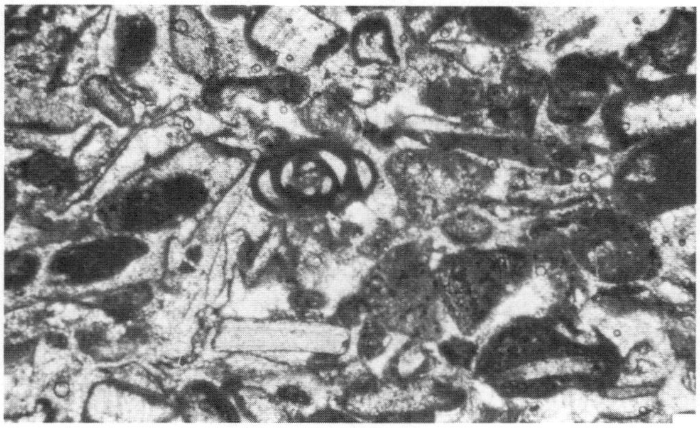

Plate 5

Alternating Skeletal Grainstone and Wackestone Facies
Plate 5. Well-rounded and well-sorted grainstone from upper portion of facies. Miliolid foraminifer near center; micritized grains are numerous. TS 2-8. Bar = 0.1 mm.

Plate 6

Rudist Bafflestone and Wackestone Facies
Plate 6. Outcrop of rudist bafflestone and wackestone facies. The lower unit, composed of reef debris, is separated by a bored surface from the bafflestone of the patch reef itself. Also visible are the underlying grainstone of the mobil grainflats and the overlying grainstone of the beach complex. Measured section 4. Scale = 15 cm.

Plate 7

Rudist Bafflestone and Wackestone Facies
Plate 7. Small conical ***Chondrodonta*** bioherm in the rudist bafflestone facies. Hammer rests on upper surface of the concave-upward mound. Measured section 5.

Plate 8

Crossbedded Skeletal Grainstone Facies
Plate 8. Fine-grained, well-sorted grainstone of the lower shoreface, overlain by large scale accretion sets of the foreshore. Measured section 7. Increments on staff = 0.5 m.

Plate 9

Crossbedded Skeletal Grainstone Facies
Plate 9. Alternating coarse and fine laminations of foreshore with upper and lower shoreface below. Lagoonal wackestone sharply overlies the case-hardened upper foreshore.

Plate 10

Burrowed Dolomitic Wackestone Facies
Plate 10. Burrowed lagoonal wackestone rests sharply on accretion sets of facies below. Intertidal and supratidal mudstone caps the sequence.

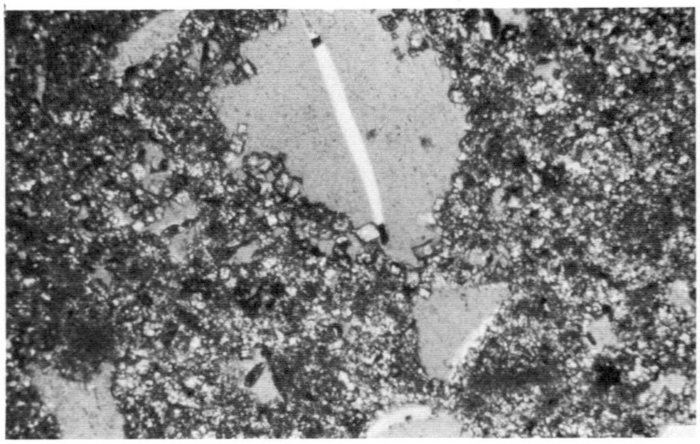

Plate 11

Burrowed Dolomitic Wackestone Facies
Plate 11. Pelleted dolomitic wackestone of the lagoonal environment. Large pore spaces are gastropod solution cavities. Dolomitic intercrystalline porosity is also present. Bar = 0.1 mm.

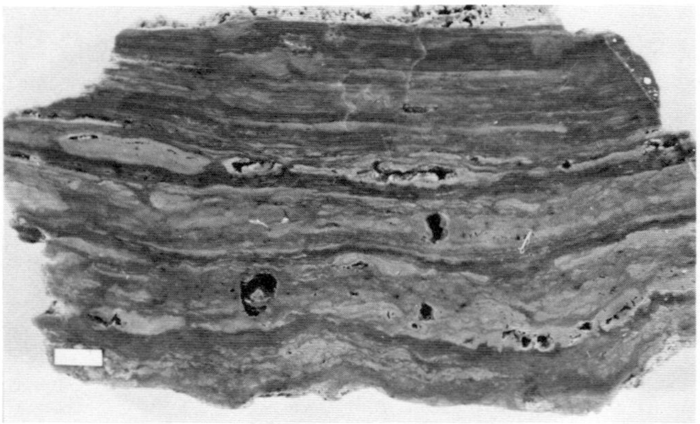

Plate 12

Burrowed and Laminated Dolomitic Mudstone Facies
Plate 12. Laminated caliche-algal crust from supratidal flat. Clasts may be mud chips eroded from mud-crack polygons. Sawed slab 7-18. Bar = 1 cm.

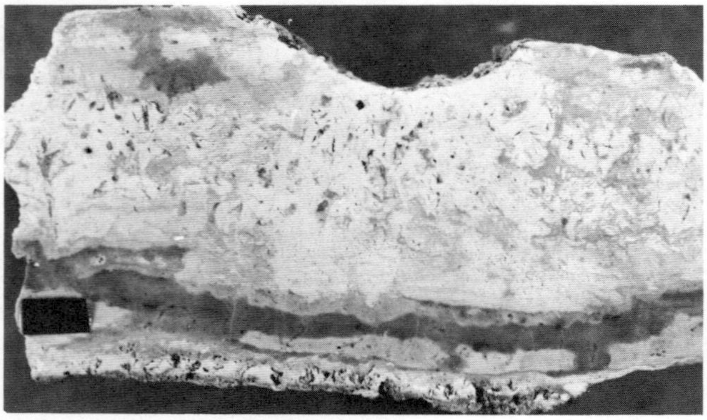

Plate 13

Burrowed and Laminated Dolomitic Mudstone Facies
Plate 13. Supratidal storm and marsh deposits. Basal skeletal lag is overlain by mud-cracked polygons and root-perforated mudstone. Root perforations resemble those of ***Scytonema***, a blue-green algae described from the Andros Island supratidal flats by Shinn and others (1969). Bar = 1 cm.

Plate 14

Burrowed and Laminated Dolomitic Mudstone Facies
Plate 14. Large-scale mud-crack polygons exposed near the top of the quarry at measured section 8.

sis occurred from the time of sediment deposition until regional exposure of the Moffat Lentil at the end of Edwards time, and "late" diagenesis occurred from the time of regional exposure, through later burial, until uplift. Table 1 summarizes the events assigned to each of these stages.

Diagenetic Events

Grain rounding

Carbonate allochems have been rounded and broken in the environment of deposition either by physical abrasion or biological degradation (pl. 15). Wave and current activity on the submerged grainflats as well as the beach complex was sufficient to winnow out the lime muds and physically abrade the soft carbonate allochems. Rounding can also occur through the chemical digestion of particles by burrowing organisms (Hanor and Marshall, 1971) or attack by boring organisms and microborers (Perkins and Halsey, 1971).

Stabilization of Mg calcite

Metastable Mg calcite stabilizes either by exsolution of Mg to form low-Mg calcite or by absorption of Mg to form dolomite (Land, 1967). This Mg calcite stabilization starts very early after exposure to meteoric water (Land, 1970). No evidence of selective early dolomitization of Mg calcite allochems was observed in the Belton Quarry section. Therefore, it is probable that stabilization was accomplished by exsolution.

Dissolution of aragonite

Aragonite is also unstable in meteoric water and will stabilize by neomorphism (inversion) to calcite or by total dissolution and reprecipitation of calcite (Land, 1967). No examples of aragonite inversion were discovered in the Belton Quarry section, so it is concluded that the majority of aragonite stabilization was by the dissolution and cavity-fill process. Land and others (1967) documented this process in the Pleistocene of Bermuda.

Fine-crystalline bladed calcite crust

In the mobile grainflat and rudist patch reef facies, micrite envelopes are commonly surrounded with a fine-crystalline (10 to 30 μm) isopachous bladed crust on the outer side only. In the example illustrated in plate 16, the micrite envelope has collapsed, and the former aragonitic allochem is filled in with coarse-crystalline equant calcite. This evidence suggests that the fine-crystalline bladed crust was precipitated before aragonite dissolution and micrite envelope collapse, which are both early meteoric events. Later, equant calcite filled the solution cavity (see medium- to coarse-crystalline equant calcite).

In a second example, an echinoid fragment (pl. 17) is surrounded by a fine-crystalline bladed isopachous crust, followed by a syntaxial overgrowth in optical continuity with the echinoid fragment. In this case, the bladed-crust precipitation preceded syntaxial cementation, which is interpreted as an early meteoric phreatic event.

Evidence favoring a marine phreatic or mixing-zone origin for the bladed crust of the mobile grainflat and rudist patch reef facies follows.

(1) Many of the individual crystals are fibrous to bladed with steep terminations (pl. 18), suggestive of Mg calcite—a marine cement (Folk, 1974). L. S. Land (personal communication) has collected similar even isopachous submarine cements from the modern Jamaican reefs (pl. 19), and similar cements have been described by Milliman (1974). Meteoric low-Mg calcite cements are commonly more equant or blocky (Land, 1970).

(2) When viewed tangentially, the crystals are triangular or pyramid shaped, also suggestive of Mg calcite cement; low-Mg calcite cements commonly have hexagonal cross sections.

(3) Precipitation of the bladed crust commonly preceded aragonite dissolution—a meteoric phreatic event (Land and others, 1967; Land, 1970)—because the crusts do not commonly appear on the internal sides of voids formerly occupied by aragonitic allochems.

(4) Bladed crusts preceded or grew in competition with syntaxial

Table 1. Diagenetic Events

Early Diagenetic Events
- grain rounding
- stabilization of Mg-calcite
- fine-crystalline bladed isopachous calcite crust
- syntaxial calcite cement
- dissolution of aragonite
- calcite solution-cavity filling cement
- *medium- to coarse-crystalline equant calcite
- calichification
- *dolomitization
- *silicification of fossils

Late Diagenetic Events
- *medium- to coarse-crystalline equant calcite
- *dolomitization
- *silicification of fossils

*occurred during both early and late stages of diagenesis

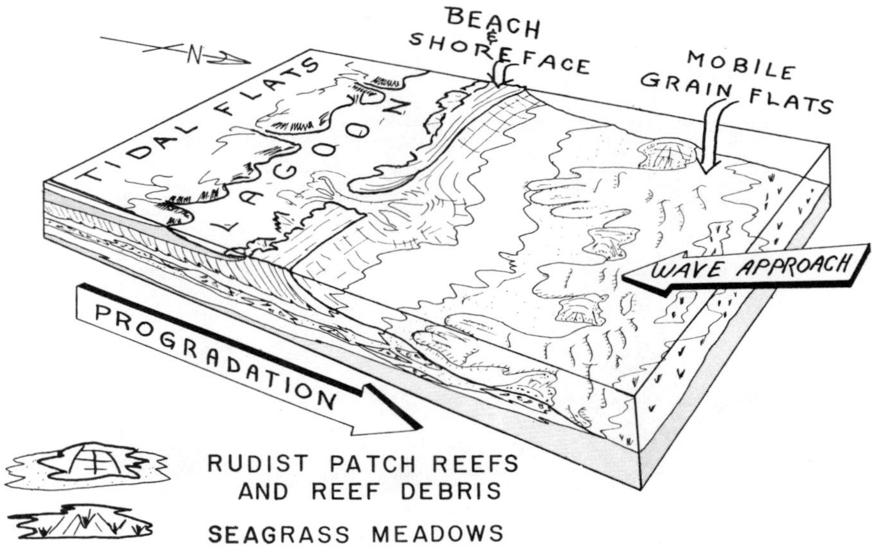

Figure 9. Depositional model developed from the Belton Quarry and Belton Dam sections.

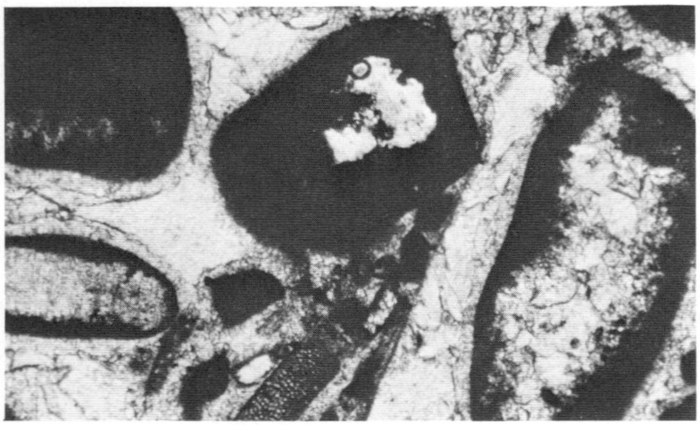

Plate 15

Diagenesis: Physical Alteration

Plate 15. Thick micrite envelopes and highly micritized well-rounded grains from mobile grainflats. Individual boring algal tubules project into grain on right side of field of view. Grain was bored first, then underwent dissolution, and equant calcite was precipitated in the void space preserving the delicate algal tubules. Bladed calcite crust surrounds allochems and coarse-crystalline equant calcite fills remaining interparticle pore space. TS 12-2B. Bar = 0.1 mm.

Plate 16

Diagenesis: Calcite Cementation

Plate 16. Thin section illustrating paragenetic sequence in mobile grainflats. The collapsed micrite envelope has an isopachous bladed crust on the outerside only, indicating that the crust was precipitated before aragonite dissolution and envelope collapse (both are meteoric phreatic events). Later coarse-crystalline equant calcite was precipitated. TS 12-2B. Bar = 0.05 mm.

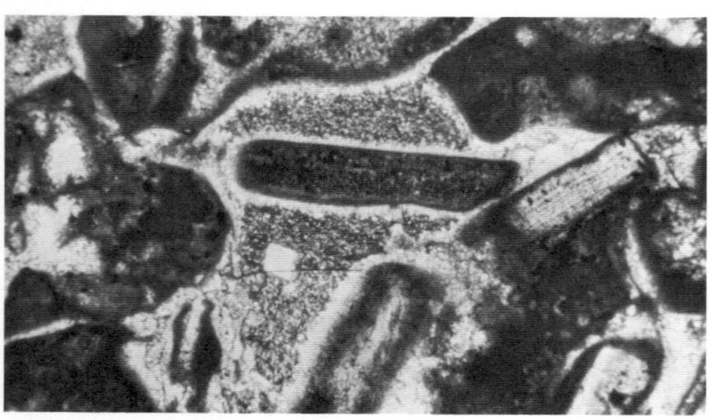

Plate 17

Diagenesis: Calcite Cementation

Plate 17. Isopachous bladed calcite crust precipitated prior to syntaxial overgrowth on an echinoid fragment. Because syntaxial cements are interpreted as meteoric phreatic, the bladed crust is probably submarine in origin. Mobile grainflats. TS 3-5. Bar = 0.1 mm. Crossed nicols.

cements, which are precipitated from solutions fresher than normal sea water (Steinen, 1974).

(5) Independent evidence shows that submarine cementation existed in the Belton Quarry section during Moffat Lentil time; the boring bivalves of the rudist patch reef facies were boring into an early lithified substrate.

In summary, the petrographic evidence suggests that the initial isopachous bladed crust in the mobile grainflats was Mg calcite. Folk (1974) suggests that the crystal habit of a carbonate precipitate is strongly influenced by the Mg/Ca ratio of the precipitating solution. The precipitation of Mg calcite requires Mg/Ca ratios greater than those of normal meteoric water. The Mg could have been supplied by seawater, in which case the crust is of marine or mixing-zone origin. However, the Mg may also have been supplied by high-Mg calcite allochems and matrix stabilizing to low-Mg calcite in the meteoric phreatic zone. The evidence favors the first hypothesis.

The calcite of the isopachous crust exhibits a different morphology and crystal habit in the shoreface and foreshore facies of the beach complex (pls. 20-22). At this stratigraphic level, the early bladed crust is generally coarser (50 to 60 μm) and approaches axial ratios of 1.5:1 (equant calcite). This change in habit could be attributed to a decreased Mg/Ca ratio in the precipitating waters, that is, in the meteoric phreatic zone. Inden (1972) found a similar decrease in crystal axial ratios in the beach facies of the Cow Creek Limestone of Central Texas.

In addition, bladed cements commonly occur on both sides of micrite envelopes in the foreshore (pl. 22) indicating precipitation after aragonite dissolution, a meteoric phreatic event. In some examples, pendulous "dripstone" cements occur within micrite envelopes, suggesting precipitation in the meteoric vadose zone (pl. 22).

Syntaxial calcite cements

Syntaxial cements are precipitated in pores and as overgrowths on echinoderm fragments in optical continuity with their substrate. In the Belton Quarry bank complex, the syntaxial cements were precipitated early and grew after or contemporaneously with the fine-crystalline bladed crusts (pl. 23). Land (1970) and Steinen (1974) discuss Pleistocene examples of syntaxial overgrowths that formed in meteoric environments.

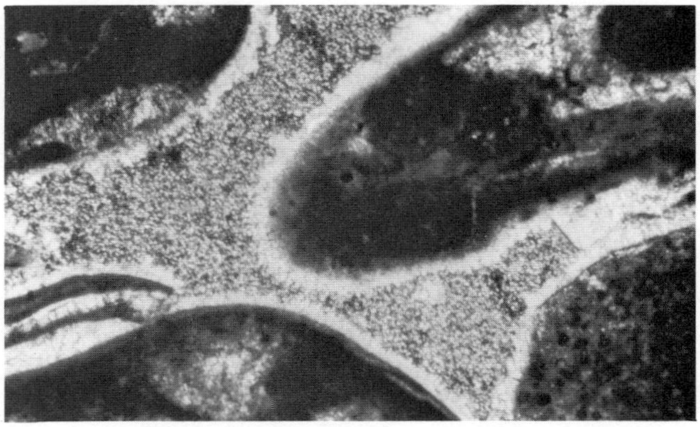

Plate 18

Diagenesis: Calcite Cementation
Plate 18. Fine-crystalline bladed crust precipitated on allochems in the mobile grainflat environment. Crystals are acicular and scalenohedrally terminated. Entire crust is of even thickness and isopachous. Compare with plate 19, TS 3-5. Crossed nicols. Bar = 0.1 mm.

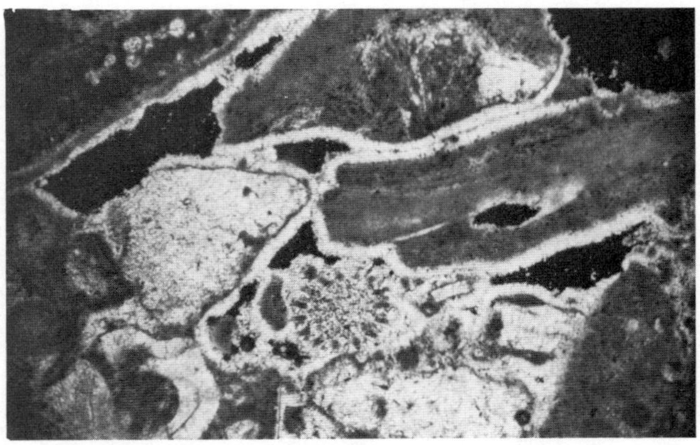

Plate 19

Diagenesis: Calcite Cementation
Plate 19. Recent example of submarine cement from Jamaica, collected by L.S. Land. The cement is similar to the Belton Quarry example shown on plate 18 in that the cement is an even isopachous bladed crust of scalenohedral calcite (probably high-Mg calcite). Bladed crust has precipitated on echinoid fragment in center prior to precipitation of any syntaxial overgrowth. Crossed nicols. Bar = 0.1 mm.

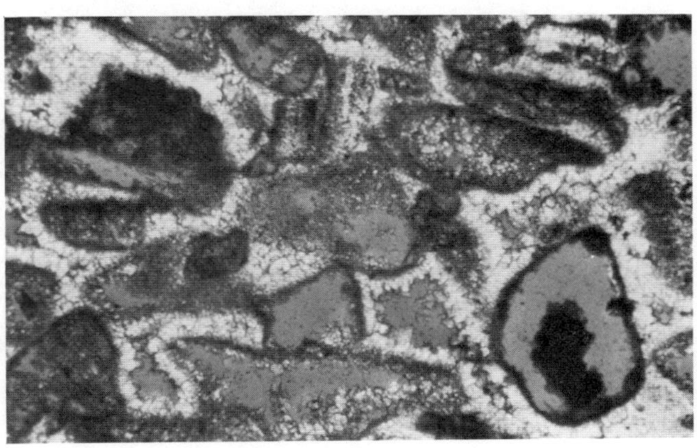

Plate 20

Diagenesis: Calcite Cementation and Calichification
Plate 20. Fine- to medium-crystalline bladed to equant calcite crust around former aragonitic allochems of the foreshore environment. Porosity is (1) secondary moldic and (2) primary intercrystalline. TS 8-7. Bar = 0.1 mm.

Medium-crystalline calcite solution-cavity fill

The central cavities of former aragonite grains (mostly fossils) are filled by a mosaic of equant calcite, generally fine-crystalline around the border, grading to medium-crystalline in the center (pls. 15, 24). Similar cement-filled solution cavities are common in Pleistocene rocks (Friedman, 1964; Bathurst, 1964; Land, 1970) and are interpreted to be the result of meteoric diagenesis.

Estimates of the vertical distribution of solution-cavity-filling cement show that it is common in all the facies with the exception of the lower and middle foreshore (pls. 20-22).

Medium- to coarse-crystalline equant calcite

Medium- to coarse-crystalline equant calcite cement commonly fills interparticle pore space of the grainstones in the Belton Quarry (pls. 15-18). This cement type is rare in the lower and middle foreshore, however (pls. 20-22).

According to the scheme of Folk (1973, 1974) and Folk and Land (1975), equant sparry calcite should precipitate in any calcite-precipitating environment with a Mg/Ca ratio less than approximately 1:1. This means that equant calcite could form early in the meteoric phreatic zone or later in the subsurface. The only petrographic evidence as to time of formation from the Belton Quarry section is that the equant calcite formed later than the fine-crystalline bladed to equant isopachous crust. All the equant cements are Fe-poor, and therefore possibly precipitated from oxidizing waters.

Loucks (1976, 1977), however, has presented qualitative evidence to distinguish early from late equant calcite stages. In the Lower Cretaceous Pearsall Formation of South Texas, early equant calcites are commonly finer than later equant calcites and more gradational in size with the early-generation isopachous cements. In addition, the contacts between crystals of early equant cements are more irregular than those of later equant cements and later equant calcite forms several large crystals in a pore space.

According to these criteria, estimates of early vs. late equant calcites in the Belton Quarry show a dominance of early equant calcites within the upper mobile grainflat facies, the rudist patch reef facies, and the lower and upper shoreface of the beach complex. Equant calcite is almost entirely absent from the middle of the foreshore facies. Late-stage equant calcite is common within the lower mobile grainflat facies only.

It is possible that this distribution of early equant calcites approximates the vertical limits of the meteoric phreatic zone during early cementation. The deeper and coarser crystalline equant calcites would therefore be of a later stage. Since most dolomitization is early, this hypothesis is supported by the greater frequency of dolomitization of early solution-cavity fill and medium-crystalline calcite cements and the lack of dolomitization of coarser crystalline equant calcites (see section on Dolomitization).

Calichification

A 0.3-m-thick caliche-algal crust (term from Ward and others, 1970) occurs at the top of the foreshore facies (pl. 25). On the outcrop, it appears case hardened and is iron stained. In this facies, it is a clotted-appearing biomicrite, and the fossil allochems are highly silicified. The texture is similar to the "clotted texture" described by James (1972) and the "secondary mud supported texture" of Read (1974).

The tightly cemented and recrystallized zone at the top of the foreshore is interpreted as paleocaliche. Moore and others (1972) reported similar subaerially formed paleocrusts from an Edwards beach complex in north-central Texas. A tightly cemented zone at the top of the beach complex probably inhibited the movement of meteoric water through the beach facies and may have been partially responsible for the absence of solution-cavity fill and equant calcite cements in much of the foreshore facies.

Late Diagenetic Features

Coarse-crystalline equant calcite cement is the most common late diagenetic feature and has been discussed previously.

Dolomitization

Dolomite is present in every lithofacies of the Belton Quarry. The percentage of dolomite was determined semiquantitatively by X-ray diffraction (fig. 10); the petrographic characteristics of the dolomites are summarized in figure 11.

Three modal sizes of dolomite crystals can be found in the lower four facies: 10 μm, 30 μm, and 70 μm. All are limpid and euhedral, but the 10-μm and 30-μm dolomites usually have replaced or displaced lime mud. The coarser size usually has replaced

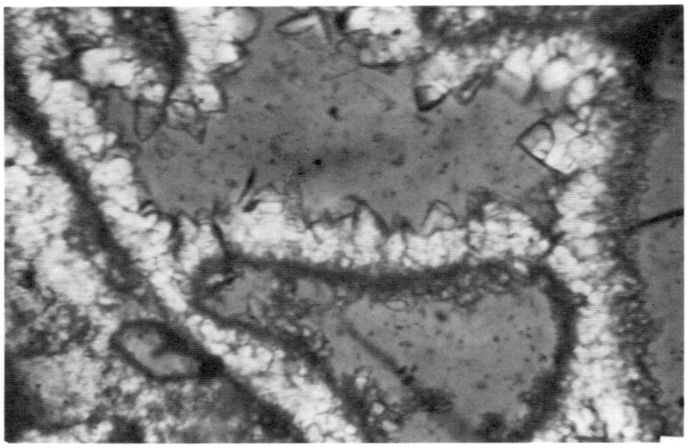

Plate 21

Diagenesis: Calcite Cementation and Calichification
Plate 21. Medium-crystalline bladed crust on both sides of micrite envelope in foreshore environment. Former aragonitic allochem has undergone dissolution. Cement inside micrite envelope is pendulous, suggesting precipitation in the vadose zone. TS 8-7. Bar = 0.05 mm.

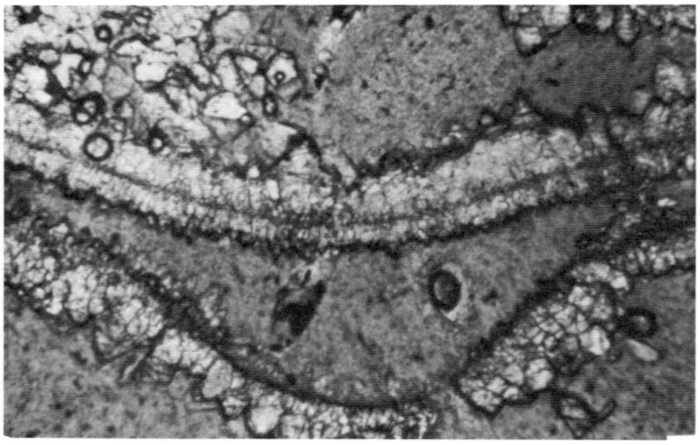

Plate 22

Diagenesis: Calcite Cementation and Calichification
Plate 22. Medium- to coarse-crystalline bladed calcite projecting into interparticle pore space. Crystals have "clean" scalenohedral terminations and show no signs of outcrop dissolution. TS 8-7. Bar = 0.05 mm.

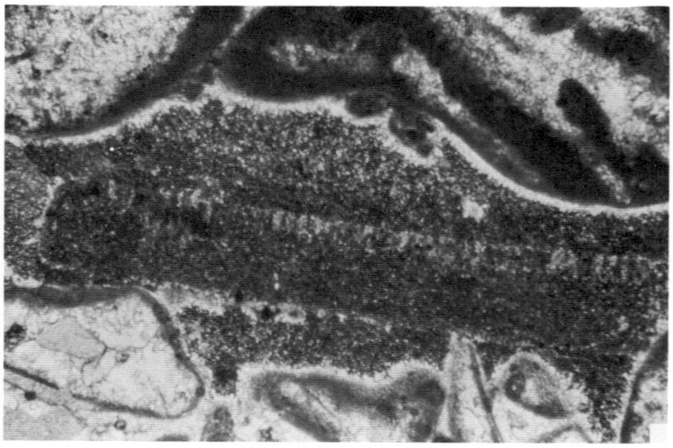

Plate 23

Diagenesis: Calcite Cementation and Calichification
Plate 23. Syntaxial overgrowth precipitated in optical continuity with echinoid spine: mobile grainflat environment. Equant calcite mosaic fills dissolved former aragonitic allochems. TS 3-5. Bar = 0.1 mm. Crossed nicols.

early medium-crystalline calcite cement but not the coarser later cements (pls. 26, 27). Land and others (1975) found a similar trimodal frequency distribution of dolomite diameters in Eocene carbonate rocks from Egypt.

In the Belton Quarry, evidence for early dolomitization includes (1) selective dolomitization of finer grained sediments, probably because of differences in porosity, permeability, organic content, or chemistry, and (2) extensive dolomitization of early calcite cements (medium-crystalline cavity fill calcite and syntaxial overgrowths), with limited dolomitization of later coarse-crystalline equant intergranular cements.

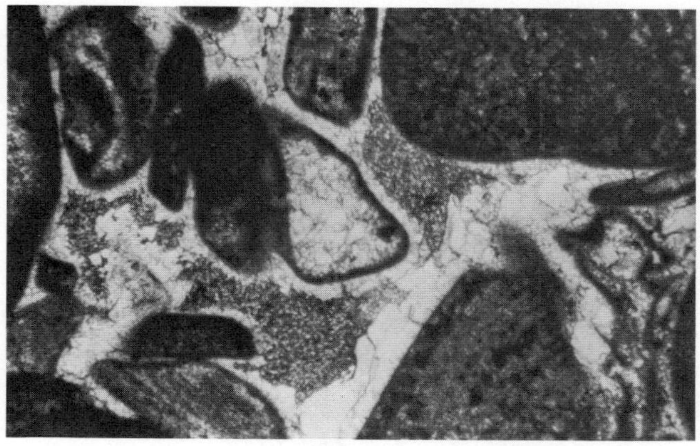

Plate 24

Diagenesis: Calcite Cementation and Calichification
Plate 24. Medium-crystalline equant calcite mosaic precipitated in former aragonitic allochem. Interparticle porespace is filled with first-generation bladed crust and second-generation equant mosaic. Mobile grainflats. TS 3-5. Bar = 0.1 mm.

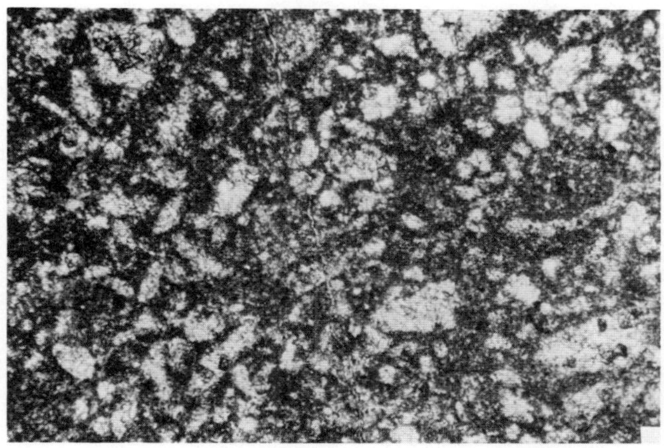

Plate 25

Diagenesis: Calcite Cementation and Calichification
Plate 25. Caliche-algal crust from top of beach foreshore. Allochems have been recrystallized to fine-crystalline equant calcite, and matrix has developed secondary mud-supported texture of Read (1974). TS 7-4TI. Bar = 0.1 mm.

Most of the dolomite within the lower four facies is clear and euhedral (limpid)—suggesting that it resulted from the mixing of fresh and saline waters (Folk and Land, 1975). (For a discussion of the fresh-water mixing hypothesis of dolomitization, see Hanshaw and others, 1971; Folk and Land, 1972, 1975; Land, 1973a, 1973b; Badiozamani, 1973.)

Zoned dolomite is common (pls. 27-28) and generally consists of three zones, with the middle zone dissolved so that the inner zone has collapsed. This zoning could be the product of alternating fresh-water and hypersaline stages of dolomitization, in which the hypersaline stage would produce the poorly ordered central zone, or an alternation of oxidizing and reducing conditions which would allow the incorporation of more Fe in the middle zone. Later, the poorly ordered or Fe-rich zone was leached out.

X-ray analysis of the dolomites revealed some poorly ordered (probably Ca-rich) dolomites in the mobile grainflat facies. Shinn and others (1965) and Brown (1975) have also reported Ca-rich poorly ordered dolomites from the Edwards of Central and north-central Texas.

The lagoon and tidal-flat facies consist of a bimodal size distribution of dolomite (pls. 29, 30): 10-μm "dirty" anhedral dolomite and 70-μm limpid euhedral pore-filling dolomite. This implies an early supratidal or hypersaline stage followed by a later fresh-water mixing stage, similar to the schizohaline dolomite formation described by Siedlecka (1972a, 1972b) and Folk and Siedlecka (1974). Again, the coarse (70 μm) zoned dolomite that is present has a dissolved middle zone.

In summary, petrologic, petrographic, and geochemical evidence suggests a multiple-stage origin for Belton Quarry dolomite. In the grassflat, mobile grainflat, rudist patch reef, and beach facies, the finer grained lime muds were the sites of initial dolomitization because of the mixing of fresh and saline waters. The source of the fresh water was meteoric lenses associated with the exposed beach complex and the subaerially exposed tidal flats. Later, early calcite cements, syntaxial overgrowths, and calcite solution-cavity fills were dolomitized by fresh-water mixing either on a local scale, or, as Nelson's (1959) regional data would suggest, a regional ground-water system developed after the entire Moffat Lentil was exposed. Intermediate stages of dolomitization due to hypersaline brines produced poorly ordered Ca-rich dolomites which enlarged the crystals of the earlier generation. Reducing waters may also have produced Fe-rich stages enlarging preexisting crystals.

In the lagoonal and tidal-flat facies, dirty 10-μm anhedral dolomite formed in the soft supratidal muds and in the shallow restricted lagoon. Later, as the result of local meteoric lenses, fresh-water influx into the lagoon (schizohaline environment), and a regional ground-water system, coarser crystalline euhedral dolomites filled moldic and intercrystalline pore space. Variations in salinity or Eh produced poorly ordered Ca-rich or Fe-rich zoned dolomite.

Diagenetic Model

Based on the diagenetic history of the Belton Quarry rocks, three stages of diagenesis are recognized: (1) early

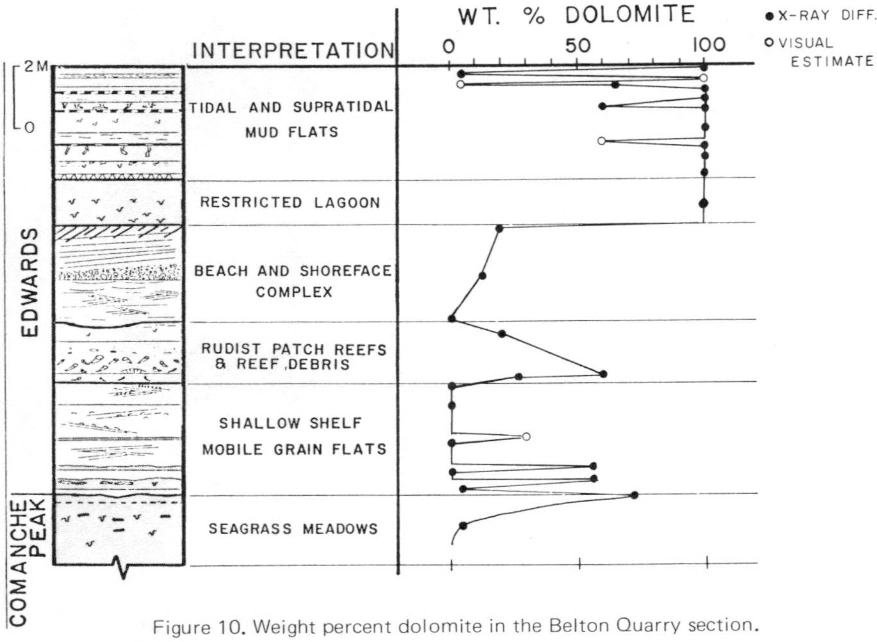

Figure 10. Weight percent dolomite in the Belton Quarry section.

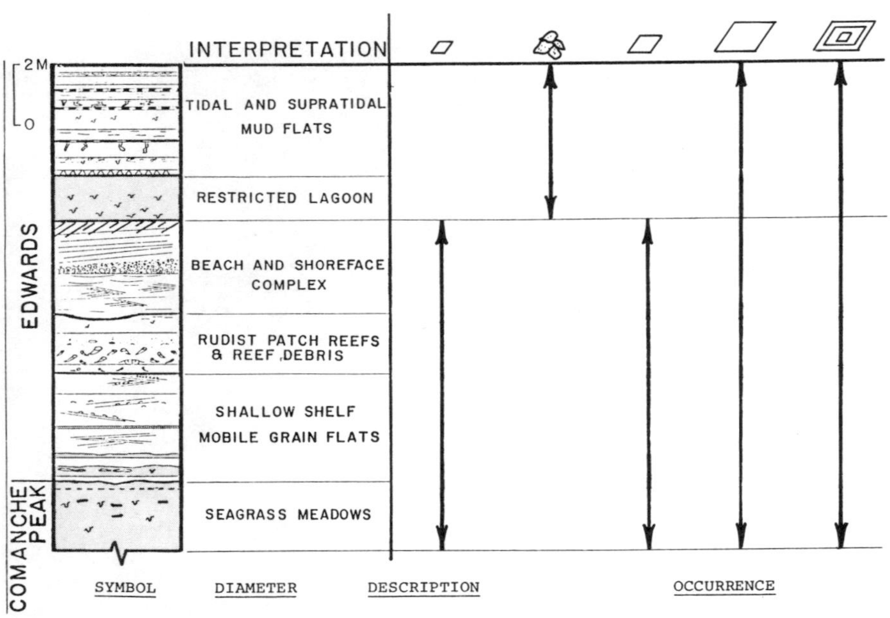

SYMBOL	DIAMETER	DESCRIPTION	OCCURRENCE
	10μm	limpid, euhedral	replaces/displaces micrite
	10μm	anhedral to subhedral "dirty", pelleted	replaces/displaces micrite
	30μm	limpid, euhedral	replaces/displaces micrite
	70μm	limpid, euhedral	replaces syntaxial cement, calcite solution-cavity-fill, and equant calcite cement; projects into moldic and intercrystalline pores
	70μm	zoned	replaces syntaxial cement, calcite solution-cavity-fill, and equant calcite cement; projects into moldic and intercrystalline pores

Figure 11. Petrographic characteristics and vertical distribution of the Belton Quarry dolomite.

marine, (2) early local meteoric, and (3) late regional meteoric.

The interpreted diagenetic environments are shown on a schematic cross section of the Belton Quarry facies tract (fig. 12). Each diagenetic environment is coded to indicate the diagenetic events associated with that environment.

The first stage of diagenesis was in the marine environment with solutions of normal-marine chemistry. The major features include rounding of grains and oolitic coatings on grains. Patchy submarine-cemented crusts formed in the quiet waters around patch reefs. Early bladed crusts of Mg calcite formed around allochems of the mobile grainflat and rudist patch reef facies. Silicification of the upper foreshore occurred, possibly because of the downward migration of silica mobilized in overlying supratidal units (see Moore and others, 1972).

The second stage of diagenesis was the local meteoric stage involving solutions of low Mg/Ca ratios in both the phreatic and vadose environments. This stage was also early and was the result of the development of fresh-water lenses below exposed islands or fresh-water tables beneath mainland tidal flats. It is a well-known principle that for every foot above sea level a water table extends, it can exist 40 feet below sea level (Tolman, 1937).

The major events were stabilization of Mg calcite, dissolution of aragonite, and precipitation of calcite-bladed crusts, syntaxial overgrowths, solution-cavity fill, and medium-crystalline equant cements. Coarse-crystalline bladed crusts were precipitated in the upper phreatic zone, whereas finer, more fibrous crusts formed lower in the phreatic zone, perhaps in response to the increased Mg content of the water.

Dolomitization of lime muds and early calcite cements occurred in the mixed-water zone in the grassflat, mobile grainflat, rudist patch reef, and beach environments. In the peritidal zone of the lagoon and tidal flat, Mg/Ca ratios and salinities were variable, and schizohaline and hypersaline dolomites formed.

The third stage of diagenesis was the late regional phreatic stage during which a regional meteoric groundwater system developed. Mg/Ca ratios were variable, but the water was normally oxidizing. This stage was probably initiated after regional exposure of the Moffat Mound at the end of Edwards (Fredericksburg) time. The major features of this stage were precipitation of coarse-crystalline

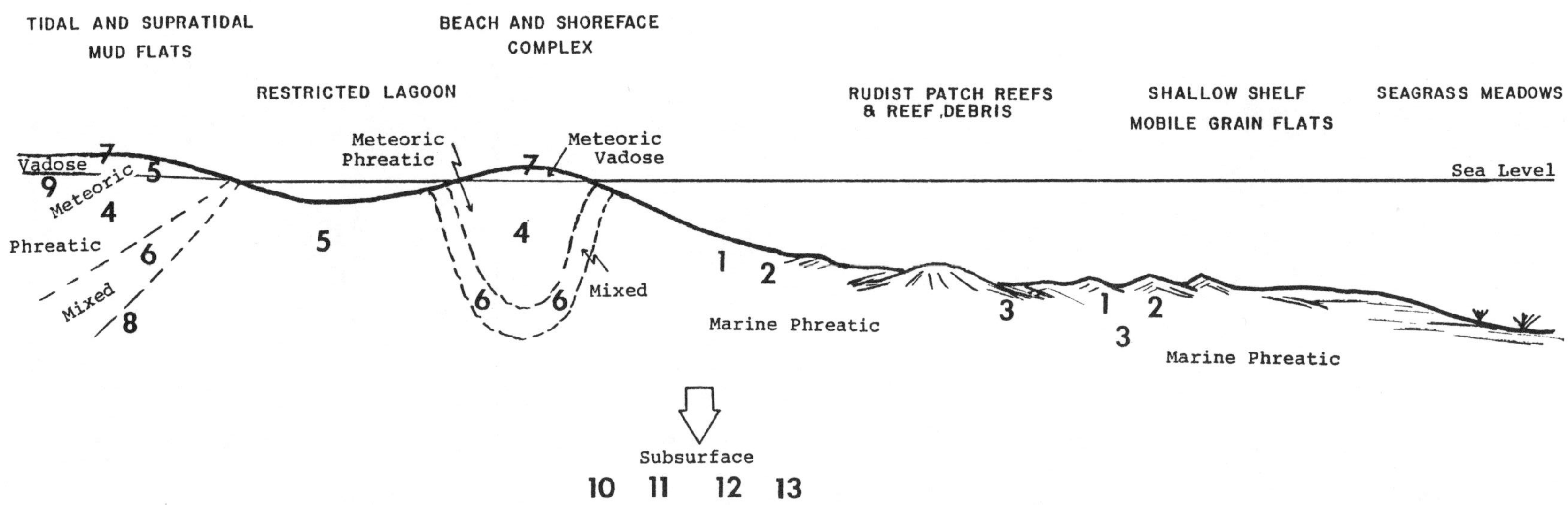

EARLY DIAGENETIC EVENTS

1 Grain rounding

2 Formation of micrite envelopes, micritization, and oolitic coating

3 Precipitation of submarine bladed Mg-calcite cement

4 Stabilization of Mg-calcite
Dissolution of aragonite
Precipitation of calcite: bladed crusts, syntaxial overgrowths, solution-cavity fill, and medium crystalline equant cements

5 Dolomitization in supratidal and schizohaline environments

6 Dolomitization in fresh water mixing zone

7 Calichification

8 Silicification of fossils

9 Chertification (?)

LATE DIAGENETIC EVENTS

10 Precipitation of coarsely crystalline equant calcite cement

11 Dolomitization by fresh water mixing

12 Silicification of fossils

13 Siliceous replacement of evaporites

Figure 12. Diagenetic model based on the Belton Quarry section.

iron-free equant calcite cements and dolomitization due to mixing of these meteoric waters with sea water or hypersaline subsurface brines.

POROSITY DEVELOPMENT

Porosity Types

Four major types of porosity developed in the Belton Quarry section: (1) primary cement-reduced interparticle, (2) secondary mollusk moldic and oomoldic, (3) secondary dolomite intercrystalline, and (4) primary intraparticle. All are fabric selective.

Primary cement-reduced interparticle porosity

Interparticle porosity accounts for up to 15 percent of the total rock mass in the grainstones of the lower and middle foreshore. It is always cement reduced by a bladed isopachous crust. (See plates 20-22.)

Secondary mollusk moldic porosity

Pore space formed by the dissolution of aragonitic pelecypod fragments is common in the lower and upper shoreface and lower and middle foreshore. Gastropod molds are common in the lagoonal and lower tidal-flat facies. Moldic porosity accounts for up to 20 percent of the rock volume in the foreshore facies and up to 10 percent of the rock volume in the lagoonal facies (gastropod molds). (See plates 11, 20, 22.)

Secondary dolomite intercrystal porosity

Pore space between dolomite crystals is common in the dolomitized mudstones as well as in grainstones where dolomite has replaced sparry-calcite cement. In the lagoonal and tidal-flat facies, dolomite intercrystal porosity can occupy up to 15 percent of the total rock mass. (See plates 27-30.)

Primary intraparticle porosity

Intraparticle porosity is commonly preserved within the original pore spaces of allochems—particularly foraminifer tests and rudist shells. It is volumetrically less than 3 percent of the rock mass in any facies. (See plate 5.)

Distribution of Porosity

The skeletal grainstones of the foreshore and shoreface are the most porous zones, with up to 35-percent (estimated) secondary mollusk moldic and primary interparticle porosity (fig. 13).

The lagoonal and tidal-flat facies have up to 20-percent (estimated) secondary moldic and dolomite intercrystal porosity, whereas the mobile grainflat facies has a total of approximately 7- to 10-percent secondary moldic, dolomite intercrystal, and primary intraparticle porosity.

The question naturally arises in an outcrop study about the effects of outcrop weathering on porosity. Several lines of evidence suggest that the moldic and interparticle porosity in the foreshore is not the result of outcrop weathering (Inden, 1972; Loucks, 1976; Mueller, 1975):

(1) the most highly porous zone is continuous along the outcrop and not related to fractures or the quarry surface;

(2) the dissolved allochems formerly were aragonitic mollusks; the fossils that were originally calcite are still intact;

(3) the early bladed crust projecting into the primary interparticle pore spaces has "clean" scalenohedral terminations and shows no signs of dissolution (pl. 21). It is unreasonable to assume that only interparticle equant calcites would be dissolved, leaving the early bladed cements intact;

(4) mollusk moldic and interparticle porosity of the Belton Quarry type is common in subsurface Lower Cretaceous grainstones.

In the lagoonal and tidal-flat facies, the mollusk moldic porosity was produced by dissolution of aragonitic allochems and is believed to have been developed in the Cretaceous. The dolomite intercrystal porosity could be Cretaceous or Recent. It is likely that the intercrystal porosity was initiated by incomplete dolomitization during Cretaceous time and has been enhanced by subsurface and possibly outcrop dissolution of the intercrystal carbonate muds.

Porosity Summary

The shoreface and foreshore facies have both high porosity and good permeability and are of primary interest. The original depositional porosity was high in these grain-supported strata. Dunham (1962) has reported that original interparticle and intraparticle porosity in carbonate grainstones can be as high as 70 percent. Physical abrasion probably reduced much of the intraparticle porosity, but interparticle pore space remained as high as 25 to 30 percent.

Submarine and meteoric phreatic diagenesis resulted in early cementation which prevented later compaction. Dissolution of metastable allochems began in the meteoric phreatic environment. As progradation continued, the foreshore became exposed in the meteoric vadose zone; aragonite dissolution probably continued, but cementation ceased. Land and others (1967) reported that diagenesis in the vadose zone typically results in high porosity because autochthonous cementation does not take place. Rather, dissolution of these overlying deposits provides carbonate for cementation in underlying sediments.

In addition, subaerial exposure of the upper foreshore led to the forma-

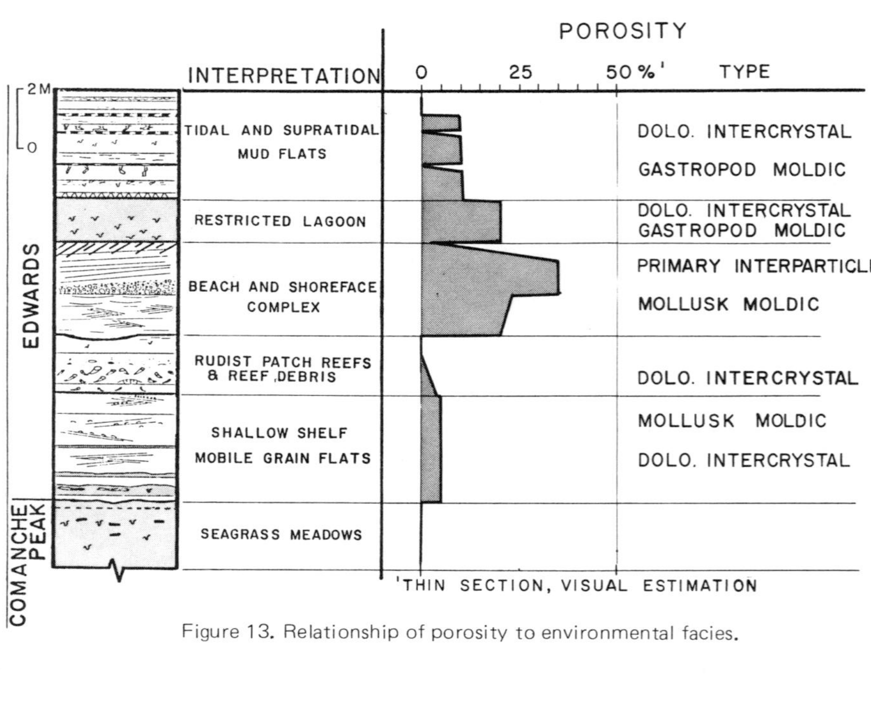

Figure 13. Relationship of porosity to environmental facies.

tion of an indurated caliche-algal crust at or near the surface and thus produced an early permeability barrier which retarded later fluid migration. This crust contributed to the lack of interparticle cementation in much of the foreshore.

Five important depositional and diagenetic elements have combined that would produce a good potential hydrocarbon reservoir in the shoreface and foreshore grainstones: (1) high initial interparticle porosity with a grain-supported texture, (2) early cementation to reduce compaction, (3) aragonite dissolution to produce moldic porosity, (4) subaerial exposure to reduce further interparticle and cavity-fill cementation, and (5) an early seal to retard fluid migration.

In the lagoonal and tidal-flat facies, dissolution of allochems, early partial dolomitization, and subsequent dissolution of calcite mud has produced high moldic and intercrystalline porosity.

CONCLUSIONS

A high-energy carbonate bank complex (the Moffat Lentil) developed on the structurally positive Belton high during deposition of the Lower Cretaceous Edwards Limestone in north-central Texas. The mobile carbonate sand bars, rudist patch reefs, and prograding beaches of the Moffat Lentil differ strikingly from the time-equivalent deeper water muds deposited in the North Texas - Tyler basin to the north and the supratidal dolomitic mudstones and evaporites of the Central Texas platform to the south.

The early stages of bank development are well exposed in a quarry near Belton, Texas, and exhibit a progradational vertical facies succession. The vertical sequence, from bottom to top, consists of (1) plant-rich skeletal wackestone of the open-marine Comanche Peak Formation, (2) alternating skeletal grainstone and burrowed wackestone deposited as mobile grain shoals, (3) rudist bafflestones associated with shallow-water patch reefs, (4) accretion-bedded skeletal grainstone of a prograding beach complex, (5) dolomitized wackestone deposited in a restricted low-energy lagoon, and (6) intertidal burrowed and supratidal laminated mudstone capping the sequence. Paleocaliche, desiccation cracks, algal boundstone, and oxidized marsh deposits suggest repeated subaerial exposure of this uppermost unit. This vertical sequence of facies also represents the lateral facies tract.

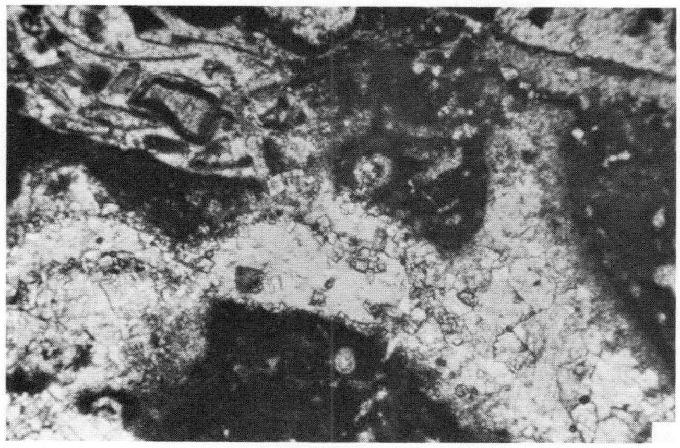

Plate 26

Diagenesis: Dolomitization

Plate 26. Medium crystalline (70 μm) dolomite replacing calcite solution-cavity fill in former aragonitic allochem, mobile grainflats. TS 12½, stained with Alizarin red-S. Bar - 0.1 mm.

Plate 27

Diagenesis: Calcite Cementation and Calichification

Plate 27. Medium-crystalline (70 μm) zoned dolomite replacing equant calcite mosaic. The middle zone of the dolomite has been dissolved and the inner core has collapsed. Rudist patch reef facies. TS 3-7T. Bar = 0.05 mm.

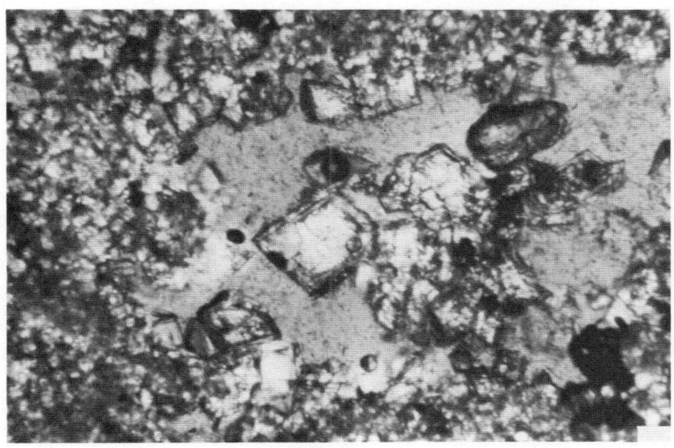

Plate 28

Diagenesis: Calcite Cementation and Calichification

Plate 28. Medium-crystalline (70 μm) dolomite from the lagoonal environment. The middle zone has undergone dissolution leaving the inner core and outer zone intact. Pelleted 5- to 10-μm dolomicrite is also present in this facies. Dolomitic intercrystalline porosity is the result of the dissolution of undolomitized micrite. TS 7-5B. Bar = 0.05 mm.

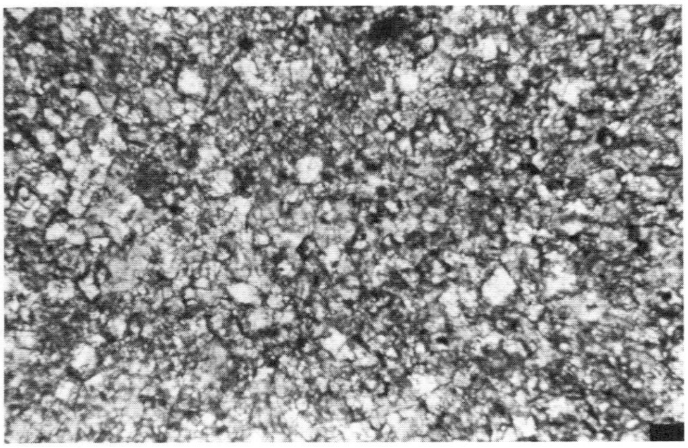

Plate 29

Diagenesis: Calcite Cementation and Calichification

Plate 29. Unfossiliferous (10 μm) pelleted "dirty" dolomite from the laminated supratidal flats. Scattered coarser dolomite rhombs also occur, and dolomitic intercrystalline pore space is present. TS 7-14. Bar = 0.05 mm.

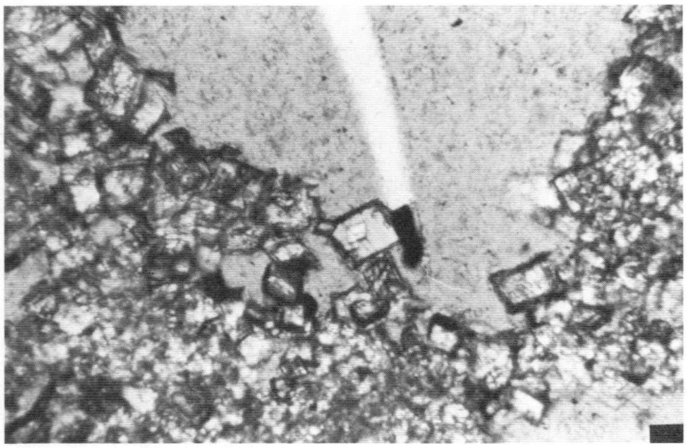

Plate 30

Diagenesis: Calcite Cementation and Calichification

Plate 30. Medium-crystalline (70 μm) euhedral limpid dolomite projecting into moldic pore space. Lagoonal facies. TS 7-5B. Bar = 0.05 mm.

During Edwards time, the carbonate complex prograded northward from the Central Texas platform into the North Texas - Tyler basin, a shallow intracratonic basin. The platform sloped gently towards the basin, and no distinct shelf break developed. Rather, scattered rudist patch reefs and mobile grain shoals occupied the basinward margins of the shelf. This type of shelf-margin buildup is similar to the knoll reef ramp of Wilson (1974).

Three stages of diagenesis are recognized based on the paragenetic sequence of diagenetic events: (1) early marine, involving solutions of normal-marine chemistry; (2) early local meteoric with solutions of low Mg/Ca ratios; and (3) late regional meteoric, with Mg/Ca ratios variable but generally low.

Most calcite cementation is early and is associated with local meteoric water tables. In this diagenetic realm, fine-crystalline bladed calcite crusts, syntaxial calcite overgrowths, calcite solution-cavity fill, and medium- to coarse-crystalline equant calcite were precipitated. Early high-Mg calcite bladed crusts were precipitated in the marine or mixed-water zones. Late-stage calcite cementation was limited to medium- to coarse-crystalline equant calcite.

Most of the Belton Quarry dolomite is of multistage fresh-water mixing origin. Dolomitization of lime muds and calcite cements occurred early, during fresh-water mixing on a local scale, and late, as a result of a regional meteoric ground-water system. Fine-crystalline anhedral dolomite in the lagoonal and tidal-flat facies is of hypersaline origin.

The shoreface and foreshore facies of the beach complex have up to 35-percent (estimated) moldic and interparticle porosity as a result of (1) high original interparticle pore space, (2) early cementation to reduce compaction, (3) aragonite dissolution to produce moldic porosity, (4) subaerial exposure to reduce further cementation, and (5) an early seal to retard fluid migration. The lagoonal facies has up to 20-percent (estimated) moldic and intercrystalline porosity because of (1) dissolution of aragonite allochems, and (2) partial dolomitization and dissolution of the original lime muds.

ACKNOWLEDGMENTS

I wish to thank A. J. Scott, to whom I am indebted for continued support. D. G. Bebout, R. L. Folk, and Keith Young offered their advice and interest throughout the study. L. S. Land generously loaned me his collection of submarine cements and offered many suggestions that improved the diagenesis section. Thanks are also due to R. G. Loucks, Bureau of Economic Geology, as well as John Boone, Tom Elliott, and Chuck Williamson, Department of Geological Sciences, The University of Texas at Austin, with whom I discussed many aspects of the project.

The University of Texas Geology Foundation provided funds for field work, and a grant from the Owens-Coates fund aided in the preparation of the manuscript for publication.

REFERENCES

Badiozamani, K., 1973, The Dorag dolomitization model—application to the Middle Ordovician of Wisconsin: Journal of Sedimentary Petrology, v. 43, p. 965-984.

Bathurst, R. G. C., 1964, The replacement of aragonite by calcite in the molluscan shell wall, *in* Imbrie, J., and Newell, N. D., eds., Approaches to paleoecology: New York, John Wiley and Sons, p. 357-376.

Brown, J. L., 1975, Paleoenvironment and diagenetic history of the Moffat Mound, Edwards Formation, Central Texas: Louisiana State Univ., Master's thesis, 148 p.

Dunham, R. J., 1962, Classification of carbonate rocks according to depositional texture, *in* Ham, W. E., ed., Classification of carbonate rocks: AAPG Memoir 1, p. 62-84.

Fisher, W. L., and Rodda, P. U., 1969, Edwards Formation (Lower Cretaceous), Texas: Dolomitization in a carbonate platform system: AAPG Bull., v. 53, p. 55-72.

Folk, R. L., 1973, Carbonate petrography in the post-Sorbian age, *in* Ginsburg, R. N., ed., Evolving concepts in sedimentology: Johns Hopkins Univ., Studies in Geology, no. 21, p. 118-158.

_______ 1974, The natural history of crystalline calcium carbonate: Effect of magnesium content and salinity: Journal of Sedimentary Petrology, v. 44, p. 40-53.

_______, and Land L. S., 1972, Mg/Ca vs. salinity; A frame of reference for crystallization of calcite, aragonite and dolomite (abstract): GSA Abstracts with Programs, v. 4, p. 508.

_______, and Land, L. S., 1975, Mg/Ca ratio and salinity: Two controls over crystallization of dolomite: AAPG Bull., v. 59, p. 60-68.

_______, and Siedlecka, A., 1974, The "schizohaline" environment: Its sedimentary and diagenetic fabrics as exemplified by Late Paleozoic rocks of Bear Island, Svalbard: Sedimentary Geology, v. 11, p. 1-15.

Friedman, G. M., 1964, Early diagenesis and lithification in carbonate sediments: Journal of Sedimentary Petrology, v. 34, p. 777-813.

Hanor, J. S., and Marshall, N. F., 1971, Mixing of sediment by organisms, *in* Perkins, B. F., ed., Trace fossils: A field guide to selected localities in Pennsylvanian, Permian, Cretaceous, and Tertiary rocks of Texas: Baton Rouge, Louisiana State Univ., School of Geosciences, Miscellaneous Pub. 71-1, p. 127-135.

Hanshaw, B. B., Back, W., and Deike, R. G., 1971, A geochemical hypothesis for dolomitization by ground water: Economic Geology, v. 66, p. 710-724.

Inden, R. F., 1972, Paleogeography, diagenesis, and paleohydrology of a Trinity Cretaceous carbonate beach sequence, Central Texas: Louisiana State Univ., Ph.D. dissertation, 264 p.

James, N., 1972, Holocene and Pleistocene calcareous crust (caliche) profiles: Criteria for subaerial exposure: Journal of Sedimentary Petrology, v. 42, p. 817-836.

Land, L. S., 1967, Diagenesis of skeletal carbonates: Journal of Sedimentary Petrology, v. 37, p. 914-930.

_______ 1970, Phreatic versus vadose meteoric diagenesis of limestones: Evidence from a fossil water table: Sedimentology, v. 14, p. 175-185.

_______ 1973a, Contemporaneous dolomitization of Middle Pleistocene reefs by meteoric water, North Jamaica: Bull. Marine Science, v. 23, p. 64-92.

_______ 1973b, Holocene meteoric dolomitization of Pleistocene limestones, North Jamaica: Sedimentology, v. 20, p. 411-424.

_______, Mackenzie, F. T., and Gould, S. J., 1967, Pleistocene history of Bermuda: GSA Bull., v. 78, p. 993-1006.

_______, Salem, M. R. I., and Morrow, D. W., 1975, Paleohydrology of ancient dolomites: Geochemical evidence: AAPG Bull., v. 59, p. 1602-1625.

Loucks, R. G., 1976, Pearsall Formation, Lower Cretaceous, South Texas: Depositional facies and carbonate diagenesis and their relationship to porosity: Univ. Texas, Austin, Ph.D. dissertation, 362 p.

_______ 1977, Porosity development and distribution in shoal-water carbonate complexes—subsurface Pearsall Formation (Lower Cretaceous), South Texas: this volume.

Milliman, J. D., 1974, Marine carbonates: Berlin, Springer, 375 p.

Moore, C. H., Jr., 1961, Stratigraphy of the Fredericksburg Division, south-central Texas: Univ. Texas, Austin, Ph.D. dissertation, 91 p.

_______ 1964, Stratigraphy of the Fredericksburg Division, south-central Texas: Univ. Texas, Austin, Bureau of Economic Geology Report of Investigations 52, 48 p.

_______, Smitherman, J. M., and Allen, S. H., 1972, Pore system evolution in a Cretaceous carbonate beach sequence, *in* Stratigraphy and sedimentology: International Geological Congress Proceedings, section 6, no. 24, p. 124-136.

Mueller, H. W., III, 1975, Centrifugal progradation of carbonate banks: A model for deposition and early diagenesis, Ft. Terrett Formation, Edwards Group, Lower Cretaceous, Central Texas: Univ. Texas, Austin, Ph.D. dissertation, 300 p.

Nelson, H. F., 1959, Deposition and alteration of the Edwards Limestone, Central Texas, *in* Lozo, F. E., Nelson, H. F., Young, K., Shelburne, O. B., and Sandidge, J. R., Symposium on Edwards Limestone in Central Texas: Univ. Texas, Austin, Pub. no. 5905, p. 21-95.

Perkins, R. D., and Halsey, S. D., 1971, Geological significance of microboring fungi and algae in Carolina shelf sediments: Journal of Sedimentary Petrology, v. 41, p. 843-853.

Read, J. F., 1974, Calcrete deposits and Quaternary sediments, Edel Province, Shark Bay, western Australia, *in* Logan, B. H., ed., Evolution and diagenesis of Quaternary carbonate sequences, Shark Bay, western Australia: AAPG Memoir 22, p. 250-282.

Rose, P. R., 1972, Edwards Group, surface and subsurface, Central Texas: Univ. Texas, Austin, Bureau of Economic Geology Report of Investigations 74, 198 p.

Shinn, E. A., Ginsburg, R. N., and Lloyd, R. M., 1965, Recent supratidal dolomite from Andros Island, Bahamas, *in* Pray, L. C., and Murray, R. C., eds., Dolomitization and limestone diagenesis: SEPM Special Pub. 13, p. 112-123.

_______, Lloyd, R. M., and Ginsburg, R. N., 1969, Anatomy of a modern tidal flat, Andros Island, Bahamas: Journal of Sedimentary Petrology, v. 39, p. 1202-1228.

Siedlecka, A., 1972a, Length-slow chalcedony and relics of sulfates—Evidences of evaporitic environments in the Upper Carboniferous and Permian beds of Bear Island, Svalbard: Journal of Sedimentary Petrology, v. 42, p. 812-816.

_______ 1972b, Sedimentary and diagenetic fabrics of some Late Paleozoic rocks of Bear Island, Svalbard (abstract): GSA Abstracts with Programs, v. 4, p. 664.

Steinen, R. P., 1974, Phreatic and vadose modifications of Pleistocene limestone: Petrographic observations from subsurface of Barbados, West Indies: AAPG Bull., v. 58, p. 1008-1024.

Tolman, C. F., 1937, Ground water: New York, McGraw-Hill, 593 p.

Tucker, D. R., 1962a, Subsurface Lower Cretaceous stratigraphy, Central Texas: Univ. Texas, Austin, Ph.D. dissertation, 137 p.

_______ 1962b, Central Texas Lower Cretaceous stratigraphy: GCAGS Transactions, v. 12, p. 89-96.

Ward, W. C., Folk, R. L., and Wilson, J. L., 1970, Blackening of eolianite and caliche adjacent to saline lakes, Isla Mujeres, Quintana Roo, Mexico: Journal of Sedimentary Petrology, v. 40, p. 548-555.

POROSITY DISTRIBUTION IN THE STUART CITY TREND, LOWER CRETACEOUS, SOUTH TEXAS

D. G. Bebout,[1] R. A. Schatzinger,[1] and R. G. Loucks[1]

ABSTRACT

The Lower Cretaceous Stuart City Trend is a complex of biogenic reefs, banks, tidal bars, channel fills, and islands that accumulated on a broad carbonate shelf encircling the Gulf of Mexico. A variety of carbonate facies were deposited in environments with a wide range of energy levels along this shelf-margin complex. Only four of these facies, however, have greater than 5 percent porosity and 5 millidarcys permeability—the algae-encrusted miliolid-coral-caprinid packstone, mollusk grainstone, rudist grainstone, and coral-stromatoporoid boundstone. Rudist grainstone is potentially the most consistent in terms of porosity and permeability, thickness, and lateral extent.

Intraparticle, interparticle, and fracture porosity are present in the thick limestone section along the Stuart City shelf margin. Intraparticle porosity, in places reaching 20 percent, is common although permeability in facies with intraparticle porosity is low. Facies with interparticle porosity of greater than 5 percent have good permeability of up to 10 millidarcys. Permeability in any facies may be enhanced by the presence of thin fractures which were noted to be common in several cores.

INTRODUCTION

General Remarks

The Stuart City Trend is a complex of biogenic reefs, banks, tidal bars, channel fills, and islands that was deposited along the basinward edge of a carbonate shelf. The shelf, which completely encircled the ancestral Gulf of Mexico (fig. 1), can be traced in outcrop from eastern Mexico to Central Texas; in the subsurface the Cretaceous shelf is known from the Yucatan Peninsula, southeastern Texas, northern Louisiana, southwestern Mississippi, south Florida, and the Bahamas. A broader geologic understanding of the Stuart City shelf-margin model, an increase in natural gas prices, and improved exploration and drilling technology have spurred a recent resurgence of interest in the productive Stuart City Trend of South Texas.

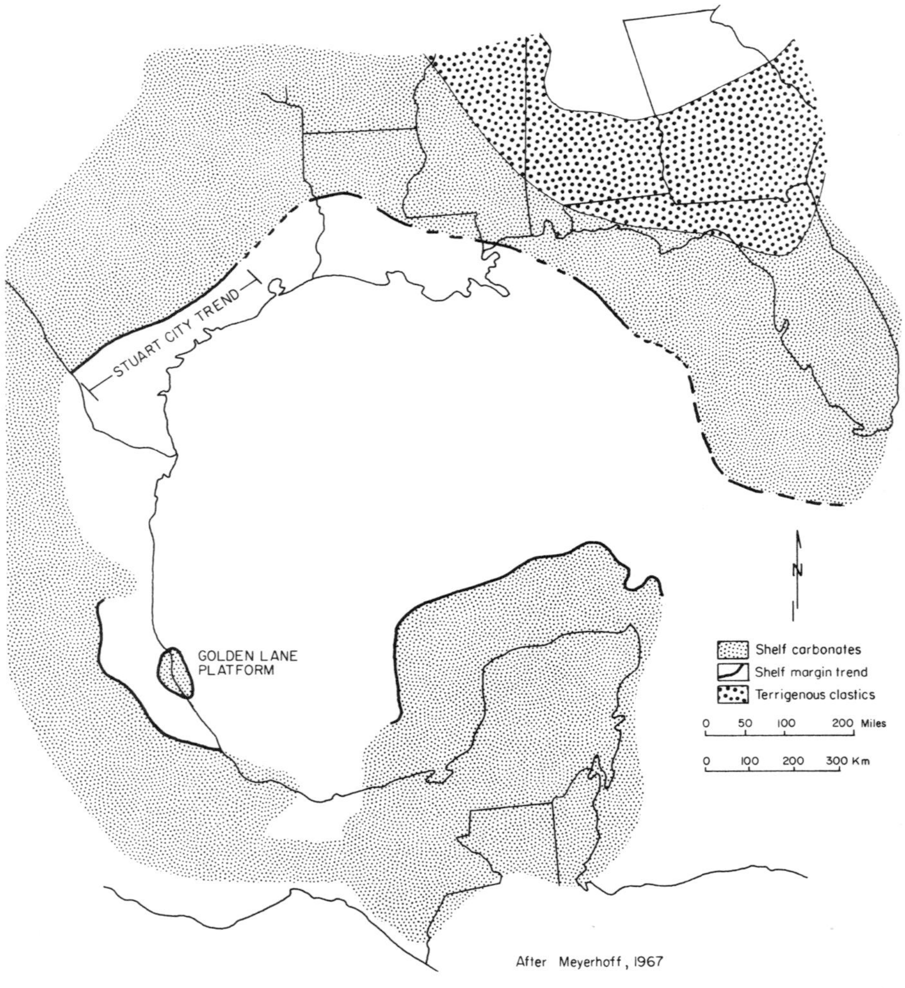

Figure 1. Distribution of Lower Cretaceous shelf margin around the Gulf of Mexico. After Meyerhoff, 1967.

Regional Setting

During Albian time, dark-colored planktonic foraminiferal carbonate muds were deposited in open-marine conditions hundreds of feet deep within the ancestral Gulf of Mexico and were separated from the 400-mile-wide carbonate shelf by shelf-margin sediments of the Stuart City Trend. The shelf-margin complex consisted mainly of rudist-coral and coral-stromatoporoid reefs and banks plus a variety of carbonate sand bodies. Other common organisms found in the shelf-edge boundstones include encrusting algae, the red algae ***Solenopora***, the benthic foraminifers ***Dictyoconus*** and ***Coskinolina***, and boring pelecypods.

The shallow-water shelf behind the shelf-edge barrier has been divided into several paleogeographic provinces (Fisher and Rodda, 1969; Rose, 1972) corresponding to broad depressions or

[1]Bureau of Economic Geology, The University of Texas at Austin.

arches in the Comanche platform (fig. 2). The main positive element, the Central Texas platform - San Marcos arch, trends southeast from near San Angelo to intersect the Stuart City Trend southeast of San Antonio. Kirschberg evaporites were deposited on top of this structure. The Central Texas platform separates the McKnight basin of West Texas and northern Mexico from the North Texas - Tyler Basin.

Objectives

Although the Stuart City Trend has been explored for a number of years, only recently has a carbonate shelf-margin model for hydrocarbon exploration been proposed (Bebout and Loucks, 1974). Bebout and Loucks summarized depositional facies and environments along the Stuart City Trend of South Texas utilizing all cores available at that time. Since then several new cores have become available and have been integrated into the shelf-margin model. The purpose of this study is to integrate the new and old data on porosity development and distribution along the Stuart City Trend in Bee and Karnes Counties (fig. 3). Facies that occur along the Stuart City Trend are described by Bebout and Loucks (1974); therefore, only the porous facies will be discussed here.

The additional insight drawn from porosity distribution as seen in cores from the Stuart City Trend will provide an improved framework for hydrocarbon exploration.

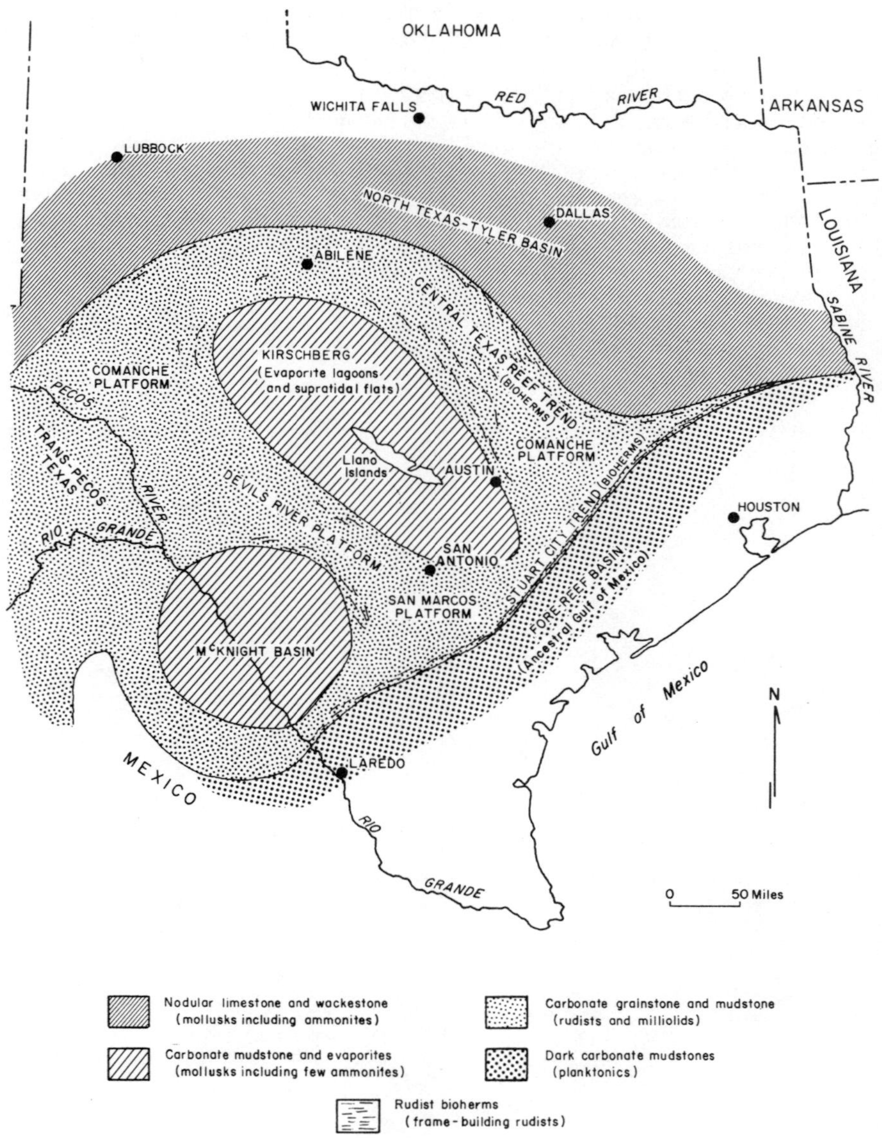

Figure 2. Paleogeography of the Lower Cretaceous of Texas. After Fisher and Rodda, 1969.

FACIES AND DEPOSITIONAL ENVIRONMENTS

General Remarks

This paper is based on a detailed study of 2,600 feet of core from 14 wells along the Stuart City Trend in Bee and Karnes Counties, Texas (fig. 3). A series of ten facies represent shelf-lagoon, stable grainflat, shelf-margin, and upper shelf-slope environments of deposition. In addition, lower shelf-slope and open-marine environments were reported by Bebout and Loucks (1974).

Environments

Environmental interpretations were based on detailed examination of core and relationship between facies. Because boundstones with very coarse fabric are difficult to identify on slabbed core surfaces, the recognition and interpretation of these facies is heavily influenced by vertical relationships with more easily recognized facies. The spatial relationship between facies of the Stuart City Trend and interpreted environments of deposition is shown by figure 4.

Open-marine conditions with water depths probably greater than 60 feet and anaerobic bottom conditions were reported by Bebout and Loucks (1974) for the planktonic foraminiferal wackestone. The planktonic foraminiferal wackestone and black limy mudstone were recorded only from the overlying Georgetown equivalent. Examination of core from the Shell #1 Leppard shows no erosional discontinuity between the light-colored Stuart City Limestone and the dark-colored Georgetown mudstone. The two facies grade one into the other through a 2-foot-thick section.

Landward, on the lower shelf slope, echinoids and mollusks were more abundant. The bottom was generally muddy and water depths probably ranged between 30 and 60 feet. Shallow water between 10 and 50 feet deep and more oxidizing substrate of the upper shelf slope were probably responsible for a greater faunal diversity. Locally, patch reefs or banks of corals and stromatoporoids were isolated from the shelf-margin banks by a muddy bottom (caprinid-coral wackestone). Other coral and stromatoporoid banks fronted the shelf-margin rudist grainstone bars and coral-caprinid reefs.

A series of stable grainflats represented by algae-encrusted miliolid-coral-caprinid packstone was protected on the seaward side by the rudist sand bodies and requieniid banks. Tidal channel debris and organic rubble were not commonly washed seaward over the shelf-margin banks onto the upper shelf slope. Evidently the main effect of storms was to wash shelf-margin bank and reef rubble into the shelf lagoon. Numerous small channels controlled tidal exchange within the mudflats. Major channels perpendicular to the shelf edge probably broke the reef trend into a series of disconnected elongate banks and islands parallel to the shelf edge. On the stable grainflats, thick algal coatings and micritized grain edges commonly formed.

Shallow-water shelf-lagoon and mudflat facies are characterized by low-energy conditions and water depths of 10 to 20 feet. Common facies are miliolid wackestone, mollusk wackestone, toucasid wackestone, and mollusk grainstone. Echinoids are relatively common and inhabited minor depressions in the shelf lagoon. Existence of small islands is evidenced by the occurrence of well-developed bird's-eye structure and palm wood. These islands were surrounded by water only a few feet deep represented by miliolid wackestone and, on the windward side, mollusk and mollusk-miliolid grainstone bodies.

Facies

Important features of the Stuart City Reef Trend facies are summarized by table 1. Those facies which commonly have 5 millidarcys or more permeability and 5 percent or more

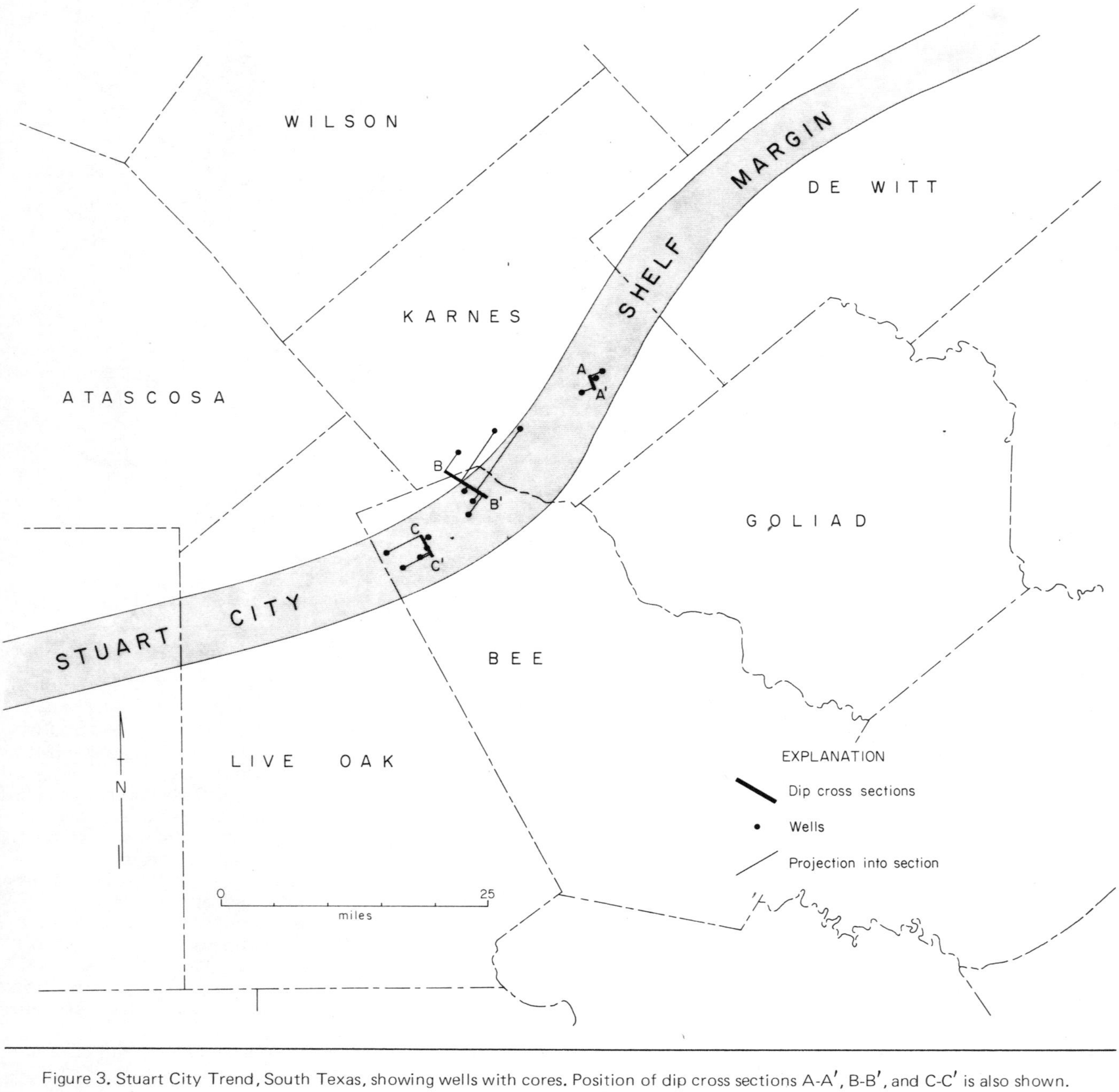

Figure 3. Stuart City Trend, South Texas, showing wells with cores. Position of dip cross sections A-A′, B-B′, and C-C′ is also shown.

Table 1. Properties of Stuart City Reef Trend facies

FACIES	ENVIRONMENT	DOMINANT FOSSILS	THICKNESS
Miliolid wackestone	Shelf-lagoon	miliolids, mollusks, *Dictyoconus*	4- 65 ft.
Mollusk wackestone		mollusks, especially nerinid gastropods	6- 74 ft.
Toucasid wackestone		toucasid rudists, mollusks, miliolids	5 ft.
Mollusk grainstone		mollusks, caprinids, *Dictyoconus,* encrusting algae	2- 45 ft.
Algae-encrusted miliolid-coral-caprinid packstone	Shelf margin stable grainflat	encrusting algae, caprinids, miliolids, *Dictyoconus*	8- 90 ft.
Rudist grainstone	Shelf margin bars, channels	caprinids, radiolitids encrusting algae	2- 70 ft.
Coral-caprinid boundstone	Shelf-margin reefs and banks	corals, caprinids, boring pelecypods	25- 40 ft.
Requieniid boundstone		requieniids, caprinids, green algae	12- 75 ft.
Coral-stromatoporoid boundstone	upper shelf slope	massive and branching corals, stromatoporoids, boring pelecypods	5- 80 ft.
Caprinid-coral wackestone		caprinids, broken corals, echinoids	20- 35 ft.
Echinoid-mollusk wackestone	lower shelf slope	echinoids, mollusks	322-405 ft.
Planktonic foraminifer wackestone	open marine	planktonic foraminifers, ostracodes	650-730 ft.

porosity (algae-encrusted miliolid-coral-caprinid packstone, mollusk grainstone, rudist grainstone, and coral-stromatoporoid boundstone) are considered to be the "porous" facies of the Stuart City Trend and are discussed in greater detail in this report.

Facies Distribution

Three dip sections across the Stuart City Trend (fig. 3) have been constructed to include the new cores (figs. 5-7). Some wells have been projected along strike into the cross sections in order to provide the best control. Porosity and permeability data, where available, have been incorporated into the cross sections.

The general progradational nature of the Stuart City cycle indicated by upward and basinward growth of the reef trend is clearly demonstrated on all cross sections. Correlation of the facies in cross sections A-A′ and B-B′, however, shows a regressive tendency of the uppermost part of the formation. The upward and shelfward building of the youngest reefal facies, the open-marine nature of the overlying Georgetown Limestone equivalent, and the lack of an erosional surface in cores at the top of the reefal section all indicate a transgressive phase which began within the top 100 feet of the Stuart City Limestone. The uppermost rudist grainstone in the Shell #1 Leppard, Bee County, grades upward through a 2-foot-thick bed of miliolid wackestone into 120 feet of black,

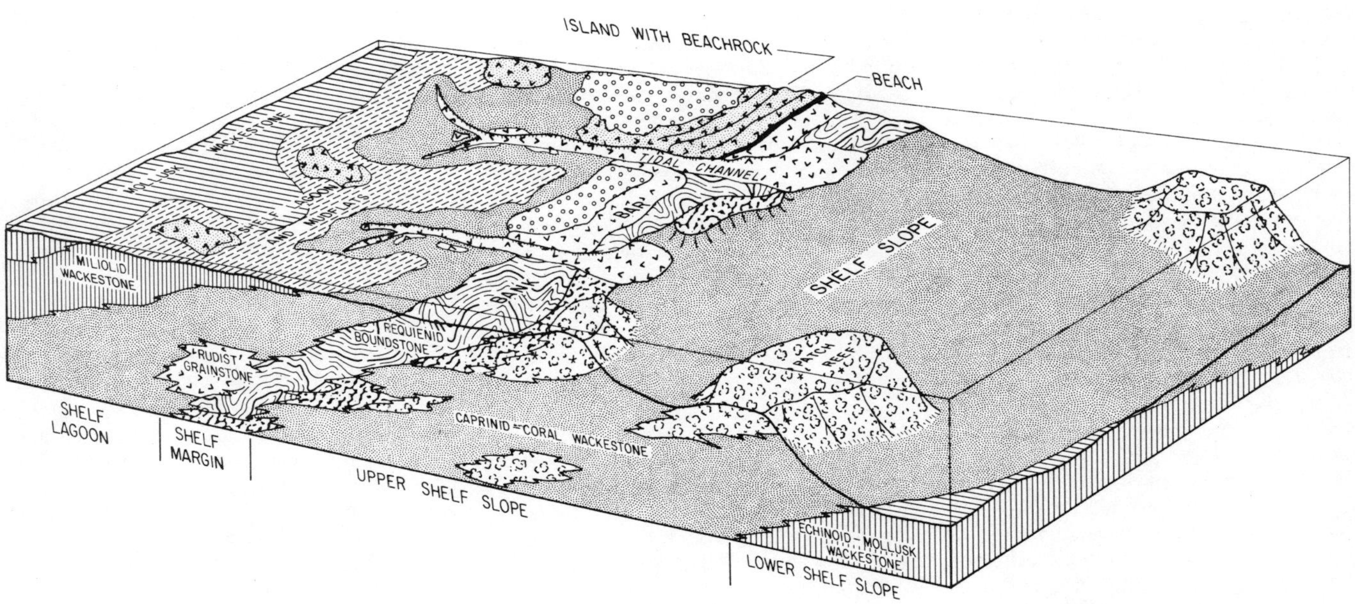

Figure 4. Facies and interpreted depositional environments across the Stuart City Trend, South Texas.

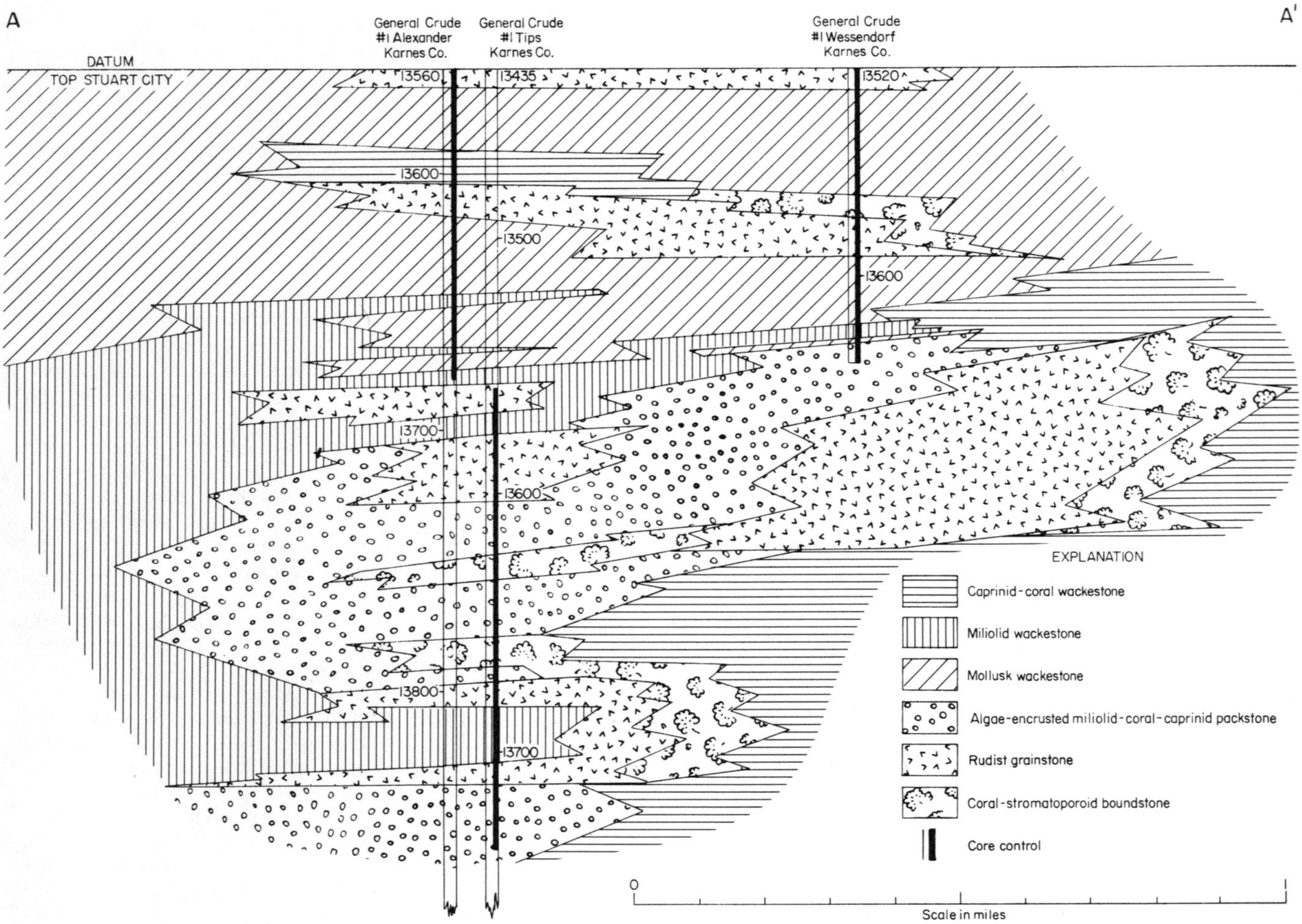

Figure 5. Facies cross section A-A'.

limy mudstone with fragments of ***Inoceramus***. Conformably above the black mudstone is at least 23 feet of planktonic foraminiferal mudstone to wackestone. This sequence is interpreted at the beginning of a transgressive phase in which the reef was "drowned," slope muds with a sparce fauna were then deposited, and, finally, the open-marine environment with its rain of planktonic foraminifers covered the entire area.

Dip section A-A' (fig. 5).—Cross section A-A' shows the projected facies relationships between three General Crude wells—#1 Tips, #1 Alexander, and #1 Wessendorf. These wells are less than a mile from one another.

Most common facies in all three cores from wells on section A-A' are interpreted to have been deposited on the shelf margin and upper shelf slope. Shelf-margin facies include algal-coated miliolid-coral-caprinid packstone and rudist grainstone. Upper shelf-slope facies include coral-stromatoporoid boundstone and caprinid-coral wackestone. Rudist grainstone commonly occurs above and/or basinward of the algal-encrusted miliolid-coral-caprinid packstone. Bitumen cement is common in pores and fractures in the rudist grainstone. The grainstone is locally bound by encrusting algae in the #1 Tips. Hardened surfaces are rare but occur within the rudist grainstone; dripstone cement is present near the base of the lower grainstone body in the #1 Tips. Both features may represent brief periods of exposure along beaches on the lagoonward side of the shelf margin.

The upper shelf-slope coral-stromatoporoid boundstone forms a seaward margin to some of the grainstone bodies. Bebout and Loucks (1974) suggested that the coral-stromatoporoid patch reefs accumulated on the upper shelf slope and were separated from the shelf-margin boundstones by upper shelf-slope muds. It is probable, however, that at least some of the upper shelf-slope patch reefs acted as basinward fringes to other shelf-margin reefs and grainstone bodies.

Cores from the upper parts of the #1 Alexander and #1 Wessendorf contain a correlative rudist grainstone. Bebout and Loucks (1974) suggested that an analogous grainstone at the top of their dip section B-B' may have represented talus from a rudist patch reef within the shelf lagoon.

In general, shelf-lagoon facies overlie shelf-margin facies and become thicker landward. Shelf-lagoon facies include mollusk wackestone which interfingers and commonly overlies the miliolid wackestone. Bird's-eye texture is common in miliolid wackestone but is not seen in the mollusk wackestone.

The most seaward extent of reef growth is basinward of the General Crude #1 Wessendorf in dip-section A-A'. The highest point of reef and bank development probably occurs no more than 1 mile southeast of the Wessendorf well.

Dip-section B-B′ (fig. 6).—Dip-section B-B′ portrays the facies relationships along the shelf edge and shelf lagoon in northwestern Bee and southwestern Karnes Counties. The General Crude #1 Wernli and #1 Strawn Estate have been projected into the section from distances twice that of the other wells but, nevertheless, do fit into the facies tract. Nearly all of the cores from the General Crude #1 Wernli and MGF # 1 Boone represent shelf-lagoon facies including mollusk wackestone, miliolid wackestone, toucasid wackestone, and mollusk grainstone. In contrast, cores from the General Crude #1 Strawn Estate, Shell #1 Ruhmann, and MGF #1 Schultz represent mostly shelf-margin and, to a lesser degree, upper shelf-slope facies. The shelf-margin rudist grainstone, algal-coated miliolid-coral-caprinid packstone, and requieniid boundstone facies are bound below and basinward by the upper shelf-slope caprinid-coral wackestone. Some caprinid-coral wackestone is interpreted to have been on the shelfward side of the shelf-margin requieniid boundstone and rudist grainstone. Such caprinid-coral muds locally separated shelf-lagoon muds from the shelf-margin banks and sand bodies. On the basis of the vertical relationship of the shelf-margin facies in the Shell #1 Ruhmann and General Crude #1 Strawn Estate, requieniid boundstone is interpreted to occur beneath and basinward of the rudist grainstone in the MGF #1 Schultz. The rapid upward and basinward growth and complex facies relationship between shelf-margin and shelf-lagoon facies is well displayed in the MGF #1 Schultz.

Shelf-lagoon mollusk grainstone in the Shell #1 Ruhmann lies directly landward of upper shelf-slope caprinid-coral wackestone in the upper part of the MFG #1 Schultz. Such a juxtaposition could represent the shelf-lagoon/upper shelf-slope transition when it occurs between discontinuous reefs and banks of the shelf margin.

Cross section C-C′ (fig. 7).—Cores from five closely spaced wells, the Shell #1 Leppard, #1 Roessler, #1 Tomasek, #1 O'Neal, and #1 Currer, arranged successively basinward, illustrate shelf-margin facies relationships in northern Bee County.

Upper shelf-slope coral-stromatoporoid boundstone is common in the lower half of the Currer, Tomasek, and Roessler. In several occurrences the coral-stromatoporoid boundstone is located directly adjacent to either rudist grainstone (Tomasek) or requieniid boundstone (Currer and O'Neal).

The lower half of the Tomasek core consists of upper shelf-slope facies which include caprinid-coral wackestone and some coral-stromatoporoid boundstone. Lower shelf-slope facies were not penetrated. Above the caprinid-coral wackestone, thick coral-caprinid boundstone and requieniid boundstone mark the shelf-margin facies.

Shelf-lagoon facies represented by mollusk wackestone and miliolid wackestone are present in the Leppard well. The restriction of shelf-lagoon facies to the top 120 feet of core from a single well in the cross section emphasizes the dominant shelf margin setting of cross section C-C′.

POROUS FACIES

Porosity and permeability data for Stuart City Reef Trend facies are compared in table 2. Porous facies include mollusk grainstone, algae-encrusted miliolid-coral-caprinid packstone, rudist grainstone, and coral-stromatoporoid boundstone. Permeabilities greater than 5 millidarcys in miliolid wackestone, mollusk wackestone, and toucasid wackestone are consistently related to fractures. Where fractures are lacking from the wackestone, permeability is less than 1 millidarcy. Porosity in the several wackestone facies is generally less than 5 percent; primary intraparticle porosity dominates, although some interparticle and moldic porosity exists.

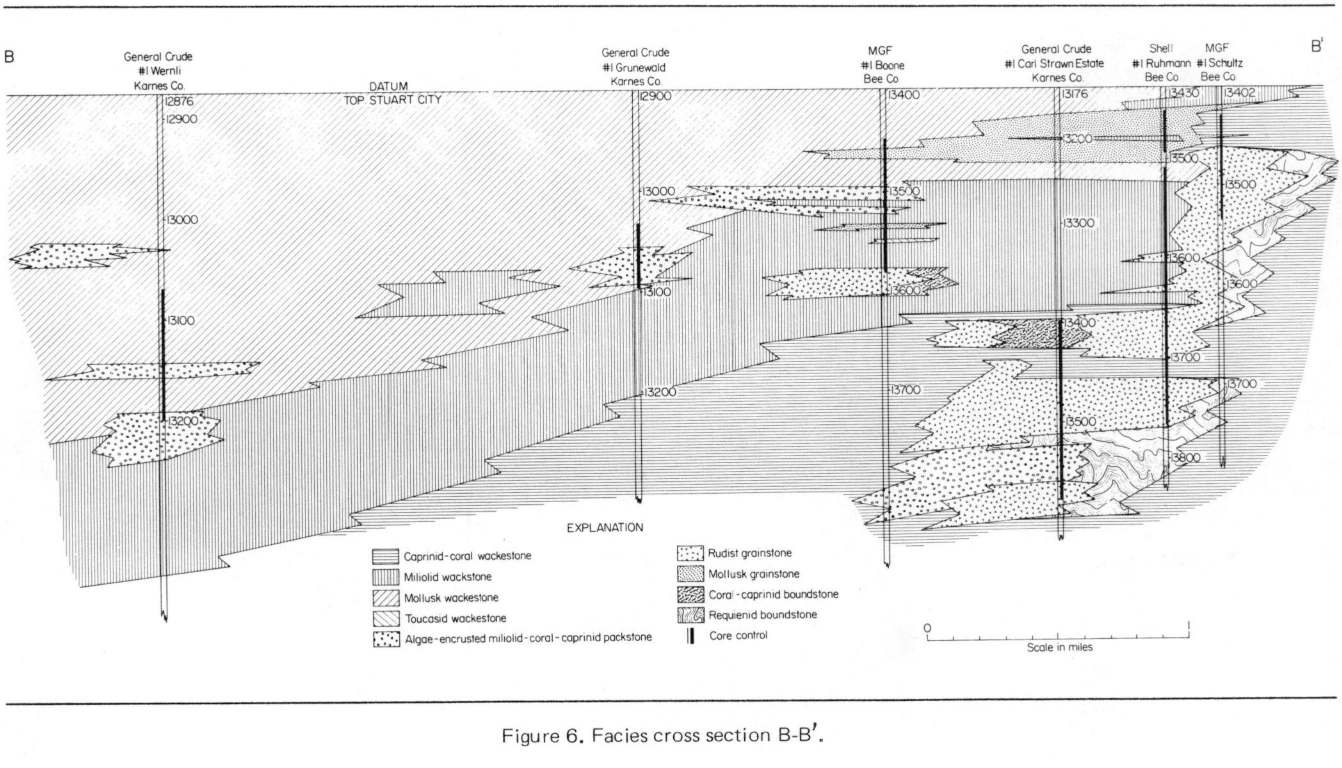

Figure 6. Facies cross section B-B′.

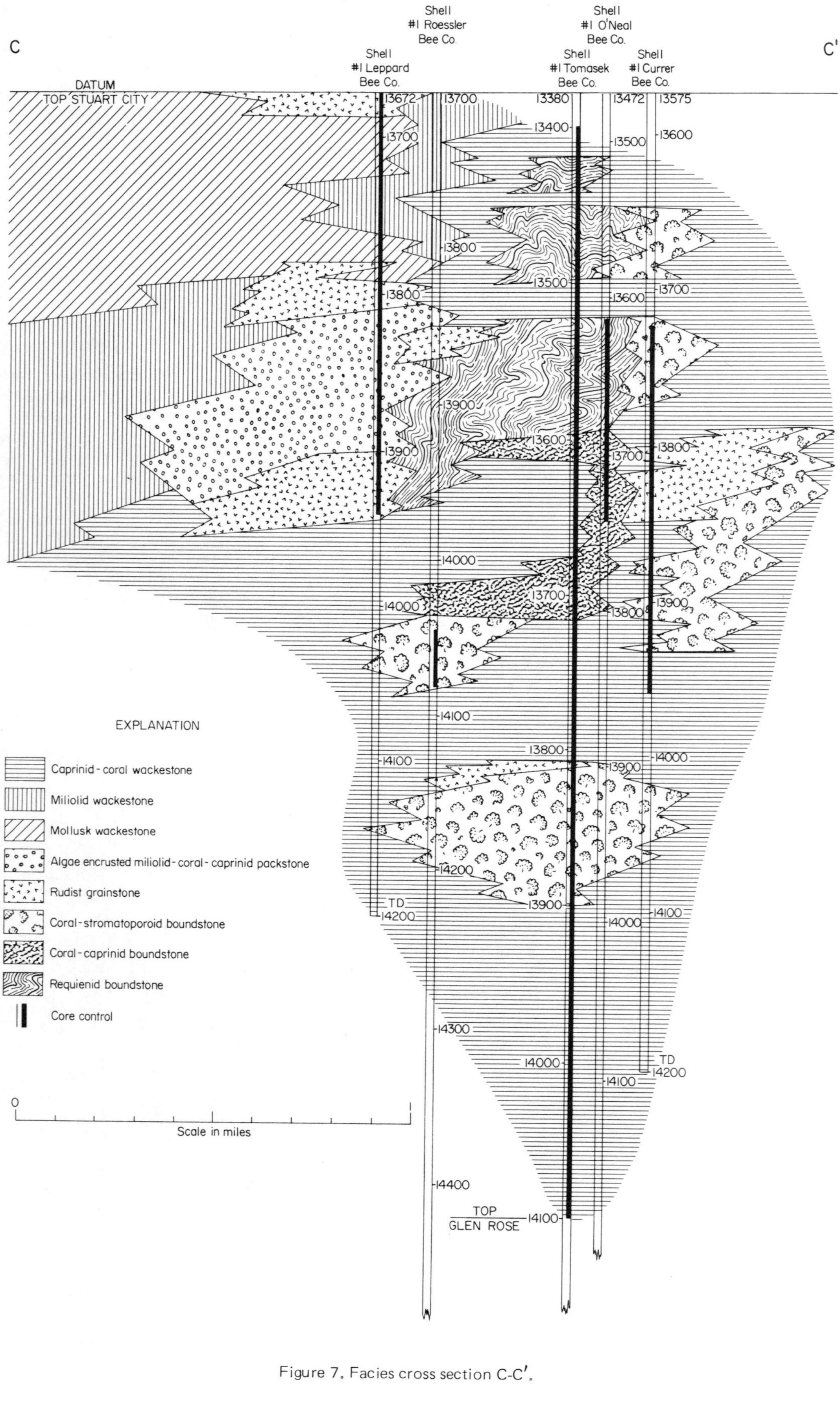

Figure 7. Facies cross section C-C'.

Figure 8. (a, b) Mollusk grainstone. Well-developed interparticle and intraparticle porosity.
(c, d) Algae-coated miliolid-coral-caprinid packstone. Algae-coated and bored requieniid shell fragments (c) and palm wood at top of core sample (d). Core slabs, xl.

Table 2. Distribution of porosity within the Stuart City Reef Trend. Symbols for type of porosity are from Choquette and Pray (1970).

FACIES	TYPE OF POROSITY	MAXIMUM POROSITY AND PERMEABILITY	COMMENTS
Miliolid wackestone	Mo, crWP	5% porosity 12 md.	high permeability only along fracture
Mollusk wackestone	FR, crWP	10% porosity 5 md	940 md in fracture Gen. Crude #1 Wernli
Toucasid wackestone	FR	4% porosity 5 md	permeability due to fracture
Mollusk grainstone	Mo, BP, WP	10% porosity 20 md	locally up to 30% intraclasts
Algae-encrusted mili-olid-coral-caprinid packstone	sxBP, Fr, V	5% porosity 15 md	
Rudist grainstone	MO, crBP, crWP	20% porosity 5 md	bitumen cement common
Coral-caprinid boundstone	crBP, WP	10% porosity 1 md	
Requieniid bound-stone	WP	5% porosity 1 md	locally up to 60% pellets
Coral-stromatopo-roid boundstone	crBP, crWP crV, GF	12% porosity 5 md	patch reefs some-times fronting grainstone belts
Caprinid-coral wackestone	WP	10% porosity 1 md	fabric commonly includes packstone

Description of Porous Facies

Mollusk grainstone (fig. 8a, 8b).

Lithology: Limestone

Color: Light to medium brown; black when oil stained

Texture: Grainstone, locally packstone. Coarse calcarenite to calcirudite layers are mainly composed of mollusk fragments, but as much as 90 percent of some layers consist of pellets (for example, Shell #1 Ruhmann). Fine-grained allochems are commonly algae bound. Locally, up to 80 percent of the grains are leached. Micritized grain edges are common. Bitumen is occasionally found in pores.

Fossils: Dominantly mollusks; occasionally corals, miliolids, *Dictyoconus* and *Solenopora*

Structures: Stylolites (up to 16 mm amplitude) and horizontal, even laminations; vertical fractures are occasionally present. Beds with poor sorting frequently consist of numerous individually well-sorted laminations.

Porosity: Up to 10 percent combined moldic, interparticle, and intraparticle porosity

Distribution: MGF Schultz and Boone; General Crude Wessendorf; Shell Ruhmann

Depositional Environment: Shelf lagoon. Moderate- to high-energy environments fringing islands and in tidal channels, with water depths up to 10 feet

Algae-encrusted miliolid-coral-caprinid packstone (figs. 8c, 8d, 9a, 9b).

Lithology: Limestone

Color: Gray-brown to black

Texture: Packstone, locally wackestone, grainstone, or boundstone. Up to 70 percent of the grains are coated; intraclasts locally constitute 10 percent of the rock and pellets range up to 30 percent. Heavy algal encrustation is diagnostic of the facies. Larger allochems tend to have thicker coatings. All miliolids, some requieniids, and some stromatoporoids are not coated. Algal oncolites are common. Occasionally bitumen is present in fractures; sulfur and galena are rare.

Fossils: Dominantly caprinids, miliolids, corals, and encrusting algae. Also present are *Dictyo-*

conus, requieniids, hippuritids, mollusks, stromatoporoids, green algae, and bryozoans. Encrusting blue-green algae occur as oncolites, coated grains, and discrete algal heads. Blue-green algal filaments are also complexly interwoven with grains and probably provided a stable substrate.

Structures: The most common structures include stylolites (up to 20 mm amplitude), bird's-eye, and geopetals. Also present are vugs, burrows, fractures, and borings in shells—especially in toucasids.

Porosity: Leached interparticle porosity up to 5 percent is the most important type of porosity (for example, General Crude Strawn Estate); also present are fracture, vug, and intraparticle porosity types. Solution enlargement of intraparticle porosity, possibly after a stage of cementation, and local solution enlargement that connected areas of vug and intraparticle porosity resulted in up to 15 millidarcys permeability.

Distribution: Shell Leppard; MGF Boone; General Crude Wessendorf, Tips, Strawn Estate, and Wernli

Depositional Environment: Shelf-margin stabilized grainflats. Thick algal encrustation, abundant algal growth as heads, and binding networks of filaments intimately mixed with the skeletal grains indicate that the substrate lay in shallow, warm, gently agitated water 1 to 5 feet deep. Carbonate beaches, bars, and spits represented by the rudist grainstone probably acted as a wave and wind baffle for the algae-stabilized flats along the seaward edge of the shelf margin. Bimodal sorting of the algae-encrusted miliolid-coral-caprinid packstone suggests that storms played an important role in mixing of the sediments (Bebout and Loucks, 1974).

Rudist grainstone (figs. 9c, 9d, 10).

Lithology: Limestone; rarely argillaceous

Color: Light brown to gray; black when oil stained

Texture: Dominantly grainstone, locally packstone and boundstone. Grains are mostly skeletal; locally 30 percent are intraclasts and up to 40 percent are lithoclasts (for example General Crude Strawn Estate); generally less than 30 percent of the grains have an algal coating. Gypsum cement is present in the uppermost grainstone from the Shell Leppard.

Fossils: Mainly caprinids, radiolitids, requieniids, and ***Dictyoconus***; also present are hippuritids, unidentified rudists, ***Solenopora***, corals, and encrusting algae. Encrusting algae are present only as grain coatings and never significantly bind the sediment.

Structures: Parallel laminae, fining upward, coarsening upward, and bimodal poorly sorted sequences. Fractures are cemented with sparry calcite, bitumen, or both. Other structures include geopetals (including an inverted geopetal in the Shell Currer), crossbedding, stylolites (up to 3 mm amplitude), and rare truncated surfaces.

Porosity: Solution-enlarged interparticle and moldic porosity are the most significant. Larger pore spaces tend to consist of cement-reduced vugs or cement-reduced intraparticle porosity. The most common type of porosity is cement-reduced intraparticle porosity, particularly within caprinid body cavities.

Distribution: Shell Currer, Tomasek, Roessler, Ruhmann, Leppard; MGF Schultz, Boone; General Crude Strawn Estate, Wessendorf, Tips, and Alexander

Depositional Environment: Shelf-margin grain belt. Vertical changes in grain size, sorting, grading, lamination, and bedding indicate that the rudist grainstone bodies accumulated as beaches, offshore bars, spits, and tidal channels in water depths ranging from inches to 10 feet (Bebout and Loucks, 1974).

Coral-stromatoporoid boundstone (fig. 11).

Lithology: Limestone

Color: Light brown to gray

Texture: Boundstone

Fossils: Dominantly branching and massive corals and stromatoporoids. Also present are caprinids, encrusting algae, ***Dictyoconus, Orbitolina***, encrusting foraminifers, boring clams, worm tubes, bryozoans, codacean and rare dasycladacean green algae, and palm wood. Stromatoporoid-encrusted wood is present in the Shell Currer.

Structures: Geopetals, graded beds (General Crude Tips), stylolites, and minor brecciation

Porosity: Cement-reduced interparticle porosity combined with cement-reduced vug porosity have associated permeability of up to 5 millidarcys. Other types of porosity that are present with lower permeability include leached framework and cement-reduced intraparticle porosity.

Distribution: Shell Currer, Tomasek, and Roessler; General Crude Tips

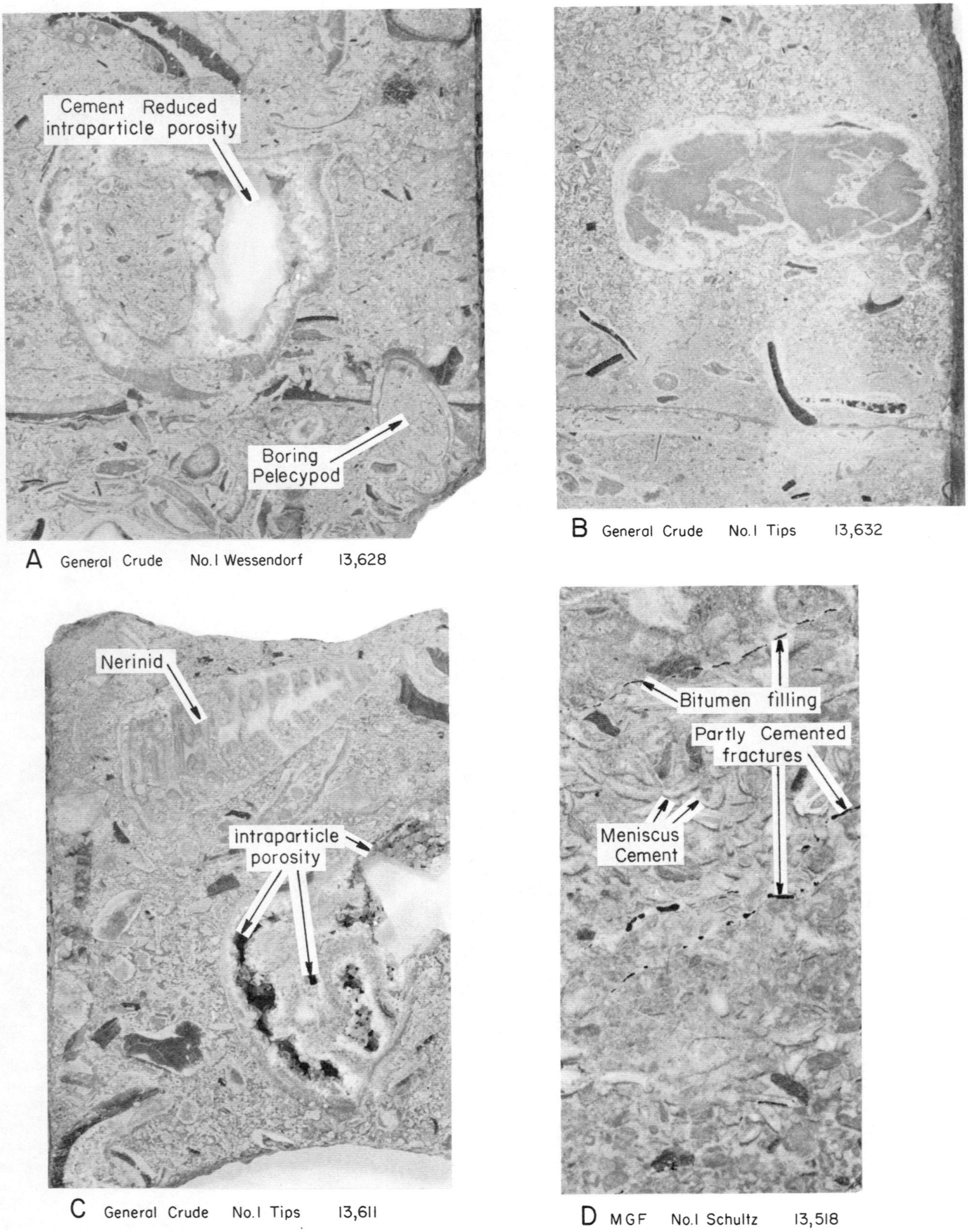

A General Crude No. I Wessendorf 13,628

B General Crude No. I Tips 13,632

C General Crude No. I Tips 13,611

D MGF No. I Schultz 13,518

Figure 9. (a, b) Algae-encrusted miliolid-coral-caprinid packstone. Cement-reduced intraparticle porosity within caprinid (a) and algae-encrusted stromatoporoid and requieniid fragments (b).
(c, d) Rudist grainstone. Large calcite rhombs and bitumen-reducing intraparticle porosity (c) and coarsening-upward sequence with meniscus cement; bitumen and sparry calcite cement in fractures (d). Core slabs, xl.

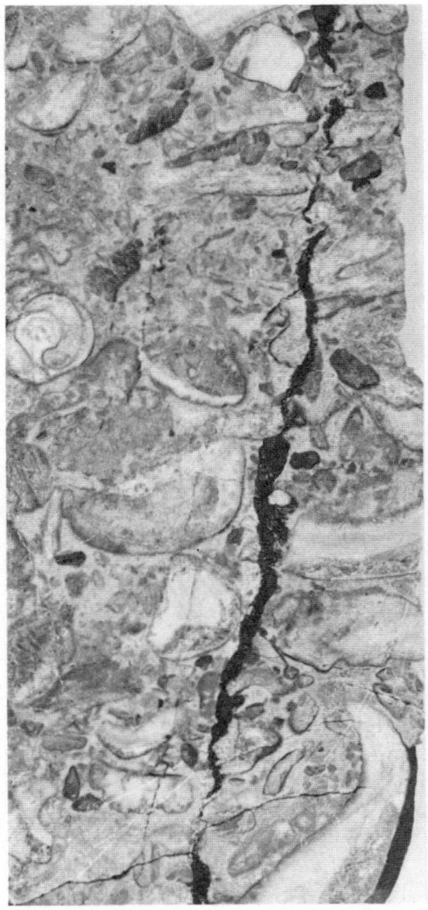

A MGF No.1 Schultz 13,468

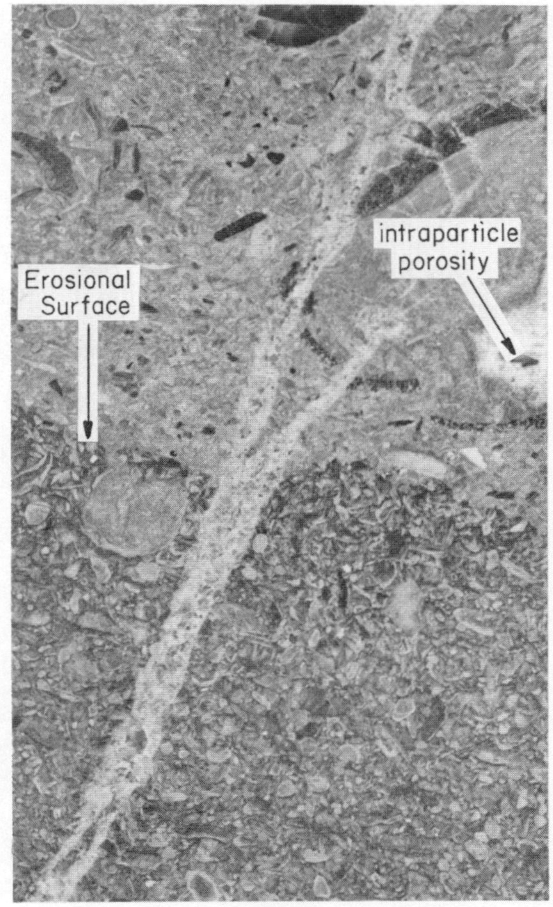

B General Crude No.1 Tips 13,677

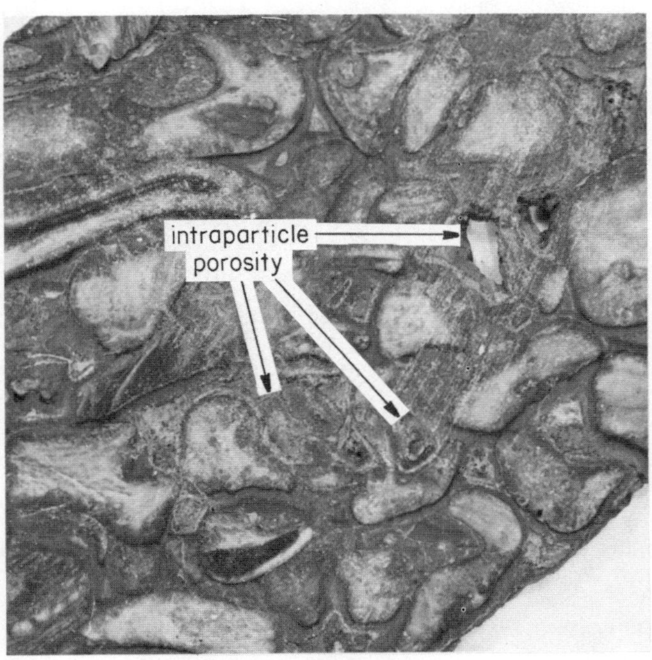

C SHELL No.1 Leppard 13,915

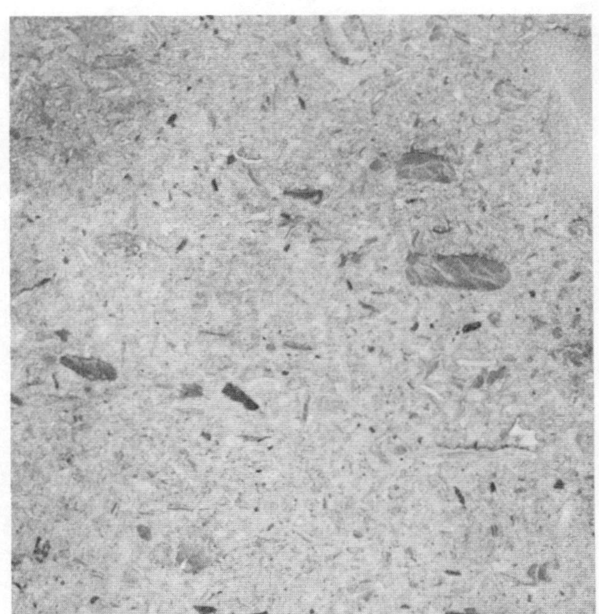

D General Crude No.1 Strawn Est. 13,576

Figure 10. Rudist grainstone. (a) Bitumen-cemented fracture in coarse grainstone. (b) Erosional surface cut by sparry calcite-cemented fracture. (c) Very coarse grainstone with meniscus cement and well-developed interparticle porosity. (d) Grainstone with fine interparticle and intraparticle porosity. Core slabs, x1.

Depositional Environment: Upper shelf-slope patch reefs. The patch reefs grew in water 10 to 30 feet deep and were separated from the shelf-edge grain belt by upper shelf-slope mud or occurred as barriers immediately basinward of the grain belt.

Porosity and Diagenesis

Approximately 40 percent porosity was present in the grainstone bodies at the time of deposition. Less than 10 percent is present today in the same grainstones. The remaining porosity results largely from incomplete cementation.

Micritization of grain rims occurred at the time of deposition while grains were still unconsolidated on the sea floor. A thin layer of isopachous cement covers micritized rims and occasionally occurs as alternating layers with geopetal cavity fill containing marine organisms. Meniscus cement occurs locally in rudist grainstone and mollusk grainstone facies. Meniscus cement also forms over micrite rims or isopachous cement.

Three variations in the cementation sequence occur after the precipitation of meniscus and isopachous cements. Each variation is responsible for the reduction of intraparticle, interparticle, and moldic porosity (fig. 12).

(1) Within the body cavities of organisms (usually rudists), a thick layer of radiaxial cement was deposited. The thick radiaxial cement is followed by equant sparry calcite which commonly filled the rest of the intraparticle space. Where the equant sparry cement did not fill the entire intraparticle space, cement-reduced primary intraparticle porosity remains (figs. 13a, 22).

(2) Interparticle pore spaces usually have the following diagenetic sequence. Thick radiaxial cement was deposited on preceding isopachous or meniscus cements (Bebout and Loucks, 1974) between grains, within fossil body cavities, and within shell walls. Coarse sparry calcite then filled all remaining pore spaces. In the General Crude #1 Tips, neomorphic radiaxial "cement" from within the shell wall of a caprinid contains original shell structure (fig. 13b); however, most neomorphic radiaxial "cement" within shell walls lacks original structure. Occasionally oil emplacement and bitumen formation followed isopachous cementation (fig. 14a). Bladed to rhombic cement only partly filled the remaining pore spaces. Interparticle pore spaces were nearly always filled by radiaxial cement, but when bitumen is present between grains, radiaxial cement is absent and cement-reduced interparticle porosity commonly remains. Interparticle porosity is uncommon where radiaxial cement is present (fig. 10c).

(3) A stage of partial to complete dissolution of grains followed the precipitation of isopachous cement and/or meniscus cement. Bladed to equant calcite lining of pores followed dissolution of the grains. Larger rhombic calcite, with bases which originated on the bladed inner rim, projects strongly into the moldic pore space (fig. 14b). A thin layer of bitumen coats the inner edge of the bladed rim and rhombic calcite. Some rhombic calcite shows evidence of corrosion, possibly by the oil (fig. 14c). A few rhombic calcite crystals projected through the bitumen coating and acted as nucleation sites for large rhombic calcite that partly filled the remaining pore spaces (figs. 14, 15a, 15b).

Distribution of Porosity

Porosity and permeability are facies controlled (figs. 16, 17). Values of greater than 5 percent porosity and 5 millidarcys permeability are limited to the four "porous" facies identified earlier in this paper—the rudist grainstone, algal-encrusted miliolid-coral-caprinid packstone, coral-stromatoporoid boundstone, and mollusk grainstone facies—which form the primarily upward- and basinward-migrating shelf-margin complex. Cross section A-A′ (fig. 16), located almost directly on the shelf margin, shows this trend of porous facies particularly well. In cross section B-B′ (fig. 17), porosity is present in the algae-encrusted miliolid-coral-caprinid packstone even landward of the shelf edge in the shelf lagoon.

Porous facies do not necessarily have permeability (figs. 18-21). Even though intraparticle porosity is the most widespread throughout the reef trend (table 2) and may range as high as 20 percent (fig. 22), associated permeability is lower than 1 millidarcy. For example, intraparticle porosity ranges up to 12 percent in rudist grainstone of the General Crude #1 Tips and permeability is less than 1 millidarcy.

Where interparticle porosity of greater than 5 percent occurs, a much better chance exists for development of permeability (figs. 19-21). High permeability not associated with greater than 5 percent porosity is produced by fractures such as in the General Crude #1 Wernli where several hundred millidarcys permeability is the result of local fractures. Although most natural fractures have only a few tens of millidarcys permeability, they may act as the only effective gas-collecting network in the Stuart City Trend. Artificial fracturing of porous but impermeable sections, such as in the General Crude #1 Tips (between 13,580 and 13,678 feet), could connect intraparticle porosity and create a significant gas-producing zone.

Thin zones with greater than 5 millidarcys permeability and 5 percent solution-enlarged (leached) interparticle porosity occur in mollusk grainstone and algal-coated miliolid-coral-caprinid packstone and rudist grainstone (figs. 20, 21). Beds with greater than 5 millidarcys permeability and 5 percent interparticle porosity possess the greatest potential for gas production.

The potentially productive facies can be ranked in terms of porosity and permeability, thickness, and extent of the reservoir section. The rudist grainstone ranks the highest. Coral-stromatoporoid boundstone with up to 12 percent vug porosity could be important reservoir facies if artificially fractured. Good permeability in the mollusk grainstone is countered by its thin, relatively rare occurrence. Mollusk grainstone ranks lower than the coral-stromatoporoid boundstone only because of the low chance of penetrating mollusk grainstone. Permeable zones in the algae-coated miliolid-coral-caprinid packstone tend to be thin and are probably discontinuous, but artificial fracturing could boost its potential to near that of the coral-stromatoporoid boundstone.

ACKNOWLEDGMENTS

We would like to express our appreciation to R. L. Nichols (Shell Oil Company), C. W. Dye (General Crude Oil Company), and A. Reid (MGF Oil Corporation) for their cooperation in obtaining cores and core analyses.

Figure 11. Coral-stromatoporoid boundstone. (a) Boring pelecypod in massive stromatoporoid. (b) Blue-green algal heads beneath caprinid with geopetal structure in the body cavity. (c) Reef debris trapped in coral growth framework. (d) Large stromatoporoid (top) and abundant binding blue-green algae. Core slabs, xl.

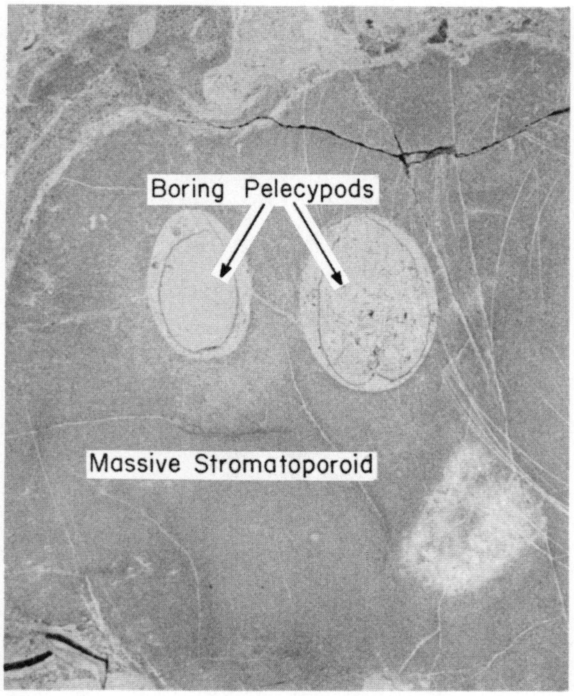

A General Crude No.1 Tips 13,627

B SHELL No.1 Currer 13,765

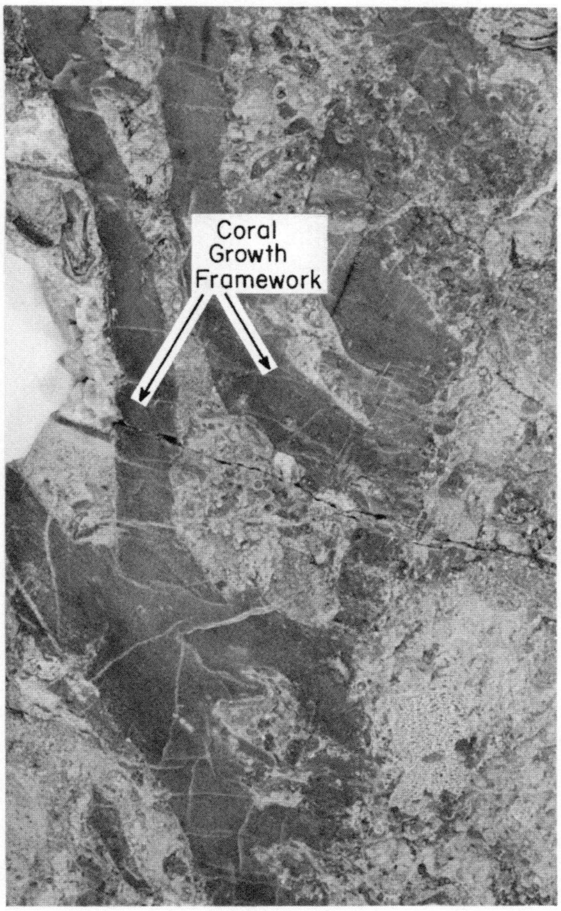

C SHELL No.1 Currer 13,737

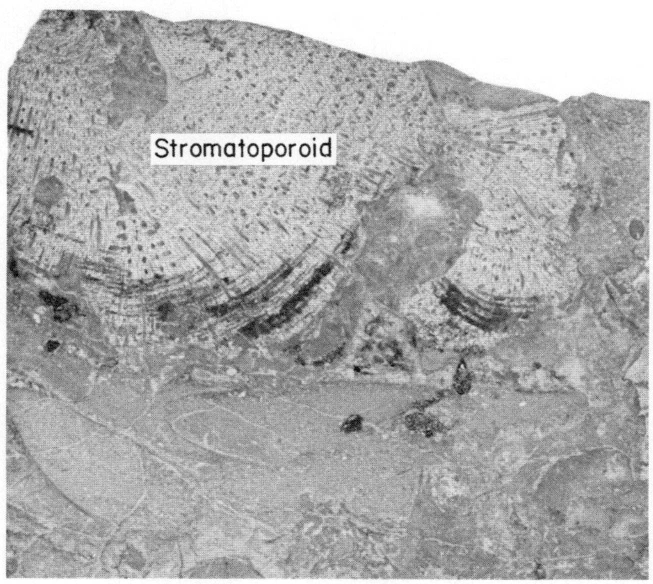

D SHELL No.1 Currer 13,728

Figure 11

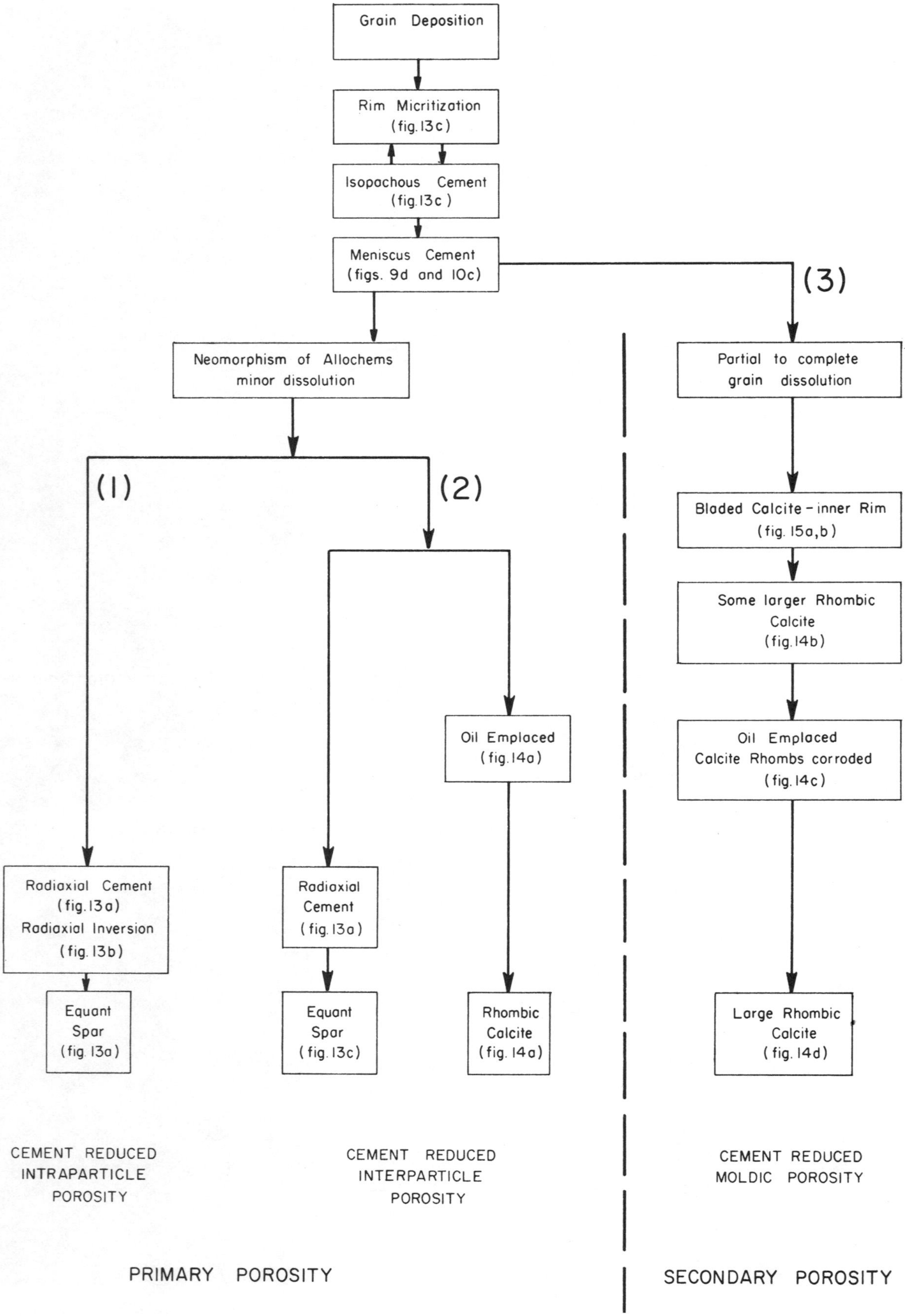

Figure 12. Diagenetic pathways leading to the development of cement-reduced intraparticle, interparticle, and moldic porosity.

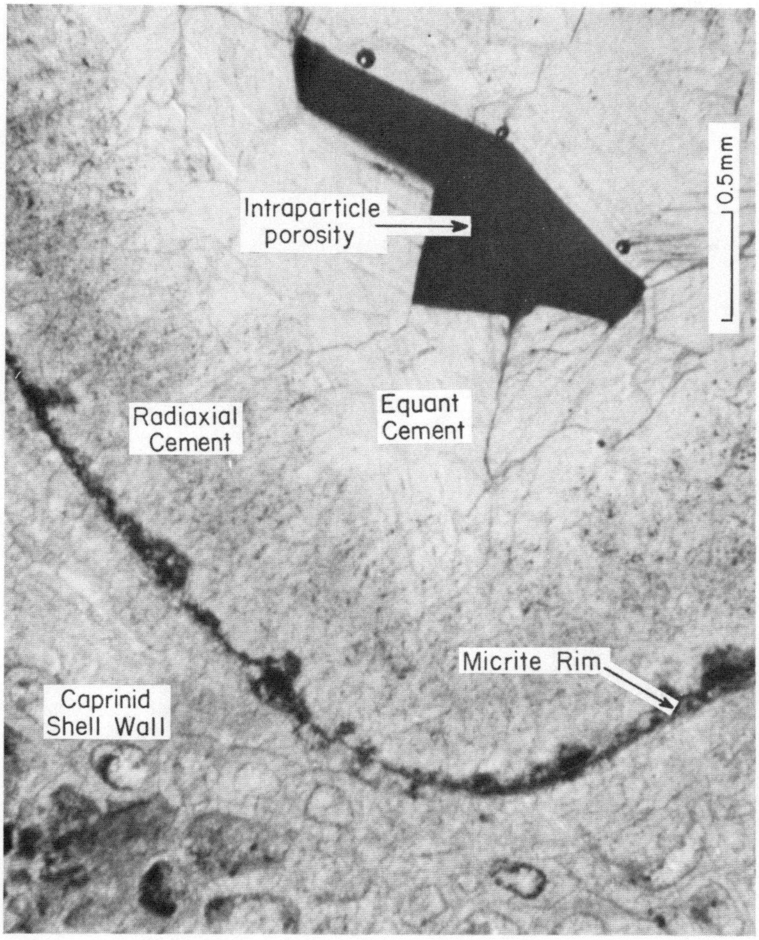

A General Crude No.1 Strawn Est. 13,483

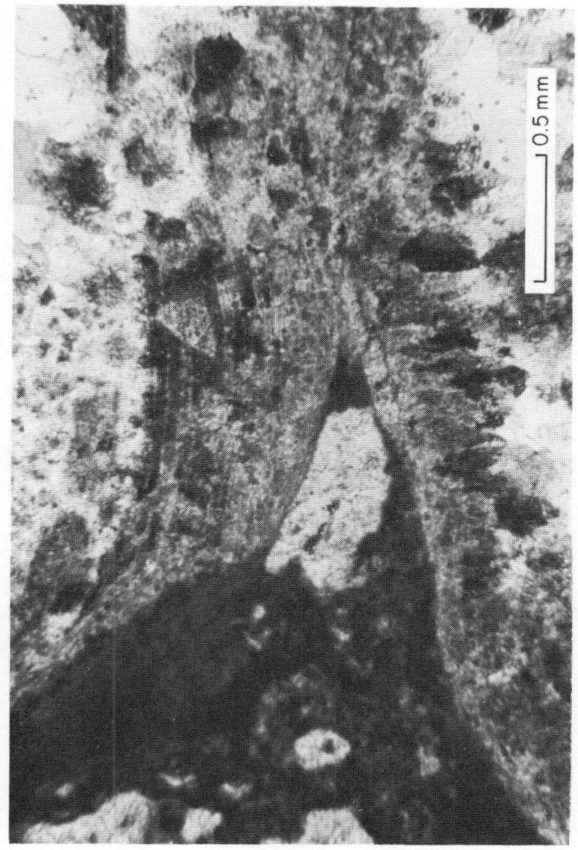

B General Crude No.1 Tips 13,623

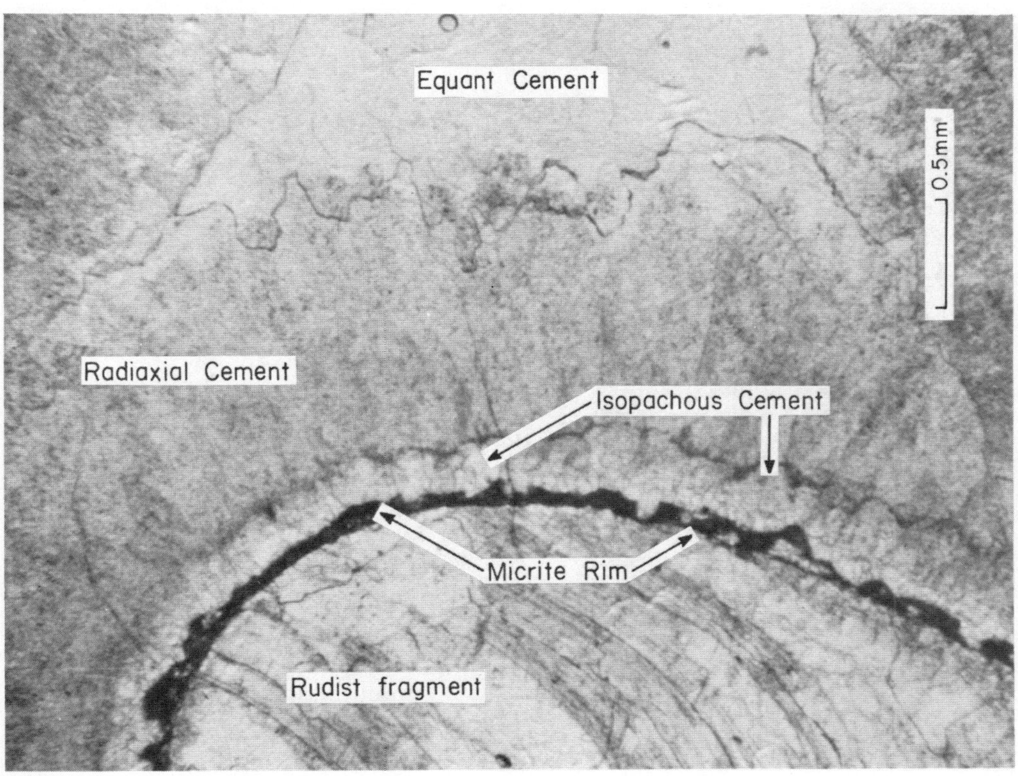

C SHELL No.1 Leppard 13,924

Figure 13

Figure 13. (a) Intraparticle porosity within a caprinid body cavity. (b) Neomorphic radiaxial "cement" in a caprinid shell wall. Growth lines from the original shell are visible in the radiaxial "cement." (c) Interparticle space completely filled by isopachous, radiaxial, and coarse equant calcite cement.

(page 250)

Figure 14. (a) Isopachous cement followed by bitumen. (b) Moldic porosity showing inner lining of bladed to equant calcite with rare, larger rhombic calcite. A thin rim of bitumen postdates this inner calcite lining. (c) Corroded rhombic calcite in moldic pore space. (d) Rhombic calcite that grew after pore space was lined with bitumen.

(page 251)

Figure 15. (a, b) Moldic porosity partly filled by inner rim of bladed to equant spar and larger rhombic calcite. (c, d) Beachrock intraclast with former square-ended aragonite crystals.

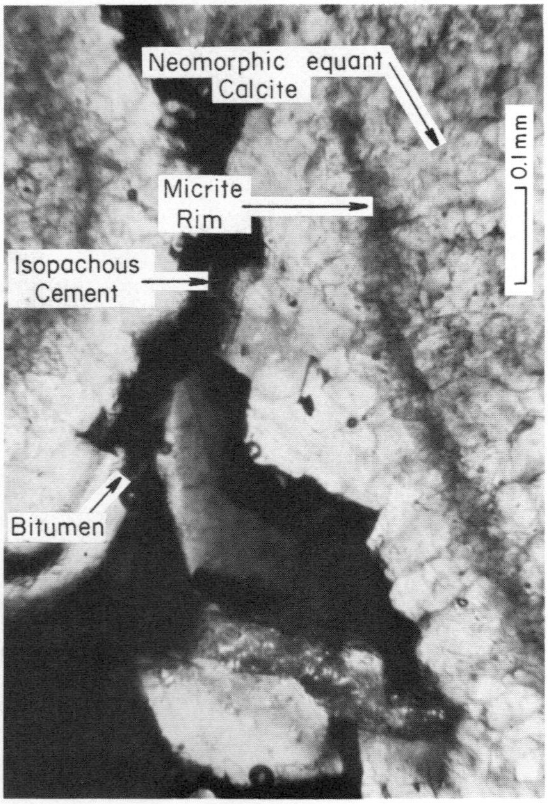

A MGF Oil Corp. No.1 Boone 13,465

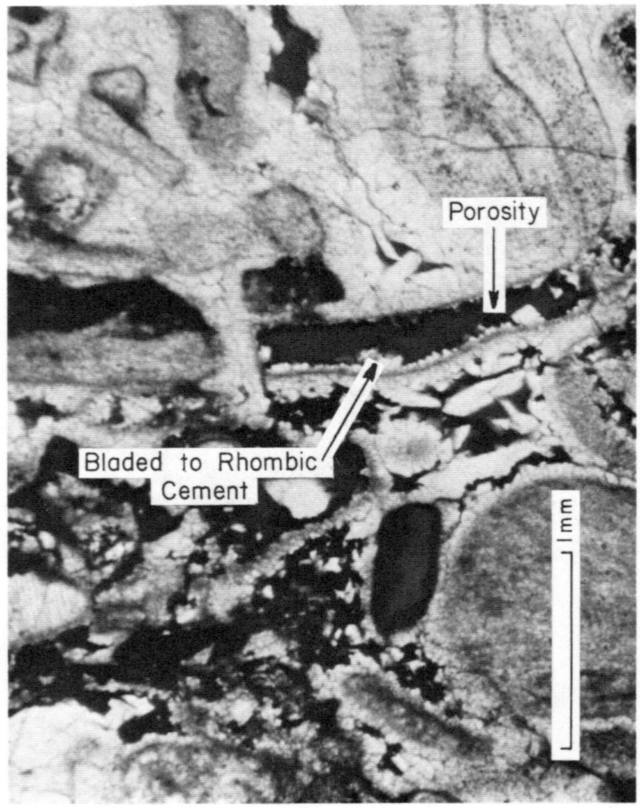

B MGF Oil Corp. No.1 Boone 13,465

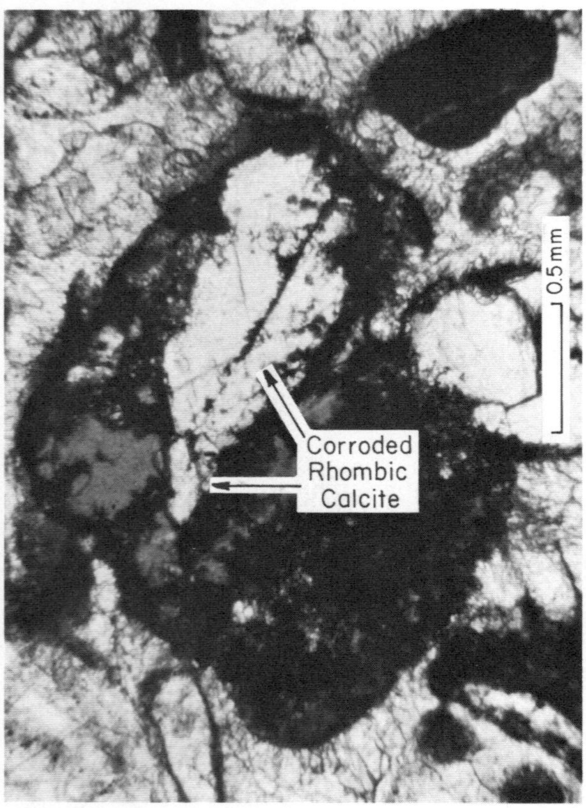

C General Crude No.1 Wessendorf 13,601

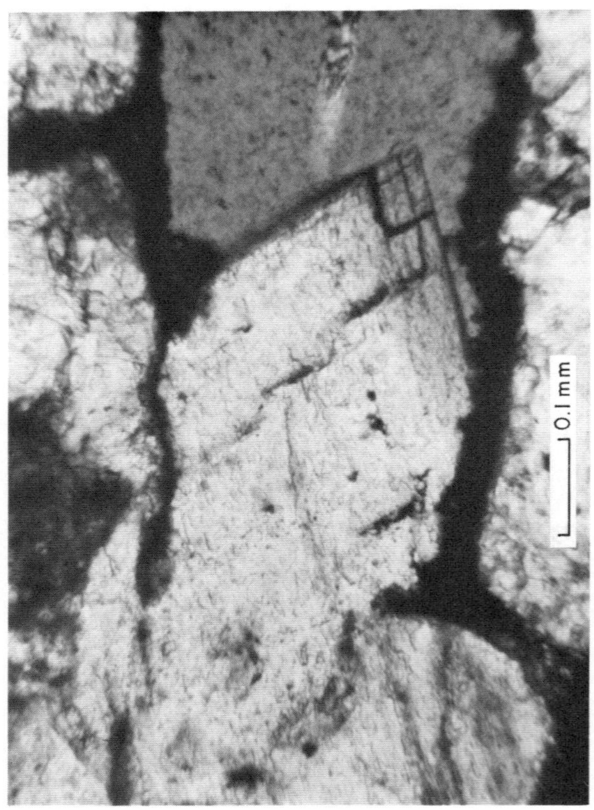

D General Crude No.1 Wessendorf 13,601

Figure 14

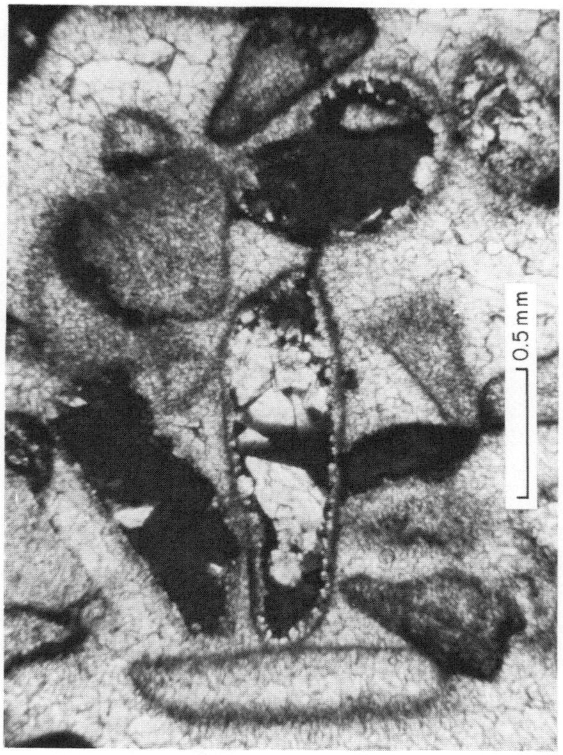

A MGF Oil Corp. #1 Boone 13,465

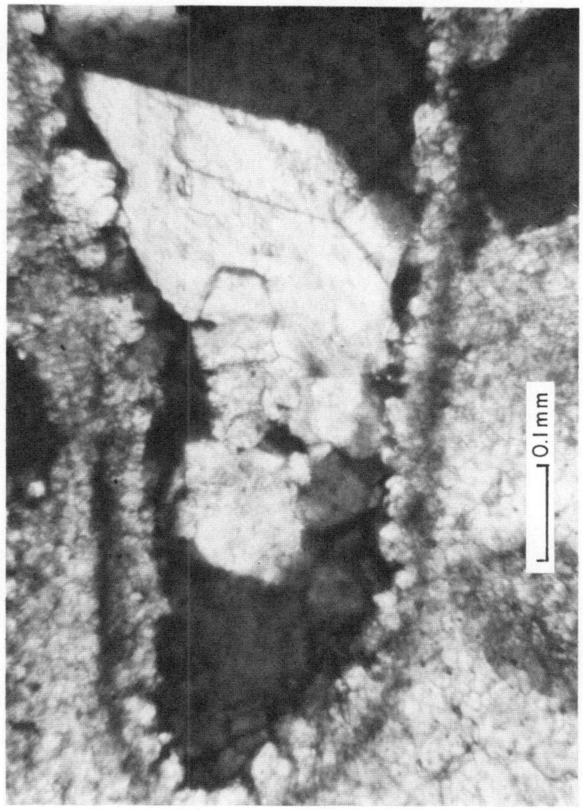

B MGF Oil Corp. #1 Boone 13,465

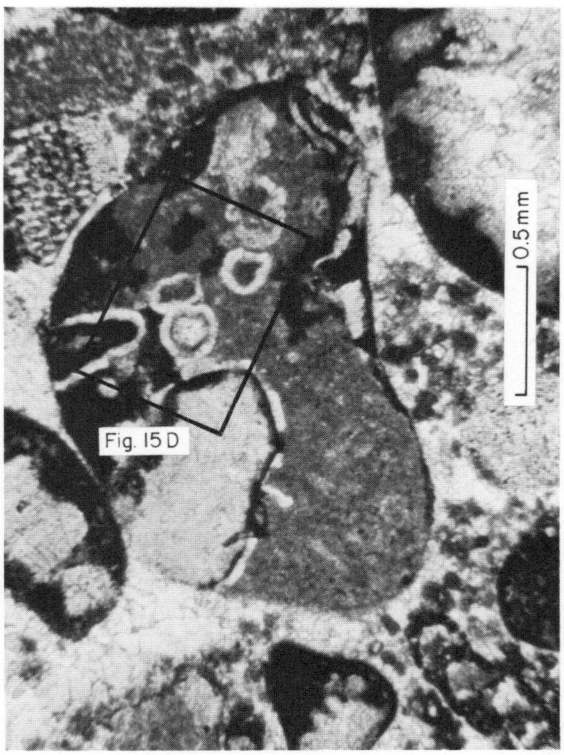

C General Crude #1 Alexander 13,564

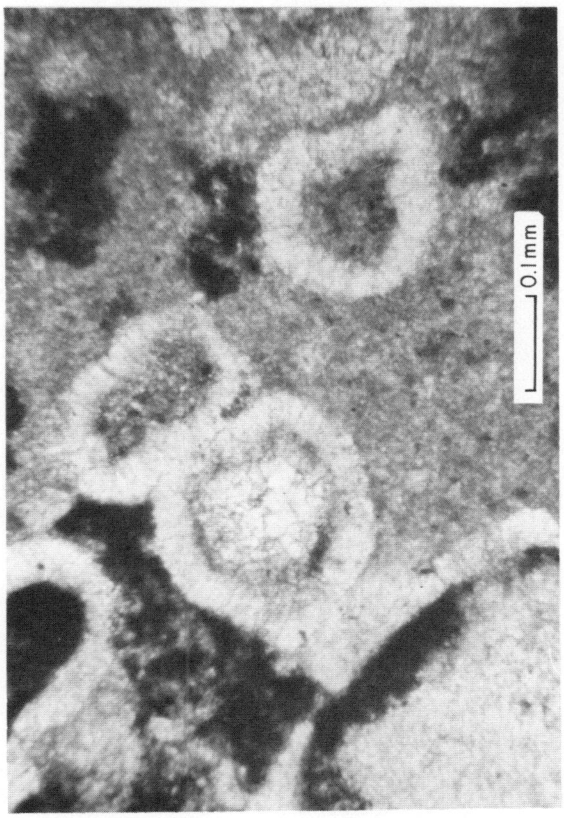

D General Crude #1 Alexander 13,564

Figure 15

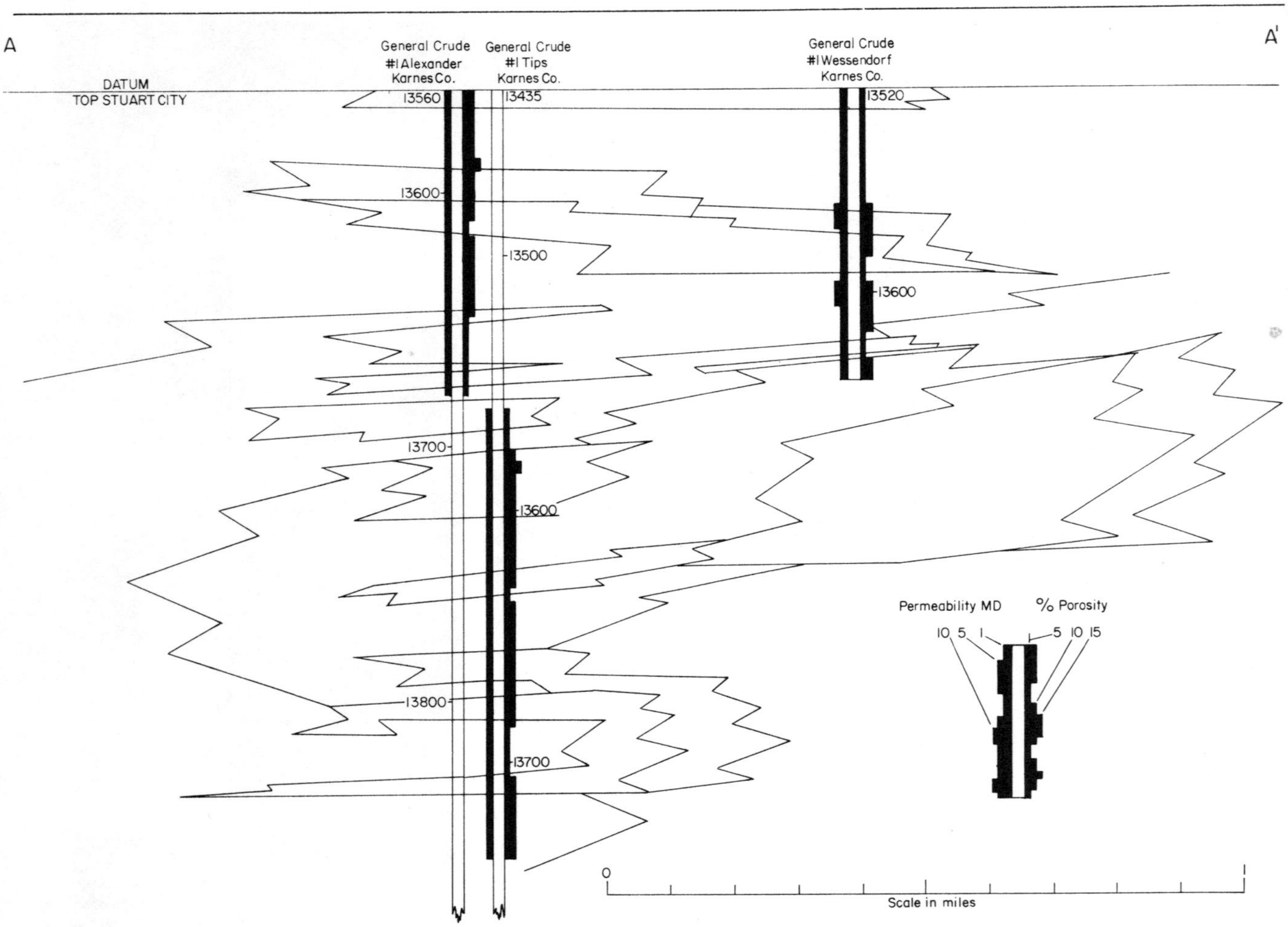

Figure 16. Porosity and permeability distribution on cross section A-A'.

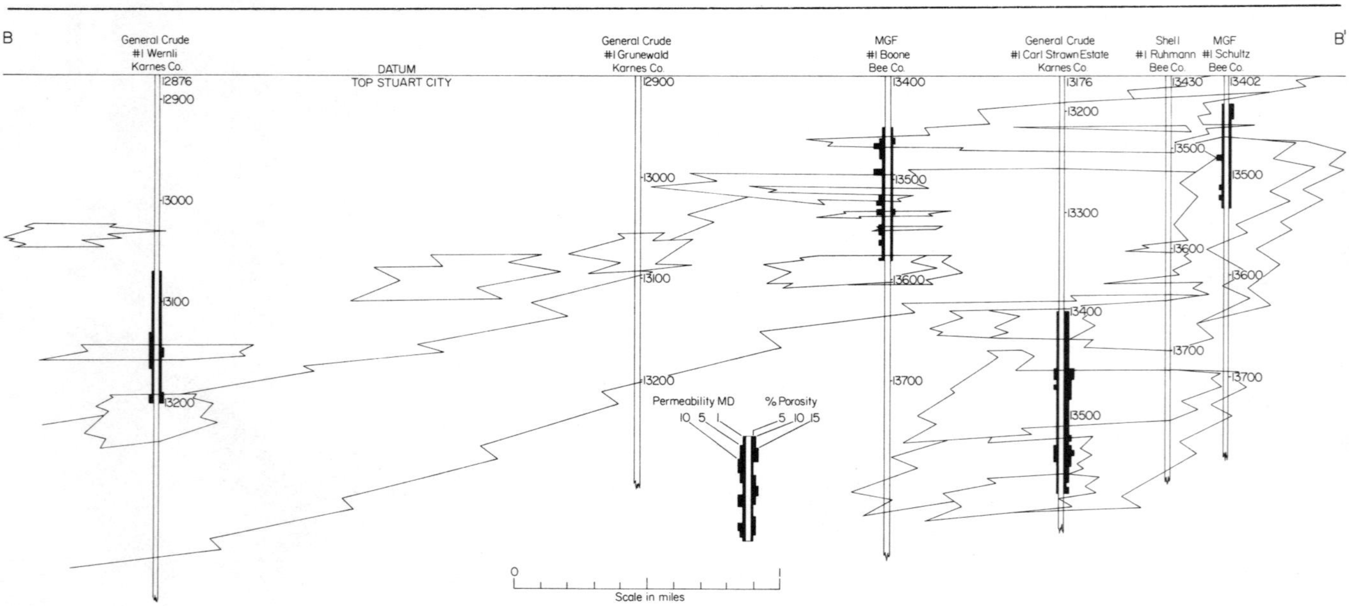

Figure 17. Porosity and permeability distribution on cross section B-B'.

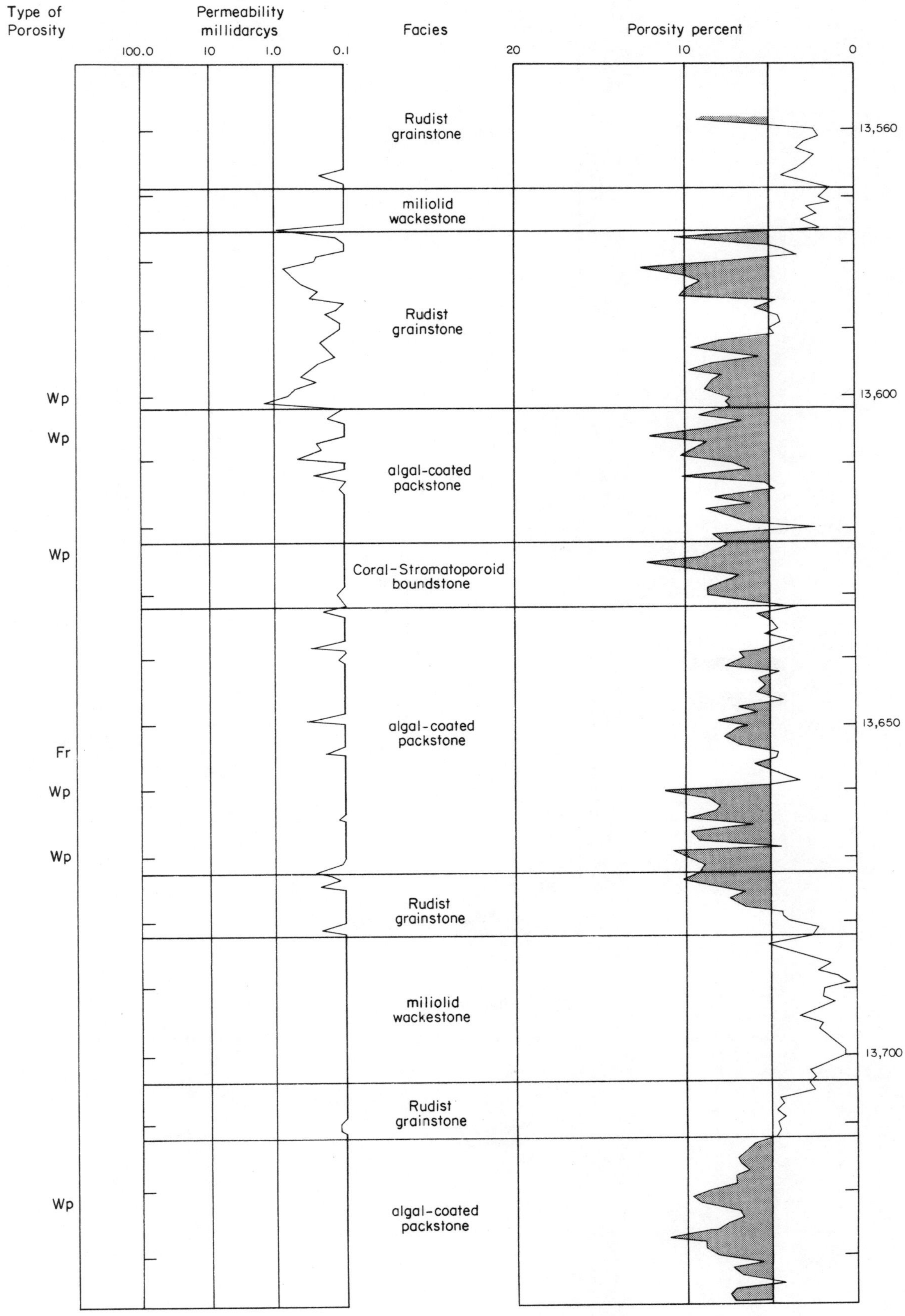

Figure 18. Relationship between porosity and permeability. General Crude Oil Company #1 Tips, Karnes County, Texas.

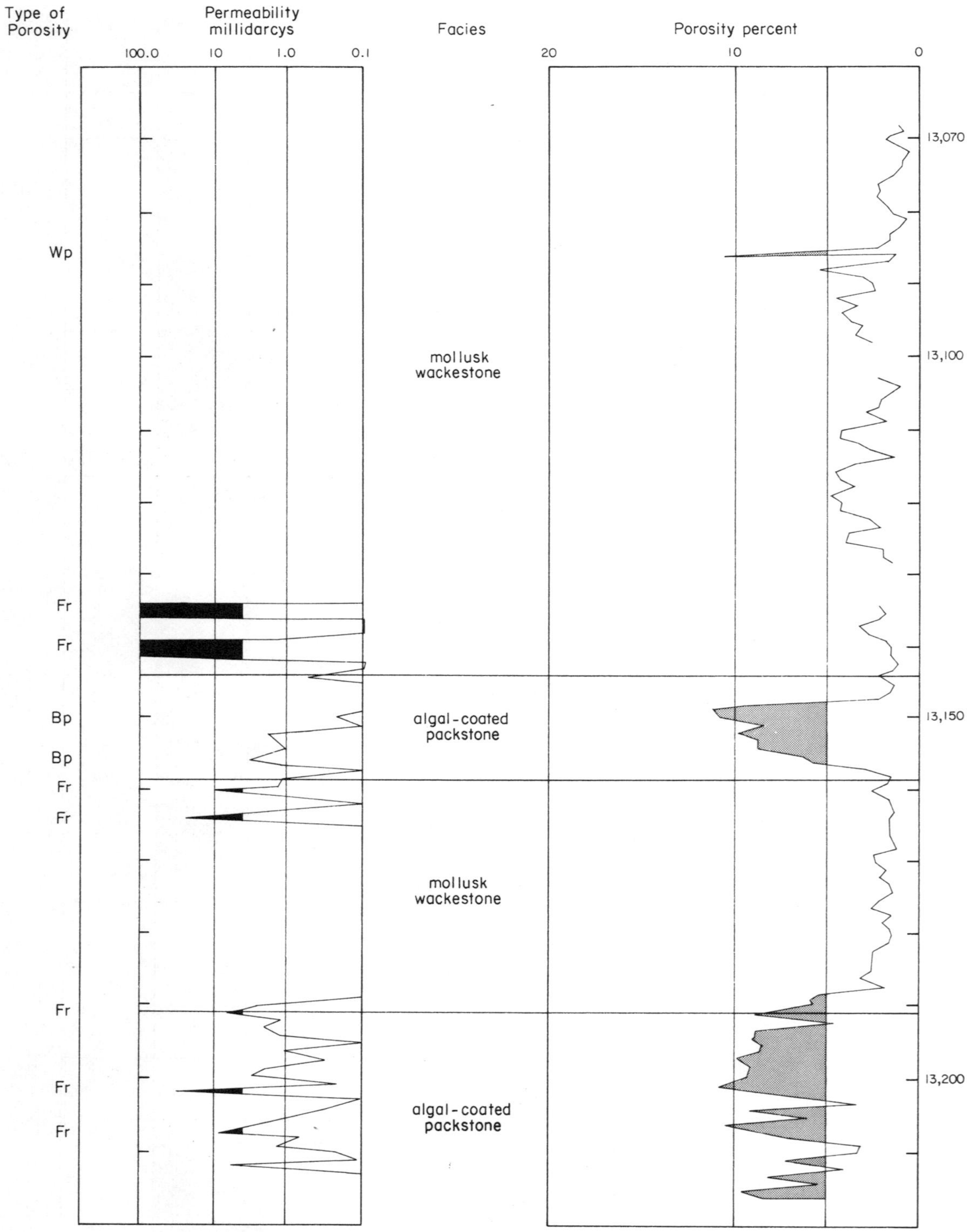

Figure 19. Relationship between porosity and permeability. General Crude Oil Company #1 Wernli, Karnes County, Texas.

Figure 20 *(page 256)*

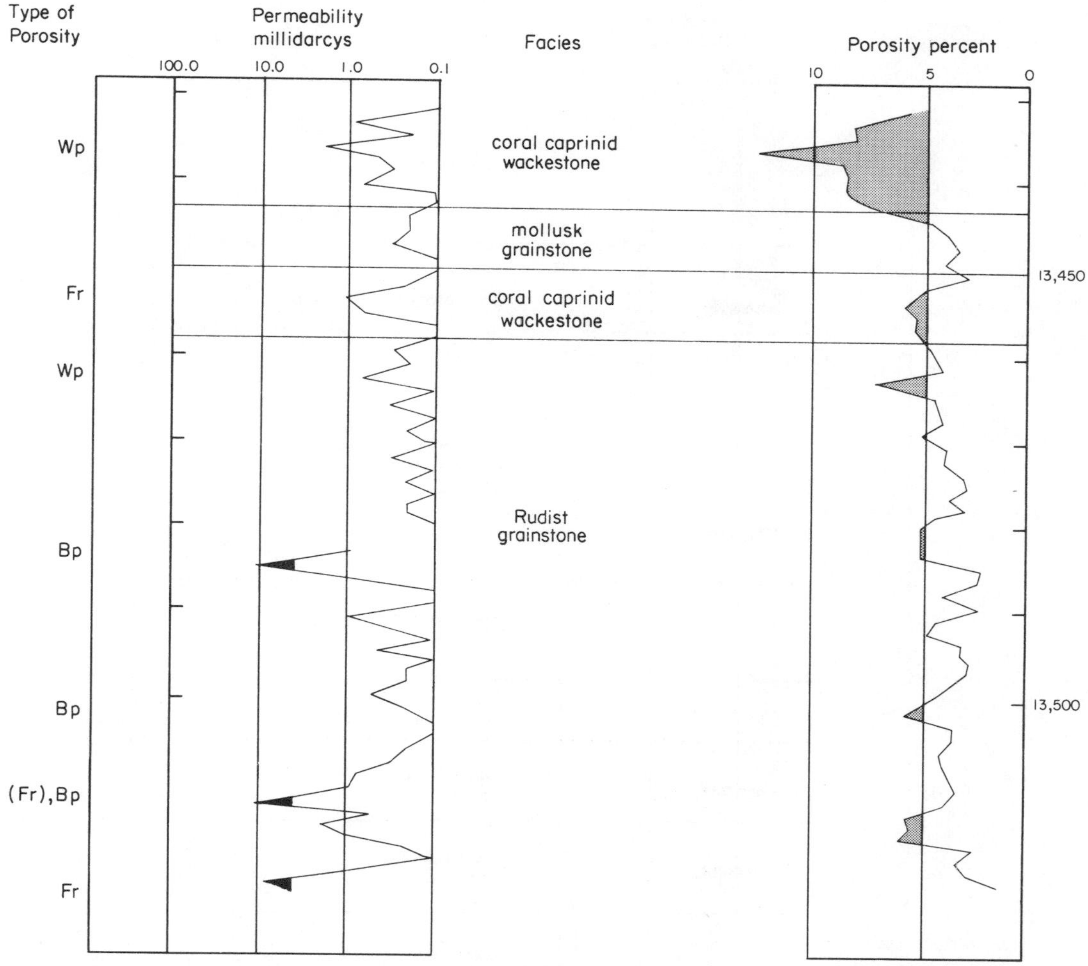

Figure 21. Relationship between porosity and permeability. MGF Oil Corporation #1 Schultz, Bee County, Texas.

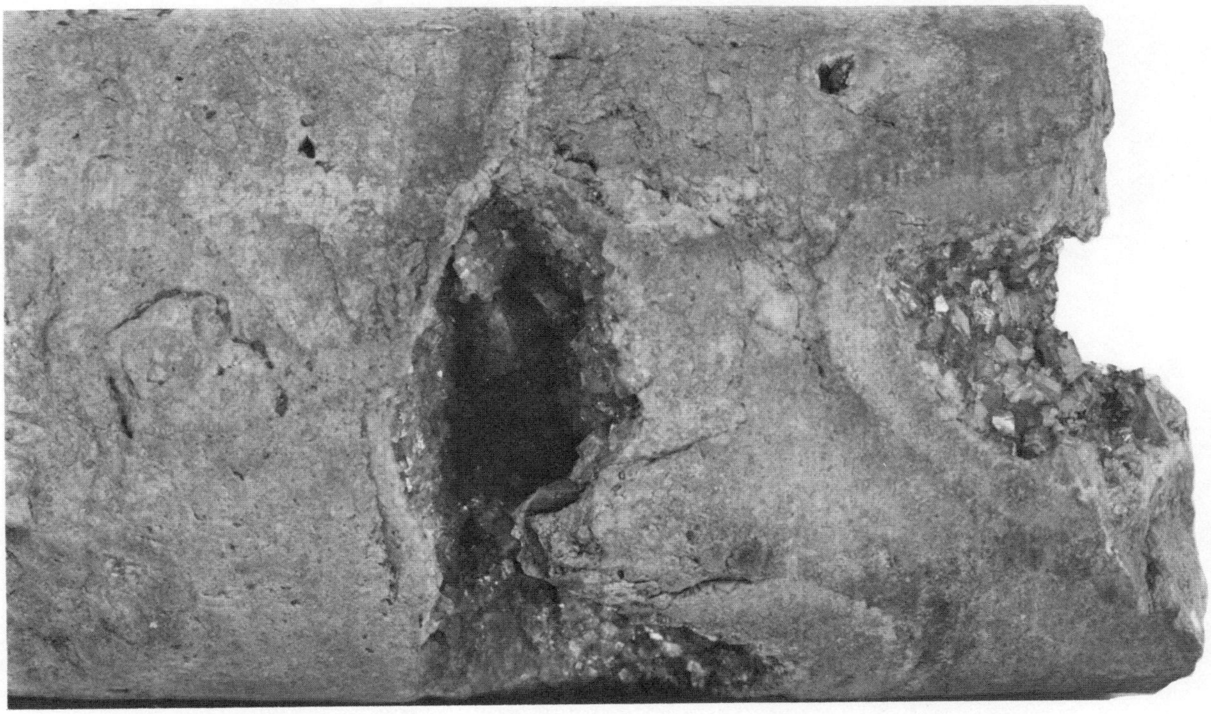

Figure 22. Cement-reduced intraparticle porosity is present in the body cavities of caprinids. Coral-caprinid boundstone from the General Crude Strawn Estate, 13,459 ft.

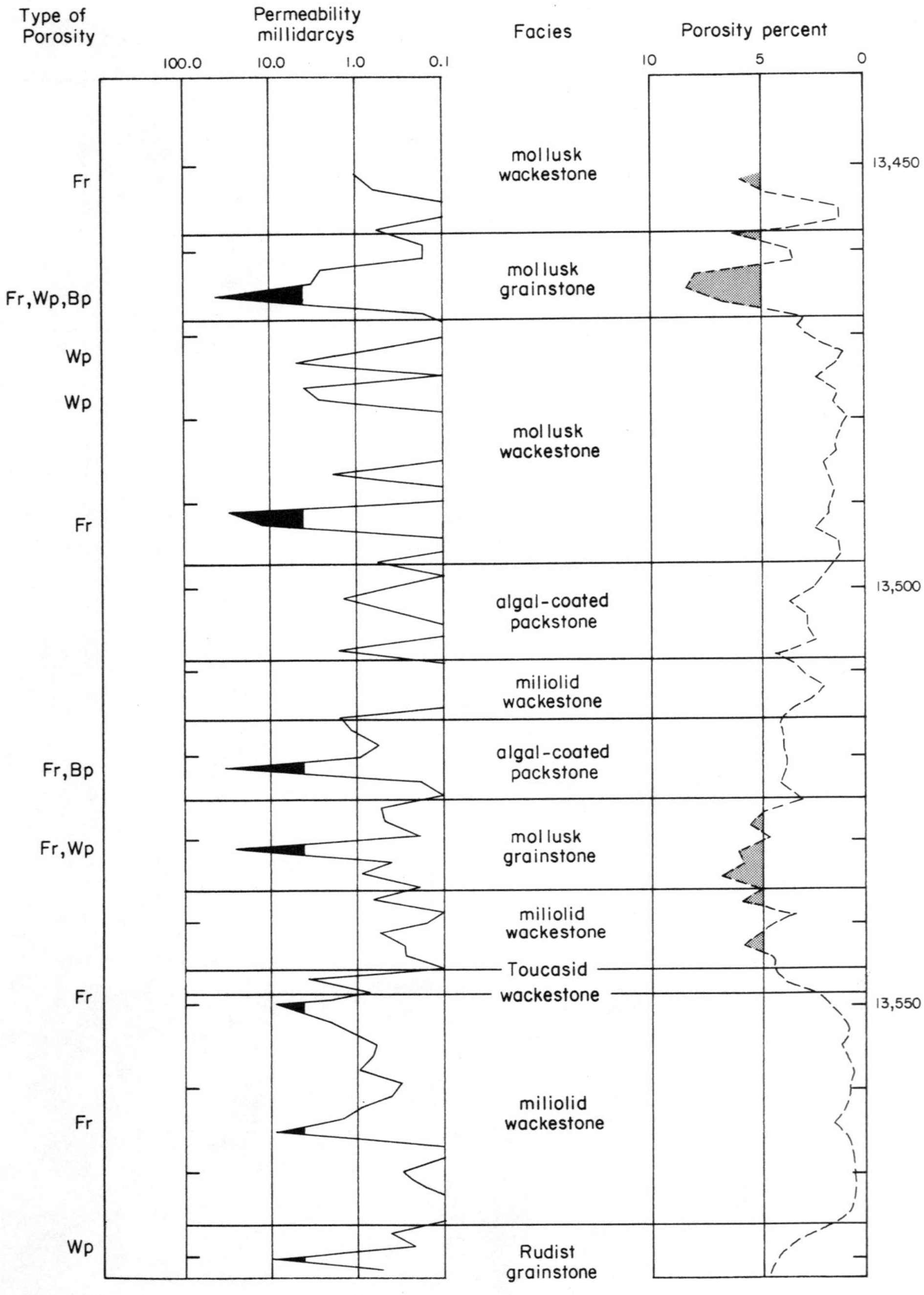

Figure 20. Relationship between porosity and permeability. MGF Oil Corporation #1 Boone, Bee County, Texas.

REFERENCES

Bebout, D. G., and Loucks, R. G., 1974, Stuart City Trend Lower Cretaceous—a carbonate shelf-margin model for hydrocarbon exploration: Univ. Texas, Austin, Bureau of Economic Geology Report of Investigations 78, 80 p.

Choquette, P. W., and Pray, L. C., 1970, Geologic nomenclature and classification of porosity in sedimentary carbonates: AAPG Bull., v. 54, p. 207-250.

Fisher, W. L., and Rodda, P. U., 1969, Edwards Formation (Lower Cretaceous), Texas: Dolomitization in a carbonate platform system: AAPG Bull., v. 53, p. 55-72.

Meyerhoff, A. A., 1967, Future hydrocarbon provinces of Gulf of Mexico-Caribbean region: GCAGS Transactions, v. 17, p. 217-260.

Rose, P.R., 1972, Edwards Group, surface and subsurface, Central Texas: Univ. Texas, Austin, Bureau of Economic Geology Report of Investigations 74, 198 p.

DIAGENETIC PATTERNS OF THE AUSTIN GROUP AND THEIR CONTROL OF PETROLEUM POTENTIAL

Peter A. Scholle[1] and Kelton Cloud[2]

The chalk of the Austin Group shows striking regional variations in porosity, permeability, and trace element and isotopic geochemistry. Porosities and permeabilities are highest across the San Marcos arch, where average values of 15 to 30 percent porosity and 0.5 to 5 md (millidarcies) matrix permeability are measured. These values decrease slightly to the north (into the northeast Texas embayment). In northern Mexico, the Austin and its equivalents have about 3 to 8 percent porosity and permeabilities of 0.01 md or less. Porosity and permeability also decrease in downdip sections of the Austin when traced from outcrop to about 4,500 m deep.

The geochemical properties follow similar trends (fig. 1). Outcrop studies show that samples from the San Marcos arch and Sabine uplift have bulk oxygen isotopic values in the range of -2.7 to -4.0 per mil (relative to PDB). In the Rio Grande embayment of South Texas and northern Mexico, these values have shifted to -5.0 to -7.0 per mil, whereas in the northeast Texas embayment they range from -3.5 to -5.0 per mil. In downdip sections near the San Marcos arch, the oxygen isotopic values shift from about -2.8 at the surface to about -8.0 at 4,500 m. Average Sr trace element values for the Austin Group on the San Marcos arch are 350 to 975 ppm, whereas in the Rio Grande embayment and the northeast Texas embayment, they range from 950 to 1,775 ppm.

All chalk undergoes both mechanical and chemical compaction (pressure solution and reprecipitation) when subjected to sufficient differential stress (Matter, 1974; Schlanger and Douglas, 1974). This stress is generally induced by addition of overburden but can also be influenced by tectonic stresses and pore-fluid pressures (Mimran, 1975; Scholle, 1977). The presence of fresh (Mg-poor) water in chalk in conjunction with elevated differential stress has been shown, both theoretically and in nature, to accelerate chemical compaction greatly (Neugebauer, 1974; Scholle, 1974, 1977). Thus, the lateral and downdip variations in the petrophysical and geochemical properties of the chalk of the Austin Group presumably reflect differences in original thickness of overburden or proximity to zones of major deformation. The noted reduction in porosity between the San Marcos arch and the Rio Grande embayment could have been produced, in the presence of Mg-poor fluids, by about 500 m difference in maximum overburden between the two areas. Greater overburden differences would have been required had marine (or other Mg-rich) pore fluids been present; less overburden difference would have been needed if differential tectonic stresses were important.

The isotopic and trace element values listed previously are compatible with these conclusions but do not uniquely distinguish among the possible explanations. The smooth shift of isotopic values, as a function of present burial depth in downdip sections and of probable paleoburial depths in lateral outcrop sections, indicates that maximum burial depth is the critical factor in porosity loss or retention. Only the ***rate*** of porosity loss is affected by water chemistry. Carbon isotopic analyses also rule out vadose diagenesis as having influenced porosity reduction in the Austin to any significant degree.

Oil production from the Austin Group is concentrated in the areas of the San Marcos arch and the Sabine uplift in a belt that is parallel to the outcrop trend and that ranges in depth from 200 to 2,000 m (fig. 2). Cumulative production from all fields in the Austin Group in Texas totals about 25 million barrels (as of January 1976). Production of oil and gas from chalks other than the Austin has been significant both on the Sabine uplift and from areas on the eastern side of the Mississippi embayment (fig. 2). Some of these reservoirs, however, may include sandy, calcarenitic, or other impure chalks.

Wells completed in the Austin have a long history of production at

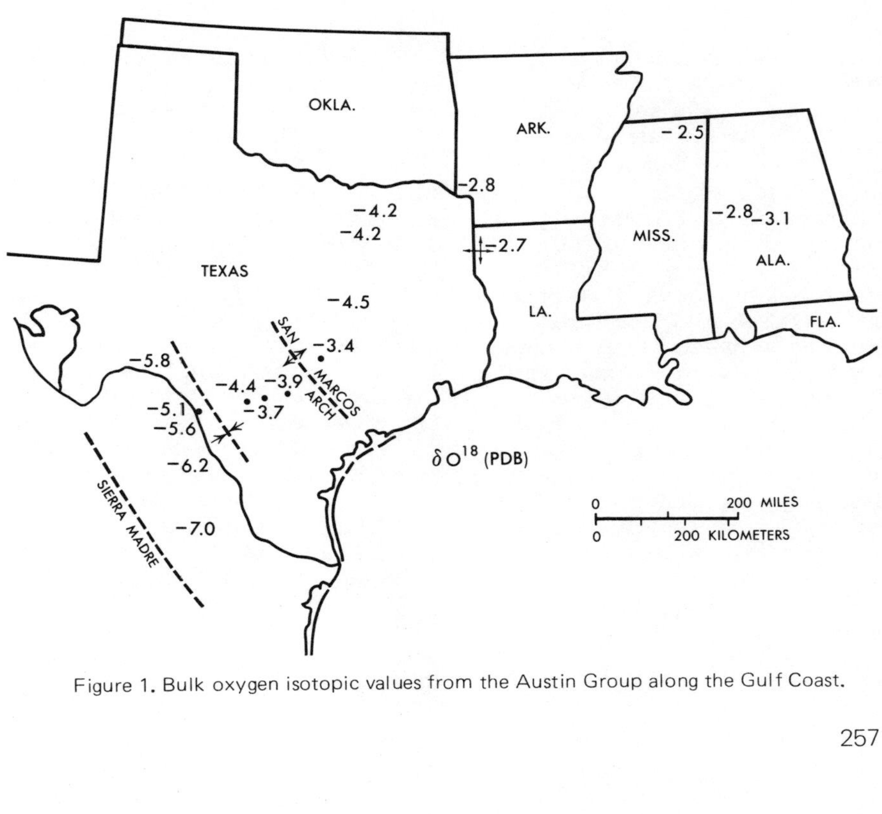

Figure 1. Bulk oxygen isotopic values from the Austin Group along the Gulf Coast.

[1]U. S. Geological Survey, Reston, Virginia

[2]Bass Enterprises Production, Fort Worth, Texas

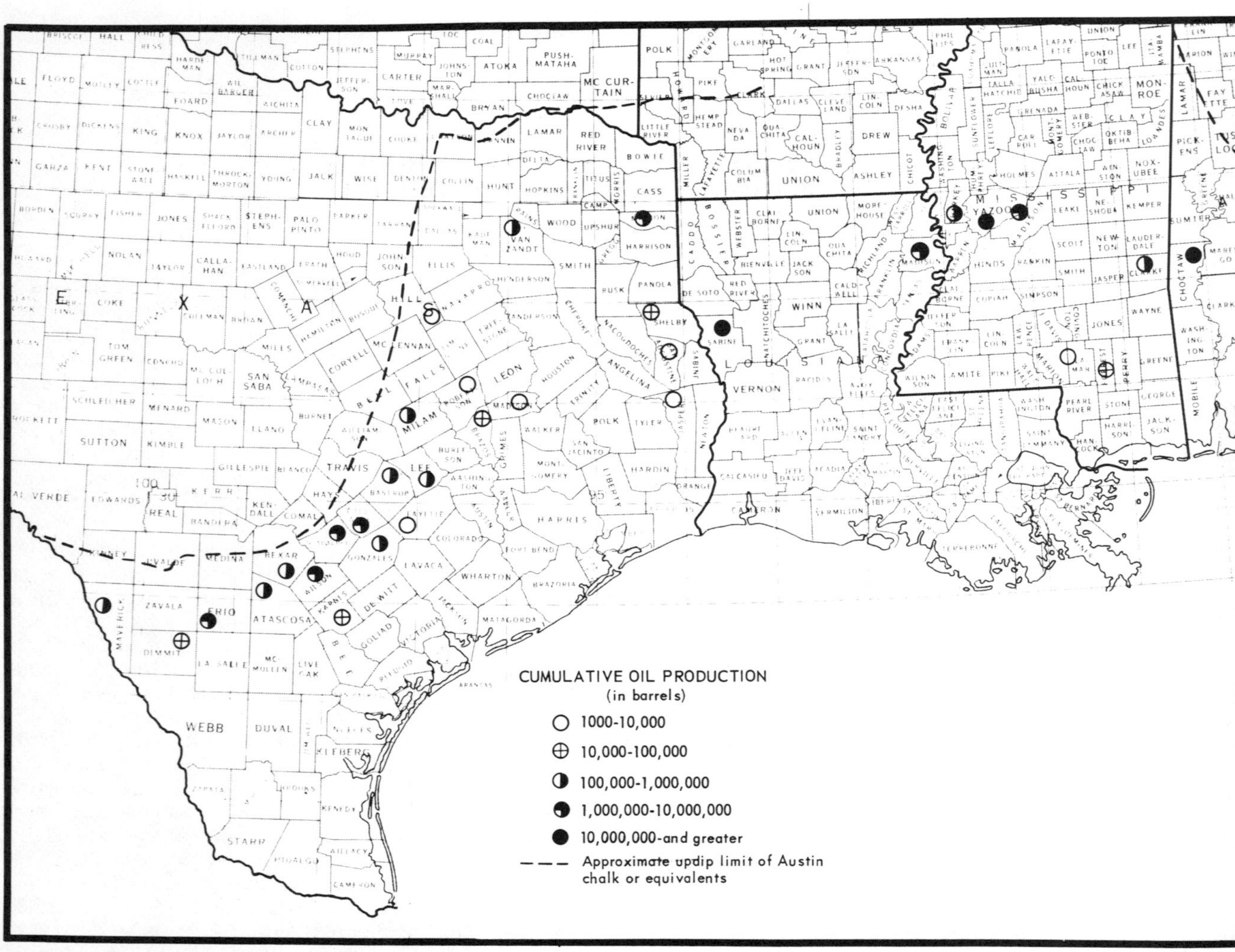

Figure 2. Oil production from the Austin Group along the Gulf Coast.

rates far lower than initial production. Indeed, the initial discovery well of the Pearsall field, drilled in 1936, was still producing at a rate of more than 200 barrels per month as of 1976 (Scott and Prestridge, 1977). Most recently drilled Austin wells have initial production rates of 200 to 500 bbls of oil per day, which decline within months to about 40 bbls per day (Long, 1976). These production histories indicate that most oil production from the Austin is from fractures. Yet, the concentration of production in areas of least diagenetic alteration, in association with the long histories of slow production, indicate that extended production is probably the result of very slow drainage of oil from the rock matrix. Artificial fracturing, a completion method used on virtually all current Austin wells, enhances both initial and long-term production by allowing shorter drainage paths through a larger number of fractures.

The best future oil and gas discoveries in the Austin and equivalent lithologies will probably be concentrated in three types of areas:

(a) where the chalks may have had any type of pore fluid but have not been deeply buried (that is, between 0 and 2,000 m);

(b) where marine pore fluids were retained and fresh water was excluded. In such areas, significant matrix porosity can be retained to as much as 3,000 m deep;

(c) where abnormally high pore-fluid pressures have reduced effective compressive stresses. Under this condition, burial depth is no longer the controlling factor in porosity loss, and porous chalks can be found at depths from 0 to greater than 4,000 m.

Other production may come from areas that have low matrix porosity but intense fracturing (as along sharp flexures or faults) or from areas of abnormal lithology (e.g., bioherms, intrusive volcanic rocks, calcarenites).

REFERENCES

Long, M., 1976, Austin chalk drilling surges in South Texas: Oil and Gas Journal, v. 74, p. 62-63.

Matter, A., 1974, Burial diagenesis of pelitic and carbonate deep-sea sediments from the Arabian Sea, *in* Whitmarsh, R. B., Weser, O. E., Ross, D. A., and others, Initial reports of the Deep Sea Drilling Project: Washington, U. S. Government Printing Office, v. 23, p. 421-469.

Mimran, Y., 1975, Fabric deformation induced in Cretaceous chalks by tectonic stresses: Tectonophysics, v. 26, p. 309-316.

Neugebauer, J., 1974, Some aspects of cementation in chalk, *in* Hsu, K. J., and Jenkyns, H. C., eds., Pelagic sediments: On land and under the sea: Blackwell, Oxford, International Association of Sedimentologists, Special Pub., no. 1, p. 149-176.

Schlanger, S. O., and Douglas, R. G., 1974, Pelagic ooze-chalk-limestone transition and its implications for marine stratigraphy, *in* Hsu, K. J., and Jenkyns, H. C., eds., Pelagic sediments: On land and under the sea: Blackwell, Oxford, International Association of Sedimentologists, Special Pub., no. 1, p. 117-148.

Scholle, P. A., 1974, Diagenesis of Upper Cretaceous chalks from England, Northern Ireland and the North Sea, *in* Hsu, K. J., and Jenkyns, H. C., eds., Pelagic sediments: On land and under the sea: Blackwell, Oxford, International Association of Sedimentologists, Special Pub., no. 1, p. 177-210.

________ 1977, Chalk diagenesis and its relation to petroleum exploration problems: Oil from chalks—A modern miracle?: AAPG Bull., v.61, no. 7.

Scott, R. J., and Prestridge, J., 1977, South Texas Austin Chalk (Cretaceous) play: AAPG Bull., v. 61, no. 2, p. 296-297.

CARBONATE FACIES DISTRIBUTION AND DIAGENESIS ASSOCIATED WITH VOLCANIC CONES—ANACACHO LIMESTONE (UPPER CRETACEOUS), ELAINE FIELD, DIMMIT COUNTY, TEXAS

P. E. Luttrell[1]

ABSTRACT

Late Cretaceous volcanic activity along the northern rim of the Rio Grande embayment resulted in the growth of a number of cones that form an arcuate trend in south-central Texas. Some of these cones grew to sea level and served as nuclei for shallow-water carbonate sedimentation. Resulting limestones are known as the Anacacho Formation. Significant hydrocarbon accumulations have been found in the limestones associated with many of the volcanoes, such as the Elaine field in Dimmit County, Texas. Facies distribution and diagenetic fabric were analyzed from core and electrical logs.

Shallow water on the flanks of the emergent volcano favored development of rudist reefs which, with other marine fauna, supplied abundant shell material for reworking into shoals. Diagenetic fabrics of the resulting grainstones include precipitation of bladed and mosaic calcite as well as limpid dolomite, thereby indicating cementation during meteoric phreatic conditions. This evidence supports the notion of subaerial exposure of the shoals as beaches, thus allowing the development of a lagoonal environment landward. Seaward of the beach, red algal ridges and a muddy sand halo developed. The muddy sand halo was supplied with shell material from high-energy areas, whereas the mud was a product of deeper water accumulation. With increasing water depth off the flanks of the volcano, these facies grade into a mud-rich open-shelf environment dominated by burrowing organisms.

After this initial carbonate buildup, the volcano subsided, and the facies onlapped the volcano as a result of the changing relative sea level. Subsidence continued until the volcano was completely submerged. Faults reflecting readjustments of strata over the plug are recorded in sediments as young as the Navarro Group.

Porosity in Elaine field carbonates occurs in areas where a fresh-water lens developed in association with subaerial exposure. In these areas, dissolution of grains and limited cementation produced excellent hydrocarbon reservoirs.

INTRODUCTION

General Statement

Elaine field is a producing oil field located about 100 miles southwest of San Antonio and 6 miles southeast of Crystal City, Texas (fig. 1). The area covers approximately 1½ square miles and the field produces from two horizons—the Anacacho Limestone associated with a volcanic cone and the sandstones of the overlying Olmos - San Miguel Formations (fig. 2). Maximum drilling usually reaches a depth of only 5,000 feet. Elaine field is one of many fields of similar size and production associated with volcanic rocks in the immediate area (fig. 3).

An arcuate trend of Late Cretaceous volcanoes in Central Texas has been recognized since R. T. Hill's (1890) investigation. Initially, the volcanoes received casual interest, probably more for their uniqueness in the area than for geologic significance. With the discovery of Thrall oil field in 1914 (Udden and Bybee, 1916), interest greatly increased, especially with the report that oil had been produced from altered igneous rock. Because "plugs" were easily outlined in the subsurface by magnetics and because they were shallow and inexpensive to drill, they became targets for hydrocarbon exploration. In 35 oil fields, production is from volcaniclastics with 14 producing from additional zones directly affected by the occurrence of the volcanic high. Total oil production to 1966 was 30,841,076 barrels (Simmons, 1967). Updip of these fields, the outcropping volcanoes and associated facies in Uvalde County contain abundant asphalt which is mined for road material; the asphalt results from loss of volatiles through inadequate seal. Higher petroleum prices have renewed interest in subsurface oil plays of the Balcones igneous province.

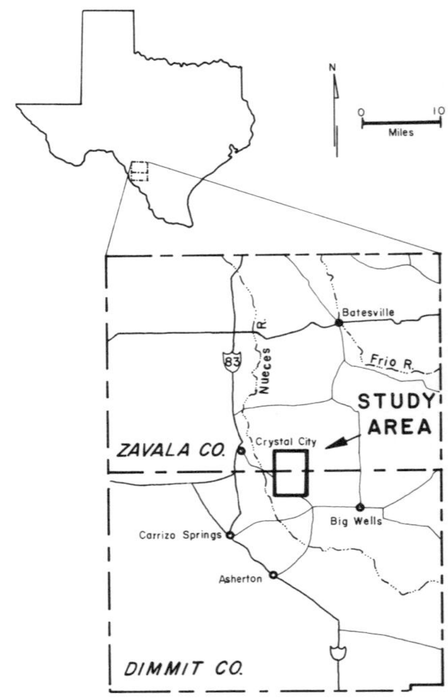

Figure 1. Location of study area.

Although most hydrocarbon production is from layered volcaniclastics, some oil is produced from carbonate grainstone bodies which discontinuously ring the volcanoes. The distribution and diagenetic fabric of the carbonate facies associated with the plug in Elaine field are the main concerns of this study. Specific objectives are outlined as follows:

1. To trace the depositional episodes resulting from development of the volcano, carbonate buildup on the flanks, and subsequent subsidence of the entire complex.
2. To delineate and describe the Anacacho Limestone facies and environments which were associated with the volcanic island.
3. To study the diagenetic fabrics of the limestone facies, to establish the sequence of cementation and porosity development, and to apply this information to petroleum exploration.

[1]Mobil Exploration and Production, Dallas, Texas

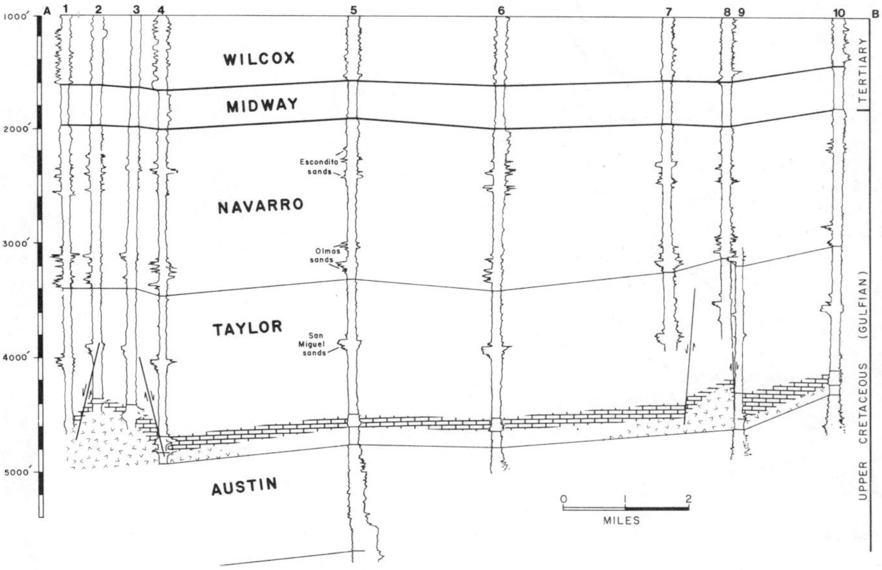

Figure 2. Stratigraphic relationship of the volcanic rocks and the Anacacho Limestone. (See Figure 3 for location of cross section.)

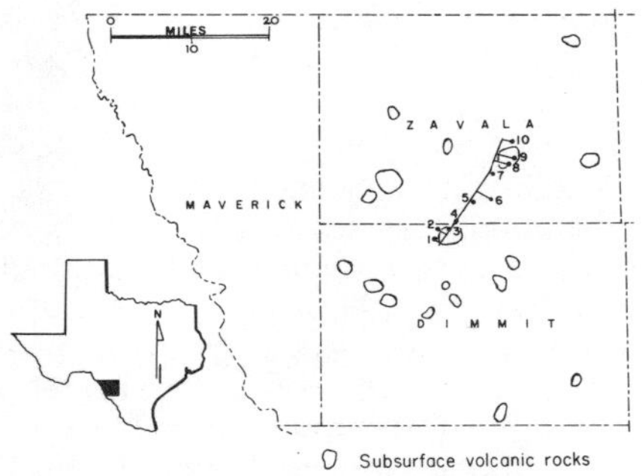

Figure 3. Distribution of subsurface volcanic plugs in the study area. 1. Shield - #1 McLean, 2. Tesoro - Crenshaw #50, 3. Tesoro - Crenshaw #58, 4. Pronto - Hagen and Dickenson #1, 5. Wier - #1 Denman, and others, 6. General Crude - #1 Donnelly, 7. Normandy - #1 Holdsworth, 8. General Crude - #5 Holdsworth, 9. General Crude - #1 Holdsworth, 10. Joceyln, and others - #1 Childress.

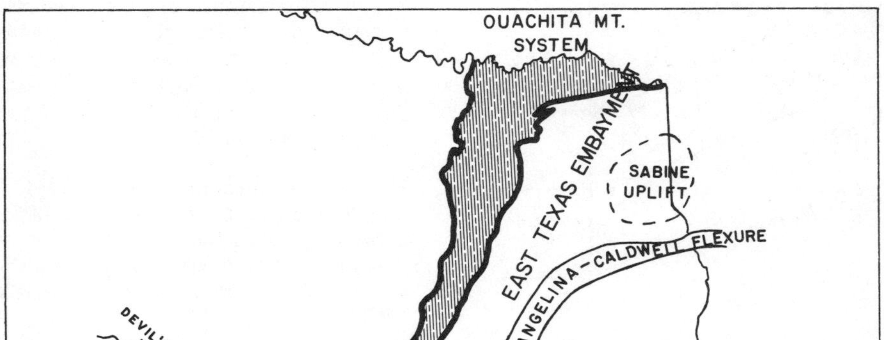

Figure 4. Index of important Late Cretaceous geologic features. After Murray (1957), Spencer (1965), and Stehli, and others (1972).

Regional Geologic Setting

The Upper Cretaceous in Texas is divided structurally into two basins, the East Texas embayment and the Rio Grande embayment. These embayments are separated by the San Marcos arch, a southeast-trending promontory of the Llano Uplift (fig. 4). Thinning of the Upper Cretaceous Austin and Eagle Ford Formations across this platform suggests that it influenced sedimentation for some time (Sellards and others, 1933). The Llano Uplift was seldom a positive structure, but rather, subsided at a substantially slower rate than the basins it divided (Loucks, 1976).

West of the San Marcos arch, Cretaceous and Tertiary carbonate terrigenous sediments were deposited in the more rapidly subsiding Rio Grande embayment. The maximum thickness of Upper Cretaceous rocks found in the Gulf Coast occurs in the Rio Grande embayment (Stephenson, 1928; Murray, 1957).

The actual shape of the Rio Grande embayment, defined by the outcrop pattern of Lower Tertiary rocks, results from structural features which are believed to have influenced sedimentation since the Late Jurassic. The Mesozoic sediments of the southeastern part of the San Marcos arch dip abruptly into the Gulf of Mexico basin. The dip on the southwest flank of the arch is considerably less and is accompanied by much less abrupt facies changes. The Balcones fault zone marks the northern boundary of the embayment, whereas the southwest side is bounded by the Salado-Tamaulipas arch, another southeast-trending positive feature. To the southeast, the embayment opens with Upper Cretaceous rocks dipping into the Gulf of Mexico basin. Continuous thickening of units throughout the entire Cretaceous attests to the influence of these tectonic elements through time (Spencer, 1965).

The northern flank of the Rio Grande embayment was the focus of volcanic activity in a narrow east-west band (fig. 5). At various localities volcanic ash is interstratified with all the formations ranging from Del Rio to Escondido with volcaniclastics common in the Austin and Taylor Groups (fig. 6). Near the end of Taylor deposition, uplift and erosion removed Upper Taylor sediments along the northern margin of the embayment. Through eastern Uvalde, Medina, and Bexar Counties, therefore, the Escondido overlies the Anacacho Limestone (Spencer, 1965). Near the end of the Late Cretaceous the appearance of

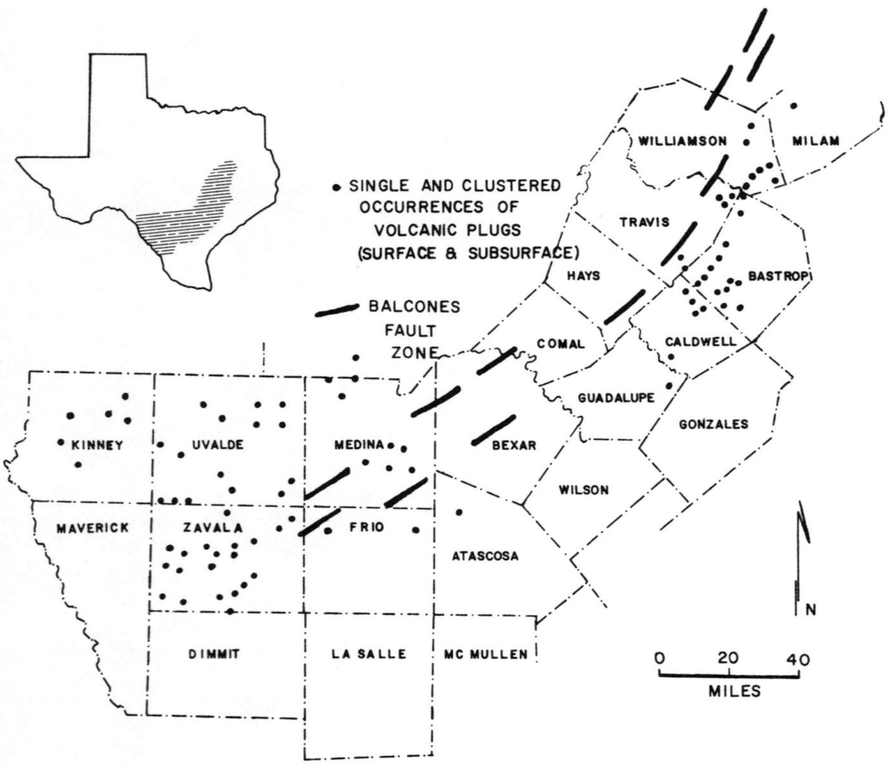

Figure 5. Distribution of surface and subsurface volcanic plugs. After Simmons (1967) and Lonsdale (1927).

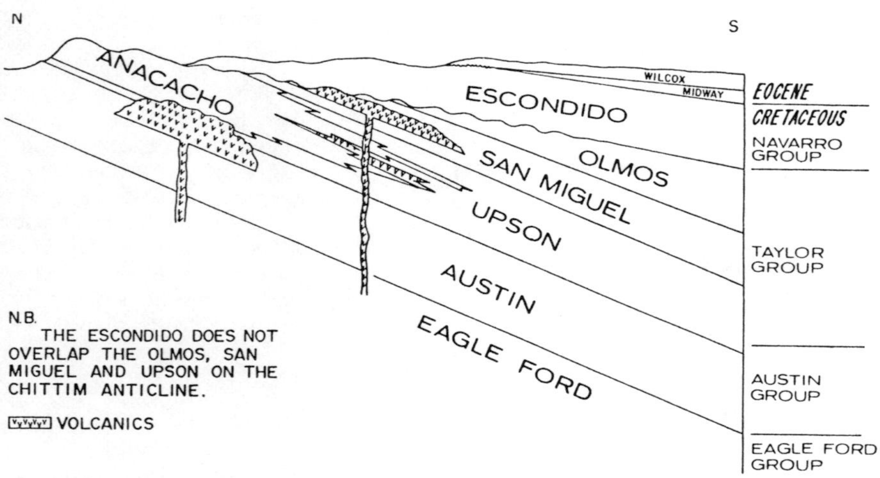

Figure 6. Generalized dip section across the northern margin of the Rio Grande embayment. From Spencer (1965).

nearshore and fluvial-deltaic facies, the Olmos and San Miguel sands, is indicative of an advancing shoreline. These formations, which outcrop on the flanks of the Chittim arch, are present in the subsurface only to the east and south of Maverick County (Spencer, 1965).

Downwarp of the Upper Cretaceous, units began with the deposition of the Navarro Group and continued into the Eocene when the embayment was filled and the depocenter shifted gradually southeastward into the Gulf Coast basin (Spencer, 1965).

CARBONATE FACIES

Five facies were delineated on the basis of allochem type and relative amount of lime mud in the carbonate rocks associated with the volcano of the Elaine field. These facies are planktonic foraminifer mudstone/wackestone, oyster-echinoid-algal grainstone, oyster-echinoid-algal packstone, oyster-echinoid green algal packstone, and rudist-algal-echinoid packstone, representing open-shelf, beach, lagoon, packstone halo, and reef environments, respectively. Volcanic material is described separately when it occurs as the most dominant constituent, that is, ash deposits, ash-flow lapilli, volcanic beach.

Cores from wells provided the primary information source for facies descriptions and environmental interpretation. Six cores (fig. 7) totaling 300 ft were examined and described. By comparing electrical logs of cored intervals with logs of uncored intervals, facies were projected laterally, and their distribution was mapped.

Tabular-form descriptions (table 1) are presented first to give a concise introduction to each facies. This method facilitates comparison of specific points. A discussion of distribution and depositional history follows this description.

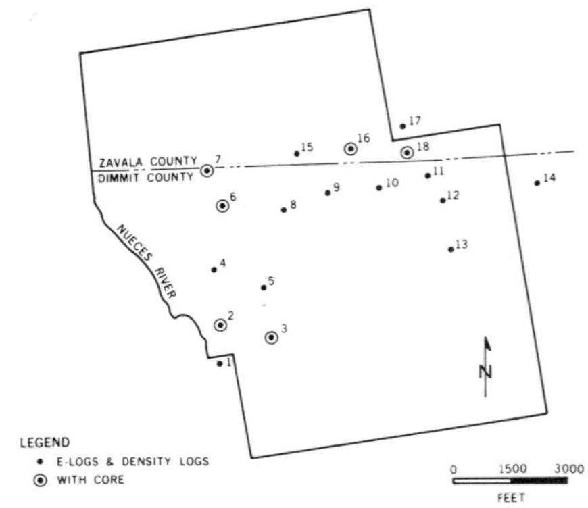

Figure 7. Location of well control and cores. (See Appendix A for list of wells.)

Volcanic Rocks

In addition to the carbonate facies previously described, several types of volcanic rocks were observed in the cores. Table 2 describes the principal varieties.

FACIES DISTRIBUTION AND DEPOSITIONAL HISTORY

Volcanic Rocks

The top of the Elaine cone (fig. 8) is irregular and elongate from the northwest to the southeast. Analysis of surface fractures in Uvalde County (Wermund and others, in press), which is north of the study area, reveals a northwest fracture trend. These trends are thought to reflect ancient basement fractures which influenced jointing in the overlying strata and probably the shape of the volcanic cone. Further, Degolyer (1915) sug-

Table 1. Facies Descriptions

CARBONATE FACIES	ALLOCHEMS DOMINANT	ALLOCHEMS MINOR	THICKNESS	STRUCTURES/DIAGNOSTIC FEATURES	ϕ
Planktonic Foraminifer Mudstone/Wackestone (PLATES I and II)	Pelagic Foraminifers (Esp. *Globigerina*) Echinoids Pelecypods	Red Algae	1-33′ in 5 cores	Streaky bedding with laminations; small lenticular burrows (5.0 mm) filled with very fine shell hash packstone; large oval echinoid burrows (4-10 cm); argillaceous; thin ash layers; abundant pyrite	0%
Oyster-Echinoid-Algal Grainstone/Packstone (PLATE III)	Oysters; Thin-shelled Mollusks; Echinoids; Red Algae	Green Algae Foraminifers Bryozoans	12-23′ in 3 cores	Large and small scale crossbeds; burrows; algal encrustation and boring of grains; geopetal fill; oil stain in core with stylolitization; dolomite	Interparticle Intraparticle Moldic 10-50%
Oyster-Echinoid-Algal Packstone/Wackestone (PLATE IV)	Oysters; Thin-shelled Mollusks; Echinoids	Red Algae Green Algae Foraminifers Bryozoans	5-7′ in 3 cores	Minor crossbedding; burrows; volcanic clasts; stylolites	Moldic 5-20%
Oyster-Echinoid Green Algal Packstone/Wackestone (PLATE V)	Whole ***Exogyra*** and ***Gryphaea***; Green Algae; Echinoids; Pellets	Red Algae Foraminifers Bryozoans	4-30′ in 4 cores	Burrows; generally lacking structures; geopetals; algal borings; abundant stylolites	Interparticle Moldic 10-30%
Rudist-Algal-Echinoid Packstone/Wackestone (PLATE VI)	Whole ***Durania*** Rudists; Mollusks; Echinoids	Red Algae Green Algae	4′ in 1 core	Burrows (4 cm long)	Intraparticle 5-10%

Table 2. Volcanic Rock Descriptions

DESCRIPTION	ALLOCHEMS	STRUCTURES	CEMENT	ENVIRONMENT
Fine-grained ash beds (PLATE II A)	None	Vague horizontal layering	Calcite	Associated with fine-grained sediments offshore. A fine ash which was airborne and water-laid.
Ash flow lapilli and lapilli tuff (PLATE VII B)	None	Graded beds	Calcite, opal cement	Forms the ash cone of the volcano. Most carbonate sediments rest on this type of deposit.
Lime mud in irregular contact with pyroclastic material. (PLATE II B)	Red algae, oysters	Slumping, microfaulting	Chlorite and calcite cement, dolomite, neomorphosed aragonite	On flanks of volcano marine cemented VRFs (possibly beachrock), some evidence of subaerial exposure, later covered by lime mud. Transgression of beach due to subsidence (?).
Fossil-bearing volcaniclastic sandstone (PLATE V B)	*Major* Gastropod, pelecypod, green and red algae, echinoids *Minor* Foraminifers, coral	Stylolitized grain boundaries, dolomite	Calcite	Beginning of burrowed packstone halo facies in volcanic material.

gested fracture control of volcanism in Mexico, and Romberg and Barnes (1954) found by gravity anomaly that the Pilot Knob volcano, Travis County, occupied an east-west fracture with a less prominent cross fracture.

The carbonate rocks rest on volcanic material, which typically is composed of graded beds of ash-flow lapilli (pl. VI). One core of the flank of the volcano exhibits large, disoriented clasts, probably representing bombs thrown with great force from the volcano. In describing various types of graded beds, Kuenen (1953) noted that many volcanic cones are composed primarily of graded pyroclastic deposits. The volcanic bombs and clasts may be sorted by air initially, but the water column ultimately is far more efficient in this process. The near-spherical shape of the clasts is typical of hot volcanic material deposited in water, causing granulation upon rapid chilling (Fuller, 1931). Kuenen also suggests grading could occur as a result of a nuée ardente eruption, similar to a turbidity flow.

Carbonate Facies and Environments

Shallow-water areas around the submarine cone were soon occupied by organisms; resulting patch reefs and other mollusks provided the shell debris necessary for shoal and fringing beach formation. Although amounts of the in situ reef material in the cores studied are small, the talus from these reefs is abundant in several cores. It has been observed that in modern reef environments talus far exceeds in situ material.

Carbonate facies with high shell content, for example, grainstones, packstones, and boundstones, are best developed along the northeastern flank of the volcanic high (fig. 9). White (1960) and Garner and Young (1976) showed that shoal and reef distribu-

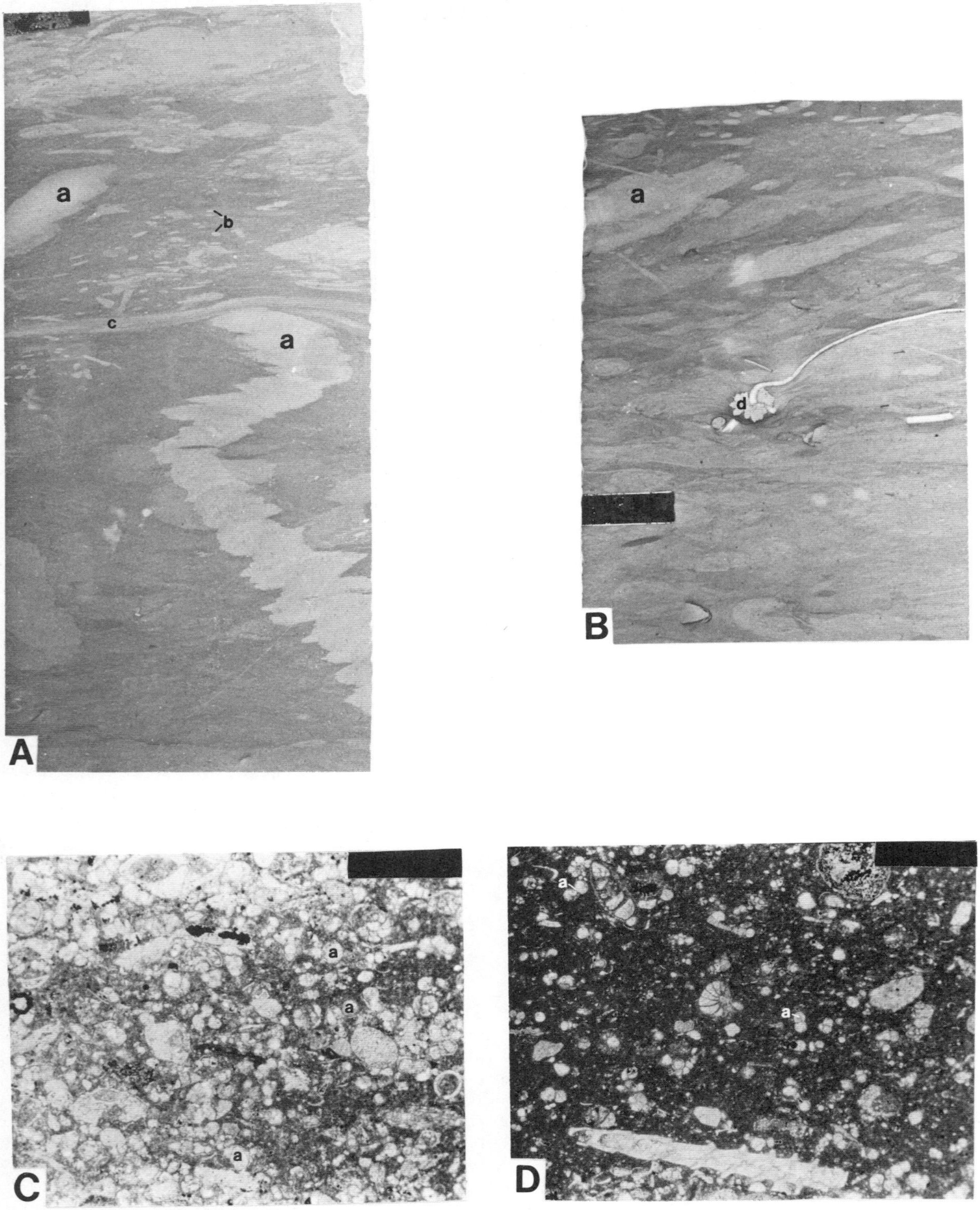

PLATE I

PLANKTONIC FORAMINIFER MUDSTONE/WACKESTONE

A, B. Typical burrowing found in this facies includes large echinoid burrows (a) and smaller worm burrows (b). Preservation of delicate laminae is shown by (c). Rare large shell material usually containing pyrite (d). Core slab for A: #50-58. Core slab for B: #50-12. Both bar scales = 2 cm. C, D. Most abundant allochem is the foraminifer, ***Globigerina*** (a). Other types are also common. Burrow fill tends to be more micritic and shell rich as in C while matrix shows markedly less shell and more clay, D. Thin section for C: #54-4. Thin section for D: 1-165-1. Bar scales for both 0.5 mm.

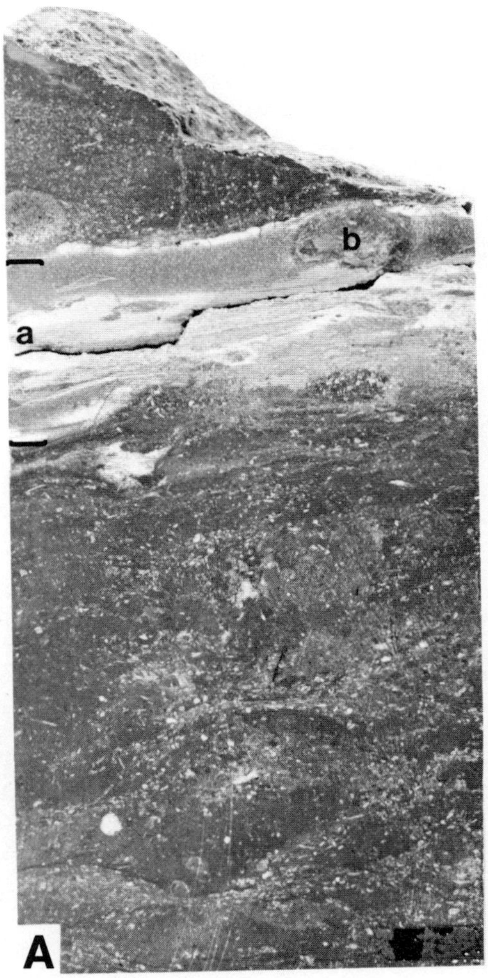

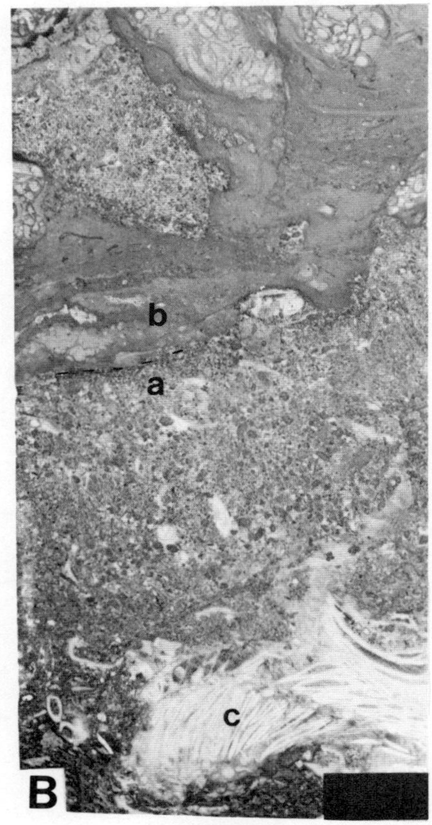

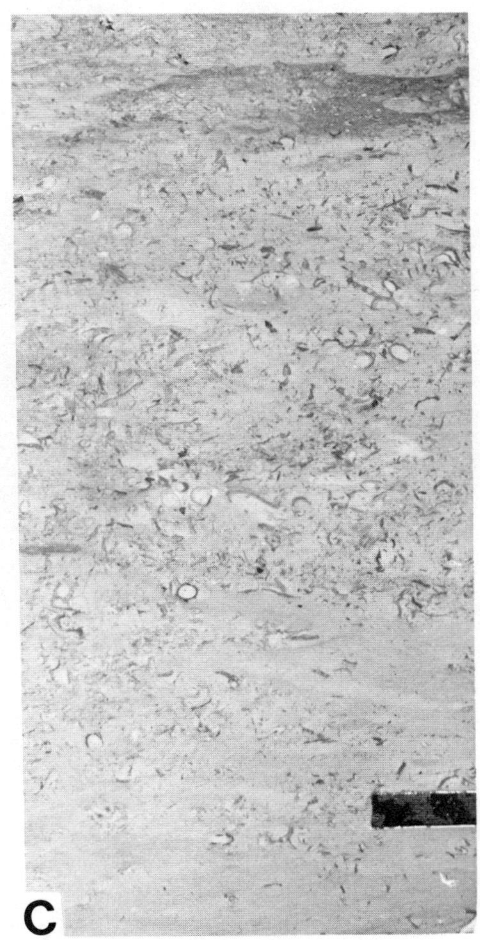

PLATE II

A. Volcanic ash layer (a) interrupting carbonate sedimentation in planktonic foraminifer wackestone facies. Ash layer contains oval echinoid burrow (b). Core slab #1-165-84. Bar scale = 2 cm.
B. Irregular boundary of aragonite-cemented volcanic fragments (possibly beachrock) (a) overlain by planktonic foraminifer wackestone (b) containing reworked fragments of lower unit. Large shell is rudist fragment (c). Core slab #54-18. Bar scale = 2 cm.
C. Sudden influx of shell debris into planktonic foraminifer mudstone. Possible mudflow or slumping. Core slab #54-15. Bar scale = 2 cm.

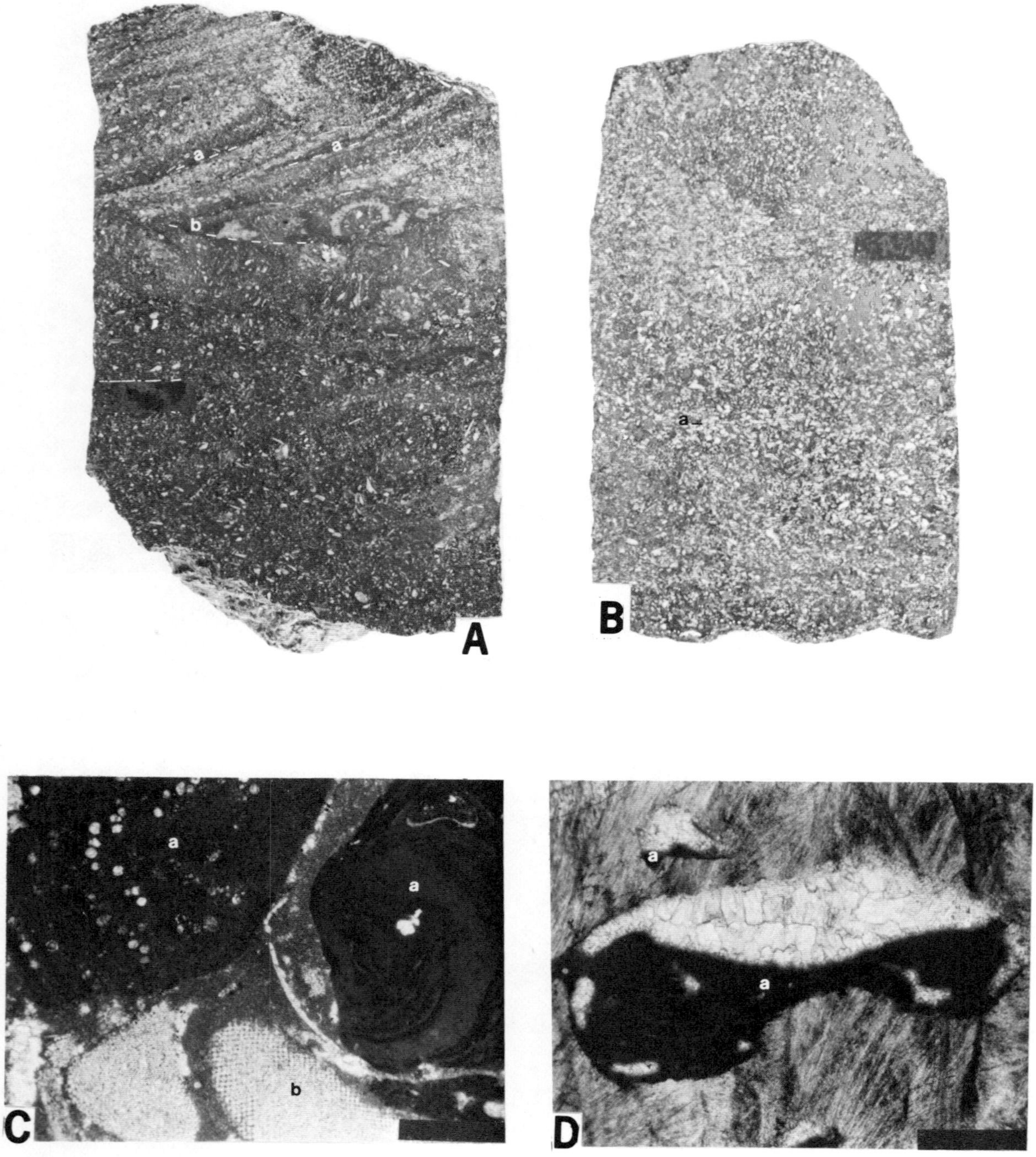

PLATE III

OYSTER-ECHINOID-ALGAL GRAINSTONE

A. Crossbedding of shoal complex (a) with erosional surface (b). Dark color is due to heavy oil stain. Core slab #165-64. Bar scale = 2 cm.

B. Well-sorted grainstone composed of rounded grains with some parallel, horizontal orientation (a). Heavily oil stained. Core slab #1-165-60. Bar scale = 2 cm.

C. Red algae, ***Archaeolithothamnion*** (a), with cellular structures preserved. This encrusting algae often binds shell in the shoal area. Echinoid plate (b). Thin section #55-16. Bar scale = 0.5 mm.

D. Geopetal fill (a) in oyster shell structure. Later, pore space filled with equant calcite cement. This type of structure is common in the grainstone shoal complex. Thin section #51-12b. Bar scale = 0.5 mm.

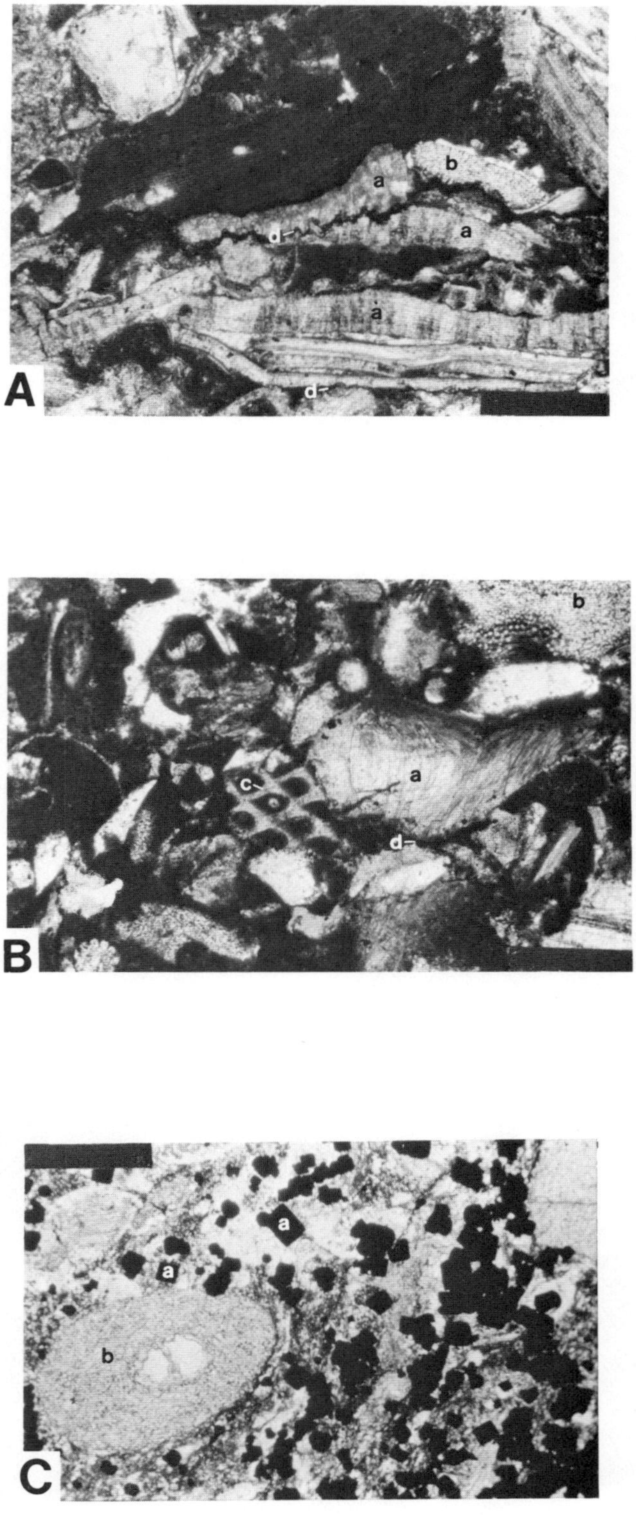

PLATE IV

OYSTER-ECHINOID-ALGAL PACKSTONE

A, B. Dominant allochems are best viewed in this facies in thin section. They include: oyster fragments (a), echinoid plates (b), ***Gryphaea*** fragments (c). Stylolites are common along grain boundaries (c). Fragments of shell vary from rounded to very angular. The occurrence of rounded shell suggests washovers from the beach-shoal area. Thin section A: #51-54. Thin section B: #51-51. Bar scales for both = 0.5 mm.

C. Abundant pyrite (a) occurs in zones, suggesting reducing conditions. Echinoid fragment of single crystal is shown (b). Thin section #1-165-84. Bar scale = 0.5 mm.

tion of the Pilot Knob volcano was also best developed to the northeast. A study of the volcano at Thrall in Williamson County by Udden and Bybee (1916) mentions a "porous shell breccia" in this same orientation. This relationship at several localities suggests conditions common to the volcanoes which would favor biologic productivity and shell reworking. A prevailing wind from the northeast would favor development of reefs as well as provide energy for reworking of talus and other shell into grain bars and beaches. Folk and Robles (1964) documented greater biologic activity and shell material on the windward side of Alacran Reef in Mexico with mud accumulating leeward of the reef. The leeward side of the Elaine field volcano shows some accumulation of shell but no occurrence of grainstones.

The windward side of the volcano has a gentle slope, whereas the opposite side is steeper. Thickness of the transgressive packstone halo facies (fig. 10) was apparently affected by these differences in slope causing the facies to move more quickly up the windward or high-energy side of the volcanic flank. Higher wave action could also plane off the windward flank of the volcano before carbonate deposition occurred. For example, on Bikini Island today the windward side of the island has a gentler slope than the leeward side (Emery and others, 1954). Other factors to consider include increased subsidence on one flank of the volcano causing slope steepening. The volcanic flanks appear to be steeper after carbonate deposition than they were at the time of deposition (figs. 11, 12). Whether this subsidence caused the slope steepening of the flank or influenced carbonate deposition is unknown; however, it is obvious that subsidence on the flanks of the volcano continued for some time, affecting sediments of the upper Taylor Group and lower Navarro Group (fig. 2). The center of the volcano probably did not experience as much subsidence as the flanks, because it is more rigid and resistant to loading by overlying strata than the loosely packed pyroclastics of the flanks.

The shallow-water facies are dynamic, and interchange occurs among all of them. Fringing shoal areas were cut by tidal channels, allowing rounded grains from the shoals to be transported into the lower energy lagoons where muds were being deposited. The shoal sands were also reworked into deeper, fringing environments by storm ebbs. The fringing

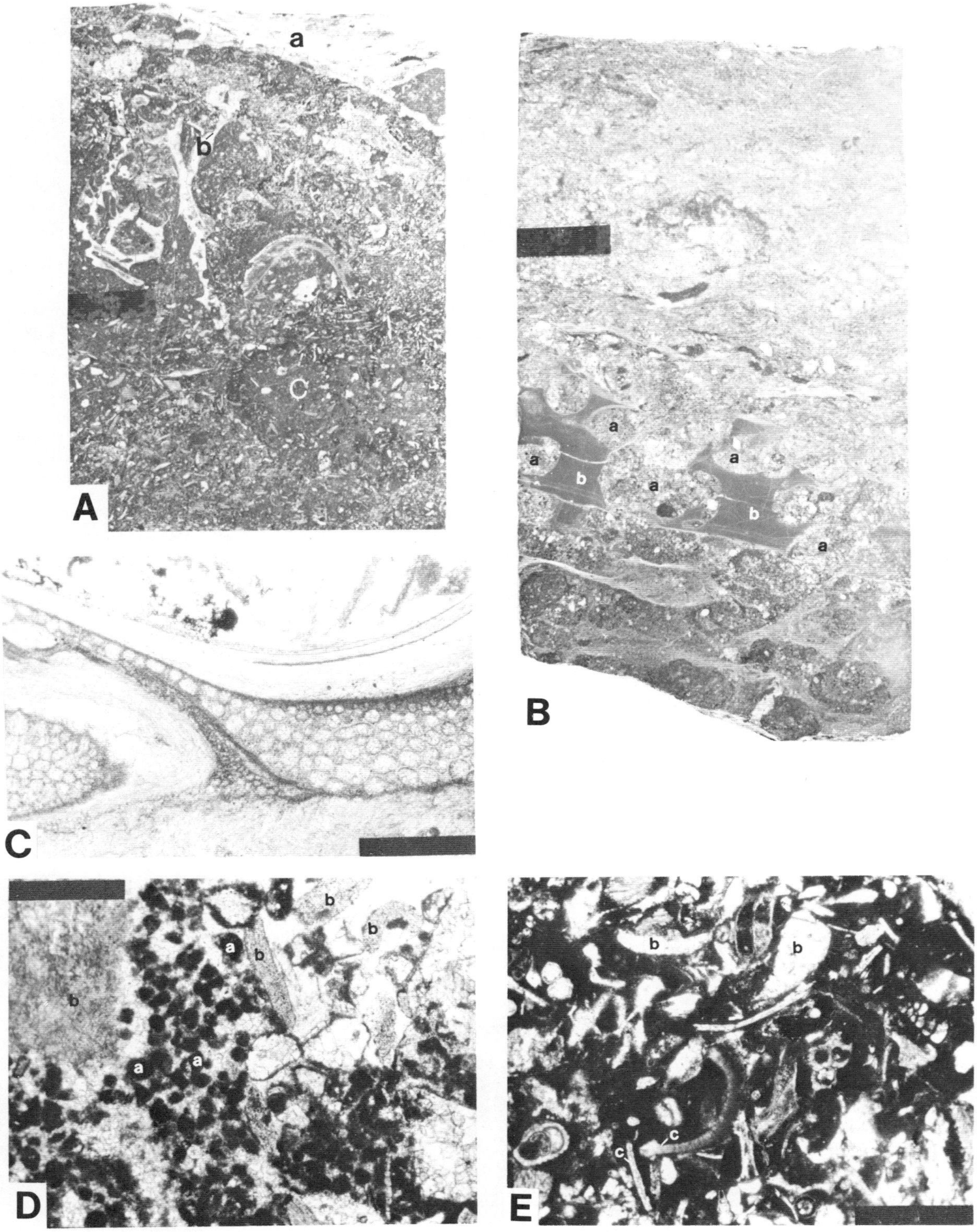

PLATE V

OYSTER-ECHINOID GREEN ALGAL PACKSTONE

A. Whole (a) and broken (b) ***Gryphaea*** and ***Exogyra*** shells. Round echinoid burrow (c). Core slab #55-36. Bar scale = 2 cm.

B. Basal part of unit. Echinoid burrows (a) in volcanic ash layer (b). Volcanic clay-rich at base becoming more calcareous upward. Core slab #52-28. Bar scale = 2 cm.

C. Detail of ***Gryphaea*** shell structure found in packstone halo. Thin section #55-26. Bar scale = 0.5 mm.

D, E. Abundant pellets (a), echinoid fragments (b), thin shelled mollusk fragments (c), pelagic foraminifers (d) are often seen in the packstone halo. Thin section in D: #55-58. Bar scale = 0.5 mm. Thin section in E: #55-12. Bar scale = 0.5 mm.

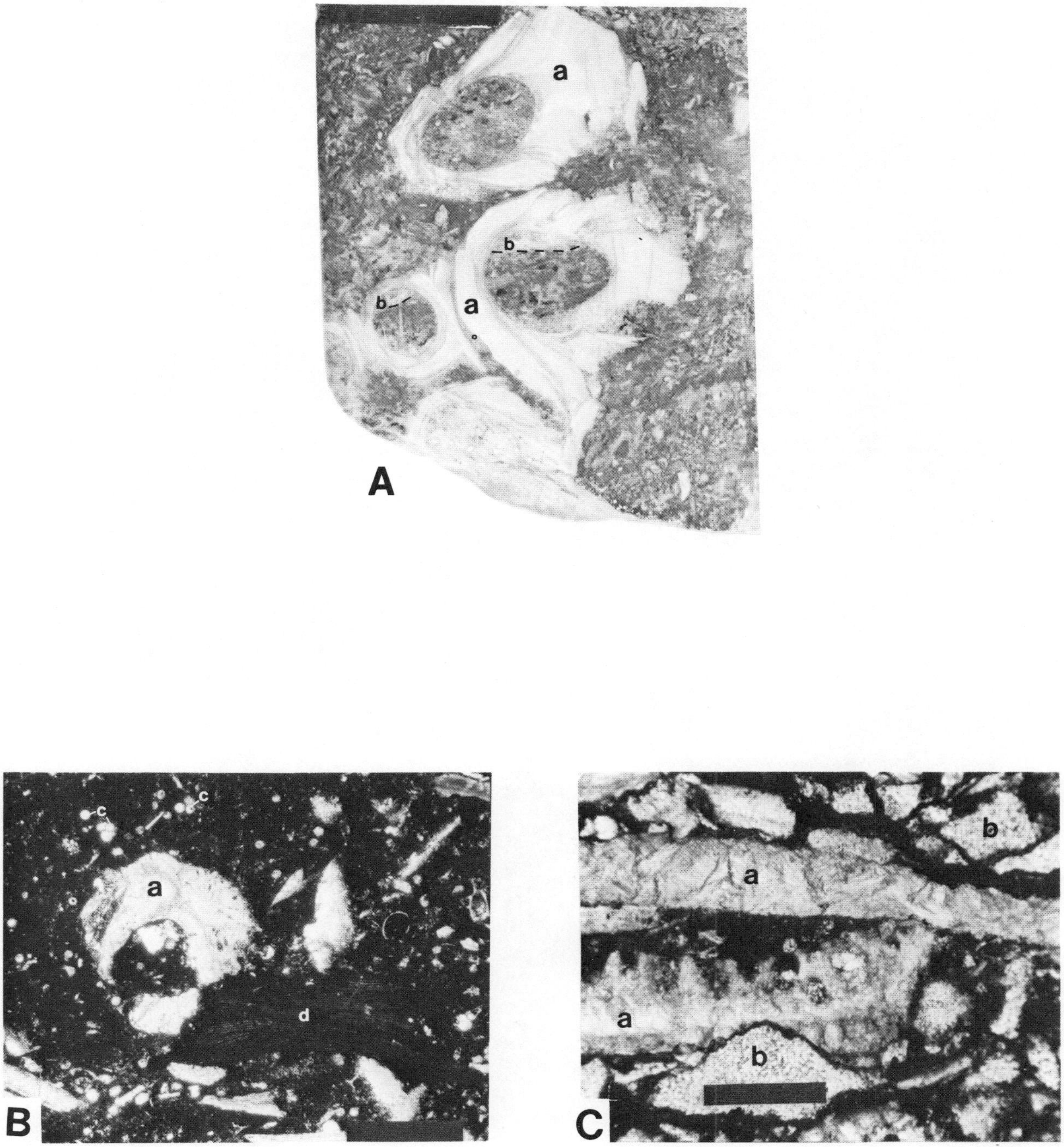

PLATE VI

RUDIST-ALGAL-ECHINOID PACKSTONE

A. Detail of rudist (***Durania***) boundstone (a) with geopetal fill (b). This rudist formed small patch reefs seaward of the grainstone shoals. Core slab #55-73. Bar scale = 2 cm.

B, C. Typical shell material associated with rudist talus; oyster fragments (a), echinoid plates (b), pelagic foraminifers (c), and red algae (d). Thin sections #55-74. Bar scales = 0.5 mm.

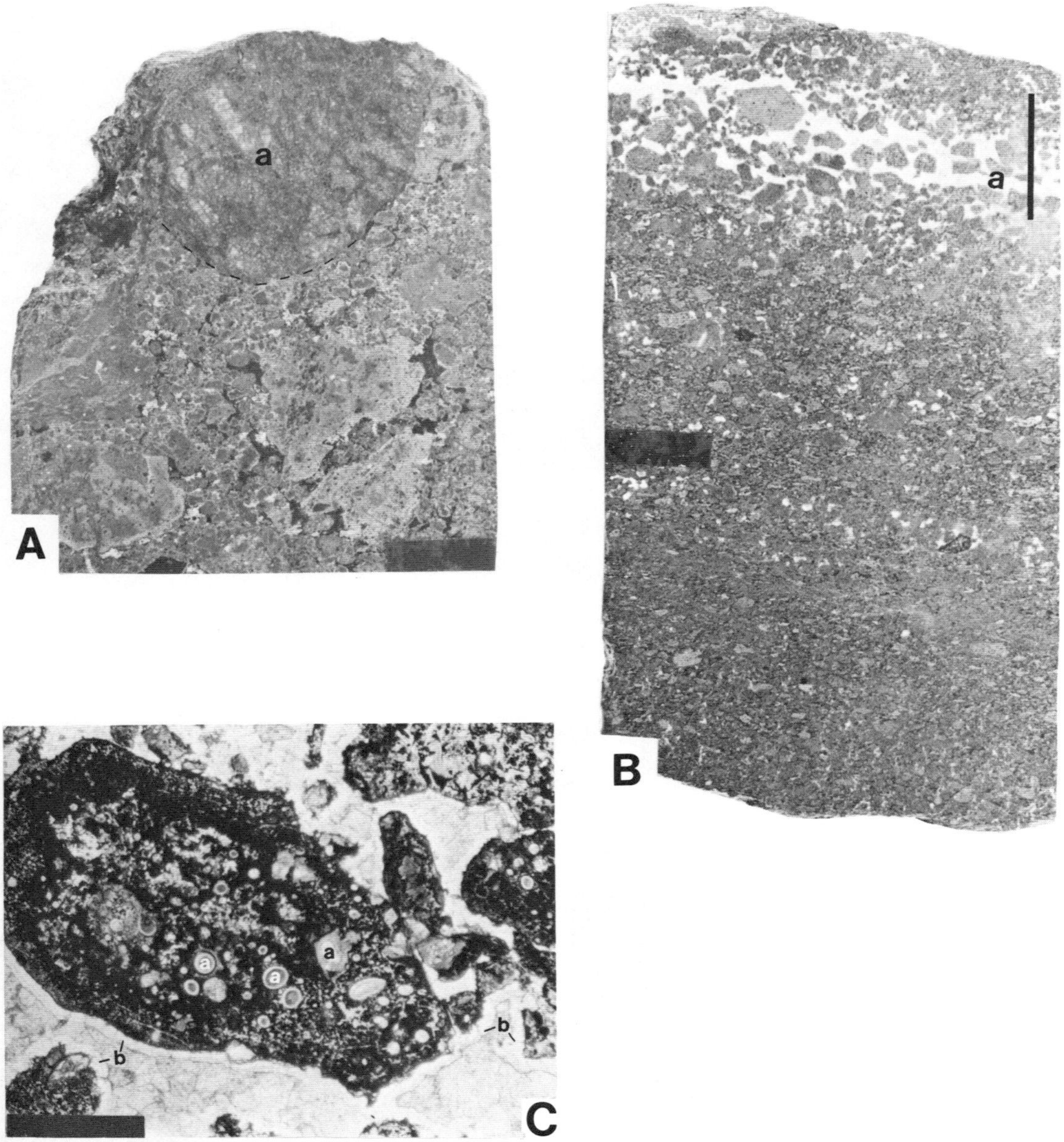

PLATE VII

A. Large rounded volcanic clasts (a) in fine-grained matrix. These large clasts could possibly be "bombs" ejected from the volcano. Core slab, unknown depth. Bar scale = 2 cm.
B. Layers of graded beds (a) which represent the structure of the majority of volcanic deposits of the study area. Core slab #55-88. Bar scale = 2 cm.
C. Thin section of a volcanic clast retaining relict gas vessicles (a). These are often partially filled by fibrous chlorite. A thin clear rim of opal cement (b) surrounds clast, followed by equant calcite cement. Thin section #52-30. Bar scale = 0.5 mm.

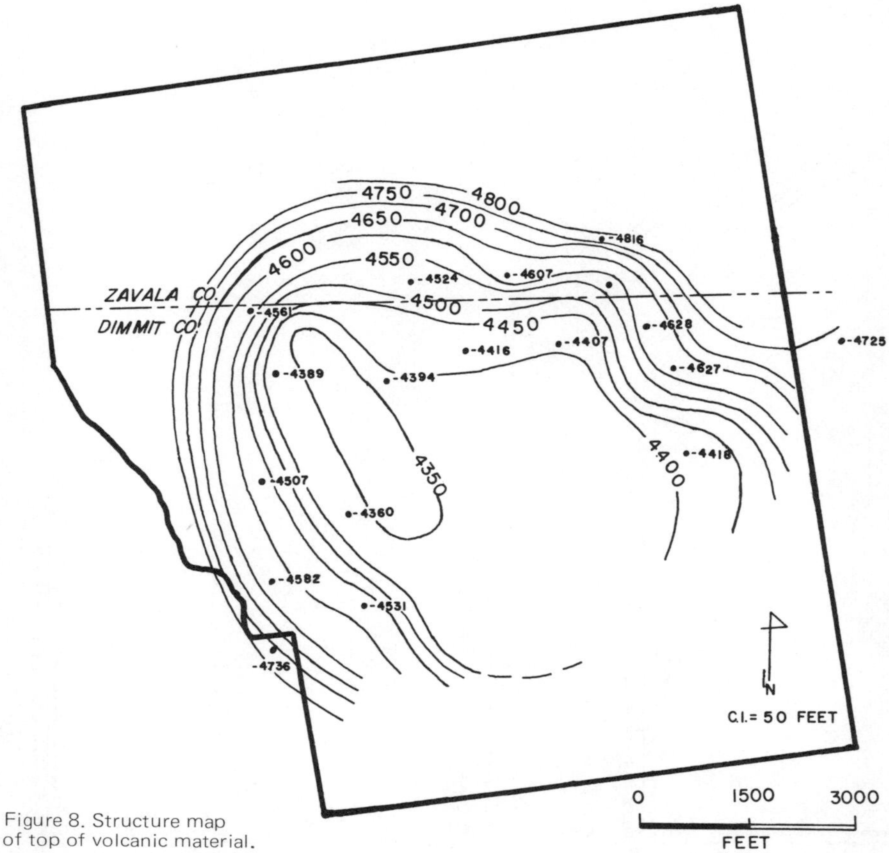

Figure 8. Structure map of top of volcanic material.

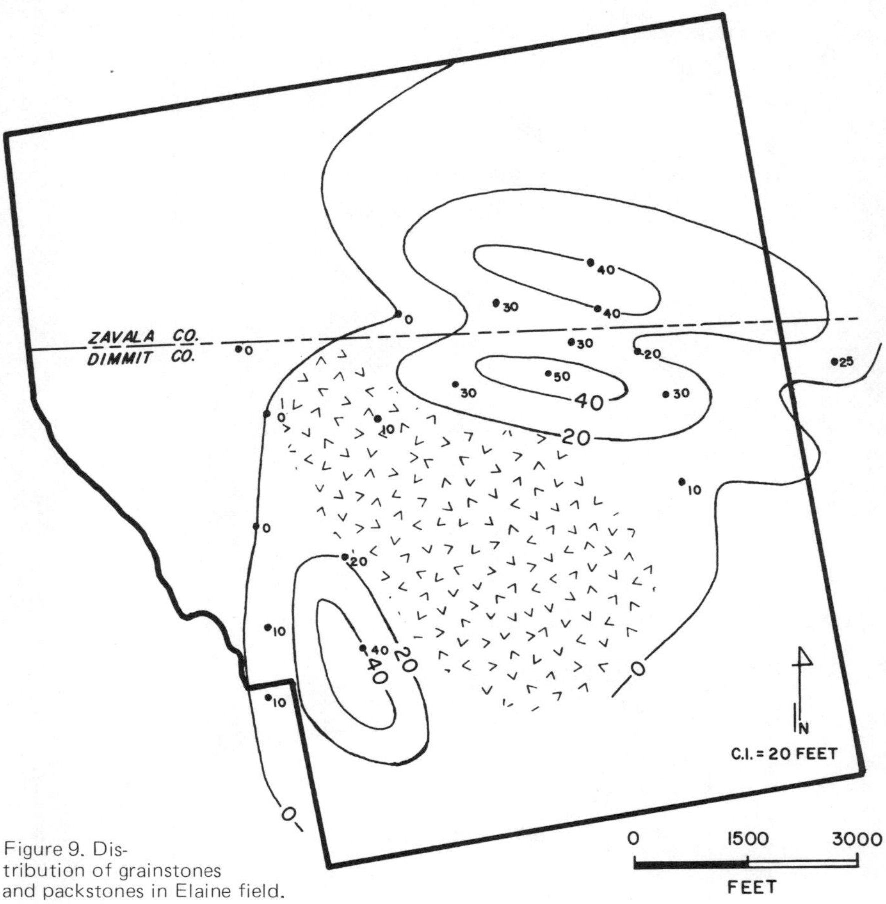

Figure 9. Distribution of grainstones and packstones in Elaine field.

"packstone halo" formed in the lower shoreface environment (fig. 13).

The low-energy lagoon environment is characterized by poor rounding and sorting of grains, with oyster, echinoid, and red algal fragments most common. The presence of pyrite suggests reducing conditions, and volcanic clasts reworked within the lagoon attest to close proximity to the volcanic island. Folk and Robles (1964) discuss the characteristics of mud-rich protected environments present today, and Scott (1968) describes the lagoonal environment along the modern Texas coast as a successful habitat for oyster growth.

Rudist talus, whole ***Exogyra*** and ***Gryphaea*** shells, and green algae are most common in the packstone halo. The abundance of this shell material decreases seaward as water deepens, mud content increases, and the facies grades into the open-shelf argillaceous planktonic foraminifer mudstone. Bioturbation is most evident in these offshore facies with delicate burrowing traces commonly preserved in the open-shelf mudstone. Pyrite is very rare in the packstone halo, but it is common in the mudstone. This variance is consistent with the interpretation of a relatively high-energy, oxygenated fringe environment grading into a low-energy, reducing environment. Sediment baffling by green algae could have contributed to mud accumulation in the packstone halo, a phenomenon often seen in modern (Ginsburg and Lowenstam, 1958) and ancient environments (Wilson, 1975).

The term "shoal complex" in this study includes four environments: beaches, red algal ridges, tidal inlets, and rudist reefs. Maximum development of the shoal complex occurs in areas of greatest current or wave energy. Red algal reef development has been reported as particularly sensitive to these conditions (Johnson, 1971). Colin (personal communication, 1976) has observed red algae growing at a 220-ft depth off the coast of Bermuda, indicating shallow water is not the exclusive environment of these organisms. Adey and MacIntyre (1973) reevaluate the shallow-water dogma for red algae. Red algal fragments can be identified in all the environments of this study, but are found most commonly encrusting and binding rounded shell material associated with the shoal area. These boundstones of ***Archaeolithothamnion*** grew parallel to and seaward of the fringing beaches. Similar distribution of the red alga ***Lithothamnion*** can be seen in modern

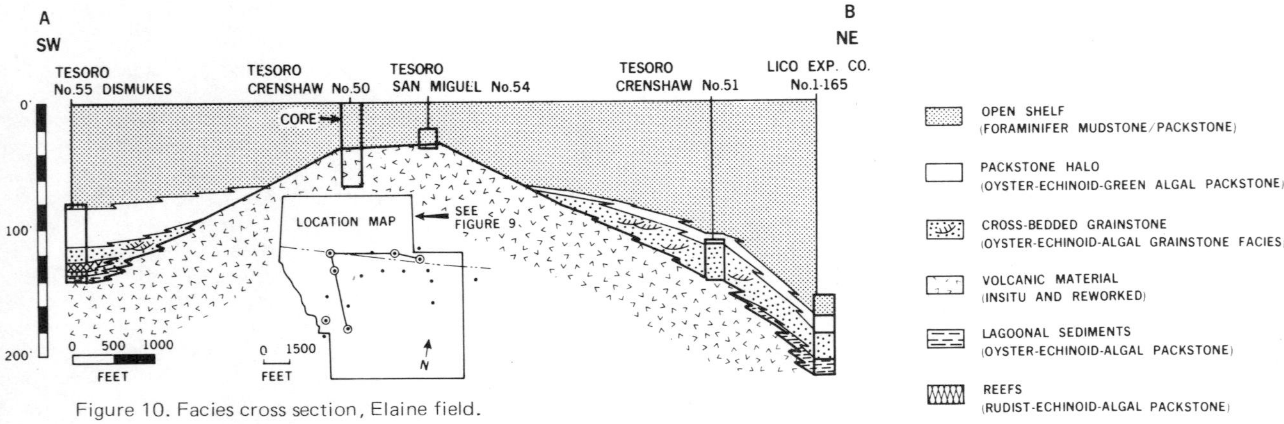

Figure 10. Facies cross section, Elaine field.

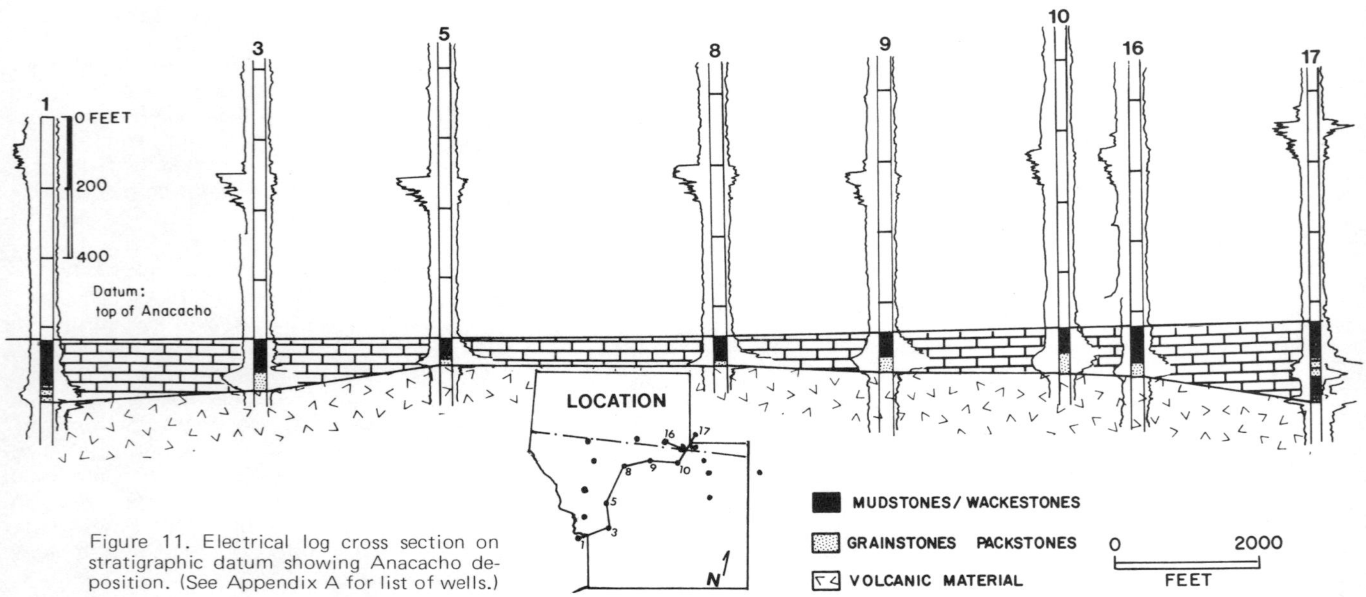

Figure 11. Electrical log cross section on stratigraphic datum showing Anacacho deposition. (See Appendix A for list of wells.)

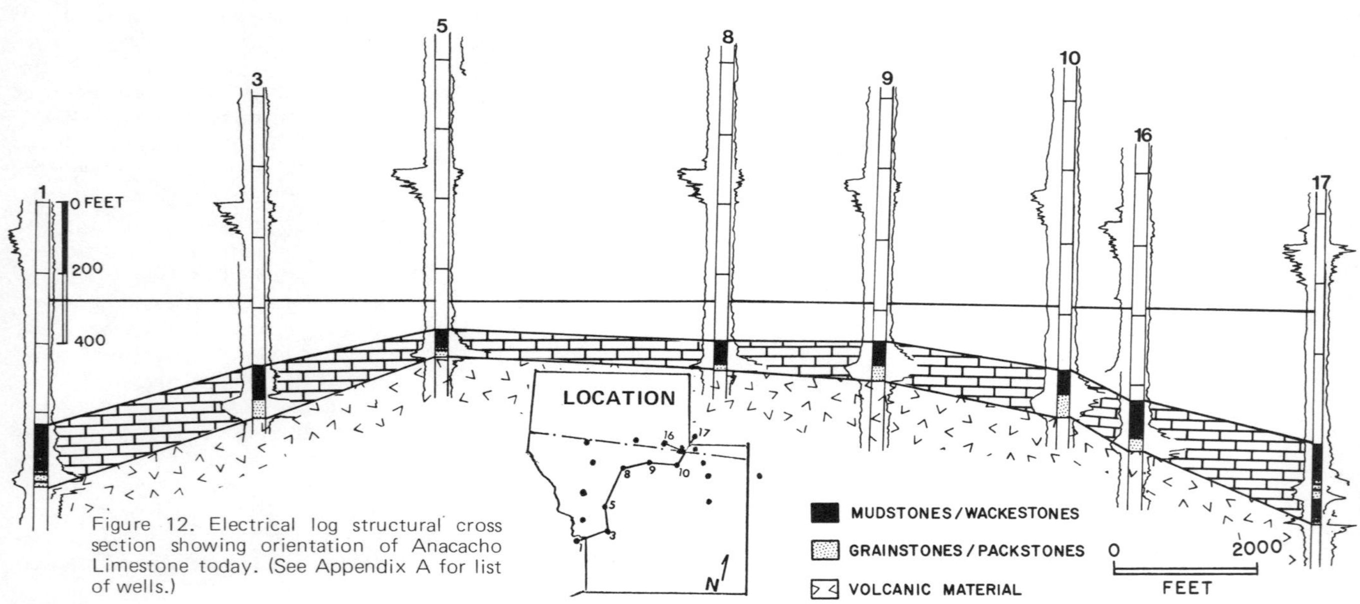

Figure 12. Electrical log structural cross section showing orientation of Anacacho Limestone today. (See Appendix A for list of wells.)

environments (fig. 14). Two possibilities could account for the distribution seen in the study area: either the red algae grew in shallow water at the same time the beaches were forming (fig. 14), or the algae encrusted the carbonate beach after it was lithified by beachrock cementation and subsided into deeper water. Interstratification of the algal boundstone with beach and upper shoreface environments and lack of evidence for beachrock cementation suggest that the algal ridges were contemporaneous with shallow-water facies.

The association of red algae with radiolitid rudists has been noted by Wilson (1975) in the middle Cretaceous of Texas and Mexico. The rudists, which form the patch reefs in the study area, were identified as ***Durania*** (pl. V) by Young (The University of Texas at Austin, personal communication, 1976). This radiolitid rudist, which had requirements similar to those of red algae, is found where conditions of good water circulation prevailed, especially the upper shoreface. Perkins (1974) found similar distribution of rudists in the Glen Rose Limestone of Central Texas. Stenzel (1971) considered robust form, simple ribbing, and rarity of frills as morphologic adaptations for habitats with vigorous wave action and water circulation. These rudists contribute large amounts of shell material which is subsequently reworked into shoals landward of the reef.

Carbonate beaches are recognized in ancient rocks by several criteria observed in modern beaches: (1) rounding of grains (Folk and Robles, 1974), (2) near-horizontal laminae within low-angle dipping beds (Wilson, 1975), (3) general lack of mud, that is, good sorting (Folk and Robles, 1964), and (4) associated bounding facies relationships. These criteria were used to recognize beach deposits within the oyster-echinoid-algal grainstone facies. Outcrops of carbonate beaches are easily identified by low-angle crossbeds, associated trough crossbeds, and lateral relationships—features difficult to observe using core samples. Diagenetic fabric, discussed in the last section of this paper, supports the conclusion that the shoals were at times subaerially exposed as beaches. When the exposure occurred, barriers developed, and a lagoonal environment was established. The barriers were cut by tidal channels to accommodate tidal exchange (fig. 15). Low-angle dipping sets of laminated sand and large ripple crossbedding characterize this foreshore facies. Tidal channels or washovers are inferred because of the occurrence of rounded sand grains in the lagoonal sediments and the presence of these inlets in modern clastic and carbonate barriers. Evidence was also found to support the presence of a beach on the volcanic

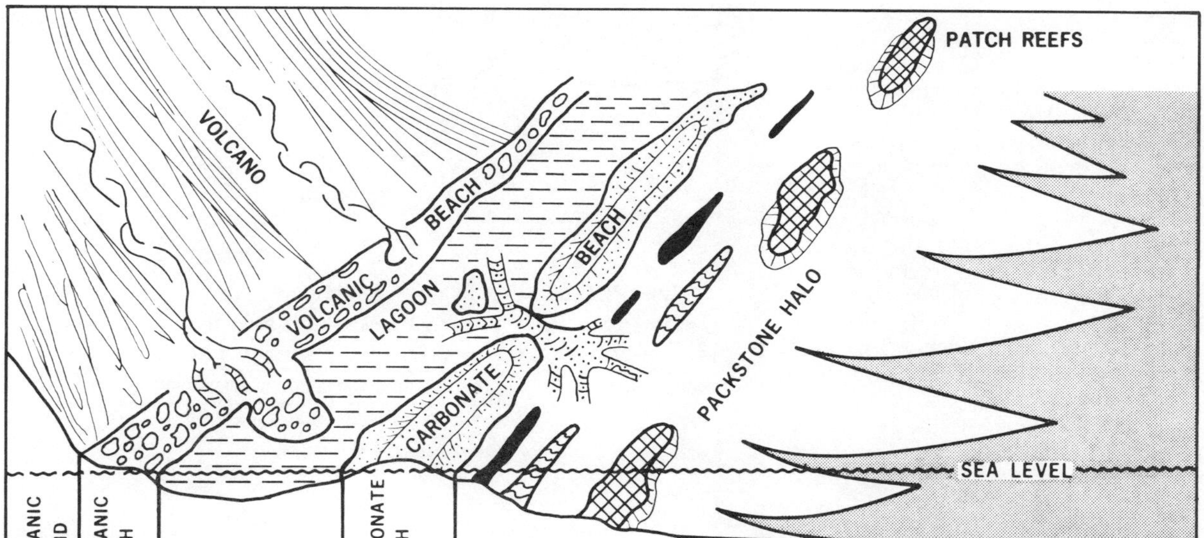

ENVIRONMENT	VOLCANIC ISLAND	VOLCANIC BEACH	LAGOON	CARBONATE BEACH	SHOREFACE	TRANSITION ZONE	OPEN SHELF
DOMINANT FEATURES		POOR SORT-ING	VOLCANIC DEBRIS, BEACH ALLOCHEMS	ALGAL BORING, ROUNDING, SORTING	RED ALGAL RIDGES, RUDIST REEFS, PACKSTONE HALO	STREAKY BEDDING, REEF DEBRIS	STREAKY BEDDING, LAMINATIONS, ARGILLACEOUS
PHYSICAL PROCESSES	EROS-ION	RE-WORK-ING	BURROWING, RIPPLES	CROSS-BEDDING, STORM DEPOSITS	CROSS-BEDDING, STORM DEPOSITS	BURROWING, STORM DEPOSITS	BURROWING, SLOW DEPOSITION, STORM DEPOSITS
TEXTURE			PACKSTONE GRAINSTONE WACKESTONE	GRAIN-STONE BOUND-STONE PACK-STONE	PACKSTONE WACKESTONE • BOUNDSTONE	WACKESTONE	MUDSTONE WACKESTONE

Figure 13. Facies tract detail.

island itself. Stratigraphic position of the deposit, as well as diagenetic fabric revealing beachrock-type cementation followed by meteoric phreatic cements, strongly suggests this environment. Because the beach deposit is composed mostly of rounded volcanic fragments with minor shell content, it is discussed in the volcanic rock section of the facies description rather than as a carbonate facies.

Following initial development of the shoal and lagoonal environments, which probably prograded somewhat basinward, subsidence of the island prompted transgression by the facies to maintain shallow-water conditions. Thus, the shoal environments were reworked back over the lagoon, preserving only a remnant of the lagoon (fig. 10). Once subsidence began, it apparently accelerated, and lagoonal deposits were completely removed near the upper flank. When the shoal complex could no longer accommodate increasing subsidence rate, the open-shelf basinal facies covered the entire sequence. The basinal facies therefore thins over the high areas of the volcanic and shoal complexes and reaches maximum thickness off the flanks (fig. 16).

Recent analogs best depicting the processes observed in this study are common in the South Pacific. The Austral Islands, for example, have similarities to the study area. Shoals, beaches, lagoons, tidal channels, and reefs seen in the South Pacific Islands are all developed as in Elaine field (fig. 17). Abundant examples of atoll formation in the Austral Islands (Devaney and Randall, 1973) illustrate the ultimate fate of the islands—a process of upbuilding and subsidence. This process is similar to that which occurred in the Elaine field area (fig. 18).

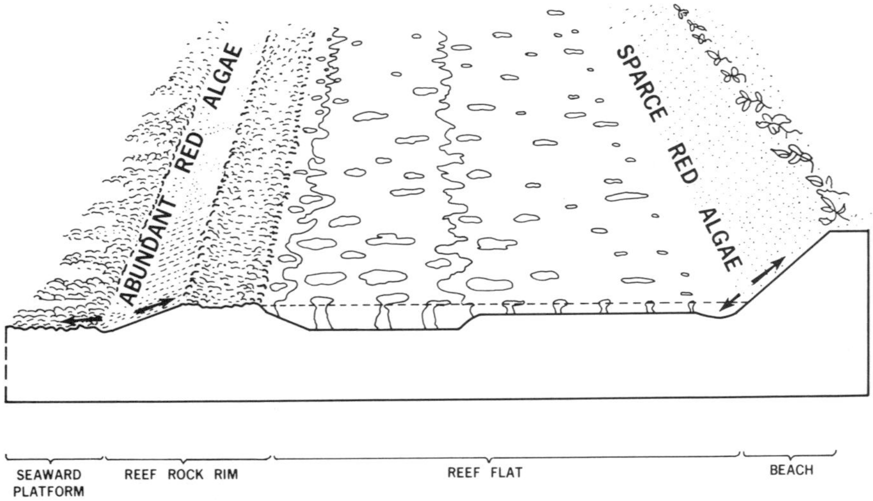

Figure 14. Distribution of red algae, ***Lithothamnion,*** in Recent environments. After Cribb (1973).

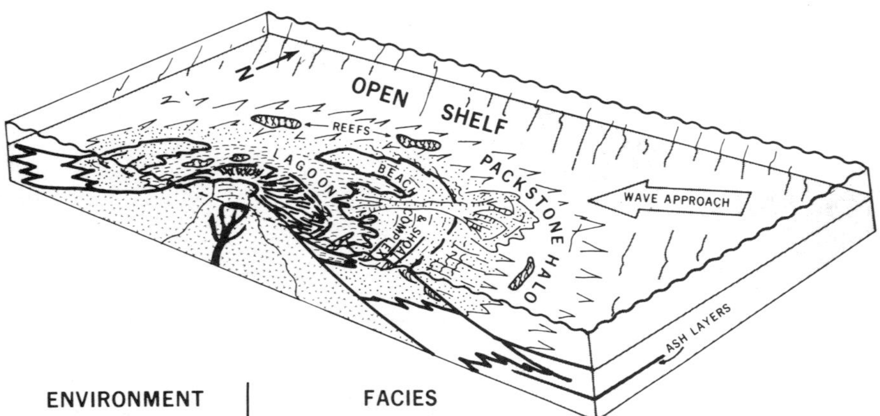

ENVIRONMENT	FACIES
LAGOON	OYSTER-ECHINOID-ALGAL PACKSTONE
BEACH/SHOAL COMPLEX	OYSTER-ECHINOID-ALGAL GRAINSTONE
PACKSTONE HALO	OYSTER-ECHINOID GREEN ALGAL PACKSTONE
REEFS	RUDIST-ECHINOID-ALGAL PACKSTONE
OPEN SHELF	FORAMINIFER MUDSTONE

Figure 15. Block diagram showing environmental reconstruction.

DIAGENESIS

General Statement

Diagenetic processes, which alter carbonate sediments mechanically, biologically, and chemically, influence retention, enhancement, and reduction of porosity. Porosity is found in varying percentages in all carbonate facies of Elaine field, but is best developed in the grainstone-packstone facies. An isopach map of all facies with greater than 5-percent porosity (fig. 19) shows a configuration very similar to the grainstone/packstone isopach map (fig. 9). Data for the porosity-distribution map was obtained from formation-density logs and direct measurement from core.

Diagenesis of the Elaine field carbonates is classified in relative terms as "early" or "late." Early diagenesis refers to alteration of carbonate sediments just after deposition but before burial. Cements formed during this period reflect specifically the environment of deposition. Late diagenesis is limited to alteration occurring with deep subsurface burial.

Diagenetic Model for Anacacho Grainstone Facies

Cements in Anacacho grainstone facies reflect different water chemistries within original pore space of carbonate grains. Folk (1974) proposed a diagenetic model based essentially on the effect of the Mg/Ca ratio and salinity of these pore fluids in controlling mineralogy and crystal shape. Several diagenetic environments affected the grainstone facies. They include marine, meteoric (local and regional), and deep subsurface (fig. 20).

Marine diagenetic environment

The effects of the marine environment on allochems include (1) abrasion (pl. VIII), (2) microborings (pl. VIII C, D, E) which ultimately form micrite envelopes (pl. VIII, A, C, D) and micritization of grains (pl. VIII

F), and (3) isopachous aragonite cement (pls. IX, X). The first two alterations are chiefly mechanical due to waves and biogenic activity. These effects are best developed in shallow water where energy is greatest and light penetration encourages growth of boring algae (Perkins and Halsey, 1971). Isopachous aragonite formed as beachrock to submarine cement under open-marine conditions. Submarine cementation is restricted in quiet-water environments because carbonate mud frequently fills pore space and inhibits aragonite crystal growth. Therefore, crystallization of aragonite in mud-rich carbonates is found most often in open structural parts of shells which are protected from mud accumulation. Marine cement, originally aragonite or high-Mg calcite, is stable in the marine waters and characteristically precipitates as blunt-tipped, fibrous, isopachous cement (pls. IX, X). Examples of this cement in Elaine field have been observed within structural pores of animals and as intergranular cement in beachrock. This cement is not as stable as other types outside the marine environment and changes to a neomorphic mosaic spar (pl. IX E). The interpretation of beachrock has been suggested because of its proximity to the marine environ-

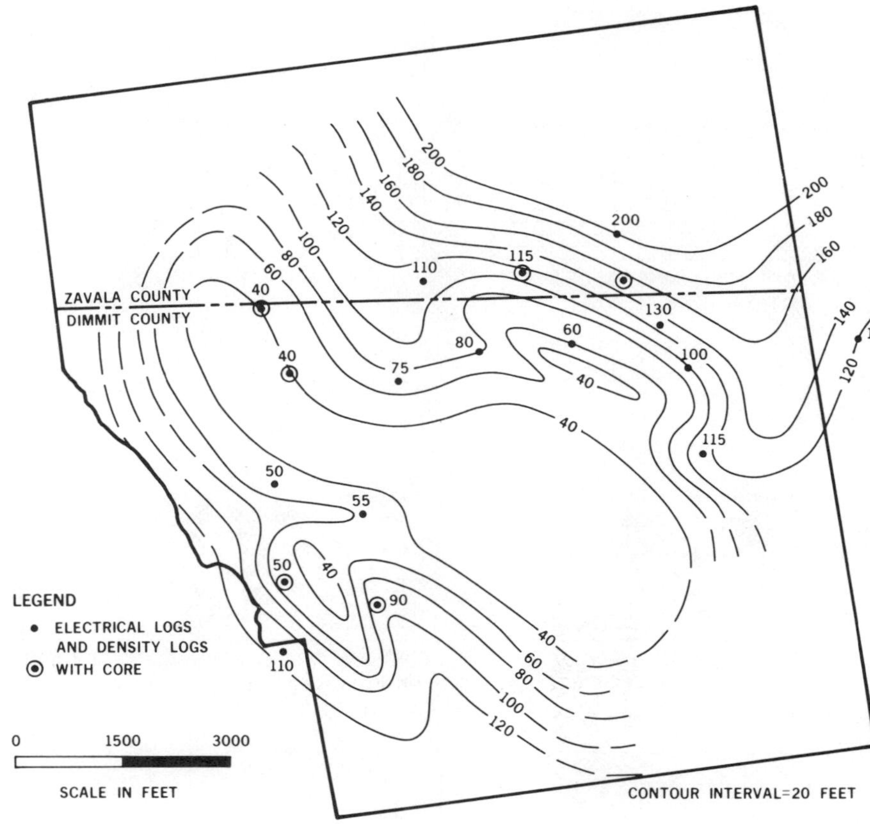

Figure 16. Isopach of mudstone/wackestone units.

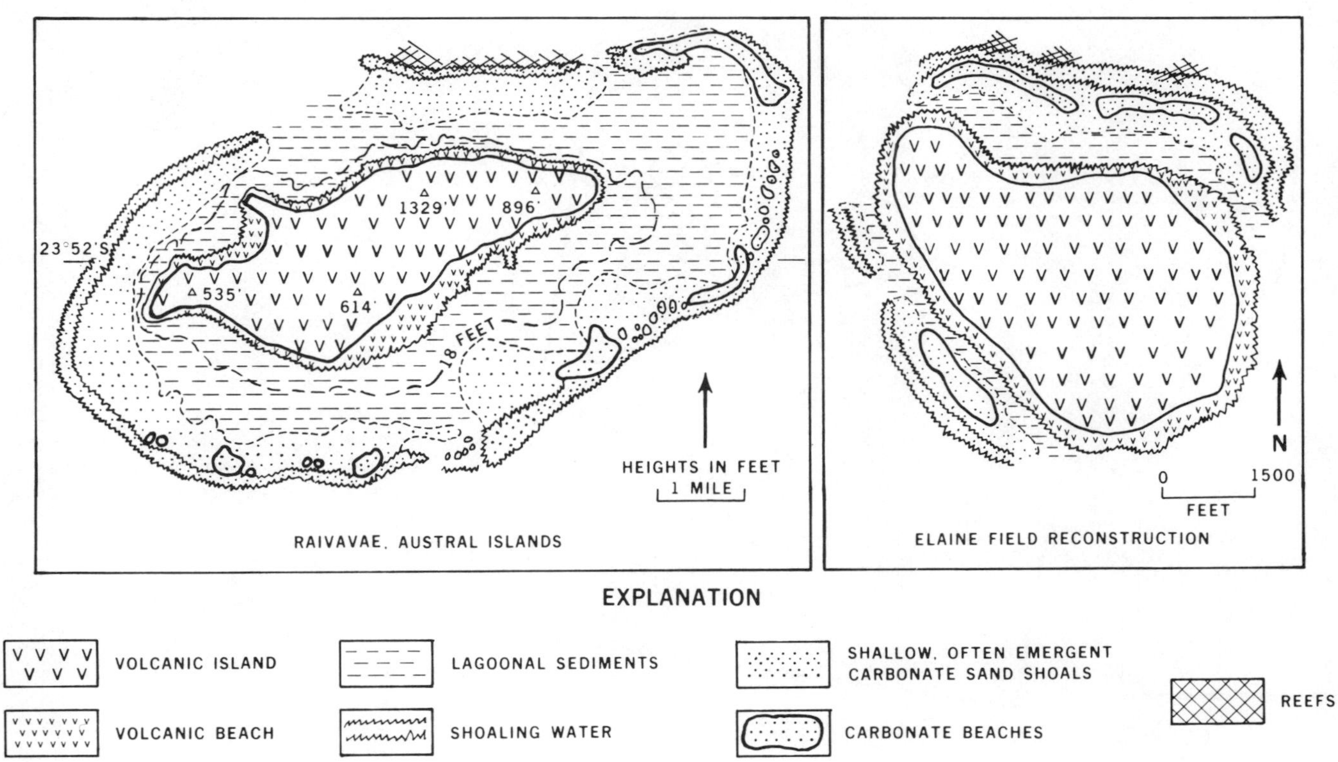

Figure 17. Comparison of environments and distribution of facies within modern island and ancient island.

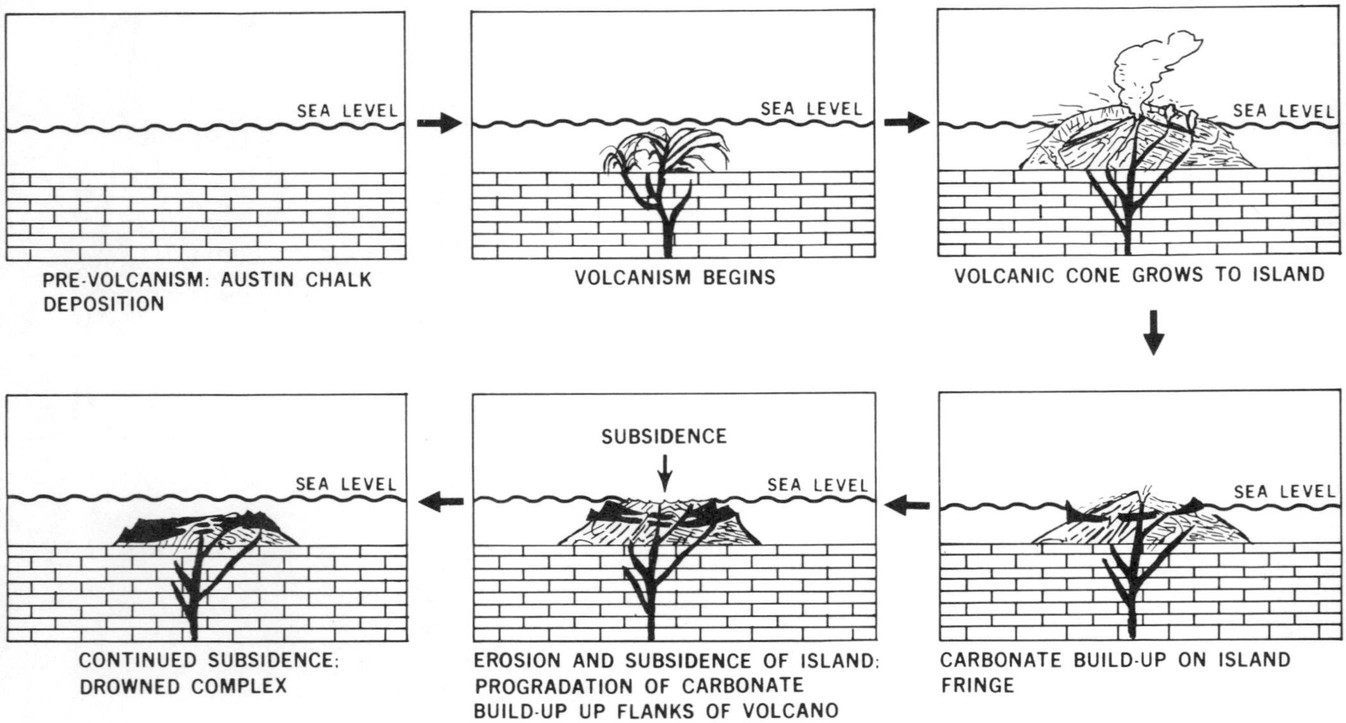

Figure 18. Proposed tectonic history and depositional development of Elaine field volcano.

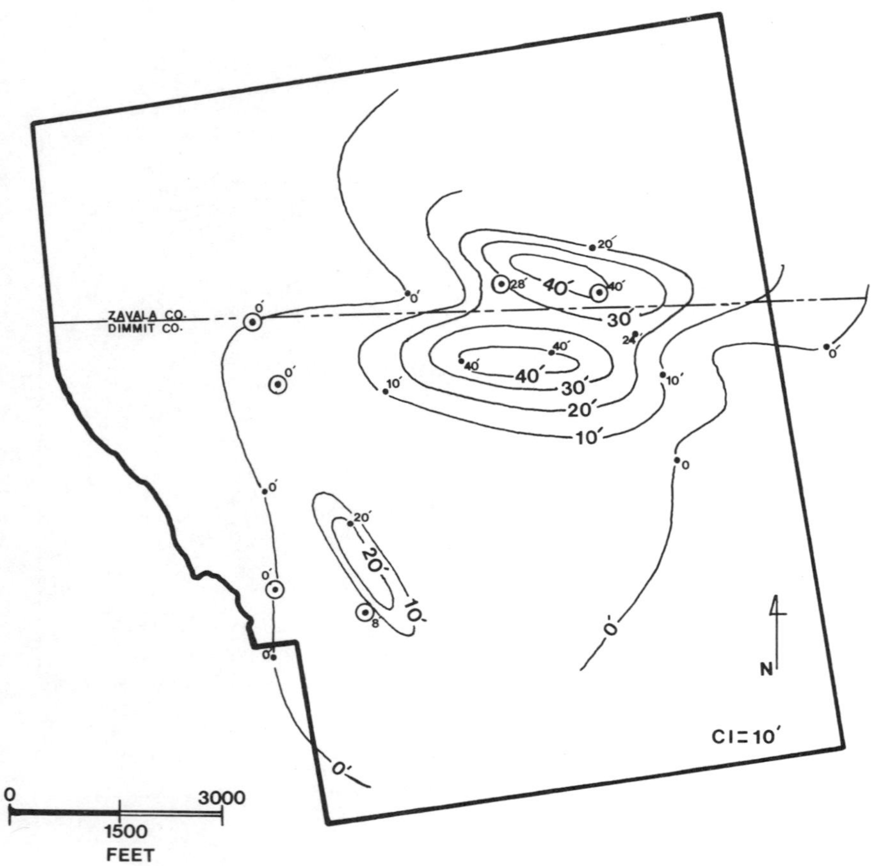

Figure 19. Isopach of facies exhibiting greater than 5 percent porosity.

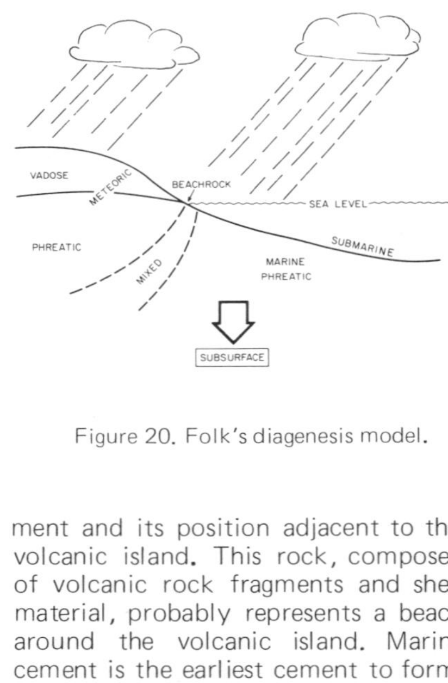

Figure 20. Folk's diagenesis model.

ment and its position adjacent to the volcanic island. This rock, composed of volcanic rock fragments and shell material, probably represents a beach around the volcanic island. Marine cement is the earliest cement to form; it predates even echinoid syntaxial cementation.

Meteoric-vadose, phreatic, and mixing-zone environments

A fresh-water lens developed with subaerial exposure of carbonate beaches. As this lens may extend beneath the surface as much as 40 times its height above sea level (Purdy, 1968), its effects are extensive relative to the surface area (fig. 21). Allochems

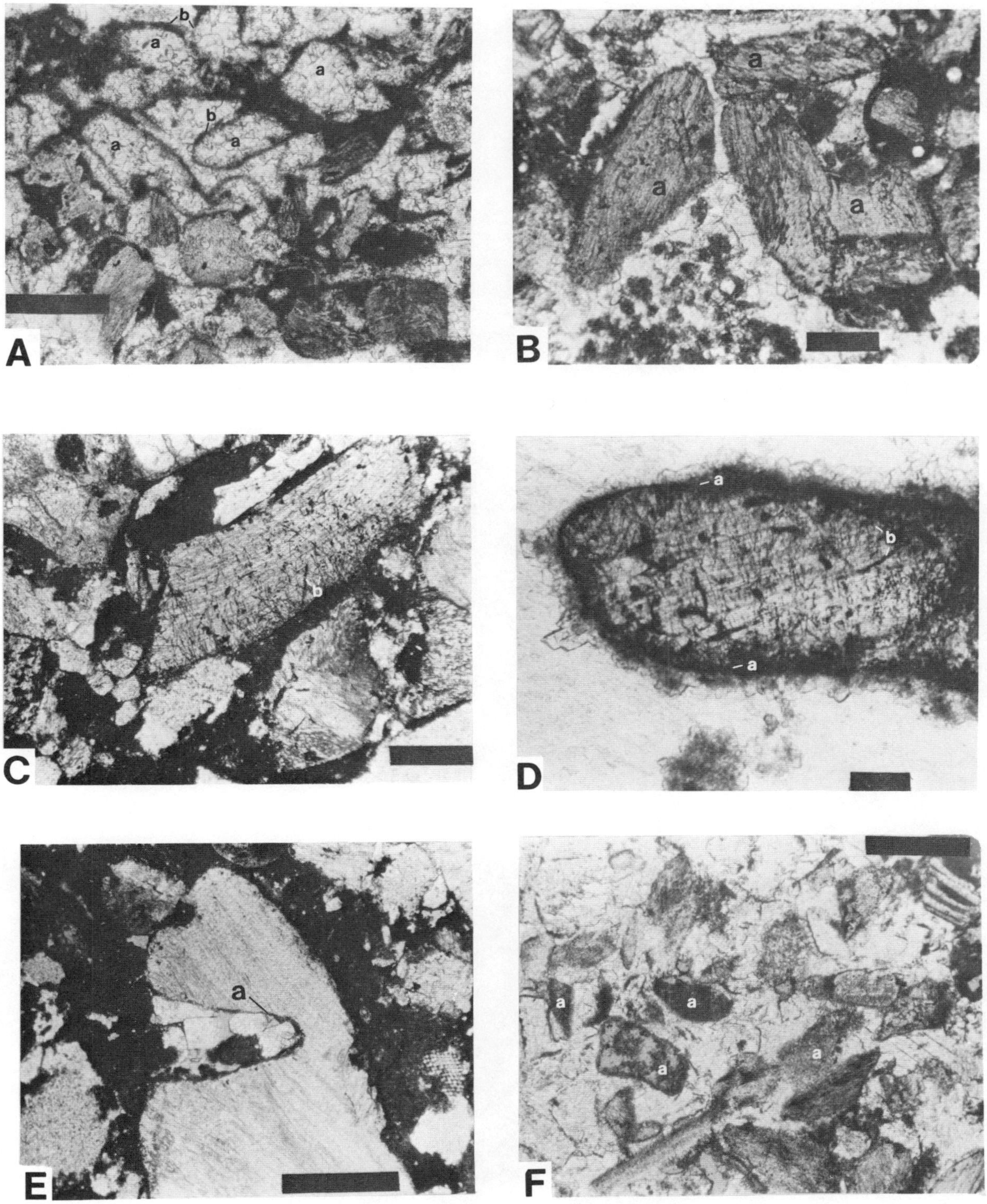

PLATE VIII

A. Oyster-echinoid-algal grainstone facies. Rounded, dissolved aragonite shells (a) filled with equant spar cement with micrite envelopes preserved (b). No porosity remaining. Thin section #51-12a. Bar scale = 0.5 mm.
B. Oyster-echinoid-algal grainstone facies. Rounding of oyster shells (a) retaining original calcite shell material. Thin section #51-41. Bar scale = 0.2 mm.
C, D. Oyster-echinoid-algal grainstone and packstone facies. Abrasion of oyster shell, development of micrite envelope (a), extensive algal microborings (b). Thin section at C: #55-61. Bar scale = 0.2 mm. Thin section at D: #1-165-58. Bar scale = 0.1 mm.
E. Oyster-echinoid green algal packstone facies. Large boring (a), possibly by sponge in oyster shell. Subsequently infilled by pellets and other grains. Thin section #55-26. Bar scale = 0.5 mm.
F. Oyster-echinoid green algal packstone facies. Rounded shell material undergoing various stages of micritization (a). Thin section #51-25. Bar scale = 0.5 mm.

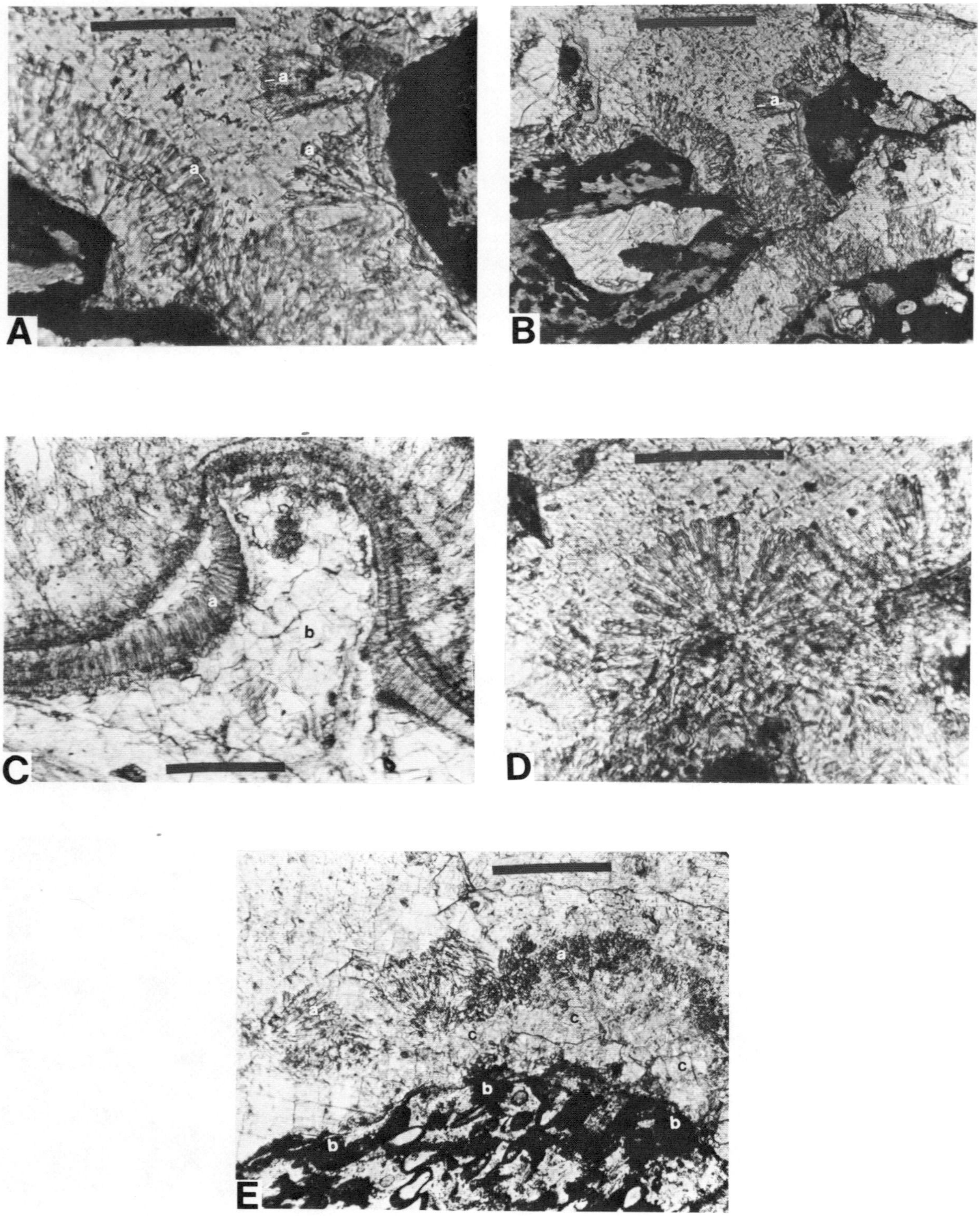

PLATE IX

A, B. Isopachous neomorphosed calcite cemented volcanic grains. The former aragonite, indicating marine environment, retains its relict blunt-tipped crystal shape (a) even though the unstable mineral has neomorphosed to more stable calcite. A is closeup of B. Thin section #54-18. Bar scale for A = 0.05 mm. Bar scale for B = 0.2 mm.

C. Former fibrous aragonite cement (a) preserved in a protected interior part of oyster shell. Later, equant calcite cement filled the remaining pore space (b). Thin section #55-16. Bar scale = 0.2 mm.

D. Close-up of excellent preservation of neomorphosed calcite after aragonite with blunt-tipped needles around pellet. Thin section #54-18. Bar scale = 0.05 mm.

E. Neomorphosed calcite cement (a) on a volcanic rock fragment (b). The basal part (c) has converted to calcite mosaic spar, obliterating the relict aragonite crystal morphology. Thin section #54-18. Bar scale = 0.2 mm.

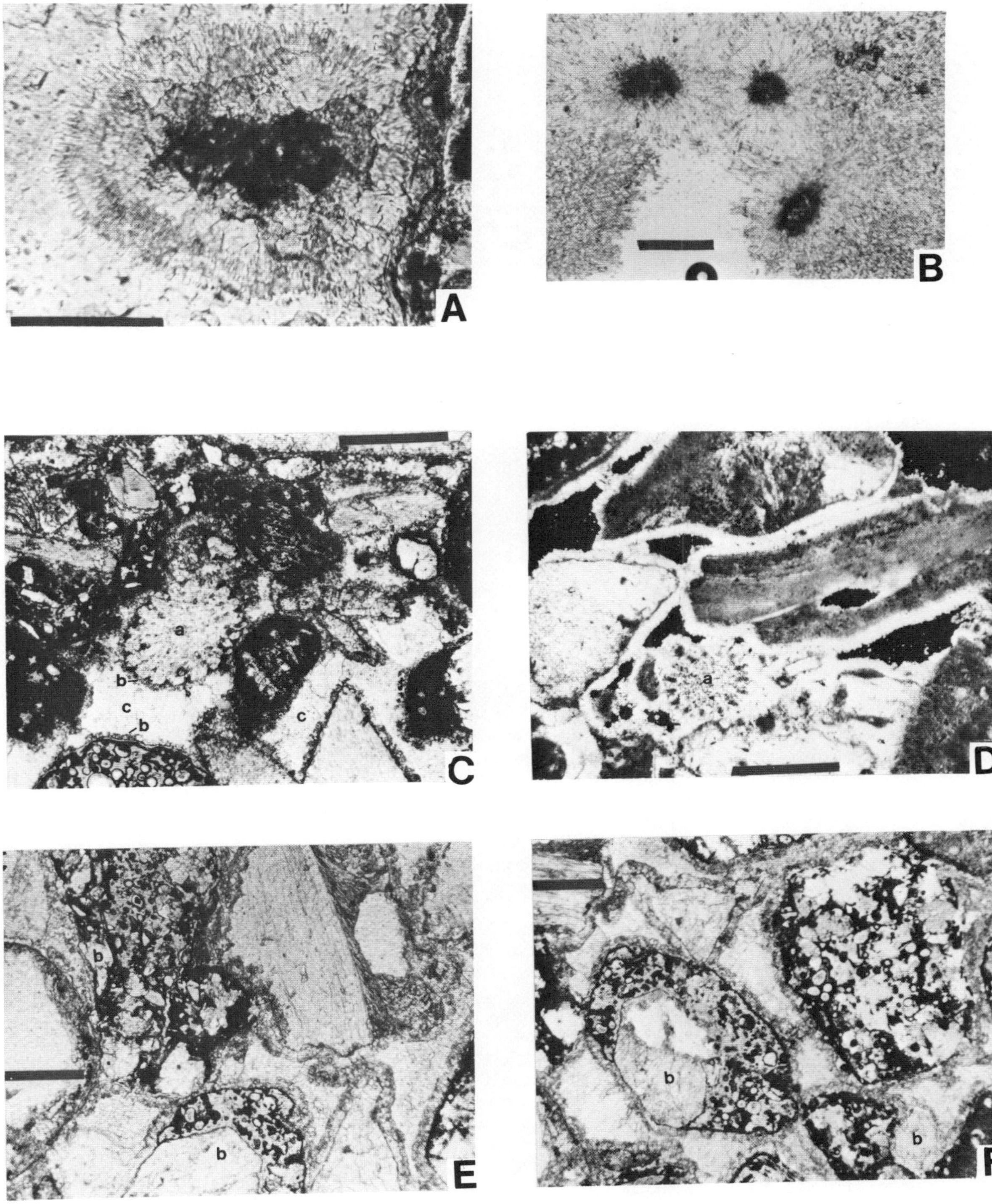

PLATE X

A. Isopachous aragonite neomorphosed to calcite blunt-tipped needle cement surrounding micrite pellet. Later cementation is equant calcite mosaic cement. Thin section #54-18. Bar scale = 0.05 mm.
B. Cementation similar to A seen in modern submarine environment in Grand Bahama Island area. Aragonite cement surrounds pellets with equant cement forming later. Thin section courtesy of L. S. Land. Bar scale = 0.2 mm.
C. Various allochems, including echinoid spine (a) cemented by fine former aragonite (b) followed by calcite mosaic cement (c). Thin section #54-18. Bar scale = 0.5 mm.
D. Cementation of allochems in modern Jamaican submarine environment. Echinoid spine (a) and other allochems cemented by aragonite. Black area is pore space. Thin section bears striking resemblance to C. Sample courtesy of L. S. Land. Bar scale = 0.5 mm.
E, F. Rounded shell and volcanic rock fragments cemented by aragonite neomorphosed to calcite. Isopachous character of cement is well demonstrated here. Volcanic rock fragments are being replaced by calcite (b). Thin section #54-18. Bar scale = 0.2 mm.

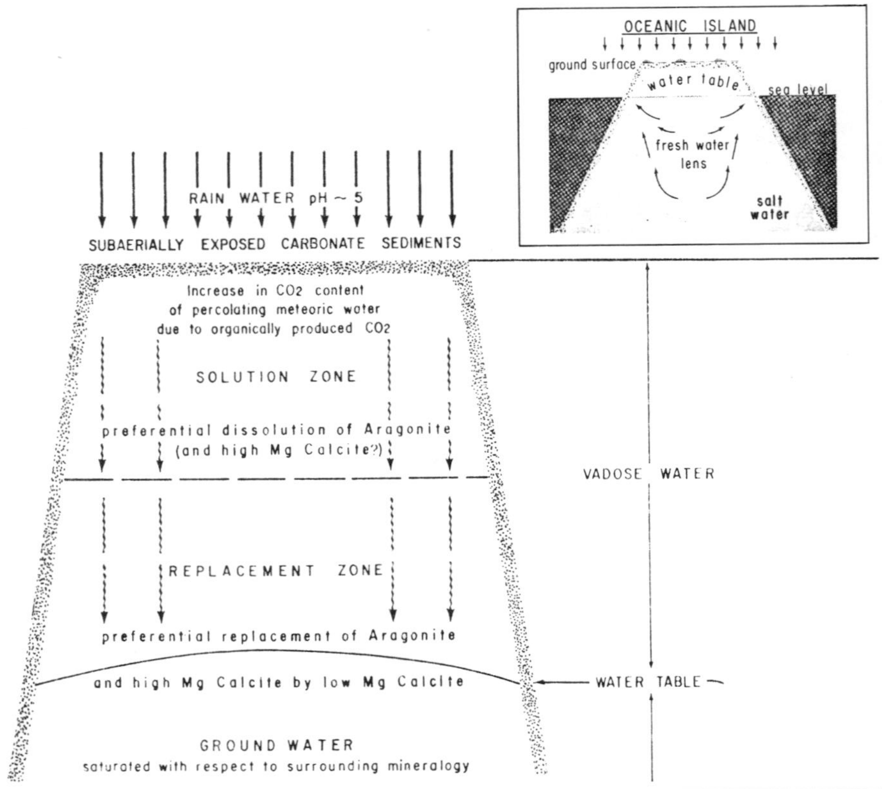

Figure 21. Purdy's subaerial diagenesis model. After Purdy (1968).

of aragonite and high-Mg calcite, which are relatively stable in the marine environment, are extensively altered when in contact with fresh water. This alteration usually occurs as complete dissolution of the allochem, which greatly enhances porosity, or as neomorphism from high- to low-Mg calcite (fig. 21). There are three meteoric subenvironments: (1) vadose, the uppermost part of the fresh-water lens through which water percolates; (2) phreatic, situated below the water table and recharged by the vadose zone; and (3) the mixing zone, a transitional area between fresh and salt waters. Pendulous and meniscus cements are characteristic of the vadose zone (Land, 1970), but neither was observed in this study.

The greatest alteration of carbonate grains occurs in the meteoric-phreatic zone because of constant saturation by fresh water. Aragonitic and high-Mg calcite allochems begin to stablize chemically relative to Mg-poor interstitial waters with Mg/Ca ratios approximately 1:3 (Folk and Land, 1975). This stabilization includes dissolution of grains and magnesium exsolution from allochems resulting in their neomorphism to low-Mg calcite. Dissolution of grains is more common than neomorphism in the Elaine field grainstones. Some dissolution is so complete that only the micrite envelopes remain to suggest the original grain shape (pl. VIII A). Green algal fragments and thin mollusk shells are most subject to dissolution because of their original aragonitic composition. This dissolution provided abundant calcium carbonate in the interstitial waters for precipitation of calcite cements. Cements associated with the meteoric phreatic zone are monocrystalline syntaxial cement (pl. XI) directed by echinoid fragments, the bladed calcite crust cement (pl. XI) with which the syntaxial cement grows in competition, and, finally, limpid dolomite (pls. XII A, F, XIII). Syntaxial cements associated with echinoid fragments are interpreted to result from phreatic waters precipitating calcite and the strong tendency of a monocrystalline substrate to nucleate a monocrystalline cement. Because this cement competes with the bladed cement (pl. XI F), it is concluded that both form at approximately the same time and under the same conditions. Bathurst (1966) offers conclusive evidence that the syntaxial cement grows faster than polycrystalline cements. The bladed crust reflects pore waters with Mg/Ca ratios near 2:1 with sidewise poisoning of the crystal lattice by magnesium only weakly executed. Folk and Land (1975) report these conditions can produce crudely fibrous, bladed, or prismatic to scalene calcite. These ratios can occur either in a fresh-water lens where magnesium expulsion from allochems has slightly increased magnesium content in the waters or in the "mixing zone," a transitional zone between fresh and marine waters (fig. 22).

With the incursion of fresh water, dolomite also may form. Folk and Land (1972) and Folk and Siedlecka (1973) have concluded that formation of dolomite occurs in waters with Mg/Ca ratios as low as 1:1, provided the crystallization rate is slow enough for the delicate ordering to form. High salinity is not a requirement; in fact, it tends to actually retard dolomite formation (Folk and Land, 1975). Limpid dolomite, a euhedral, clear form (Folk and Siedlecka, 1973), protrudes into the secondary pore space, which resulted from dissolution of fossils, and into primary pore space, which remained after precipitation of the bladed calcite crust (pls. XI A and XII F). These relationships suggest that, after the establishment of a fresh-water lense, grains were dissolved, providing a source of magnesium which slightly altered calcite cementation to form a bladed crystal morphology. Further freshening of the waters finally allowed the precipitation of the euhedral dolomite which did not completely fill the remaining pore space. The best porosity in Elain field is in these areas of limpid dolomite precipitation. This relationship can also be observed where "dirty" dolomite formed in the structural pores of the red alga ***Archeolithothamnion*** (pl. VIII B), an allochem well known for its high-Mg calcite skeleton. Waters trapped in structural pores do not circulate freely and stagnate in a very localized environment termed "micro-sabkha" (Folk and Siedlecka, 1975). The "dirty" dolomite, then, is formed in hypersaline conditions. Surrounding these rhombs are clear zones of limpid dolomite. The relationship clearly suggests a history of waters which are becoming increasingly fresher.

Regional ground-water system and deep subsurface environments

Final cementation of equant calcite (pls. IX, X, XI, XII) filled remaining pore space in some of the Elaine field grainstones. It is difficult to document when this cementation occurred, but either a regional ground-

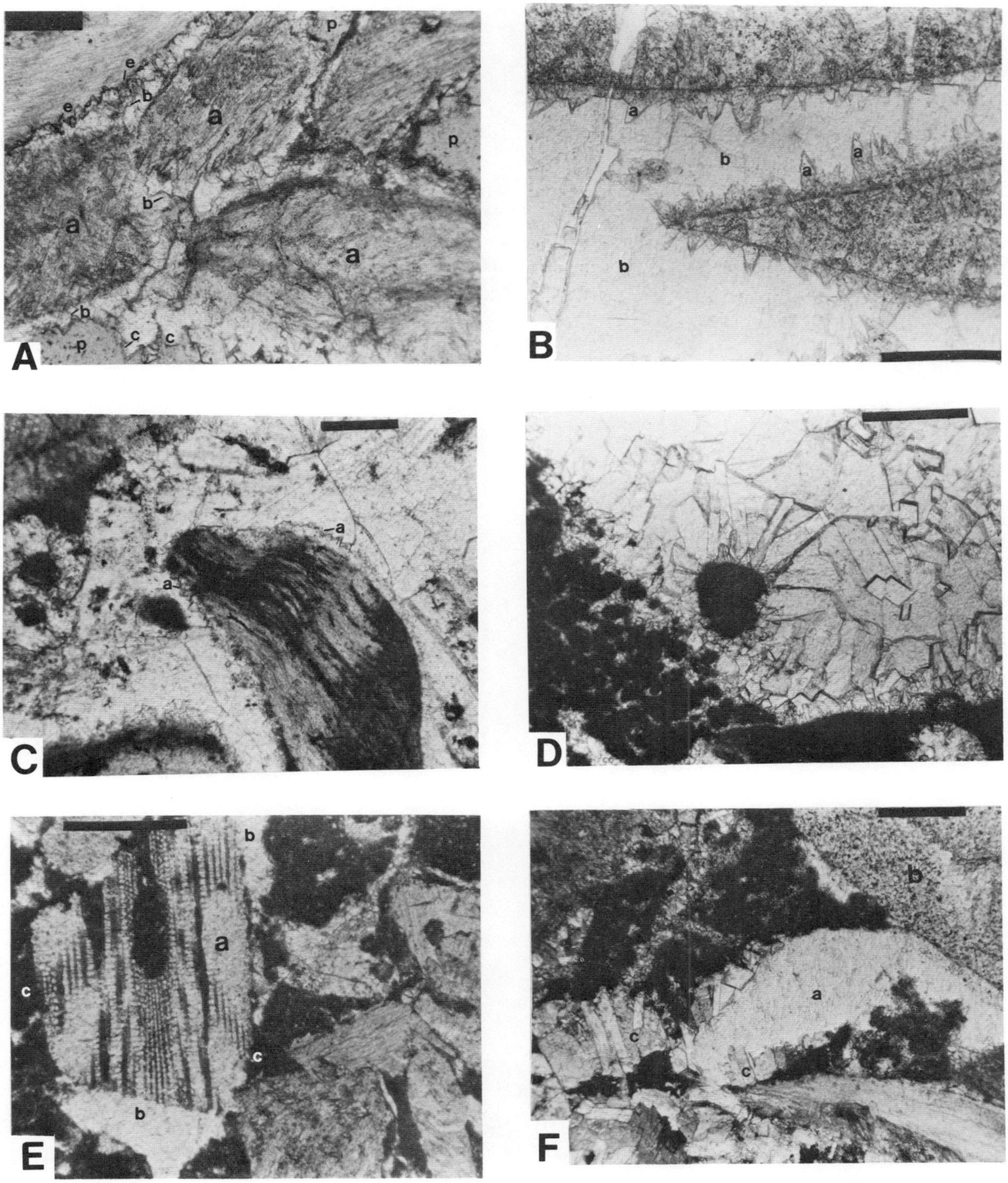

PLATE XI

A. Oyster-echinoid-algal grainstone facies. Rounded oyster fragments (a) cemented by bladed phreatic meteoric cement (b) followed by limpid dolomite rhombs (c). Pore space (p) is preserved in spite of cementation and subsequent stylolitization of grain boundaries (e). Thin section #51-17b. Bar scale = 0.2 mm.
B. Possible volcanic beach facies. Parts of rudist structure have been dissolved and subsequent porosity filled by bladed phreatic meteoric cement (a) and later calcite spar cement (b). Thin section #54-18. Bar scale = 0.5 mm.
C. Oyster-echinoid-algal grainstone facies. Bladed cement (a) on rounded oyster fragments, no porosity remaining. Thin section #51-12b. Bar scale = 0.2 mm.
D. Oyster-echinoid green algal packstone facies. Geopetal fill of pellets with void space filled by equant cement. The pellets probably influenced the nucleation of the cement causing it to grow initially somewhat elongated. Thin section #52-18. Bar scale = 0.2 mm.
E. Oyster-echinoid-algal grainstone facies. Echinoid fragment (a) with syntaxial cement (b). Cement is inhibited where allochem is surrounded by micrite (c). Thin section #51-6. Bar scale = 0.5 mm.
F. Oyster-echinoid green algal packstone facies. Syntaxial cement (a) on echinoid fragment (b) "competing" with equant cement (c). Equant cement is best developed away from the echinoid fragment. Thin section #55-32. Bar scale = 0.2 mm.

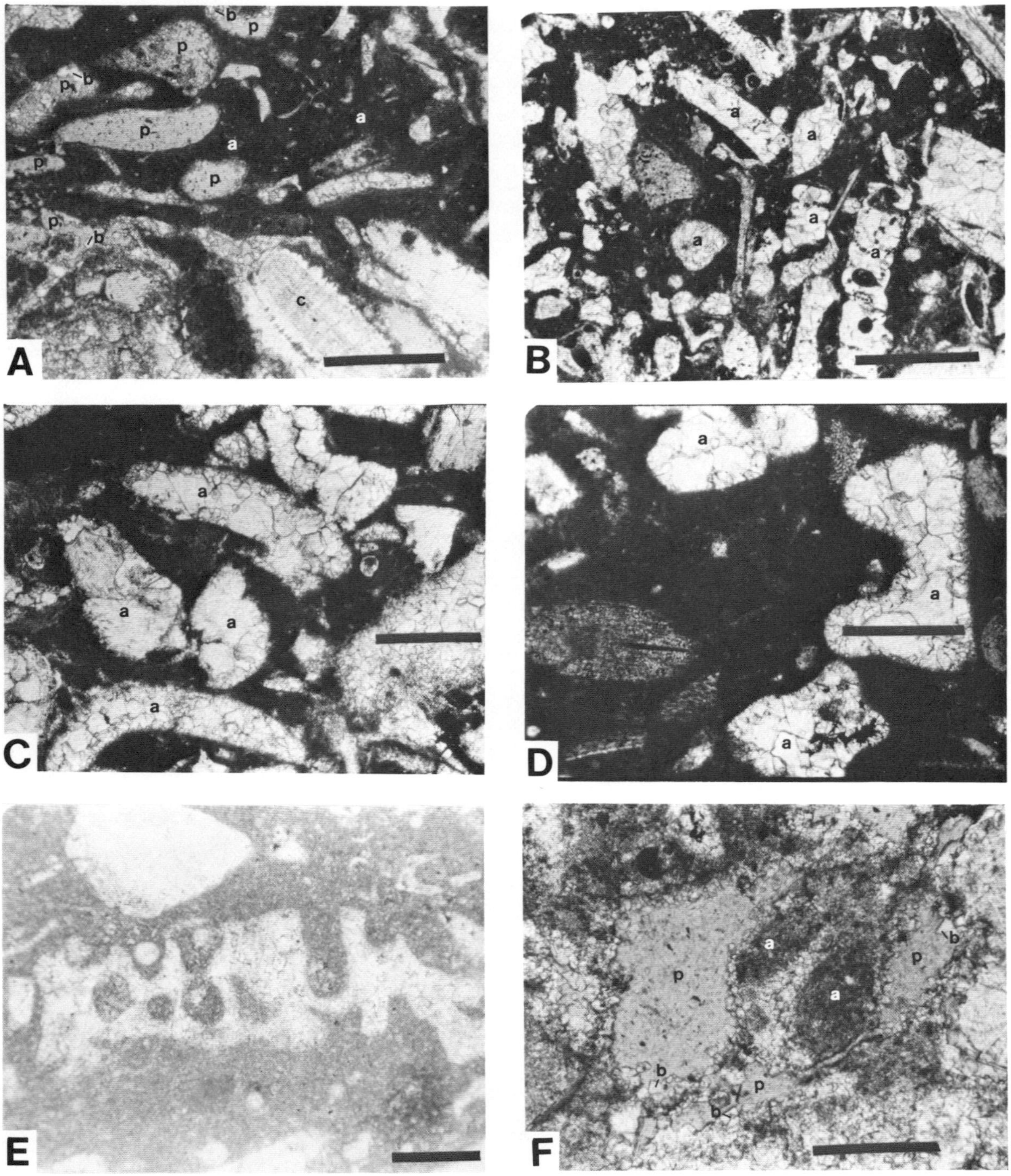

PLATE XII

A. Oyster-echinoid-algal boundstone. Dark matrix (a) is red algae heavily stained by oil. Aragonite grains are dissolved out and pore space remains (p). Very-fine-crystalline limpid dolomite reduces porosity slightly (b). Non-aragonitic shell (c) remains intact. Thin section #51-17b. Bar scale = 0.5 mm.

B. Oyster-echinoid-algal grainstone facies. Neomorphosed, aragonitic green algae fragments (a). Now seen as equant calcite cement. Algal pores are preserved. Thin section #55-69. Bar scale = 0.5 mm.

C, D. Oyster-echinoid green algal packstone facies. Rounded, dissolved aragonite shells of mollusks and green algae (a), filled with fine- to coarse-crystalline, equant calcite cement in the packstone halo environment. No shell structure is preserved with dissolution. Thin section C: #1-165-35. Thin section D: #52-9. Bar scales for both = 0.5 mm.

E. Oyster-echinoid green algal packstone facies. Neomorphosed green algal fragments preserving shell structure. Cement is equant calcite. Thin section #52-2. Bar scale = 0.2 mm.

F. Oyster-echinoid-algal grainstone. Micritization of grains (a), dissolved aragonite shell represented by preserved pore space (p) partially reduced by fresh-water limpid dolomite (b). Thin section #51-12A. Bar scale = 0.2 mm.

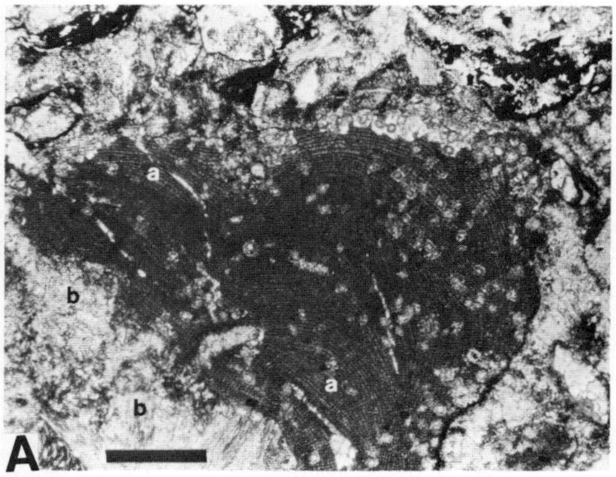

PLATE XIII

A. Red algae (a) encrusting oyster fragment (b). Small dolomite rhombs can be seen in the cellular structure of the algae. These can be attributed to stabilization of high-Mg calcite red algae which, in the presence of fresh water, could form dolomite, or to "micro-sabkha" conditions precipitating dolomite in the pores of the algae. Thin section #54-18. Bar scale = 0.2 mm.
B. Detail of A shows dolomite to be zoned. The "dirty" central rhombs probably formed as discussed above while the clear outer zones are limpid dolomite, forming in the later, fresher water conditions. As the meteoric water became fresher, the dolomite growth became slower resulting in the clear zones. Bar scale = 0.2 mm.
C. Volcanic rock fragment filled with calcite spar (a) and large limpid dolomite crystals (b). Thin section #54-18. Bar scale = 0.2 mm.

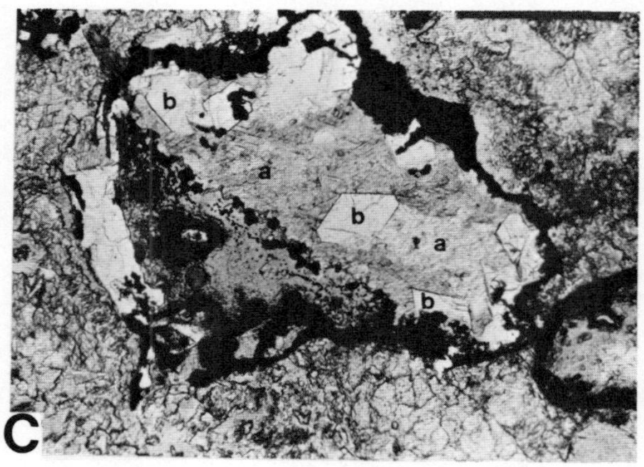

water system or subsurface burial could produce equant calcite. Differences in crystal size and boundaries suggest two separate episodes of formation. Grain stylolitization (pls. IV, VI, XI) is attributed to deep subsurface burial and compaction.

In conclusion, at least three stages of diagenesis altered the grainstones of Elaine field by reducing and enhancing porosity. The best preserved porosity occurs in arèas subject to meteoric-phreatic diagenesis.

CONCLUSIONS

1. An arcuate trend of Late Cretaceous volcanoes occurs in Central Texas; these volcanoes provided nuclei for shallow-water carbonate sedimentation and accumulation of the Anacacho Limestone.

2. Carbonates associated with one of these volcanoes, located in the subsurface of Dimmit and Zavala Counties, were studied using cores and electrical log data to delineate facies and environments within the island complex. Five facies were recognized: (1) the planktonic foraminifer mudstone/wackestone (open shelf),

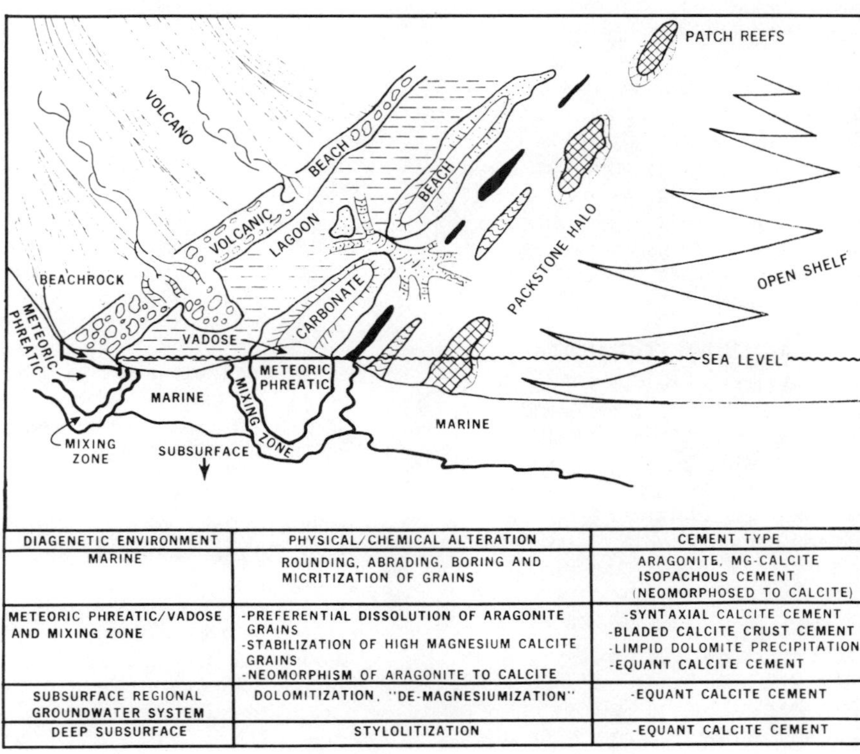

DIAGENETIC ENVIRONMENT	PHYSICAL/CHEMICAL ALTERATION	CEMENT TYPE
MARINE	ROUNDING, ABRADING, BORING AND MICRITIZATION OF GRAINS	ARAGONITE, MG-CALCITE ISOPACHOUS CEMENT (NEOMORPHOSED TO CALCITE)
METEORIC PHREATIC/VADOSE AND MIXING ZONE	-PREFERENTIAL DISSOLUTION OF ARAGONITE GRAINS -STABILIZATION OF HIGH MAGNESIUM CALCITE GRAINS -NEOMORPHISM OF ARAGONITE TO CALCITE	-SYNTAXIAL CALCITE CEMENT -BLADED CALCITE CRUST CEMENT -LIMPID DOLOMITE PRECIPITATION -EQUANT CALCITE CEMENT
SUBSURFACE REGIONAL GROUNDWATER SYSTEM	DOLOMITIZATION, "DE-MAGNESIUMIZATION"	-EQUANT CALCITE CEMENT
DEEP SUBSURFACE	STYLOLITIZATION	-EQUANT CALCITE CEMENT

Figure 22. Summary and relationship of environments to diagenesis in Elaine field.

(2) oyster-echinoid green algal packstone (packstone halo), (3) rudist-algal-echinoid packstone/boundstone (reef), (4) oyster-echinoid-algal grainstone (beach), and (5) oyster-echinoid-algal packstone/wackestone (lagoon).

3. The carbonate buildup developed and prograded for a time until subsidence of the volcano changed relative sea level. The facies then prograded up the flanks of the volcano until continued subsidence drowned the entire complex which was then covered with fine-grained sediments.

4. Analogous facies and environments can be seen in other complexes of this ancient trend of volcanoes in Texas and in modern seas. Modern analogs can be observed in the Austral Island trend in the South Pacific Ocean.

5. Diagenesis of the Elaine field grainstones includes at least three stages. The first stage occurred under marine conditions resulting in grain breakage, abrasion, and biogenic boring. Porosity was reduced in this stage by these processes as well as by marine cementation by blunt-tipped, fibrous aragonite or high-Mg calcite. The second stage was more extensive in its alteration of the grainstones. This stage included the development of a local fresh-water lens which dissolved unstable allochems and greatly enhanced porosity. Cementation by syntaxial calcite cement, bladed calcite crust, and limpid dolomite then reduced this porosity. All zones with porosity remaining show evidence that alteration in this diagenetic environment has occurred. In some cases, the regional groundwater system and deep subsurface burial destroyed this porosity through late equant calcite precipitation and stylolitization.

ACKNOWLEDGMENTS

I would like to thank A. J. Scott, D. G. Bebout, R. G. Loucks,and Keith Young for their critique and discussion of this manuscript. Also I am grateful to H. A. Semler and D. A. Urbanec who, courtesy of Tesoro Oil Company, provided me with the necessary core and electrical logs for the study. L. S. Land was kind enough to lend me his slides of recent carbonate cements.

This study was supported financially by the Bureau of Economic Geology, and by the Geology Foundation and the Owen-Coates Fund of the Department of Geological Sciences, The University of Texas at Austin.

REFERENCES

Adey, W., and MacIntyre, I., 1973, Crustose coralline algae: A re-evaluation in the geological sciences: GSA, v. 84, p. 883-904.

Bathurst, R. G. C., 1966, Boring algae, micrite envelopes and lithification of molluscan biosparites: Journal of Geology, v. 5, p. 15-32.

Cribb, A. B., 1973, The algae of the Great Barrier Reefs: *in* Biology and geology of coral reefs; v. 2, Biol. I: New York, Academic Press, p. 47-72.

DeGolyer, E., 1915, The effect of igneous intrusions on the accumulation of oil in the Tampico Tuxpam Region, Mexico: Economic Geology, v. 10, p. 651-662.

Devaney, D. M., and Randall, J. E., 1973, Investigations of ***Acanthaster planci*** in southeastern Polynesia during 1970-71: Atoll Research Bull. no. 169, 38 p.

Emery, K. D., Tracey, J. I., Jr., Ladd, H. S., 1954, Geology of Bikini and nearby atolls: USGS Professional Paper 260-A, 265 p.

Folk, R. L., 1974, The natural history of crystalline calcium carbonate: Effect of magnesium content and salinity: Journal of Sedimentary Petrology, v. 44, p. 40-53.

_______, and Robles, R., 1964, Carbonate sands of Isla Perez, Alacran Reef Complex, Yucatan: Journal of Geology, v. 72, p. 255-292.

_______, and Land, L. S., 1972, Mg/Ca vs. salinity; A frame of reference for crystallization of calcite, aragonite and dolomite (abstract): GSA Abstracts with Programs v. 4, p. 508.

_______ 1975, Mg/Ca ratio and salinity: Two controls over crystallization of dolomite: AAPG Bull., v. 59, p. 60-68.

_______, and Siedlecka, Anna, 1974, The "schizohaline" environment: Its sedimentary and diagenetic fabrics as exemplified by late Paleozoic rocks of Bear Island, Svalbard: Sedimentary Geology, v. 11, p. 1-15.

Fuller, R. E., 1931, The aqueous chilling of lava on the Columbia River Plateau: American Journal of Science p. 281.

Garner, L.E., and Young, Keith, 1976, Environmental geology of the Austin area: Univ. Texas, Austin, Bureau of Economic Geology Report of Investigations 86, 39 p.

Ginsburg, R. N., and Lowenstam, H. A., 1958, The influence of marine bottom communities on the depositional environment of sediments: Journal of Geology, v. 66, p. 310-318.

Johnson, J. H., 1971, An introduction to the study of organic limestones (revised edition): Quarterly of the Colorado School of Mines, v. 66, no. 2, 185 p.

Land, L. S., 1970, Phreatic versus vadose meteoric diagenesis of limestones: Evidence from a fossil water table: Sedimentology, v. 14, p. 175-185.

Lonsdale, J. T., 1927, The igneous rocks of the Balcones fault region of Texas: Univ. Texas, Austin, Bull. 2744, 178 p.

Loucks, R. G., 1976, Pearsall Formation, Lower Cretaceous, South Texas: Depositional facies and carbonate diagenesis and their relationship to porosity: Univ. Texas, Austin, unpublished Ph.D. dissertation, 362 p.

Murray, G. E., 1957, Geologic occurrence of hydrocarbons in Gulf Coastal province of the United States: GCAGS, v. 7, p. 253-299.

Perkins, R. D., and Halsey, S. D., 1971, Geological significance of microboring fungi and algae in Carolina shelf sediments: Journal of Sedimentary Petrology, v. 41, p. 843-853.

Perkins, R. F., 1974, Paleoecology of a rudist reef complex in the Comanche Cretaceous Glen Rose Limestone of Central Texas: Geoscience and Man, v. VIII, April 1, p. 131-173.

Purdy, E. G., 1968, Carbonate diagenesis: an environmental survey: Geol. Romana, v. 7, p. 183-228.

Romberg, G., and Barnes, V. E., 1954, A geological and geophysical study of Pilot Knob, (south), Travis County, Texas: Geophysics, v. 19, p. 438-454.

Scott, A. J., 1968, Environmental factors controlling oyster shell deposits, Texas coast: *in* Brown, L. F. (ed.), Proceedings, fourth forum on geology of industrial minerals: Univ. Texas, Austin, Bureau of Economic Geology, p. 129-150.

Sellards, E. H., Adkins, W. S., and Plummer, F. B., 1933, The geology of Texas, v. 1 - Stratigraphy: Univ. Texas Bull. 3232, 1007 p.

Simmons, K. A., 1967, A primer on "serpentine plugs" in South Texas, *in* Ellis, W. G., (ed.), Contributions to the geology of South Texas: San Antonio, Texas, South Texas Geological Society, p. 125-132.

Spencer, A. B., 1965, Upper Cretaceous asphalt deposits of the Rio Grande embayment: Corpus Christi Geological Society Guidebook, 67 p.
Stehli, F. G., Creath, W. B., Upshaw, C. F., Forgotson, J. M., 1972, Depositional history of Gulfian Cretaceous of East Texas embayment: AAPG Bull., v. 56, no. 1, p. 38-67.
Stenzel, H. B., 1971, Oysters, *in* Moore, R. C., (ed.), Treatise on invertebrate paleontology: Lawrence, Kansas, Kansas Univ. Press and GSA, p. N953-N1224.
Stephenson, L. W., 1928, Correlation of the Upper Cretaceous or Gulf series of the Gulf Coast Plain: American Journal of Science, 5th series, v. 16, p. 485-496.
Udden, J. A., and Bybee, H. P., 1916, The Thrall oil field: Univ. Texas, Austin, Bull. 1919, p. 1-78.
Wermund, E. G., Cepeda, J. C., and Luttrell, P. E., (in press), Fracture zones of the southern Edwards Plateau, Part 1. Regional distribution of fractures and their relation to tectonics and caves: Univ. Texas, Austin, Bureau of Economic Geology, 29 p.
White, R. H., Jr., 1960, Petrology and depositional patterns in the upper Austin Group, Pilot Knob area, Travis County, Texas: Univ. Texas, Austin, Master's thesis, 133 p.
Wilson, J. L., 1975, Carbonate facies in geologic history: New York, Springer-Verlag, 471 p.

APPENDIX A: LIST OF WELLS

Dimmit County

1. Fred W. Shield
   Marrs McLean #1
2. Tesoro
   San Miguel #52
3. Tesoro
   E. L. Dismukes #55
4. Tesoro
   E. L. Dismukes #57
5. Tesoro
   E. L. Dismukes #47
6. Tesoro
   G. C. Crenshaw #50
7. Tesoro
   San Miguel #54
8. Tesoro
   G. C. Crenshaw #79
9. Tesoro
   G. C. Crenshaw #77
10. Tesoro
    G. C. Crenshaw #58
11. Pet. Corp. Tex.
    Lico 2-165
12. Tesoro
    C. F. Groos #59
13. Tesoro
    C. F. Groos #78
14. Pet. Corp. Tex.
    Lico 1-164

Zavala County

15. Tesoro
    G. C. Crenshaw #81
16. Tesoro
    G. C. Crenshaw #51
17. Pronto Drilling Co.
    Hagen Dickenson #1
18. Pet. Corp. Tex.
    Lico 1-165

A LOWER CRETACEOUS SHELF MARGIN IN NORTHERN MEXICO

James Lee Wilson[1] and Giampaolo Pialli[2]

ABSTRACT

In seven sections in the Monterrey-Saltillo area of Mexico the 800- to 500-m-thick Lower Cretaceous Cupido Limestone and the equally thick underlying Taraises Formation of shale and black lime mudstone have been studied petrographically. These units apparently represent in part complimentary facies of carbonate bank and basinal environments. If used for paleotectonic interpretation, isopach maps should include both formations. The formations record a marine transgression in earliest Cretaceous time over a positive element in central Mexico which furnished sands and muds to the east. This transgression was followed in Barremian time by eastward progradation of a carbonate bank out from the positive element. The fully expanded late Cupido bank is overlain by the transgressive La Peña black shale and limestone of late Aptian age. The bank, as developed around Saltillo, consists almost wholly of cyclic grainstone and tidal-flat sediments showing progressive-upward shoaling. To the east around Monterrey, a bank edge appears, marked by more than 100 m of rudists and corals. This facies migrates eastward and rises in the section as the bank expands. The downslope facies in this area (Taraises Formation) is thick and well developed, contains lithoclastic conglomerates in black micritic matrix, and also has tumbled remains of corals and rudists. An eastern edgeline of the bank occurs at Saddle Mountain in Monterrey, and at the Sierra Minas Viejas 50 km northeast of the city. Still farther east and south of Monterrey in the Sierra de la Silla, basinal micritic limestone occurs through the total Lower Cretaceous section.

The extension of this trend northeast into Texas is difficult to follow. From outcrop studies around Monterrey-Saltillo and north at Sierra de la Gavia and Bustamante, it is possible to predict that the bank margin has a gentle slope over some tens of kilometers. Initial porosity and brecciation, vuggy and cavernous secondary voids, and dolomitization in the bank edge are encouraging signs for subsurface reservoir development.

[1]Rice University, Houston, Texas

[2]University of Perugi, Italy

INTRODUCTION

This paper reports preliminary results of a study of a Lower Cretaceous bank in northern Mexico from Sabinas Hidalgo to Monterrey and Saltillo (fig. 1). The rock units studied are the Cupido and Taraises Formations and their basinal equivalent, the lower Tamaulipas Limestone. The top of the sequence is marked by a widespread upper Aptian dark shale termed La Peña, and the base, by paleontological markers indicating the top of the Jurassic (fig. 2).

Detailed study of the Sierra Madre outcrops of the bank may be economically important. The shelf margin of the Cupido bank is known to extend northward into Texas and occurs beneath the Albian Deep Edwards trend; this Early Cretaceous formation produces gas near Nuevo Laredo on the Mexican side of the border.

What is the biologic and petrographic character of the Early Cretaceous bank? How steep was its seaward slope? How rapidly do the facies change across the bank margin? What diagenetic processes may be expected to influence significant porosity and permeability? Does late Paleozoic - early Mesozoic paleogeography influence Cretaceous facies? How different or similar is the Lower Cretaceous from the Middle Cretaceous El Abra - Deep Edwards? Recent work in northern Mexico offers a point of departure for attempts to answer some of the above questions. Work by Imlay (1938, 1943) and Humphrey (1949) and an unpublished stratigraphic synthesis by W. Humphrey and T. Diaz prepared in 1956 were valuable guides for locating sections and for establishing ammonite control. A Ph. D. dissertation by S. Charleston (1974) over the vast area of Coahuila northwest of Nuevo León enabled a more precise understanding of regional facies of the Lower Cretaceous. Petrographic-stratigraphic studies of the Cupido - Lower Tamaulipas by Bishop (1970, 1972) gave clues to the basinal facies, and A. Guzman's (1974) diagenetic study of the Arteaga - Los Chorros section near Saltillo aided our interpretation of the bank interior. A. Weidie and W. C. Ward of the University of New Orleans, Jack Conklin and Clyde Moore of Louisiana State University, and Ian Evans of the University of Houston aided in locating sections. The writer is very grateful to Warren Cooper of Gulf Research and Development Company for assistance with coccolith zonation of the Taraises and to Mary Pace of the same organization for assistance with the drafting of figures 3, 4, and 5.

Work at Rice University has proceeded for several years, with sporadic field and petrographic studies by numerous classes and by several weeks of sustained field work by G. Pialli (University of Perugia, Italy). Samples at 5-foot intervals have been collected, polished and peeled, described petrographically, and studied paleontologically at the following sections (fig. 3).

(1) Los Chorros - Arteaga Canyon, 15 miles south of Saltillo, Coahuila (Weidie and others, 1959).

(2) Huasteca Canyon, 15 miles east of Monterrey at Santa Caterina, on the northern side of the anticline (Vokes, 1963).

(3) Potrero Garcia along the arroyo at the Garcia Caverns, Sierra de Fraile (Wall and others, 1961).

(4) Potrero Chico, Sierra de Fraile (Wall and others, 1961), west of Hidalgo village and cement plant.

(5) Potrero Minas Viejas, 3 miles east of Hidalgo village, 20 miles north of Monterrey (Wall and others, 1961).

These sections plus Bishop's studies in Sierra Picachos (1970, 1972) and reconnaissance at Sierra de la Silla (La Boca - Presa Rodrigo Gomez), San Isidrio west of Horsetail Falls (Cola de Caballo), Santa Rosa Canyon west of Linares (Weidie and others, 1959), and unpublished work by Conklin at Bustamante north of Monterrey have enabled construction of an east-west cross section of the bank and the tracing of its eastern edge for about 100 miles north-south.

The stratigraphic section as determined from earlier work referred to above is shown on figure 2. Charleston (1974) and Humphrey and Diaz (1956) have demonstrated the great complexity and thickness of the Lower Cretaceous in the Sabinas basin. The simple couplet of a Lower Cretaceous argillaceous sequence (Taraises) over-

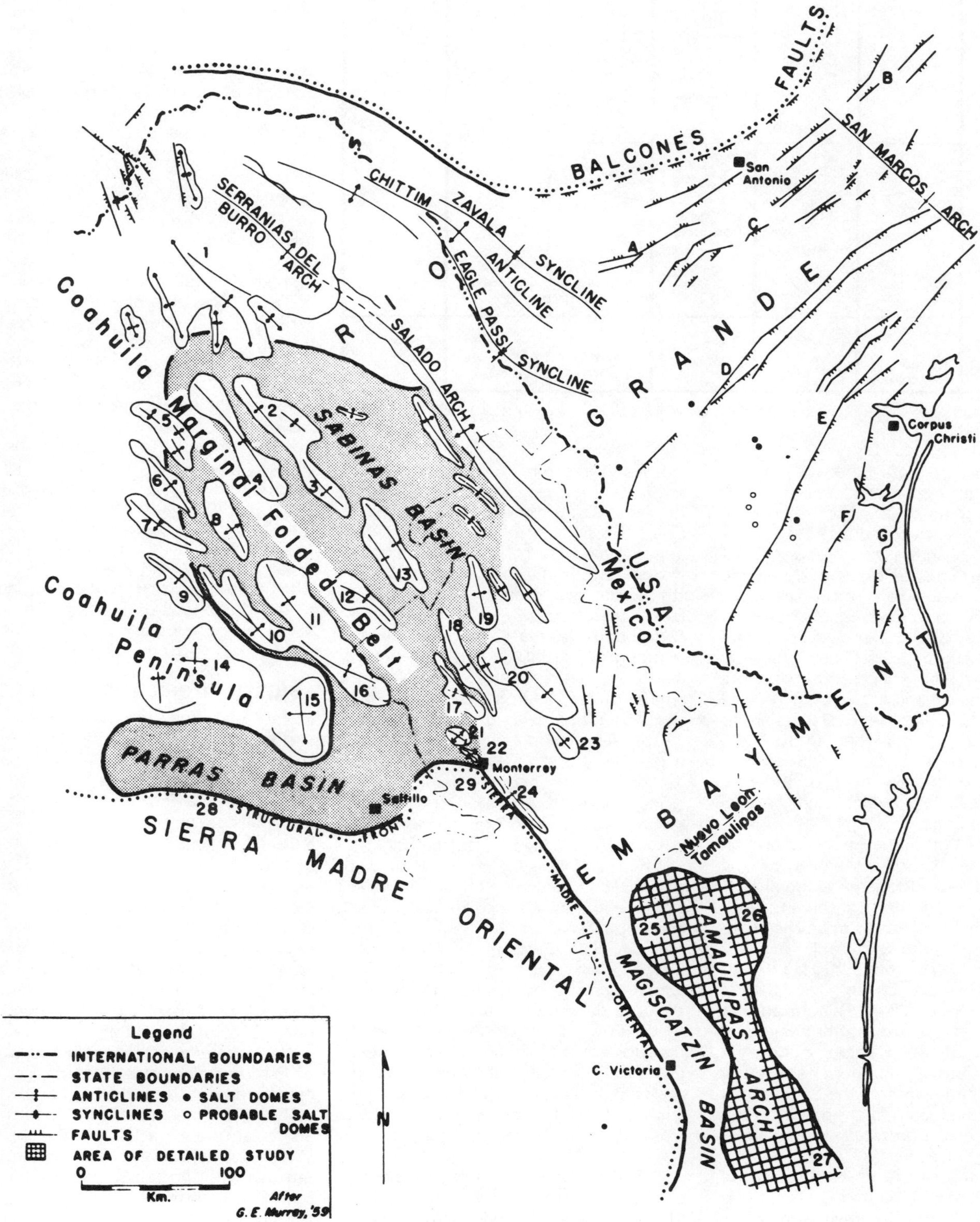

Figure 1. Regional tectonic framework and index map of northeastern Mexico. From Wall and others, 1961.

Stage	Sierra Madre west of Linares (Galeana)	Sierra Madre south of Saltillo	Sierra de Fraile north of Monterrey	Sabinas basin (Charleston, 1974)	Eastern Mexico basin	South Texas Subsurface
Upper Aptian	La Peña >100 ft	La Peña (20 ft)	La Peña <100 ft	La Peña 100-300 ft	Otates-La Peña 300 ft	Bexar shale
Aptian-Barremian	Cupido Lower Tamaulipas	Cupido Limestone	"Cupidito" shale Cupido Limestone	Cupido Patula Arkose La Mula shale La Virgen Evap.	Lower Tamaulipas Limestone	Cow Creek Limestone Hammett shale Sligo
Hauterivian	Taraises Galeana	? Sandstone "La Casita" of Authors ?	Shaly Limestone of Taraises Formation	Barril Viejo Marls		
Valanginian	Taraises		?	Menchaca Limestone		
Berriasian				Cupido Taraises		
Tithonian Kimeridgian	La Casita Limestone and shale	La Casita	La Casita shale	La Casita	La Casita Shale and Limestone	Cotton Valley

Figure 2. Stratigraphic sections and correlation chart for Lower Cretaceous of Mexico and Texas.

lain by a carbonate bank (Cupido) in the Monterrey-Saltillo area becomes more complex to the north and west where coarse terrigenous clastics entered the Sabinas basin pouring out from the flanks of the early Mesozoic Coahuila Peninsula uplift. The same situation prevailed southwest of Monterrey where the Galeana sandstone tongue appears. The Taraises is generally the argillaceous offshore equivalent of these westerly derived clastics. It ranges from Berriasian to Hauterivian age and from 800 to 1,600 ft thick. Identical facies relationships are seen in the Hosston, of the East Texas subsurface, which grades downdip to the Sligo.

The Cupido Limestone is laterally equivalent to and overlies the westerly derived clastic sediments. It is thus a diachronous carbonate unit, and much of it formed as an eastward-prograding carbonate bank (fig. 4). Thickness ranges from 2,500 ft (700 m) at Arteaga - Los Chorros to 2,900 ft (900 m) at Minas Viejas. Paleontological evidence of the time transgression of the base of the Cupido is not yet abundant but is consistent. The lowest beds of the Cupido in the Saltillo area are presumed Barremian, based on the occurrence of ***Choffatella***, and may be Aptian at Minas Viejas.

The top of the Lower Cretaceous (Nuevo León - Durango series) is marked by a well-known and widespread marker of dark shale and argillaceous limestone, commonly with some thin stringers of dark chert—the La Peña (Otates of the Mexican Gulf Coast subsurface). This unit is the upper shale of a triplet of Aptian age known both in the subsurface of Texas (the Bexar Shale) and in the northern Mexican outcrops. The unit bears the ***Dufrenoya texana*** ammonite fauna (Humphrey, 1949), the first well-developed globigerinids, and ***Nannoconus truitti***, and is just below the zone of ***Colomiella mexicana*** (Trejo, 1975). The middle limestone of the triplet is the Cow Creek of Central Texas outcrops and the James Limestone of the East Texas subsurface. Where developed in Mexico, this unit has been termed informally the "Cupidito." The lower shale of the triplet is the Hammett of Central Texas and Pine Island of East Texas and Louisiana. It has no name in Mexico. The Pearsall Formation of South Texas embraces this triplet which seems to be developed best over the Cupido shelf margin. Thus, many of the Lower Cretaceous formations in the subsurface of Texas are represented in the outcrops in Mexico. This uniformity is the result of a gently curving shelf margin which stretches from Tampico at least as far as East Texas.

The Mesozoic tectonic framework of northern Mexico is outlined by Murray (1961), Wall and others (1961), and McBride and others (1974) (fig. 1). It was first worked out by Böse (1923) and Kellum and others (1936). The southern edge of the Coahuila Peninsula stretches east to a point north of Monterrey. This early Mesozoic landmass is separated from the southward-trending Tamaulipas arch along the eastern margin of Mexico by the pronounced Sabinas basin filled with very thick Cretaceous sediments. Both the Coahuila Peninsula and the Tamaulipas arch are formed by roots of the late Paleozoic Ouachita-Marathon orogeny and were uplifted lands during the early Mesozoic. No Jurassic sediments appear on them. The Paleozoic to Cretaceous paleostructure appears to have controlled both the style and trend of Laramide folding. The pronounced right-angle bend of the Sierra Madre Oriental at Monterrey and the obvious crowding of the folds against the Coahuila Peninsula to the north give evidence of this folding. The offset between the Salado-Burro arch and the Tamaulipas arch is probably also controlled by pre-Mesozoic structure, possibly a basement strike-slip fault.

The objective of this report is to describe the facies and stratigraphic age relationships of the Neocomian strata of the Monterrey-Saltillo area Cupido and Taraises Formations, and to interpret the nature of this major carbonate bank and shelf margin. The work so far has identified a carbonate bank with a well-defined complex of facies which prograded eastward into the Sabinas basin north of Monterrey.

BIOSTRATIGRAPHIC MARKERS

The La Peña Formation (restricted to the unit above the "Cupidito Limestone") contains an upper Aptian ***Dufrenoya texana*** fauna (described by Humphrey, 1949) from the Sierra de los Muertos between Monterrey and Saltillo. A detailed microfauna study of the Aptian-Albian boundary was published by Trejo (1975) using sections in Huasteca Canyon and Canyon La Boca in the southern Sierra de la Silla. The tintinnids ***Colomiella recta*** and ***Colomiella mexicana*** are considered to mark lowermost Albian and to occur within or just above the upper beds of the La Peña.

Choffatella descipiens and species of ***Pseudocyclammina*** are large agglutinative foraminifers which exist principally in the normal-marine limestone and grainstone facies of the bank. These forms range from Barremian into Aptian. ***Choffatella*** present within 300 ft of the base of the Cupido at Arteaga Canyon is significant because it indicates that, here, the whole 2,500 ft (700 m) of the formation is post Neocomian.

The Taraises facies contains ammonites which are useful in correlation. Imlay (1938, 1940) described a large fauna of Hauterivian-Valanginian age and an older one belonging to the Berriasian stage. The writers collected ammonites of the younger Neocomian fauna, 200 ft (160 m) from the top of the upper Taraises, in Minas Viejas in unit 3 of the Taraises of Charleston (1975). Vokes (1963) reports the same fauna from both top and base of the 950 ft (290 m) thick Taraises in Huasteca Canyon. Humphrey and Diaz (1956) report some ammonites of this fauna at Potrero Garcia about 600 to 1,000 ft (200 to 300 m) below the massive base of the Cupido. This abundant and widespread fauna consists of the following genera: ***Olcostephanus, Distiloceras, Leopoldia, Neocomites, Acanthodiscus, Bochianites, Crioceras, Thurmannites, Maderia, Mexicanoceras,*** and ***Rogersites.*** It is generally considered to represent early Hauterivian and late Valanginian by Imlay (1938). The lower 215 ft (66 m) of the Taraises of Humphrey and Diaz (1956) at Potrero Garcia contain ***Subthurmannia*** (?) and ***Thurmannites***, Valanginian-Berriasian according to these authors.

No well-recognized tintinnid zones have been observed in the Barremian strata, although ***Nannoconus truitti***

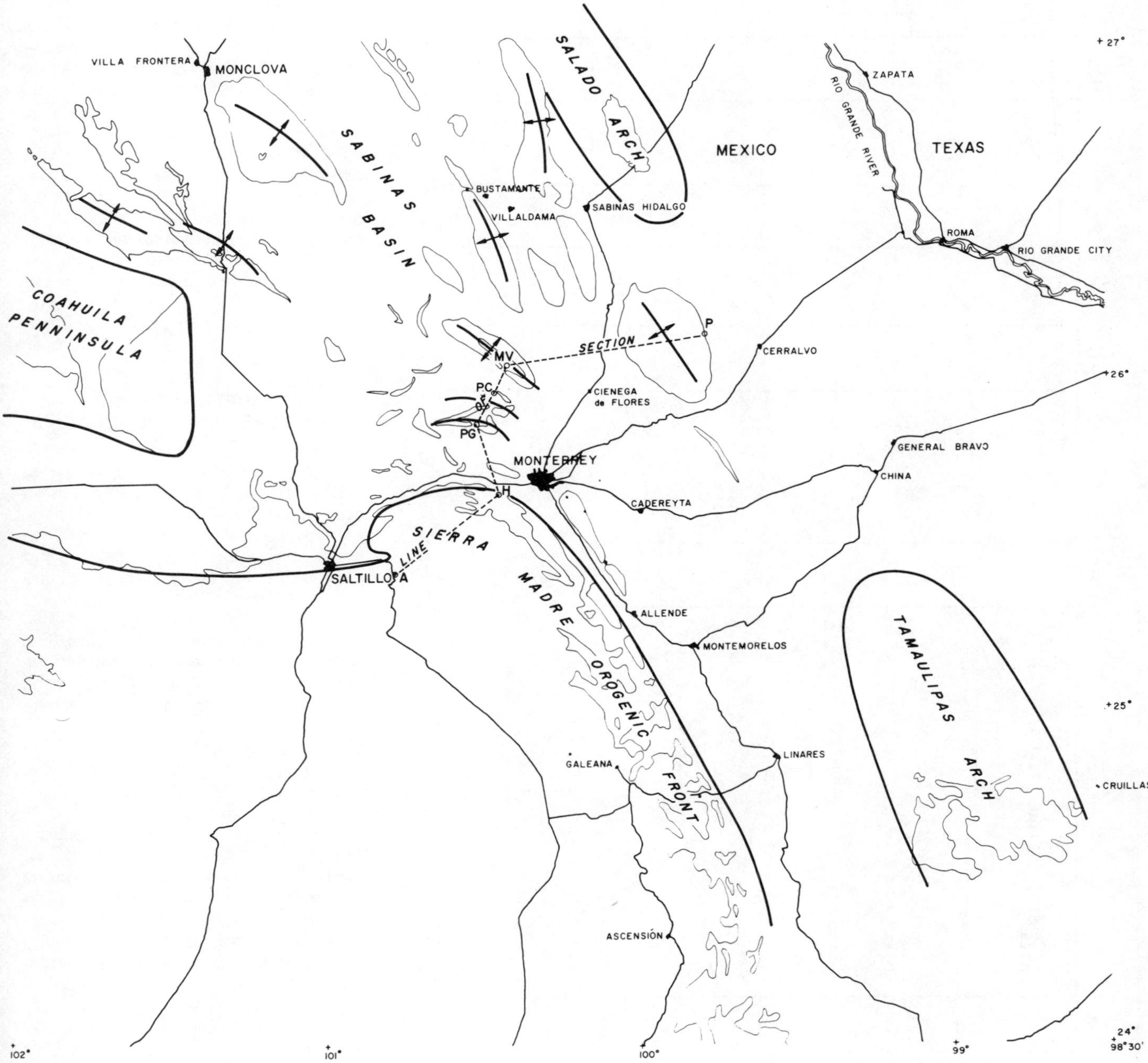

Figure 3. Line of cross section and location of measured sections (A, H, PG, PC, MV, and Picachos, P) on map of Laramide tectonic elements in Cretaceous of northeastern Mexico.

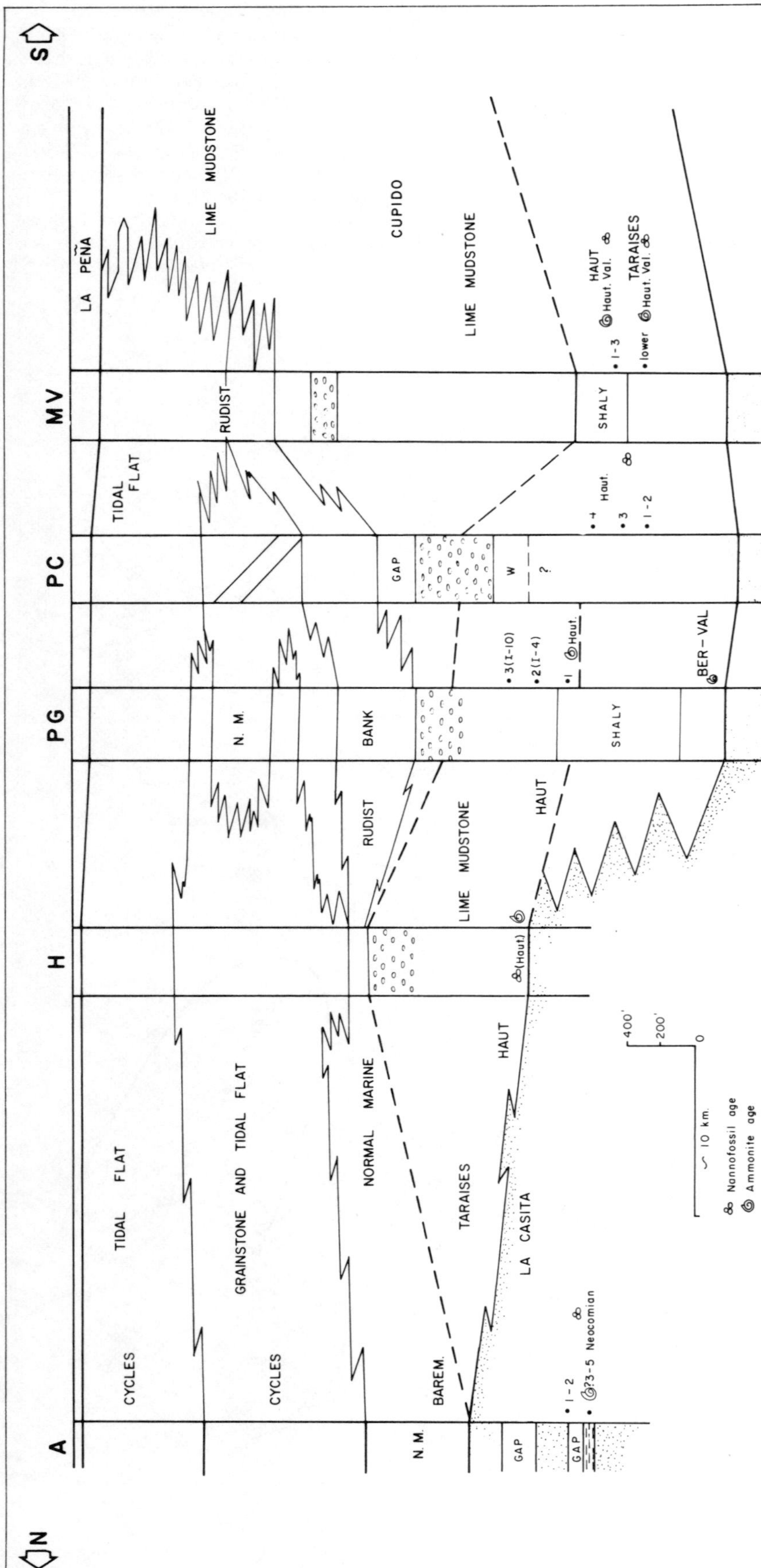

Figure 4. Cross section of Lower Cretaceous (Taraises-Cupido shelf edge) from Saltillo to Minas Viejas, north of Monterrey. Position of localities is indicated on figure 3. Dashed line is position of Taraises-Cupido boundary as mapped in field. Open circle pattern indicates lithoclastic limestone interpreted as forereef slope deposit. Dot pattern is quartzose sandstone.

and *Nannoconus wassalli* appear in the upper Aptian Cupido (Trejo, 1975). Nannoconids and coccolithophores are useful in zoning the shaly intervals of the Taraises. They indicate that the La Casita sandstone facies west of Monterrey in Los Chorros - Arteaga Canyons is in part Berriasian and younger Neocomian and, thus, that the Taraises shale and limestone in the mountains north of Monterrey may be equivalent in time to sandstone in the Sierra Madre foldbelt.

Work by Bonet (1956), Trejo (1975), and others on microfauna in the lower Tamaulipas and basinal Cupido Limestones of La Boca de Sierra de la Silla and Sierra de las Cruillas lists tintinnids and ammonites as correlation markers. These include, according to Bonet and Trejo, *Colomiella mexicana* and *Colomiella recta* which are typical of the Aptian-Albian boundary. Distinctive tintinnids appear rare to absent in the Neocomian to Aptian beds, but *Microcalamoides* is considered a Barremian marker by Bonet. *Calpionella alpina* is useful as a marker for Tithonian-Berriasian age. The large fauna recognized by Bonet in the Neocomian has not yet been seen in our sections.

DESCRIPTION OF BASIC ROCK TYPES AND MICROFACIES

The Taraises and Cupido contain a limited number of distinctive strata which aid in environmental interpretations. As with most limestones, petrographic work is necessary to understand these rocks fully. Cut and polished plaquettes, acetate peels, and thin sections of samples at regular 5 ft (1.5 m) intervals were used in the five sections listed in the Introduction. The basic rock types and microfacies are listed below in sequence from the basin to the center of the carbonate bank.

(1) Dark even and medium-bedded lime mudstone with fine bioclastic debris scattered throughout. The regular bedding is caused by thin shale layers. Some burrow mottling is seen. The rare bioclasts are fragments of mollusks and echinoderms including swimming crinoids and ophiuroids and pelagic microorganisms. Ammonites are present in argillaceous beds, usually a few hundred feet below the Cupido. The facies is generally mapped as Taraises where it is sufficiently shaly to weather in slopes rather than crags. Interpretation: Downslope mudstones accumulated at the base of the shelf margin (Wilson, 1975).

(2) Dark, thick- to medium-bedded lime mudstone and wackestone with

abundant lithoclasts of varying size (floatstone, lithoclastic wackestone). The clasts are jumbled cobbles of mudstone, rolled coral heads, and rudists. Rare graded bedding is observable. Weathering results in rounded to well bedded strata. Interpretation: Coarse material accumulated on a gentle slope of a carbonate shelf and in quiet water (Wilson, 1975). These massive beds are generally mapped as Taraises in Huasteca Canyon and Sierra de Fraile. They are not present at Arteaga near Saltillo and are included within the lower member of the Cupido at Minas Viejas (Charleston, 1974).

(3) Medium- to thick-bedded, gray, normal marine wackestone. Bioclasts are larger fragments and whole shells of echinoderms and mollusks in dark micrite. Dolomite is fairly common at the Arteaga - Los Chorros section. Interbedded grainier strata contain ***Choffatella*** and other foraminifers. The strata are included in the Cupido. They are common in its base at Arteaga - Los Chorros and above the basal rudist bank at sections around Monterrey.

(4) Rudist bank. Massive strata from 150 to 400 ft (50 to 120 m) thick are considered the base of the Cupido in sections around Monterrey. The rock is difficult to interpret because it is highly brecciated. It appears to have been originally a coarse shell hash and conglomerate with rolled and algally coated thick-shelled requieniid rudists and coral lumps (usually branched ***Cladophyllia***), spongiomorph stromatoporoids, and red algae. No in situ growth of reef-frame organisms is discernible although the organisms are common in reefs of the Mesozoic. Secondary brecciation resulted from Tertiary tectonic activity. Calcite veins, at least in part late diagenetic, are present throughout the rock. This facies is considered to represent accumulation of detritus at a shelf margin (Wilson, 1975).

(5) Three microfacies are present in the grainstone and packstone strata forming much of the middle Cupido. These include: (a) coated foraminifer and dasycladacean grains with common ***Choffatella,*** pseudocyclamminids, and the codiacean ***Cayeuxia.*** (b) Typical bahamith lumps, aggregates, and peloids. (c) Well-formed, multicoated ooids. These lime sands were partially cemented early and, at a later time, pore centers were in-filled with secondary phreatic-zone anhedral, blocky calcite. These facies may be interbedded with bioclastic normal-marine wackestone and with intertidal deposits in cyclic sequences. Interpretation: Shoal sands behind the rudist bank margin, probably deposited in lenses and sand bars in and out of water and in passes through lagoonal water of varying salinity (Wilson, 1975).

(6) Bank-interior facies have been studied and environmentally analyzed by Ekdale and others (1975). Although the samples analyzed for this work represented only the upper 400 ft (130 m) of Cupido at the Arteaga - Los Chorros section, the eastward progradation of the bank in the upper Cupido makes the component microfacies and sedimentary structures easily recognizable in the upper several hundred feet as far east as Minas Viejas. Ekdale and others (1975) identified the following petrographic microfacies.

(A) Rudist and/or oyster packstone with many miliolid foraminifers.

(B) Black pelleted bioturbated wackestone.

(C) Light-colored wackestone with abundant miliolids and dasycladacean algae.

(D) Rudist and/or oyster packstone with abundant codiacean and dasycladacean algae.

(E) Light-colored pelleted bioturbated wackestone.

(F) Black pelleted wackestone with abundant miliolids and dasycladaceans.

(G) Superficial oolite with coated pellets, foraminifer tests, and shell fragments.

(H) Finely laminated black lime mudstone.

(I) Black laminated lithoclastic packstone with abundant miliolids.

(J) Homogeneous lime mudstone with bird's-eye structure and miliolids.

(K) Homogeneous lime mudstone with bird's-eye structure.

(L) Homogeneous lime mudstone with abundant miliolids.

(M) Algal stromatolites.

These microfacies were based on cluster analysis of 15 petrographic characters, including various sedimentary structures, types of grains, and Dunham textural terms. Ordination of the basic data points out differences and gradational relationships between the samples and enables eight primary facies to be identified. These alternate with each other to form a number of imperfectly developed sedimentary cycles. The eight groups are:

(1) Bioturbated pelleted wackestone (BEF)

(2) Miliolid and codiacean-rich wackestone (C)

(3) Shelly packstone (A&D)

(4) Superficial oolite (G)

(5) Finely laminated lime mudstone (H)

(6) Miliolid lithoclastic packstone (I)

(7) Algal stromatolite (M)

(8) Homogeneous lime mudstone (JKL)

In terms of Holocene sedimentary environments, facies 1 may be interpreted as shallow bay of open-marine salinity; facies 2 as restricted marine bay; facies 3 and 4 as winnowed shoals and bars within such bays. Facies 5 and 6 are intertidal mudflats and ponds with an exposure index of about 50 percent; facies 7 and 8 are high intertidal, with the stromatolites of facies 7 marking mean high tide. The cyclic arrangement of these rock types, generally characteristic of the platform environment, is unmistakable although not as regular in the Cupido bank. An upward-shoaling sequence is imperfectly indicated with variable and alternating normal-marine bay and shoal deposits followed by intertidal to supratidal homogeneous mudstone, stromatolites, or traces of evaporite solution breccia.

Further study will probably differentiate between the biostromes of "oysters" (which are probably lithiotids) and requieniid rudists. Brecciated beds also may occur above and within primary facies 5 to 8 and indicate supratidal evaporite ponds or sabkha deposits from which gypsum and anhydrite have been leached. North of the area studied, the upper Cupido facies includes extensive evaporites, the La Virgen Formation.

THE WEST TO EAST DEVELOPMENT OF THE CUPIDO BANK

The progradation of the Cupido bank facies from the Arteaga - Los Chorros area (A) to Huasteca Canyon (H) and northeast through Sierra de Fraile (Potrero Garcia and Potrero Chico) to Minas Viejas (MV) is shown on figure 4. If this section is continued farther east to Sierra de Picachos (Bishop, 1970, 1972) or to the mountains at Sabinas Hidalgo, the total Cupido appears in a lime mudstone facies indicative of lower slope to basinal environments. The section is termed the lower Tamaulipas Limestone farther east and south of the Sierra de Picachos area.

The cross section (fig. 4) indicates an outpouring of sandy clastics from the west during the beginning of the Cretaceous. A clay unit within the sandstone and dolomite, about 750 ft (230 m) below Cupido Limestone, contains Neocomian nannofossils at

Arteaga. Here, beds 300 to 750 ft (90 to 230 m) above the base of the Cupido bear the Barremian foraminifer ***Choffatella***. Previously, this sandstone unit has been termed La Casita and assumed to be Jurassic. Terrigenous influx was carried far east, beyond Sierra de Fraile and Minas Viejas where the Lower Cretaceous argillaceous limestones formed the Taraises. This influx of clastics was gradually overcome by marine transgression to the west because, by Barremian time, normal-marine lime wackestone was deposited as far west as Arteaga - Los Chorros directly above the Neocomian sandstone. The growth of the carbonate bank began here in Barremian time. Age-equivalent downslope rhythmic-bedded lime mudstone and shale were deposited eastward in the Sierra Fraile and Huasteca areas. The Neocomian beds in the eastern sections are relatively thick compared to those in Huasteca and Arteaga sections. The strata contain pelagic microorganisms and the Neocomian ammonite faunas described by Imlay (1938). These beds are termed Taraises where shaly. In Huasteca Canyon and Sierra de Fraile, the base of the carbonate bank (approximately Barremian age) is the convenient mapping unit of Taraises and Cupido because of the resistant, massive nature of the limestone. The beds mapped as Taraises are thick- to medium-bedded limestone strata alternating with thin shaly intervals. Many beds of limestone consist of large cobbles to sand-sized grains of lime mudstone and fossiliferous wackestone debris with some rounded individual rudists and corals. The matrix for this material is a dark micrite containing very fine bioclastic debris. It seems clear that this sediment is a downslope deposit of bank-derived sediment, including vast quantities of shelf-derived lime mud mixed with pelagic microorganisms.

The overlying rudist bank consists primarily of rolled and coated requieniids mixed with clumps of corals, stromatoporoids, and red algae. The fauna indicates an open-marine assemblage but lacks the radiolitid and caprinid rudists seen in Albian and Cenomanian bank margins. No reef framework patches have been recognized to date. Extensive brecciation throughout the massive rock makes its sedimentary interpretation difficult. The bank is thicker in Sierra de Fraile than elsewhere. Because of facies progression, it rises eastward relative to the upper Aptian top of the La Peña. In Minas Viejas the beds equivalent in age to the bank, as developed in Huasteca and Sierra de Fraile, are dark, thin, medium-bedded lime mudstone. The bank at Minas Viejas lacks the abundant corals and stromatoporoids, and bedding is better developed because of the presence of individual mounds and flank beds.

Most of the Cupido from Arteaga to Potrero Garcia consists of grainstones, packstones, and pelloidal-bioclastic wackestones in repetitive sequences alternating with some intertidal laminites. This grainy facies gives way basinward to more normal marine wackestone in the Sierra de Fraile sections. More detailed petrography is needed to clarify relationships between Potrero Garcia, Chico, and Minas Viejas at this level. Individual cyclic units petrographically defined can probably be traced between these sections. The primary evidence indicates that the surface of the bank is dropping into deeper, quieter water toward its eastern edge at and beyond Minas Viejas.

The uppermost beds of the Cupido in all sections consist of the above discussed upperward-shoaling cycles and include many beds of restricted-marine environment and prominent biostromes of oysters or lithiotids and requieniids. Laminites also become relatively more important in these upper Cupido strata. This very shoal and intertidal facies progrades beyond the grainstone and is 550 ft (170 m) thick at Minas Viejas. It apparently disappears a little distance to the east of Minas Viejas and is present only in the uppermost beds at Sierra Picachos or at Sabinas Hidalgo. It is overlain abruptly by the transgressive La Peña which thickens eastward.

NATURE OF SHELF MARGIN

In exploration, predicting the degree of slope on shelf margins becomes important because the slope, in turn, controls breadth and regularity of facies belts which might contain reservoirs. Thickening of the La Peña (as restricted by Humphrey, 1949) from Potrero Chico through Minas Viejas to Sierra de Picachos indicates that the slope of the prograding margin of the Cupido bank was very gentle. Assuming that the top of the transgressive La Peña is essentially a time line, the difference in thickness of the formation from an offbank to an onbank position should indicate the relief on the shelf margin. Even if the La Peña (restricted) thickens by the addition of shaly basal Albian beds, as believed by Humphrey and Diaz (1956), the original relief on the bank is described. The La Peña is 200 ft (70 m) at Sierra de Sabinas, 100 ft (30 m) at Picachos, and about 60 ft (20 m) at Potrero Chico and northern Minas Viejas. This slope is less than 10 ft per mile if a plannar surface is assumed and palinspastic reconstruction of the distance is made. The thinning of the rudist bank at the base of the Cupido between Potrero Chico (400 ft, 125 m) and Minas Viejas (150 ft, 50 m) perhaps gives another indication of the early Cupido slope. This slope is approximately 40 ft per mile or less than one-half degree dip when the shortening by folding is eliminated.

Two observations indicate, however, that the slope may not have been regular and was locally steeper. Persistent limestone cobbles below the rudist bank indicate downslope movement of blocks off the bank; their presence hints at some relief, but perhaps only 1 or 2 degrees. At the Cerro de Labrador near Galeanna, a facies change from bank to basin may be seen on the east side of the mountain and from the Saddle Prongs at the north end of Sierra de la Silla, Monterrey, to the San Roque section, a distance of only 4 miles (7 km). The shelf has disappeared by facies change into basinal sediments (T. Diaz, personal communication). The difference in total thickness between the Early Cretaceous at the shelf margin and the well-developed basinal facies in eastern Mexico is about 1,200 ft (350 m). Charleston (1974) conjectures that this figure could approximate the depth of the basin at Aptian time, but distances between control points are great, and the general slope would have been slight.

CHARACTER OF THE BANK EDGE

The width of the prograding rudist bank (30 miles) was somewhat greater than the width of the shelf margin, and the high-energy edge of the bank was even narrower. Probably the lime-sand facies behind the rudist bank was the high point on the profile and had the most agitated water. In many places, the rudist accumulations contain micritic matrix. The configuration of the shelf margin is that of Wilson's type II (1974).

The bank edge is remarkably straight (fig. 5). It trends from Cerro de Labrador at Galeana where it may be seen in the mountain profile (T. Diaz, personal communication) past Cola de Caballo (Horsetail Falls) to the north end of Saddle Mountain above Monterrey. It passes between Minas Viejas - Bustamante and Sabinas

Hidalgo and thus has a strike almost due north. It appears to cross the Salado-Burro arch without being affected by the Tamaulipas early Mesozoic uplift (Charleston, 1974). In so doing, the Cupido bank edge cuts directly across the Sabinas basin. It is as if Cupido and Taraises deposition had buried all of the Jurassic paleostructure so that the bank edge has an even, curvilinear trend around the Gulf into Texas, crossing the border near Laredo.

The following are some differences between the well-known El Abra - Edwards facies of Albian age and the Cupido bank.

(1) There is a difference in the character of the rudists. Although the Cupido front contains the normal Cretaceous reefy fauna of stromatoporoids, spongiomorphs, corals, and red algae, it is dominated by thick-shelled, coated, large rudists which are mainly requieniids. The largest seen are about 6 inches across. The stalk-like large caprinids of the Albian are rare. Individual clumps or piles of caprinid rudists have not been observed. All the requieniids seem to have been moved

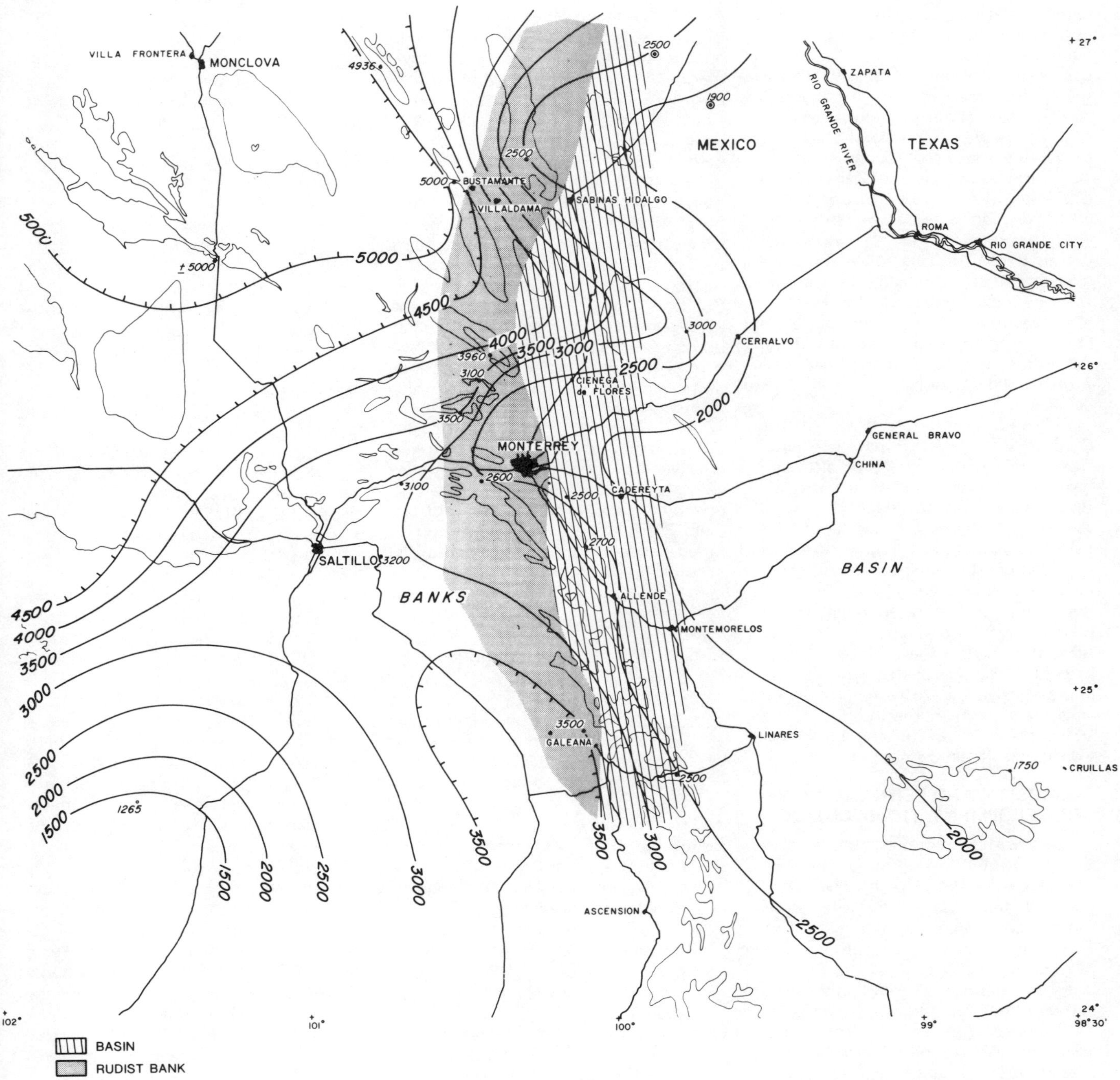

Figure 5. Thickness map (in ft) of Lower Cretaceous Neocomian through Aptian in northern Mexico (Cupido and Taraises Formations). Thick areas indicated by hachured contours outline Sabinas basin and Galeana region where sandy clastic influx occurs from the west. The rudist shelf-edge bank at the base of massive Cupido is superimposed on isopachs.

and perhaps piled mechanically; mostly they are arranged haphazardly, and in a few places they are bedded.

(2) The slope appears gentle, dropping on average only half a degree at the top of the Cupido bank. The El Abra bank edges are much steeper, perhaps as much as 5 to 10 degrees, although the faulting at the edges of the banks makes exact determination difficult. The greater and sharper vertical relief of the El Abra results probably in part from the ability of the rudists to form patches of framework. In both banks, lithoclastic limestone blocks accumulated downslope.

(3) As far as it can be presently traced, the Cupido shelf margin in Mexico is straight to curvilinear and simple as well as gentle. No separate isolated banks like those that occur in the El Abra are known in the Early Cretaceous of Mexico. The Lower Cretaceous Cupido acts as an underlying wide platform below the individual Albian banks, bearing much the same relationship to them as the Leonardian to the Guadalupian of the Permian reef complex, the Anisian Serla Dolomite to the Ladinian banks of the dolomites, and the Cooking Lake platform to the Devonian Leduc reefs of Alberta.

(4) The bank-interior facies of the Cupido differs somewhat in detail from that in the El Abra, although basically the two facies are similar. Both contain upward-shoaling cycles a few meters thick and have abundant tidal-flat laminations and biostromes of requieniid rudists. The Cupido lagoonal facies has fewer miliolid grainstones and more abundant shelly beds of oysters or lithiotids. In addition, thin beds of evaporite solution breccia at the Arteaga Canyon section and extensive Early Cretaceous evaporites are known in the interior of the Valles platform and in the La Virgen Formation of the Salinas basin.

POROSITY AND PERMEABILITY DEVELOPMENT IN THE CUPIDO

Several favorable rock characteristics for potential reservoir development exist in the Cupido. The rudist bank at the base of the Cupido is highly brecciated in the Sierra de Fraile and Bustamante sections. The brecciation appears to have formed by original deposition and increased through later tectonic stress applied to an unbedded mass of carbonate. Extensive veining and mineralization is present in the basal unit in some places.

Conklin (personal communication) has pointed out the extensive dolomitization common in the basal rudist bank. This lower dolomite is also present in the basal normal-marine, bedded Cupido Limestone in the Arteaga - Los Chorros section. The reason for apparently more dolomitization in the lower Cupido is not known at present. Dolomite and extensive brecciation create potential reservoirs.

Guzman (1974) has presented evidence for early generation of cement in the lime grainstones of the Cupido bank at the Arteaga - Los Chorros section. Some of the cements appear to have formed in shallow-marine to intertidal (splash zone) environments. Much of the cementation took place in the vadose-phreatic zone as the lime sand shoals of the bank were intermittently exposed during their formation. How much marine (first generation) cement and how much vadose-phreatic (second-generation) cement was early remains to be demonstrated by comparison with subsurface samples of the Cupido. The above processes, plus late-generation cementation and porosity occlusion under subsurface burial conditions, have possibly destroyed all original pore space in the grainstone of the Cupido bank.

The fact that the top of the Cupido bank is buried under a transgressive dark-colored pelagic argillaceous unit (La Peña) indicates the possibility for stratigraphic entrapment if dolomitization or inhibition of late-generation calcite cementation occurred in the upper Cupido.

REFERENCES

Bishop, W. F., 1970, Stratigraphy of Sierra de Picachos: AAPG Bull., v. 54, p. 1245-1270.

_______ 1972, Petrography and origin of Cretaceous limestones, Sierra de Picachos and vicinity, Nuevo León, Mexico: Journal of Sedimentary Petrology, v. 42, p. 270-286.

Bonet, F., 1956, Zonification microfaunistica de las Calizas Cretacicas del este de Mexico: XX Congreso Geologico Internacional, 102 p.

Böse, E., 1923, Vestiges of an ancient continent in northeast Mexico. American Journal of Science, 5th Series, v. 206, p. 127-136, 196-214, 310-337.

Charleston, S., 1974, Stratigraphy, tectonics, and hydrocarbon potential of the Lower Cretaceous, Coahuila series, Coahuila, Mexico: Univ. of Michigan Ph.D. dissertation, 268 p.

Ekdale, A., Ekdale, S. R., and Wilson, J. L., 1976, Numerical analysis of carbonate microfacies in the Cupido Limestone (Neocomian-Aptian), Coahuila, Mexico: Journal of Sedimentary Petrology, v. 46, no. 2, p. 362-368.

Guzman, A.E., 1974, Diagenesis de la Caliza Cupido del Cretacico Inferior, Coahuila, Mexico: Rev. Inst. Mexico Petroleo, v. 6, p. 20-40.

Humphrey, W. E., 1949, Geology of the Sierra de los Muertos area, Mexico: GSA Bull., v. 69, p. 89-176.

_______, and Diaz, G. T., 1956, Jurassic and Lower Cretaceous stratigraphy and tectonics of northeast Mexico: unpublished MS.

Imlay, R. W., 1938, Ammonites of the Taraises Formation of northern Mexico: GSA Bull., v. 49, p. 539-602.

_______ 1940, Neocomian faunas of northern Mexico: GSA Bull., v. 51, p. 117-190.

Kellum, L. B., Imlay, R. W., and Kanes, W. G., 1936, Evolution of the Coahuila Peninsula, Mexico, part 1: GSA Bull., v. 47, p. 969-1008.

McBride, E. F., Weidie, J. A., Wolleben, J. A., and Laudon, R. C., 1974, Stratigraphy and structure of the Parras and La Popa basins, northeastern Mexico: GSA Bull., v. 84, p. 1603-1622.

Trejo, M., 1975, Zonificacion del Limite Aptiano-Albiano de Mexico: Rev. Inst. Mexico Petroleo, v. 7, p. 6-29.

Vokes, H. E., 1963, Geology of the Cañon de la Huasteca area in the Sierra Madre Oriental, Nuevo León, Mexico: Tulane Studies in Geology, v. 1, p. 125-148.

Wall, J. R., Murray, G. E., and Diaz, G. T., 1961, Intrusive gypsum in Coahuila marginal fold belt: AAPG Bull., v. 45, p. 1504-1522.

_______ 1961, Geology of the Monterrey area, Nuevo León, Mexico: GCAGS Transactions, v. XI, p. 57-71.

Weidie, A. E., Wolleben, J. A., and Murray, G. E., 1959, Preliminary report on the structure of the Parras basin in the vicinity of Saltillo, Coahuila: South Texas Geological Society Guidebook, Southeastern Coahuila and Western Nuevo León.

Wilson, J. L., 1974, Characteristics of carbonate platform margins: AAPG Bull., v. 58, p. 810-824.

_______ 1975, Carbonate facies in geologic history: New York, Springer-Verlag, 471 p.

STUDY OF THE COCCOLITHS AND *NANNOCONUS* FROM THE TARAISES–CUPIDO SHELF MARGIN, NORTHERN MEXICO

Janine Barrier[1]

Samples from the following outcrops have been studied: Arteaga Canyon, Minas Viejas, Potrero Chico, and Potrero Garcia (tables 1, 2, 3). The sections are arranged across a carbonate shelf margin demonstrated by the stratigraphy of the Cupido and Taraises Formations in the Monterrey-Saltillo area of northern Mexico (Wilson, 1977). The age of the Cupido (shelf limestone) is given by benthonic foraminifers, but the biostratigraphy of the offshore argillaceous Taraises can be ascertained only with rare ammonites and nannofossils. The sampled intervals studied are indicated on the cross section by Wilson (1977). The samples, collected from soft, marly sediments, were treated by washing and centrifugation (Noel, 1965-1970). The study has been done entirely by means of a Zeiss Photomicroscope II.

Most of the samples contain a moderately abundant, poorly preserved nannoflora; some are barren, but a few contain enough calcareous nannoplankton to allow rather interesting biostratigraphic conclusions.

The distribution of this nannoplankton is figured in three range charts (tables 1-3) and is illustrated on plates 1 through 3: (1) Arteaga - Los Chorros Cañon, near Saltillo, Coahuila; (2) Potrero Chico, near Hidalgo, Nuevo León, Sierra de Fraile, north of Monterrey; and (3) Minas Viejas, near Hidalgo, Nuevo León, Potrero Minas Viejas, north of Monterrey.

The samples from Potrero Garcia of Sierra de Fraile are barren and are not figured.

The Arteaga Canyon samples disclosed a poor, rather badly preserved nannoflora. Among the coccoliths, a few forms are already present in the Jurassic, but others, such as ***Parhabdolithus asper, Cretarhabdus sp., Cruciellipsis chiasta,*** and ***Micrantholithus hoschulzi,*** are Cretaceous. Nannoconids are abundant, but they are restricted to a few species; their association is rather similar to that of the Berriasian (Derès, 1972). The Arteaga Canyon samples are considered Berriasian in age.

[1]c/o Rice University, Houston, Texas

Table 1. Occurrence of coccoliths and *Nannoconus* from the Arteaga Canyon section; sample numbers refer to those of Wilson (this volume, fig. 4).

ARTEAGA CANYON	SAMPLE NUMBERS			
	2	3	4	5
Parhabdolithus asper		—	—	
Parhabdolithus embergeri		—	—	—
Cretarhabdus sp.		—	—	
Cruciellipsis chiasta		—	—	—
Diazomalithus lehmani		—	—	
Cyclagelosphaera margereli	—	—	—	—
Ellipsagelosphaera communis		—	—	—
Ellipsagelosphaera britannica?		—	—	
Watznaueria barnesae	—	—	—	—
Watznaueria biporta		—	—	
Micrantholithus hoschulzi		—	—	
Intermediate forms Micrantholithus - Braarudosphaera		—	—	
Nannoconus steinmanni		—	—	
Nannoconus globulus		—	—	

Table 2. Occurrence of coccoliths and *Nannoconus* from the Potrero Chico section; sample numbers refer to those of Wilson (this volume, fig. 4).

POTRERO CHICO	SAMPLE NUMBERS				
	0	1	2	3	3a
Calcicalathina oblongata				—	—
Parhabdolithus asper	—			—	
Parhabdolithus embergeri	—	—		—	—
Cretarhabdus sp.	—				—
Cruciellipsis chiasta	—				—
Diazomalithus lehmani	—		—	—	—
Cyclagelosphaera margereli	—			—	—
Ellipsagelosphaera communis	—		—	—	—
Ellipsagelosphaera britannica					—
Chiastoplacolithus quadratus?	—			—	—
Watznaueria barnesae	—	—	—	—	—
Watznaueria biporta	—			—	—
Micrantholithus hoschulzi	—			—	—
Braarudosphaera bigelowi	—			—	—
Nannoconus steinmanni	—		—	—	—
Nannoconus globulus					—
Nannoconus minutus					—
Nannoconus truitti?				—	—

Table 3. Occurrence of coccoliths and *Nannoconus* from the Minas Viejas section; sample numbers refer to those of Wilson (this volume, fig. 4).

MINAS VIEJAS	SAMPLE NUMBERS					
	1	2	3	4	5	6
Calcicalathina oblongata	—	—	—	—	—	
Parhabdolithus asper				—	—	—
Parhabdolithus embergeri	—		—	—	—	—
Cretarhabdus conicus			—	—	—	
Cretarhabdus crenulatus?			—	—		
Cruciellipsis chiasta	—		—	—	—	
Cruciellipsis cuvillieri			—	—	—	—
Diazomalithus lehmani		—	—	—	—	—
Cyclagelosphaera margereli			—		—	—
Ellipsagelosphaera communis		—		—	—	
Ellipsagelosphaera britannica					—	
Chiastoplacolithus quadratus?			—	—		
Watznaueria barnesae	—	—	—	—	—	—
Watznaueria biporta	—		—		—	—
Micrantholithus hoschulzi			—	—	—	—
Braarudosphaera bigelowi	—		—	—	—	
Nannoconus steinmanni	—		—	—	—	—
Nannoconus globulus			—			—
Nannoconus aff. bronnimanni			—			
Nannoconus aff. kamptneri			—			
Nannoconus truitti?	—		—	—		—

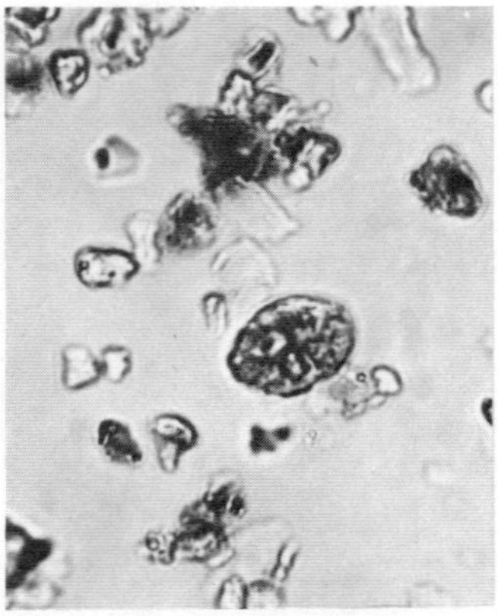

Figure 1
PC3 Transmitted light, 1000 x.

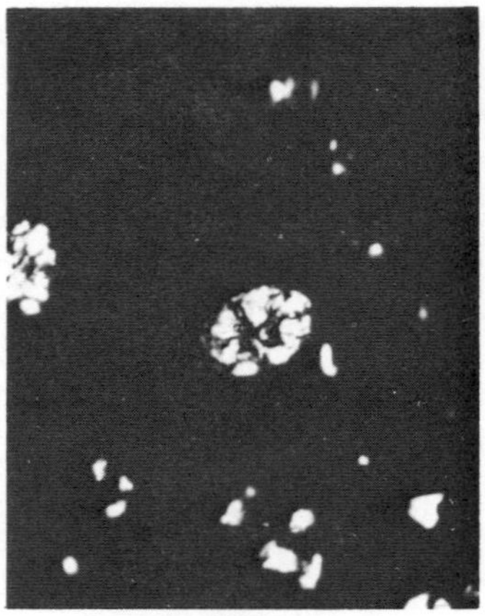

Figure 2
Cross-polarized light, 1000 x, same specimen as Figure 1.

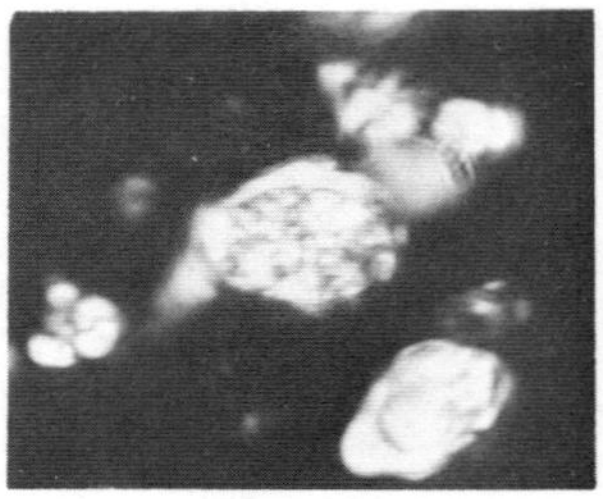

Figure 3
MV3 Cross-polarized light, 1200 x.

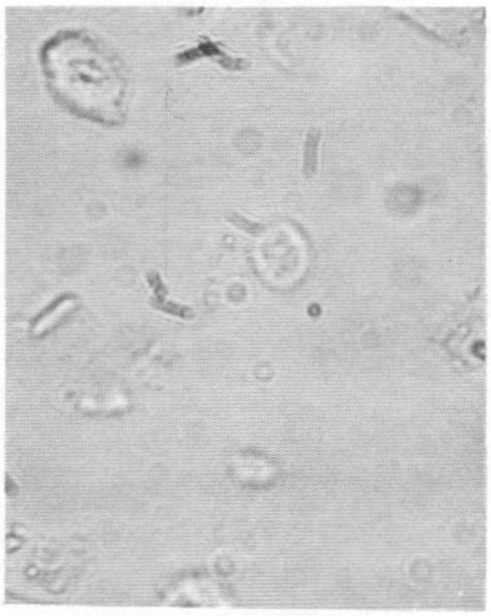

Figure 4
A3 Transmitted light, 1000 x.

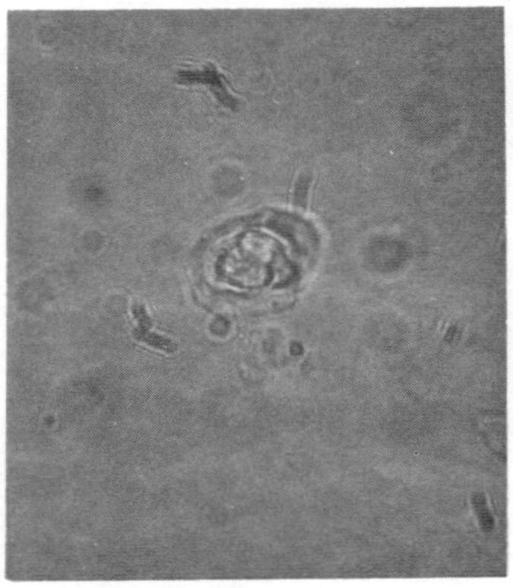

Figure 5
MV3 Transmitted light, 1200 x.

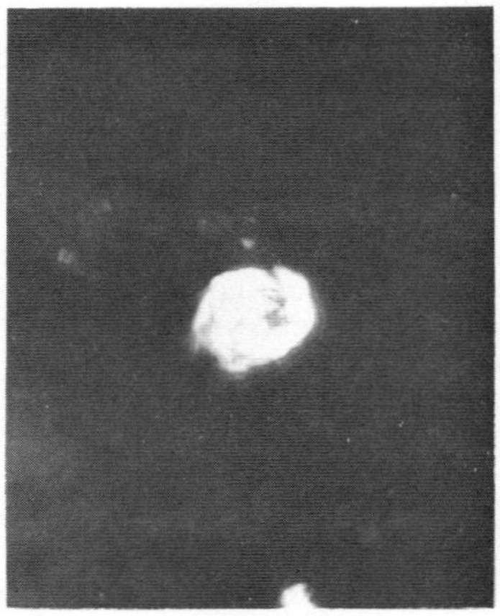

Figure 6
Cross-polarized light, 1200 x, same specimen as Figure 5.

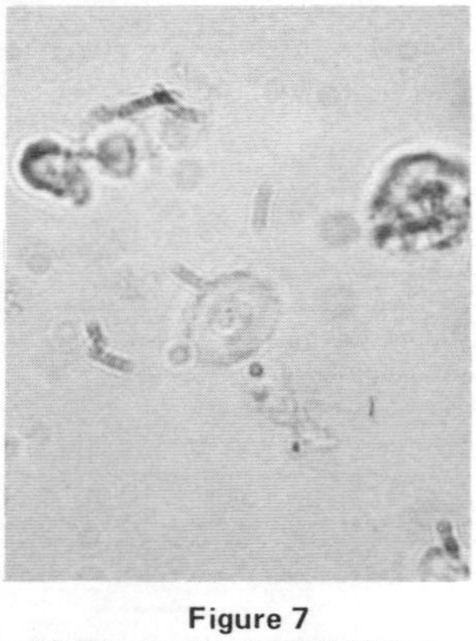

Figure 7
A3 Transmitted light, 1200 x.

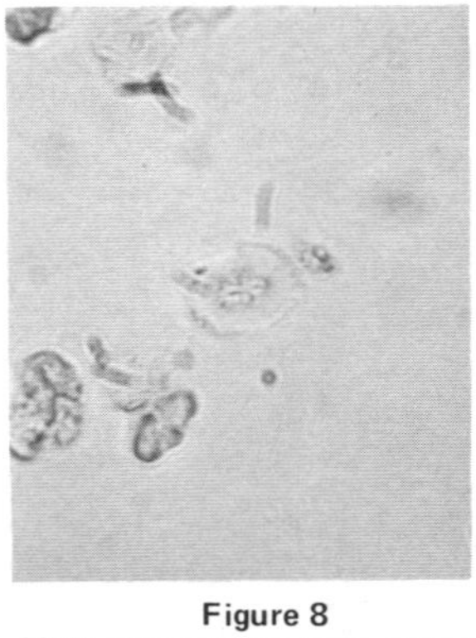

Figure 8
MV4 Transmitted light, 1000 x.

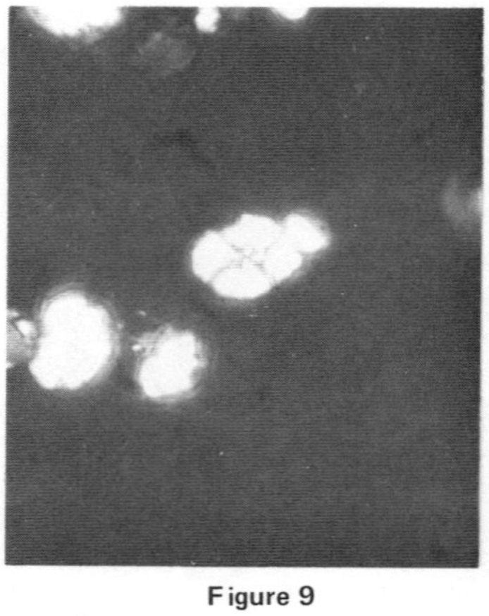

Figure 9
Cross-polarized light, 1000 x, same specimen as Figure 8.

PLATE 1

Figures 1-3. ***Calcicalathina oblongata*** (Worsley, 1971) Thierstein, 1973. Figure 4. ***Parhabdolithus asper*** (Stradner, 1963) Manivit, 1971. Figures 5-6. ***Parhabdolithus embergeri*** (Noël, 1958) Stradner, 1963. Figure 7. ***Cruciellipsis chiasta*** (Worsley, 1971) Thierstein, 1972. Figures 8-9. ***Cruciellipsis cuvillieri*** (Manivit, 1966) Thierstein, 1971.

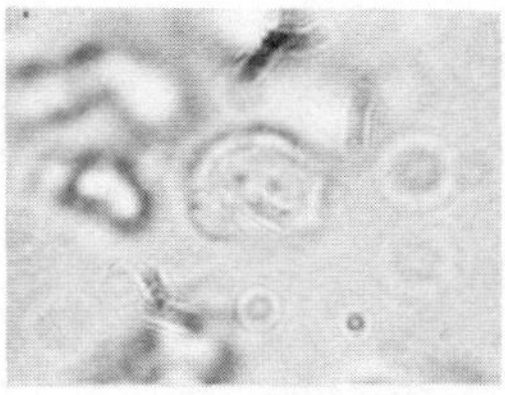

Figure 10
MV3 Transmitted light, 1000 x.

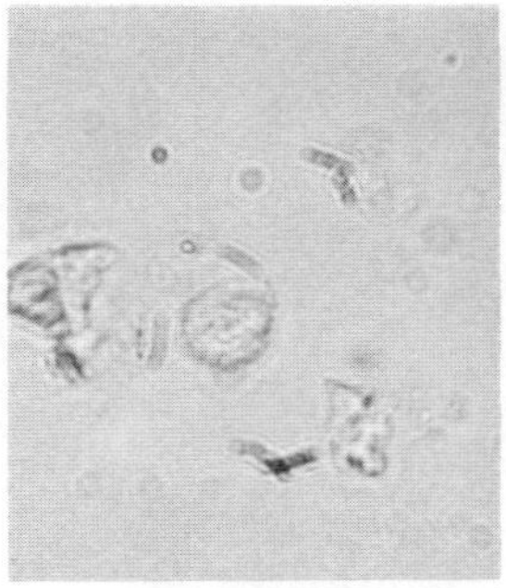

Figure 11
MV3 Transmitted light, 1200 x.

Figure 12
MV3 Cross-polarized light, 1200 x.

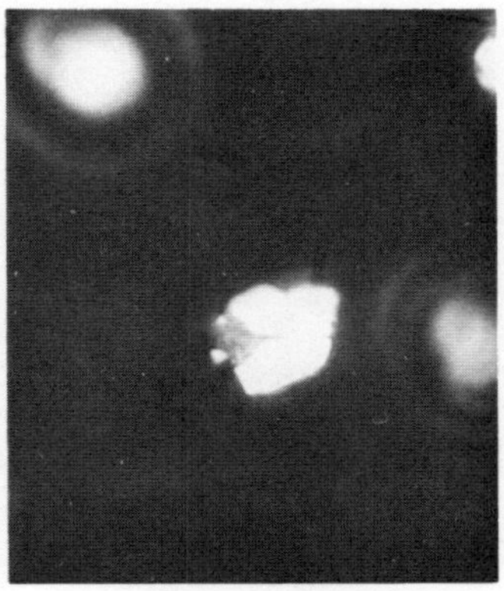

Figure 13
MV4 Cross-polarized light, 1000 x.

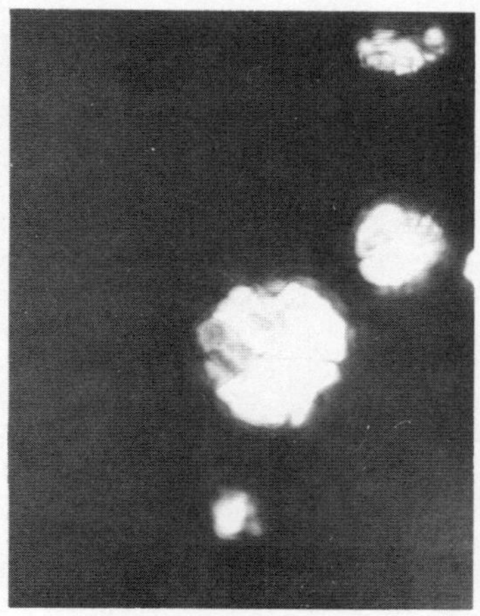

Figure 14
MV4 Cross-polarized light, 1200 x.

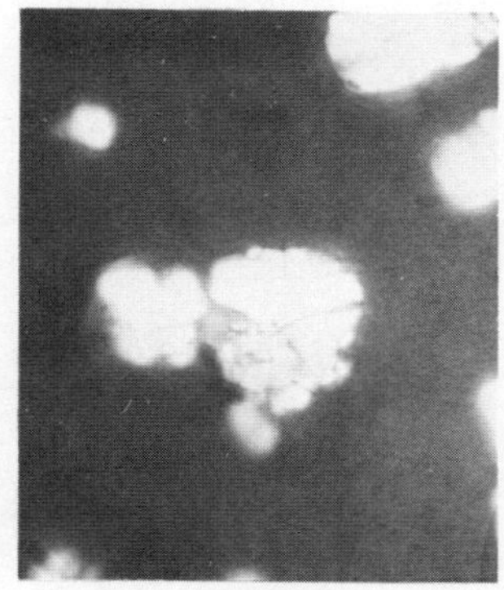

Figure 15
MV3 Cross-polarized light, 1200 x.

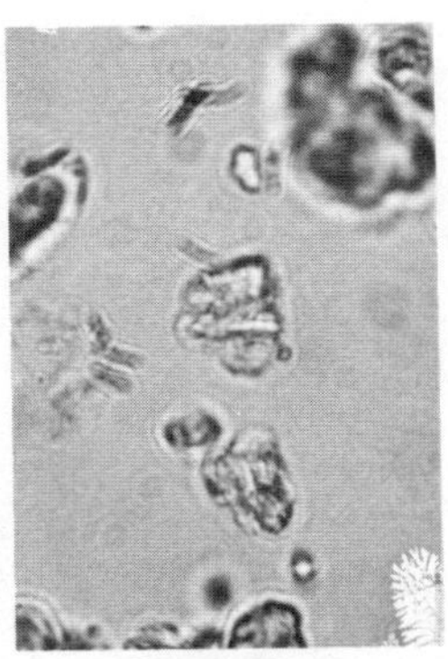

Figure 16
MV3 Transmitted light, 1000 x.

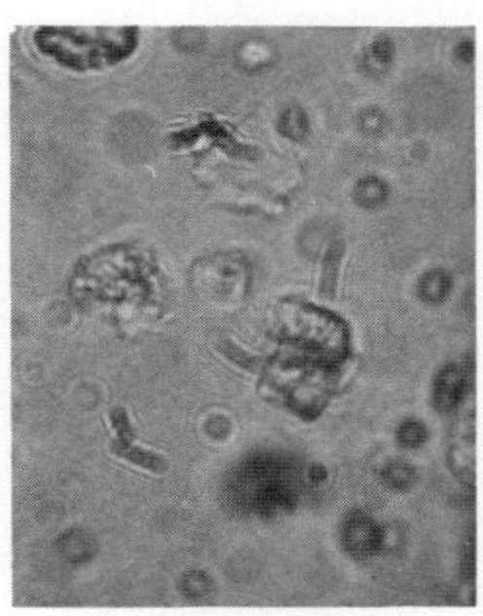

Figure 17
MV3 Transmitted light, 1000 x.

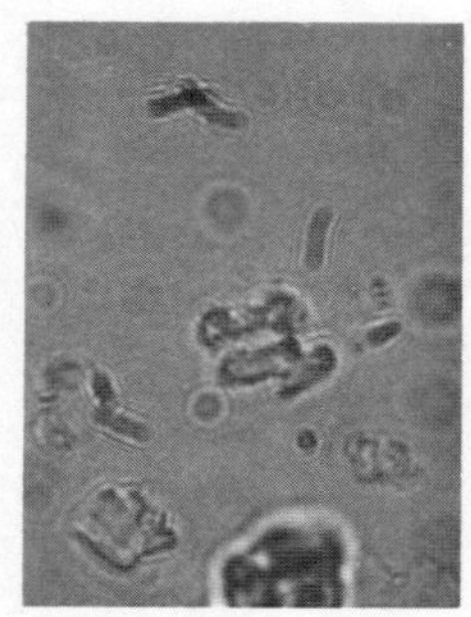

Figure 18
MV4 Transmitted light, 1000 x.

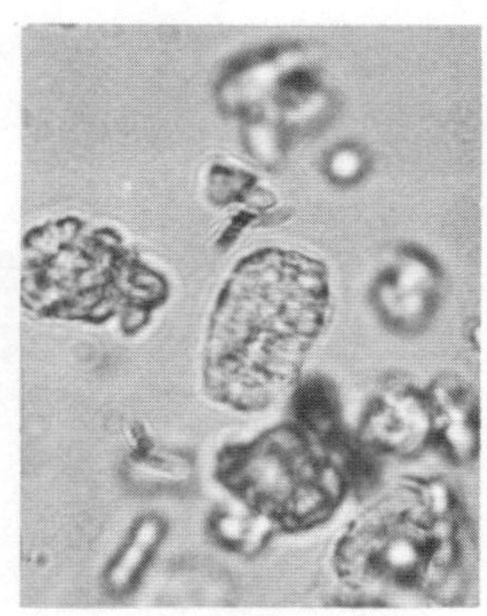

Figure 19
MV3 Transmitted light, 1000 x.

PLATE 2

Figure 10. ***Ellipsagelosphaera britannica***? (Stradner, 1963) Perch-Nielsen, 1968. Figure 11. ***Watznaueria barnesae*** (Black, 1969) Perch-Nielsen, 1968. Figure 12. ***Micrantholithus hoschulzi*** (Reinhardt, 1966a) Thierstein, 1971. Figure 13. ***Braarudosphaera bigelowi*** (Gran and Braarud, 1935) Deflandre, 1947. Figures 14-15. Intermediate forms between ***Micrantholithus hoschulzi*** and ***Braarudosphaera bigelowi***. Figure 16. ***Nannoconus*** aff. ***brönnimanni*** Trejo, 1956. Figures 17-18. ***Nannoconus truitti***? Brönnimann, 1955. Figure 19. ***Nannoconus steinmanni*** Kamptner, 1931.

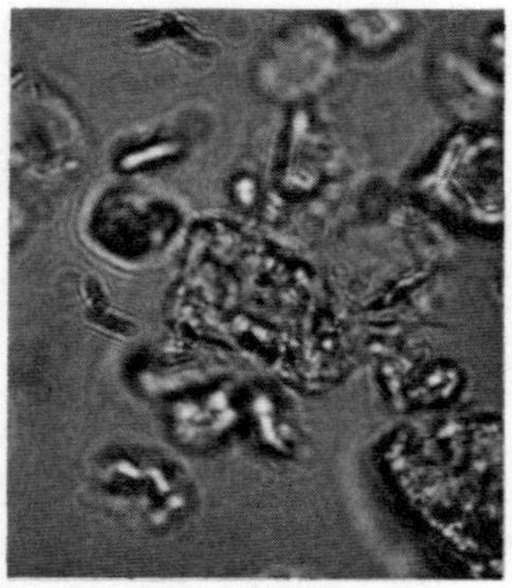

Figure 20
MV3 Transmitted light, 1000 x.

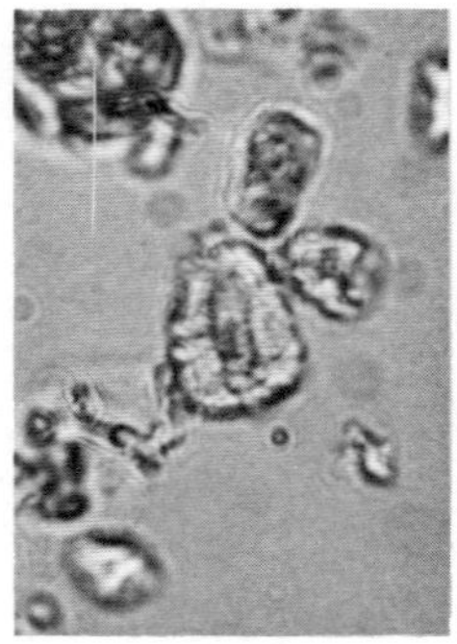

Figure 22
MV3 Transmitted light, 1000 x.

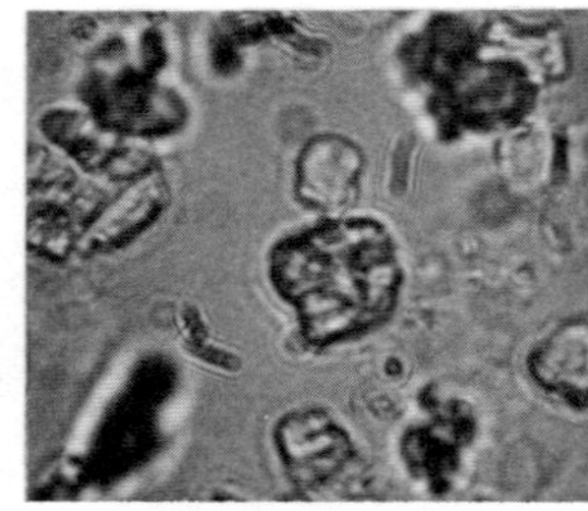

Figure 21
MV4 Transmitted light, 1000 x.

PLATE 3

Figures 20-21. *Nannoconus* aff. *kamptneri* Brönnimann, 1955. Figure 22. *Nannoconus globulus* Brönnimann, 1955.

The Potrero Chico samples show a moderately abundant nannoflora. Most of the coccolith species occur throughout the Lower Cretaceous, except for *Calcicalthina oblongata,* a west Atlantic form which ranges from Valanginian to Albian. Their scarcity and the fact that the *Nannoconus* do not show any of the typical Hauterivian forms lead to the conclusion that the Potrero Chico samples very likely belong to the top of the "*Calcicalathina oblongata* zone" (upper Valanginian to lowermost Hauterivian).

The Minas Viejas samples contain the coccoliths *Calcicalathina oblongata* and *Cruciellipsis cuvillieri,* which have a stratigraphic range of Valanginian to Hauterivian. More information, however, is provided by the nannoconids. The association of *Nannoconus steinmanni, Nannoconus globulus, Nannoconus minutus,* and a few *Nannoconus aff Kamptneri* is similar to the Hauterivian association described by Derès (1972). Though the type coccolith is absent, those samples seem to belong to the "*Lithraphidites bollii* zone" of lowermost Hauterivian to lowermost Barremian.

Paleoecology.—The presence of nannoconids and abundance of *Braarudosphaera, Micrantholithus, Watznaueria,* and *Cyclagelosphaera* point to a shallow-water, nearshore environment.

Notes on Micrantholithus.—Thierstein (1973) remarks that "intermediate forms between *Micrantholithus hoschulzi* and typical *Braarudosphaera bigelowi* have been observed from lowermost Berriasian to Cenomanian; when their outer margin of the single elements is distinctly convex they are here included in *Braarudosphaera bigelowi*" (1973, p. 44). Several of these intermediate forms were found in this study, and, following Thierstein's indications, they have been attributed to either *Micrantholithus* or *Braarudosphaera.*

A similar convergence of species exists between *Micrantholithus hoschulzi* and *Micrantholithus obtusus.* In the studied samples, most *Micrantholithus* are intermediate forms, their elements having commonly a slightly indented edge. Sometimes only one or two of their elements show any indentation; the others have a straight edge. I prefer to consider the forms which have an obviously concave outer edge as *Micrantholithus obtusus* (Barrier, in press). For this reason, all the *Micrantholithus* found in the studied samples are assigned to *Micrantholithus hoschulzi.*

ACKNOWLEDGMENTS

I thank W. W. Cooper of Gulf Oil Corporation, Technical Services Laboratory, in Houston, for a helpful discussion. Philippe Taugourdeau provided facilities for the preparation of photographs. M. Neumann and J. Taugourdeau also provided assistance with preparation of photographs.

REFERENCES

Barrier, J., in press, Nannofossils calcaires des Marnes de l'Aptien inférieur type: Bédoulien de Cassis - La Bédoule, Bull. Muséum National d'Histoire Naturelle.

Derès, F., 1972, Echelle biostratigraphique des Nannoconidés dans le Cretacé: Publication interne. SNPA.

Thierstein, H., 1971, Tentative Lower Cretaceous nannoplankton zonation: Eclogae Geologicae Helvetiae 64/3, p. 455-488.

Thierstein, H., 1973, Lower Cretaceous nannoplankton biostratigraphy: Abh. Geol. Bundesanst. Wien 29.

Wilson, J. L., and Pialli, G., 1977, A Lower Cretaceous shelf margin in northern Mexico: this volume.

Worsley, T., 1971, Calcareous nannofossil zonation of Upper Jurassic and Lower Cretaceous sediments from the western Atlantic: Proc. II, Plankt. Conf. Roma, 1970-2, p. 1301-1322 (Ed Tecnosienzia).

INITIATION OF LOWER CRETACEOUS REEFS IN SABINAS BASIN, NORTHEAST MEXICO

Colin L. Stabler[1] and Benjamin Marquez D.[2]

ABSTRACT

Coral-algal and rudist reefs were initiated in the center of the subsiding Sabinas basin during the Hauterivian.

It is speculated that the reefs may have grown along a hinge line or perhaps over salt or shale anticlines. The reefs, together with their associated lagoons and sabkhas, mark the beginning of a vast carbonate-evaporite platform sequence, equivalent to the Sligo, which prograded eastward and built out over nearly all of the former basin.

INTRODUCTION

This paper reports some field observations of the earliest reef developments in the Lower Cretaceous Coahuila series and briefly reviews some speculations about their origin. The reefs are of interest in trying to understand how carbonate platforms evolve in sedimentary basins.

The Sabinas basin developed in the Upper Jurassic when an embayment (Humphrey, 1956) formed between two land masses, the Coahuila Massif (Coahuila Peninsula of Humphrey) and the Burro Massif (Tamaulipas Peninsula of Humphrey). Coarse-grained terrigenous sediments were deposited in the embayment flanking the massifs, and graded into calcareous shales towards the center (La Casita Group).

The same general paleogeography persisted in the basal Lower Cretaceous and until Hauterivian time. Wedges of coarse-grained clastic sediments (Hosston and its equivalent, San Marcos, Formations) spilled off the massifs, passing into argillaceous sediments (Menchaca and Barril Viejo Formations). This sequence of sediments is called the Durango Group[3] and was probably deposited on a sedimentary "ramp," the surface of which curves around in the form of an embayment.

In contrast, the sequence of facies which makes up the overlying Nuevo León Group[3] (Sligo equivalent) includes sabkha evaporite facies (La Virgen Formation), lagoonal and reef complex carbonates (Cupido Formation), and open-marine limestones (lower Tamaulipas Formation). The reef complex was deposited on a platform and prograded eastward into the open sea, spreading over nearly all of the former gulf, building up a platform margin of approximately 500 m relief. Details of the facies and diagenesis of this carbonate platform are discussed in two other papers presented at this symposium, (Wilson and Pialli, 1977, Conklin and Moore, 1977).

In this paper, the initial platform facies are described and possible mechanisms responsible for the change from ramp to platform are suggested.

FACIES AND DEPOSITIONAL ENVIRONMENTS OF THE PADILLA FORMATION (LOWER SLIGO)

During Hauterivian time, depositional environments underwent a significant change with the development of reefs in a broad tract across the central part of the Gulf of Sabinas (fig. 1). Sheltered behind the reef tract, an extensive open lagoon developed. These two environments are represented by members of the Padilla Formation: the Borregos Member (Marquez and Stabler, in preparation), and the Oballos Member (Charleston, 1973), respectively.

Reef Facies

Reef-type carbonates as early as Hauterivian in age are known in only two outcrops: Potrero Oballos and Potrero Pajaros Azules. (In the latter, the reef lithology is nearly completely dolomitized.) Despite this lack of outcrops, a similar reef facies has been encountered in this formation in several surrounding wildcat wells (fig. 1).

In both reef outcrops it has been possible to identify a vertical sequence of macrofossils and, assuming progradation, to postulate how the reef developed. Thus, it appears that the reefs started to grow over stromatoporoid mounds. Corals ramified by calcareous algae formed the reef framework on the seaward side, whereas *Caprinids* and *Toucasia* grew on the sheltered side. Periodic destruction is indicated by beds of rounded Caprinid shells. Other hard-shelled organisms living in the reef environment include pelecypods, gastropods, echinoderms, foraminifers, and bryozoans.

Lagoon Facies

Bioturbated mudstones and skeletal wackestones were deposited in a broad area behind the reef tract. This area appears to have been a relatively quiet water area, protected from wave action, except for short periods of storms during which thin crossbedded calcarenites were laid down.

The fauna include small *Toucasia,* miliolids, textularias, small stromatoporoid colonies, and large gastropods. This assemblage suggests that the salinity of the lagoonal waters was not significantly different from that of the open sea, implying ample circulation through the reef tract.

Farther from the reef tract, however, indications of more restricted conditions become apparent in the lagoonal sediments. Calcite nodules (replaced sulphate?) occur in burrows in the calcilutites. Microdolomite crystals start to appear in the calcilutites, primarily concentrated in the burrows. In one horizon at Potrero Padilla, pseudomorphs of calcite after celestite were also discovered in the burrows. At the same locality, a sequence of nodular calcite (after anhydrite) occurs at the base of the lagoonal sequence. Additional evidence of evaporite deposition is an anhydrite bed in some wells near the top of the formation and a collapse breccia at Potrero Agua Chiquita. Finally, a bed of microdolomitic desiccation breccia was observed at Potrero Menchaca.

These observations suggest that sabkhas existed in the western part of the basin. However, closer to the margins of the massifs, as at Potrero La Mula, terrigenous sandy facies were deposited. To the southwest, at Potrero La Gavia, arkosic sandstones of the Patula Arkose were deposited in a deltaic environment (Krutak, 1956). Flanking the delta, at Potrero Barril Viejo, sandy bioclastic, oolitic and oncolitic, graded and crossbedded

[1]Shell Oil Company, New Orleans. Formerly with Illing Associates Ltd., London.

[2]Petroleos Mexicanos, Reynosa.

[3]For more details of the local stratigraphy see Charleston (1973) and Marquez and Stabler (in preparation).

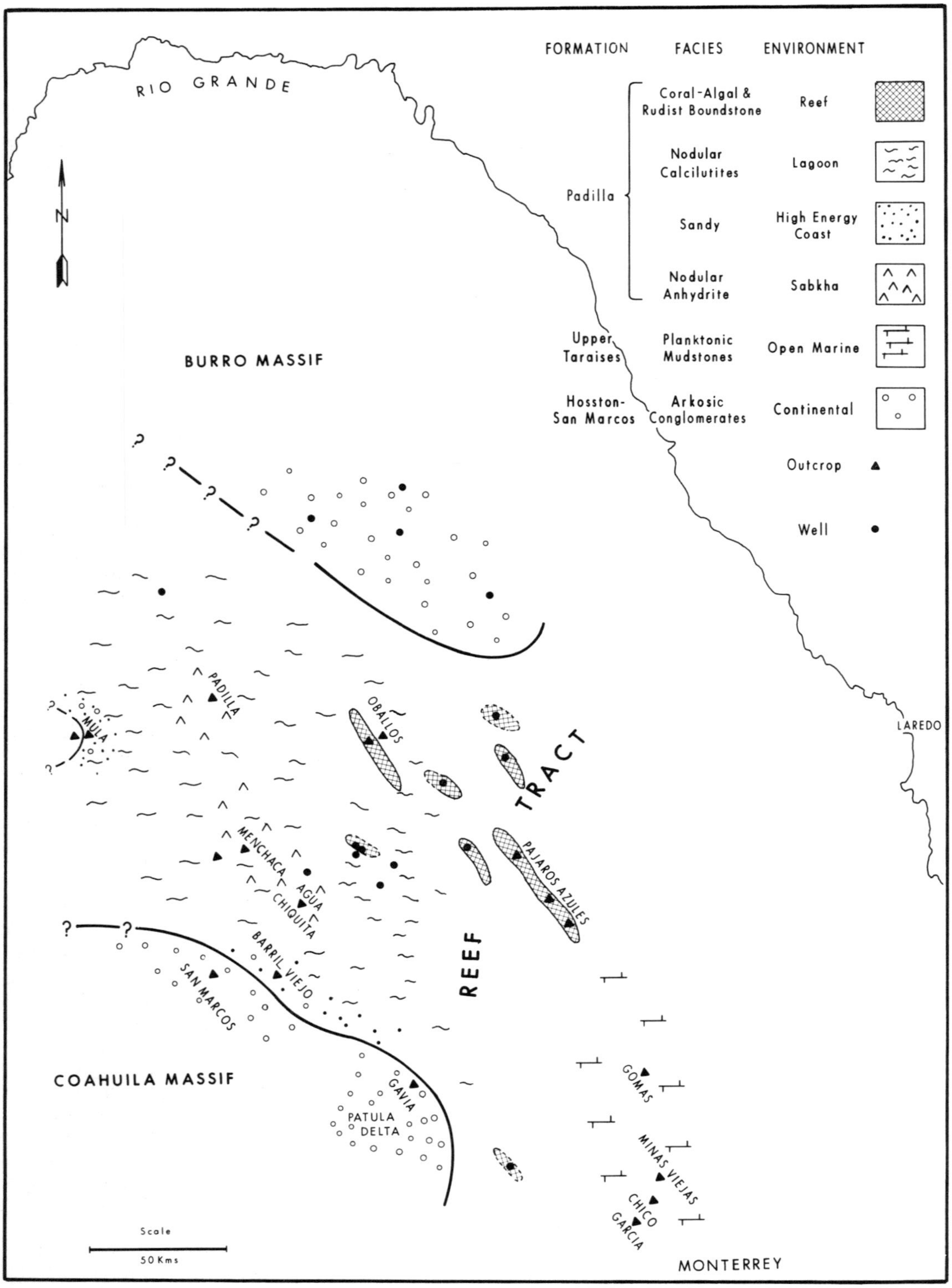

Figure 1. Paleogeography of the Padilla Formation, Sabinas Basin.

carbonate grainstones were laid down in a beach or bar environment during Hauterivian time.

POSSIBLE MECHANISMS FOR REEF INITIATION

The appearance of reefs in the middle of a subsiding basin is an enigma because there is no regional evidence for a lowering of sea level or a change in terrestrial sediment supply. The following mechanisms are suggested.

Hinge Line

A facies change in the basal Cretaceous sediments, from shales (Menchaca and Barril Viejo Formations) to calcareous shales (Taraises Formation) has been mapped by previous workers within the broad area of the reef tract in the overlying Padilla Formation. This facies change could have reflected a change in subsidence rates across a hinge line. Slight adjustments along this line may have allowed sites to become colonized by hard-shelled organisms (shell bank, stromatoporoid mound) which could then have become colonized by corals.

Buried Paleo-highs

A few areas of broad mountain topography, reflecting little tectonic disturbance, are known to occur in the Sabinas basin. They are postulated as buried Jurassic paleotopographic highs. Accommodation around their margins could have formed sufficient topographic relief on the Hauterivian sea floor to serve as a foundation for reef colonization. This mechanism is not favored, however, as most of the known reef facies are not closely related to the gently dipping paleotopographic high areas.

Salt Anticlines

The Sabinas basin was deformed in the Laramide Orogeny into a series of narrow elongated anticlines. Some workers believe that the Jurassic salt and anhydrite, known to underlie a large part of the basin, took an active role in the deformation (Humphrey, 1956; Weidie and Martinez, 1970), and it is possible that salt anticlines and diapirs existed before the Laramide. Recent work immediately to the north of Potrero Pajaros Azules (Vinet, 1976) disclosed evidence that suggests that some uplift occurred during the Albian (Aurora Formation, equivalent to Lower Glen Rose). Local thinning over anticlines in the Aptian (La Pena Formation, equivalent to Pine Island) indicates some movement at this earlier date. Therefore, it seems reasonable to postulate that slight salt rolling took place even earlier, during the Hauterivian, which brought the sediment floor of the basin up to a depth suitable for coral colonization and the development of reefs.

The main disadvantage of the salt anticline mechanism is that salt is not known to have occurred in the Oballos and Pajaros Azules area. In fact, the high gravity anomalies tend to confirm that the sulphates exposed at the surface continue as anhydrite, rather than salt, to depth in the cores of the anticlines.

Shale Anticlines

A similar mechanism that could have been responsible for early local uplifts, thinning, and facies changes is the upwelling of the thick shale sequence at the top of the Jurassic/ base of the Cretaceous (La Casita Group and Menchaca Formation) which underlies most of the Sabinas basin. Such upwelling could have started in the Hauterivian and localized sites of reefing throughout a broad tract where marine conditions were optimum.

All known occurrences of reef and lagoon facies of the Padilla Formation, whether from outcrops or wildcat wells, are located on or close to the crests of the elongated anticlines. If we favor the mechanism of shale anticlines as the best tentative working model for reef development, then it is possible that more normal conditions could have persisted off the crests of the anticlines. The variation from massive reef on the west flank to patch reef on the east flank of Potrero Oballos suggests that these normal conditions may have existed. Another example occurs in the subsurface near Monclova, where developmental drilling has encountered nonreef facies on the western flank of an anticline. The paleogeographic reconstruction shown in figure 1 has been drawn with this model in mind.

It is further possible that the sabkhas observed in the western part of the basin are restricted to the anticlines and represent periods when shale upwelling brought the sediment surface into the supratidal environment.

Thus, the development of the Sligo carbonate platform in the center of the Sabinas basin is tentatively postulated as being due to reef colonization on the crests of shale anticlines that started to develop in the Hauterivian because of overloading of the underlying argillaceous sequence. A vast amount of outcrop exposed in the Sabinas basin awaits further study. Hopefully, future workers will be encouraged to continue investigations of the evolution of this intriguing sedimentary basin.

CONCLUSIONS

1. Coral-algal and rudist reefs, together with associated lagoon and sabkhas, were initiated in the Hauterivian.
2. The reefs grew in the center of the subsiding Sabinas basin, possibly along a hinge line, over buried paleotopographic highs, or over salt or shale anticlines.
3. The appearance of these reefs represents the first step in the development of a vast, prograding carbonate-evaporite platform sequence, equivalent to the Sligo.

ACKNOWLEDGMENTS

The authors wish to thank the Exploration Management of Petroleos Mexicanos for permission to publish these results, and also L. V. Illing of Illing Associated Ltd., under whose auspices the study was carried out.

REFERENCES

Charleston, S., 1973, Stratigraphy, tectonics and hydrocarbon potential of the Lower Cretaceous, Coahuila series, Coahuila, Mexico: Univ. Michigan, Ph.D. thesis.

Conklin, J., and Moore, C., 1977, Paleoenvironmental analysis of the Lower Cretaceous Cupido Formation, northeast Mexico: this volume.

Humphrey, W.E., 1956, Tectonic framework of northeastern Mexico: GCAGS Transactions, v. VI, p. 25.

Krutak, P.R., 1965, Source areas of the Patula Arkose (Lower Cretaceous) Coahuila: Journal of Sedimentary Petrology, v. 35, p. 512-517.

Marquez, B., and Stabler, C.L., in preparation, Razonamiento de la estratigrafia de la Serie Coahuila: Bol. Assoc. Mex. Geol.

Vinet, M.J., 1976, Geology of Sierra de Baluartes and Sierra de Pajaros Azules: Univ. New Orleans, Master's thesis.

Weidie, A.E., and Martinez, J.D., 1970, Evidence for evaporite diapirism in northeastern Mexico: AAPG Bull., v. 54, no. 4, p. 655-661.

Wilson, J.L., and Pialli, G., 1977, An Early Cretaceous carbonate bank in the Monterrey-Saltillo area, northern Mexico: this volume.

PALEOENVIRONMENTAL ANALYSIS OF THE LOWER CRETACEOUS CUPIDO FORMATION, NORTHEAST MEXICO

Jack Conklin[1] and Clyde Moore[2]

ABSTRACT

Limestones of the Lower Cretaceous Cupido Formation (late Neocomian to early Aptian) were deposited during the general Mesozoic transgression of a warm, clear-water, epeiric seaway over northeastern Mexico.

The development of the massive Cupido carbonate platform, with the Coahuila Peninsula acting as a nucleus, was primarily due to the sudden proliferation of rudist bivalves in the Lower Cretaceous. Rudists, mostly requieniids and caprinids, along with dendroid corals, red algae, and stromatoporoids constructed ecologic reefs in the higher energy zone as waves touched bottom along a shallow carbonate ramp (formed by the underlying Taraises Formation). The reefs trapped bioclastic debris and baffled the wave energy to produce lower energy backreef environments. The accumulation of carbonate material in the reef complex kept up with basinal subsidence and sea-level rise and, in approximately 9 million years, formed a platform with a shelf margin of considerable submarine topographic relief.

The vertical sequence of lithologies found at several localities indicates localized regression of the sea as the reef and its laterally coexisting environmental facies prograded over basinal lime muds. Six laterally coexisting facies are listed below in the order in which they are commonly found in a vertical measured section:

6. Lagoon, restricted-lagoon, and sabkha facies
5. Near-reef tidal-flat facies
4. Near-reef shoal facies
3. Organic reef facies
2. Forereef slope talus facies
1. Basinal facies.

INTRODUCTION

The Lower Cretaceous Cupido Formation in northeastern Mexico consists of up to approximately 800 m of carbonate platform type deposits. Its massive limestone beds resist erosion and form impressive ridges in many of the breached anticlines in the study area (figs. 1, 5). Of the fourteen localities studied, eight were sampled and measured in detail. The wide areal distribution of well-exposed outcrops makes the Cupido Formation an excellent subject for a regional study of the relationship between lithofacies and platform development of an ancient carbonate sequence. Such a study is

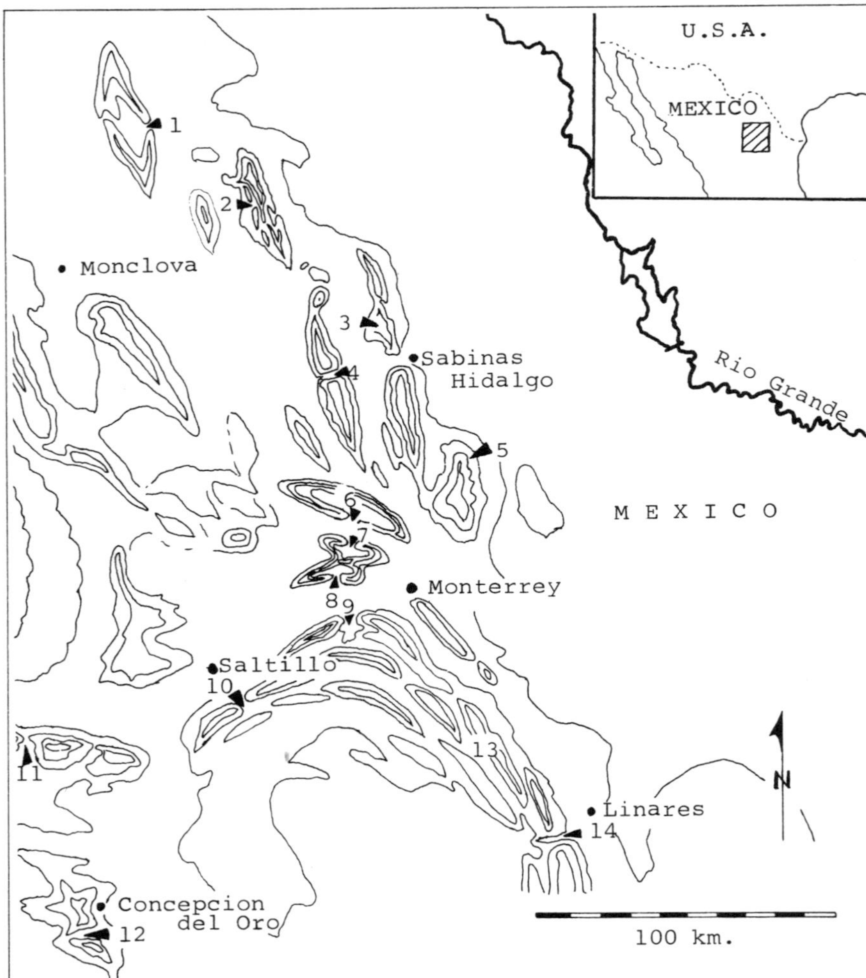

1. Potrero Oballos
2. Pajaros Azules Canyon
3. Sierra Sabinas*
4. Bustamante Canyon
5. Sierra Picachos*
6. Potrero Minas Viejas*
7. Potrero Chico
8. Potrero Garcia*
9. Huasteca Canyon
10. Arteaga Canyon
11. La Casita Canyon*
12. Astillero Canyon
13. Sierra Madre Oriental*
14. Santa Rosa Canyon

(* unmeasured section referred to in text)

Figure 1. Location Map

[1] Cities Service Oil Company, Tulsa, Oklahoma.

[2] Department of Geology, Louisiana State University, Baton Rouge, Louisiana.

helpful in understanding similar Lower Cretaceous platforms found in the deep subsurface of Texas, Louisiana, Florida, and, possibly, off the eastern coast of the United States. The Cupido Formation is an important exploration target for Petroleos Mexicanos in the region near Monclova, Coahuila. Its stratigraphic equivalent, the Sligo Formation, has been productive in eastern Texas and most notably at Black Lake field in northwestern Louisiana.

The Cretaceous carbonate platforms and offshore banks owe much of their development to the sudden proliferation of rudist bivalves (pl. 4). They evolved from the early Mesozoic ***Diceras*** of the Pachydont group, and flourished along an equatorial belt from 40° N. to 20° S. (Wilson, 1975). The rudists found in the Cupido Formation are dominantly of the requieniid and caprinid groups along with some monopleurids. Although the rudists became extinct at the end of the Mesozoic era, their paleoecology has been thoroughly studied on the Comanchean platform of Texas (Perkins, 1974) and in the Tampico area of Mexico (Coogan and others, 1972). These studies of Lower Cretaceous paleoecology, along with some parallel studies of the carbonate petrography of rudist sequences (Petta, 1976), allow an interpretation of the paleodepositional patterns of the Cupido Formation.

STRATIGRAPHY

Previous Investigations

The Cupido Formation was proposed by Imlay (1937) for 1,415 ft of resistant ridge-forming limestones exposed along the northern wall of the Canyon del Mimbre, in the center of the Sierra de Parras (60 km southwest of Parras, Coahuila). Humphrey (1949) measured the Cupido Formation on the north flank of the Sierra de los Muertos (near Monterrey), reporting 2,244 ft of "thin to thick bedded and massive limestones, dolomitic limestones, and dolomites with intercalated shales." He redefined the Cupido Formation to include the overlying lower limestone unit of Imlay's (1936) La Peña Formation. The upper member of the La Peña Formation is usually less than 30 m thick and is an excellent stratigraphic marker between the Cupido and the overlying Aurora Formation (fig. 2). The thin-bedded black shales and marls of the La Peña Formation are easily eroded and form a distinct gully between the two massive limestone formations. The La Peña Formation contains a late Aptian (Gargansian) age ammonite fauna, ***Dufrenoya texana*** (Humphrey, 1949), which is also used to mark the top of the Coahuila series (Imlay, 1944).

At most localities studied in this report, the Cupido Formation is conformably underlain by the Taraises Formation. Gradational contact between the two formations is often difficult to spot in the field. Humphrey and Diaz (1956) placed the top of the Durango group, which is also the top of the Taraises Formation (fig. 2), at the upper limit of the ammonite assemblage, ***Maderia, Mexicanoceras, Leopoldia***, and ***Acanthodiscus*** (Charleston, 1974). Therefore, the ammonite assemblages of the underlying and overlying formations restrict the Cupido Formation to late Neocomian - early Aptian time (fig. 2). This makes the Cupido Formation roughly equivalent to the Sligo Formation in the lower Trinity division of Texas. The "Sligo Trend" is considered to be a deeply buried reef complex that formed on the shelf margin around the coast of the Gulf of Mexico during the marine transgression of the Lower Cretaceous.

South of the study area in the Tampico region of Mexico, the late Neocomian - early Aptian age limestone is called the "Lower Tamaulipas" Formation, as proposed by L. W. Stephenson (1921?). Muir (1936) studied this chert-bearing black limestone and found that it had pelagic fossils which indicated that it was the bathyal facies equivalent of the Cupido Formation.

In his studies of the Lower Cretaceous formations in the extreme northeastern part of Mexico, Smith (1970) found evidence for a slow transgression in the lower Cupido Formation, a slight regression in the middle Cupido, and a more rapid transgression in the uppermost Cupido Formation.

Recently, the entire Coahuila series in the northern part of the states of Coahuila and Nuevo León have been studied by Charleston (1974). He combined a detailed field study with a petrographic and computer analysis of the Cupido and its laterally coexisting terrigenous facies: the Patula Arkose, the evaporites of the La Virgen Formation, and the argillaceous strata of the La Mula Formation (fig. 6). These formations along with the Cupido Formation make up the Nuevo León Group of Imlay (1944).

J. L. Wilson and students at Rice University are currently engaged in a detailed, ongoing study of the Cupido Formation in northeastern Mexico. They are interested in the carbonate platform development and the biostratigraphic relationships between the Cupido and the Taraises Formations.

REGIONAL GEOLOGIC SETTING

The Cupido Formation in the study area was deposited during the gradual transgression of the Mesozoic Seaway over northeastern Mexico. The boundaries for early and middle Mesozoic deposition were controlled by several broad low-relief landmasses: the Texas Craton, the Mesa Central, the Tamaulipas Peninsula, and the Coahuila Peninsula.

These positive features are thought to represent the remains of a post-Permian orogeny that intensely deformed the Paleozoic sediments and caused thick clastic, red bed wedges to be deposited, possibly from the block faulting of Triassic to Middle Jurassic time (Humphrey, 1956).

The ancient geomorphology was further affected some time in the Middle Jurassic by the block faulting of the Nevadian Orogeny, with later, possibly related pulsations along the borders of the positive elements during the Lower Cretaceous (Humphrey, 1956). The low areas between the landmasses became the sites of deposition and continued subsidence throughout the Mesozoic Era.

A major area of Mesozoic deposition was in a long, narrow trough referred to in this paper as the Mexican Seaway (fig. 3). Sedimentation also took place in an embayment between the Coahuila and Tamaulipas Peninsulas known as the Sabinas Gulf and east of the Tamaulipas Peninsula in the early Gulf Coast Geosyncline (fig. 3). Evidence of these very important Upper Jurassic paleogeographic elements can still be seen today on satellite photographs (fig. 4). As Humphrey (1956) eloquently stated, "The structural patterns in northeastern Mexico are largely the result of deformation by Laramide stresses acting on discrete areas of Mesozoic sedimentation. The character and extent of tectonic types in those areas have to do with thickness and nature of the sedimentary sections involved and with the relative strength and the margins of contemporaneous Mesozoic positive elements. Thus, as present-day geomorphology in northeastern Mexico more or less faithfully represents the distribution and character of the structural patterns, so the character of the structural deformation of the Mesozoic sediments reflects the nature and disposition of the sedimentary areas. All

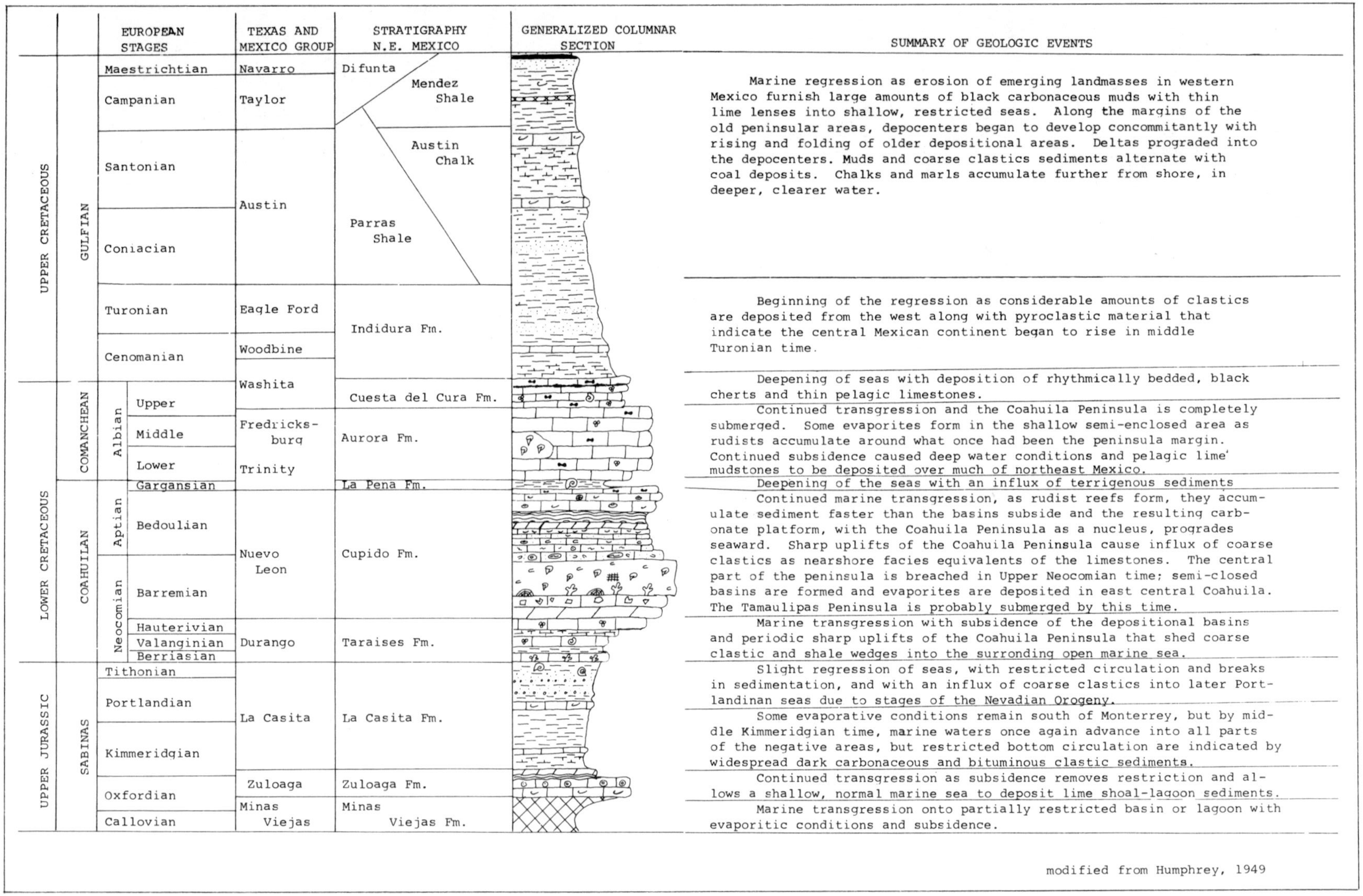

Figure 2. Stratigraphic Correlation Chart, and Short History of Sedimentation.

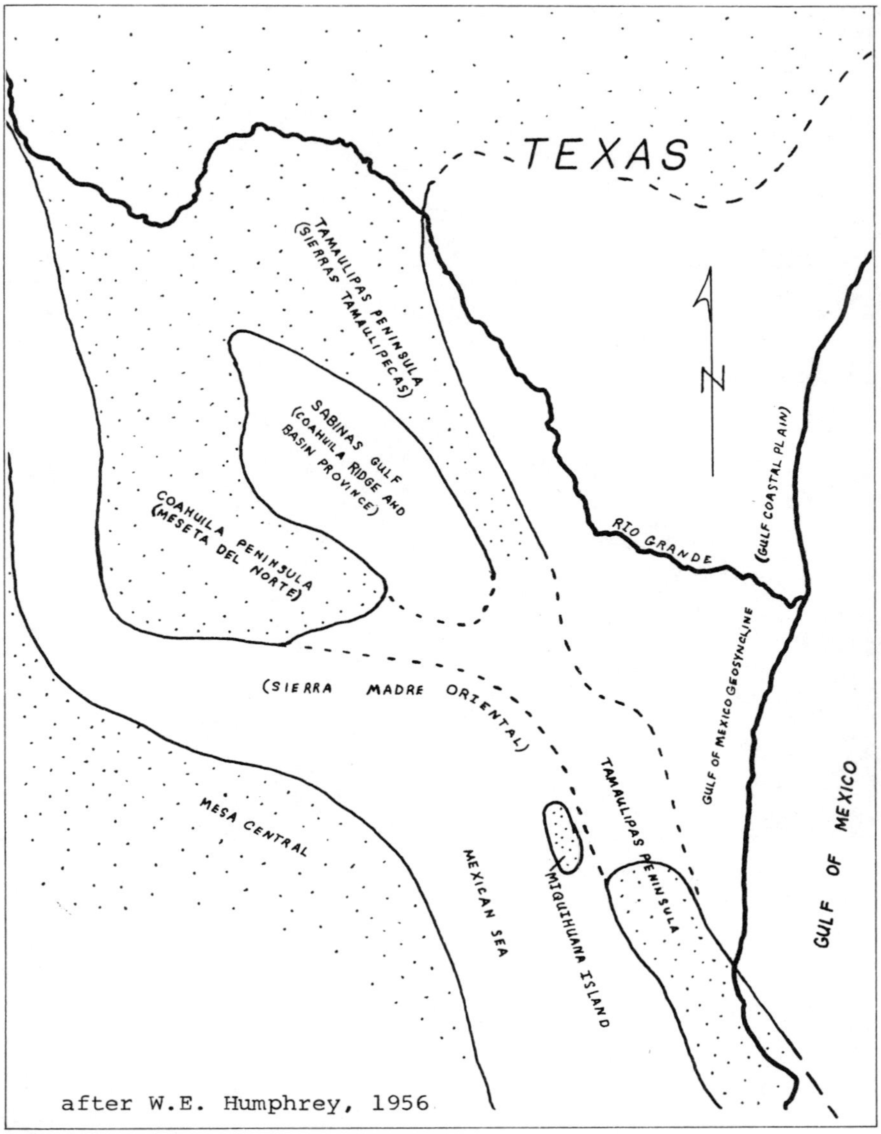

Figure 3. Relationships between Early Upper Jurassic paleogeography and modern geomorphologic provinces.

are intimately related to the fundamental landforms of early Upper Jurassic paleogeography and to the response of those land margins during their progressive destruction through Upper Jurassic and Lower Cretaceous time."

Little is known about the ancient landmasses that correspond to the present topographic province known as the "Mesa Central" (fig. 3) because it is now covered by Tertiary volcanics. Humphrey (1956) thought that it was once an extensive welt of Paleozoic sediments, which probably became a gently shelving "hinterland" for Mesozoic sediments. Because the "Mesa Central" is west of the study area, the western boundary of the Cupido Formation (fig. 3) is speculative.

The Tamaulipas Peninsula (fig. 3), which coincides with present Sierra Tamaulipas trend, may have had little effect on the deposition of the Cupido Formation. It was probably submerged by the end of the Upper Jurassic (Humphrey, 1956). Therefore, the ancient Gulf of Mexico and the Sabinas Gulf probably merged by Early Cretaceous time. Bishop (1966) found the Lower Cretaceous Cupido Formation at the Sierra Picachos locality to be dominantly pelagic, "basinal" limestones. This locality coincides with the trend of the Tamaulipas Peninsula (figs. 1, 3). Smith (1970) drew the northern strand line for the Early Cretaceous seaway near the present Big Bend area of West Texas, and indicated that the northern part of the Tamaulipas Peninsula was totally submerged. The famous Faja Del Oro and the Sierra Del Abra are massive (Albian) rudist buildups that formed on submerged topographic highs possibly associated with the southern extension of the Tamaulipas Peninsula (Wilson, 1975).

A positive feature termed the "Miquihuana Island" (fig. 3) was emergent during the Upper Jurassic (Oivanki, 1973). It possibly continued to be an emergent feature or a slightly submerged positive feature into the Lower Cretaceous. Enos (1974) showed that it became the site of the "Valles Platform," a major carbonate accumulation during Albian time. It is possible that the base of the Valles platform might have Neocomian and Aptian fauna similar to the Cupido Formation. This platform is south of the study area.

In 1923, Emil Böse found what he thought were "the vestiges of an ancient continent" in northeastern Mexico. This was later confirmed by the work of Kellum and others (1936) who called the landmass the Coahuila Peninsula. This feature was to have a profound influence on the depositional facies of the Cupido Formation.

The Coahuila Peninsula was connected to the Texas Craton during the Jurassic and the very early Cretaceous (early Neocomian). Charleston (1974) shows, however, that the middle part of the peninsula was later breached during late Neocomian time, leaving an isolated landmass that remained emergent until Albian time (fig. 3). The ancient Coahuila Peninsula coincides with the present-day topographic province called the "Meseta del Norte" (Humphrey, 1956).

The trend of the Mexican Seaway ("Mexican Geosyncline" of Humphrey, 1956) during the Early Cretaceous period coincides with the trend of the present-day Sierra Madre Oriental. In northeast Mexico, en echelon folds of the sierras extend from west of Ciudad Victoria, Tamaulipas, to just south of Monterrey, Nuevo León, where the mountain range curves sharply westward to the vicinity of Saltillo, Coahuila (Humphrey, 1956). The sierras then wrap around the southern and southwestern part of the Coahuila Peninsula and, in the States of Durango and Chihuahua, trend north-northwest where they eventually cease to exist.

The Sierra Madre Oriental contains more than 5,000 m of continental and marine sediments ranging from Triassic to Upper Cretaceous in age. Two localized depocenters within the Mexican Seaway, the Parras basin

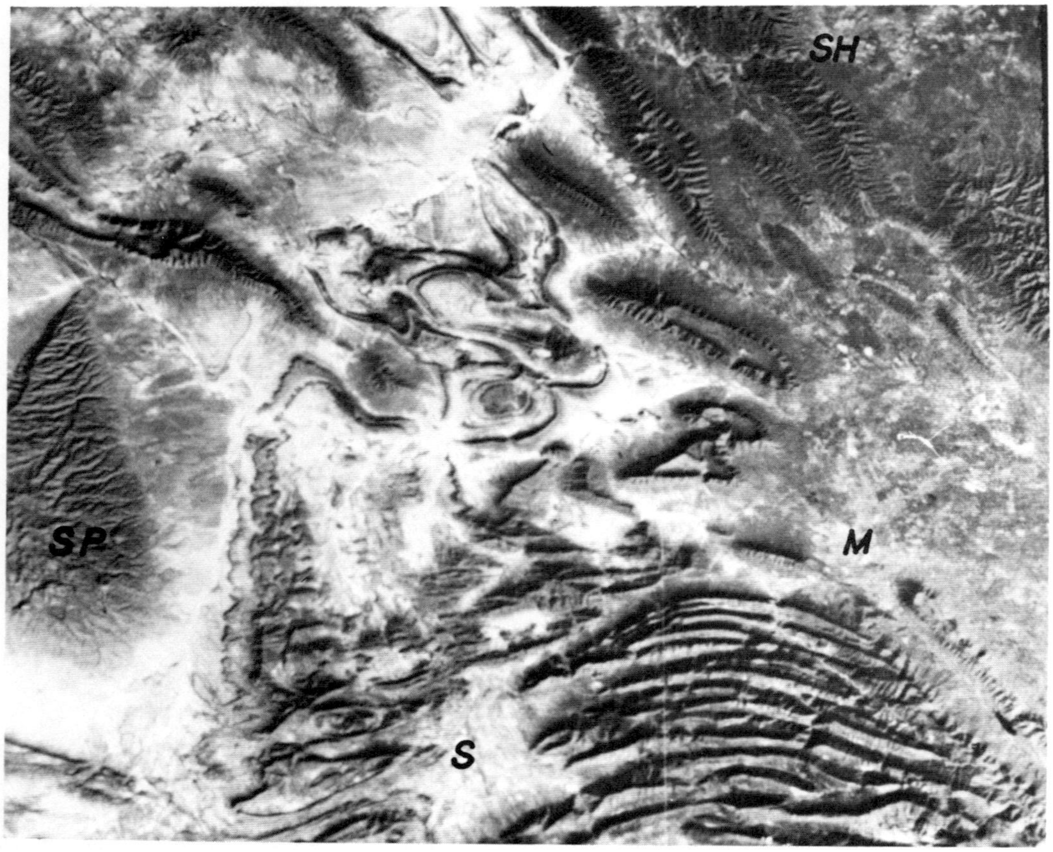

Figure 4. Satellite photograph of much of the study area. S = Saltillo, M = Monterrey, SH = Sabinas Hidalgo, SP = Sierra de la Paila (probably coincides with the southern tip of the ancient Coahuila Peninsula).

and the Magiscatzin basin, rapidly subsided during the Upper Cretaceous and contain thick deltaic sequences (Humphrey, 1956).

The deformation of the Ouachita Geosyncline created the Coahuila and Tamaulipas Peninsulas (fig. 3), and the Sabinas basin was the resulting graben-like trough or intermountain basin between the two highlands. Most of the Mesozoic history in the area involves the gradual infilling of the Sabinas basin and the gradual submergence of the Tamaulipas and Coahuila Peninsulas.

The first cycle of basin infill probably began some time in the Upper Triassic or Lower Jurassic Periods with the deposition of clastic red beds similar to those of the Huizachal Group found farther south in the Ciudad Victoria region of Mexico. However, there are no outcrops of these red beds in the study area.

The second phase of basin infill was marked by evaporite deposition of the Minas Viejas Formation. During the Middle or Upper Jurassic, the Sabinas basin was a restricted embayment with the Coahuila Peninsula on the west, the Texas Craton to the north, and the Tamaulipas Peninsula on the east and northeast. The restriction was sufficient to cause the very thick sequence of evaporites, the Minas Viejas Formation, to be deposited in the Sabinas basin.

Most of the study area is within the Sabinas basin. Some of the best outcrop localities are in the breached anticlines known locally as "potreros" (a hidden valley or a steep-walled, box canyon suitable for use as a corral). Some of these potreros are associated with diapirism of the Minas Viejas evaporites during the Laramide Orogeny (fig. 5).

The water depths of the early Sabinas basin during the deposition of the Minas Viejas evaporites are debatable. Oivanki (1973) found that the limestones of the overlying Zuloaga Formation contain widespread, shallow-water, lagoon and tidal-flat facies. This indicates that much of the Sabinas basin had filled to near sea level by Late Jurassic time, and that thick accumulation of the overlying sediments was made possible only by regional subsidence and sea-level rise.

During the uppermost Jurassic Period, the Nevadian Orogeny caused uplift of the Coahuila Peninsula and deposition of argillaceous and clastic sediments in the Sabinas basin (Humphrey, 1956). Intermittent pulses of this orogeny continued during the Early Cretaceous transgression and the deposition of the Taraises and Cupido limestones. Figure 6 shows the thick clastic wedges and the laterally co-existing carbonates of the Coahuila series.

The thickness distribution of the Coahuila series is shown on figure 7. Its apparent similarity to the isopach pattern of the Cupido Formation (fig. 8) shows that the Cupido limestone is the main constituent of the Coahuila series. The regional extent and the contour pattern of figure 8 indicate that the Cupido Formation represents a major carbonate platform. It is thought to have evolved from the underlying Taraises Formation as waves touched bottom on a gently sloping carbonate ramp, causing three

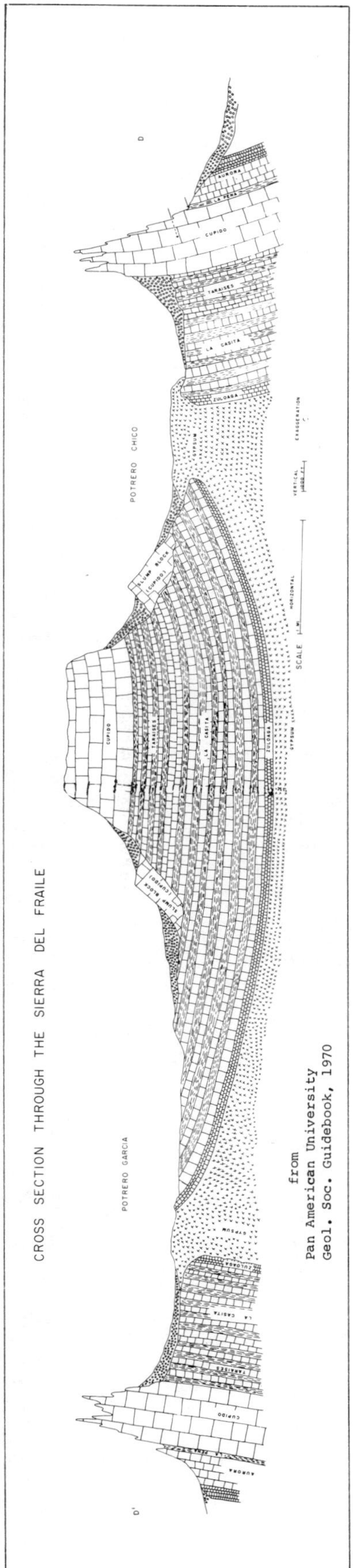

Figure 5. Potreros that resulted from Diapirism of the Minas Viejas Evaporites.

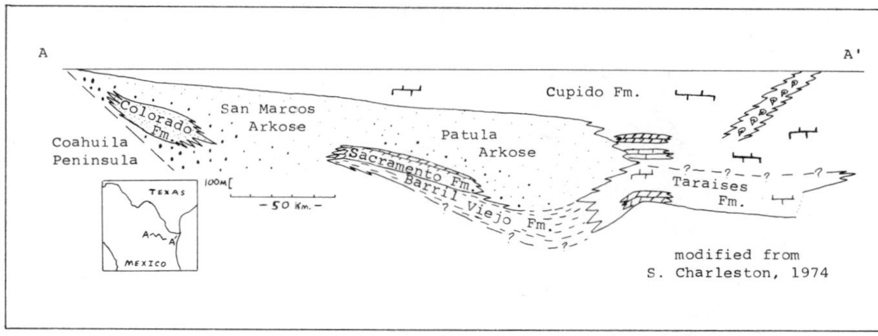

Figure 6. Cross Section of the Formations in the Coahuila Series.

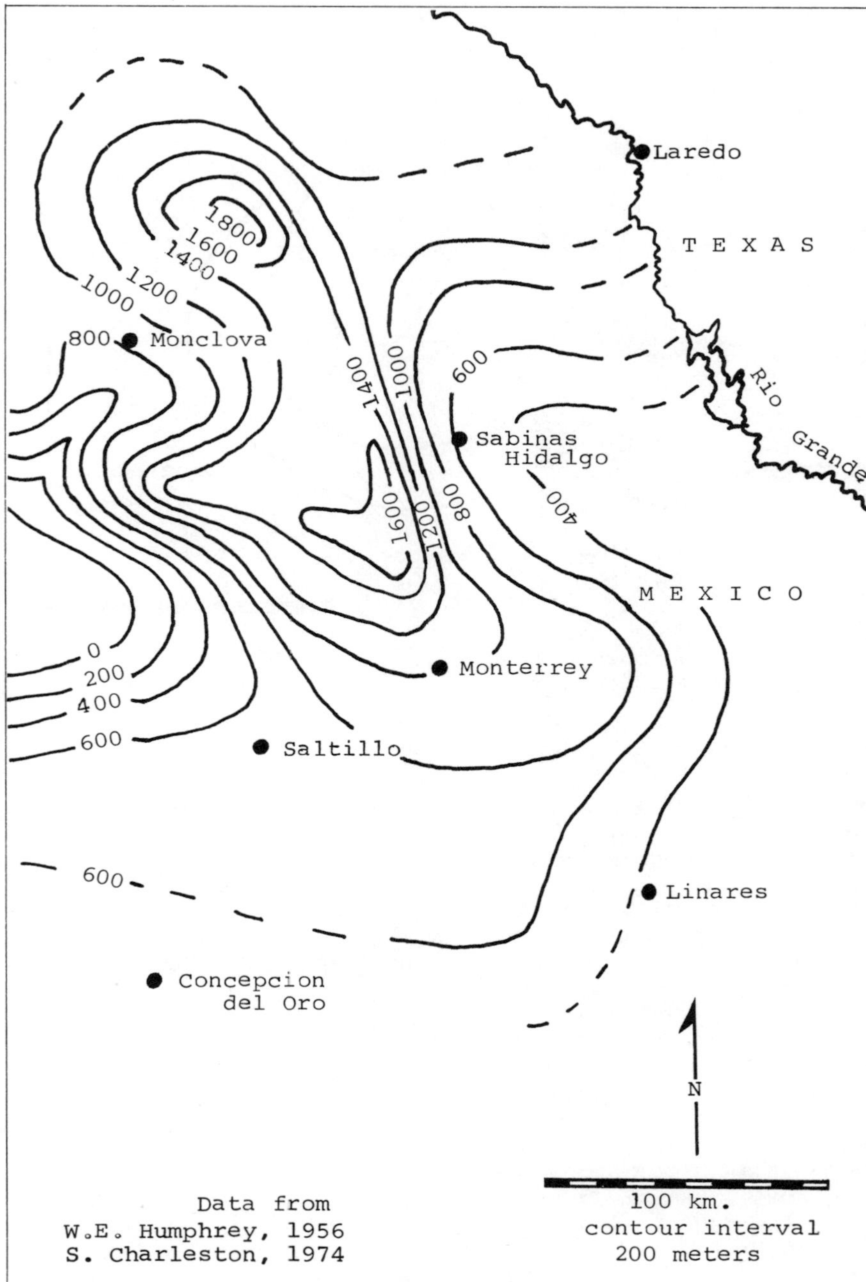

Figure 7. Isopach map of the Coahuila Series.

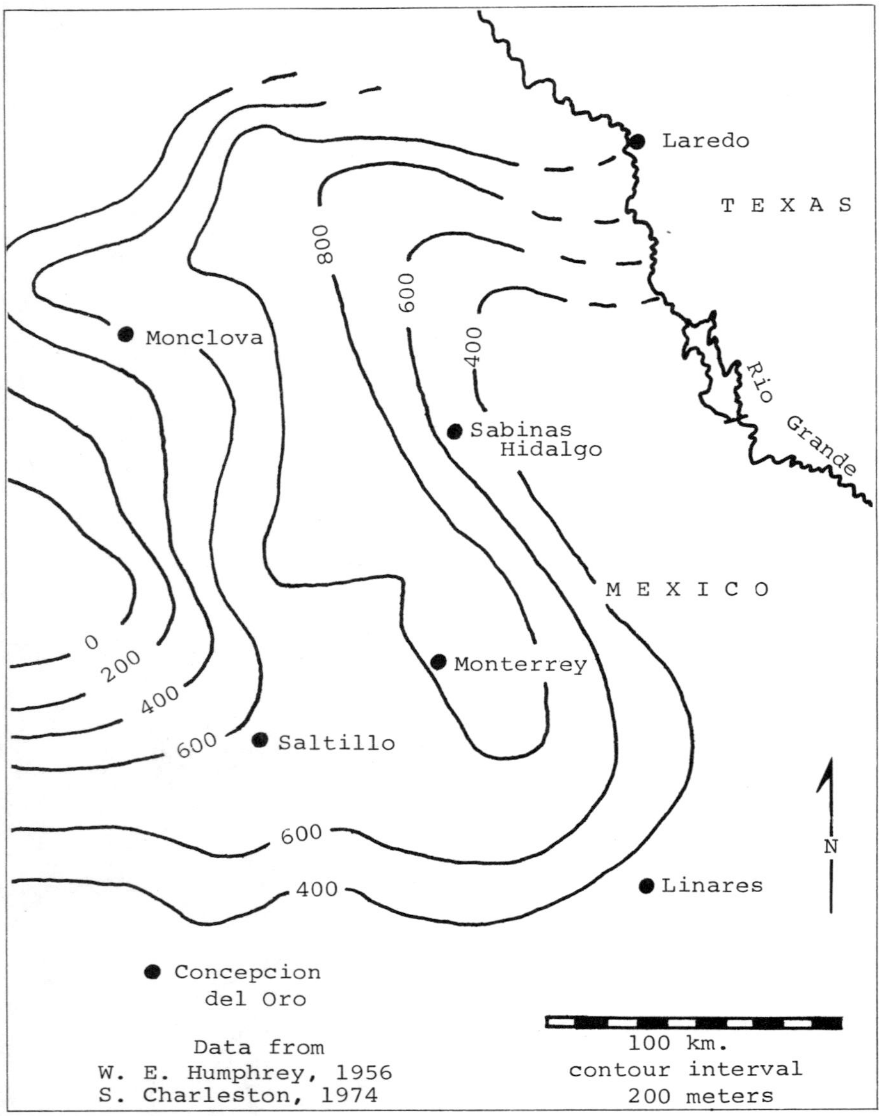

Figure 8. Isopach map of the Cupido Formation.

major energy and carbonate accumulation zones to develop. Irwin (1965) utilized the ramp concept to describe the development of a carbonate platform in the Willaston basin. The greatest biologic productivity and differential carbonate accumulation are indicated by the sharp decrease in thicknesses eastward into the early Gulf of Mexico. The area east of Monterrey is considered to be a "starved basin" where accumulation rates could not keep up with subsidence or platform growth. The isopach pattern suggests, but cannot prove, considerable submarine topographic relief. The platform margin probably trends north-northwest, from Monterrey to slightly west of Sabinas Hidalgo. The trend of the platform margin south of Monterrey is still uncertain, but it is thought to arc westward and pass south of Saltillo, possibly extending around to the western side of the Coahuila Peninsula.

LITHOFACIES AND PALEOENVIRONMENTS

The characteristic outcrop-weathering profile of the Cupido Formation and the distribution of lithologies within the formation are closely related. This relationship was found to be consistent at several widely spaced localities; it forms the basis for dividing the Cupido Formation into six different units (fig. 9). These six units are systematically described in the normal stratigraphic sequence commonly found in outcrops. The descriptions include: (1) the localities where the units are exposed, which are used to determine their regional distribution and trends; (2) the localities where the units occur in the vertical measured sections, to determine transgressive or regressive sequences; and (3) an analysis of the units' lithofacies, to postulate environments of deposition. A synthesis of this information would yield a plan-view facies map and a model for the paleoenvironments of the Cupido Formation.

Unit A

The best examples of the basal unit of the Cupido Formation are found at localities along the eastern edge of the study area (fig. 1). The localities are the Sierra Sabinas, the Sierra Picachos, and the Santa Rosa Canyon area in the Sierra Madre Oriental. This unit ranges from 350 to 500 m thick at these localities and conformably overlies the Taraises Formation with a gradational contact that is often difficult to observe in the field. The contact is usually picked where thinly bedded black limestones with interbedded shales (often containing ammonites) grade upward into more resistant, less argillaceous, black limestones.

Very thick beds and undulatory bedding planes are typical of the unit at the Sierra Sabinas locality (pl. 1). The unit also has some very thin shaly interbeds, stylolites, and black chert nodules. Some bedded chert (approximately 40 cm thick) was found at the Santa Rosa Canyon section. The limestones are characteristically gray to dark-gray lime mudstones with rare echinoids and ammonites. Bishop (1972) found that the Cupido Formation at the Sierra Picachos locality averaged only 4 percent allochems consisting principally of "ostracods, echinoid fragments, foraminifers, intraclasts, and pellets." His scanning electron microscope photographs show the remains of nannoplankton in the overlying Aurora Formation; they are suspected to occur in the Cupido Formation as well. Bishop also found the Cupido Formation to be strongly dolomitized and stated that, "the dolomite occurs as distinct rhombs and constitutes fifty-four percent of the rock by volume." The finely crystalline dolomite commonly occurs as euhedral rhombs that average 0.04 mm in length (Bishop, 1972). Similar dolomite crystals were found at the Astillero Canyon section. The relict inclusions or ghosts of micrite within the rhombs suggest replacement of the original lime mud during diagenesis (Bishop, 1972). However, many samples from other similar limestones show no dolomite. Doubly terminated euhedral quartz crystals (0.5 mm in length) and pyrite crystals occur in some of the lime mudstones.

Although there are some peloids present, burrows were not observed. The lack of burrows at the Sierra Sabinas section, the Santa Rosa Canyon section, and possibly the Sierra Picachos section suggests that the basin floor was below the oxygenation level. This indication of relative depth, along with the lack of any recognizable reef-derived debris or shallow-water type sediments, and the dominance of pelagic fauna suggest that the limestones of unit A represent a "basinal lithofacies."

The sediments in this bathyal to outer neritic environmental zone are thought to have formed seaward of the developing carbonate platform in what might well be called a "starved basin." In this area the relatively slow deposition of pelagic microfossils from the open-marine water column, along with the influx of some very fine sediments from the platform, could not keep pace with either regional subsidence or the more rapid growth of the carbonate platform. The difference in formation thickness between the platform and the basin can be more than 400 m depending on the distance from the platform.

Other basinal pelagic lime mudstones were recognized that had slightly different characteristics which indicated they were deposited in shallower waters and closer to the platform.

The 360-m-thick section of the Cupido Formation at Astillero Canyon consists predominantly of lime mudstone-wackestone lithofacies with a deep-water pelagic fossil assemblage, but it is also heavily burrowed and pelleted in places, has gastropods and pelecypods, and has at least two thin beds of oolitic lime grainstones and packstones. Several beds with aggregates of algae-coated peloids (like "grapestones") were also observed. This suite of lithofacies would indicate that the Astillero section was above the oxygenation level and probably shallow enough to be affected by storm waves that could transport oolite-sized material from the nearby shelf into the basin. Nelson (1962) suggested that sea bottoms normally experience moderate wave action to depths of 100 ft and occasionally much greater (400 ft) (Heckel, 1974).

The third type of basinal limestones is found even nearer the shelf margin. The best examples are the pelagic lime mudstones at Potrero

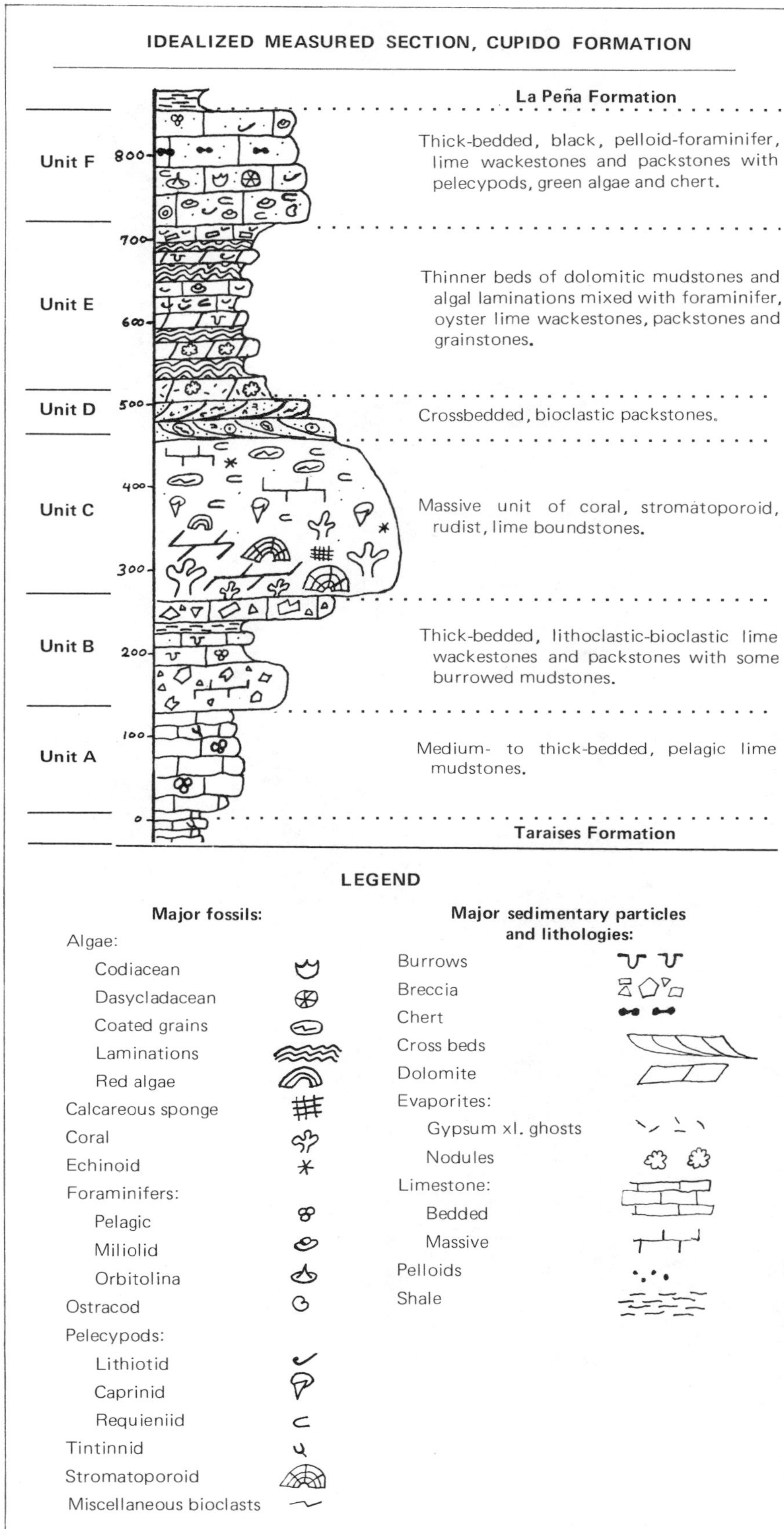

Figure 9. Idealized Measured Section, Cupido Formation.

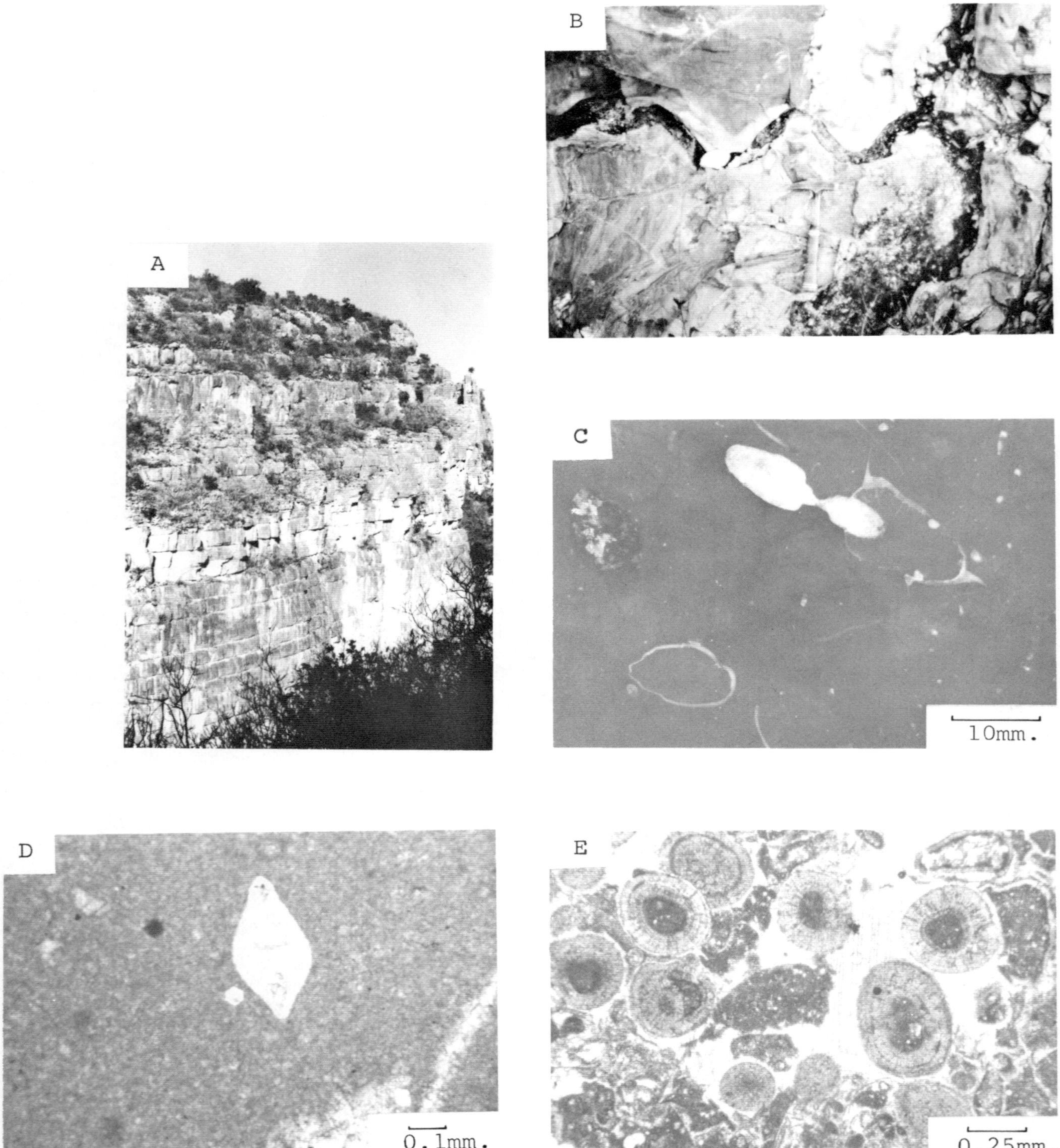

PLATE 1

A. Very thick-bedded, basinal lime mudstones of the Cupido Formation. Sierra Sabinas locality. B. Undulatory bedding planes with very thin interbedded shaley layer. Sierra Sabinas locality. C. Polished slab: black, basinal lime mudstone-wackestone, with cross section of ammonite and possible brachiopod. Santa Rosa Canyon. D. Thin section: basinal, lime mudstone with unidentified foraminifers alongside the hexagonal cross section of a small, authigenic quartz crystal. Santa Rosa Canyon locality. E. Thin section: oolitic lime packstone, with well-formed concentric laminations around skeletal fragments and pelloid nuclei; much of the mud matrix has been recrystallized to a poikilotopic sparry calcite. The oolites are thought to have been transported from the shelf into the basin. Astillero Canyon locality.

Minas Viejas, Potrero Chico, and at Bustamante Canyon. Charleston (1974) reports 600 m of these pelagic lime mudstones at Potrero Minas Viejas. They are commonly massive to very thick bedded and have black chert nodules, well-developed stylolites, planktonic foraminifers, radiolarians, ostracods, and calcispherules (Charleston, 1974). This lithofacies is directly overlain by what is thought to be coarse debris derived from the platform and, therefore, might correspond to Wilson's "Basin Margin," standard facies belt.

Unit B

The limestones of unit B are thickest and best exposed at Potrero Chico (approximately 170 m thick), but some are also found at Bustamante Canyon, Potrero Minas Viejas, and possibly at Huasteca Canyon (approximately 10 to 20 m thick). The unit conformably overlies the basinal lime mudstones.

Unit B has a lithofacies consisting of thick- to very-thick-bedded stylolitic, bioclastic-lithoclastic, lime wackestones (pl. 2). The bioclasts include: stromatoporoids, solitary and colonial corals, well-preserved sponges and spicules, encrusting red algae, echinoid fragments, brachiopods (*Kingena* cf.), caprinids, requieniids, bryozoa, ostracods, and several types of benthonic foraminifers. The lime mud matrix of the wackestone contains pelagic foraminifers, tintinnids (?), and calcareous silt, probably from the shelf. There are some bedding planes (5 to 30 cm thick) near the middle of the unit of black pelleted lime mudstone interbedded with burrowed, nodular, very thinly bedded marly lime mudstones and some thicker mudstone beds (greater than 1 m thick). Stratigraphically below the massive cliff-forming unit at Potrero Chico are approximately 12 m of very coarse-grained, angular, lithoclastic-bioclastic lime packstones and probable grainstones (pl. 2).

This suite of lithofacies is interpreted to be "reef-derived submarine talus deposits" on the neritic, gently inclined slope environment of the platform margin. The massiveness of some of the lower parts of the unit and the way some clasts are "floating" in the mud matrix suggest slumping or periodic (storm induced?) mudflows mixed with periods of normal, open-marine sedimentation as represented by the burrowed, pelagic lime mudstones.

Unit C

Unit C is a very massive unit that lies directly over the submarine talus deposits. It is quite noticeable and very well exposed at Pajaros Azules Canyon, Bustamante Canyon, Potrero Chico, Potrero Garcia, Potrero Minas Viejas, and Huasteca Canyon. The maximum thickness found for unit C was approximately 230 m at the Potrero Chico section, and the minimum thickness was 47 m (Charleston, 1974) at the Potrero Minas Viejas section. This unit is commonly found in the lower half or near the base of the Cupido Formation and has a characteristic massive, unbedded nature that is resistant to erosion and forms vertical escarpments, often more than 300 m high (pl. 3). This unit is commonly highly fractured and partially or completely dolomitized.

The unit has four important lithofacies. They are: (1) coral-stromatoporoid lime boundstones, (2) rudist lime boundstones, (3) bioclastic lime packstones, and (4) dolomitized equivalents of these boundstones and packstones.

At Bustamante Canyon and Pajaros Azules Canyon, unit C has been extensively dolomitized, but at these outcrops the "ghosts" of very large, in situ, dendroid corals of the coral-stromatoporoid boundstone lithofacies are still recognizable (pl. 3). This lithofacies is 60 m thick at Bustamante Canyon, and 38 m thick at Pajaros Azules Canyon. Along with the dendroid corals are several types of smaller colonial corals. Associated with the corals are well-preserved stromatoporoids, coralline red algae, bryozoa, occasional caprinids, and the hexactinellid spicule pattern of calcareous sponges.

According to Heckel's (1974) classification of reefs, this type of boundstone lithofacies would indicate an "organic framework reef." His primary criteria for this type of reef include: (1) evidence for wave resistance, and control over the surrounding environment, and (2) evidence for contemporaneous reef talus that consists of organically encrusted sediment and large pieces of colonial organisms. Heckel's criteria would appear to be well met by the previously described boundstones and the lithoclasts of the forereef slope talus facies.

Overlying the coral-stromatoporoid lime boundstone lithofacies is the rudist lime boundstone lithofacies. This lithofacies is only 17 m thick at Bustamante Canyon and only 9 m thick at Pajaros Azules Canyon; yet it apparently predominates at the reef units of the other shelf-margin localities (for example, Potrero Chico). The lithofacies consists of caprinid- and requieniid-type rudists with abundant red algae, encrusting foraminifers, and a few corals, stromatoporoids, and thick-shelled oysters. Most of the rudists are recumbent and their primary effect on the environment was to baffle the wave energy and trap sediment. The caprinids stood erect during life and could form a mutually supportive framework (Petta, 1976), but the requieniids are thought to be unattached forms (Perkins, 1974) and probably did not construct wave-resistant structures by themselves without the aid of "binding" organisms. The requieniids are usually thought to have preferred the lower energy lagoon and other backreef environments; their abundance on the shelf margin is unusual. The rudists are particularly well preserved and exposed at the Potrero Minas Viejas (pl. 3) where the reef is notable because it contains a great quantity of lime mud and because it is not highly fractured and dolomitized as in most of the other localities.

Many of the requieniid shells retain parts of their original texture; some even have a slight greenish color and a characteristic layer of black, carbonaceous (?) film in their outer shell walls. Petta (1975) inferred that the original mineralogy of the outer prismatic layer of the requieniids was calcite and, therefore, stable in the marine environment. The inner, lamellar layer of the requieniid is thought to be originally aragonite and was relatively unstable so that it recrystallized to calcite. The caprinids were probably originally composed of aragonite; they also have been inverted or dissolved to molds, then refilled with calcite (Petta, 1976).

In the Potrera Minas Viejas reef several well-preserved, articulated, requieniids apparently attached to each other somewhat like modern oysters (pl. 4). They, and neighboring rudists, have similar alignments of their internal sediments (geopetal structures) in both their central shell cavities and in the borings in their shell walls. This would indicate that the requieniids along with the "toppled" caprinids were bored and infilled after they had been transported only a short distance, and/or at least some of the rudists are "in place." In either case, the preserved, intact nature of the rudists indicates that something stabilized the reef against the relatively high-energy waves on the shelf margin.

The reef at Potrero Minas Viejas has many caprinids and requieniids that are extensively "bound" by a mucilaginous green alga which re-

PLATE 2

A. The forereef slope talus facies (Unit B) is found stratigraphically below (to the right of, in photo A) the massive cliff-forming limestones of Unit C. Potrero Minas Viejas locality. B. Bioclastic-lithoclastic, lime wackestone found near the base of Unit B at Potrero Chico. This lithology is thought to represent sediment transport from the shelf margin reefs onto the forereef slope and into the basin. C. Large lithoclasts near the top of Unit B at Potrero Chico. Coin is 30 mm in diameter. D. Closeup photograph of reef debris, showing abundant caprinid-type rudists. Potrero Chico locality. E. Polished slab: light-colored, angular lithoclasts with stromatoporoids, calcareous sponges (?), benthonic foraminifers and other bioclasts floating in a black mud matrix. Potrero Chico locality. F. Thin section: photomicrograph of the mud matrix near the base of Unit B, showing cross section of a planktonic (?) foraminifer (Planomalinidae c.f.). Potrero Chico locality.

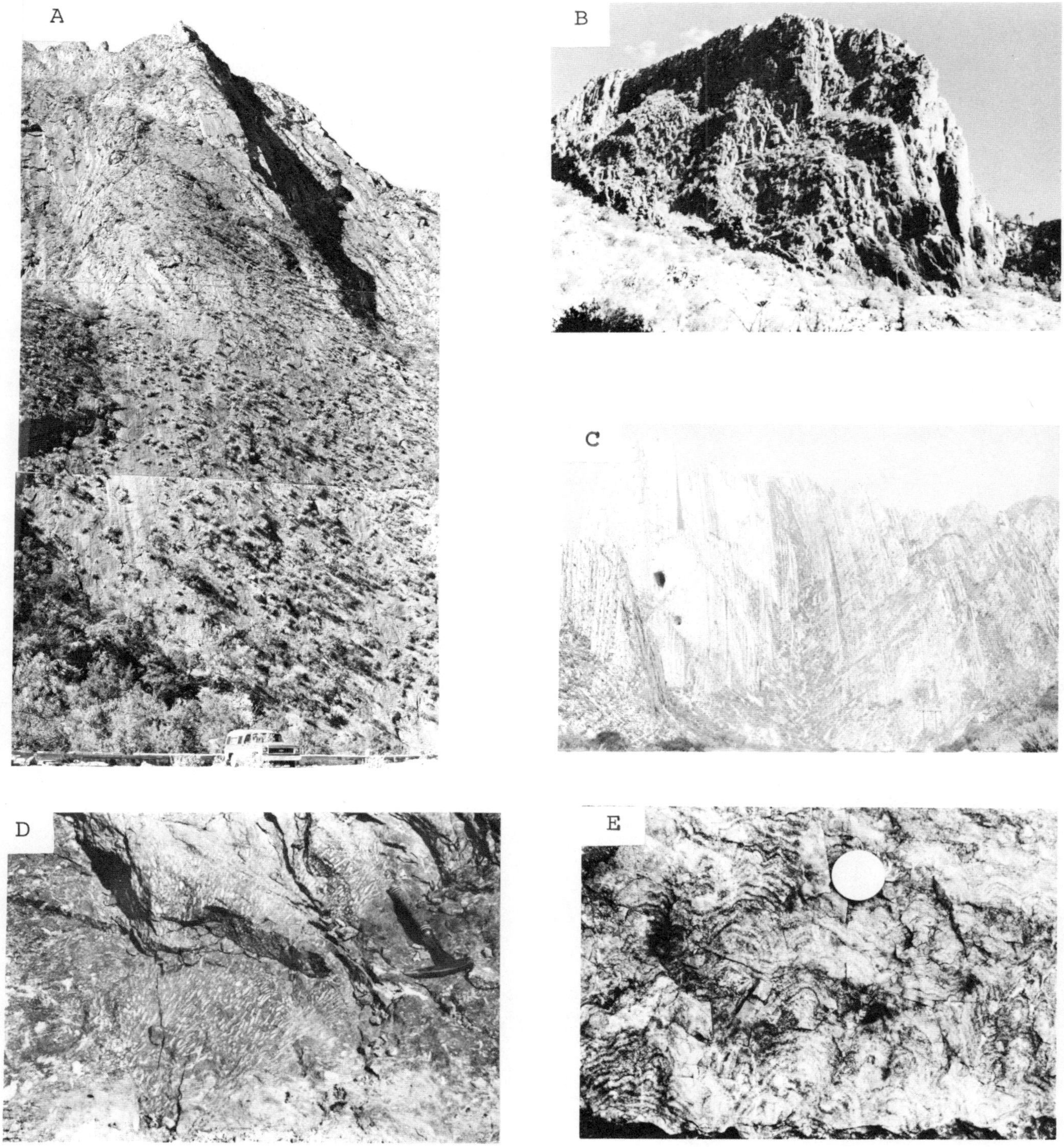

PLATE 3

A. Massive, vertically dipping, cliff-forming limestones of Unit C at Potrero Chico. The unbedded nature of this unit is a characteristic feature of the organic reef facies. However, a smaller, but very similar cliff in the overlying units D and E consists of very large, brecciated blocks cemented together after falling into the canyon or possibly as cavern infill during the late Cenozoic. Very large caverns are found at the Potrero Garcia locality. B. Dolomitized, vertically dipping, organic reef near the base of the Cupido Formation at Pajaros Azules Canyon. In the Monclova region of Coahuila, Mexico. This mappable unit is called the "Padilla Dolomite" by Petroleos Mexicanos geologists. C. The Cupido Formation at Huasteca Canyon near Monterrey. The massive, unbedded, reefal unit appears in the background directly below the label ***C***. The bedded units to the right of the large cavern are stratigraphically upsection and are thought to represent backreef sediments. D. Very large, dendroid, colonial coral thought to be in growth position in the now dolomitized, organic reef of Unit C at Bustamante Canyon. E. Convolute laminations of stromatoporoids in the organic reef of Unit C at Pajaros Azules Canyon. Coin is 30 mm in diameter.

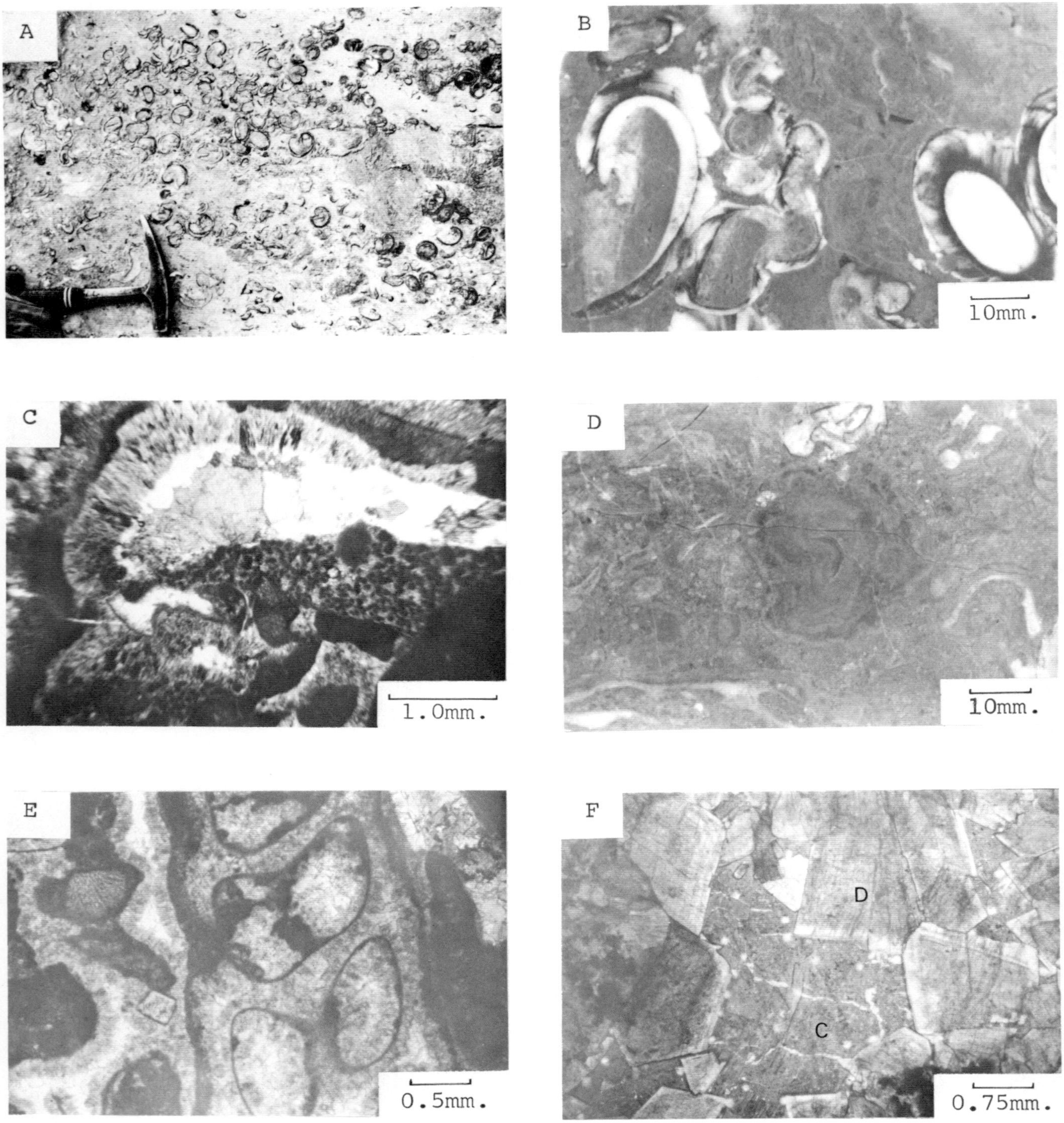

PLATE 4

A. Abundant rudists (mostly requieniids with some caprinids) in the organic reef facies of Unit C at Potrero Minas Viejas. B. Polished slab from the area shown in photo A: requieniids with traces of mucilaginous algal binding. Up is to the left. C. Thin section, polarized light: Isopachous, fibrous cement inside reef framework void, with pelleted internal sediment. Scanning electron microscope analysis showed high strontium values which indicate that this was originally an early marine cement. Unit C, Potrero Chico locality. D. Large oncolite thought to have formed on a calcarenite plain slightly behind the main reef buttress zone. Unit C, Potrero Chico. E. Thin section, plane light: Isopachous, fibrous cement coating the grains of the calcarenite plain. The large grain in the center of the photograph is a high spired gastropod; the shell has been micritized and leached, leaving a "micrite envelope," which was then cemented with the fibrous marine cement. A final stage of pore-fill cement is represented by the blocky calcspar cement within the isopachous, fibrous cement. F. Stained thin section, plane light: Coarse-grained, euhedral dolomite crystals (D) with undulatory extinction patterns surrounding the darker (stained) calcite pore-fill cement (C). Small white dots were bubbles that formed during the staining process. This sample is from the "Padilla Dolomite" of the reef at Bustamante Canyon.

sembles ***Girvanella*** (?). The alga is difficult to see (pl. 4), but appears to be similar to the type which forms oncolites and algally coated grains at the Potrero Chico section (pl. 4). There must also be an associated endolithic alga because many of the rudists have been completely "micritized" with only the general outline of their shells and a central spar-filled void remaining.

Other "binding" organisms found among the rudists (and also in the coral-stromatoporoid facies) include: a relatively rare encrusting foraminifer similar to ***Coscinophragma*** described by Petta (1976) and a common red alga that resembles ***Parachaetetes lobatum***, first described from the Pipe Creek reef of Central Texas by Johnson (1968). Petta (1976) thought that this type of red algae "did not have an encrusting habit, but attached itself to the sides of framework voids within the reef." He further stated that, "this algae may have indirectly strengthened the framework as it grew in irregular to lobate branching masses that aided in trapping sediment within framework voids."

The stabilization of the shelf-margin reef by the organic "binding" of the encrusting foraminifers and algae was probably aided by early, inorganic submarine cementation in the intergranular pore spaces and by early lithification of the internal sediments, as illustrated by Petta (1976) for the Pipe Creek reef and by Goreau and Land (1974) for the Jamaican reefs. The apparent anomalous occurrence of carbonate mud and silt found within this shelf margin "high energy zone" is probably due to a combination of: (1) the baffling action of the bottom-dwelling rudists, (2) the trapping of the mud and fine internal sediments by the encrusting organisms, and (3) early "marine" cementation.

The third major lithofacies produced within the reef environment is the bioclastic lime packstone-grainstone. It is best exposed at the Potrero Chico locality and at the Bustamante Canyon locality. This lithofacies consists of dark-gray to cream-colored, poorly sorted, fine- to very-coarse-grained reef detritus and backreef material that was trapped as it was being transported across the reef. The large grains (approximately 3 cm in diameter) are commonly algally coated rudist fragments, but many of these grains are unidentifiable because they are highly recrystallized, and some are apparently "micritized" by endolithic algae (pl. 4). A few oncolites mixed in with the larger algae-coated grains, indicating that the grains actually rolled about on the reef. The smaller grains (approximately 1 mm) are pelloids, which appear to be some type of micritized foraminifer rather than faecal material. The other bioclasts include: miliolids, ***Textularia***, dasycladacean algae, fragments of red algae, fragments of small branching corals (like ***Cladophyllia***), ostracods, echinoid spines, and gastropods.

Within the grainstone portion of the lithofacies, the intergranular voids commonly have an isopachous fringe of fibrous calcite cement (pl. 4). Preliminary results from microprobe analysis indicate that the fibrous cements have relatively high strontium values (7,500 ppm) and probably represent early "marine" cements. Petta (1975) found similar cements from the Pipe Creek reef of Central Texas. Unfortunately, at Potrero Chico most of the early fibrous cement has been destroyed by a replacement of a blocky, cloudy white, calcite cement.

The bioclastic lime packstone-grainstone lithofacies is thought to have been deposited on a calcareous "sand plain" that probably existed slightly behind the major framework reef. This sand plain would be similar to that illustrated by Roberts (1974) for the modern, low-energy shelf margin of Grand Cayman, or the Holocene "sand cover" shown by Hine and Neuman (1977) for Little Bahama Bank. Some of the bioclastic lime packstones and grainstones of the sand plain become internal sediment in the reef-framework cavities, and the sand plain itself is eventually incorporated into the massive reefal unit during progradation (fig. 13).

The dolomites are a particularly important reef lithofacies. The dolomites are best exposed at Bustamante Canyon and at Pajaras Azules Canyon starting near the base of the reef and extending up into the overlying unit. The dolomites consist of black to gray to white, medium- to very-coarse-grained (up to 2 mm), euhedral to anhedral crystals (dolospar) that replace matrix, allochems, and previous cements. Sometimes the original texture of the limestone is completely destroyed and only "ghosts" of fossils remain. Under crossed nicols, the dolospar commonly displays an undulatory extinction pattern which is indicative of strain caused by the inclusions of impurities in the crystal lattice (pl. 4). Folk and Asserto (1974) applied the term "baroque dolomite" to similar dolomites found in the Triassic of Lombardy, Italy. However, the Italian baroque dolomites were white and filled only pore space (an early post-depositional cement), while the dolomites of the Cupido Formation are characteristically gray and show definite replacement fabrics which indicate that the dolomite formed after the primary fibrous marine cements and after some grains had been neomorphosed. Some dolomite was also observed filling fractures, possibly as a replacement of an earlier blocky calcite vein fill.

Another (much scarcer) type of dolomite was observed to be exceptionally clear and to have unit extinction. This type of dolomite was called "limpid dolomite" by Folk and Land (1975) and is thought to have formed from relatively fresh (phreatic-meteoric) waters.

At the Potrero Chico reef section, there are angular, coarsely crystalline, dolomite blocks, isolated within the surrounding nondolomitized reef sediments. Nearby brecciated blocks appear to have cosertal boundaries, and the dolomitized blocks may have been dolomitized "in place" because their shapes seem to be controlled by fracture patterns. If this is actually the case, then it is consistent with an observation that the forereef slope talus facies lacks detrital, dolomitic lithoclasts. This lack of lithoclasts suggests that the dolomitization was due to late, post-depositional processes, possibly related to deep burial and/or diapirism.

The large amount of coral, stromatoporoid, and caprinid debris found in unit B at Potrero Chico indicates that a framework of some type may have existed, although it is now unrecognizable on the outcrop. A look at modern reefs provides an explanation. According to Moore (1975), "Borers such as Clionids affect the destruction of this frame and the generation of new secondary pore space. This pore space along with primary constructional pore space is filled with sediments termed ***internal sediments*** that can have their origin in a variety of ways including pelagic components from the water column, debris from the boring activity of sponges or from sediment traction load being transported across the reef. These internal sediments can then be cemented ***in situ*** shortly after deposition becoming an integral part of the reef rock. Because the reef rock is continuously in contact with marine conditions the processes of internal sedimentation, boring, and cementation can continue for long periods of time leading ultimately to a rock that has little or no original frame left . . .

The recognition of these processes in ancient sequences then, is one of the basic keys to the recognition of ancient ecologic reefing conditions."

Several of the samples (pl. 4) from the Potrero Chico reefal section show some evidence of being "destroyed reef rock." Petta (1976), Bein (1976), and Aguayo (1976) found similar diagenetic textures and described submarine cementation, organic boring, and internal sedimentation as being contemporaneous with rudist framework accretion in Cretaceous fringing reef complexes of Central Texas, Israel, and the El Abra of Mexico, respectively.

Unit D

Unit D is a very-thick-bedded unit which directly overlies the more resistant massive "organic framework reef" facies in vertical stratigraphic sections. Unit D was found at Potrero Chico, Huasteca Canyon, Potrero Minas Viejas, Bustamante Canyon, and Pajaro Azules Canyon. This unit is approximately 30 to 45 m thick. The contact with overlying unit is usually extremely gradational. The best exposure of unit D is probably at Bustamante Canyon. Even though the unit is extensively dolomitized, small scale crossbedding is still visible there (pl. 5).

The lithofacies believed to be characteristic of the unit are:

(1) dark-brown, pelloidal, oncolitic, lime packstones with abundant mucilaginous algae coatings on coarse-grained coral, rudist, and gastropod fragments mixed with miliolid and other benthonic foraminifers, dasycladacean algae, and micritized grains;

(2) tan- to gray-colored, benthonic foraminifers, dasycladacean algae, lime grainstones with ostracods, micrite envelopes, pelloids, and some oolites;

(3) dark-colored, dolomitized, bioclastic lime packstones and grainstones with "ghosts" of rudist fragments in mosaic of coarse-grained subhedral dolomite.

The inferred environment for these lithofacies consists of relatively shallow, high-energy, near-reef shoaling conditions slightly behind the submerged reef trend. Wilson (1975) would consider these to be the "winnowed platform edge sands."

Unit E

Unit E is thin to thick bedded and is much less resistant than its neighboring units. It is usually found as an approximately 200-m-thick erosional reentrant near the middle of the Cupido Formation. This unit is well exposed at Pajaros Azules Canyon, Bustamante Canyon, Potrero Chico, and Huasteca Canyon. Unit E is possibly exposed as a thin unit at Potrero Minas Viejas. It is transitional into the underlying "near-reef shoal facies."

Unit E is notable for its highly variable lithofacies. The following is a summary of the four major types of limestones thought to be characteristic of this unit:

(1) Black miliolid, pelloid, lime wackestones, packstones, and grainstones with abundant and often dolomitized burrows, dasycladacean and codiacean algae, some pelecypods, biserial foraminifers *(Textularia?)* ostracods, and small gastropods. The grainstones of this group may contain dolomitic "rip-up" clasts.

(2) Black pelecypod lime wackestones and packstones with bored, often fragmented, thin-shelled pelecypods, and some small (.05 mm) dolomite crystals, miliolids, and pelloids.

(3) Light brown dolomicrite with "popcorn" evaporite nodules (now replaced by calcite) and "ghosts" of gypsum crystals.

(4) Gray smooth to slightly convolute, algae laminations. Some well-preserved algae laminations have no dolomite. Others are dolomitized and have calcite-replaced evaporite nodules, rare desiccation features, and associated solution breccias.

Owing to their geographic location in relation to the Coahuila landmass (more than 100 km away?) and to their stratigraphic position in relation to the underlying reef and near-reef shoals, these four types of limestones are thought to have been deposited in shallow subtidal to high intertidal waters, slightly behind the main reef trend. The occurrence of the dolomicrites, evaporite nodules, and desiccation features of the algae laminations, along with the rare occurrence of layered evaporites and solution breccias, indicates that some of these limestones were probably formed in the supratidal zone. The occurrence of dolomitized rip-up clasts suggests reworking of these supratidal sediments by storms or localized shoreline retreat.

The four major types of limestones found in this unit along with their many variations are repeated many times throughout the unit. This repetition suggests generally "shoaling upward cycles," but no definite cyclic pattern could be recognized. These beds are considered to be a "near-reef lagoon and tidal-flat" facies that formed on top of the near-reef shoals.

Unit F

Unit F has thick to very thick beds that are more resistant than the underlying "near-reef, lagoon and tidal-flat" facies. At Pajaros Azules Canyon, Bustamante Canyon, Potrero Chico, Potrero Minas Viejas, Potrero Garcia, and Huasteca Canyon, this unit is found at the very top of the Cupido Formation and averages approximately 100 m thick. Further to the west, the entire 680-m-thick Arteaga Canyon section, along with the entire 330-m-thick Potrero Oballos section of the Cupido Formation, is considered to belong to unit F.

The lithofacies found at the localities along the Huasteca Canyon to Pajaros Azules Canyon trend are characteristically very thickly bedded, black, pelecypod, miliolid lime wackestones and packstones with requieniids, dasycladacean and codeacean algae, pelloids, ***Choffatella, Orbitulina, Textularia(?)***, and some black chert nodules.

The lithofacies is interpreted to be the result of normal-marine sedimentation in a subtidal, backreef lagoonal environment. The lithofacies of unit F at the Arteaga Canyon and Potrero Oballos sections are more variable than the ones just described because these sections are much closer to the remnant of the Coahuila Peninsula. The characteristic lithofacies for unit F at the base of these localities include: dolomitic lime mudstones; burrowed, pellet, lime wackestones; and bioclastic lime packstones and grainstones with intraclasts, pelecypod fragments, miliolids, ***Choffatella***, gastropods, dasycladacean algae, ostracods, and some superficial oolites. Higher in the section, these beds alternate with dolomitic algal laminations, dolomicrites containing evaporite nodules, and zones of solution breccia. The tops of these sections also have rudist (requieniids and some monopleurids) biostromes. The top of the Cupido Formation at the Potrero Oballos section has a caprinid organic reef, 5 to 18 m thick with some associated corals (pl. 7). It is thought to be a large patch reef on the shelf interior, well behind the platform-margin reefs.

Unit F at the Arteaga and Potrero Oballos sections is also thought to represent a lagoonal environment. The tidal-flat cycles and solution breccias in the middle part of the formation suggest that supratidal conditions

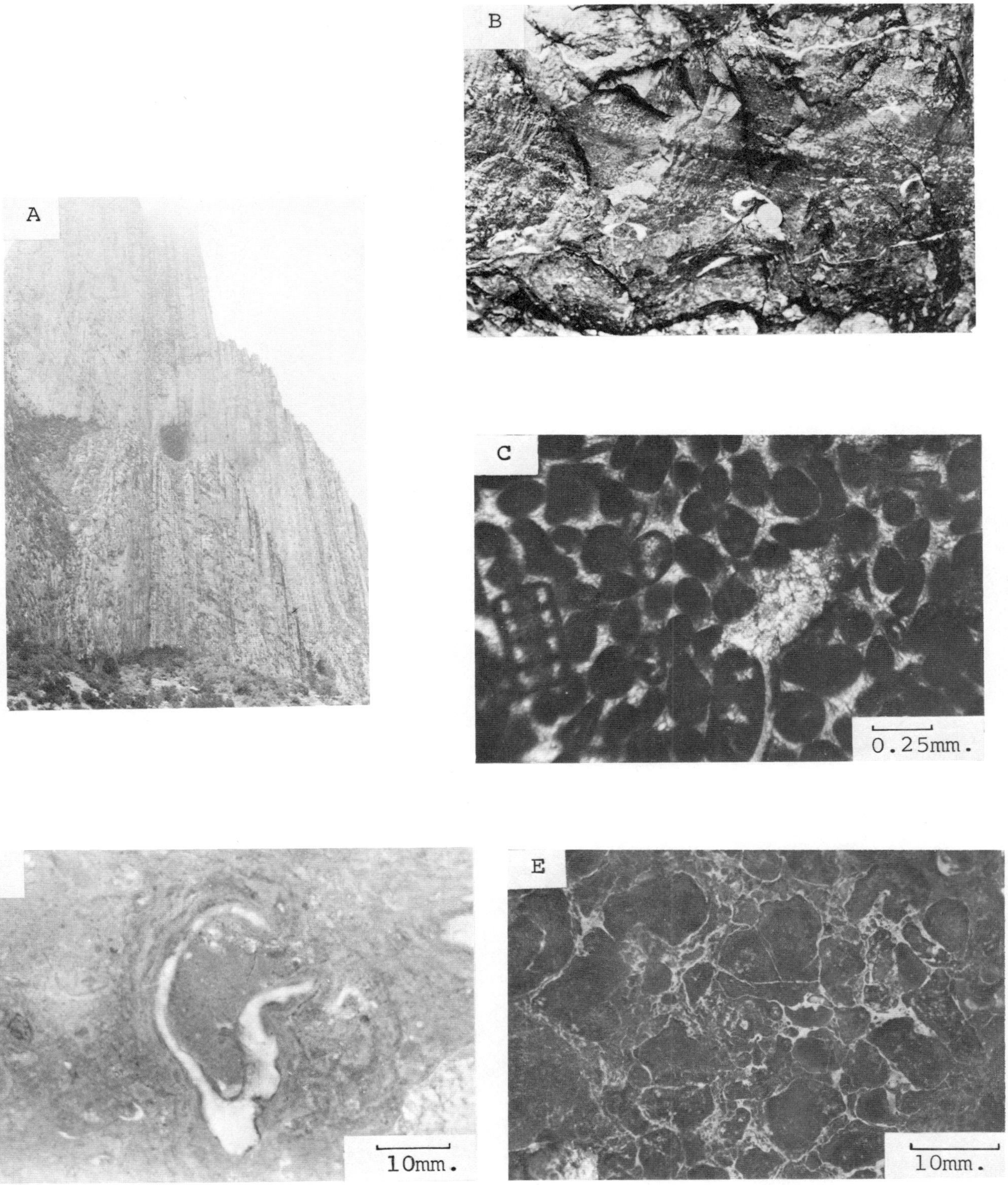

PLATE 5

A. Vertically dipping limestone beds at Huasteca Canyon. The massive organic reef Unit C is located directly below the label A. The cavern and the beds to the right belong to the overlying "near-reef shoal" facies of Unit D. B. Dolomitized crossbedded calcarenites directly above the organic reef at Bustamante Canyon. Coin is 30 mm in diameter. C. Thin section, plane light: foraminifer, pelloid lime grainstone with some dasycladacean algae and micrite envelopes. The grainstone has equant crusts overlain by blocky calcspar cements, along with possible keystone vugs. Unit D at Huasteca Canyon. D. Polished slab: algal coated bioclastic lime packstone with blue-green (?) algae encrusting a caprinid shell. The original aragonite shell structure of the caprinid has recrystallized to calcite and appears to have undergone some micritization. Unit D, Potrero Chico. E. Polished slab: oncolitic lime packstone from Unit D, Pajaros Azules Canyon.

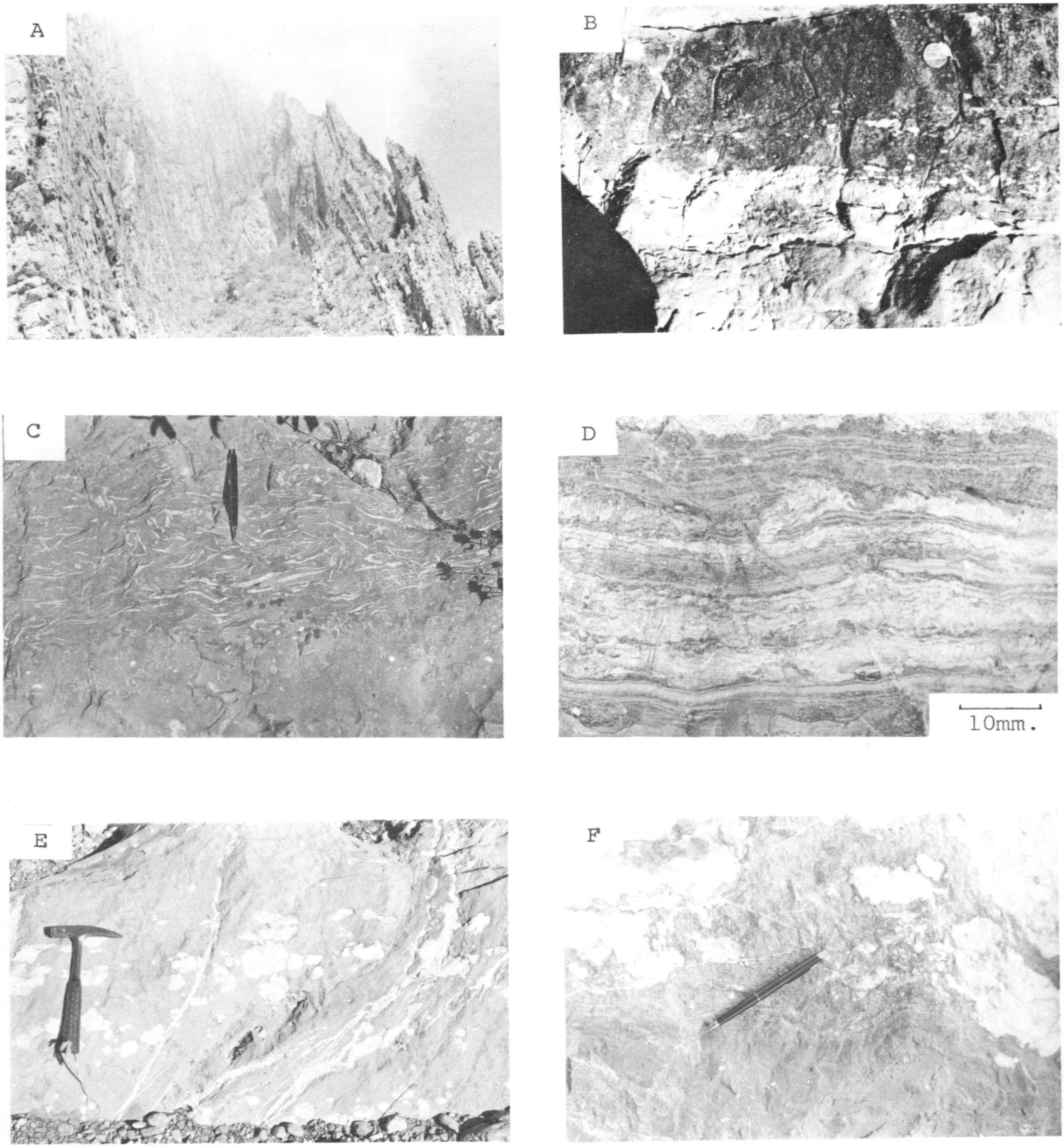

PLATE 6

A. Indentation of the outcrop in Unit E at Potrero Chico because of the relatively less resistant tidal-flat facies. Vertically dipping beds; stratigraphic up is to the right. B. Flat pebble conglomerate composed of dolomitic rip-up clasts indicating a localized transgression and start of a new shoaling upward cycle. This could possibly represent a storm deposit or shifting of a broad tidal channel and strand line over a supratidal or high intertidal algal mat area. Unit E, Potrero Chico. Coin is 25 mm in diameter. C. Pelecypod (lithiotids?) lime wackestone-packstone, with miliolid foraminifers, pellets and a few requieniids; probably developed in a subtidal environment. Unit E, Potrero Chico. D. Slightly convoluted algal laminations; these laminations are not dessicated or dolomitized, and are thought to have formed in the lower intertidal zone. Unit E, Potrero Chico. E. "Popcorn" evaporite nodules (now replaced by calcite) in a light brown, finely crystalline dolomitic mudstone. These evaporite nodules are thought to have formed in the subsurface below the supratidal and high intertidal zones. Unit E, Potrero Chico. F. Closer view of evaporite nodules with trace of algal laminations still visible. Unit E, Potrero Chico.

PLATE 7

A. Vertically dipping, very thick to thin beds of the lagoon-tidal flat-restricted lagoon facies. Unit F, Arteaga Canyon, southeast of Saltillo. B. Caprinid patch reef near the top of the Cupido Formation. Stratigraphic up is to the right. The overlying La Peña Formation is in the gulley and the resistant limestones on the far right of the photograph belong to the Aurora Formation. Unit F, eastern side of Potrero Oballos. The reef is better developed on the western side of the potrero. C. Orbitolina lime wackestone of the lagoonal facies found stratigraphically below the reef at Potrero Oballos. Coin is 30 mm in diameter. D. Polished slab: pellet, pelecypod lime wackestone of the subtidal, lagoonal facies. Unit F, Arteaga Canyon. E. Solution breccia due to dissolution of tidal-flat evaporites and collapse of surrounding beds. Unit F, Potrero Oballos. This solution breccia might represent a tongue of the La Virgen Evaporites that form a mappable unit to the west of the Potrero Oballos locality. F. Oolitic, miliolid lime packstone with aggregate grains, dasycladacean algae, small gastropods, Dicyclina, and pelecypod fragments. This packstone is thought to have formed in tidal channels and deltas near the sabkha. Unit F, Arteaga Canyon.

occurred at these localities and that a sabkha possibly could have extended back to the landmass. Ekdale and others (1976), through a numerical analysis of the upper 400 ft of the Arteaga Canyon section, found that general upward-shoaling cycles were present, but that "no perfect cyclothems existed." They found that the possibilities for variations "are numerous and are controlled by shifting of tidal channels, bays, bars, and mud flats during continuous deposition."

DEVELOPMENT OF A PLATFORM MODEL

The Cupido platform probably does not have a good modern analog because the primary biologic component of the platform, the rudists, became extinct at the end of the Cretaceous period. However, the hermatypic hexacorals, foraminifers, and algae found at several localities are comparable to Recent forms found on the Florida Keys, Bahama Banks, and other modern reefs. Black (1930), Field (1930), Grimsdale (1950?), and Newell (1953) recognized the similarity between Recent Bahama Bank sediments and lithofacies in ancient limestones (Wilson, 1975).

The concept that most carbonate platforms contain similar patterns of environmental belts and that each belt contains characteristic lithofacies has proven to be a useful exploration tool. An example of this concept is the model constructed by Irwin (1965). This model consists of three primary energy zones for the Mississippian carbonates of the Williston basin. Wilson (1975), on the basis of 40 examples throughout the geologic record, has expanded the earlier models into a single model with nine "standard facies belts." He emphasized that, although it is an idealized pattern and that no one carbonate platform should necessarily include all nine facies belts, the overall patterns of the belts are remarkably consistent even in a variety of tectonic settings. Six such facies belts are recognizable on the Cupido platform. These six facies belts were based on environmental interpretations given to the characteristic lithofacies in each of the six primary units (fig. 9). They are (1) the basinal facies (unit A), (2) the forereef slope talus facies (unit B), (3) the organic framework reef facies (unit C), (4) the near-reef shoaling facies (unit D), (5) the near-reef tidal-flat facies (unit E), and (6) the lagoon-restricted lagoon-sabkha facies (unit F).

A measured section of the Cupido Formation at one of the platform-margin localities (localities 1 through 9) shows that deeper water environments are overlain by increasingly shallower water environments. Figure 10 is a cross section across the carbonate platform which shows, with three measured sections, that there is a

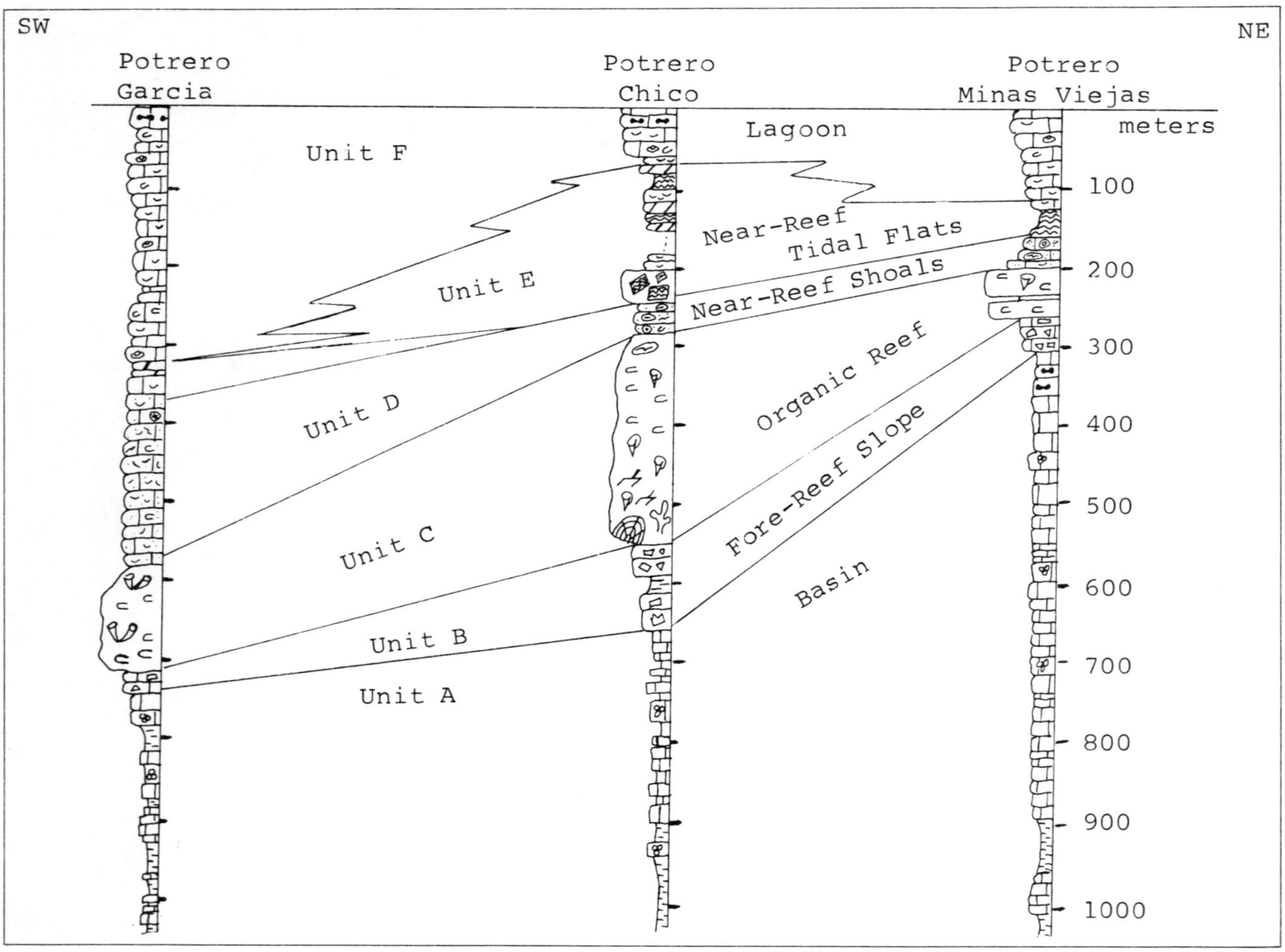

Figure 10. Cross section showing vertical and horizontal facies relationships.

definite pattern to this trend. The lateral continuity of most of the environmental facies and the order in which they occur are indications that the Cupido platform prograded seaward (towards the east) over the deeper water basinal sediments. This progradation strongly suggests that the rates of biologic productivity and resulting carbonate accumulation on the platform were greater than the combined effects of sea-level rise, regional subsidence, and off-platform sediment transport.

However, the rates of these platform-controlling processes must have fluctuated. Figure 11 is a hypothetical model for platform progradation based on the cross section shown in figure 10. The platform probably evolved from the underlying Taraises carbonate ramp. During the early stages of platform growth (time interval T_1) progradation exceeded subsidence and/or sea-level rise. However, during the interval T_2, the vectors of progradation and subsidence and/or sea-level rise are almost equal, showing a near balance between the processes. This balance would allow nearly vertical reef growth. The reef unit at Potrero Chico is anomalously thick (220 m) and might be the result of such a balance. In interval T_3, progradation far exceeded subsidence and/or sea-level rise. This interval is thought to correspond to a possible temporary standstill of the generally rising Cretaceous sea level, or perhaps to a slight regression. Smith (1970) noted evidence for such a regression in the middle Cupido Formation in the Big Bend area of Texas. The extensive "near-reef tidal-flat facies" (unit E) along the platform margin and the very thin reef at Potrero Minas Viejas (and the La Virgen Evaporites on the shelf interior) are thought to have formed under these conditions. Finally, in interval T_4 the processes might have again balanced. This hypothesis is based on the presence of lagoonal facies at the very top of the Cupido Formation at Potrero Minas Viejas, Bustamante Canyon, and other platform-margin locations.

Platform growth was ended when a very rapid transgression occurred in Gargansian time. The growth rates of the rudists and other reef organisms probably could not match this rate of relative sea-level rise when coupled with an argillaceous influx that also occurred in Gargansian time. This event can be traced regionally by the sharp contact with the overlying La Peña Formation. The La Peña Formation has deep-water assemblages of fossils and thin beds of black shales and limestones. Isopach maps of the La Peña Formation show that its fine terrigenous sediments "draped" over the Cupido platform margin and the thickest accumulation coincides with the platform foreslope and inner basin.

This "drape effect" also implies that the Cupido platform had considerable submarine topographic relief. Thickness variations from the platform into the inner basin suggest a relief of approximately 400 m (Charleston, 1974), but this value along with calculated slope angles is uncertain because of the possibility of preferential subsidence under the load of the thickened reef trend. The forereef slope talus facies (unit B) also implies submarine topographic relief. A possible localized submarine escarpment was constructed by the framework organic reef and could have been approximately 50 m high.

CONCLUSIONS

A synthesis of the information gathered from the 14 localities where the Cupido Formation was studied can be collected and organized into two remaining diagrams.

Figure 12 shows a model of the environments of deposition for the

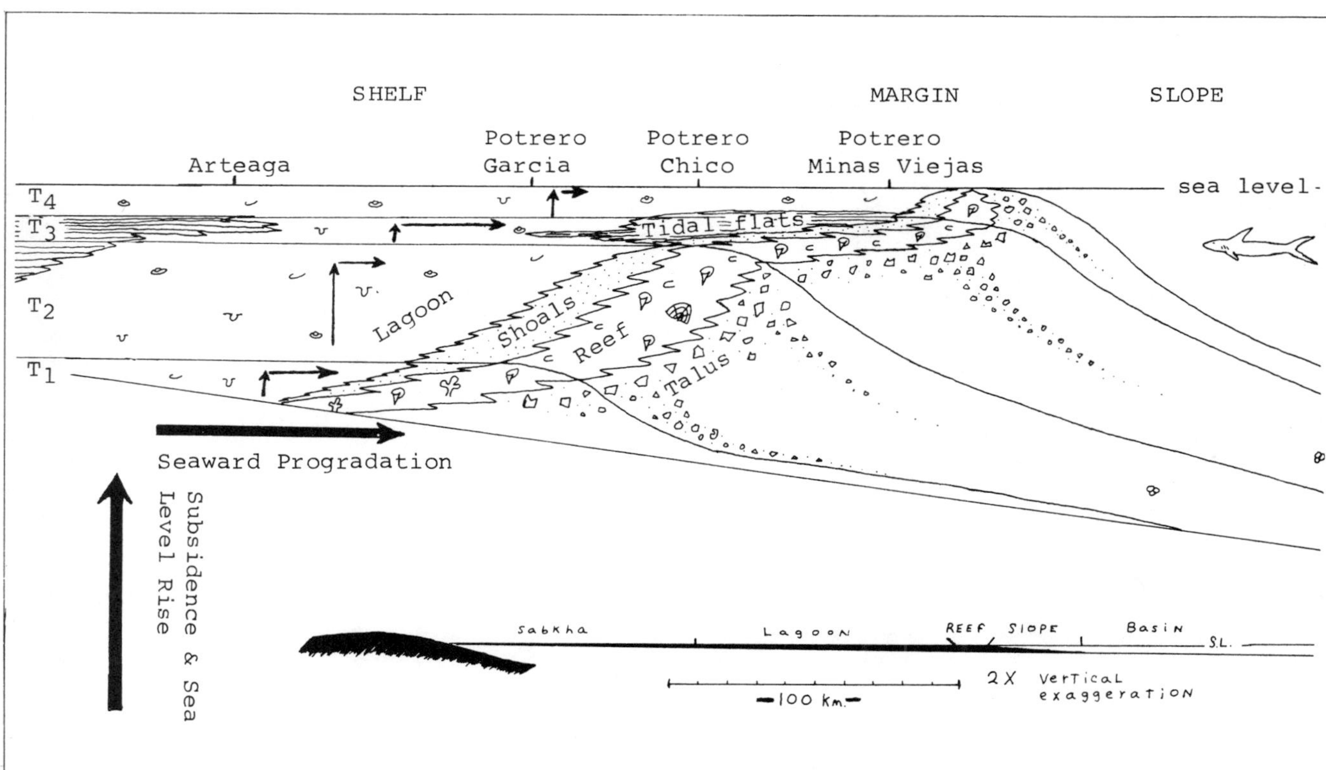

Figure 11. Hypothesized Model for Platform Progradation, Considering Relative Rates of Accumulation, Subsidence, Sea Level Change, and Time.

MODEL OF THE CUPIDO PLATFORM

EAST — WEST

Diagrammatic cross section (not to scale)

Facies Number	1.	2.	3.	4.	5.	6.
Facies & Environments	Basin & open marine (neritic) (below wave base)	Fore-reef slope talus (neritic) (below normal wave base?)	Organic reef & intrareef bioclastic accumulation (inner neritic, above wave base)	Near-reef shoals (inner neritic to littoral)	Near-reef tidal flats (littoral to supratidal)	Lagoon to restricted lagoon and sabkha (inner neritic to supra-tidal)
Lithology	Black to dark grey, pelagic, lime mudstones & wackestones: often with burrows, pellets, and black chert nodules. Some of the micrite has been replaced by 0.04mm. euhedral dolomite crystals.	Black to dark grey litho-clastic-bioclastic, lime wackestones. Exotic block "floating" in pelagic lime mudstones & wackestones. Possibly episodic mudflows alternating with normal pelagic sedimentation and deposition of shales.	Dark grey to light colored coral, stromatoporoid, and rudist lime boundstone, and intrareef algal coated, bioclastic lime packstones. Base of reef is often thoroughly dolomitized with coarse grained dolospar.	Brown to grey and tan colored, algal coated bioclastic packstones; oolitic- oncolitic pack-stones and grainstones; foram, pellet lime packstones.	Dark brown to grey colored cycles of interbedded: algal laminations; dolomitic lime mudstones with psuedomorphs of evaporites. Pelecypod, miliolid lime wackestones & packstones and zones of solution breccia.	Black to light grey colored, burrowed (sometimes cherty) pelloidal, mollusc miliolid, lime wackestone-mudstones, with some rudist bioherms and biostromes. Some localities have cyclic interbedded algal laminations & dolomicrites & intraclastic, psuedo oolitic lime grainstones & packstones.
Biologic Components	1 Pelagic Foraminifera 2 Tintinids 3 Ammonites 4 Ostracods 5 Echinoids 6 Radiolarians (?) 7 Calcispheres (?)	1 Pelagic Foraminifera 2 Tintinids 3 Ostracods 4 Bioclasts from reef 5 Brachiopods 6 Calcareous sponges (?)	1 Corals 2 Stromatoporoids 3 Caprinids 4 Requienids 5 Encrusting algae 6 Encrusting Foraminifera 7 Echinoids 8 Calcareous sponges 9 Bryzoan 10 Boring clams	1 Rudist fragments 2 Dasycladacean algae 3 Codiacean fragments 4 Orbitolina 5 Miliolids 6 Lithiotid fragments 7 Coral fragments 8 Pelloids	1 Algal laminations (blue-green algae?) 2 Lithiotids 3 Miliolids 4 Small gastropods 5 Pelloids 6 Dasycladacean algae	1 Requienids 2 Lithiotids (?) 3 Miliolids 4 Orbitolina 5 Ostracods 6 Caprinids (rare) 7 Dasycladacean algae 8 Codiaceaen algae 9 Choffatella 10 Small gastropods
Bedding & Sedimentary Structures	Very thick bedded, with undulatory bedding planes and loading into thin, black shale beds.	Massive to very thick beds, randomly alternating with thick beds and shale laminations.	Massive, no beds visible Reef thickness variable, up to 215 meters thick.	Medium to very thick bedded, with some low amplitude cross bedding.	Thick to thin bedded, with zones of solution breccia and thin zones of flat pebble conglomerate. Abundant slightly convoluted algal laminations	Thick to thin beds with some small ripples and cross beds. Stromatolitic units have some rare birdseye fabric & desiccation cracks. Some solution breccias.
Localities where examples are best observed	1 Sierra Sabinas 2 Astillero Canyon 3 Sierra Picachos 4 Santa Rosa Canyon	1 Potrero Chico 2 Huasteca Canyon 3 Bustamante Canyon 4 Pajaro Azules Canyon	1 Potrero Chico 2 Huasteca Canyon 3 Bustamante Canyon 4 Pajaro Azules Canyon	1 Potrero Chico 2 Bustamante Canyon 3 Huasteca Canyon 4 Pajaro Azules Canyon	1 Potrero Chico 2 Bustamante Canyon 3 Huasteca Canyon 4 Pajaros Azules Canyon	1 Arteaga Canyon 2 Potrero Oballos 3 Huasteca Canyon 4 Potrero Chico

Figure 12. Model of the Cupido Platform.

Cupido Formation, and figure 13 shows where these environments might have been located geographically during early Aptian time.

ACKNOWLEDGMENTS

This report is condensed from a Master of Science thesis to be completed by Jack Conklin at Louisiana State University in 1977. The thesis work was supported by a teaching assistantship from the Department of Geology at L.S.U. and by a grant from the American Association of Petroleum Geologists. The authors would like to express their gratitude to James L. Wilson and Colin Stabler who critically reviewed this manuscript and gave much helpful advice and information on the Cupido Formation. Special thanks are extended to James Cavanaugh, Hugo Conklin, and Bill Ward, who, on different occasions, accompanied the authors in the field.

REFERENCES

Aguayo, J. E. C., 1976, Sedimentary environments and diagenesis of the El Abra Limestone at its type locality, eastern Mexico (abstract): AAPG, Annual Meeting, New Orleans, p. 644.

Bien, A., 1976, Rudistid fringing reefs of Cretaceous shallow carbonate platform of Israel: AAPG Bull., v. 60, p. 258-272.

Bishop, B. A., 1970, Stratigraphy of the Sierra de Picachos and vicinity Nuevo León, Mexico: AAPG Bull., v. 54, p. 1245-1269.

_______ 1972, Petrography and origin of Cretaceous limestones, Sierra de Picachos vicinity, Nuevo León, Mexico: Journal of Sedimentary Petrology, v. 42, p. 270-286.

Böse, E., 1923, Vestiges of an ancient continent in northeast Mexico: American Journal of Science, v. 206, p. 127-136.

Charleston, S., 1974, Stratigraphy, tectonics, and hydrocarbon potential of the Lower Cretaceous, Coahuila series, Coahuila, Mexico: Univ. Michigan, Ph.D. dissertation, 268 p.

Coogan, A. H., Bebout, D. G., Maggio, C., 1972, Depositional environments and geologic history of Golden Lane and Poyo Rico Trend, Mexico, an alternative view: AAPG Bull., v. 56, p. 1419-1447.

Dunham, R. J., 1962, Classification of carbonate rocks according to depositional texture: AAPG Memoir 1, p. 108-121.

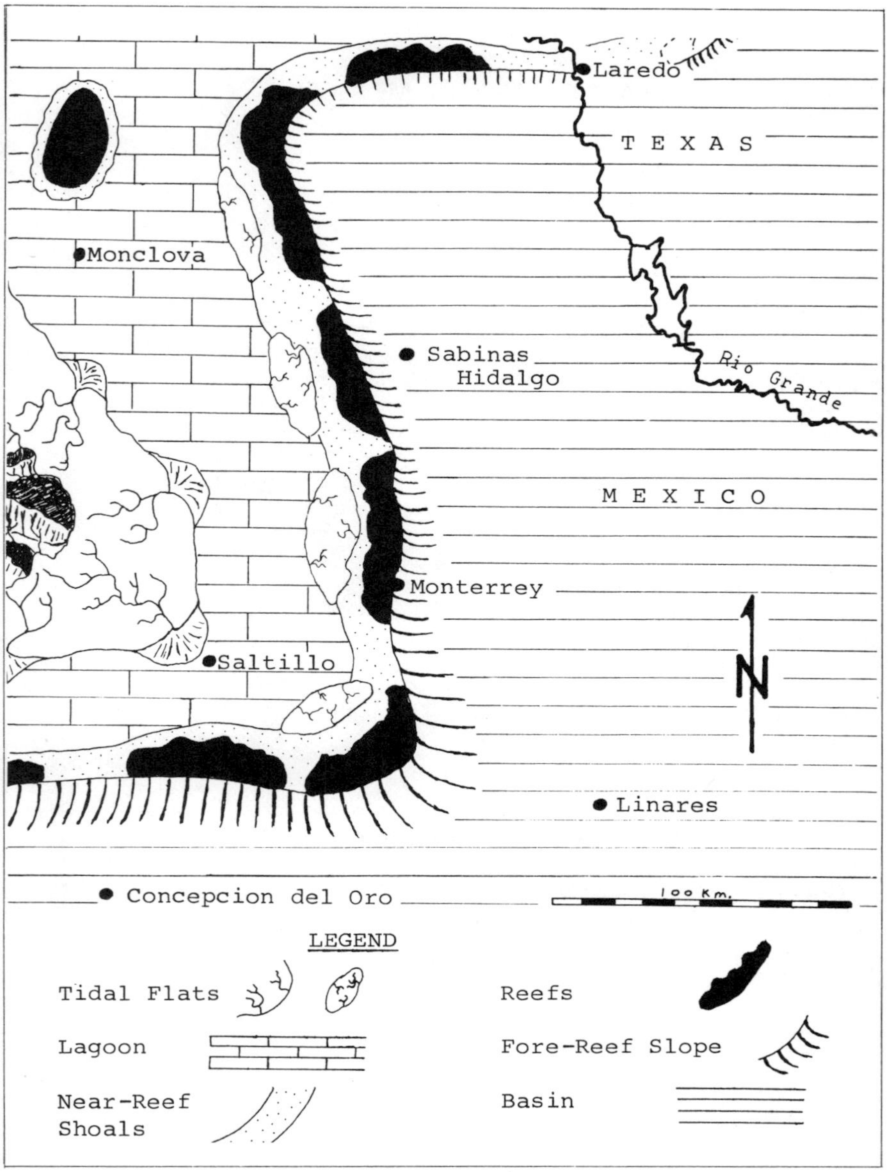

Figure 13. Idealized paleogeography and environments of the Cupido Platform during early Aptian time.

Ekdale, A. A., Ekdale, S. F. and Wilson, J. L., 1976, Numerical analysis of carbonate microfacies in the Cupido limestone (Neocomian-Aptian) Coahuila, Mexico: Journal of Sedimentary Petrology, v. 46, p. 362-368.

Enos, P., 1974, Reefs, platforms, and basins of Middle Cretaceous in northeast Mexico: AAPG Bull., v. 58, p. 800-809.

Folk, R. L., and Asserto, R., 1974, Giant aragonite rays and baroque white dolomite in tepee-fillings, Triassic of Lombardy, Italy: AAPG and SEPM Annual Meeting abstracts, v. 1, p. 34-35.

_______, and Land, L. S., 1975, Mg/Ca ratio and salinity: two controls over crystallization of dolomites; AAPG Bull., v. 59, p. 60-68.

Goreau, T. F., and Land, L. S., 1974, Fore-reef morphology and depositional processes, north Jamaica, *in* Laporte, L. F., ed., Reefs in time and space: Tulsa, Oklahoma, SEPM Special Pub. 18, p. 77-89.

Heckel, P. H., 1974, Carbonate buildups in the geologic records, *in* Laporte, L. F., ed., Reefs in time and space: SEPM, Special Pub. 18, p. 90-154.

Hine, A. C., and Neuman, A. C., 1977, Shallow carbonate-bank-margin growth and structure, Little Bahama Bank, Bahamas: AAPG Bull., v. 60, p. 89-176.

Humphrey, W. E., 1949, Geology of the Sierra de los Muertos area, Mexico (with descriptions of Aptian Cephalopods from the La Peña Formation): AAPG Bull., v. 60, p. 89-176.

_______ 1956, Tectonic framework of northeast Mexico: GCAGS, v. 60, p. 25-35.

_______ 1961, Notes on the geology of northeast Mexico: South Texas Geological Societies Guidebook 11, p. 6-60.

Imlay, R. W., 1936, Geology of the western part of the Sierra de Parras, GSA Bull., v. 47 p. 1723-1744.

_______ 1944, Cretaceous Formations of Central America and Mexico: AAPG Bull., v. 29, p. 1416-1469.

Irwin, M. L., 1965, General theory of epeiric clear water sedimentation: AAPG Bull., v. 49, p. 445-460.

Kellum, L. B., 1936, Evolution of the Coahuila peninsula, Mexico, part III, geology of the mountains west of the Laguna district: GSA Bull., v. 47, p. 969-1008.

Moore, C. H., 1975, Ancient carbonates: AAPG Field Seminar Guidebook, 93 p.

Muir, J. M., 1936, Geology of the Tampico region, Mexico: AAPG Special Pub., 280 p.

Nelson, H. F., Brown, C. W., and Brineman, J. H., 1962, Skeletal limestone classification: AAPG Memoir 1, p. 224-252.

Oivanki, S. M., 1973, Paleoenvironmental analysis of the Zuloaga Formation, Mexico: Louisiana State Univ., Master's thesis, 138 p.

Perkins, B. F., 1974, Paleoecology of a rudist reef complex in the Comanche Cretaceous Glen Rose Limestone of Central Texas: Louisiana State Univ., Geoscience and Man, v. 8, p. 131-173.

Petta, T. J., 1976, Diagenesis and paleohydrology of a rudist reef complex (Cretaceous), Bandera County, Texas: Louisiana State Univ., Ph.D. dissertation, 212 p.

Roberts, H. H., 1974, Variability of reefs with regard to changes in wave power around an island: Proceedings of Second International Reef Symposium 2.

Smith, C. I., 1970, Lower Cretaceous stratigraphy, northern Coahuila, Mexico: Univ. Texas, Austin, Bureau of Economic Geology, Report of Investigations 65, 101 p.

Wilson, J. L., 1975, Carbonate facies in geologic history: New York, Springer Verlag, 471 p.

DIAGENESIS OF A GIANT: POZA RICA TREND, MEXICO

Paul Enos[1]

ABSTRACT

The Poza Rica trend, Veracruz, Mexico, has been Mexico's largest petroleum producer since the 1930's. The productive Tamabra Limestone of this trend consists of rudist-fragment packstone, wackestone, and grainstone; breccia; and pelagic wackestone. The breccia and rudist limestones are interpreted as mass-flow deposits from the 1,000-m-high, east-adjacent escarpment of the Golden Lane "atoll" (Enos, 1977).

The major stages in porosity development which can be identified despite local complexities are: (1) primary, sedimentary porosity, (2) early cementation, (3) matrix lithification (probably overlaps cementation), (4) skeletal mold and vug formation, (5) fracturing (overlaps 4), (6) dolomitization (limited), and (7) calcite cementation. Porosity at each stage varied with depositional texture (table 1). Primary porosity, which was accompanied by effective permeability only in low-matrix packstone, grainstone, and low-matrix breccia with porous clasts, was almost completely destroyed in early stages of diagenesis. The slightly greater porosity retention in the more matrix-free rocks is probably significant for later porosity development.

Productive porosity is almost entirely skeletal molds in the rudist limestones. Favorable porosity is developed in some nonproductive wells. Production is determined by a basinward stratigraphic pinch-out of the coarser grained limestones coincident with a broad plunging anticline; good porosity is developed over a wider area, although not basinward of the pinch-out of the coarser grained rocks. Leaching is ascribed to meteoric waters descending westward from cavernous limestone during intermittent periods of subaerial exposure of the Golden Lane Escarpment. By this model, impermeable pelagic carbonates deposited during and immediately following deposition of the Tamabra Limestone confined the water until it reached areas where the pelagic carbonates were missing and emerged as submarine springs. Dolomitization may have resulted from brines seeping downward along the same routes from evaporite lagoons within the Golden Lane "atoll" or from mixing of fresh water and sea water within the aquifer. An average δO^{18} of -2.5 in the dolomites suggests low-salinity waters.

Table 1. Evolution of porosity; preliminary data from thin-section point counts.

	DIAGENETIC STAGE*	DEPOSITIONAL TEXTURE**				
		Grainstone	Packstone	Wackestone	Breccia	Breccia Clasts
(1)	Sedimentation (primary) BP, WP, SH, BC (matrix)	28.7	32.8	45.1	36	56
(2)	Early cementation cr BP, WP, SH, BC (matrix)	1.0	14.6	42.6	17	48
(3)	Matrix lithification cr BP, WP, SH, BC (neomatrix)	0.3	0.7	0.1	–	0.4
(4)	Leaching MO, VUG, cr BP	22.2	17.7	5.5	–	9
(5)	Fracture MO, WP, SH, BC, FR		Increases	1 percent;	isolated samples	
(6) & (7)	Present porosity Cr (MO, VUG, FR, BP)	14.2	10.4	4.0	2.6	4
	Number of samples	5	19	23	3	2

*Porosity types indicated by terminology of Choquette and Pray (1970) **Porosity values are mean percentages

[1]State University of New York at Binghamton

MIDDLE CRETACEOUS ROCKS OF MEXICO AND TEXAS

Keith Young[1]

ABSTRACT

Middle Cretaceous (late Albian to Turonian) rocks in Texas are mostly terrigenous or nodular and micritic. In Mexico, the terrigenous deposits are replaced by limestones, which are more reefoid to the south than to the north. South into Jalisco and Colima there are many reefoid outcrops, and east into Queretaro and Guerrero similar rocks intertongue with marls and terrigenous rocks.

The Belton high, San Marcos platform, Del Carmen trend, Sierra de la Párra, Coahuila Peninsula, and Miquihuana platform all contribute to facies changes around their margins and to thinning of nonreefoid rocks over these structurally higher areas.

[1]Department of Geological Sciences, The University of Texas at Austin

INTRODUCTION

General

Far southwest in Jalisco and adjacent Colima, Mexico, there are Cenomanian (Paso del Rio and, perhaps, Soyatlan de Adentro) and Turonian (Huescalapa) biolithites and reefs (fig. 1). Between these and their time equivalents in northern Mexico and Texas—the Del Rio, Buda, and Eagle Ford Formations—lie almost 1,000 km of limestone represented by the Cuesta del Cura (lower part) and the San Felipe Limestone; in places these Turonian and upper Cenomanian strata are replaced by the flyschoid Caracol or the more shaly Agua Nueva Formations.

Turonian and Campanian reefoid rocks of the Cuautla and Tamazopo

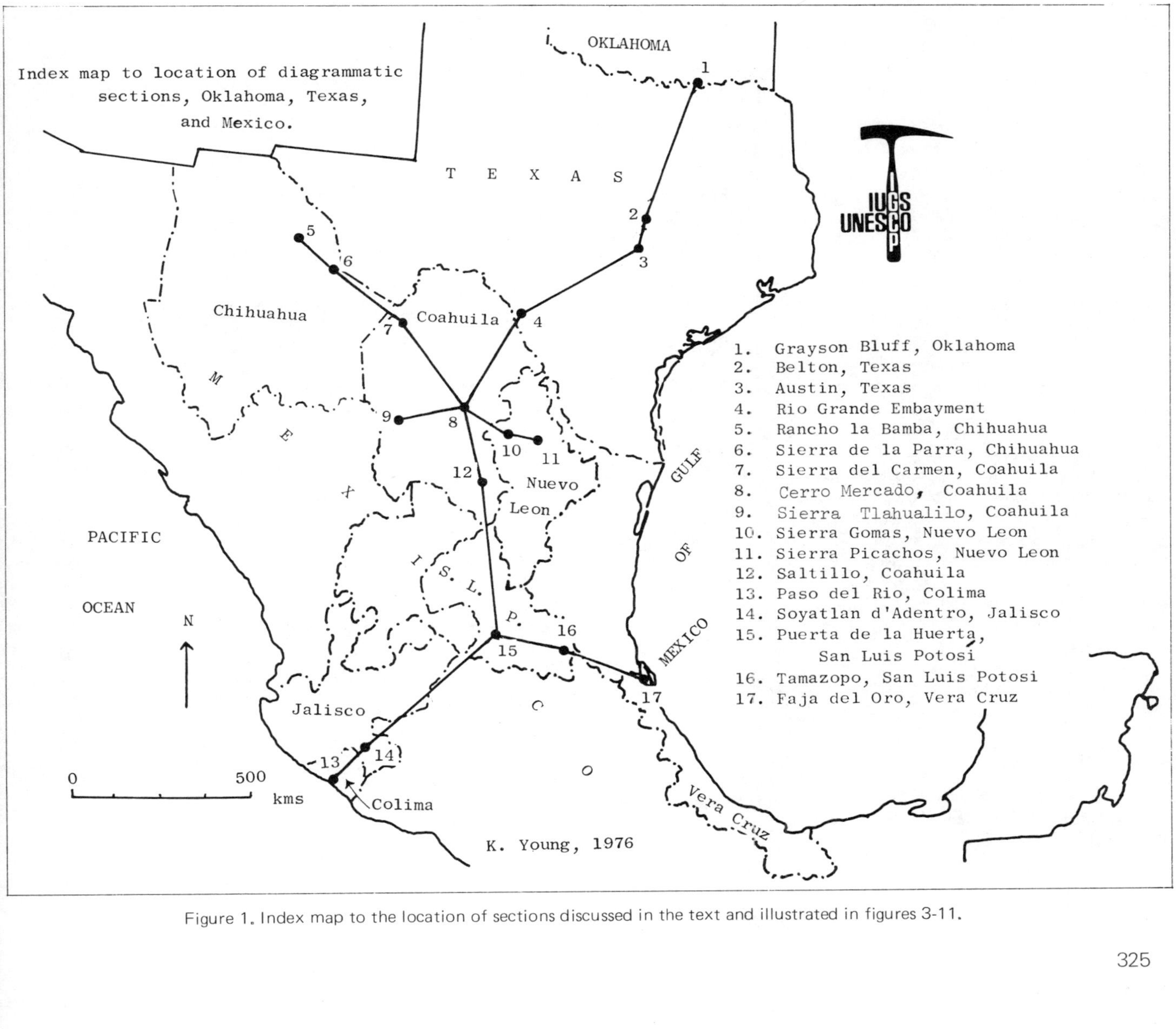

Figure 1. Index map to the location of sections discussed in the text and illustrated in figures 3-11.

Formations, respectively, overlie the Miquihuana platform (fig. 2). In south-central Coahuila and eastern Durango, the entire Cenomanian and Turonian section, including the lower Cenomanian, is represented by the Indidura Formation. Cenomanian evaporites, thin-bedded limestones of deeper water deposition, and a fine-grained, terrigenous lithofacies, perhaps of the outer part of the delta-platform are reported in the mountains south of Colima.

Problems of Correlation

Many of the localities of Cretaceous rocks occur in isolated mountains which were produced by block faulting and protrude through Cenozoic bolson deposits (Coahuila, Chihuahua, Durango, Zacatecas, and San Luis Potosí); other outcrops of Cretaceous rocks are inliers in faulted Cenozoic volcanics (Jalisco, Nayarit, Sinaloa, Colima, and Michoacan). Single outcrops are often separated from the next nearest outcrop by tens or even hundreds of kilometers. Therefore, many of these isolated sections must be tied together biostratigraphically. Equivalent deposits extend on east of Michoacan into Queretaro, Guerrero, Mexico, D. F., Pueblo, and into southern Vera Cruz. From Colima to Vera Cruz, Cenomanian and Turonian rocks are usually included in the Lower Cretaceous on the geologic map of Mexico (Salas and others, 1968; Córdoba and others, 1976), but the rocks sometimes are classified as part of the "Mesozoic undifferentiated."

Although extremely rare south and east of Coahuila and northern Zacatecas, ammonites have been the best tools for correlation. In their absence, the overlapping ranges of ***Hedbergella, Ticinella, Rotalipora, Praeglobotruncana, Globigerellinoides***, and other planktonic foraminifers are important (Pessagno, 1969). Palmer (1928) originally dated the deposits in Jalisco and Colima by comparing the rudists with those of Sicily (Douvillé, 1910). However, rudists and ammonites do not often occur in the same strata in the same area, and foraminifers still are the best tool for tying ammonite facies to rudist facies where surface mapping cannot meet such objectives.

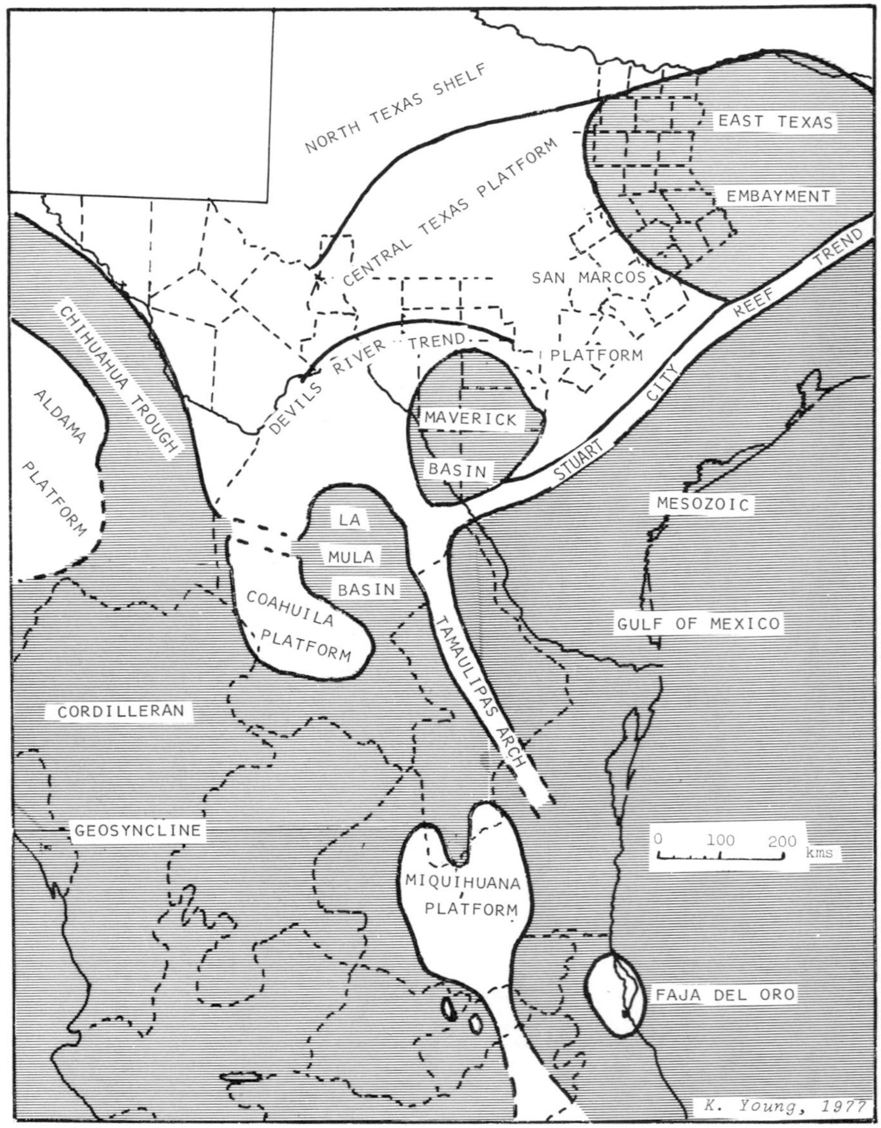

Figure 2. Paleogeography of Texas and northern Mexico during the early part of the Upper Cretaceous. The Miquihuana platform is only approximately outlined; Imlay (1943) and Carrasco (1970) suggested that this feature may represent an archipelago of islands.

RELATIONS OF MIDDLE CRETACEOUS STRATA FROM NORTH TO SOUTH

North-Central Texas

Adkins and Lozo (1951) and Lozo (1951) have delineated the relationships of Cenomanian and Turonian formations from Central and North Texas. In Grayson County, sandstones and shales of formations of the Woodbine Group conformably overlie the Grayson Formation and are overlain conformably by claystones and marls of the Eagle Ford Group (figs. 1, 3, 4). Several fossils from the upper part of the Grayson Formation of this area occur elsewhere only in the Buda Limestone. Furthermore, ammonites known from the middle (Modlin) limestone member of the Grayson Formation are also restricted to the Buda Limestone elsewhere. According to Lozo (1951), that part of the Grayson Formation overlying this limestone member is probably equivalent to the Maness Shale in the subsurface (Bailey and others, 1945).

Ammonites common to the upper part of the Grayson Formation of North Texas and the Buda Limestone are ***Budaiceras hyatti*** (Shattuck), ***B. elegantior*** (Lasswitz), and ***Faraudiella roemeri*** (Lasswitz). The same three species also occur in the middle (Modlin) limestone member of the Grayson Formation but have not been recovered lower in that formation. The foraminifers (Pessagno, 1969) are not zoned as closely as are the ammonites and are not as diagnostic; however, evidence from foraminifers does not contradict the ammonite correlations.

Over the Belton high, minor disconformities represent

nondeposition between the Edwards and Georgetown Formations and between the Buda and Pepper (Woodbine) Formations. The latter disconformity extends north to at least beyond the Johnson and Hill county line, where Winston and Scott (1922) described rolled boulders of Buda Limestone between the Grayson and Woodbine.

At Austin, Travis County, the disconformity between the Pepper (Woodbine) and the Buda Limestone is still apparent, although here, for convenience, the 1 m of Pepper Shale is mapped as part of the Eagle Ford Group. There is a minor disconformity at Austin between the Del Rio and Buda Formations (Martin, 1967) and a considerable disconformity between the Eagle Ford and Austin Groups (Adkins and Lozo, 1951). These disconformities are not found to the south in the Rio Grande embayment.

Coahuila

Middle Cretaceous outcrops north of Monclova, Coahuila, occur only in the isolated mountains south and west of Muzquiz and on the outcrops to the west of the Maverick basin in the tectonically higher areas of the Devils River and Del Carmen trends. A few kilometers southwest of Monclova at Cerro Mercado, the equivalents of the upper part of the Salmon Peak Formation occur in the lower part of an unnamed, thin-bedded limestone without chert (fig. 5), and equivalents of the lower part of the Salmon Peak Formation occur in the marly beds with limestone interbeds of the Sombrerotillo Formation. The thin-bedded limestone above the Sombrerotillo extends upward to include the equivalents of the Del Rio and Buda Formations (fig. 6). The Del Rio equivalent has yellow to orange beds of limestone similar to those at Rancho la Bamba, Chihuahua, but lacks the small ammonites. This limestone could be called Cuesta del Cura Formation, but it almost completely lacks the black chert typical of this formation. Humphrey (1956a, 1956b) applied the name "Monclova Limestone" to carbonates of Washita age in this area, but the context of the description indicates that this name should be applied to the limestone bank at the top of the Sierra de la Gloria southeast of Monclova, a limestone that now appears to represent only the upper part of the lower Cenomanian and is probably an isolated bank. At Cerro Mercado (fig. 5), this unnamed limestone is overlain by the early upper Cenomanian Agua Nueva Shale with ***Eucalycoceras***.

Both ***Graysonites adkinsi*** Young, from limestones of Del Rio age, and species of ***Budaiceras***, from limestones of Buda age, have been recovered from the areas of Monclova and Cuatro Cienegas and permit an approximate

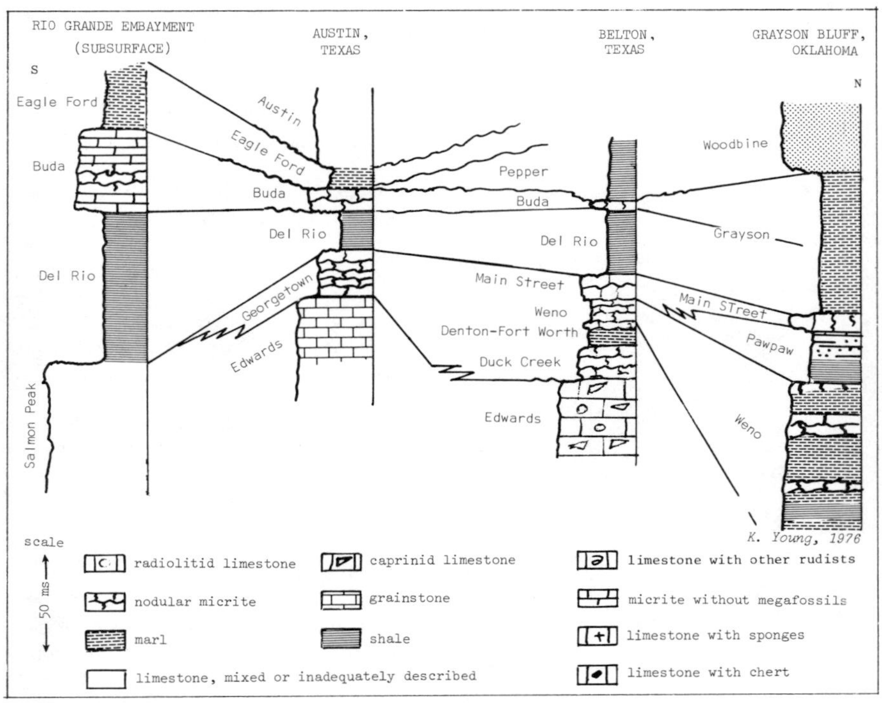

Figure 3

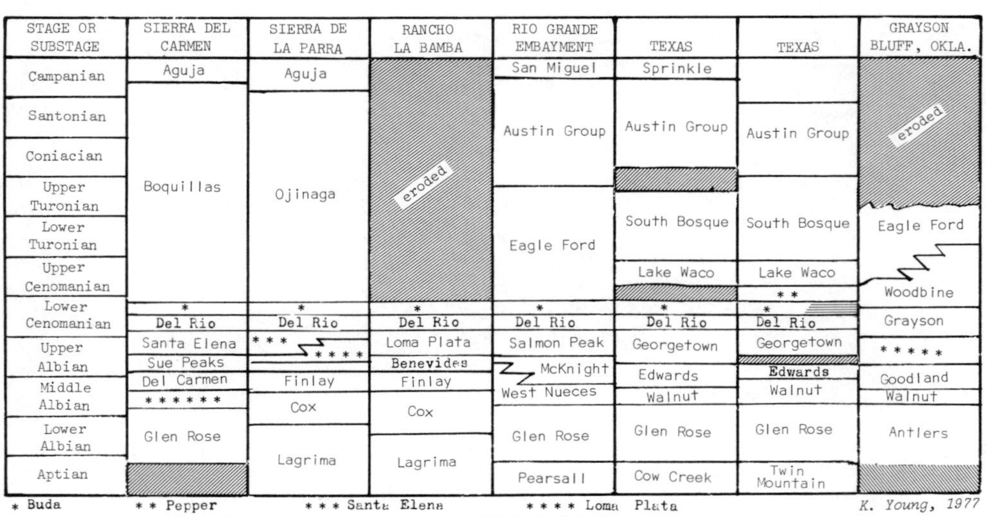

Figure 4

Figure 3. Correlation and facies of representative Middle Cretaceous rocks from Grayson Bluff, Oklahoma, to the Rio Grande embayment, Texas.

Figure 4. Correlation of formations of the middle of the Cretaceous for northern Coahuila, northeastern Chihuahua, and Texas. The Campanian and Aptian Formations are not complete.

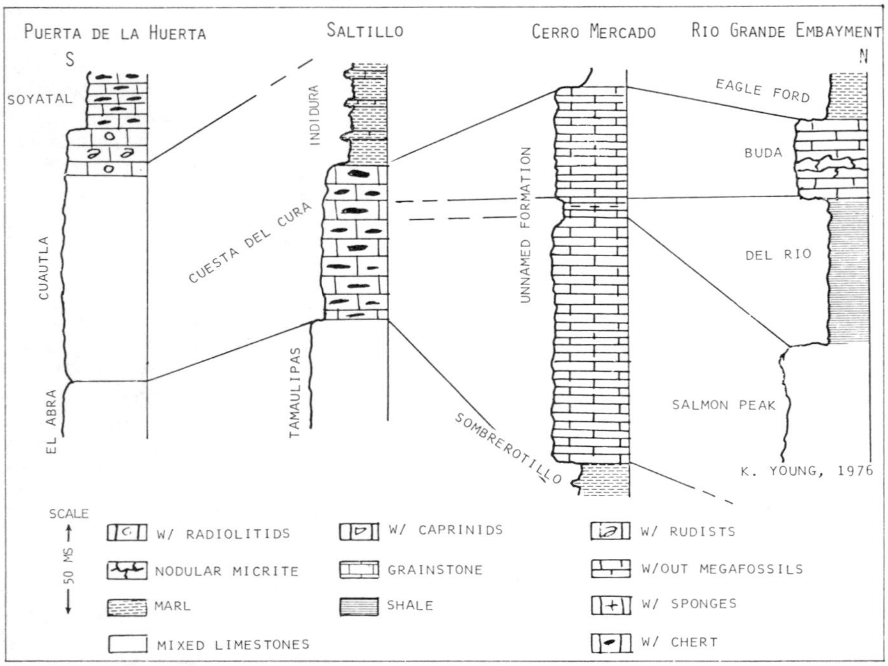

Figure 5. Correlation of Middle Cretaceous rocks and their facies from the Rio Grande embayment, Texas, through Coahuila, to the Puerta de la Huerta, San Luis Potosí.

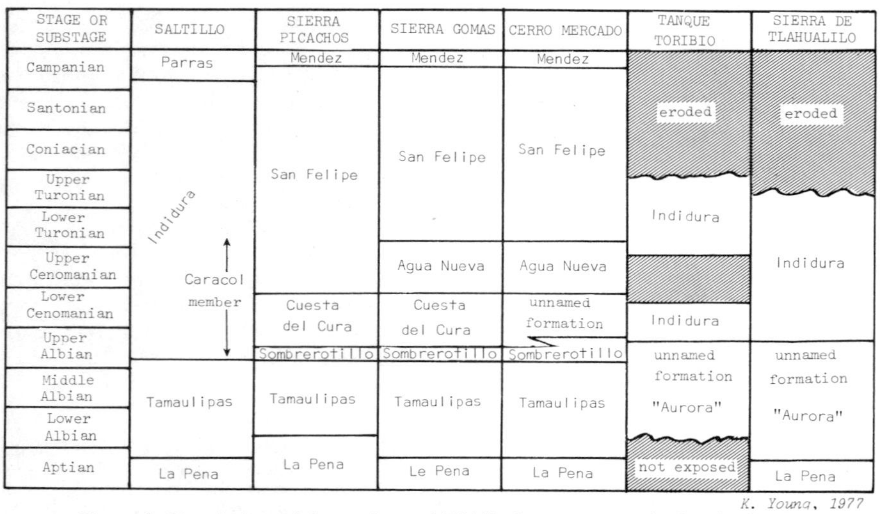

Figure 6. Correlation of formations of Middle Cretaceous rocks for the States of Coahuila and Nuevo León, Mexico. The Campanian and Aptian Formations are not complete.

correlation. At Cerro Mercado the Sombrerotillo Formation contains many specimens of ***Boeseites romeri*** (Haas), early upper Albian, and also many specimens of ***Androivites*** and ***Manuaniceras*** similar to those described by Collignon (1936) and Besairie (1936); this fauna has not been studied.

Near Saltillo, Coahuila, 150 km south, the situation has again greatly changed (fig. 5). Overlying the Tamaulipas Limestone here is the Cuesta del Cura, which contains the equivalents of the Sombrerotillo, Salmon Peak, Del Rio, and Buda Formations. The Indidura Formation overlies the Cuesta del Cura. At its base, the Indidura contains the Caracol Member which some have considered more flyschoid.

Zacatecas and San Luis Potosí

In northeastern Zacatecas, the lower Cenomanian deposits—the middle part of the Cuesta del Cura Formation—contain the siliceous ammonites described by Böse (1923) that represent the lower part of Dubourdieu's (1956) ***schneegansi*** zone of northern Africa. Dubourdieu's scheme and correlations made by this author of the lower Cenomanian of northern Mexico and Texas appear in figure 7. Some of the lowermost Cenomanian silicified ammonites include ***Tetragonites zacatecanus*** Böse, ***Turrilites multipunctatus*** Böse, and ***Anisoceras camachoense*** Böse.

That part of the American section above typical Albian Mortoniceridae and below ***Graysonites adkinsi*** correlates with subzone I of Dubourdieu. The remainder of the Del Rio Formation correlates generally with Dubourdieu's subzone II, and the Buda Limestone, with his subzone III. Böse's (1923) fauna from Zacatecas is above ***Mortoniceras*** but below ***Graysonites***, although some of his species range up into beds bearing ***Graysonites*** at Rancho la Bamba, Chihuahua.

From Saltillo, Coahuila, south into San Luis Potosí, along the western margin of the Miquihuana platform (Valles platform) (fig. 2), the picture becomes less clear. Thin-bedded limestones similar to Cuesta del Cura are common, but some Cuesta del Cura strata may even extend down to below the upper Albian. Many exposures of thin-bedded limestone with black chert occur between San Roberto and Matehuala, but not all of them have been correctly dated. South of Matehuala at Charco Blanco, there is a westward salient of the Miquihuana platform where reef and reef-talus deposits extend farther west. These deposits contain large species of ***Durania***, advanced ***Rotalipora***, ***Praeglobotruncana*** (?), miliolids, and a foraminifer similar to ***Edomia***; they are at least as young as Cenomanian and perhaps Turonian—probably Cuautla Formation.

Some 40 to 60 km east of San Luis Potosí, from just west of Santa Catarina and west through the Valle de Fantasmos to Puerta de la Huerta, Cserna and Bello-Barradas (1963) have described a section of the Cuautla Formation (their El Doctor, at least in part) containing many rudists and overlain by mostly unfossiliferous, siliceous beds of the Soyatal Formation (fig. 8). The Cuautla and Soyatal Formations also occur to the south of Mexico City in Guerrero and Morelos (Bonet, 1971; Fries, 1960; Bauman, 1958). West of Santa Catarina in San Luis Potosí, one of the diagnostic fossils is ***Hippurites resectus mexicanus*** (Barcena); it occurs in the Cuautla Formation and in a limestone bed in the Soyatal Formation (Cserna and Bello-Barradas, 1963). *Hippurites resectus* Defrance is an index to the lower part of the upper Turonian in Europe. *Radiolites* spp. (similar to those described by Bauman in Bartlett, 1977) and *Bayleoidea* (?) sp. are also present. The *Toucasia* sp. mentioned by Cserna and Bello-Barradas is probably *Bayleoidea* sp.

RELATIONS OF MIDDLE CRETACEOUS STRATA FROM WEST TO EAST

Three lines of sections from west to east illustrate the broad relations of Middle Cretaceous strata (figs. 9, 10, 11). One of these (fig. 9) is in the extreme north of Mexico, extending from Rancho la Bamba, northeastern Chihuahua (Haenggi, 1966), to Cerro Mercado southwest of Monclova, northern Coahuila. The second line of sections (fig. 10) extends from the Sierra de Tlahualilo, western Coahuila (Kellum and Robinson, 1963), east to the Sierra de Picachos, Nuevo León (Bishop, 1970). The third line of sections (fig. 11) is several hundred kilometers to the south, extending from Paso del Rio, south of Colima, Colima (Palmer, 1928), eastward to the Faja del Oro, Vera Cruz.

Rancho la Bamba to Cerro Mercado

At Rancho la Bamba, the largely upper Albian Loma Plata Limestone is thin to medium bedded and has many marly interbeds (fig. 9). It is a micrite overlain by the Del Rio Claystone, of which only the lower part (subzone II of Dubourdieu (1956) *schneegansi* zone) is exposed. Fossils from the Del Rio at this locality include *Sciponoceras* sp. cf. *S. Baculoides* (Mantell), *Turrilities bosquensis* Adkins, *T. multipunctatus* (Böse), *Scaphites bosquensis* Böse, *Otoscaphites subevolutus* (Böse), *Eoscaphites* sp. cf. *E. tenuicostatus* (Pervinquiere), *Scaphites* sp. cf. *S. hugardianus* d'Orbigny, *Ficheuria* sp. aff. *F. pernoni* Dubourdieu, *Prionocycloides* sp. cf. *P. proratum* (Pervinquiere), and *Graysonites* (?) sp. juv. The Del Rio is overlain by the sponge facies of the Buda Limestone described by Reaser (1970).

Considerable changes of facies take place between Rancho la Bamba and the Canyon of the Rio Concho through the Sierra de la Párra, northeastern Chihuahua (Gries, 1970). At the Sierra de la Párra the entire upper Albian and lower Cenomanian sections are much thinner. The mudstone and wackestone facies of the Buda Limestone are overlain by the upper Cenomanian black shales of the lower part of the Ojinaga Formation and are underlain by a thinned Del Rio Claystone (3.5 m). The Del Rio Claystone, from which no fossils were recovered, sharply overlies the Santa Elena (fig. 9), the uppermost limestone bed of which contains fragments of large rudists (up to 12 cm wide and 1 m long) of the dimensions of *Immanitas* Palmer (1928); they are probably *Kimbleia occidentalis* (Conrad). The Santa Elena at lower horizons also contains *Mexicaprina, Radiolites,* and many specimens of *Toucasia,* and, in turn, overlies the

Africa (Dubourdieu, 1956)			Texas (Young)	
Zone	Subzones	Description	Zone	Formation
schneegansi	III	Typical Mantelliceras of the saxbyi group	Budaiceras hyatti	Buda Limestone
	II	Compressed mentellicerids	Graysonites lozoi	Del Rio Claystone
			Graysonites adkinsi	
	I	Below Mantelliceras, but above Mortoniceras	Plesioturrilites brazoensis	Georgetown Limestone
UPPER ALBIAN				

Figure 7. Correlation of Dubourdieu's (1956) zonation of the lower Cenomanian of North Africa with that of northern Mexico and Texas.

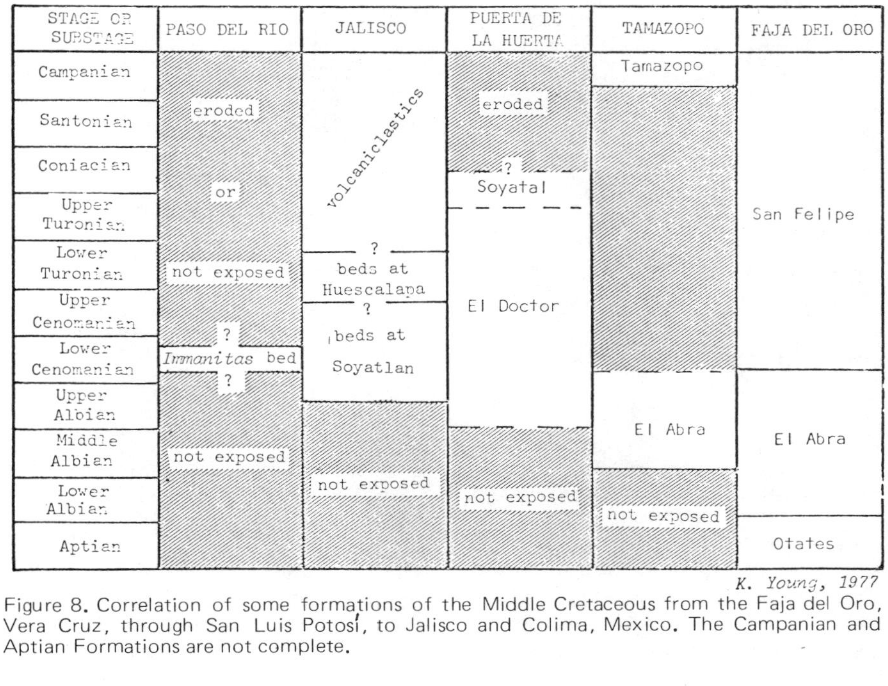

Figure 8. Correlation of some formations of the Middle Cretaceous from the Faja del Oro, Vera Cruz, through San Luis Potosí, to Jalisco and Colima, Mexico. The Campanian and Aptian Formations are not complete.

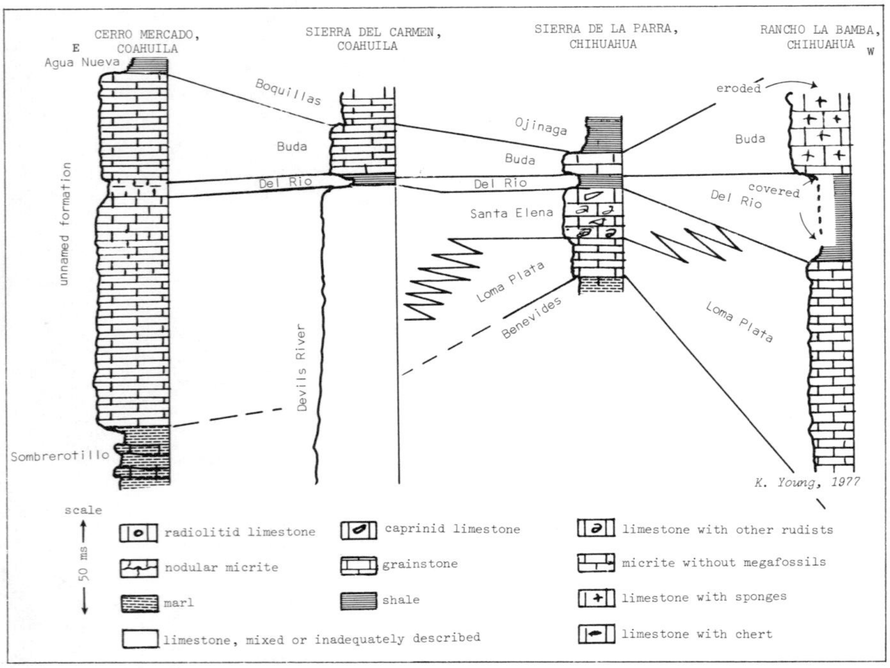

Figure 9. Correlation of some Middle Cretaceous sections and facies from Chihuahua and Coahuila.

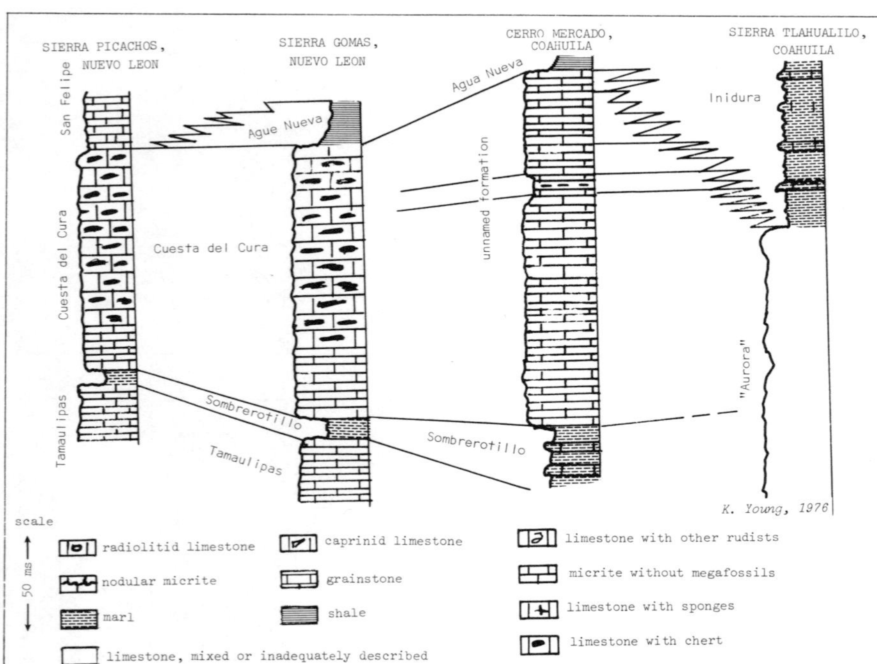

Figure 10. Correlation of some sections and facies of the Middle Cretaceous of Coahuila and Nuevo León, Mexico.

Loma Plata. The Santa Elena is a lateral facies of the upper part of the Loma Plata Limestone at Rancho la Bamba (fig. 9). The Benevides Formation separates the Loma Plata from the underlying Finlay Limestone.

The Del Rio Claystone is thinner or nonexistent to the east, across the Sierra del Carmen in western Coahuila. Where the Del Rio is not present, a coarser grained Buda Limestone rests with erosional contact on the Devils River or Santa Elena Limestone.

Farther southeast at Cerro Mercado near Monclova, Coahuila, the Buda Limestone, Del Rio Claystone, and underlying Salmon Peak Limestone are represented by a thin-bedded mudstone to wackestone with little chert, showing the transition of the northwestern facies toward that of the Cuesta del Cura farther east and south (fig. 10). The black Agua Nueva Shale, with ***Eucalycoceras***, occurs just above the unnamed limestone but still represents the lower part of the upper Cenomanian. The Sombrerotillo Formation overlies the mudstone of the Tamaulipas Limestone.

Sierra de Tlahualilo to the Sierra de Picachos

In the Sierra de Tlahualilo (fig. 10), the entire Cenomanian and Turonian sections, and, perhaps, the latest Albian are represented by the alternating nodular limestone and marl facies of the Indidura Formation (Kellum and Mintz, 1962; Kellum and Robinson, 1963). East of the Sierra de Tlahualilo on the southern tip of the Coahuila platform, as at Tanque Toríbio (Jones, 1938), there is a disconformity within the Indidura Formation, and the latest Cenomanian ***Kanabiceras septemseriatum*** (Cragin) rests on eroded beds of some part of the Del Rio (lower Cenomanian) containing *Ilymatogyra arietina* (Römer). The Indidura is underlain by strata usually called "Aurora" even though the name "Aurora" is no longer used at its type locality more than 500 km to the northwest.

East of Cerro Mercado across a Late Cretaceous basin now represented by the Difunta (Maestrichtian) Formation, the Sombrerotillo Formation is only 3 or 4 m thick and is overlain by the wavy-bedded black limestone and chert of the Cuesta del Cura. The Cuesta del Cura occurs along the front of the Sierra Madre Oriental for several hundred kilometers to the south. Chert is less abundant in the lower part of the Cuesta del Cura, which is underlain by the thin Sombrerotillo Marl and then by the Tamaulipas Limestone. In the Sierra Picachos (fig. 10) (Bishop, 1970), the section is much like that in the Sierra Gomas (Fuentes, 1964) except that the Agua Nueva is replaced by the lower part of the thin-bedded black micritic limestone with shale interbeds of the San Felipe Formation which, in turn, overlies the Cuesta del Cura Limestone. Lenses of coarse-grained grainstone of undetermined origin occur in the upper part of the Cuesta del Cura along the western flank of the Sierra Gomas. They contain large rudists similar to those in the top part of the Santa Elena Limestone in the Sierra de la Párra, Chihuahua.

Paso del Rio to the Faja del Oro

Several hundred kilometers farther south, the middle Cretaceous facies greatly change. In Colima the westernmost outcrop of Cenomanian consists of a single cycle of carbonate deposition and terminates in a reefal section. This outcrop is at the old Paso del Rio Armería, north of Tecomán at the western edge of the village of Periquillos. The limestone appears to be a xenolithic remnant in a large intrusion of diorite and includes a rudist fauna of ***Immanitas, Palus, Apricardia, Horiopleura, Sabinia***, and a large species of ***Durania*** that indicates rocks in this outcrop cannot be older than Cenomanian—the age that Palmer (1928) first gave to these beds (fig. 11). The relation of this bed to those at Soyatlan de Adentro, containing species of ***Caprinuloidea, Coalcomana, Planocaprina, Chaperia***, and ***Horiopleura***, is uncertain. Most of the rudists at Soyatlan are fragments in downslope debris or mudflows and, consequently, many may be reworked.

A considerable thickness of cyclic lime mudstones and wackestones, containing ***Bayleoidea*** and ***Apricardia***, culminates in beds with several species of ***Radiolites*** and caprinids at Huescalapa, Jalisco; they are capped by intertidal grainstones. These rocks also contain ***Rotalipora, Rhapydionina***(?), and ***Edomia***(?); on the basis of the rudists, Palmer considered them Turonian. Further east in eastern Querétaro, rudist rocks of Turonian age are known as the Cuautla Formation, and rocks partly equivalent to and younger than those at Huescalapa are known as the Soyatal Formation. The Soyatal Formation is largely unfossiliferous and is probably mostly equivalent to Upper Cretaceous volcaniclastic rocks of Jalisco (Palmer, 1928). In Querétaro, ***Radiolites*** sp. and ***Hippurites resectus mexicanus*** are common to the Cuautla Formation. ***Hippurites*** has not been recovered from Turonian rocks at Huescalapa.

East of San Luis Potosí, reefoid rocks occur on both sides of the Miquihuana platform. On the west, Albian rocks have not yet been authenticated, but they probably underlie the Cuautla Formation; the Cuautla Formation is overlain by sparsely fossiliferous rocks of the siliceous Soyatal Formation. Here the Cuautla Formation contains ***Hippurites resectus mexicanus*** (Barcena) and abundant specimens of ***Radiolites*** (similar to those described by Bauman in Bartlett, 1977). Cserna and Bello-Barradas (1963) also found ***Hippurites resectus mexicanus*** in a limestone bed in the Soyatal. The Cuautla and Soyatal of the Miquihuana platform are probably upper Turonian.

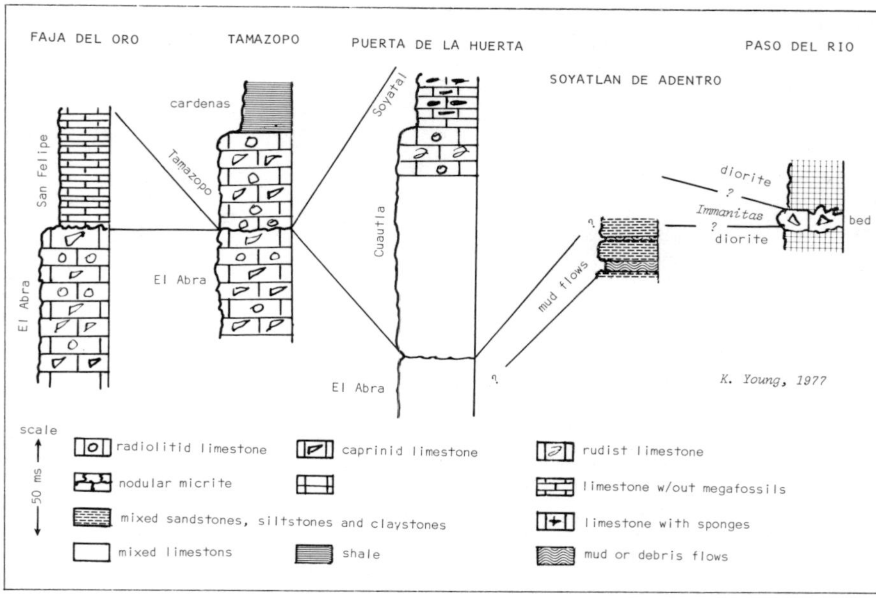

Figure 11. Correlation of some sections and facies of the middle of the Cretaceous from Colima, Jalisco, San Luis Potosí, and Vera Cruz, Mexico.

Between Valles and Rio Verde, at the eastern edge of the Miquihuana platform, tan-colored grainstones of the Tamazopo Formation contain ***Thyrastylon*** sp., ***Tampsia*** sp., ***Sauvagesia*** sp., ***Biradiolites aguilerae*** Böse, ***Durania*** sp., ***Distefanella*** sp., large bryozoans, and many other fossils. These fossils are at least as young as Campanian and are probably Maestrichtian. The Tamazopo lies unconformably on the gray, commonly rudist-bearing strata of the Tamabra and El Abra Limestones (Nigra, 1951), with the upper Cenomanian, Turonian, and most of the Senonian represented by the intervening disconformity. At Valles, just east of the Miquihuana platform, the magnitude of this unconformity is considerably reduced, and at the Faja del Oro, along the Gulf of Mexico, the unconformity is even less significant. Here, the overlying San Felipe Limestone, probably upper Cenomanian at its base, rests disconformably on El Abra Limestone, probably lower Cenomanian at its top.

SUMMARY

This brief review summarizes the present knowledge of the distribution of Middle Cretaceous rocks of Texas and northern Mexico. Much of the area lacks sufficient data for the construction of facies maps. Ongoing research by academic and industrial groups continues to add knowledge to the Cretaceous stratigraphy of this area and will modify the picture presented here.

REFERENCES

Adkins, W. S., and Lozo, F. E., 1951, Stratigraphy of the Woodbine and Eagle Ford Waco area, Texas: Southern Methodist Univ. Press, Fondren Scientific Series, no. 4, p. 101-161.

Alencaster, G., 1971, Rudistas del Cretacico superior de Chiapas: Instituto Geología (México), Paleont. Mexicana, no. 34, 91 p.

Bailey, T. L., Evans, F. G., and Adkins, W. S., 1945, Revision of stratigraphy of part of Cretaceous in Tyler basin, northeast Texas: AAPG Bull., v. 29, p. 170-186.

Bartlett, S. G., (in preparation), Carbonate rocks of the area around San Francisco, Municipio Zaragoza, State of San Luis Potosí: Univ. Texas, Austin, Master's thesis.

Bauman, C. F., Jr., 1958, Dos radiolítidos nuevos de la región de Cuernavaca, Morelos: Instituto Geología (México), Paleont. Mexicana, no. 3, 9 p.

Besairie, H., 1936, Les Manuaniceras de l'Albien moyen (niveau supérieur), *in* Besairie, H., La

Geologie du Nord-Ouest: Memoires de l'Académie Malgache, v. 21, p. 188-190.

Bishop, B. S., 1970, Stratigraphy of Sierra de Picachos and vicinity, Nuevo León, Mexico: AAPG, v. 54, p. 1245-1270.

Bonet, F., 1971, Espeleología de la Región de Cacahuamilpa, Gro.: Instituto Geología (México), bol. no. 90, 98 p.

Böse, E., 1923, Algunas faunas Cretácicas de Zacatecas, Durango, y Guerrero: Instituto Geología (México), bol. no. 42, 219 p.

Carrasco, B., 1970, La Formación El Abra (Formación El Doctor) en la Platforma Valles - San Luis Potosí: Instituto Mexicano del Petrolero, Revista, v. 2, no. 3, p. 97-99.

Collignon, M., 1936, Les Oxytropidoceras de l'Albien moyen (niveau supérieur) de la Province d'Analalava, *in* Besairie, H., La Géologie du Nord-Ouest: Memoires de l'Académie Malgache, v. 21, p. 176-187.

Coogan, A. H., 1973, Nuevos rudistas del Albiano y Cenomaniano de México y del sur de Texas: Instituto Mexicano del Petrolero, Revista, v. 5, no. 2, p. 51-82.

Cordoba, D. A., 1976, Carta Geológica de la Republica Mexicana, 4th ed.: Comité de la Carta Geológica de México, México, D.F., escala 1:2,000,000.

Cserna, E. G., and Bello-Barradas, A., 1963, Geología de la parte central de la Sierra de Alvarez, Municipio de Zaragoza, Estado de San Luis Potosí: Instituto Geología (Mexico), Bol. no. 71, pte. 2, p. 23-63.

Douvillé, H., 1910, Études sur les rudistes: rudistes de Sicile, d'Algérie, d'Egypte, du Liban, et de la Perse: Société Géologique de France, Paléontologie, Memoir 41, 84 p.

Dubourdieu, G., 1956, Étude Géologique de la Région de l'Ouenza (Confins Algéro-Tunisiens): Service Carte Géol. de l'Algérie, n.s., Bull. 10, 659 p.

Fries, C., Jr., 1960, Geología del Estado de Morelos y de partes adyacentes de México y Guerrero, Región Central Meridional de México: Instituto Geología (México), Bol. no. 60, 236 p.

Fuentes, R. P., 1964, Stratigraphy of Sierra Santa Clara and Sierra Gomas, Nuevo León, Mexico: Univ. Texas, Austin, Master's thesis, 216 p.

Gries, J. C., 1970, Geology of the Sierra de la Parra area, northeast Chihuahua, Mexico: Univ. Texas, Austin, Master's thesis, 151 p.

Haenggi, W. T., 1966, Geology of the El Cuervo area, northeastern Chihuahua, Mexico: Univ. Texas, Austin, Master's thesis, 401 p.

Humphrey, W. E., 1956, Notes on the geology of northeast Mexico: supplement to Volk, H. W., Jr., chairman, Corpus Christi Geological Society, annual field trip, 44 p.

________ 1956, Bosquejo geológico de la región que cruzara la excursión C-5: p. 8-60.

Imlay, R. W., 1943, Evidence for Upper Jurassic landmass in eastern Mexico: AAPG Bull., v. 27, p. 524-529.

Jones, T. S., 1938, Geology of the Sierra de la Peña and paleontology of the Indidura Formation, Coahuila, Mexico: GSA Bull., v. 49, p. 61-150.

Kellum, L. B., and Robinson, W. I., 1962, Cenomanian ammonites from the Sierra de Tlahualilo, Coahuila, Mexico: Univ. Michigan Contributions from the Museum of Paleontology, v. 13, no. 10, p. 267-287.

__________ 1963, Geology of the west-central part of the Sierra de Tlahualilo, Coahuila, Mexico: Michigan Academy of Sciences, Arts, and Letters, papers, 1962, v. 48, p. 223-261.

Lozo, F. E., 1951, Stratigraphic notes on the Maness (Comanche Cretaceous) shale: *in* Woodbine and adjacent strata: Southern Methodist Univ. Press, Fondren Scientific Series, no. 4, p. 65-91.

Martin, K. J., 1967, Stratigraphy of the Buda Limestone, south-central Texas, *in* Hendricks, L., ed., Comanchean (Lower Cretaceous) stratigraphy and paleontology of Texas: SEPM Permian Basin Section, Pub. 67-8, p. 287-299.

Mullerried, F. K. G., 1930, El llamado ***Hippurites mexicana*** Barcena: Instituto Biología (México), Anales, v. 1, no. 1, p. 63-70.

Nigra, J. O., 1951, El Cretácico medio de México, con especial referencia a la facies de caliza arrecifal del Albiano-Cenomaniano en la Conobahia de Tampico-Tuxpan: Asociación Mexicana de Geólogos Petroleros, v. 3, p. 107-175.

Palmer, R. H., 1928, The Rudistids of southern Mexico: San Francisco, California Academy of Sciences Occasional Papers, 14, 137 p.

Perkins, B. F., 1969, Rudist faunas in the Comanche Cretaceous of Texas, *in* Shreveport Geological Society, Guidebook, 1969 spring field trip, p. 121-137.

Pessagno, E. A., Jr., 1969, Upper Cretaceous stratigraphy of the western Gulf Coast area of Mexico, Texas, and Arkansas: GSA Memoir 111, 139 p.

Powell, J. D., 1963, Cenomanian-Turonian (Cretaceous) ammonites from Trans-Pecos Texas and northeastern Chihuahua, Mexico: Journal of Paleontology, v. 37, p. 309-322.

________ 1963, Turonian (Cretaceous) ammonites from northeastern Chihuahua, Mexico: Journal of Paleontology, v. 37, p. 1217-1232.

__________ 1970, Early Upper Cretaceous faunal zones southwest of the Diablo-Coahuila platform, *in* Campbell, D. H., Reaser, D. F., and Jones, B. R., Geology of the southern Quitman Mountains area, Trans-Pecos Texas: SEPM, Permian Basin Section, Pub. 70-12, p. 96-99.

Reaser, D. F., 1970, Preliminary study of the Buda Limestone in foothills of the southern Quitman Mountains, Trans-Pecos Texas, *in* Campbell, D. H., Reaser, D. F., and Jones, B. R., Geology of the southern Quitman Mountains area, Trans-Pecos Texas: SEPM, Permian Basin Section, Pub. 70-12, p. 76-81.

Salas, G. P., and others, Carta Geológica de la Republica de México: Comité de la Carta Geológica de México, México, D.F., 2 sheets, escala 1:2,000,000.

Stephenson, L. W., 1944, Fossils from limestone of Buda age in Denton County, Texas: AAPG Bull., v. 28, p. 1538-1541.

Taff, J. A., and Leverett, S., 1893, Report on the Cretaceous area north of the Colorado River: Texas Geological Survey, Annual Report, v. 4, pt. 1, p. 293-354.

Tucker, D. R., 1962, Subsurface Lower Cretaceous stratigraphy, Central Texas, *in* Stapp, W. L., ed., Contributions to geology of South Texas: South Texas Geological Society, p. 177-216.

Wilson, B. W., Hernández, J. P., and Meave-Torrescano, E., 1955, Un banco calizo del Cretácico en la parte oriental del Estado de Querétaro, Mexico: Sociedad Geológica Mexicana Bol., v. 18, no. 1.

Winton, W. M., and Scott, G., 1922, The geology of Johnson County: Univ. Texas, Austin, Bull. 2229, 68 p.